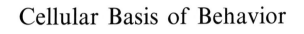

Cellular Basis of Behavior

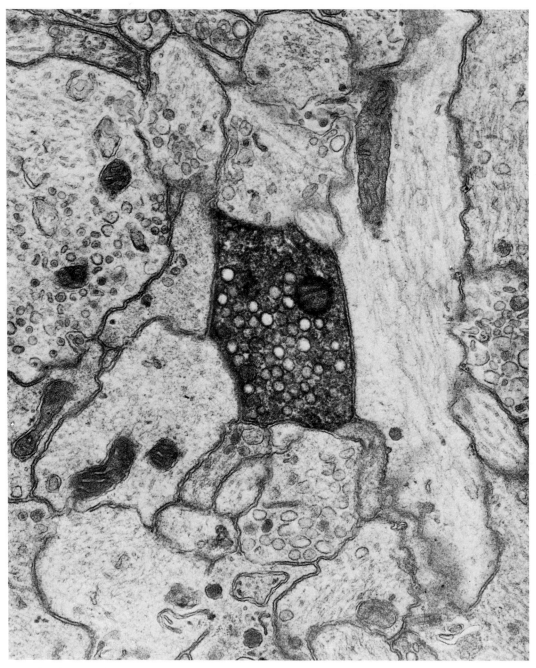

An electron micrograph of a darkened profile of a presynaptic terminal from a sensory neuron injected with a marker substance (×40,000). An action potential in the sensory neuron produces a synaptic potential in a motor neuron. This synapse undergoes plastic changes associated with habituation and sensitization. [From experiments by C. Bailey, V. Castellucci, E. R. Kandel, and E. B. Thompson.]

Cellular Basis of Behavior

AN INTRODUCTION TO BEHAVIORAL NEUROBIOLOGY

Eric R. Kandel

COLUMBIA UNIVERSITY

W. H. Freeman and Company

SAN FRANCISCO

A series of Books in Psychology

EDITORS: *Richard C. Atkinson*
Jonathan Freedman
Gardner Lindzey
Richard F. Thompson

Library of Congress Cataloging in Publication Data

Kandel, Eric R.
     Cellular basis of behavior.

     Bibliography: p.
     Includes index.
     1. Neurobiology. 2. Psychology, Physiological.
I. Title. [DNLM: 1. Behavior. 2. Neurophysiology.
WL102 K156c]
QP360.K37   591.1'88   76-8277
ISBN 0-7167-0523-0
ISBN 0-7167-0522-2 pbk.

Printed in the United States of America

9 8 7 6 5 4 3 2 1

To W. Alden Spencer

Contents

Preface   xv

PART I   STRATEGIES IN THE STUDY OF BEHAVIOR

1   The Study of Behavior: The Interface Between Psychology
   and Biology   3

   Development of Behavioral Paradigms   4
      Experimental Study of Learned Behavior   5
      Naturalistic Study of Instinctive Behavior   14
      Biological Constraints on Learning   18
      Types of Learning Not Dependent on Reinforcement   20
   Reappraisal of the Study of Behavior   21
   Intervening Variables, Brain Processes, and Behavior   22
   The Convergence of Psychology and Neurobiology   25
   Summary and Perspective   25
   Selected Reading   27

2   The Neurobiological Analysis of Behavior: The Use
   of Invertebrates   29

   Toward an Operational Definition of Behavior   30
      Classification of Behavior   32
      Intervening Variables and the Classification of Behavior   34
   Behavioral Advantages of Studying Invertebrates   35
   Neurobiological Advantages of Studying Invertebrates   36
   Goals of a Cellular Approach   37
   Strategy and Limitations in the Use of Invertebrates   38
   Higher Invertebrates and the Evolution of Vertebrate Behavior   40
   Summary and Perspective   41
   Selected Reading   43

3   The Choice Among Invertebrates: Cellular and Genetic Approaches
    to Behavior   45

        Studies of Invertebrate Behavior   46
        Single-Cell Studies of Invertebrate Nervous Systems   48
        Genetic Analysis of Invertebrate Nervous Systems   59
        Is There One Ideal Organism for Studying Behavior?   61
        Summary and Perspective   64
        Selected Reading   65

4   The Behavioral Biology of a Single Invertebrate: The Marine
    Snail *Aplysia*   67

        The Molluscs   69
        Studies of *Aplysia*   73
        Natural History and Biology of *Aplysia*   74
        Nervous System of *Aplysia*   77
            General Plan   77
            Fine Structure of the Neuron   82
        Behavior of *Aplysia*   83
            Elementary Behavior: Reflex Acts   84
                Neuroendocrine Reflexes   84
                Defensive Reflex of the Mantle Organs   84
            Elementary Behavior: Fixed Acts   84
                Inking and Egg-Laying   84
        Complex Behavior   85
            Locomotion   85
            Escape   87
            Homeostatic Adjustments and Respiratory Pumping   87
            Feeding   89
        Higher-Order Behavior: Group Sex   90
        Summary and Perspective   91
        Selected Reading   92

PART II   STRATEGIES IN THE STUDY OF NERVE CELLS

5   Introduction to the Study of Neurons   95

        Cellular Elements of the Nervous System   96
        Electrical Signals in Nerve Cells   98
            Membrane and Action Potentials   98
            Synaptic Transmission   102
            Triggering of Action Potentials by Synaptic Potentials   105
        Neurons and Reflex Behavior   105
        The Model Neuron   109
        Passive Properties of the Membrane   110
            Resistance   111
            Current–Voltage Relationship   114
            Capacitance   115
            Time Constant and Space Constant   120
        Active Properties of the Membrane   123
        Passive Properties and Neuronal Integration   126
        Passive Properties and Synaptic Transmission   130

Summary and Perspective   132
Selected Reading   135

6   Cellular Mechanisms of Neuronal Function   137

Ionic Mechanisms of Membrane and Action Potentials   138
Membrane Potential   140
Action Potential   143
The Sodium–Potassium Pump   145
Testing the Ionic Hypothesis   145
The Ionic Hypothesis and the Equivalent Circuit   146
Quantifying the Ionic Hypothesis   153
Postsynaptic Mechanisms of Chemical Transmission   163
Synaptic Potentials Due to Increased Ionic Conductance   165
Excitatory Postsynaptic Potentials   165
Inhibitory Postsynaptic Potentials   171
Synaptic Potentials Due to Decreased Ionic Conductance   174
Excitatory Postsynaptic Potentials   175
Inhibitory Postsynaptic Potentials   177
Presynaptic Mechanisms of Chemical Transmission   179
Quantal Release of Transmitter   184
Calcium Influx and Quantal Release   197
Factors Controlling the Membrane Potential of the
Presynaptic Terminals   198
Chemical Nature of Transmitters and Receptors   202
Summary and Perspective   204
Selected Reading   209

7   Identified Neurons   211

Identification of Neurons   215
Identified Neurons in Gastropods   221
Identified Neurons in *Aplysia*   225
Abdominal Ganglion   225
Is the Entire Structure Invariant?   229
Functional Organization   229
Buccal and Pleural Ganglia   231
Identified Neurons in Other Opisthobranchs   234
Identified Neurons in Other Invertebrates   235
Diversity Among Identifiable Neurons   238
Firing Patterns   238
Mechanisms for Generating Membrane Potentials   239
Mechanisms for Generating Action Potentials   246
Mechanisms for Generating Afterpotentials   251
Sites for Impulse Initiation   251
Signaling Function of Autoactive Rhythms   256
Mechanisms of Various Autoactive Rhythms   256
Beating Cells   256
Bursting Cells   260
Protein Synthesis   268
Transmitter Biochemistry and Pharmacology   270
Morphology   276

Summary and Perspective   278
Selected Reading   280

8   Patterns of Synaptic Connection   281

Elementary Divergent Aggregates   282
Electrophysiological Criteria for Direct and Common
   Connections   289
      Chemically Mediated Direct Connections   289
      Chemically Mediated Common Connections   292
      Electrically Mediated Connections   294
A Multiaction Cholinergic Neuron in the Abdominal Ganglion
   of *Aplysia*   295
      Opposite Actions Mediated to Different Follower Neurons   295
      Dual Actions Mediated to a Single Follower Neuron   304
         Excitatory-Inhibitory Actions   305
         Dual Inhibitory Actions   311
      Electrical Synaptic Actions   313
Multiaction Cholinergic Neurons in the Buccal Ganglia   317
A Multiaction Cholinergic Neuron in the Left Pleural Ganglion   322
Pharmacological Properties of the Three Types of ACh Receptors   324
Multiaction Cholinergic Neurons in Other Invertebrates   324
Multiaction Serotonergic and Dopaminergic Neurons   327
Multiaction Neurons in Vertebrates   328
Elementary Convergent Aggregates   330
Higher-Order Patterns of Interconnection   330
      Convergent Connections   331
      Feedback Connections by Follower Neurons   333
      Feed-Forward Connections   334
         Feed-Forward Substitution   334
         Feed-Forward Summation   338
Summary and Perspective   340
Selected Reading   342

PART III   NERVE CELLS AND BEHAVIOR

9   The Neuronal Organization of Elementary Behavior   345

The Search for Units of Behavior   346
A Somatic-Motor Reflex Act: Defensive Withdrawal of the Mantle
   Organs   350
      The Behavior   350
      The Neuronal Circuit   353
         Motor Neurons   354
         Quantitative Contributions of Individual Motor
            Neurons   362
         Sensory Neurons   365
         Quantitative Contributions of Individual Sensory
            Neurons   372
      Quantitative Analysis of a Behavior   374
A Visceral-Motor Fixed Act: Respiratory Pumping   378
      The Behavior   378
      The Neuronal Circuit   380

Multiple Control of a Single Effector System   384
A Glandular Fixed Act: Inking   385
   The Behavior   386
   The Neuronal Circuit   387
         Properties of the Neuronal Circuit and Properties
            of the Behavior   393
         The Central Program for Inking   397
         Two Ways of Triggering the Central Program
            for Inking   399
A Neuroendocrine Fixed Act: Egg-Laying   406
   The Behavior   407
   The Neuronal Circuit   408
Common Features of Fixed Acts: Inking and Egg-Laying   415
A Comparison of Reflex and Fixed Acts   415
Summary and Perspective   416
Selected Reading   418

10   The Neuronal Organization of Complex Behavior   419

Circulatory Fixed-Action Patterns   419
   Behavior   419
   Circulatory System of *Aplysia*   421
   Neuronal Circuits   421
      Motor Neurons   421
      Interneurons   428
   Centrally Programmed Increase of Heart Rate   429
   Hormonal Control of Heart Rate   436
   Centrally Programmed Decrease of Heart Rate   436
   Command Elements and Central Programs   438
   Integration of Fixed Acts into a Fixed-Action Pattern   441
   Coordination of Antagonist Fixed Acts   443
Defensive Swimming and Feeding   446
   Defensive Swimming of *Tritonia*   449
   Swimming of the Leech   455
   Feeding   457
   Behavioral Choice   470
Summary and Perspective   471
Selected Reading   473

11   Neuronal Plasticity: Possible Mechanisms for Behavioral
      Modification   475

Neurophysiological Hypotheses of Behavioral Modification   475
   Plastic Change Hypothesis   476
   Dynamic Change Hypothesis   478
Effects of Usage at Vertebrate Synapses   480
Effects of Usage at Invertebrate Synapses   491
Plasticity of Electrical and Chemical Synapses   491
Neural Analogs of Learning   494
   Classical Conditioning   496
   Instrumental Conditioning   506
   Habituation   518
   Heuristic Value of Neural Analogs   524

Biochemical Aspects of Neuronal Plasticity   527
Neuronal Control of Macromolecular Synthesis   528
Macromolecular Synthesis and Short-Term Functioning
of Neurons   528
Neuronal Control of Small-Molecule Synthesis   529
Morphological Aspects of Neuronal Plasticity   531
Summary and Perspective   534
Selected Reading   536

12   Short-Term Behavioral Modification   537

General Features of Habituation and Dishabituation   538
Habituation and Dishabituation of the Gill-Withdrawal Reflex   542
Neural Analysis of Habituation of the Gill-Withdrawal Reflex   546
Localization of Functional Change   546
Motor Fatigue and Receptor Adaptation   546
Central Synaptic Changes   548
Change in Input Resistance versus Synaptic Drive   555
Postsynaptic Inhibitory Buildup versus Excitatory
Synaptic Depression   555
Presynaptic Inhibition versus Homosynaptic
Depression   557
Mechanism of Habituation   564
Synaptic Depression and the Level of Transmitter
Release   564
Postsynaptic versus Presynaptic Mechanisms   565
Locus and Mechanism of Dishabituation of the Gill-Withdrawal
Reflex   575
A Cellular Electrophysiological Model of Short-Term Habituation
and Dishabituation   583
A Mollecular Model of Dishabituation   585
Cellular Analyses of Habituation in Other Invertebrates   591
Escape Swimming of the Crayfish   591
Escape Response of the Cockroach   595
Cellular Analysis of Habituation in Vertebrates   596
Higher-Order Features of Habituation   600
Summary and Perspective   601
Selected Reading   603

13   Relationships Between Short-Term and Long-Term Behavioral
Modifications   605

Short-Term and Long-Term Memory   606
The Transition from Short-Term to Long-Term
Habituation   609
Neural Studies of the Transition from Short-Term to
Long-Term Habituation   611
Transmitter Release and the Acquisition of Long-Term
Habituation   616
Comparison of Short-Term and Long-Term Habituation   617
Relationship of Synaptic and Behavioral Changes   620
Relationship of Long-Term Habituation to Complex
Learning   620

Habituation, Dishabituation, and Sensitization   625
Sensitization and Arousal   635
Short-Term and Long-Term Sensitization   636
Sensitization and Associative Classical Conditioning   636
Summary and Perspective   640
Selected Reading   642

14   Cellular Studies of Behavior in Perspective   643

Complementarity of Cellular Studies of Behavior in *Aplysia* and
Other Invertebrates   643
Neural Control of Locomotion   644
Command Cells   646
Organization of Sensory and Motor Systems   647
Locus and Mechanism for Habituation   648
Restrictions and Limitations of a Simple-Systems Approach   648
Methodological Limitations   648
Closeness of Fit   650
Generality of Findings   650
Selected Reading   652

15   Implications for the Study of Abnormal Behavior   653

Abnormal Behavior Produced by Stress   654
Motivational State and Stress   657
Behavioral Modification and Development   660
Selected Reading   662

Appendix   663

TABLE I   Identified neurons and cell clusters in the abdominal
ganglion of *Aplysia*   664
TABLE II   Known connections made by interneurons located in the
abdominal ganglion of *Aplysia*   666

Bibliography   667

Index   711

Preface

This book is an introduction to the principles of cellular biology that underlie simple forms of behavior and behavioral modification. My aim throughout has been to make this book usable by undergraduate and graduate students—even those having little background in behavior, neurobiology, or invertebrate zoology—as well as to provide an overview of the neurobiology of behavior for scientists in other fields. I therefore do not assume that the reader is familiar with cellular neurobiology or experimental psychology. I have introduced most topics by outlining the historical development of the neurobiological and behavioral concepts necessary for understanding the main arguments. An historical perspective is particularly important in the study of behavior. Problems dealt with today were formulated many years ago, and some traditional terms, such as "behavior," "arousal," "motivation," and "learning," were initially only vaguely defined. Over the years they have taken on new and often surplus meanings. Many key terms cannot be properly understood without knowing how they were first used and how their present usage has evolved.

The most interesting aspect of any field is not the collection of facts it presents, but the clusters of findings generated in the course of testing a few crucial ideas. Neurobiological studies of behavior are often embedded in one or another theoretical framework. To appreciate the logic of these studies it is essential to know even those theories that have been abandoned. Indeed, some previously discarded theories may later be resurrected in another form. Therefore, in introducing the neurobiology of behavior I think it is important not only to make the key ideas explicit, but also to show why some theories continue to prompt useful new experiments while others have failed to withstand critical testing and have been abandoned.

In addition to defining key concepts and explaining their history, I have felt it important to bring the reader closer to the original experiments through often detailed description and abundant illustration. Because students entering neurobiology have particular difficulty learning to interpret electrophysiological recordings, I have included many illustrations designed to help readers develop the necessary skill of analyzing this type of data.

The principles considered in this book have been derived from experiments on different invertebrate animals, but the key issues are illustrated with studies of a single invertebrate, the marine snail *Aplysia californica*. I focus on *Aplysia* because the ideas I want to develop become most clear when viewed from the perspective of a single animal. I hasten to emphasize that the findings from studies of *Aplysia* are not unique; similar ideas have developed from studies of other higher invertebrates that are technically advantageous for studying the nervous system at the cellular level.

Until this work in invertebrates, the distance between the study of the physiological mechanisms of behavior and traditional psychology was often considered the widest between any related scientific disciplines. But the accelerated progress in cell biology during the past ten years has led to concepts and methods that have made it possible to analyze behavior in a more fundamental way than previously. Most important have been improvements in the physiological, biochemical, and morphological techniques used to study individual nerve cells and the interaction between groups of cells. As a result, a number of neurobiologists have been encouraged to apply these techniques to the study of behavior and the modifications produced by learning. For this purpose particular experimental animals are needed, namely, species with interesting behavior but with nervous systems simple enough to be studied effectively with cellular techniques. The search for suitable experimental animals has been guided by developments in comparative psychology. From these we have learned that despite great variation in the capabilities of different animals the basic response patterns essential for survival—feeding, defense, and sexual behavior—are universal. Aspects of these behavioral responses are often quite similar in simple invertebrate and complex vertebrate animals. Even behavioral modification, at least in its elementary forms, seems to follow similar principles in widely different species. As a result, studies of invertebrate animals can be used to further our understanding of the behavior of vertebrates.

The nervous systems of certain higher invertebrates, such as crayfish, leeches, locusts, roaches, and snails, are particularly amenable to cellular analysis. Compared with the billions of neurons in the brains of vertebrates, the cells of the central nervous systems of these invertebrates are large in size and few in number, perhaps 10,000 to 100,000. It is therefore possible to trace *at the level of individual cells* not only the sensory information entering the nervous system and the motor action emerging from it, but also the entire sequence of intervening events that underlie a behavioral response. By combining psychological techniques that demonstrate the behavioral capabilities of these simple animals with cellular techniques that analyze the biological mechanisms, it is becoming possible to clarify some relationships between behavior and its neural mechanisms. Although we are still far from explaining the complex behavior and learning of vertebrates, we are beginning to understand elementary forms of behavior and behavior modification in invertebrates that may have parallels in vertebrates.

Because recent advances in the understanding of the cellular mechanisms of behavior are based both upon cellular neurobiology and comparative psychology, effective work in this area requires familiarity with both disciplines. These fields represent two aspects of a common study and their convergence marks the emergence of a more fruitful science.

An effective approach to the cellular biology of behavior can best be achieved by concentrating on a specific organism. As early as 1880, T. H. Huxley pointed out how much can be learned about biology by studying one animal in depth. In the preface to his book *The Crayfish: An Introduction to the Study of Zoology* (pp. v–vi) Huxley wrote:

> In writing a book about Crayfish it has not been my intention to compose a zoological monograph on that group of animals . . . nor has it been my ambition to write a treatise upon our English crayfish. . . .
>
> What I have had in view is a much humbler, though perhaps in the present state of science, not less useful object. I have desired, in fact, to show how the careful study of one of the common and most insignificant animals leads up, step by step, from everyday knowledge to the wide generalization and the most difficult problems of zoology; and indeed, of biological science in general.

Here Huxley prescribes an antidote to the opportunism of biological research. Initially at least, problems in a given area of biology must be examined in whatever animals are most suitable to the methodologies or research tools of the time. As in other areas of biology, the key principles of neurobiology have been derived from work on a variety of organisms: action-potential generation from studies of the squid giant axon, synaptic transmission from studies of various vertebrate and invertebrate nerve-muscle synapses, visual integration from studies of the cat cortex, and motor coordination from studies of arthropods and monkeys. This approach leads to a rapid discovery of basic ideas. But, in certain areas, it leaves in its wake a mosaic of knowledge—a collection of disparate findings that cannot readily be pieced together. This fragmentation of knowledge is evident today in the neurobiology of behavior. For instance, much of what is known about learning in invertebrates comes from studies of octopods, but little is known about the cellular physiology of the octopod brain. On the other hand, the behavior of invertebrates whose brains have been explored at the cellular level, such as crayfish, leeches, and opisthobranch molluscs, has until recently been ignored. Even when behavioral and cellular approaches have been combined, different animals have been used to study different behaviors, so that escape has been analyzed in one, feeding in another, sexual behavior in still another. The result is that the neural controls underlying one behavior get examined independently of the controls underlying other behaviors, and the relationships *between* behavioral responses are not examined.

In other areas of biology the fragmentation of research has sometimes been overcome by viewing a given problem from the perspective of a single organism or group of closely related organisms. This was accomplished for classical transmissional genetics in the 1920s by focusing on *Drosophila,* and for molecular genetics in the 1950s by concentrating on bacteriophages. I hope this book will encourage the development of a similar holistic approach to the study of the neurobiological mechanisms of behavior.

Aplysia has a central nervous system consisting of only nine central ganglia. One of these, the abdominal ganglion, offers many technical advantages for cellular studies. Some nerve cell bodies in this ganglion are very large, reaching a diameter of almost one millimeter, a volume of up to 70×10^{-9} liter, and a protein content of 10 μg. Thus the biophysical properties of these cells, their interconnections, and cellular biochemistry can be studied in detail. In addition, some cells have characteristic features and functions that allow them to be recognized reliably in every individual of the species. The ability to identify cells as unique individuals makes it possible to pinpoint the cells involved in a specific behavior and to determine the exact quantitative contribution of each of these cells to the behavior. Because the nerve cells are large, the cellular components of the behavior of this animal can be analyzed with a variety of techniques: physiological, biochemical, and morphological. Furthermore, although the abdominal ganglion contains only 2,000 cells, it carries out many of the same functions as a brain. It generates a variety of behavioral responses, involving all four classes of effector systems controlled by the nervous system:

somatic-motor, visceral-motor, neuroglandular, and neuroendocrine. This makes it possible to examine the ways in which different behavioral responses are integrated.

Some of the advantages offered by *Aplysia* are shared by other animals, and the work on *Aplysia* has been greatly influenced by parallel studies of other opisthobranch molluscs and non-molluscan invertebrates, as well as by work on vertebrates. Whenever appropriate I have situated studies of *Aplysia* within the larger neurobiological and behavioral context from which they derive.

Many of the ideas developed in this book have been sharpened in discussions with the several colleagues with whom it has been my good fortune to work for the past 10 years in a common research group. We first joined forces at the New York University School of Medicine, and more recently have continued to work together as the Division of Neurobiology and Behavior at the College of Physicians and Surgeons, Columbia University. In particular, I have learned much from a long and rewarding association with Alden Spencer, with whom I collaborated on my first serious research effort 18 years ago; from James Schwartz, a friend from college and medical school days, and with whom I have had the privilege of collaborating recently; and from Irving Kupfermann, a former student who has certainly taught me as much as I have taught him. Each of them combines a fine sense of his own discipline—mammalian neurophysiology, neurochemistry, and psychology, respectively—with a broad interest in other areas of neural science. I am fortunate to have matured scientifically in their company. I was also fortunate to have begun my work on *Aplysia* by collaborating with Ladislav Tauc, one of the pioneers in work on this organism. Spending the year 1962–63 in his laboratory was an unparalleled introduction to research on *Aplysia*.

I have benefited greatly from detailed comments on the manuscript by John Koester and three other young colleagues in our group, John Byrne, Vincent Castellucci, and Thomas Carew, as well as from the pointed questions raised by two graduate students, Wayne Hening and Abraham Susswein. The manuscript was also read by Mitchell Glickstein, Richard Thompson, Jeffrey Wine, and Philip Zeigler; individual chapters were read by Michael V. L. Bennett and Gerald Fischbach. Each of them spared me errors in fact and helped clarify numerous ambiguities.

The normally arduous task of putting a book together was considerably lightened by Kathrin Hilten, who prepared the preliminary versions of all the illustrations with her usual skill, thoughtfulness, and good cheer, and by Sally Muir, who edited and typed the manuscript and, with the help of Penny Bailey, proofread the galleys. The final work was edited by Howard Beckman at W. H. Freeman and Company with a fine sense for the logic of argument and clarity of expression, both of which he demanded of me, sometimes unsuccessfully. As in my other attempts at writing, I have enjoyed the advice of Denise Kandel. Her humorous comments on the text and its author helped give both a more balanced perspective.

The original research described in this book was supported by a Research Scientist Award from the National Institute of Mental Health, and by research grants MH-19795 from the National Institute of Mental Health and NS-09631 from the National Institute of Neurological and Communicative Disorders and Stroke. I am grateful for this continued support.

May, 1976   *Eric R. Kandel*
                                                        COLUMBIA UNIVERSITY

Cellular Basis of Behavior

PART I

STRATEGIES IN
THE STUDY
OF BEHAVIOR

Psychology and neurobiology now share a common interest in the study of behavior. Independent progress in both fields had led each to the point where it can profit from advances in the other. For example, attempts by psychologists to classify behavior into basic units, such as *reflex patterns* and *fixed-action patterns*, have been confounded by the finding that many types of behavior do not fit exclusively into one or the other category. A detailed knowledge of the neuronal architecture of different behavioral responses may provide the criteria for a better classification. Biological studies can also clarify some previously intractable questions in the psychology of learning: Where and how is memory stored? Do short- and long-term memory represent different processes, with different loci and different mechanisms, or phases of a single process? How do the mechanisms of simple nonassociative learning compare to those of more complex associative learning? These questions, central to psychology, cannot be solved with behavioral techniques alone; they require a combined behavioral and neurobiological approach.

At the same time that psychologists are turning to neurobiology, neurobiologists are turning to the study of behavior. Many neurobiologists are now looking beyond the function of individual cells, which they are beginning to understand, to the analysis of systems of cells. A particularly attractive system is the interconnected group of cells that mediates a complete behavioral act.

At first thought, the prospect of studying a behavioral act on the cellular level seems awesome. The brain of man is made up of 1,000,000,000,000 cells, and the connections between these cells are many times more numerous still. To specify the complete neural circuit mediating a behavior in such a complex brain is difficult. Fortunately, this task can be simplified in a number of ways. One simplification is to select for study very simple types of behavior; another is to use higher invertebrates, such as molluscs, worms, crayfish, and lobsters. The central nervous systems of higher invertebrates contain only 100,000 cells, so that interconnections can be examined more thoroughly than in vertebrates. In addition, nerve cells in invertebrates are often large, which makes them particularly suitable for cellular analyses.

Part I introduces the reader to a strategy for studying the cellular mechanisms of behavior based on the use of these simple animal systems. The first chapter delineates the advances in psychology that have led to the current interest in and need for neural analyses of behavior. Chapter 2 considers why neural analyses of behavior are most likely to be informative if carried out on the cellular level. Chapter 3 reviews the advantages of certain invertebrates for a cellular biological approach to behavior and of other invertebrates for a genetic approach. The differences between the two approaches and their respective technical requirements are outlined. This part of the book concludes with a brief chapter on *Aplysia,* an invertebrate animal that has been particularly useful for a cellular biological study of behavior.

CHAPTER 1

The Study of Behavior:
The Interface Between Psychology
and Biology

In the last century psychology and biology have been characterized by alternate movements toward and away from each other. The first convergence of the two disciplines dates to the work of Charles Darwin in the middle of the nineteenth century. Darwin's discovery of a behavioral continuum between animals and man suggested to him that human behavior might profitably be studied by examining its simpler analogs in lower forms of animals. This notion stimulated the comparative study of behavior and led to the development of animal models designed to relate the nervous system to behavior. This new perspective brought about the acceptance among psychologists of the idea that all behavior, even the most complex, derives from the nervous system. Acceptance of this idea was exemplified in William James's *Principles of Psychology,* published in 1890. James's influential book generated a widespread optimism about the capability of neurological science to relate brain mechanisms to behavior. However, this optimism proved premature. Thirty years later studies of brain function had succeeded only in localizing sensory and motor functions in various regions of the brain. And in the period 1925 to 1955 even those localizations were challenged (for example see Lashley, 1929). To many, the neural mechanisms of behavior seemed intractable to empirical analysis, and experimentation was often replaced by elaborate theorizing. As a result, psychology and the neural sciences drifted apart.

Rejection of neural models extended to many schools of psychology, from introspectionism, based on verbal reports of subjective experience, to behaviorism, based on objective description of observable acts. For example, when Sigmund Freud (1895) first explored the relation of unconscious psychological processes to behavior at the beginning of this century, he adopted a neural model of behavior, but soon abandoned this model for a purely mentalistic one. Thirty years later B. F. Skinner rejected neurological theorizing in favor of empirical behaviorism when he began his studies of conditioning.

For a time this separation was healthy for psychology. It permitted the development of systematic descriptions of behavior that were not contingent on still vague correlations with neural mechanisms. Skinner argued that separation of the study of behavior from the neural sciences was in fact scientifically necessary.

> What is generally not understood by those interested in establishing neurological bases [of behavior] is that a rigorous description at the level of behavior is necessary for the demonstration of a neurological correlate. The discovery of neurological facts may proceed independently of a science of behavior if the facts are directly observed as structural and functional changes in tissue; but before such a fact may be shown to account for a change in behavior, both must be quantitatively described and shown to correspond in all their properties.[1]

But the hope persisted that biological studies of the brain would someday allow a new level of insight into behavior. This goal is now within reach. As a result of progress in the last two decades, the nervous system can now be studied at the cellular level. Cellular studies have yielded remarkable insights about the functioning and interconnection of nerve cells. These studies are beginning to show how various regions of the brain, including the cerebral cortex of mammals, are organized. In so doing, cellular studies have resolved conceptual problems that eluded earlier investigators using more global techniques. The cellular approach therefore provides a completely new basis for a biological study of behavioral mechanisms.

In this chapter I shall outline some of the recent history of experimental psychology and delineate the problems and forces that have fostered a new rapprochement between psychology and biology. This historical review will also provide the background in psychology we shall need in later chapters.

DEVELOPMENT OF BEHAVIORAL PARADIGMS

The application of biological techniques to the study of behavior has its roots in the nineteenth century, particularly in the later writings of Charles Darwin

[1]B. F. Skinner, 1938, p. 422.

(1860, 1871, 1873). Darwin appreciated that behavior has a biological basis and argued that certain types of behavior—instincts—are inherited and therefore subject to the same pressures of natural selection as morphological characters (Darwin, 1884). In his chapter on "The comparisons of the mental powers of man and the lower animals" in *The Descent of Man* (1871), and again in *The Expression of Emotions in Man and Animals* (1873), Darwin argued that since man evolved from lower animals, his behavior must have parallels in these lower forms.

> The difference in mind between man and the higher animals, great as it is, certainly is one of degree and not of kind. We have seen that the senses and intuitions, the various emotions and faculties, such as love, memory, attention, curiosity, imitation, reason, etc., of which man boasts, may be found in an incipient, or even sometimes in a well-developed condition in lower animals.[2]

This radical position, emphasizing behavioral as well as structural continuity within and between species, helped dissociate psychology from philosophy and promote contact with experimental biology. Darwin's writings directly led to two major developments that have since characterized experimental psychology: experimental study of learning and naturalistic study of behavior.

Experimental Study of Learned Behavior

Ten years after the publication of *The Expression of Emotions in Man and Animals,* Darwin's colleague zoologist George Romanes published *Animal Intelligence* (1883), the first modern text on comparative psychology. In this work and in *Mental Evolution in Man* (1888) Romanes sought to support the controversial notion of evolution by demonstrating similarities in the behavior of man and other animals. Although Romanes did some experimental work, his approach was basically anecdotal. He assumed (as did Darwin at times) that the experience of all animals was sufficiently like man's to permit all behavior to be described in subjective terms. Therefore, on the basis of introspection and empathy, he ascribed feeling, reason, and free will to the actions of invertebrates.

> If we observe an ant or a bee apparently exhibiting sympathy or rage, we must either conclude that some psychological state resembling that of sympathy or rage is present, or else refuse to think about the subject at all; from the observable facts there is no other inference open.[3]

[2]C. Darwin, 1871, p. 126.
[3]G. J. Romanes, 1883, p. 10.

Within three decades of Darwin's work—and due in no small part to Romanes's efforts at developing interest in animal psychology—the introspectionist tradition was replaced by a more objective approach. One of the first challenges to anthropomorphism came from Jean Henri Fabre (1897–1907).[4] Fabre rejected the arbitrariness of subjective interpretations of behavior and substituted careful observations. By this means he discovered the stereotypic sequences of insect behavior later called *fixed-action patterns.* Impressed with the rigidity of insect behavior, Fabre conceived of them as mindless biological machines. Fabre's approach, followed by C. Lloyd Morgan (1894), John Lubbock (1882, 1888), Jacques Loeb (1918), Max Verworn (1889), Herbert Spencer Jennings (1906), and John B. Watson (1913), was to dominate modern animal psychology and eventually human psychology. The new outlook was expressed best by H. S. Jennings (1906) as he faced the problem that had previously confronted Romanes and the introspectionists: the study of behavior in lower animals.

In describing the behavior of lower organisms we have used in the present work, so far as possible, objective terms—those having no implication of psychic or subjective qualities. We have looked at organisms as masses of matter, and have attempted to determine the laws of their movements. In ourselves we find movements and reactions resembling in some respect those of the lower organisms. But in ourselves there is the very interesting additional fact that these movements, reactions, and physiological states are often accompanied by subjective states—states of consciousness. . . .

The peculiarity of subjective states is that they can be perceived only by the one person directly experiencing them—by the subject. . . . But observation and experiment are the only direct means of studying behavior in the lower organisms. We can reason concerning their behavior, and through reasoning by analogy we may perhaps conclude that they also have conscious states. But reasoning by analogy, when it is afterward tested by observation and experiment, has often shown itself fallacious, so that where it cannot be tested, we must distrust its conclusiveness. Moreover, in different men it leads to different conclusions, so that it does not result in admitted certainty. Hence it seems important to keep the results of observation and experiment distinct from those of reasoning by analogy, so that we may know what is really established. . . .

While this exclusive use of objective terms has great advantages, it has one possible disadvantage. It seems to make an absolute gulf between the behavior of the lower organisms on the one hand, and that of man and higher animals on the other. . . .

Does such a gulf actually exist, or does it lie only in our manner of speech? We can best get evidence on this question by comparing the objective features

[4]For selected translations see Fabre, 1921.

of behavior in lower and in higher organisms. In any animal outside of man, and even in man outside of self, the existence of perception, choice, desire, memory, emotion, intelligence, reasoning, etc., is judged from certain objective facts—certain things which the organisms do. Do we find in the lower organisms objective phenomena of a similar character, so that the same psychic names would be applied to them if found in higher organisms? Do the objective factors in the behavior of lower organisms follow laws that are similar to the laws of psychic states? Only by comparing the objective factors can we determine whether there is continuity or a gulf between the behavior of lower and higher organisms (including man), for it is only these factors that we know.[5]

Loeb and Watson were the most radical advocates of an objective psychology. Loeb, a botanist by training, regarded behavior as a series of highly stereotypic, almost mechanical movements, comparable to the tropisms of plants. For Loeb, behavior was a series of responses directly elicited and controlled by sensory stimuli (Loeb, 1918).

Watson, a student of Jennings, was equally committed to objectifying psychology. The introspectionists had often placed little value on overt behavioral responses, seeing them only as a means of gaining insight into internal, conscious states. Watson rejected internal states and claimed that behavioral responses were the actual, and only, data of psychology. In his famous manifesto of 1913, *Psychology as the Behaviorist Views It*, Watson argued that psychology should be redefined and restricted to the study of observable behavior.

> I believe I can write a psychology . . . and never go back upon our definition: never use the terms consciousness, mental states, mind, content, introspectively verifiable, imagery, and the like. . . . It can be done in terms of stimulus and response; in terms of habit formation, habit integrations and the like.
>
> In a system of psychology completely worked out, given the response, the stimuli can be predicted; given the stimuli, the response can be predicted.[6]

The behaviorist approach, first developed by Jennings, was dramatically extended by Edward L. Thorndike (1898) and Ivan Pavlov (1906, 1927), who developed two model systems, or paradigms, for examining the learning abilities of animals: *instrumental* and *classical conditioning*. Under the influence of Thorndike and Pavlov, psychologists became less concerned with behavior and more concerned with learning, or how behavior is modified. Most researchers in the area of learning lost interest in comparing the learning capabilities of different species and focused their attention instead on demonstrating

[5] pp. 328–335.
[6] p. 167.

general laws of learning in selected species. These two changes of focus proved critical for the development of modern psychology, particularly in America.

The systematic analysis of learning dates from important contributions by Thorndike (1898), who devised the first reliable techniques for measuring learning in animals. A hungry cat was placed in a "problem box" and food was placed outside the box. The cat could get to the food only by pulling a latch that opened a door. Thorndike noted that the cat made a variety of movements and responses before pulling the latch; this he called *trial-and-error learning.* Thorndike found that after several trials in the box the cat made fewer erroneous responses before finding the latch, so that with each succeeding trial there was a gradual decrease in the time it took to get the food. This decrease in time provided a measure of learning (Figure 1-1).

Thorndike also discovered the importance of *reinforcement.* He found that learning occurred more quickly if only correct, rather than all, performance was rewarded. This finding, which Thorndike called the *law of effect,* is the basis of the instrumental (operant) conditioning paradigm, according to which an animal's behavior is instrumental in producing a reward. In addition, Thorndike and Woodworth (1901a, b, c) illustrated *transfer of learning,* in which training in one skill enhances performance in another.

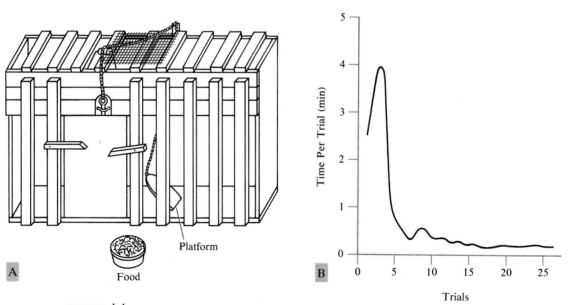

FIGURE 1-1

One of Thorndike's several problem boxes **(A)** and a mean learning curve obtained from 13 cats **(B).** The cat must depress the platform to raise a bolt and then pull one of the two latches to open the door. [Modified from Thorndike, 1898.]

Another experimental approach to learning was developed independently by Pavlov. While studying digestive reflexes of dogs, Pavlov noted that salivation could be triggered by an apparently indirect stimulus, for example, the sight of an approaching attendant. A major requirement for this type of learning was that the otherwise ineffective or *conditioned stimulus* (CS), the research attendant, be associated or paired with an effective or *unconditioned stimulus* (US) such as food. After repeated pairing within a definite time sequence, Pavlov found that the conditioned stimulus was able to elicit a *conditioned response* (CR). If after conditioning had occurred the unconditioned stimulus was withheld, the conditioned stimulus alone elicited the response. However, repeated presentation of the conditioned stimulus without the unconditioned stimulus led to extinction, the gradual decline of the ability of the conditioned stimulus to elicit the conditioned response (Pavlov, 1906, 1927).

Pavlov's work thus led to a second important learning paradigm, *classical conditioning*. Like Thorndike's instrumental conditioning, classical conditioning involved the association of two events. But whereas in instrumental conditioning the association is between an animal's response and the reinforcing stimulus, in classical conditioning the association is between two stimuli: the conditioned and the unconditioned (reinforcing) stimulus. Unlike instrumental conditioning, in which an animal's correct response leads directly to the presentation of a stimulus (reinforcement), in classical conditioning the experimenter presents the animal with a second, unconditioned stimulus at some fixed interval after the presentation of the conditioned stimulus. The presentation of the US in classical conditioning is therefore not linked to a response; conditioning occurs even when the animal's unconditioned response is prevented by temporary muscular paralysis. Also, the nature of reinforcement can differ in classical and instrumental conditioning even when the same unconditioned stimulus is used. For example, with shock as an aversive (negative) reinforcing stimulus, classically conditioned responses, such as heart rate, are reinforced by the occurrence of the shock, but instrumentally conditioned responses, such as escape or avoidance, are reinforced by the termination of the shock. There are also differences in extinction. Instrumentally conditioned responses resist extinction if during training the reinforcing stimulus is withheld on some trials (partial reinforcement); classically conditioned responses do not show this effect. Finally, classical conditioning is generally more effective for conditioning visceral responses than is instrumental conditioning. These and other distinctions have led Miller and Konorski (1928), Skinner (1938), and others to propose a two-process learning theory based on the idea that each paradigm produces a different type of learning.[7]

---

[7]For a recent review see Rescorla and Solomon, 1967.

Other psychologists have argued that the distinctions between the two types of learning are superficial compared to their common features. Both give rise to conditioning as a result of a specific association and in each case extinction occurs when reinforcement is withheld. In addition, the stimuli that can serve as reinforcement in instrumental conditioning (such as food, water, or shock) can also serve as unconditioned stimuli in classical conditioning. Also, it is often difficult to arrange a stimulus condition for the one type of conditioning without the possibility that the other may occur.[8] The similarity between classical and instrumental conditioning and the difficulty in disentangling them experimentally have led some theorists to assert that the two paradigms do not represent two types of learning, but different aspects of a common learning process.

Unfortunately, the controversy between theorists advocating two learning processes and those advocating only one cannot be readily resolved because the relationships between classical and operant conditioning are difficult to analyze with behavioral techniques. As we will see later the inability to distinguish behaviorally between one-process and two-process theories is also found in other learning paradigms.

The independent discovery of two interrelated learning paradigms based on a common principle of association had an extraordinary impact on psychology. Instrumental and classical conditioning offered reliable rules expressed in simple paradigmatic forms for altering behavior and thereby established a basis for a completely mechanistic explanation of acquired behavior. Moreover, the principle of *learning by association,* on which the paradigms were based, seemed inherently plausible because it has long been familiar to Western

---

[8]For example, in instrumental conditioning the experimenter often does specifically present a stimulus (comparable to the CS) in developing an association between response and reinforcement. Yet various stimuli, called discriminative stimuli, are always present in the environment (the lever, the key, or the latch that the animal has to work to gain reward). An invariant discriminative stimulus can gain control over the instrumentally conditioned response much as the invariant CS gains control over the classically conditioned response. Both paradigms therefore share a common sequence: Stimulus$_1$ (a CS or instrumental discriminative stimulus) followed by Response (CR or instrumentally conditioned response) followed by Stimulus$_2$ (US or instrumental reinforcing stimulus). In most classical conditioning experiments the CR can enhance or modify the reinforcing properties of the US. The pairing of a light (CS) with food (US) will give rise to classical conditioning, that is, the light will elicit salivation, a response previously given only to the food. But the salivation also affects the US: it makes the food taste better and easier to digest. Salivation therefore serves as a potential instrumental response for a reinforcing stimulus, food. Each time the animal salivates, it is reinforced with food. As a result, in each instance of classical conditioning there is a possibility for instrumental conditioning (see Terrace, 1973).

thought (Herrnstein and Boring, 1965). Aristotle had already suggested that memory (the retention and recall of learned information) requires the association of ideas. This notion was elaborated and formalized by Locke and the British empiricists, the forerunners of modern psychology. The classical and instrumental paradigms went one important step further, however. They combined the older concept of learning by association with the modern concept of *reflex act*. As described by Sherrington (1906), the reflex act is largely shaped by environmental contingencies, such as the intensity and pattern of sensory stimuli. Following the early behaviorists, Thorndike and Pavlov rejected the unobservable psychological construct of 'idea' in favor of an observable behavioral construct, the reflex act. This marked a decisive shift toward a behaviorist concept of learning. To Thorndike and Pavlov the stuff to be learned was no longer the association between ideas but that between stimuli and behavior. This shift made the study of learning amenable to experimental analysis; responses could be measured objectively and the parameters relating a stimulus to a response could be specified and even modified.

Pavlov (1927), and later Grether (1938), also described two nonassociative behavioral modifications: habituation and sensitization. *Habituation* refers to a decrease in the amplitude of a progressive reflex response when the response is repeatedly elicited. When an animal is exposed to a novel or unanticipated stimulus, it reacts to it with a characteristic set of somatic and autonomic responses called the orienting reflex. For example, the somatic response to a novel sound consists of a turning of the animal's head to the source of the stimulus; the autonomic response consists of a change in heart and respiratory rate that enhances the animal's readiness to act. When the stimulus is presented repeatedly the animal gradually stops responding. Responsiveness returns only after some time has elapsed in which the stimulus is not presented. Because of its similarity to the extinction of a conditioned response in the absence of reinforcement, Pavlov called the decrease in the unconditioned reflex response *extinction*. Earlier, however, the Peckhams (1887) had found a decrease in the escape response of spiders following repeated vibratory stimuli unrelated to classical conditioning. Pieron (1913), Dodge (1923a, b), Humphrey (1933), and Harris (1943) found a similar decrease in a variety of defensive reflexes. To distinguish this type of decrease in reflex strength from the extinction of classical conditioning, Dodge and Humphrey referred to it as *habituation*. These workers also found that a sudden or strong stimulus immediately restored the reflex responsiveness of habituated responses. This restoration they called "dehabituation," now called *dishabituation*. Subsequently, other features were added and the paradigm was refined (see Chapter 11, pp. 518–519, and Chapter 12, pp. 538–541).

Sensitization refers to an enhancement of the response to a stimulus as a result of the presentation of another (usually strong or noxious) stimulus where enhancement does not depend on pairing of the stimuli. A form of sensitization (pseudoconditioning) was first discovered by Grether (1938). Following a series of strong unconditioned stimuli presented alone, another (previously ineffective) stimulus was able to produce a response similar to the unconditioned response to the strong unconditioned stimulus (for details see Table 11-2 and pp. 496–498).[9]

Because habituation and sensitization were not dependent on specific associations between stimuli, psychologists treated them as adventitious and confounding variables of classical conditioning that were basically much less interesting and important than associative learning. Consequently, nonassociative paradigms were not systematically explored. By contrast, associative learning rapidly found widespread experimental use and provided the foundation for popular theories of learning. Psychologists were impressed with the associative-learning paradigms because they led to general laws about conditions under which environmental stimuli modify behavior. Experimental arrangements were purposely designed to be unnatural and to reduce the effectiveness of any previous experience or biological capabilities of test animals. The very arbitrariness of the experimental arrangements used by Pavlov and Thorndike seemed to guarantee the generality of the learning principles that emerged. Associative learning was thought to be so powerful that learning could occur with any combination of variables. Thus, in 1928 Pavlov wrote:

> If our hypothesis as to the origin of the conditioned reflex is correct, it follows that any natural phenomenon chosen at will may be converted into a conditioned stimulus . . . any visual stimulus, any desired sound, any odor, and the stimulation of any part of the skin, whether by mechanical means or by the application of heat or cold. . . .[10]

---

[9]Two formally similar processes, nonassociative classical conditioning (pseudo-conditioning) and nonassociative alpha conditioning, are often both subsumed under sensitization. The distinction between them depends upon whether the facilitation brings about a new response to a CS that was not present before the presentation of the US (pseudoconditioning) or whether the US enhances a previously existing (alpha) response to CS (alpha conditioning). Both processes are nonassociative (quasiconditioning) in not being dependent on a specific pairing sequence, but they can often be distinguished behaviorally (see Hilgard and Marquis, 1940; and Table 11-2 here).

[10]Pavlov, 1928, p. 86.

Believing that there was basically only one variety of learning, associative conditioning, psychologists felt justified in thinking that further analysis of associative conditioning would lead to an understanding of most variants of learning in all animals, including man.

This position was enunciated by Edward Tolman in his presidential address to the American Psychological Association in 1938:

> Let me close now, with a final confession of faith. I believe that everything important in psychology (except perhaps such matters as the building up of a super-ego, that is everything save such matters as involve society and words) can be investigated in essence through the continued experimental and theoretical analysis of the determiners of rat behavior at a choice point in the maze.[11]

During the period from 1920 to 1960 the heavy emphasis in American psychology on associative learning led to a somewhat restricted view of behavior. Also, there was an almost exclusive reliance on three laboratory animals: the rat, pigeon, and dog. Although useful in focusing early explorations of the associative-learning paradigms, in the long run this narrow focus led to a neglect of comparative psychology (Beach, 1950). There were exceptions to this trend. Theodore Schneirla and his students at the American Museum of Natural History emphasized differences in the level of organization of animals at different phylogenetic points.[12] Schneirla found that different organisms solve the same learning task in different ways. Although the ant and the rat can both learn to run a maze with surprisingly similar learning curves (Figure 1-2), the behavioral processes of the two are different. The ant learns the maze slowly, by stages. It first learns individual choices at each point in the maze and only gradually, in later runs, integrates the knowledge of each correct decision. The rat integrates the learning of choice points on the very first run and soon anticipates distant parts of the maze. Moreover, the rat learns how to learn. In learning to run one maze the rat also learns how to run other mazes better. For the insect each maze is a completely new problem; there is no transfer of learning from one maze to another.

But despite the occasional voice in American psychology calling for comparative studies of behavior, the major impetus for moving behavior off the narrow track of learning curves came from outside psychology—from the vigorous European tradition of natural history.

---

[11]Page 34.
[12]See Schneirla, 1934, 1953a, b, c; and Beach, 1950.

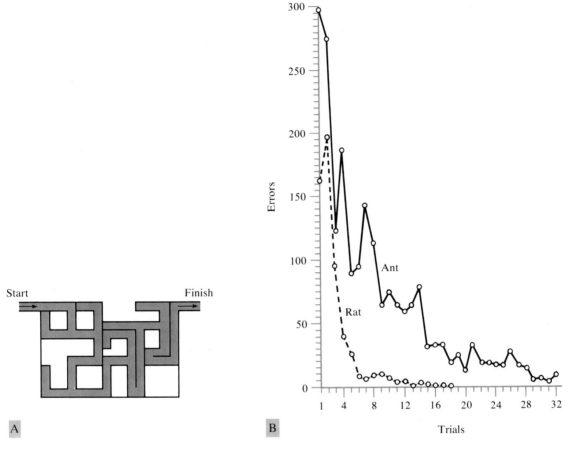

FIGURE 1-2.
The maze shown was used by Theodore Schneirla to compare the learning capabilities of the ant and the rat; the graph compares the progress of the two. Both species achieved a high level of performance, but the rat learned the maze more rapidly than the ant. [From Schneirla, 1953a.]

Naturalistic Study of Instinctive Behavior

Among animals in nature, stimulus-guided, modifiable reflex acts account for only a part of behavior. Animals also show purposeful responses that are less obviously stimulus directed and may even be generated in the absence of an evident directing stimulus. Such behavior is called instinctive, and includes stereotypic motor responses found in every member of the same sex in a species. Many of these responses are not readily modified by experience. Instinctive behavior equips an animal with adaptive responses that often appear complete at their first elicitation. A honey-bee, for example, inherits not only structures—wings and wing muscles for flight—it also inherits the behavioral capability to seek out flowers and collect pollen and nectar. This class of behavior is not learned, and is clearly an advantage for animals with short life-spans and little or no postnatal parental care (Manning, 1972).

The distinction between reflexive and instinctive behavior was defined by the ethologists, a group of biologists interested in comparative studies and in the classification of behavior into basic units.[13] Jennings had already appreciated instinctive behavior in his dispute with the more mechanistically minded Loeb. Loeb often worked with sessile animals and emphasized the importance of stimuli in the control of behavior. Jennings (1906) argued that stimuli have only limited control. Even in simple animals the same stimulus can produce a variety of responses in different individuals because physiological states vary considerably. Moreover, a given action is elicited by only a few of the many stimuli that an animal's sense organs may perceive at any given time. Each animal, Jennings argued, responds to certain *representative stimuli.*

> This reaction to representative stimuli is evidently of the greatest value from the biological standpoint. It enables organisms to flee from injury even before the injury occurs or to go toward a beneficial agent that is at a distance. Such reactions reach an immense development in higher animals.

> Most of our own reactions, for example, are to such representative stimuli. Only as we react to actual physical pain or pleasure do we share with lower organisms the fundamental reaction to direct injury or benefit. Practically all our reactions to things seen or heard are such reactions to representative stimuli. While such behavior plays a much larger part in higher than in lower organisms, the existence of reactions to representative stimuli even in the lower organisms considered in the present work is an evident fact.[14]

Jacob von Uexküll, a contemporary of Jennings, also emphasized that behavior is a response to only a fraction of all perceptible stimuli (Uexküll, 1957). Uexküll proposed that the notion of responsiveness implied selection of stimuli; this concept was the origin of the idea of *sign stimulus* (Tinbergen, 1951).

Jennings also stressed that to study behavior it is necessary to understand the *action systems* of an organism, its coordinated way of acting. This view was

---

[13]As used by ethologists, the term "instinctive behavior" is limited to highly stereotypic movements (fixed-action patterns) released by specific stimuli. This term should not be confused with the term "instinct" used at the beginning of the century by psychologists, who, following William James (1890), applied the term to most types of behavior.

[14]Jennings, 1906, p. 297.

the basis for Konrad Lorenz's notion of *fixed-action pattern,* which describes an animal's relatively stereotyped and species-specific motor responses (Lorenz, 1937, 1950; Lorenz and Tinbergen, 1938).

In focusing on apparently unlearned responses, Jennings was greatly influenced by the work of two naturalists, Charles Whitman (1899, 1919) of the University of Chicago and Oskar Heinroth (1911) of the Berlin Aquarium, the two founders of modern ethology. Whitman and Heinroth independently discovered that the sterotypic responses of birds could be used as taxonomic criteria for phylogenetic classification, just as comparative anatomists use morphological characters. Whitman, an evolutionary biologist, interbred closely related species of pigeons and studied the inheritance of morphological characters. He found that the nest building and sexual displays of different species varied in predictable ways and were as taxonomically useful as bones, teeth, or feathers. Heinroth reached a similar conclusion while studying ducks. These discoveries drew the attention of biologists to the usefulness of behavioral characters in the study of evolution (see for example Mayr, 1942, 1963; Roe and Simpson, 1958).

The work of Whitman and Heinroth was elaborated by Whitman's student, Wallace Craig, who noted that many complex behavioral sequences included two components: (1) a reflex *steering* (or *appetitive*) component, consisting of variable searching or orienting movements, followed and terminated by (2) a stereotypic, fixed-action *consummatory* component.

Craig (1918) and subsequently Lorenz (1935, 1937, 1965) found that orienting behavior is readily influenced by learning, whereas fixed-action patterns tend to be invariant and less influenced by learning. The finding that many forms of complex behavior have both types of components helped reconcile the disparate views of behavior held by experimental psychologists—who emphasized the variability of reflex behavior and its modifiability by reinforcement—and those held by ethologists—who emphasized the stereotypy of fixed-action patterns. For example, the greylag goose always uses the same technique for retrieving an egg that has rolled out of its nest. The goose puts its bill beyond the egg and makes side-to-side movements of its head to center the bill on the egg; it then rolls the egg back under its breast by drawing in its head. The variable side-to-side orienting movements of the head are reflexively determined by and continually adjusted to the size and shape of the stimulus (the egg), but the retrieval movement is an all-or-none fixed act and is independent of the size of the stimulus. The bird uses the identical motion to retrieve a giant egg, a small light cube, or a normal-sized egg. Although retrieval is fairly efficient for the greylag goose, which has a broad-based bill, the same technique is used by other ground-nesting birds with narrow bills, such as gulls. These animals have difficulty retrieving eggs with their bills because the egg tends to slip away; retrieval would be simpler were they to use their wings or their webbed feet instead of their bill. But birds show no ability to modify this fixed act or substitute another for it (Tinbergen, 1951).

The discovery of fixed-action patterns is important because it defines a new unit of behavior. In addition, the distinction between reflex patterns and fixed-action patterns suggests that a large body of behavior might be classified into two fundamental units (Lorenz and Tinbergen, 1938). The amplitude of many simple reflex actions, such as head orienting or flexion reflex, tends to be a smoothly graded function of the intensity of the eliciting stimulus. On the other hand, fixed-action patterns, such as head tucking, are usually not directly related to the intensity of the stimulus; the stimulus triggers a relatively stereotypic, all-or-none response. The form, sequence, and intensity of the fixed-action pattern are often relatively independent of the intensity and pattern of the stimulus. The control exerted by stimuli on a reflex pattern is analagous to the control that a driver exerts over the acceleration of a car, whereas the control of stimuli over fixed-action patterns is analogous to the control an operator has on the ringing of a telephone (Morris, 1957). The harder the driver presses the accelerator pedal the faster the car moves, but no matter how urgently the operator places a call, the telephone bell rings at the same rate and intensity.

Moreover, the shape and pattern of stimuli eliciting a reflex pattern tend to be less specific (and sometimes less complex) than that of stimuli eliciting a fixed-action pattern. Finally, reflex patterns invariably require a discernible stimulus, whereas fixed-action patterns can occasionally occur spontaneously in the absence of any evident stimulus, presumably in response to an internal command or trigger. The centering movements of the head in the egg-retrieval response of the greylag goose are due to the irregular rolling of the egg; the goose does not make these movements if the egg is replaced by a cylinder that rolls smoothly. By contrast, once triggered, bill retrieval by the greylag goose continues to completion even if the egg is removed. At the extreme, fixed-action patterns may occur in the absence of any obvious external stimulus, as *vacuum activity.* Thus, birds go through the motion of snapping at and swallowing flies even when flies are absent.

Because expression of a fixed-action pattern is stereotypic and independent of the strength and pattern of the stimulus, Lorenz argued that the sequence of movements is determined by a central nervous system machinery (later called a *central program*) that does not require proprioceptive (or environmental) feedback for maintaining the timing and sequence of the behavior (Lorenz and Tinbergen, 1938; Lorenz, 1965). Because fixed-action patterns are also species-specific, and found in every individual of the same sex in the species, Lorenz further proposed that they are inherited. In reality, most fixed-action patterns are not completely stereotyped. Although the relationship and sequence of their components tend to be invariant, fixed-action patterns often vary in intensity and completeness. Moreover, in some cases components of a fixed-action pattern are not inherited and are dependent on environmental factors for their expression. But despite many exceptions to these criteria and much criticism of fixed-action patterns as discrete behavioral units (see for example, Lehrman, 1953; Hinde, 1970, pp. 19–23; Barlow, 1968) both the basic concept and Lorenz's three corollaries—central program, stereotypy, and genetic determinism—have remained useful (Manning, 1972; Hinde, 1970; see also pp. 346–350, here).

Biological Constraints on Learning

Ethological thinking liberalized the study of behavior by emphasizing its adaptive significance and the importance of unlearned (instinctive) responses. The biological perspective of ethologists also influenced the study of learning. Perhaps the best example of this influence is the recent widespread interest in experimental psychology in new learning paradigms derived from research into the biological advantages and limits of learning.

Ethologists have long argued that an organism brings to any task a specified sensory and motor apparatus that is the product of evolution. These specializations make certain stimuli more readily perceived than others, and certain environmental contingencies more important than others for survival (Seligman and Hager, 1972; Garcia *et al.*, 1974). As a result, an animal learns some tasks readily and others not at all. Ethologists therefore argue that learning cannot be only a general and arbitrary associative process, as experimental psychologists had earlier maintained. Each species is constrained by specific biological realities.

The most instructive examples of biological constraints on learning have come from the analysis of *sickness-induced taste aversion* (also called "bait shy" behavior) by John Garcia and his colleagues (Garcia, McGowan, and Green, 1972; Garcia *et al.*, 1974; see also Barnett, 1963). These remarkable studies have led to a better understanding of the function of reinforcement in classical and instrumental conditioning. Wild rats who have survived a poisoning attempt become "bait shy"; using taste and smell, they avoid poisonous bait that once made them sick, sometimes without avoiding the place where the poisoning occurred. In its outline this phenomenon resembles classical conditioning: a US (the ingested poisoned food) giving rise to a UR (malaise) is paired with a CS (the smell and taste of poisoned food) to produce a CR (avoidance of food). But sickness-induced taste aversion differs from classical conditioning in two ways.

In classical conditioning the optimum interval between the conditioned stimulus and the reinforcement (the CS-US interval) is a fraction of a second. If reinforcement is delayed even slightly, conditioning becomes less effective and ultimately will fail. In sickness-induced taste aversion behavior the aversive consequence (poisoning) can follow the CS (food) by hours. In addition, the learned aversion tends to be specific to the particular taste and smell of the food; it is difficult to elicit aversion for other conditioned stimuli, e.g., auditory or visual stimuli or other foods. Sickness-induced taste aversion suggests that the brain of the rat, a nocturnal animal that relies heavily on taste and smell for finding food, is designed to learn to associate illness and food-related stimuli, such as smell or taste, but not illness and auditory or visual stimuli (Seligman and Hager, 1972; Garcia et al., 1974). By contrast, birds are largely visual feeders. They readily learn to associate visual stimuli, but not taste, with illness (Wilcoxon, Dragoin, and Kral, 1971).

Thus, Garcia's studies show that animals readily learn to associate events that have survival value. Compared to the learning of arbitrary associations, learning of biologically significant associations is easier to achieve and is more resistant to extinction; it also tolerates great delay in reinforcement. Since illness often occurs long after eating, a learning process that tolerates a long delay in reinforcement is a considerable selective advantage (Garcia et al., 1972).

There are also biological constraints on instrumental conditioning. Traditionally it had been thought that birds pecking for grain and rats pressing a bar for food manifested similar instrumental responses. However, Brown and Jenkins (1968) found that these responses are biologically different. Pressing a bar to get food is part of an experimental set up only, and rats will not press when not reinforced with food. But pecking is essential to the pigeon's feeding. Pigeons are biologically so predisposed to peck that under certain experimental conditions (called autoshaping) pecking is continued even when it is not reinforced with food and even when it deprives the animal of grain. Furthermore, instrumental responses to which an animal is predisposed are highly resistant to extinction (Williams and Williams, 1969). Behavior that is maintained in the absence of reinforcement and is resistant to extinction is not the usual result of instrumental conditioning. As with experimental classical conditioning, the selection of a response for experimental instrumental conditioning cannot be arbitrary, as psychologists once thought—the effectiveness of reinforcement depends on the biological significance of the response for the animal being tested.

Types of Learning Not Dependent on Reinforcement

Further exploration of behavior stimulated new interest in three previously discovered learning paradigms: latent learning, sensory preconditioning, and imprinting. Latent learning and sensory preconditioning had been discovered in laboratory studies, and imprinting had been observed by naturalists. But none of these could be easily explained in terms of traditional stimulus-response psychology, since reinforcement did not seem to play a part in the learning process.

Latent learning is learning without an evident reward, as when animals learn about their environment through exploration. Rats allowed to explore a maze for several days without reward later make fewer errors and complete the maze in a shorter time when presented with a food reward than do control animals with no previous exposure to the maze (Blodgett, 1929). A rat can learn the entire maze by exploration without reward; reward merely induces the animal to learn more quickly (Maier, 1932). Latent learning is common in nature and is evident in the exploratory behavior of homing animals (Thorpe, 1943, 1944a, b). Many insects learn the geography of the home area in detail and can repeatedly return to the same spot using the relative position of the sun or nearby landmarks as guideposts (von Frisch, 1950).

Sensory preconditioning is learning in which a sensory stimulus, e.g., a tone, elicits a conditioned response without prior pairing with an unconditioned stimulus (Brogden, 1939). If the tone is first paired with another sensory stimulus, light, which is then paired with an unconditioned stimulus, shock, the tone will prove capable of eliciting a conditioned response even though it was never paired with the shock.

Imprinting, commonly encountered in birds that move about as soon as they are hatched, is an attachment or following response to any prominent moving object, normally the mother, to which the animal is exposed immediately after birth (Heinroth, 1911; Lorenz, 1935; Immelmann, 1972). Attachment is not necessarily to a fellow member of the species; a young goose reared in isolation by a human will follow him. Imprinting illustrates a potentially close relation between development and certain types of learned behavior, suggesting that the two processes may share some common features. Thus, imprinting is rapidly acquired but only during a critical early period in development; once acquired, it generally does not disappear. In some species imprinting is important in establishing the identity of future sex partners. When a bird imprinted upon and reared by a foster mother of another species becomes sexually mature, it will court and attempt to mate with members of the foster mother's species (Immelmann, 1972).

REAPPRAISAL OF THE STUDY OF BEHAVIOR

The competitive claims emanating from experimental and naturalistic studies generated a healthy conflict within psychology and led eventually to a reappraisal of the study of behavior. Generally, experimental psychologists emphasized learning to the neglect of other behavior. They emphasized the role of experiential factors in determining behavior and ignored genetic ones. Finally, learning was equated with associative conditioning, to the exclusion of nonassociative behavioral modification. By equating learning with associative conditioning, psychologists lost sight of the fact, emphasized by the logical positivists, that scientific concepts are merely the agreed upon operations by which phenomena are observed. This point was explained by Bridgman:

> The new attitude toward a concept is entirely different. We may illustrate it by considering the concept of length: what do we mean by the length of an object? We evidently know what we mean by length if we can tell what the length of any and every object is, and for the physicist nothing more is required. To find the length of an object, we have to perform certain operations. The concept of length is therefore fixed when the operations by which length is measured are fixed; that is, the concept of length involves as much and nothing more than a set of operations; the concept is synonymous with the corresponding set of operations.[15]

Early on, psychologists established good operational definitions for instrumental and classical conditioning. The discovery and application of conditioning techniques was a major advance; certain types of learning came under experimental control and some important principles of behavioral modification were derived. However, due partly to the importance placed by Western philosophical thought on learning by association, the paradigms for conditioning eventually came to be regarded as the *only* paradigms for learning (see for example Skinner, 1938; Miller, 1967). Behavioral modifications that fell outside these paradigms were considered not merely something other than learning, but uninteresting and unimportant as well. Generally, psychologists failed to appreciate how arbitrary this exclusion was; they failed to appreciate that the stimulus sequences of nonassociative paradigms produced behavioral modifications that are similar to associative learning. Because of their concern with the specificity of association between stimuli or between stimulus and response, experimental psychologists often ignored other aspects of learning, such as time course and biological importance. Classical conditioning of an

[15]P. Bridgman, 1927, p. 5.

arbitrary response retained for only a few minutes or hours was considered a more interesting example of learning than biologically important instances of sensitization lasting months (for examples of this position see Miller, 1967; Hilgard and Marquis, 1940). This created a factitious hierarchy of values in which special value was attached to demonstrating associative conditioning and little or no value was attached to demonstrating nonassociative behavioral modification.

In contrast to experimental psychologists, ethologists held a more comprehensive view of behavior and learning. They regarded behavior as a family of adaptive processes, each interesting in its own right for the selective advantage it offered to an animal. Ethologists studied the overall behavior patterns of a wide variety of species and in so doing defined new units of behavior (fixed-action patterns). To ethologists, learning was only a subset of an animal's behavior. Even in studying learning, the ethologists cast a wider net. They defined learning broadly, as *any performance altered by experience, exclusive of maturation, fatigue, or injury* (see for example Thorpe, 1956). Ethologists therefore explored habituation, sensitization, and imprinting, and in so doing, discovered new and surprising features. Habituation was thought by many experimental psychologists to be a short-lived and fairly trivial behavior modification. But ethologists found that in nature habituation is often long lasting and can have a critical role in various adaptive processes, including stimulus recognition and selective perception (Hinde, 1954a, b; Thorpe, 1956; and see p. 538ff. here). Finally, ethologists explored several central issues in comparative biology that were ignored by the experimental psychologists, such as the comparative behavior of closely related species, the evolution of behavior, and the role of behavior in speciation.

The influence between ethology and psychology has not been all one way. Because ethological work was carried out largely in the field, it tended to be descriptive rather than experimental, at times relying on anecdotal evidence (Lehrman, 1953). The looseness of methodology in ethology partially justified psychologists in their belief that there was not much scientifically valuable work in ethological studies. Contact with the rigorous methodology of experimental psychology and its concern with scientific evidence improved research in ethology.

INTERVENING VARIABLES, BRAIN PROCESSES, AND BEHAVIOR

The gradual resolution of the conflict between ethology and psychology has introduced new vigor and confidence into psychology. Students of behavior now feel that they have a good descriptive understanding of the several forms of behavior and classes of learning. As a result, both ethologists and behaviorists are increasingly ready to address themselves to a new range of questions concerning mechanisms of behavior: How does a stimulus give rise to one response rather than another? What are the fundamental differences between classes of behavior and classes of learning? Although Watson and the early behaviorists were willing to limit their studies to descriptions of stimulus-response relationships, modern behaviorists, following Edward Chase Tolman (1932), increasingly wanted to know how one behavioral process relates to another. Since these questions reflect internal (often brain) processes, not stimulus-response relationships, Tolman began to speculate about internal processes in behavioral terms. Tolman shared the belief of the early behaviorists that stimuli initiate behavior and that the responses that constitute behavior must be objectively observed and operationally described. But for Tolman observable stimulus-response relations represented only the tip of the iceberg. The actual determinants of behavior, which he thought psychologists should explore, are the internal (and therefore not directly observable) factors interposed between the stimulus and the response. Tolman called these *intervening variables.*

To fill the gaps in their understanding of stimulus-response relationships, psychologists following Tolman invoked a number of intervening variables, such as hunger, thirst, motivation, learning strategies, and memory storage. Intervening variables cannot be observed; they are inferred from stimulus-response relationships. For example, hunger cannot be observed but can be quantitatively related to observable behavior, to the amount of food consumed and how rapidly it is ingested. Thus, an unobserved variable can be defined, quantified, and manipulated if it can be clearly related to a stimulus–response relation. Only by exploring the brain of the organism, however, can the process whereby food deprivation leads to food consumption be directly investigated.

Many intervening variables are designed to represent biological processes; some psychologists have even argued that every intervening variable should be thought of in neurological terms (Krech, 1950). As learning theory became more dependent on these variables the need to explore brain processes increased, because intervening variables simply cannot be studied directly with behavioral techniques. Behavioral techniques alone cannot resolve such questions as: What are the mechanisms of the learning process? Do classical and instrumental conditioning represent the same or different learning processes? Are associative and nonassociative processes related? What is the nature of memory storage? How are short-term and long-term memory related? Are there one, two, or more stages to memory storage?

Even though intervening variables (such as learning processes, memory storage, and motivational state) are often vaguely formulated, they are conceptually useful particularly in the study of human behavior. As a result, many psychologists now believe that the utility and rigor of these constructs would be enhanced if the biological processes they are usually designed to represent were understood. Biological studies are also becoming of great interest to ethologists because they may answer a number of key questions. Do fixed-action patterns always involve central programs in the brain? How do inborn, developmental, and experiential factors interact in determining the final form of a behavior? Do ontogenetic and learning processes share common mechanisms as is suggested by the similarity of imprinting to adult learning? These questions can only be answered by studying the nervous system directly.

This prescription is of course not new. Psychologists never denied the importance of the brain. A segment of psychology, physiological psychology, has continually worked on furthering biological understanding of behavior (see Lashley, 1929). Many who abandoned biology did so not from choice but because they thought that biology was not likely to inform psychology in the near future. Ernest Hilgard pointed out in 1956:

> This relative lack of interest by learning theorists in the nervous system, particularly in the period 1930 to 1950, is again a problem for the historian. . . . Possibly one influence was a certain disillusionment about progress in neuroanatomy and neurophysiology after the breakdown of the localization doctrine under many lines of evidence, including Head's studies of aphasia and Lashley's cortical ablation studies. If nobody really knew anything about how the brain worked, what help could present neurophysiology be to the learning theorists?[16]

If past attempts to explain the neural mechanisms of behavior were premature, what is different now? One difference is the changes that have occurred within psychology. Freed from neurological constraints, psychologists have developed a rigorous science of behavior much as Skinner (1938) predicted. The combined efforts of ethology and experimental psychology have also enlarged the view of behavior and of its ability to be modified. Various combinations of stimuli, responses, and reinforcement have been found that produce reliable, biologically significant effects that can be summarized in a small number of paradigms. Insight into the neural mechanisms of any one of these paradigms now becomes an issue of considerable scientific interest. In addition, the range of animals to which these paradigms apply, and in which they can therefore be explored, has been considerably expanded to include lower vertebrates and invertebrates.

[16]E. Hilgard, 1956, p. 452.

But the key change has occurred within neural science. The development of powerful analytic techniques makes it possible to study the cellular biology of behavior—how different interconnected groups of cells control different behavioral responses. By combining cellular methods with appropriate experimental animals—and there are many of these now—it has become possible to explore the principles underlying different classes of behavior as well as those underlying different types of behavioral modifications. The opportunity therefore exists for exploring a range of previously inaccessible behavioral problems on a mechanistic level.

THE CONVERGENCE OF PSYCHOLOGY AND NEUROBIOLOGY

The new convergence between psychology and neural science has not resulted in a massive fusion of the two fields—that should not be expected and may not even be desirable. What has happened is more modest. Segments of several independent research traditions—the comparative study of behavior, studies of the mechanisms of behavior, and cellular neurobiology—have converged. As a result, some psychologists previously interested in behavior or its evolution are extending their work to the analysis of neural mechanisms. In turn, some cellular neurobiologists who previously studied cellular and synaptic processes are extending their interests to interconnected groups of cells and to the question of how different patterns of connections lead to the expression of behavior. Both groups appreciate that behavioral techniques are necessary but insufficient for analyzing the mechanisms of behavior, just as cellular techniques are necessary but insufficient for understanding the role of nerve cells in the functioning of the animal. As physiological and behavioral analyses approach one another they can be seen as two aspects of a unified study. The convergence of the two fields marks the beginning of a more effective science.

For those involved in the convergence the initial task is clear: to find experimental animals in which the various key questions of behavior can be effectively analyzed on the cellular level.

SUMMARY AND PERSPECTIVE

The recent convergence of cellular neurobiology and comparative behavior has its roots in the early history of experimental psychology. Following Darwin's discovery of a behavioral as well as morphological continuity among species, two traditions of psychology developed: the laboratory analysis of learned behavior and the naturalistic study of instinctive behavior.

Laboratory studies began with anecdotal descriptions of animal behavior. This was soon replaced by more objective techniques and culminated in the discovery by Thorndike, Pavlov, and Grether of instrumental conditioning, classical conditioning, habituation, and sensitization. These paradigms allowed psychologists to compare the learning capabilities of various animals. In instrumental conditioning the animal's behavior is instrumental in producing a reward. In classical conditioning a previously ineffective (conditioned) stimulus becomes effective in producing a response as a result of being repeatedly paired with an unconditioned stimulus. In habituation an initially effective, novel stimulus becomes progressively less effective as it is repeated and loses its novelty. In sensitization a powerful unconditioned stimulus leads to elicitation of a response to a previously ineffective stimulus that is not dependent on precise pairing of two stimuli.

Following Thorndike and Pavlov, experimental psychologists (most of whom were American) focused primarily on reflex behavior and on associative learning. The study of behavior was expanded by ethologists, a group of biologists (many of whom were European) interested in the naturalistic study of behavior. Jennings, Whitman, Heinroth, Craig, Tinbergen, and Lorenz studied a variety of naturally occurring adaptive responses and described fixed-action patterns, new units of behavior parallel to reflex acts. These stereotypic behavioral sequences are not controlled by a stimulus, but are simply triggered. Ethologists and psychologists also observed three types of behavioral modification that could not easily be explained by the theory of associative conditioning, because these types lacked obvious reinforcement: latent learning, sensory preconditioning, and imprinting. The search for new learning paradigms was further extended by Garcia's study of bait-shyness, which showed that animals are biologically predisposed to associate certain types of stimuli and to ignore others. The existence of many types of behavioral modification raised questions about their interrelation. These questions have intensified the interest in exploring the neural mechanisms of behavior and behavioral modification.

SELECTED READING

Herrnstein, R. J., and E. G. Boring, eds. 1965. *A Source Book in the History of Psychology.* Harvard University Press. A selection of primary literature with introductory essays.

Manning, A. 1972. *Introduction to Animal Behavior,* second edition. Reading: Addison-Wesley. A good introduction to ethology; small, well written, thoughtful.

Seligman, M. E. P., and J. L. Hager. 1972. *Biological Boundaries of Learning.* New York: Appleton-Century-Crofts. A collection of original papers on bait-shyness and other constraints on learning, with introductory essays.

Thorpe, W. H. 1956. *Learning and Instinct in Animals.* Harvard University Press. A classic in comparative behavior. This book brought to the attention of laboratory psychologists the adaptive importance of habituation and did much to bridge the distance between the thinking of ethologists and experimental psychologists.

CHAPTER 2

The Neurobiological Analysis of Behavior: The Use of Invertebrates

The convergence of psychological and neurobiological research in the problem of behavior has made it necessary to find experimental animals suitable to a combined approach. Organisms are needed in which the neural elements controlling specific behavior can be delineated in adequate physiological and morphological detail. Experimentation with several animals has shown that higher invertebrates are particularly well suited to neurobiological studies of behavior.

Here I believe it is important to restate the obvious in order to answer the question of how the study of invertebrates can possibly illuminate our understanding of the behavior of man. All animals are faced with the universal problems of reproduction, adaptation and survival. An important assumption of biology is that phylogenetically diverse organisms share similar sets of solutions to these problems. Since in the end we are concerned with identifying biological principles applicable to human behavior, the invertebrate is simply a convenient, but necessary, substitute for people. Although a solution found in invertebrates may not be the only mechanism for a given biological problem, the solution is likely to be a common mechanism that might be found as well in vertebrates, including man.

Invertebrates are useful precisely because both their behavior and nervous systems are relatively simple. In this chapter I shall discuss these two advantages of invertebrates as well as the goals and limitations of an experimental approach to the neurobiological study of behavior using invertebrates.

TOWARD AN OPERATIONAL DEFINITION OF BEHAVIOR

To understand why the simple behavior of invertebrates is advantageous, we must first define what is meant by behavior. The word "behavior" is used in various ways by psychologists. It is used to describe activities ranging in complexity from the coordinated movement of a limb to the writing of a sonnet. Clearly the two usages are not comparable and one would not seek out the same animal to study both processes. In an attempt to restrict its meaning, some argue that the word "behavior" should be limited to thinking and other complex mental functions, and not used to describe overt activity. Most psychologists disagree with this view. They argue that one can establish operationally a hierarchy of behavior; in that hierarchy limb movement is simply an example of an elementary behavior, and creative expression is a reflection of a very high order of behavior.

There are great problems, however, in defining a hierarchy of behavior that is applicable throughout phylogeny. In addition to observable components, the psychic life of man (and perhaps of other animals) also contains conscious but nonobservable components (thinking, feeling, planning) and even unconscious components (conflicts, strivings). From many perspectives the nonobservable components are the most fascinating aspects of mental life. But how does one include unobservable psychic processes in a definition of behavior?

The advance into modern psychology, culminating in behaviorism and its recent extensions, came with the realization that, for a time anyway, scientific inquiry could explore only the *observable* aspects of mentation. This position does not deny the importance of nonobservable mental processes in the mechanism of behavior. But it restricts the study of these processes to observable indexes. Most experimental psychologists agree therefore that psychic events that cannot be observed or inferred from outside the organism cannot, as yet, be studied experimentally. For example, one cannot measure perception; one can only measure the ability to discriminate, measured against some standard. The private experience is reduced to the reportable operation of discrimination.

> If I try to describe to you my sensation of the color red, I find it cannot be done. I can point to a red object, I can say that red is somewhat like orange or purple, and very different from green and blue, quite a gay, stimulating color, very nice for a tie, but a little too gay for a professor's overcoat—I can put red in many relations but I cannot describe the sensation itself. I should have the same difficulty in trying to describe my feelings of pleasantness.[1]

In such a restricted scientific sense, behavior refers only to observable mental life. Thus, psychologists with quite different orientations, such as

[1]R. S. Woodworth, 1948, p. 106.

B. F. Skinner and Donald Hebb, define behavior in similar terms. Skinner writes: "Behavior is what an organism is doing—or more accurately what it is observed by another organism to be doing. . . . By behavior, then, I mean the movement of an organism or its parts in a frame of reference provided by the organism itself or by various external objects."[2] In a similar vein Hebb states that behavior is "the publicly observable activity of muscles or glands of external secretion as manifested in movements of parts of the body or in the appearance of tears, sweat, saliva, and so forth. Talking is behavior; so is a smile, a grimace, eye watering, trembling, blushing (which is produced by muscular changes in blood vessels), changing one's posture or following the words of a printed page with the eyes."[3]

Although the definition of behavior as observable movement derives from traditional behaviorism, there is a great difference between the view of mental life held by John Watson and that held by most current students of behavior. Watson and the early behaviorists argued that observable behavior is synonymous with mental life (see p. 7). But the work of introspectionists, ethologists, and modern cognitive psychologists, as well as the demands of common sense, make it necessary to abandon this position. Watson narrowly defined a larger reality, psychic life, in terms of the scientific techniques available for studying it. By so doing he denied the existence of consciousness and unconscious cerebration, of feelings, motivation, ideas, sensations, and memories, merely because he could not study them objectively.

The growing reaction within experimental psychology against traditional, Watsonian behaviorism, often called *neobehaviorism,* began with Tolman. Tolman correctly emphasized the need to include cognitive processes—knowing, reasoning and problem solving—within a behaviorist framework. In his autobiography he writes:

> [I began to have] a growing belief that a really useful Behaviorism would not be a mere "muscle twitchism" such as Watson's. It soon appeared to me that "responses" as significant for psychology are defined not by their physiological detail but rather by the sort of rearrangements between organism and environment . . . which they achieve. . . . [There was] the further notion that purpose and cognition are essential descriptive ingredients of such nonphysiologically defined behavior . . . I also spent considerable effort in trying to translate some of the familiar prebehavioral concepts such as 'sensation', 'emotion', 'ideas' and 'consciousness' into these new nonphysiological behavioral terms.[4]

[2]1938, p. 6.
[3]1958, p. 2.
[4]Cited by D. Kretch in the introduction to Tolman's *Purposive Behavior in Animals and Man,* 1967, Meredith, p. xiii.

Following Tolman, modern psychology has been concerned with internal as well as external events. But it has studied internal events only insofar as they can be examined by means of objective techniques applied to external events. This neobehaviorist position combines the objectivity demanded by Watson with a concern for the fuller range of mental life. According to this view, observable behavior is not the *all* of mental life (as Watson thought it was), but represents the only way now available to examine or infer *any* aspect of mental life, and the most satisfactory way to begin a classification of behavior.

Classification of Behavior

Following Skinner and Hebb, I will define behavior to include *all observable processes by which an animal responds to perceived changes in the internal state of its body or in the external world.* An animal may respond to a danger signal with escape, a simple withdrawal movement, or by freezing. All three are considered behavior, although each represents a different complexity of behavior.

Even among researchers who think of behavior as observable psychological processes, there are some who resist classifying as behavior the simple movements of a muscle group or the secretion of a gland. Nevertheless, much has been learned about the logic of complex behavior from the study of just such simple responses. The main principles of classical conditioning derive originally from studies of salivary behavior in dogs, behavior defined in terms of changes in the rate of saliva secretion by the parotid gland.

The definition of behavior advocated by Skinner and Hebb separates unobservable psychological processes from observable ones. But even this more limited definition includes under one term simple movements as well as integrated chains or patterns of actions. How can behavior be classified according to complexity? A distinction between elementary and complex reflex behavior was first proposed almost concurrently by Sherrington and Pavlov. This distinction was later refined by Hebb (1958), who distinguished simple behavioral acts from the pattern of separate acts that make up an integrated behavioral sequence. Hebb called the latter "organized behavior." Although organized behavior usually does not consist of smoothly integrated units, nevertheless I find it useful, for analytic purposes, to follow Hebb and explicitly divide organized behavior into elementary, operationally defined units. I will refer to organized behavior as *complex behavior* or *behavioral patterns,* and to the subunits as *elementary behavior* or *behavioral acts.* Elementary behavior consists of an isolated response of a single effector organ. Complex behavior consists of a sequence of responses or the activity of several different effector organs.

Sherrington, using physiological techniques, and Pavlov, using behavioral ones, both regarded complex *reflex patterns* as being made up of elementary *reflex acts. Fixed-action patterns* (p. 16ff.) can also be divided into elementary components, which I will refer to as *fixed acts.*[5]

Both complex and elementary behavior are *lower-order behavior.* Advanced behavior that falls beyond the definition of complex behavior—ranging from courtship behavior, in which each behavioral sequence triggers the next, to simple forms of communication—will be referred to as *higher-order behavior* (Table 2-1). I would emphasize that the lines between elementary and complex

TABLE 2-1.
Classification of behavior with some selected examples.

Lower-Order	Higher-Order
Elementary (acts)	Courtship
Reflex acts	Nest-building
Fixed acts	Communication
Complex (patterns)	
Reflex patterns	
Fixed-action patterns	

behavior and between complex and higher-order behavior are not fixed. Moreover, the idea that a given behavior is made up of discrete elementary units of behavior, while conceptually attractive, must in each instance be demonstrated experimentally. At the moment, higher-order behavior is not readily amenable to cellular neurobiological analysis. I will therefore only be concerned with lower-order behavior. I will focus in particular on elementary behavior and on some simple examples of complex behavior.

From the above definition of elementary behavior as simple movements, it follows that the distinction between behavioral and physiological analyses of an organism is sometimes arbitrary and often more a question of perspective or emphasis than of substance. In general, physiological analysis is concerned with the functioning of a part isolated from the whole, whereas behavioral analysis is concerned with the function of the part in the context of the whole animal.

[5]For a discussion of fixed acts see Russell *et al.,* 1954; Barlow, 1968.

Intervening Variables and the Classification of Behavior

Observable acts are not the only criteria for classifying behavior. Tolman and others have tried to classify behavior on the basis of intervening variables, particularly cognitive ones, such as goal and purpose. Despite their interest, however, these inferred variables are not as reliable a means of classification as observable acts. If, for example, behavior were classified according to *purpose,* as the purpose became more complex it would become increasingly difficult for independent observers to agree upon the cluster of responses belonging to a particular category. Consider the difficulty of classifying behavior according to the purpose of defense. The simplest purpose would be withdrawal, manifested in the observable response of a single limb. Next would come defensive action, combinations of movements that lead to escape. More complex still would be the recognition of potential danger and the expression of fear. Although nonobservable (intervening) variables are often theoretically useful— for example, *strategy* is now commonly used by cognitive psychologists to explain human learning and memory—they are not practical for a classification of behavior. They are inferences and are therefore open to greater disagreement than are observations. By basing classification on observable details, one gains objectivity—and avoids commitment to a particular theoretical bias— while retaining the freedom to explore specific intervening variables that seem useful.

A compromise between the classification of behavior in terms of either movement or goals is classification in terms of *outcome.* This objective index has been used in instrumental conditioning where learning is scored on the basis of, for example, the *rate* at which a rat presses a bar for food or runs a maze, irrespective of the movements by which the task is accomplished (Skinner, 1938). Thus, a rat may use either of his two hind paws, his two front paws, or his head to press the bar. However, because outcomes are independent of movement, they are more effectively used to rate learning than to classify behavior. In particular, they are not useful for classifying behavior as a basis for neurobiological studies. The fact that outcomes afford a good description of learning but a poor description of behavior (at least for studying the neural mechanisms of behavior) illustrates that there is no single way to classify behavior; the classification is in part determined by the problem at hand.

Since cellular studies of behavior are most effectively carried out by classifying behavior according to movement, how likely are these studies to enlighten our understanding of intervening variables, such as purpose or strategy? The very effort to analyze a behavioral variable in cellular terms demands that the variable be highly specified. In turn, the better the specification, that is, the more the variable can be directly inferred from measurable changes in responses, the greater is the likelihood that it can be studied systematically. Postulating an intervening variable necessarily implies that some neural transformation has as yet not been specified. Specifying the transformation provides an important validation and refinement of the variable. As I will show, studies of three intervening variables—learning, memory storage, and motivation—are most likely to be rewarding in the near future. These variables are moderately well-defined operationally and testable at the cellular level.

BEHAVIORAL ADVANTAGES OF STUDYING INVERTEBRATES

Both reflex and fixed acts are easily identified in invertebrates, and the combination of these elementary units into complex behavioral patterns can usually be easily recognized. For example, the escape behavior of the nudibranch mollusc *Tritonia* consists of a series of elementary behavioral responses: first a local then a general body withdrawal, followed by a stereotypic swimming sequence consisting of alternating dorsal and ventral body flexion (Willows, 1968; and Figure 10-24, p. 450).

In vertebrates the relationship of behavioral components is often not obvious, so that a reductive analysis is difficult. In contrast to vertebrates, invertebrates perceive less and do less, and their behavioral patterns are often quite stereotyped. As a result, even suprisingly complex higher-order behavior involving decision and choice can sometimes be analyzed into its components. For example, a male cricket can emit six types of songs: a song of recognition, to announce his presence; a song of calling, to establish contact with the female at a distance; a song of courtship at close range; a song of courtship interception, for when the female moves away; a song of rivalry, for repelling another male; and a song following copulation, for maintaining contact after mating (Alexander, 1962). Each song is a highly stereotypic fixed-action pattern consisting of a fixed number of trills and chirps temporally separated in characteristic ways and which are produced by rubbing the wings together. The conditions under which each is released are by no means simple. There are several variables: the cricket's physiological state, where he is, and who is nearby. If a female is near, he will produce a courtship song; if a rival is near, he will produce a rivalry song; if a predatory bird is near, he will not sing but will simply try to escape. Despite their variety and complexity, each complex (sign) stimulus can be defined and analyzed. The lower-order behavior of higher invertebrates typically occurs over a defined, usually short, time period. This facilitates the search for neural events that are temporally correlated with and causally related to a behavior.

NEUROBIOLOGICAL ADVANTAGES OF STUDYING INVERTEBRATES

In addition to their behavioral advantages, invertebrates offer a second advantage: their nervous systems are amenable to cellular studies of the neural mechanisms of behavior and learning. The relationship of the nervous system to behavior has been studied in two ways. One is through gross localization of the regional substrates of behavior; that is, areas of the brain are related to specific organismic functions. Gross localization is aimed at determining where in the nervous system a particular behavior is controlled. This has been done with global techniques, such as permanent or temporary (reversible) ablation of a part of the brain, stimulation of brain areas, or recording the summed electrical activity of large populations of neurons, as in the electroencephalogram.

Another way to study the relationship of the nervous system to a particular behavior is through cellular localization of the neural mechanisms of the behavior. Here one needs to understand the chain of cellular events underlying the behavior, from the sensory transduction of the stimulus to the motor response. For this type of analysis the properties of the relevant neurons and their interconnections need to be known. This information is not obtainable with global techniques because these do not permit one to establish a causal relationship between regional processes and cellular events. Although global techniques are an important preliminary step to cellular studies, they cannot provide information about the relationship of individual cells to a given behavior. For an understanding of the neuronal mechanisms of a behavior, the parts of the brain responsible for the behavior need to be examined one cell at a time. This is not to imply that the neuronal mechanism of a particular behavior will necessarily be found in single cells. The mechanism will more likely arise out of the interactions of many cells. However, a detailed analysis of the component cells, either singly or in combinations, is required. This can be done using microelectrodes, the electron microscope, and biochemical techniques that yield direct information about the relationships between cell properties and behavior, and about changes in cell function and learning.

The importance of distinguishing between global methods for studying localization and cellular methods for studying mechanisms was appreciated at

the beginning of this century by Santiago Ramón y Cajal, the Spanish anat-
omist. Cajal's thinking about behavior and learning provided much of the
foundation for the current attack on the cellular study of behavior.

> No matter how excellent, every physiological teaching on the working of the
> brain based on localization leaves us ignorant of the mechanism of mental activ-
> ity. These actions are certainly accompanied by molecular modifications in the
> nervous cells and preceded by complex changes in the relationship between
> neurons. To understand mental activity it is necessary to understand these mo-
> lecular modifications and changes in neuronal relationships. One must know, of
> course, the complete and exact histology of cerebral centers and their tracks.
> But that is not enough; it will be necessary to know the energetic transformations
> of the nervous system which accompany perception and thought, consciousness
> and emotion.[6]

The mammalian brain has many cells (about 10^{12}), which are typically
small. The nervous system of higher invertebrates has far fewer cells (about
10^5 to 10^6), many of which are large and identifiable. Although 10^5 is still a
large number, invertebrates have other features advantageous for studying
the mechanisms of behavior, features that are unparalleled in vertebrates. In
invertebrates connections between indivdiual cells can be mapped and specific
cells can be related to the sensory and motor periphery. Certain invertebrates—
the gastropod molluscs *Aplysia, Tritonia, Helisoma, Limax,* and *Helix;* the
annelid and nematode worms; and many types of insects—can be reared in the
laboratory to provide relatively pure genetic stocks. Also, some of these
(*Drosophila* and nematode worms) have a generation time short enough to
permit isolation of varieties of behavioral mutants.

GOALS OF A CELLULAR APPROACH

Because cellular studies of behavior in invertebrates can combine cellular and
behavioral techniques, they can illuminate both psychology and neurobiology.
Cellular studies can inform psychology by providing an understanding of the
neuronal architecture of a given behavior, the rules that relate different types
of neural architecture to different classes of behavior, and the mechanisms for
integrating units of behavior. What is the difference in the functional organization of the reflex and fixed acts? How does a single effector system take part in
both? Cellular studies can also examine the loci and the mechanisms of be-

---

[6]Cajal, 1911, p. 8.

havioral modification. These analyses are not only interesting in their own right, they also provide new means for examining the interrelationships between types of behavioral modifications: between short-term memory (lasting minutes and hours) and long-term memory (lasting days), habituation and dishabituation, dishabituation and sensitization, and between sensitization and classical conditioning. Until recently these relationships had been inferred from studies using behavioral techniques that cannot establish underlying mechanisms (see p. 22ff.). For instance, the relationship of short-term to long-term memory is poorly understood. An examination of their neuronal mechanisms could reveal whether these two types of memory have different mechanisms or whether they represent stages in a common cellular mechanism with a variable time course.

In addition to leading to a better understanding of behavioral mechanisms, studies of cellular mechanisms and neuronal circuits of behavior can advance understanding of basic problems in neurobiology by specifying the relative contributions of heredity, development, and experience to neuronal functions. Do behavioral modifications involve the growth of new nerve cells or the sprouting of new connections, or do they involve changes in the functional effectiveness of previously existing connections? If we could answer this question in any one animal, we could broaden our understanding of the functional organization of nervous systems in general. Also, behavior typically involves the action not of a single neuron but of groups of neurons. In nervous systems, as in other complex biological systems, the dynamic functioning of the system as a whole may be difficult to infer from an examination of its isolated parts. By combining behavioral and cellular neurobiological approaches, the functional interaction of individual cells may be better understood in the context of the behavioral system to which each cell belongs.

STRATEGY AND LIMITATIONS IN THE USE OF INVERTEBRATES

It could be argued that an understanding of the behavior of an animal requires a total understanding of the animal's nervous system. But the immediate strategy advocated by a number of neural scientists, and to which I am committed, is more restricted in its goal. It seems to me impossible at present to obtain a complete understanding of any nervous system capable of complex behavior. Even a nervous system as relatively simple as that of the leech has a large number of neurons (approximately 10^5). To reduce the study of an invertebrate nervous system to more manageable proportions, it is necessary to focus on specific instances of behavior and to analyze only the parts of the nervous system relevant to that behavior. Indeed, it may not even be necessary to work out the details of an entire ganglion, although this is now technically possible. But it is necessary to delineate in complete cellular detail the neural circuits of a number of interrelated behavioral responses.

Although this strategy is new in the study behavior, the strategy itself is not new. Advances in experimental biology often depend on studies of selected simple organisms in which a family of interrelated problems can best be studied. The crucial task in such an approach is to define the problems clearly enough to be able to select appropriate experimental organisms. The history of modern genetics offers several examples of how the selection of experimental animals follows naturally from a clear statement of a problem. For example, early in this history geneticists decided to focus on the transmission of genetic information. This decision led to the selection of experimental organisms—at first the fruitfly *Drosophila* and later microorganisms and viruses—that offered appropriate technical advantages, such as short generation time and the ability to grow under defined culture conditions.

In the neurobiological analysis of behavior we need to focus on the mechanisms of elementary behavioral responses and their modification. This requires experimental animals in which a relatively complete understanding of the neuronal architecture of the behavior can be achieved. Certain invertebrates seem particularly advantageous for this purpose.

It is my assumption, and that of others working in this area, that some of the biological mechanisms of behavior and learning found in any animal are likely to exist in all animals. This strategy is not universally accepted. The neural organizations of vertebrates and invertebrates are structurally different in detail and may also differ in principle. A comparative and reductive approach to behavior may therefore be misleading. Although some features are common to all phyla, the study of one phylum or even studies using representatives of all major invertebrate phyla will not reveal *all* the important principles of neural organization that have evolved. Some principles will only be found among higher mammals and some will be found only in man. For example, man's capacities of abstract thinking and language are not found in invertebrates, and these capacities clearly require unique types of neural organization.

The relevant question, however, is not whether some types of behavior and their neural organization are peculiar to the mammalian brain—obviously some are—but whether any types of behavior are general enough to suggest common patterns of neural organization. Here the answer seems clear: elementary perception, motor coordination, motivation, learning, and memory are found in higher invertebrates as well as in vertebrates. Invertebrates have solved many biological problems that vertebrates face. For example, one particularly successful set of solutions is the ability to modify behavior. This ability has been achieved independently by many invertebrate and vertebrate lines, and the formal similarity of some of the solutions suggests that the neuronal mechanisms of behavioral modification in different species are likely to share common features.

The existence of behavioral capabilities common to higher invertebrates and vertebrates does not necessarily mean that identical neural mechanisms are involved. It does suggest that the mechanisms may be general and should be explored fully using those experimental animals in which they can be studied most effectively. I therefore believe that a complete and rigorous cellular analysis of a learning task, no matter how simple the animal or the task, is likely to prove more instructive than incomplete or less direct studies of learning in animals whose nervous systems are so complex as to preclude cellular analysis at this time.

However, I would here sound a cautionary note. The fact that different brains accomplish the same end need not imply that they do so by the same means. The eyes of both cephalopods and man are image-forming, but in cephalopods they develop directly from the ectoderm, and in man only indirectly by way of the neural tube. The cephalopod eye aims its photoreceptors toward the light, the vertebrate eye aims them away from the light. Evolution is opportunistic and will seek the most appropriate solutions with whatever biological materials are at hand. If the materials are different, the mechanisms may be different, despite convergent evolution (Simpson, 1949). Although the nervous systems of higher invertebrates and vertebrates differ in the details of their structural plans, the basic biological materials—the neuronal building blocks— are surprisingly similar. So are some principles of neuronal interconnection. These biological facts are the basis of the optimism of those who study the neural mechanisms of learning in invertebrates. Nevertheless these arguments make clear that any ideas about the generality of mechanisms based on findings in invertebrates will ultimately have to be checked in studies of vertebrates.

HIGHER INVERTEBRATES AND
THE EVOLUTION OF VERTEBRATE BEHAVIOR

It is important to distinguish between studies of the general mechanisms of behavior and comparative studies of behavior designed to examine phylogenetic relationships. Any invertebrate, independent of its phylogenetic position, can be used to study the general mechanisms of behavior common to all animals. However, Darwin, Romanes, and some early students of invertebrate behavior thought that invertebrates were lower than vertebrates on a "phylogenetic scale" and therefore could be used to study the evolution of vertebrate behavior.

The older notion of a phylogenetic scale is no longer tenable. Animals cannot be arranged in continuous *scala naturae,* with sponges at the bottom, man on top, and a hierarchy of invertebrates and vertebrates in between. Present evidence on animal evolution indicates that there has been extensive divergence of evolutionary lines, leading to parallel evolution as well as the extinction of many intermediate forms (Simpson, 1949). Therefore, we cannot trace the evolution of a behavior in "higher" animals by studying the same behavior in "lower" forms of life (Hodos and Campbell, 1969). It is now clear that even where analagous behaviors are involved, studies in invertebrates cannot clarify the evolution of vertebrate behavior. A particular escape response in vertebrates has not simply evolved from an escape response in a higher invertebrate. *Aplysia,* crayfish, leech, lobster, and other invertebrates used in neurobiological studies are not descendants of animals that led directly to vertebrates (see Figure 3-1, p. 49). The evolution of behavior can be studied only in animals specifically selected for their presumed phylogenetic relationships (see Kandel, *The Behavioral Biology of Aplysia*).

SUMMARY AND PERSPECTIVE

Most invertebrates have limited behavioral and learning capabilities. Why should they be used to study these functions? The main reasons are the relative simplicity of the behavior of invertebrates and the small number of cells constituting their nervous systems. To understand why invertebrates are appropriate for this research, one needs to define what is meant by behavior and how its neural mechanisms are best studied.

As now used by most psychologists, the term behavior refers only to a restricted aspect of mental life, that which is observable. Within that domain behavior includes all processes by which an animal responds to perceived changes in the internal state of its body or in the external world. An animal may respond to a danger signal with an escape response, a simple withdrawal movement, or complete inactivity and only an increase in heart rate. All three responses are equally defined as behavior. The definition of behavior in terms of observable response does not exclude the importance for mental life of nonobservable conscious mental processes, such as thinking or feeling, or even unconscious processes, such as conflict. But the definition restricts scientific study of these processes to their observable manifestations.

Elementary and complex aspects of two major classes of observable behavior, reflex and fixed-action patterns, occur in invertebrates as well as in vertebrates. But in vertebrates the relationship between elementary and complex behavioral responses is often not obvious, so that a reductive analysis is difficult. Invertebrates perceive less and do less than vertebrates, and their behavioral patterns are often quite stereotypic. The behavioral and structural simplicity of invertebrates brings certain psychological issues into clear focus. As a result, in invertebrates the elementary behavioral responses that make up a complex behavior can often be discovered and analyzed. Moreover, the lower-order behavior of higher invertebrates typically occurs over a defined, usually short, time period. This facilitates the search for neural events that are temporally correlated with and causally related to a behavior. Finally, invertebrates manifest some kinds of lower-order behavior common to all animals having well-differentiated central nervous systems. Lower-order behavior is also the most likely to be well understood at the cellular level in the near future. Before higher-order behavior can be effectively examined it will be useful to understand the principles that underlie lower-order behavior.

The relation of the nervous system to behavior has been studied in two ways. The first is global localization of areas of the brain correlated with specific behavioral functions. The other is cellular localization of the mechanisms of behavior in which the chain of cellular events underlying the behavior is specified, from the stimulus to the motor response. For this type of analysis it is necessary to know the properties of those neurons and their connections relevant to the behavior. The mammalian brain has many cells, and these are typically small. By contrast, the nervous systems of invertebrates have fewer cells by several orders of magnitude, and many of these cells are large and identifiable. As a result, connections between cells in invertebrates can be mapped on a cell-to-cell basis and individual cells can be related to sensory and motor structures. Several types of invertebrates can be cultured in the laboratory and some have a short generation time, so that isogenic lines and behavioral mutants can be reared and analyzed. These advantages are difficult to find in vertebrates.

There are, however, also limitations to a simple-systems approach to behavior using invertebrates. The most obvious of these is that only certain types of behavior can be studied. Even though some features are common to all animals, the study of invertebrate phyla is not likely to reveal all the important principles of organization that nervous systems have evolved. Some features of design will be found only among higher mammals, and some are likely to be unique to man. Even where common principles seem to be involved, it will still be necessary to demonstrate by cellular studies in vertebrates that these principles are indeed common. But for the present, complete and rigorous analyses of various classes of behavior and learning in invertebrates, no matter how simple the behavior, are likely to prove more informative than incomplete or indirect studies of behavior and learning in vertebrates.

SELECTED READING

Kennedy, D., A. L. Selverston, and M. P. Remler. 1969. Analysis of restricted neural networks. *Science,* 164: 1488–1496. An analysis of the neural circuitry underlying an escape behavior in the crayfish, combining morphological and physiological techniques.

Nicholls, J. G., and D. Van Essen. 1974. The nervous system of the leech. *Scientific American,* 230:38–48. A summary of the recent work in the leech relating electrophysiological studies of single identified nerve cells to behavior.

Willows, A. O. D. 1971. Giant brain cells in mollusks. *Scientific American,* 224:68–75. A description of the cellular analyses of escape-swimming, a fixed-action pattern, in the nudibranch mollusc *Tritonia.*

CHAPTER 3

The Choice Among Invertebrates:
Cellular and Genetic Approaches to Behavior

The neuronal mechanisms underlying behavior can be analyzed with either genetic or cellular biological techniques. Desirable features for these analyses include: (1) Well-delineated, simple, yet interesting behavioral responses that can undergo short- and long-term modifications. (2) Short generation time from birth to reproductive maturity so that genetically pure strains are readily obtainable. These strains allow analysis both of the way genes control the development of the nervous system and of how genes control behavior. (3) Nerve cells that are few in number, large, and identifiable. (4) Connections between neurons that can be mapped on a cell-to-cell basis using both electrophysiological and morphological techniques. These features were combined into a specification of the neural scientist's ideal organism by Chip Quinn, a student of *Drosophila* genetics:

> The organism should have no more than three genes, a generation time of twelve hours, be able to play the cello or at least recite classical Greek, and learn these tasks with a nervous system containing only ten large, differently colored, and therefore easily recognizable neurons.[1]

The search for suitable invertebrates having, or at least approaching, these features involves some intriguing aspects in the history of both psychology and neural science. In this chapter I shall first trace the development of interest in

[1]Personal communication.

invertebrate behavior in more detail and then describe the studies that led to the selection of certain animals for cellular biological and genetic studies.

STUDIES OF INVERTEBRATE BEHAVIOR

The struggle for an objective psychology was in part fought, and won, on the basis of studies of invertebrates. Darwin and Romanes worked with earthworms, jellyfish, starfish, and sea urchins. The next generation of behaviorists, including Lubbock, Loeb, Verworn, and Jennings, were all invertebrate biologists. Between 1930 and 1960, however, interest in invertebrates declined among experimental psychologists (see Beach, 1950). Associative learning proved difficult to demonstrate in many invertebrates. In the search for the behavioral laws of associative learning, most psychologists focused on vertebrates (see p. 13).

Ethologists, however, continued to study invertebrates. Being naturalists, they had (and have) a great curiosity about all animals, including simple animals. Following Darwin's example, they were interested in the taxonomic possibilities of behavior, for which invertebrates proved particularly useful. Moreover, ethologists were more concerned with the adaptive significance of behavior than with learning processes. They found in invertebrates elementary examples of universal instinctive behavior—feeding, sex, defense, social behavior, and communication. Certain invertebrates, particularly insects, were used extensively to study fixed-action patterns, social behavior, and communication.

Two biologically oriented American psychologists, Karl Lashley and his student Theodore Schneirla, also appreciated quite early the usefulness of invertebrates for studying behavior. In his presidential address to the American Psychological Association in 1938, Lashley began a review of instinctive behavior by describing how the planarian *Microstoma* captures its prey. "Some of the most remarkable observations in the literature of comparative psychology are reported in the study of *Microstoma*. . . . Here, in a length of half a millimeter, are encompassed all the major problems of dynamic psychology."[2] The usefulness of invertebrates for studying general behavioral processes is perhaps best exemplified in the study of bees by several generations of naturalists, culminating in von Frisch's description of the dance language used by bees to communicate to each other precise information about the location of food (von Frisch, 1950).

[2]1938, p. 445.

Studies of invertebrate behavior and learning were codified by W. H. Thorpe in several influential reviews (1943, 1944a, b, and 1956) in which he integrated the findings of ethology with those of experimental psychology and established criteria for demonstrating learning in invertebrates. B. Boycott and J. Z. Young (1950 and 1955) applied these learning paradigms to cephalopod molluscs and found that they were capable of extraordinary degrees of learning, comparable in many ways to vertebrates.

The greatest impetus to the study of invertebrate behavior came from two developments in neural science. One was the early attempt by neurophysiologists to apply the techniques of single-cell recording to the study of the invertebrate nervous system. A number of neurophysiologists—E. D. Adrian, J. Z. Young, H. K. Hartline, C. Ladd Prosser, Kenneth Roeder, Theodore H. Bullock, and Cornelis Wiersma—approached the invertebrate nervous system with questions that had intrigued ethologists. They were particularly interested in knowing how the nervous system could generate rhythmic behavior in the absence of obvious stimuli. Sherrington's work illustrated how a stimulus could elicit and guide reflex responses only. The identification of fixed-action patterns posed a new set of questions. How are these patterns generated in the nervous system? What triggers their occurrence? Is there a central nervous system program that determines their temporal sequence? Since fixed-action patterns can occur spontaneously, their existence in invertebrates made such animals ideal for analyzing these questions on the cellular level. This approach proved immediately rewarding. By recording from single cells in invertebrate central nervous systems, neurophysiologists soon developed preliminary answers to their own and to ethologists' questions about the nature of spontaneous behavior. These studies held out the first hope for relating cellular processes to behavior. In addition, neurophysiologists developed a family of experimental animals that proved immensely useful for studying impulse conduction in axons and synaptic transmission.

The other development came from studies of the transmission of hereditary characteristics. Much work in this area was carried out on the fruit fly, *Drosophila;* many mutant strains were produced, including some in which behavior had been altered. Ethologists had long emphasized the possibility that certain types of behavior, e.g., fixed-action patterns, were largely inborn. To test these ideas, genetic control of behavior of the sort displayed by *Drosophila* mutants was clearly useful. In addition, mutations proved to be powerful tools for analyzing the relationship between the development of the nervous system and behavior.

I will describe how cellular neurophysiological techniques were first applied to the nervous system of invertebrates and then review the development of interest in the behavioral genetics of *Drosophila* and nematode worms. Figure 3-1 illustrates the postulated evolutionary relationships among the major higher invertebrates, a few of which I will consider below, and their relationship with the vertebrates, on the one hand, and the major lower invertebrates, on the other.

SINGLE-CELL STUDIES OF INVERTEBRATE NERVOUS SYSTEMS

In the course of his studies of the vertebrate nervous system, Edgar Adrian was the first to record the electrical activity of single nerve cells (Adrian and Zotterman, 1926; Adrian and Bronk, 1928). He did so by developing surgical techniques for isolating axons, which made it possible to stimulate and record from them extracellularly. It was perhaps only to be expected, therefore, that

Edgar Adrian, circa 1961
(Courtesy of Professor
Detlev Bronk)

Adrian should also be the first to record from invertebrate nerve cells. In a short abstract on the nervous system of the caterpillar, presented to the British Physiological Society in 1930, Adrian assured his colleagues: "The neurones of the caterpillar do not differ greatly from those of the vertebrate in the general nature of their activity: in particular, the motor discharge shows a remarkable likeness to that in the motor fibers of a vertebrate. . . ."[3]

[3]1930, p. 34.

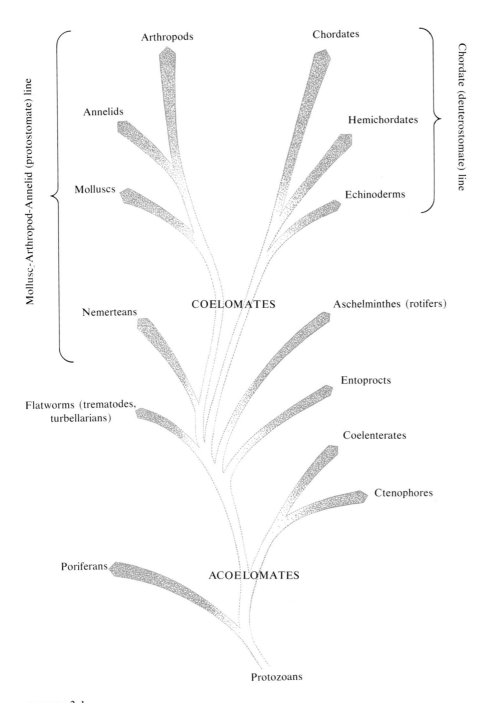

FIGURE 3-1.
A possible phylogeny of the metazoa, illustrating the two great lines of evolution, the interrelationships between phyla within each line, and the relationships between vertebrates and invertebrates. The deuterstome line includes the chordates; the protostome line includes the higher invertebrates: molluscs, annelids, and arthropods. [Adapted from Villee *et al.*, 1963.]

Using single-cell recordings, Adrian also found that some neurons seemed to be spontaneously active. Sensory stimuli did not impinge upon a neural tabula rasa, but modified existing patterns of spontaneous activity. Adrian was particularly interested in respiratory rhythms. To examine whether these were autoactive, it was necessary to disconnect the neural grouping controlling respiration from the rest of the brain. Whereas this could not be readily accomplished in the mammalian brain, Adrian succeeded in a lower vertebrate, the goldfish, and in some invertebrates, namely caterpillars and beetles (Adrian, 1930, 1932a, 1937). The neurons so isolated were indeed spontaneously active.

> One of the activities of the central nervous system which is particularly easy to study is that concerned with respiration. The movements are fairly simple and they are repeated rhythmically at convenient intervals . . . ; no external stimulation is needed to evoke them. A group of nerve cells in the brain stem enters periodically into the active state: as shown by the motor discharge, the activity rises and falls, to zero in some neurones and to a low level in others, and the cycle is constantly repeated.
>
> There are two possible explanations of a periodic activity of this sort. One that it is spontaneous, that the cells of the respiratory centre tend to beat like the heart; the other that it is reflex, that each movement produces a sensory discharge which determines the next movement and so on. In a sense both explanations are correct. There is no doubt at all that sensory discharges are responsible for the normal rhythm of breathing. The root contains a number of afferent endings from the vagus which discharge periodically when the tissues are stretched and have the effect of cutting short the movements of inspiration. The frequency of respiration is governed by these and can be altered by periodical inflation of the lungs. The normal movements are reflexly controlled as are all the movements of the body, but respiration is not a reflex if by reflex we mean a reaction which would not take place at all without the afferent discharges.[4]

Adrian thus found that the cells of the nervous systems of invertebrates not only resemble those of vertebrates but that the invertebrate brain could be used to study problems that could not yet be tackled in the mammalian brain. In the most remarkable example, Adrian compared the gross electrical activity of the optic ganglion of the water beetle *Dytiscus* to his own electroencephalogram, recorded from the visual cortex, and found that they showed comparable responses to light (Figure 3-2).

In 1934 C. Ladd Prosser recorded from single cells in the nervous system of the crayfish (Prosser 1934a, b). As Donald Kennedy (1971) has pointed out, Prosser's strategy of applying cellular techniques to invertebrate ganglia to

[4]Adrian, 1932b, p. 78.

Eyes shut   Open   Shut

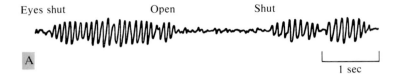

A |_____|
                                                                1 sec

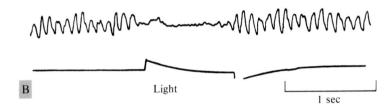

B   Light |_____|
                                                    1 sec

FIGURE 3-2.
Comparison of a human (Edgar Adrian's) electroencephalogram (A) with the
electrical activity of the optic ganglion of a water beetle *Dytiscus* (B). In both
instances the slow waves of the electrical activity are disrupted (desynchronized)
when light impinges on the retina. In man this is done by having the subject
open his previously closed eyes. In *Dytiscus* light is shown on the retina. [From
Adrian and Matthews, 1934; Adrian, 1937.]

study general problems in neural organization explains the recent interest of
many workers in the invertebrate nervous system. Prosser wrote:

> The peripheral sensory and motor units of vertebrates have been more ame-
> nable to study by electrical methods than have the units in the central nervous
> system. The principal difficulty encountered in studying electrically the interac-
> tion between the central neurons is the confusion resulting from the large num-
> bers of neurons in any given region of the central nervous system. Characteristic
> of invertebrate nervous systems is their more diffuse nature and the fact that
> each ganglionic center contains relatively few cells, when compared, for example,
> with one segment of the spinal cord of the cat. Very few attempts have been
> made to apply to invertebrate ganglionated systems the electrical methods used
> in studying the peripheral elements in classical vertebrate preparations. The
> present series of papers is the result of an effort to elucidate some of the problems
> of the central nervous system by studying action potentials in the relatively
> simple nervous system of the crayfish.[5]

Prosser's work on the crayfish confirmed Adrian's findings on spontaneous
activity in insect ganglia (Prosser, 1934a, b, 1936) and also showed that invertebrates were suitable for studying synaptic transmission (Prosser, 1935, 1946).

[5]1934a, p. 185.

It marked the beginning of comparative synaptic physiology, later developed more extensively by T. H. Bullock (1947) and H. Grundfest (1957).

In 1932 Keefer Hartline, a student of Adrian's, began his classic work on the eye of the horseshoe crab (*Limulus polyphermus*), demonstrating that knowledge of the elementary sensory properties of invertebrates can be used to understand human perception (Hartline and Graham, 1932; Hartline, 1934). In one of the earliest laboratory studies of psychological processes, Weber (1834) measured the ability of humans to detect small differences (difference threshold) in length, weight, and tone. These experiments were continued by Fechner (1860), who proposed that sensation is proportional to the logarithm of the stimulus strength. In the course of his studies on single sensory fibers Adrian found that cells signal changes in intensity of the stimulus by changes in their frequency of firing. This provided an opportunity of testing Fechner's law in single sensory cells. Using the eye of the horseshoe crab, Hartline and Graham (1932) recorded from single fibers that were directly connected to visual receptors and found, as predicted by Fechner, that the frequency of the cell's discharge was proportional to the logarithm of the stimulus intensity (Figure 3-3). Hartline and Graham also found that Fechner's law stems from the basic properties of the receptor.[6]

Because of the pioneering work of Adrian, Prosser, and Hartline, crustaceans and insects were the favored experimental animals for workers on the invertebrate nervous system from 1930 to 1965. Neurophysiological research in arthropods was restricted, however, to studies of individual axons surgically isolated. The new technique of *intra*cellular recording, introduced in 1946, could not be successfully applied in the central nervous systems of these animals until recently. Nevertheless, this line of research in arthropods resulted in four key findings. (1) Elucidation by Hartline and Ratliff of the role of lateral inhibition in certain psychophysical processes, such as Mach bands. (2) The discovery by Wiersma of cells in the visual system of crayfish that respond to complex features of a stimulus. (3) The discovery by Wiersma, Kennedy, and their colleagues of command neurons, single cells that control behavioral sequence. (4) Elucidation by Wilson of central control of rhythmic locomotor patterns (insect flight).[7]

---

[6]Psychophysical and sensory receptor processes are not exclusively logarithmic, however. Other mathematical relationships exist for different sensory modalities or even different subgroups of the same modality.

[7]For a discussion of these findings in arthropods see Chapter 14. For a review of the work on the eye of *Limulus* see Hartline and Ratliff, 1957; for insect flight see Wilson, 1968. For command elements see Wiersma, 1952; Wiersma and Ikeda, 1964; Kennedy, 1971; and Kennedy and Davis, 1976.

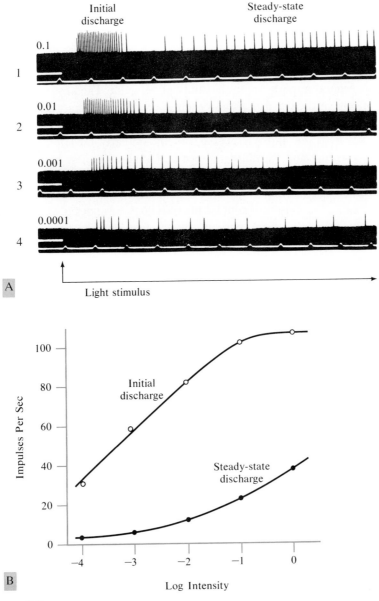

FIGURE 3-3.
Cellular analysis of Fechner's psychophysical law. [From Hartline and Graham, 1932.]

A. Selected records illustrating extracellularly recorded action potentials from a single optic nerve fiber in *Limulus*. The eye was dark-adapted. The frequency of the action potentials varies as a function of stimulus intensity. Intensity is indicated in arbitrary units (1 unit = 630,000 meter candles on the surface of the eye). The marks on the white line below the action potential record represent 0.2 sec intervals.

B. Graph of the data illustrated in *A*, showing the Fechner relationship between impulse frequency (impulses per sec) and the logarithm of the intensity of stimulating light (intensity in arbitrary units, as in *A*). Top curve: maximal frequency of discharge during the initial phase of firing (first 3.5 seconds). Bottom curve: frequency of discharge 3.5 seconds after onset of illumination, when a steady-state firing frequency is reached.

As with Hartline's earlier confirmation of Fechner's law in *Limulus*, these insights into the functional organization of invertebrate nervous systems proved of general importance to the study of behavior. The abstracting capabilities of cells in the crayfish visual system found by Wiersma were similar to those later found in the frog by Lettvin and his colleagues (1959) and in the cat and monkey by Hubel and Wiesel (1963, 1970). Similarly, the lateral inhibition found by Hartline and Ratliff in the eye of *Limulus* has now been found in several vertebrate sensory and motor systems.

During this early period of modern invertebrate research the value of annelid worms and molluscs for cellular neurobiology was not fully appreciated. The advantages of the leech nervous system first came to the attention of modern neurobiologists through the magnificent drawings of its central ganglia made in 1891 by Gustaf Retzius. Seventy years later these drawings caught the attention of Stephen Kuffler and David Potter, who were searching for a simple experimental preparation in which to study glial cells, the supporting cells of the nervous system (Kuffler and Potter, 1964).

Retzius's drawings illustrated that the segmental ganglia of the leech are divided into regions by an orderly arrangement of giant glial cells. Each cluster of nerve cells is imbedded in one large glial cell. Using these features to advantage, Kuffler and his colleagues described, for the first time, some of the functional properties of the glial cells (Kuffler and Nicholls, 1966). Other advantages of the leech nervous system were appreciated by John Nicholls: the ordered arrangement into highly repetitive segmental ganglia, the lack of a peripheral nerve net, the small number of cells in each ganglion, the even smaller number involved in a particular function, and the invariance of many cells and their connections. Together with his associates, Nicholls has used the leech nervous system to great advantage in exploring the relation of cellular properties to behavioral function, the specificity of neurons and their connections, and the regrowth of connections following injury (see Nicholls and Van Essen, 1974).

The history of molluscan neurobiology follows the discovery, neglect, and then the rediscovery of the giant fiber system of the squid. The giant axons of squid are so large (1 mm in diameter) that for many years they were thought to be blood vessels. The nerve cell bodies that give rise to them were described by Williams in 1907 in a privately distributed paper, which was overlooked until the cells (and subsequently William's description) were rediscovered by J. Z. Young. Of this discovery Young writes:

> I went to Naples in 1928 and took up work with the late Professor Enrico Sereni on regeneration of nerves of octopus. In the course of this I noticed a

J. Z. Young, circa 1950
(Courtesy of
Professor Young)

small yellow spot on the stellate ganglion. I made sections of this purely out of curiosity and found the epistellar body.

Looking for the epistellar body in squids, I failed to find it. The ganglia contained numerous large spaces, which for three or four years I thought were veins, and did not follow them. One day it struck me that the blood in them looked very curious, and examining more closely I found that they were nerve fibers. Moreover, they arose from the giant cell lobe in the position of the epistellar body of octopods. I think this must have been in 1932. It was not until 1936 that I finally established that they were nerve fibers. Working at Woods Hole, I could show that stimulating a nerve produced a twitch, provided the giant fibre was intact. At the same time Gerard, Bronk, and Hartline undertook to help me to record impulses from them [see Bronk, Young, and Gerard, 1936]. This proved not altogether easy. Every time they stimulated the nerve electrically, the oscilloscope waggled so much that they couldn't decide whether it was an impulse. One day in despair I said to Hartline, let's stimulate this nerve with oxalate. We put a crystal on the end and joined the system to a loudspeaker. Out came a wonderful buzz, and that was it.

It was only at about this stage that I came across Williams' book on the squid, in which, if you look closely, he describes the giant fibers in the stellar nerves. They are referred to only very vaguely, and of course he supposed that the whole system consisted of a single unit throughout.[8]

[8]Personal communication, July 2, 1971.

Angelique Arvanitaki, circa 1941
(Courtesy of Professor
Arvanitaki-Chalazonitis)

Young's contribution is basic to current developments in molluscan neurobiology. He not only discovered the giant fibers but also immediately appreciated their value for behavioral and biophysical studies, and urged physiologists to exploit their advantages. In 1939 he wrote that the third-order giant axon was "readily available for the study of its optical, chemical, and physical properties and not difficult to isolate completely." Young's claim was amply confirmed when Curtis and Cole, as well as Hodgkin, Huxley, and Katz used the giant axon for their studies of the ionic mechanisms of the action potential (see Chapters 5 and 6).

Only one conceptual step was needed to move from giant nerve fibers to giant nerve cells. This step was taken by Angelique Arvanitaki. As a graduate student working at the marine biological station in Tamaris, France, Arvanitaki studied repetitive firing in the giant axons of the cuttlefish *Sepia*. Adrian and Prosser had shown that small groups of isolated neurons were capable of spontaneous activity, but they were still uncertain whether this property was intrinsic to individual cells or resulted from reverberating activity among several interconnected neurons. Arvanitaki (1938, 1939) discovered that in low-calcium solutions isolated single fibers of *Sepia* underwent graded, local oscillatory potentials that grew large and generated rhythmically recurrent firing (Figure 3-4*A*). To see whether neurons in the central nervous system fired spontaneously under normal circumstances, she began to search for an animal that had "comfortably large nerve cells."[9]

[9]Personal communication, 1971.

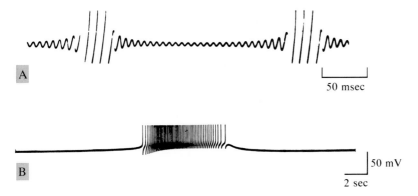

FIGURE 3-4.
Comparison of spontaneous oscillatory activity and action potentials recorded extracellularly in the isolated giant axon of the cuttlefish *Sepia* and intracellularly in the giant cell body of an identified neuron (R15) in *Aplysia.* **A.** The membrane potential of the giant axon of *Sepia* was made to oscillate by adding 10 percent sodium citrate to the physiological bathing solution, thereby removing calcium. The oscillations increased in amplitude until the threshold for spike generation was reached. Regularly occurring bursts of spikes are separated by oscillatory potentials. [From Arvanitaki, 1939.] **B.** Intracellular recording of characteristic spontaneous oscillatory activity in cell R15 of the abdominal ganglion of *Aplysia.* Spontaneous bursts of action potentials of the sort illustrated in the figure recur in a fairly regular fashion every 20 or 40 seconds. [From Arvanitaki and Chalazonitis, 1964.]

It had been known at least since the last quarter of the nineteenth century that many gastropod molluscs have gigantic nerve cells (Ihering, 1877), and that these are especially well developed in opisthobranchs (Mazzarelli, 1893). Arvanitaki examined the nervous systems of several gastropods and introduced into cellular neurophysiology the land snail *Helix* and the marine snail *Aplysia* (Arvanitaki and Cardot, 1941a, b, c, d). She found the abdominal ganglion of *Aplysia* particularly suited to the examination of spontaneous activity, since many of its cells had characteristic firing patterns: regular beating, regular bursting, and irregular bursting (Figure 3-4*B*). Interest in this ganglion was heightened when Ladislav Tauc and his colleagues found that *Aplysia* is useful for studying the biophysical properties of neurons, central synaptic transmission, and transmitter pharmacology (Tauc, 1958; Tauc and Gerschenfeld, 1961; for review see Tauc, 1966).

As early as 1942 Arvanitaki and Tchou found that certain cells in the abdominal ganglion of *Aplysia* can be repeatedly identified in every member of the species. This work was extended by Coggeshall, Kandel, Kupfermann,

Ladislav Tauc, 1973
(Courtesy of L. Tauc)

and Waziri (1966; see also Frazier *et al.*, 1967). More recently, Gardner (1971) mapped identified cells in the buccal ganglion and Kehoe (1972b) mapped identified clusters of cells in the pleural ganglia of *Aplysia*. Maps of identified cells have also been obtained for ganglia of other opisthobranchs: for *Tritonia* (Dorsett, 1967: Willows, 1968; Willows, Dorsett, and Hoyle, 1973a), *Navanax* (Levitan, Tauc, and Segundo, 1970), the land snail *Helix* (Kerkut, French and Walker, 1970), and *Helisoma* (Kater, 1974), as well as for central ganglia of the lobster (Otsuka, Kravitz, and Potter, 1967), the crayfish (Kennedy, Selverston, and Remler, 1969), the leech (Nicholls and Baylor, 1968; Stuart, 1970), the cockroach (Cohen and Jacklett, 1967) and the locust (Burrows and Hoyle, 1973).

The identification of neurons marked a new departure in the study of behavior, making it possible to relate unique cells to specific behavior and facilitating the study of the mechanisms that underlie behavioral modification. In addition, cell identification may help achieve the ethologists' hope of using behavioral criteria in taxonomy (see p. 16). So far this task has been hampered by technical and interpretive difficulties. Repeated attempts have been made to establish homologies of behavior and to trace behavior in different but closely related species to common ancestors, much as comparative anatomists have done with homologous structures. But behavior is more difficult than structure to treat comparatively because it is typically transient and may be variable and hard to quantify (Atz, 1970). Moreover, the current techniques for studying behavior are more limited than those for studying structure. The discovery that particular cells mediate behavior has raised the possibility that the difficulties of establishing homologies might be overcome by using the neural architecture of behavior as the basic taxonomic characteristic.

GENETIC ANALYSIS OF INVERTEBRATE NERVOUS SYSTEMS

I have so far described only one aspect of the current interest in the behavior of higher invertebrates, the cellular physiological and biochemical studies. A second aspect is the interest in genetic analyses of behavior (behavioral genetics), as exemplified in studies of the fruit fly *Drosophila* and the nematode worm *Caenorhabditis elegans.*

Modern genetics originated with the rediscovery of Gregor Mendel's theory (1866) that hereditary characteristics are determined by elementary units transmitted in a predictable fashion from one generation to another. Each unit, or gene, serves two functions. It transmits hereditary information from one generation to another so as to allow each succeeding generation to have a physical copy of the unit, and it dictates information about structure, function, color, and other biological features to all cells that carry the copy. Because of the dual function of the gene there have been two approaches to the study of genetics. One is concerned with identifying the physical substance, the genetic material, passed between generations; microorganisms have proved immensely useful for these studies. The second approach is concerned with identifying the rules by which the gene action is expressed in biological characters. Here a variety of organisms have been useful. One of the most instructive has been *Drosophila.*

The genetic potentialities of *Drosophila* were first fully realized by Thomas Hunt Morgan. Morgan's work (1927) illustrated the many advantages that *Drosophila* has over other animals for genetic studies. In both plants and animals genes are arranged linearly along threads, called chromosomes, located in the nucleus of each cell. Germ cells contain one copy (haploid number) of each chromosome, whereas the body (somatic) cells contain two copies (diploid), one from each parent. Animals vary in the number of chromosomes per cell. *Drosophila* has only four chromosomes in its germ cells. This contrasts with seven chromosomes in the germ cells of the peas used by Mendel, 15 in the cricket, 17 in *Aplysia*, and 23 in man. *Drosophila* are very small and can be reared by the thousands in the laboratory. Also, their generation time is relatively short: two weeks. Because of these advantageous features, Morgan and his colleagues discovered a great number of hereditary variants (mutants) that differed from the normal (wild type) in wing size, eye size, body color, and eye color. Each mutant was initially recognized as a single deviant in a population of thousands of individuals.

Morgan inferred that in each case deviation reflected a rare, spontaneously occurring change (mutation) in the structure of the gene controlling that character. By cross breeding mutants he was able to probe the mechanisms for hereditary transmission. He determined that some combinations of traits are inherited together because they are physically carried together (linked) on the same chromosome. From these types of data Morgan developed genetic maps of the four *Drosophila* chromosomes indicating the location of the mutant genes (Morgan, 1927). Soon, H. J. Muller (1927) found that X-rays dramatically increased the frequency of mutation in *Drosophila.* Subsequently, several chemical substances were found that also increase the mutation rate. It thus became possible to produce a large variety of mutants having morphological and biochemical abnormalities (Strickberger, 1968) as well as behavioral abnormalities (Benzer, 1967; 1973). Some mutants are disturbed in their ability to locomote: they are sluggish, or hyperkinetic, nonclimbing, or flightless. Other mutants are disturbed in their sexual behavior. A behavioral mutant called *stuck* cannot disengage after copulation; another called *coitus interruptus* disengages prematurely and produces no offspring.

In recent years Seymour Benzer (1973) has further increased the interest in *Drosophila* by using behavioral mutants to analyze and trace the development of the nervous system in relation to behavior. Previously, molecular biologists had used mutants to analyze the biochemical pathways of cellular function in bacteria by examining the consequences of removing one reaction at a time. Benzer is now using single-gene mutants to analyze how the neural circuits that control specific behavior are altered. This permits him to identify the genetic component of a behavior and to relate the behavior both to a particular gene and to the anatomical site in the nervous system where the gene exerts its action.

By these means one might be able to examine what parts of the neural circuit of a behavior are specified by single genes. Is the development of all the motor cells or all the sensory cells specified? Are patterns of interconnection specified? Once a particular single-gene mutation producing an abnormality in behavior has been ascribed to a particular anatomical locus, one can then determine whether this site is the locus where the mutant gene acts, or whether the locus is elsewhere and the gene produces its action by the presence (or absence) of a diffusible substance (a toxin or a metabolite). This analysis is made possible by another attractive feature of *Drosophila,* the ability to grow *mosaics* (composite organisms) in which all parts of the animal are normal except one, namely the anatomical site where the mutant gene is presumed to exert its action. This is equivalent to grafting a small piece of mutant tissue on a normal fly. By means of mosaics, Benzer and his colleagues have found that the eye is the locus of action for certain nonphototactic mutants, mutants that do not respond to light.

Finally, Benzer has refined an older technique, *fate mapping*, in which anatomical parts of the differentiated adult are related to their sites of origin in the blastoderm (Benzer, 1973). According to Benzer, fate mapping can "unroll the fantastically complex adult fly, in which sense organs, nerve cells, and muscles are completely interwoven, back in time to the blastoderm, a stage at which the different structures have not yet come together." This is another method by which mutant behavior can be localized to the site of expression.

Using these techniques, Benzer hopes to be able to relate a gene sequentially to its initial expression in the blastoderm and then to behavior in the adult. Mutant analysis and the use of mosaics and fate mapping are therefore extraordinarily powerful tools for analyzing behavior.

Genetic studies are also being carried out on a soil nematode, *Caenorhabditis elegans,* introduced into neurobiological studies by Sidney Brenner (1973). Since the worms are self-fertilizing hermaphrodites, mutants give rise to progeny that are homozygous at almost all gene loci. The generation time is very short (2.5 days) and each parent gives rise to 250 progeny. This worm is very small (60 μm by 1 mm), so that up to 10^5 individuals can be maintained in a small petri dish. Moreover, all the organ systems, including the nervous system, have a fixed number of cells (see p. 220). The nervous system is remarkably simple, containing only 300 cells. It is therefore possible, with electron microscopy, to reconstruct the three-dimensional morphology of its nervous system, including the position, size, shape, branching patterns, and interconnections of cells. This kind of reconstruction permits detection even at the synaptic level of morphological changes that mutant genes might produce. Many behavior mutants have been identified. Some of these have altered motor activity, others have sensory deficits, particularly in their chemical senses. Examination by electronmicroscopy of serial sections of some of these mutants has revealed abnormalities in the morphology of what are presumed to be motor and sensory cells.

Similar techniques are also being used by Cyrus Levinthal and his colleagues to study the development of connections between cells in isogenic lines of the crustacean *Daphnia,* the water flea (Macagno, Lopresti, and Levinthal, 1973; Lopresti, Macagno, and Levinthal, 1973).

IS THERE ONE IDEAL ORGANISM FOR STUDYING BEHAVIOR?

In view of the obvious advantages for the study of behavior of combining cellular biological and genetic techniques in the same animal, it would be desirable to concentrate on one organism. At the moment, however, no single invertebrate appears to be ideally suitable for the study of behavior. Indeed, some of the desired characteristics are in theory mutually exclusive. *Drosophila* and nematode worms, with favorable genetic properties for mutant isolation, are physically very small. With the exception of a few motor cells in *Drosophila* that may reach 15 μm, the nerve cells are generally less than 5 μm in diameter. Small cells are not convenient for physiological and biochemical studies of single cells. Conversely, animals useful for cellular studies are usually larger and have longer generation times; they are therefore not ideal for genetic studies. There is as yet no perfect subject—no animal is both genetically tractable and well-suited to cellular studies. Although it greatly compounds the task, the lack of an ideal experimental animal may even have some advantages. Overreliance on a single animal is bound to produce work that is to some degree provincial and overlooks important dimensions of behavior.[10] Work on several invertebrates, each having advantages for solving particular behavioral problems, is more likely to produce general principles relating neuronal function to behavior.

Moreover, for certain behavioral problems the disadvantages of sacrificing some aspects of either cellular or genetic analyses are outweighed by the advantage that accrues from a compromise based on a combined approach. The best example is the cricket. Some tropical crickets of the genus *Teleogryllus* have a six-week generation time. Although not ideal, this generation time permits certain types of genetic analyses in an animal that is one thousand times larger than *Drosophila* and in which some single-cell investigations are possible. Bentley (1971; Bentley and Hoy, 1974) has studied the calling song by which males attract receptive females (see p. 35). The song consists of a chirp, followed by several trills. The trills are produced by wing movements driven by wing-opener and closer motor neurons located in two thoracic ganglia. The species *T. oceanicus* has two pulses per trill; *T. commodus* has 14 pulses per trill. The differences are reflected in the firing pattern of the motor neurons. Mating of the two species followed by backcrossing with the parental strain leads to hybrid animals having combinations of parental characteristics: 3,4,5 and 6 pulses per trill; the trills are again reflected in the discharge pattern of motor neurons. Thus, genetic changes producing even a slight change in the motor output of homologous neurons produce different patterns of behavior.

So far, most of the genetic variants in crickets have been derived from natural, not laboratory-reared populations. Since variants in natural populations usually have many genetic differences, these differences are difficult to analyze. Indeed, the genetic differences involved in the calling song involve many genetic loci on different chromosomes (Bentley, 1971).

---

[10]See p. 13 for the history of research on the white rat.

In the long run it will prove possible to develop single-gene mutations in crickets or to do neurobiological analyses in a genetically favorable animal, such as *Drosophila*. It may even be possible to do occasional genetic analyses in animals favorable for single-cell studies, such as the gastropod molluscs *Helisoma* and *Aplysia*. Nevertheless, compromising the choice of research animals risks the danger that neither the genetics nor the cellular biology of the animal will be optimally understood. Most investigators have therefore chosen animals that are heavily advantageous for either a cellular biological or a genetic analysis. Furthermore, the two approaches have proven complementary; usually what can be accomplished by one cannot be done by the other.

In the remaining chapters I will focus on studies of behavior based on analyses of single nerve cells. In addition to reflecting my experience and bias, this decision reflects the fact that presently the cellular biological approach to the study of behavior is more advanced in reaching its objectives than is the genetic approach. This is because studies of the genetics of behavior have only recently used single-gene mutations to analyze cellular mechanisms. Much work has necessarily been spent on developing the techniques—such as rapid screening of mutants, development of mosaics, fate mapping, and three-dimensional reconstructions—required to analyze single-gene mutations. In addition, most behavioral mutants are developmental mutants. Their abnormal behavior reflects a failure in the nervous system to develop normally and not a failure of the normally developed nervous system to function correctly (Brenner, 1973). As a result, genetic studies of behavior have been more concerned with the question: What aspects of the neural circuit of a behavior are coded by single genes? These studies have therefore been concerned less with the regulatory processes that specify how the functioning nervous system generates behavior and how learning modifies cellular function (for an interesting exception, see Quinn, Harris, and Benzer, 1974).

In contrast to genetic studies, cellular studies of behavior have required fewer technical innovations. These studies have been able to use the powerful and versatile methods developed by workers in cellular biology and psychology. As a result, some principles have already emerged from cellular studies of behavior in insects, crustaceans, annelid worms, and gastropod molluscs.

The selection of a specific animal for cellular biological studies is somewhat arbitrary. Insects and cephalopod molluscs are the most interesting behaviorally, but they are also the most difficult to use for cellular studies. Gastropod molluscs, whose behavior is perhaps least varied, are technically the most advantageous. The cell bodies are large and, to a greater degree than is generally the case in other invertebrates, permit faithful recordings of most synaptic and other subthreshold potentials generated in the neuropil (see Chapter 3). Lobsters, crayfish, and annelid worms fall in between. Useful insights have come from studies of each group of animals. Nevertheless, the difficulty of carrying out cellular analysis in a given animal has invariably proven more limiting than the lack of interesting behavior. For this reason gastropod molluscs, even though their behavior is limited, are presently particularly good subjects for neurobiological analyses simply because they afford the greatest advantages for studying individual nerve cells.

SUMMARY AND PERSPECTIVE

The modern struggle for an objective psychology was in part won on the basis of successful studies of the behavior of invertebrates. Darwin and Romanes worked on earthworms, jellyfish, starfish, and sea urchins. Many behaviorists of the next generation—Lubbock, Loeb, Verworn, and Jennings—also worked with invertebrates. In the period 1930 to 1960, however, interest in invertebrates declined among experimental psychologists. Associative learning was difficult to demonstrate in many invertebrates. Instead, psychologists focused on vertebrates in their search for the general principles of behavior underlying associative learning. But ethologists continued to study invertebrates; working in the tradition of natural history, and continuing the work of Darwin, they studied behavior comparatively. They were also more concerned with the adaptive significance of behavior than with specific instances of associative learning. Their interest in feeding, defense, and sexual and social behavior led them to find in invertebrates many elementary examples of universal behavior. Certain invertebrates, particularly insects, were used extensively to study fixed-action patterns and simple forms of nonassociative learning, such as habituation. Developments in the ethology and psychology of invertebrates were codified and integrated in 1944 and again in 1956 by W. H. Thorpe in several major papers in which he established criteria for demonstrating learning in invertebrates.

Another impetus to the study of invertebrate behavior came from neural science. Early workers on the physiology of single nerve cells of invertebrates (Adrian, Prosser, and Hartline) were interested in arthropods and insects. Only later, after the work of Young and Arvanitaki, were the advantages of molluscs appreciated. The value of using invertebrates to study behavior was significantly enhanced by the discovery that in certain invertebrate ganglia some cells were invariant in every animal of the species. Furthermore, some identified cells within species were shown to have invariably the same behavioral function, making it possible to relate specific behavior to specific nerve cells.

Certain invertebrates—in particular the soil nematode *Caenorhabditis elegans* and the fruit fly *Drosophila*—attracted the interest of neurobiologists because of their usefulness for genetic studies. These animals reproduce rapidly and can be made to generate behavioral mutants that facilitate the study of the neuroanatomical basis of behavior. However, animals useful for genetic studies are often not suitable for cellular studies; similarly, animals that are appropriate for studies of single nerve cells are not ideal for genetic studies. In addition, the objectives of the two approaches differ. Cellular studies are chiefly directed toward analyzing how the nervous system controls behavior and how these control mechanisms are modified by learning. The genetic approach is primarily concerned with analyzing how genes determine the structure and development of the nervous system. Students of the mechanisms of behavior have usually had to choose between one approach and the other. Of the two, the cellular study of behavior is at the moment more advanced and its objectives more immediately relevant to understanding how behavior is generated. This book will therefore focus on the cellular biological approach.

SELECTED READING

Bentley, D., and R. R. Hoy. 1974. The neurobiology of cricket song. *Scientific American*, 231(2):34–44. An introduction to cricket neurobiology, illustrating the usefulness of genetic analysis of behavior.

Benzer, S. 1973. Genetic dissection of behavior. *Scientific American*, 229(6): 24–37. A review of recent studies of *Drosophila* behavior, illustrating the power of fate mapping, mosaics, and genetic dissection of the nervous system.

Brenner, S. 1973. The genetics of behavior. *British Medical Bulletin*, 29:269–271. An outline of the strategies used to study the soil nematode *Caenorhabditis elegans*.

Kennedy, D. 1971. Fifteenth Bowditch lecture: crayfish interneurons. *Physiologist*, 14:5–30. Surveys work on the crayfish nervous system and emphasizes the importance of command fibers.

Kennedy, D., and W. Davis. 1976. The organization of invertebrate motor systems. In "Cellular Biology of Neurons" (Vol. 1, Sect. 1, *Handbook of Physiology*), E. R. Kandel, ed. Williams and Wilkins. Outlines the organizational principles of invertebrate motor systems, based largely on work on arthropods.

Nicholls, J. G., and D. Van Essen. 1974. The nervous system of the leech. *Scientific American,* 230(1):38–48. An introduction to the neurobiology of the leech.

Roeder, K. D. 1963. *Nerve Cells and Insect Behavior.* Harvard University Press. An early description of how cellular techniques have been used to study insect behavior.

Thorpe, W. H. 1956. *Learning and Instinct in Animals.* Harvard University Press. A very influential book that illustrates the continuity between invertebrate and vertebrate learning.

Willows, A. O. D. 1973. Learning in gastropod mollusks. In *Invertebrate Learning,* W. C. Corning, J. A. Dyal and A. O. D. Willows, eds. Plenum Press. Reviews gastropod behavior, emphasizing cellular studies.

CHAPTER 4

The Behavioral Biology of a
Single Invertebrate:
The Marine Snail *Aplysia*

The sometimes complex and fascinating behavior of crustaceans, annelid worms, and gastropod molluscs is not as apparent as the dramatic variety of behavior of insects. Consequently, neurobiologists interested in behavior were first attracted to insects.[1] However, the neurobiological study of behavior requires experimental animals whose nervous systems can be conveniently studied at a cellular level. As it became clear that the limiting factor in the neural analysis of behavior is the difficulty of applying cellular techniques, annelids, crustaceans, and gastropods began to be selected for cellular studies of behavior because of the technical advantages of their nervous systems. This selection was also based on two assumptions. First, all ganglionated central nervous systems mediate interesting behavior. Second, the principles of the neuronal organization of behavior are likely to be general, so that detailed insights into the neuronal architecture of any one behavior are likely to be useful for analyzing other behavior.

The selection of *Aplysia* represents an extreme case of this pragmatic strategy. Despite its limited behavior, even compared to other invertebrates, *Aplysia* was selected by neurobiologists because of the great analytic advantages of its central nervous system and because considerable information on the electrophysiological properties of its neurons had accumulated as a result of the pioneering work of Arvanitaki, Tauc, and Gerschenfeld.

---

[1]For example, see von Frisch, 1950; Wilson, 1968; Hoyle, 1973.

The central nervous system of *Aplysia* is remarkably well suited for studies of single cells. Almost all its nerve-cell bodies are large enough to permit the insertion of microelectrodes for intracellular recording. Many of the cells can be individually identified (see Chapter 7) so that identical neurons can be examined in any number of individual animals under a variety of conditions. Single cell bodies can be dissected for biochemical studies. Radioactive chemical substances or dyes can be injected into the cell body and their movement throughout the neuron monitored morphologically and biochemically. Unlike arthropods, the synaptic area in the neuropil of *Aplysia* and other opisthobranch molluscs is relatively close electrically to the cell body. As a result, synaptic connections can often be monitored electrically, with minimal attenuation, by obtaining intracellular recordings in the cell body of the neuron (see Chapter 8).

The ability to trace connections from cell to cell makes it possible to relate types of neuronal circuits and types of behavior. In addition, the abdominal ganglion, on which much of the behavioral work in *Aplysia* has been carried out, functions to some degree as a small brain, controlling several behavioral responses. This permits one to examine the ways in which a central nervous system establishes temporal sequences between synergistic responses and selects from among competing responses. Finally, *Aplysia* can be cultured in the laboratory from a fertilized egg to the reproductive adult within a generation time of 19 weeks, so that genetically controlled strains can be raised. Such inbred strains are more homogenous than animals in the wild and therefore reduce variability for behavioral and neurobiological studies. In addition, these animals can be used for analyzing the development of the nervous system and behavior.

These technical advantages made *Aplysia* attractive for the exploration of the rules governing the patterns of nervous connections and the relationship between different patterns of connections and behavior. But since little was known about the behavior of *Aplysia*, considerable effort first had to be invested in the study of its behavioral capabilities. This chapter is a brief introduction to what is now known about the biology, including behavior, of *Aplysia*.[2] The next two chapters take up in detail the cellular biology of neurons.

Here, and in later chapters, I shall discuss some aspects of the physiology of effector organs. This perhaps needs explaining. Conditioned by everyday talk, one is inclined to think of behavior in terms of reflexes, such as withdrawal, or fixed-action patterns, such as escape, or perhaps a higher-order of behavior, such as playing tennis. Reduced to their observable essentials, however, all

[2]This subject is treated in greater detail in Kandel, *The Behavioral Biology of Aplysia,* forthcoming.

these examples consist of a series of somatic muscular movements, usually accompanied by altered activity in visceral muscles or glands—changes in heart rate, salivation, stomach contractions. As behaviorists have emphasized, the contraction of muscles, changes in visceral tone, and secretions of glands are the basic elements of observable behavior. They are the only aspects of mental processes that can be studied directly (see Chapter 2, p. 30). Analysis of the behavioral strategies and thought processes of another person (or animal) is strictly *inferential;* only changes in muscular and glandular activity can be directly observed and measured. The neural control of behavior therefore reduces to the neural control of effector organs.

THE MOLLUSCS

The phylum Mollusca, to which the genus *Aplysia* belongs, is the second largest in the animal kingdom. In the number and variety of its species, it is surpassed only by the arthropods. The embryonic development of molluscs and arthropods is quite similar, and the development of both resembles that of a third invertebrate phylum, the annelids. These three phyla are therefore thought to be derived from a common ancestor. Embryological similarity has given rise to the idea that the animal kingdom evolved along two lines: the molluscan-arthropod-annelid line, which includes flatworms and nemertines, and the chordate line, which includes echinoderms and hemichordates (see Figure 3-1, p. 49).

Molluscs are highly adaptable from an evolutionary point of view. This is evident in the variety of body structures and in the flexible plans of their nervous systems. The adaptability of molluscs has been well described by Nellie B. Eales, one of the major students of *Aplysia* biology.

> Arthropods and molluscs have reached the top of the invertebrate scale of organisms. They both brought in new ideas in structure and function, the arthropods being segmented, with numerous paired appendages and a secreted external armour, all in one piece, which necessitates growth in a series of jumps in the short interval between the casting of the old exoskeleton and the laying down of the new one. In the molluscs the plan is quite different, and on the whole, more adaptable. There is no segmentation, no wasteful repetition of organs, but rather concentration on the production of efficient units, e.g., foot, mantle, pallial [visceral mass] complex. The body remains plastic, so that molluscs can explore the possibilities of the organs they have evolved. Their protective shell is not an armour which binds and limits the form of the soft parts, but a covering that grows with the animal, and it can be enlarged or reduced or even enclosed to suit the requirements of the individual. The mucous skin makes moisture a ne-

cessity, so that their range is more limited than that of arthropods. The majority of the species are marine; about one in eight inhabit fresh water, rather less than half are terrestrial, and none can fly. Most move slowly but surely, but the cephalopods, by learning to modify the organs they inherited, invented jet propulsion and are the swiftest moving of the marine animals.[3]

Molluscs range in complexity from simple parasitic forms to cephalopods—squid, cuttlefish, and octopods, the most intelligent of invertebrate animals. Indeed, the learning capabilities of cephalopod molluscs rival those of vertebrates. In spite of great heterogeneity, all molluscs are built on a simple and general body plan (Figure 4-1). The plan, which is thought to derive from an ancestral (archimolluscan) type, consists of four main parts: (1) the *head*, containing the mouth, tentacles, and eyes; (2) the *foot*, a ventral muscular surface used for locomotion; (3) the *visceral mass*, a domelike structure containing the heart, kidney, and reproductive organs; and (4) the *mantle*, a wide skirt that covers and protects the visceral mass. The mantle also secretes the shell. The area between the mantle and the visceral mass forms a sheltered space, the *mantle cavity*. The mantle cavity is primarily a respiratory chamber containing the gill; it also houses the terminations of the digestive and renal systems, which are discharged into it. In some gastropods a fold of the mantle forms a siphon that channels water into or out of the mantle cavity.

Molluscs are classified into three relatively small and primitive minor classes, Monoplacophora, Amphineura, and Scaphopoda, and three larger and more advanced major classes, Bivalvia, Gastropoda, and Cephalopoda (Figure 4-1). The nervous systems of the molluscan classes vary from the simple cord-like systems of the monoplacophorans and amphineurans, which are comparable in structure to the nervous system of flatworms, to the complex brain of cephalopods, which compares in both size and capability to that of the lower vertebrates. As is true for other higher invertebrates and vertebrates, increased complexity of the molluscan nervous system typically involves an increase in size, a shortening of the connectives and commissures, and a concentration and ultimate fusion of individual ganglionic masses into a more massive brain (Figure 4-2).

Aplysia and the other molluscs that we shall consider in later chapters belong to the class Gastropoda (belly-footed). This class contains three subclasses: (1) the prosobranchs, or fore-gilled snails; (2) the opisthobranchs, or hind-gilled snails (including the neurobiologically useful *Aplysia, Navanax, Pleurobranchaea, Tritonia, Melibe,* and *Anisodoris*); and (3) the pulmonates, or lung

[3]Eales, 1950, p. 53.

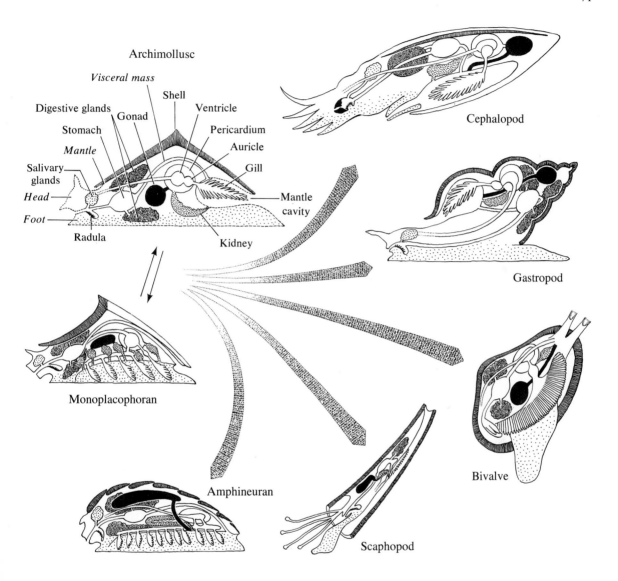

FIGURE 4-1.
The phylum Mollusca includes the second-widest range of species in the animal kingdom. Yet all members of the phylum share a common anatomical plan and seem to be descendants of common ancestors. The body plan (or morphotype) of the phylum is presented here as the archimollusc. Monoplacophorans, the most primitive molluscan class, of which there is only one living genus, *Neopilinia,* have hard parts that resemble those thought to have been present in the earliest molluscs. *Neopilinia* embodies many of the features specified in the theoretical archimollusc; they also have some signs of possible segmentation, not found in more advanced molluscs. The other five molluscan classes (gastropods, amphineurans, scaphopods, bivalve gastropods, and cephalopods) can also be derived from the hypothetical archimollusc. Homologous anatomical parts are identified by similar shading. [Modified from Wells, 1968.]

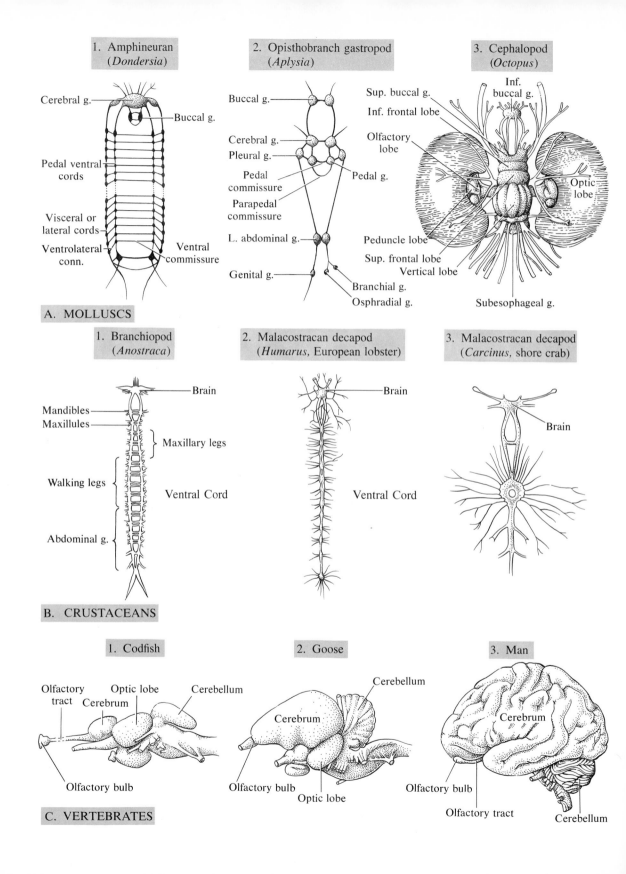

1. Amphineuran
(*Dondersia*)

Cerebral g.

Buccal g.

Pedal ventral cords

Visceral or lateral cords

Ventrolateral conn.

Ventral commissure

A. MOLLUSCS

2. Opisthobranch gastropod
(*Aplysia*)

Buccal g.

Cerebral g.

Pleural g.

Pedal commissure

Parapedal commissure

Pedal g.

L. abdominal g.

Genital g.

Branchial g.

Osphradial g.

3. Cephalopod
(*Octopus*)

Inf. buccal g.

Sup. buccal g.

Inf. frontal lobe

Olfactory lobe

Optic lobe

Peduncle lobe

Sup. frontal lobe

Vertical lobe

Subesophageal g.

1. Branchiopod
(*Anostraca*)

Brain

Mandibles

Maxillules

Maxillary legs

Walking legs

Ventral Cord

Abdominal g.

B. CRUSTACEANS

2. Malacostracan decapod
(*Humarus*, European lobster)

Brain

Ventral Cord

3. Malacostracan decapod
(*Carcinus*, shore crab)

Brain

1. Codfish

Olfactory tract

Cerebrum

Optic lobe

Cerebellum

Olfactory bulb

2. Goose

Cerebellum

Cerebrum

Olfactory bulb

Optic lobe

3. Man

Cerebellum

Cerebrum

Olfactory bulb

Olfactory tract

Cerebellum

C. VERTEBRATES

FIGURE 4-2.
Progression of brain complexity in animals belonging to the molluscan-arthropod and chordate lines. In each phylum there are tendencies for (1) a progressive increase in size of the nervous system; (2) shortening of connectives and commissures, thereby bringing ganglion masses closer together; and (3) fusion of ganglia into a complex brain (usually in the head region). **A.** Molluscs. [Modified from Bullock and Horridge, 1965.] **B.** Crustaceans. [Ibid.] **C.** Vertebrates. [From Truex, 1959.] The drawings are not to scale.

carrying snails (*Helix, Limax* and *Helisoma*). Gastropods have diverged extensively. Among the 35,000 existing species there are marine species adapted to all types of ocean life, species that have invaded fresh water, and species (pulmonate snails and slugs) that have become land dwellers.

STUDIES OF *APLYSIA*

For at least 2,000 years before Angelique Arvanitaki found specimens of an *Aplysia* species on the beach of Tamaris in 1937, naturalists had known and studied a large, strange looking marine snail, presumably *Aplysia,* and described its ability to secrete jets of dark purple ink. *Aplysia* was first seriously studied at the end of the nineteenth century by biologists interested in tracing the origins of opisthobranchs from their presumed prosobranch ancestry. Because *Aplysia* has anatomical features that are intermediate between those of more primitive prosobranchs and those of more advanced opisthobranchs, it provided an interesting transitional form in the study of gastropod evolution. Subsequently, physiologists were drawn to *Aplysia* because of its size, its abundance, and the ease with which it could be dissected. The large foot of *Aplysia* was used in studies of locomotion, and the accessibility of its nervous system and the wide distribution of its ganglia were convenient for studying neural control of the heart and other visceral functions.

After the turn of the century interest in *Aplysia* declined. Between 1920 and 1940 *Aplysia* was primarily used for studies of muscle tone and neuromuscular transmission, problems for which it was not particularly advantageous. It was not until Arvanitaki pointed out the advantages of the large nerve cells and distributed ganglia for electrophysiological studies of single cells that interest in this animal again emerged, this time focusing on the nervous system. Recently, history has come half turn. There are an increasing number of studies of the nervous system of *Aplysia* and relatively few of its organ physiology and behavior—studies that are now much needed.

NATURAL HISTORY AND BIOLOGY OF *APLYSIA*

The opisthobranch genus *Aplysia* is one of the 10 known genera of the family Aplysiidae, belonging to the order Anaspidea. It includes five subgenera and 35 identified species.[4] The subgenus now most commonly used by physiologists is *Neoaplysia,* the only known species of which is *Aplysia californica,* found along the California coast. Three species belonging to the subgenus *Varria,* *A. brasiliania, A. dactylomela,* and *A. willcoxi,* are also found off the coasts of North America, as is the species *A. vaccaria* of the subgenus *Aplysia.* The three most common European species are *A. (Pruvotaplysia) punctata, A. (Varria) fasciata,* and *A. (Aplysia) depilans.*

Aplysia is an herbivore that feeds on seaweed. It lives in two types of shore environments: the littoral or intertidal zone, which is submerged and uncovered almost daily, and the sublittoral zone, which is never exposed to air. Animals living in the littoral zone are often exposed to the air for several hours a day and are buffeted about by the tides. These animals must withstand wide and rapid changes in temperature, humidity, salinity, pressure, mechanical stimulation, and wind, and survive repeated exposure to air and immersion in the sea (Yonge, 1949; Hardy, 1959; Kupfermann and Carew, 1974). By contrast, sublittoral animals are rarely exposed to air.

Aplysia has a one-year life cycle with maximum growth in the late summer, followed by death soon after spawning in the autumn. During spawning the animal lays large egg masses containing 10^5 to 10^6 eggs. These hatch about 12 days after they are laid, and free-swimming larvae (veligers) emerge (Figure 4-3). Although the veliger is a developing animal, it is an independent organism that performs all necessary life functions except for reproduction, namely, feeding, locomotion, and escape. The veliger metamorphoses after 33 to 37 days and undergoes a radical change into a *juvenile,* progressively taking on the form, diet, behavior pattern, and life style of the adult (Figure 4-3; and Kriegstein, Castellucci, and Kandel, 1974). The juvenile reaches reproductive maturity three months after metamorphosis.[5]

Adult *Aplysia* have the typical features of the molluscan body plan: head, foot, mantle, and visceral mass (Figure 4-4). The *head* contains a mouth, two small eyes, and anterior and posterior tentacles or feelers. Both tentacles are chemosensory; the anterior ones participate in feeding, the posterior ones aid in sensing food at a distance.

[4]For details see Eales, 1960.
[5]For details see Kandel, *Behavioral Biology of Aplysia,* forthcoming.

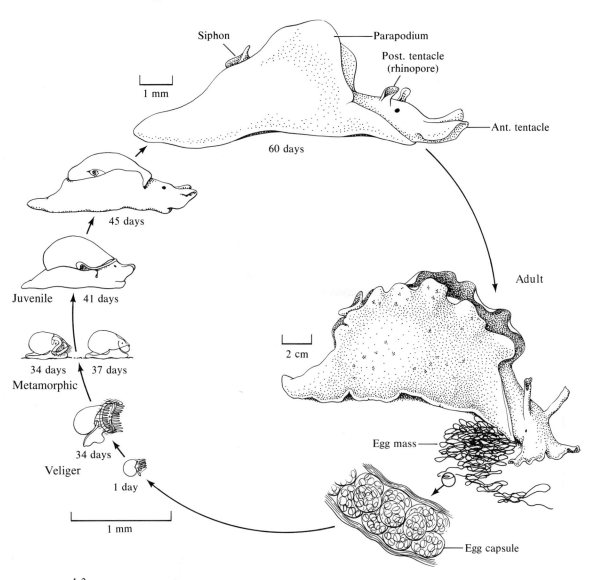

Siphon

Parapodium

Post. tentacle
(rhinopore)

1 mm

60 days

Ant. tentacle

Adult

45 days

Juvenile   41 days

34 days   37 days
Metamorphic

2 cm

34 days

Veliger

1 day

Egg mass

1 mm

Egg capsule

FIGURE 4-3.
Life cycle of *Aplysia californica*. Post-hatching development can be divided into four major phases: the veliger or planktonic (days 1–34), the metamorphic (days 34–37), the juvenile (day 37 to sexual maturity at approximately day 120), and the adult. External features at each of these phases are illustrated in the drawings (drawings are to the same scale, with the exception of the 60-day juvenile and the adult). The adult deposits egg masses consisting of a long string of egg capsules, each containing 15 to 20 eggs. [After Kriegstein *et al.*, 1974.]

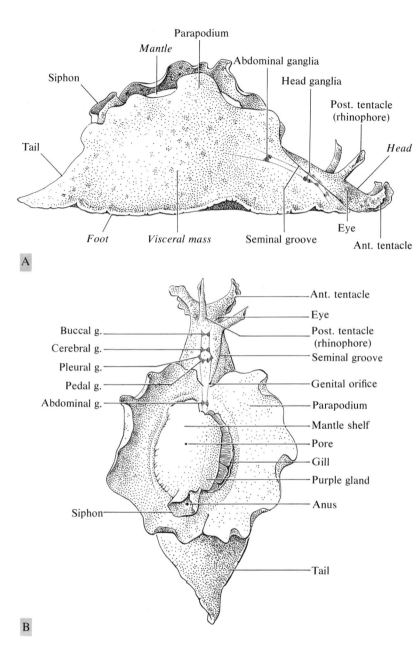

FIGURE 4-4.
Aplysia californica. **A.** Side view illustrating head, mantle, visceral mass, and foot. The major ganglia of the nervous system are superimposed in the relative position they have inside the animal. **B.** Dorsal view. The parapodia have been retracted and the mantle is partially exposed. The outline of the nervous system has again been super-imposed to indicate the relative positions of the various ganglia.

The *foot* is long and extends along the whole length of the animal from the mouth to the short tail or metapodium. The parapodia, two lateral extensions of the foot, are used in some species for swimming.

The *mantle* consists of the mantle shelf; its posterior extension, the siphon; and the mantle cavity, containing the gill, osphradium, and opaline gland, which releases a whitish secretion of unknown function. The osphradium is a sense organ that tests osmolarity and perhaps other chemical properties of the seawater bathing the mantle cavity. On the edge of the mantle shelf is the purple gland, which secretes ink when the animal is perturbed.. The mantle contains the shell, which is relatively small in the adult.

The *visceral mass* contains the heart, most of the gastrointestinal tract, and the urinary and reproductive systems.

NERVOUS SYSTEM OF *APLYSIA*

General Plan

The central nervous system of *Aplysia* consists of four paired head ganglia—cerebral, buccal, pleural, and pedal—that form a ring around the esophagus, and a single unpaired abdominal ganglion below the esophagus (Figures 4-2 and 4-5; and Eales, 1921). The cerebral, pedal, and buccal ganglia on each side are united to their symmetrical mate by nerve fiber tracts called *commisures* and to other ganglia by fiber tracts called *connectives*. The cerebral ganglia innervate the tentacles and eyes, the buccals innervate the buccal musculature and the rostral portion of the intestinal tract, and the pleuro-pedals innervate the foot and the parapodia. The pleurals give rise to the pleuroabdominal connectives that connect to the abdominal ganglion. This ganglion is made up of five fused ganglia that lie close together and are interconnected by a commisure that is usually obscured in the adult animal. The head ganglia innervate the head and foot and control the major somatic functions of *Aplysia*: feeding, copulation, and locomotion. The abdominal ganglion innervates the mantle and the visceral hump and controls the major visceral functions: circulation, respiration, excretion, and reproduction.

The nervous system is avascular, like that of all molluscs except cephalopods, but the sheath of the abdominal ganglion is highly vascularized. Numerous fine terminal processes from the neurosecretory cells in the ganglion appear to

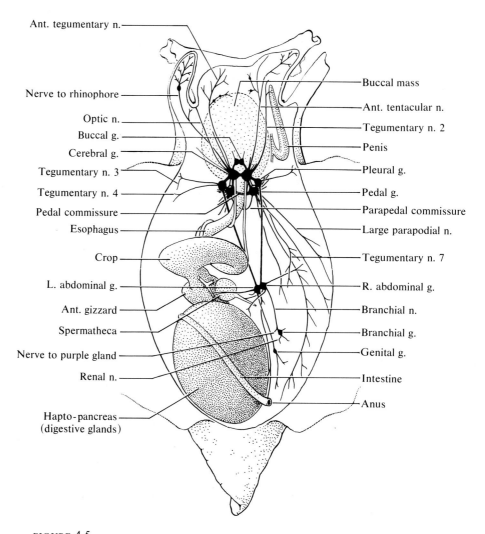

Ant. tegumentary n.

Nerve to rhinophore

Optic n.

Buccal g.

Cerebral g.

Tegumentary n. 3

Tegumentary n. 4

Pedal commissure

Esophagus

Crop

L. abdominal g.

Ant. gizzard

Spermatheca

Nerve to purple gland

Renal n.

Hapto-pancreas
(digestive glands)

Buccal mass

Ant. tentacular n.

Tegumentary n. 2

Penis

Pleural g.

Pedal g.

Parapedal commissure

Large parapodial n.

Tegumentary n. 7

R. abdominal g.

Branchial n.

Branchial g.

Genital g.

Intestine

Anus

FIGURE 4-5.
The general plan of the nervous system of *Aplysia* in relation to internal body organs. [Modified from Eales, 1921.]

release their secretory products into the sheath (Figure 4-6). Thus, in addition to its structural role, the sheath is also a neuroendocrine organ: it stores neurosecretory products and releases them into the blood stream.

Each ganglion is a discrete collection of neurons, containing about 2,000 nerve cells and many more glial (nonneural support) cells. As in all nervous systems, invertebrate ganglia are divided into two zones: cellular (the cell bodies) and fibrous (the axons). In vertebrates the two zones are intermingled

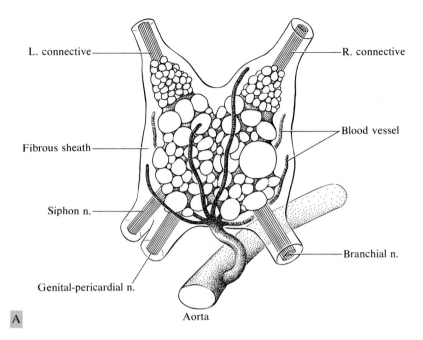

L. connective

R. connective

Fibrous sheath

Blood vessel

Siphon n.

Branchial n.

Genital-pericardial n.

Aorta

A

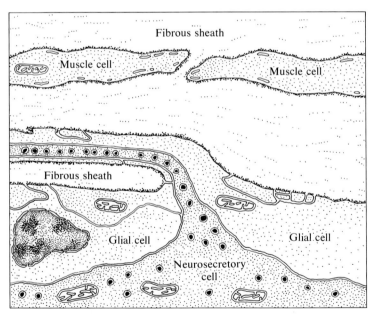

Fibrous sheath

Muscle cell

Muscle cell

Fibrous sheath

Glial cell

Glial cell

Neurosecretory
cell

B

FIGURE 4-6.
The sheath of the abdominal ganglion. **A.** Drawing illustrating the vascularization of
the sheath by branches from the aorta. The branches do not penetrate into the ganglion.
B. Schematic drawing illustrating termination in the sheath of a neurosecretory cell
process. [Based on Coggeshall, 1967, and unpublished observations.]

(axons enter into the cellular zone), in invertebrates the zones are separated. The cell bodies of invertebrate neurons are gathered in a rind around the outside of the ganglion. This cellular region is devoid of axons and synapses (Figure 4-7*A*). Each neuron sends its axon into a central region (Figure 4-7*B*), the *neuropil.* The most important feature of the neuropils is the *synaptic field,* made up of the connections between neurons lying within the ganglion and between neurons in one ganglion and axons from other ganglia or the periphery. The neuropil also contains fibers of passage that run through it without making synaptic connections. The fact that invertebrate ganglia are divided into a peripheral region of cell bodies and a central synaptic field is of great experimental advantage. The location of the cell bodies on the outer surface of the ganglion often makes it possible to see individual cells in the intact ganglion under the dissecting microscope.

In addition to the ganglia of the central nervous system, the nervous system of *Aplysia* and other molluscs contains small groups of isolated neurons in the periphery. Some of these peripheral neurons are motor neurons. Effector organs are therefore sometimes dually innervated. Some can be independently activated over central or peripheral pathways; others are activated only conjointly by activity in both pathways (for details see Kandel, *Behavioral Biology of Aplysia*).

Although they differ in detail, invertebrate and vertebrate neurons share common structural features. The major difference is in the position of the receptive pole of the neuron, the dendritic tree. With some exceptions, vertebrate neurons are typically *multipolar;* the cell body has an axon emanating from one pole and one or more dendritic arborizations emanating from other points; synaptic input is distributed over the cell body and the dendritic trees. Invertebrate neurons (and most vertebrate sensory neurons) are typically *monopolar;* the cell body gives rise to an axon only, and the dendritic branches emanate from the axon, usually the proximal part. The cell body has no synapses on it; all the synapses are distributed along the dendrites (Figure 4-8). As we shall see later, the similarities between vertebrate and invertebrate neurons can be expressed in a general model of neurons (see Figure 5-6, p. 109). Never-

FIGURE 4-7.
Structure of an invertebrate ganglion. **A.** Schematic drawing of a cross section of the abdominal ganglion indicating relation of peripheral cell bodies to central neuropil, where the synaptic contacts between cellular processes are made. [From Frazier *et al.,* 1967.] **B.** Dendritic ramification of an identified cell (L7) as revealed by intracellular injection of cobalt chloride and subsequent precipitation in ammonium sulfide. The dendritic processes arise from the axon and spread in the neuropil region of the ganglion. [From Winlow and Kandel, 1976.]

81

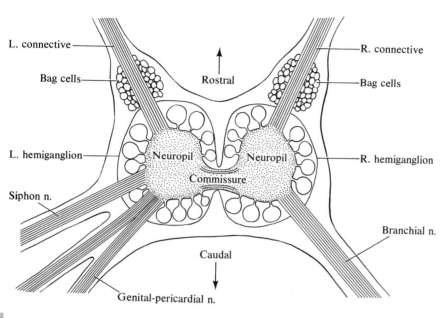

A

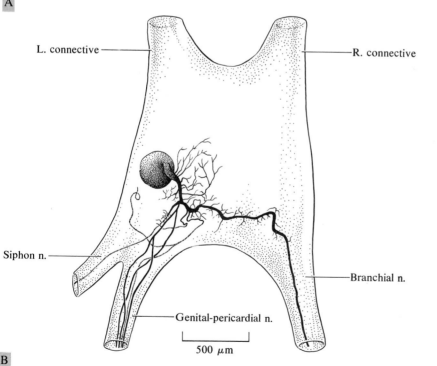

B

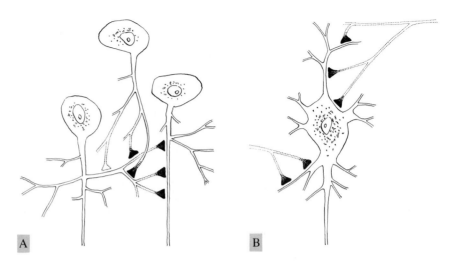

FIGURE 4-8.

Comparison of invertebrate and vertebrate central neurons. **A.** Monopolar cells characteristic of *Aplysia* and other invertebrate ganglia. Dendritic processes emanate from the main axon. Some of these processes receive synapses from presynaptic cells, others form presynaptic terminals on other cells. **B.** Typical bipolar vertebrate cell with dendritic ramification at both apical and basilar (axonal) pole of the neuron. Each of these dendritic processes receives synaptic input from other cells, but with rare exceptions they do not form presynaptic terminals.

theless, there is a range of neuronal subtypes, as evidenced in variations in the size and configuration of the cell body, the extent of the dendritic tree, the size of the main dendritic branches, the distribution of the synapses upon them, and the relation of the synapses to the trigger zones.

Fine Structure of the Neuron

The intracellular medium or cytoplasm of neurons in the abdominal ganglion contains the machinery for cellular function: nucleus, mitochondria, granular and smooth endoplasmic reticulum, free ribosomes, microtubules, neurofilaments, Golgi complexes, granules and vesicles, multivesicular bodies, and round pigment granules that resemble lysosomes in their fine structure.

Although there are differences, the fine structures of the synapses in *Aplysia* generally resemble those of vertebrates (Coggeshall, 1967). Presynaptic terminals filled with granules or vesicles abut relatively clear, postsynaptic pro-

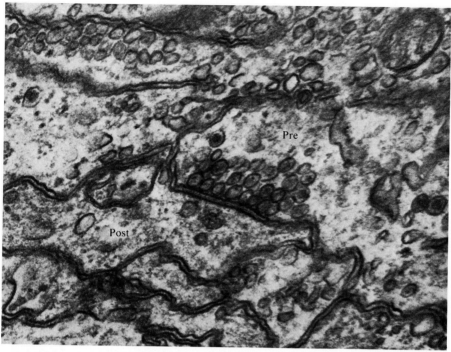

FIGURE 4-9.
Electron micrograph of a synapse in the neuropil, showing clearly defined pre- and postsynaptic contacts. The vesicles are closely applied to the presynaptic membrane. The intercellular space—widened at the synapse—is filled with a granular material that appears in this picture as an indistinct intermediate line. Very narrow layers of electron-dense cytoplasm are closely applied to the internal surfaces of opposed synaptic membranes. × 66,000. [From Coggeshall, 1967.]

cesses (Figure 4-9). In some cases the intercellular gap between the pre- and post-synaptic processes (the synaptic cleft) is slightly widened and filled with a granular material. Sometimes a narrow band of electron-dense cytoplasm seems to be closely applied to the cytoplasmic side of both pre- and post-synaptic elements. Regrettably little is known about the structure of the invertebrate neuropil and the specific morphology of the synapses of identified cells (for a beginning, see Graubard, 1973; Thompson *et al.*, 1976).

BEHAVIOR OF *APLYSIA*

The behavior of *Aplysia* ranges in complexity from elementary reflex and fixed acts, through complex reflex and fixed-action patterns, to higher-order social behavior, e.g., chain copulation. I shall describe here only those responses that have been well studied. This sketchy description does not provide a complete inventory of the behavioral repertoire of the animal.[6]

Elementary Behavior: Reflex Acts

Neuroendocrine reflexes.   When placed in a reduced salt (hypoosmotic) solution, animals maintain their weight and salt concentration (isotonicity). Reductions in salt concentration (osmolarity) are sensed by an osmoreceptor organ, the osphradium, which inhibits the release of a hormone (the equivalent of vertebrate antidiuretic hormone) that regulates the balance between salt and water (Stinnakre and Tauc, 1969; Kupfermann and Weiss, 1974, 1976).

Defensive reflex of the mantle organs.   The external organs of the mantle region (siphon, mantle shelf, and gill) participate in a defensive withdrawal reflex (Pinsker *et al.,* 1970). A weak tactile stimulus applied to the siphon or mantle shelf causes the gill to contract and withdraw into the mantle cavity and the siphon to withdraw beyond the parapodia (Figure 4-10*A*). With repeated stimulation, this defensive reflex undergoes habituation, both short-term (lasting hours) and long-term (lasting weeks) (see Chapters 12 and 13); a strong stimulus to another part of the body immediately restores the response (dishabituation). The degree of modification produced by a given training procedure seems to be determined by the motivational state of the animal. Fed animals show less habituation than hungry animals.

Elementary Behavior: Fixed Acts

Inking and egg-laying.   In addition to the withdrawal reflex of gill, siphon, and mantle shelf, the organs of the mantle region of *Aplysia californica* also generate two types of all-or-none fixed acts: inking and egg-laying (see Chapter 9, pp. 385 and 406). When *Aplysia californica* is exposed to a strong or highly noxious stimulus, it not only withdraws its siphon and gill but also secretes a massive cloud of dark purple ink from its ink gland in an all-or-none manner (Figure 4-10*B*). The inking response is lacking in all species of the subgenus *Aplysia* (which includes the species *A. vaccaria, A. juliana,* and *A. depilans*). Reproductively mature *Aplysia* periodically release large egg masses in an apparent all-or-none fashion (see Figure 4-3).

[6]For a more detailed description see Kandel, *Behavioral Biology of Aplysia,* forthcoming.

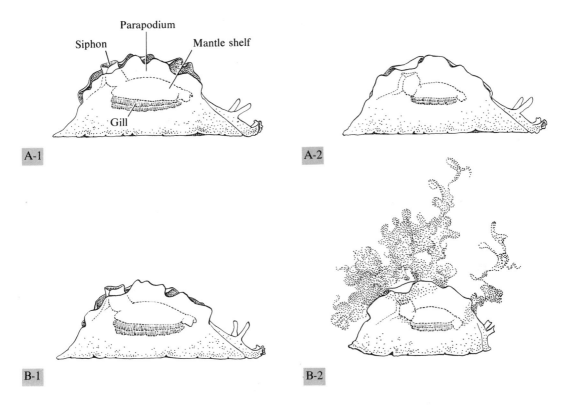

FIGURE 4-10.
Examples of elementary behavior in *Aplysia californica*. **A.** Gill and siphon withdrawal, a reflex act produced by a weak tactile stimulus to the siphon. **B.** Head withdrawal and inking, a fixed act produced by a strong stimulus to the head.

Complex Behavior

Locomotion. All *Aplysia* species move by crawling. Some animals reach speeds of 80 meters per hour for limited periods. Figure 4-11 illustrates the movements involved in one full step. The animal moves by lifting its head (*B*) and releasing suction in order to raise the anterior margin of its foot (*C*), which is then extended for a distance equal to half the length of the animal (*D* and *E*). As the anterior edge of the foot is brought down and attached to produce an arch, the site of release and extension moves to the posterior portion of the foot. The animal then contracts its anterior end to take up the slack produced by the continued extension at the posterior end of the body (*F*). The anterior and then the middle part of the foot attach sequentially, and the posterior portion of the foot is released and brought forward by a distance equal to that moved by the head (*G*). Locomotion has a circadian (daily) rhythm; animals locomote more during the day than at night (Kupfermann, 1968; Strumwasser, 1967a, 1971). In addition to crawling, some species swim, e.g., *A. brasiliana* and *A. fasciata* and *A. juliana*.

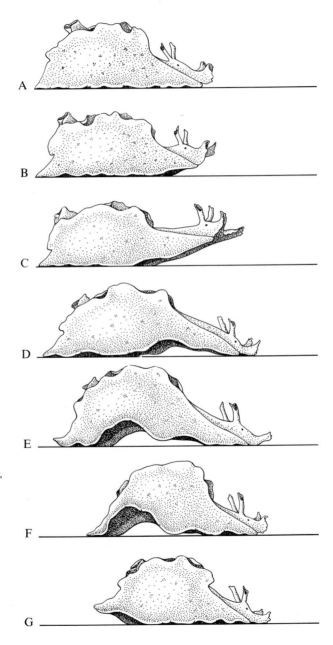

FIGURE 4-11.
The locomotor pattern of *Aplysia*, showing one full step (see p. 85 for details).

Escape.   When a particular species of starfish or other predator (*Navanax, Pleurobranchaea*) contacts the animal, it quickly withdraws the contacted part (turning from the site of contact) and initiates a series of rapid pedal waves accompanied by parapodial flapping designed to move the animal away from the predator (Figure 4-12).

Homeostatic adjustments and respiratory pumping.   Heart rate and cardiac output undergo increases that are centrally mediated and have the properties of visceromotor fixed-action patterns. Heart rate also undergoes sudden de-

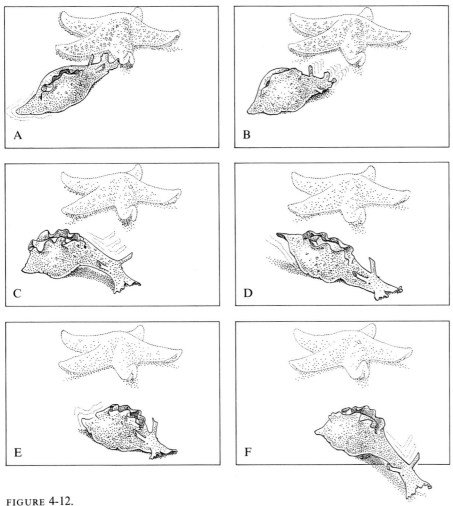

FIGURE 4-12.
Escape response of *Aplysia californica* to starfish *Astrometis sertulifera*. On contact with the starfish (**A**), the animal withdraws (**B**), turns away from the starfish (**C**), and escapes with rapid pedal waves (**D–F**). [From Dieringer and Koester, unpublished.]

creases accompanied by respiratory pumping movements, brisk contractions of the gill and parapodia. The pumping movements cause fresh seawater to circulate through the mantle cavity, thereby aerating the gill, cleansing the mantle cavity of debris, and expelling secreted ink (Figure 4-13; also see Chapter 9, pp. 378–383).

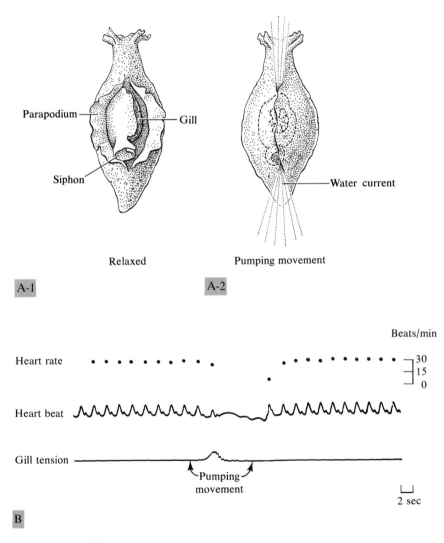

FIGURE 4-13.
Respiratory pumping movements in *Aplysia californica*. **A.** The pumping movement that circulates seawater through the mantle cavity. **B.** This movement accompanied by spontaneous gill contraction (tension record) and a decrease in heart rate.

Feeding. *Aplysia* is a grazing herbivore that consumes great quantities of seaweed. Feeding consists of two separate behavioral components: an appetitive, orienting response and an ingestive response (Figure 4-14). The appetitive response is characterized by arousal and back-and-forth orienting head movements that tend to center the mouth over the source of food. This phase is followed by a stereotypic (fixed-action pattern) ingestive response, during which the animal opens its mouth, grasps the food, pulls until it tears away the food, and then swallows. Eating behavior is under extensive motivational control: it shows satiation, circadian rhythmicity, and arousal (see Chapter 10, p. 457 ff). Feeding can also be modified. Noxious stimulation results in prolonged avoidance of food and repeated presentation of a nonnutritive tactile stimulus leads to habituation of the orienting response (Kupfermann, 1974a).

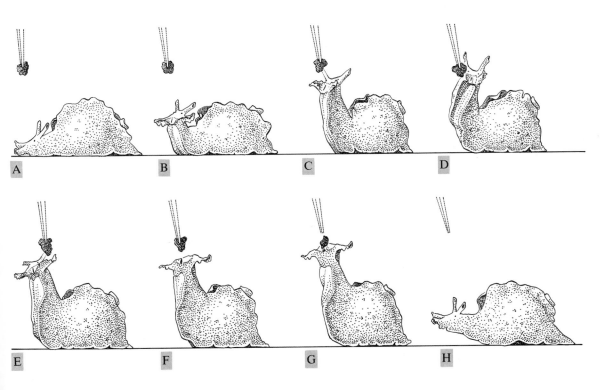

FIGURE 4-14.
Orienting (appetitive) and feeding (consummatory) responses of *Aplysia.* **A-F.** The orienting component of feeding behavior, a reflex pattern. **G.** The ingestive component, a fixed-action pattern. **H.** Return to resting state.

Higher-Order Behavior: Group Sex

The sexual behavior of *Aplysia* is the animal's most complex social behavior. *Aplysia* is hermaphroditic; one individual can serve as male or female partner for any other, but one animal cannot copulate with or fertilize itself. Often animals form coupling chains in which three to 10 animals participate (Figure 4-15). When two individuals couple, one animal serves as the female and attaches itself to a convenient crawling surface. The other animal acts as the male, crawls over the visceral hump of the first, and positions itself so that its penile aperture is level with the vaginal aperture of the first animal. The male then grasps the mantle of the female with the anterior part of its foot while the posterior part embraces the tail of the female; the female holds on to the body of the male by opening its parapodial lobes. The male then thrusts its penis into the vaginal aperture. If three or more animals are involved, the third animal attaches itself behind the second (the male) in the same way described. In this manner several animals may copulate *ensemble*. The first animal in the chain serves only as a female, the last animal in the chain serves only as a male, and all animals in between are simultaneously males and females. Occasionally the first and last animals also couple, so that the animals form a complete circle.

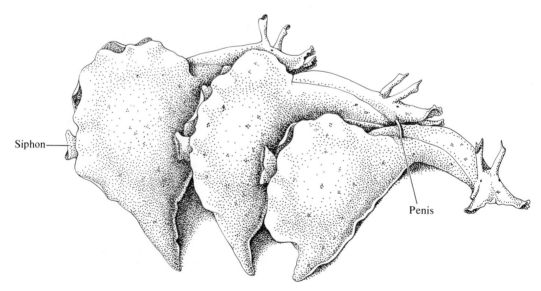

FIGURE 4-15.
Group sex in *Aplysia*. Formation of a copulatory chain by three animals. [Modified from Eales, 1921.]

SUMMARY AND PERSPECTIVE

The molluscs form one of the great invertebrate phyla, second in number of species only to the arthropods. Although the six classes of molluscs vary greatly in size and complexity, they share many basic features, as evidenced by a common anatomical plan, consisting of head, foot, mantle, and visceral hump. The head and foot are innervated by the head ganglia (cerebral, buccal, pleural, and pedal), which control primarily somatic functions: locomotion, feeding, and copulation. The mantle and visceral hump are innervated by the abdominal ganglion, which controls primarily visceromotor functions: respiration, circulation, reproduction, and excretion.

The genus *Aplysia* is a gastropod belonging to the subclass *opisthobranchia* (hind-gilled snails), the order *Anaspidea,* and the family *Aplysiidae.* The genus includes five subgenera and 35 species.

Aplysia lives in one of two shore environments: the exposed sublittoral zone and the submerged littoral zone. The animals usually reproduce in the summer, depositing long egg cords containing about a million ova. The fertilized eggs hatch into free swimming veliger larvae, which metamorphose into juvenile forms after 35 days. Animals can be reared in the laboratory from fertilized eggs to reproductive maturity with a generation time of 19 weeks.

The nervous system of *Aplysia* is avascular, like that of many other invertebrates. However, the ganglionic sheath is usually highly vascularized and serves as a neuroendocrine organ that stores and releases the hormones of neuroendocrine cells.

Each invertebrate ganglion is a discrete collection of nerve cells. For example, the abdominal ganglion contains about 2,000 neurons and many more glial cells. The cell bodies are gathered in a cortex on the outside of the ganglion. This region lacks axons and synapses. Each neuron sends an axon into the central region, or neuropil, which consists of the synaptic connections between axonal processes of the neurons and the fibers that run through the ganglion without making synaptic connections.

The neurons of invertebrates and vertebrates share a common plan. The major difference is in the position of the receptive pole of the neuron. Vertebrate neurons typically are multipolar, and the synaptic input is distributed over the cell body and the dendritic trees. Invertebrate neurons are typically monopolar. The cell body has no synapses on it; all the synapses are distributed along the dendrites which emanate from the proximal part of the axon.

Aplysia was initially selected for behavioral studies because of the experimental advantages offered by its nervous system. Although little was known about its behavior until recently, in less than a decade a substantial body of information on several types of behavior has accumulated: (1) *elementary reflex acts*—defensive reflexes of the mantle organs and neuroendocrine reflexes; (2) *elementary fixed acts*—inking and egg-laying; (3) *complex behavior patterns*—locomotion, escape, respiratory pumping, and feeding; and (4) *higher-order behavior*—group sex.

These behaviors have become focal points for research. They have been used to examine how different patterns of neuronal interconnections mediate different classes of behavior, how the nervous system integrates and coordinates different response patterns, and how motivational states and behavioral modifications alter the expression of these responses. As a result, there are now useful neurobiological insights into most of these questions.

Substantial progress has also been made in understanding the nature of several sensory receptors that transmit information from the animal's environment to its central nervous system, especially the eyes, the statocyst, the osphradium, and some mechanoreceptors. However, much work is still needed to explore fully the behavioral and learning capabilities of *Aplysia*.

SELECTED READING

Eales, N. B. 1921. *Aplysia.* Liverpool Marine Biology Committee, Memoir 24, *Proceedings and Transactions of the Liverpool Biological Society*, 35:183–266. A small classic. The first review of the biology of *Aplysia* to appear in English.

Kandel, E. R. Forthcoming. *Behavioral Biology of Aplysia.* W. H. Freeman.

Willows, A. O. D. 1973. Learning in gastropod mollusks. In *Invertebrate Learning*, edited by W. C. Corning, J. A. Dyal, and A. O. D. Willows. Plenum Press. An up-to-date overview of gastropod biology.

PART II

STRATEGIES IN THE STUDY OF NERVE CELLS

To understand the biological basis of behavior one must understand the structure and function of nerve cells, the basic building blocks and signaling units of the nervous system. Chapters 5 and 6 introduce the reader to the functioning of nerve cells. The methods for studying various electrical signals are reviewed and the significance for behavior of the results is discussed. The ionic hypothesis, which explains how the various electrical signals are produced, is outlined in detail. Based on the ionic hypothesis, a model of the nerve cell is presented that contains features common to most neurons.

Superimposed upon this model there are differences between neurons, even in a single ganglion. Chapter 7 examines this diversity, which includes differences in the mechanisms of the membrane potential, the action potential, and the presence or absence of autoactivity. These differences make it possible to distinguish one cell from another and to identify certain cells as unique individuals that occur in every member of the species. In addition to having distinctive properties, identified cells also make invariant connections. In Chapter 8 some of the principles underlying the patterns of interconnection among cells are discussed.

The emphasis throughout this part is on cellular morphology and transmitter biochemistry as well as on the electrophysiological aspects of neuronal signaling. These chapters set the stage for considering how the architecture of different types of neuronal circuits is related to different types of behavior.

CHAPTER 5

Introduction to
the Study of Neurons

Study of the cellular mechanisms of behavior requires familiarity with the biology of neurons and particularly with their signaling capabilities. The electrical signals of nerve cells are the primary means by which information is conveyed from one region of the nervous system to another (hormones are another means). All sensory stimuli that an animal receives, both from its environment and from within its body, are transformed into electrical signals. These signals constitute a neural representation of sensory stimuli and are conveyed to different regions of the central nervous system. The central nervous system in turn uses electrical signals to send out motor commands to muscles and glands to produce behavior.

The first part of this chapter is intended for readers unfamiliar with the principles of signaling in nerve cells; it presents an elementary overview of neuronal functioning in relation to a simple reflex act. In the second part of this chapter and in the chapter that follows, the arguments and experiments underlying the principles of cellular neurobiology are set forth in slightly more detail, with emphasis on some of their quantitative aspects. This detail is important for understanding the analytic studies, to be considered later, that relate the biophysical properties of specific cells to various classes of behavior and behavioral modifications. Heavy reliance is placed on studies with molluscan axons and synapses (the squid giant axon and giant synapse) since these are the basis of much of our current understanding of neuronal functioning. However, important insights have also come from studies of certain vertebrate synapses (in particular the synapse between motor neurons and skeletal muscle in the frog), and some of this work is also reviewed.

CELLULAR ELEMENTS OF THE NERVOUS SYSTEM

The nervous system consists of two classes of cells: *glial* (or neuroglial) *cells* and *nerve cells* (or neurons). In the central nervous system of both vertebrates and invertebrates the glial cells are eight or nine times more numerous than nerve cells. The glial cells serve, in part, as supporting elements, a role played by connective tissue cells in other parts of the body. A type of glial cell, the Schwann cell, forms the fatty insulating (myelin) sheath around certain (myelinated) axons. Other types of glial cells seem to segregate groups of neurons from one another. Glial cells may have additional, perhaps nutritive, functions. Although the membrane potential of the glial cell can be altered by changes in the external potassium concentration, produced by impulses in nerve cells, the glia do not seem important for signaling neural information (see Kuffler and Nicholls, 1966). The signaling unit of the nervous system is the nerve cell.

A typical nerve cell has four morphological regions (Figure 5-1): a cell body, dendrites, axon, and the presynaptic terminal (of the axon). Each region has a distinctive function. The *cell body*, or soma, contains the organelles necessary for cellular function: the nucleus, ribosomes, rough endoplasmic reticulum, the Golgi apparatus, and mitochondria. The cell body makes almost all of the neuron's macromolecules.[1] Some of these remain in the cell body; others are transported out of the cell body to other parts of the neuron. Many of the transported macromolecules are incorporated into mitochondria, vesicles, smooth endoplasmic reticulum, and lysosomes. The cell body gives rise to the *axon*, a tubular process that sometimes extends over a considerable distance. Near its ending the axon divides into many fine branches, each of which has a specialized terminal region called the *presynaptic terminal*. The terminal contacts the receptive surface of other cells, transmitting (by chemical or electrical means to be discussed below) information about the activity of the neuron to other neurons or to effector organs. The receptive surfaces, or *dendrites,* typically consist of arborizing processes that extend from the cell body

---

[1]A few types of macromolecules are made in mitochondria all along the neuron.

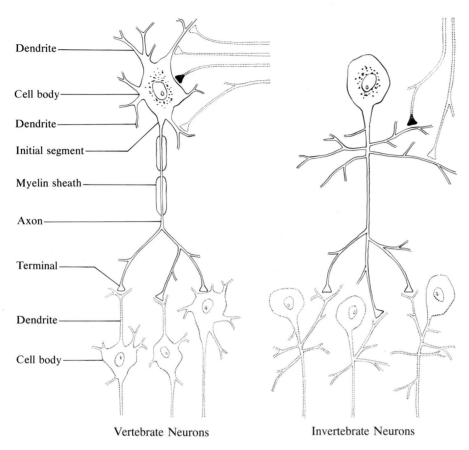

Dendrite

Cell body

Dendrite

Initial segment

Myelin sheath

Axon

Terminal

Dendrite

Cell body

Vertebrate Neurons   Invertebrate Neurons

FIGURE 5-1.
The structure of nerve cells in vertebrates and invertebrates.

(in vertebrates) or the initial segment of the axon (in both vertebrates and invertebrates).

The point of contact between two neurons is known as the *synapse*. It is formed by the presynaptic terminal of one cell (the *presynaptic cell*) and the receptive (dendritic) surface of the other cell (the *postsynaptic cell*). In vertebrates a small number of synapses (less than five percent) contact the cell-body membrane; in invertebrates there are usually no synapses on the cell body (Figure 5-1). (In Figure 5-4 and subsequent illustrations synapses occur on the cell body of invertebrate neurons. This is not a realistic depiction and is done only for illustrative convenience.)

ELECTRICAL SIGNALS IN NERVE CELLS

Nerve cells are capable of producing a variety of electrical signals. All of these are caused by changes in current flow across nerve cell membranes. A change in current flow in turn leads to a change in the membrane potential of the cell. In recent years there has been remarkable progress in understanding how the various changes in membrane potentials used for signaling are produced. This progress is the result of the development of three techniques: (1) intracellular recording, which allows direct measurement of the membrane potential; (2) intracellular stimulation, by which the membrane potential is systematically changed by injecting current into the cell; and (3) voltage clamping, by which the membrane potential is held (clamped) at any preset voltage level, allowing the current flowing through the membrane at that membrane potential to be measured. Voltage clamping is technically the most difficult of these methods but it is also the most powerful for examining the various currents that produce the voltage responses of the membrane.

Research during the last 35 years has led to a unified theory of the neuron's electrical properties that is the basis for our current understanding of electrical signaling by nerve cells. This theory was derived from two sources: (1) the study of the ionic mechanisms of the membrane and action potentials by Alan Hodgkin, Andrew Huxley, and Bernard Katz (1949, 1952; for review see Hodgkin, 1964); and (2) the analysis of chemical synaptic transmission by Katz (1966, 1969) and his colleagues, Paul Fatt, Jose del Castillo, and Ricardo Miledi, and by J. C. Eccles (1964) working with John Coombs and Paul Fatt. Before considering this theory in detail in the next chapter, I shall describe the types of signals generated by nerve cells and discuss the importance of each for behavior.

Membrane and Action Potentials

Intracellular recordings have revealed that each nerve cell is *polarized.* It maintains across its membrane an electrical potential, the *membrane* (or *resting*) *potential,* which makes the inside of the cell electrically negative in relation to the outside (Figure 5-2). The membrane potential (abbreviated V_m) is usually about -60 mV.[2] The existence of a membrane potential and its approximate value in axons and muscles had been known since the beginning of this century

[2]By convention, the membrane potential is defined as the inside potential minus the outside potential. The *resting* membrane potential in different nerve cells ranges from -40 to -70 mV. Muscle cells can have a resting potential of -90 mV.

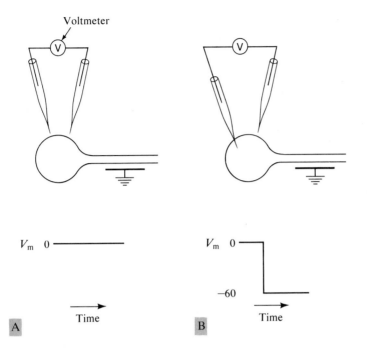

FIGURE 5-2.
Intracellular recording of the membrane potential from cell body. The resting potential, V_m, is generated across the membrane of the cell and can be measured by placing one electrode inside the cell and another on the outside. These intracellular recordings are obtained with glass micro-electrodes filled with a concentrated solution of potassium chloride and connected by means of a chlorided silver wire connected to a voltmeter (V). **A.** With both microelectrodes in the extracell-ular space, no potential is recorded by the voltmeter connected between the microelectrodes. **B.** When one electrode tip is pushed into the cell a sudden potential shift of -60 mV is recorded between the electrodes inside and outside the cell. This is the resting membrane potential of the cell. In most experiments (see Figure 5-3 and later figures) the extracellular electrode is not another glass microelectrode but a large chlorided silver wire connected to ground ($\equiv$) and placed in the bathing solution that surrounds the cells. This is equivalent to placing an electrode just outside the cell because the resistance between the electrode and the outside of the cell is negligible. Ground is defined as zero potential.

as a result of some indirect but ingenious measurements made by Julius Bernstein (1902). But to measure the membrane potential directly, an electrode had to be placed on each side of the membrane, one inside the cell and another outside it (Figure 5-2). Until 1946 this difficult task had been accomplished only for the enormous squid giant axon (1 mm in diameter and several centi-meters long). Hodgkin and Huxley (1939) and Curtis and Cole (1942) suc-ceeded independently in threading its length with a fine glass cannula electrode (see Figure 6-4, p. 154). This technique is not generally applicable to other nerve cells, however. In 1946 Graham and Gerard developed a glass capillary microelectrode with a tip less than a micrometer (1 μm $= 10^{-6}$ m) in diameter. The tip of this electrode is so small that it can be inserted into a variety of nerve cells with only minimal damage to the cell. The glass capillary is filled with a concentrated salt solution (usually 3 M potassium chloride) to conduct electricity and connected to an amplifier by means of an unpolarizable metal electrode (usually a chlorided silver wire). The change in potential is then displayed on a suitable voltmeter: an oscilloscope or a pen writer (Figure 5-2).

To change the membrane potential systematically, a second microelectrode is placed inside the cell. Current is passed between this electrode and an extracellular electrode by means of a stimulator, or current generator (Figure 5-3A). Since current flows from positive to negative, if the intracellular electrode is made positive and the extracellular electrode negative, current can be made to flow outward across the cell membrane. As a result, the membrane potential will become *depolarized* (the polarized state of the membrane is decreased by making the inside of the cell less negative with respect to the outside). Alternatively, if the intracellular electrode is made negative and the extracellular electrode positive, current can be made to flow inward and the membrane will become *hyperpolarized* (the polarized state of the membrane is increased by making the inside more negative with respect to the outside). As we shall see later, inward and outward currents are normally produced by changes in the permeability of the membrane to various ions.

A weak inward (hyperpolarizing) current pulse injected into the cell body (Figure 5-3B) produces a small hyperpolarization that decreases gradually with distance (along the axon or dendrites) over several millimeters. Similarly, a weak outward (depolarizing) current pulse causes only a small depolarization that decreases with distance. These small changes in potential are called *electrotonic potentials,* and are conducted along the neuron by a process called *electrotonic propagation* (or conduction). Electrotonic propagation is *passive:* the signal loses amplitude progressively as it is conducted along the neuron (see Figure 5-11, p. 124).

If the outward current pulse is increased and the membrane potential is made less negative by a critical amount, the membrane will suddenly generate, in an explosive manner, a large, brief (1 to 10 msec) active response, called the *action potential* or *spike* (Figure 5-3B). The membrane potential at which the action potential is triggered is called the *threshold.* The action potential is generated in an all-or-none fashion; increases in current strength above the threshold do not increase the amplitude of the action potential or change its shape (Figure 5-3B). Hodgkin and Huxley (1939, 1945) and Curtis and Cole

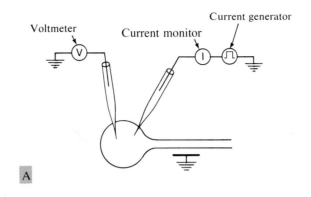

A

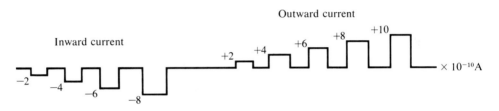

Outward current

Inward current

+2   +4   +6   +8   +10

$\times 10^{-10}$A

−2   −4   −6   −8

B–1

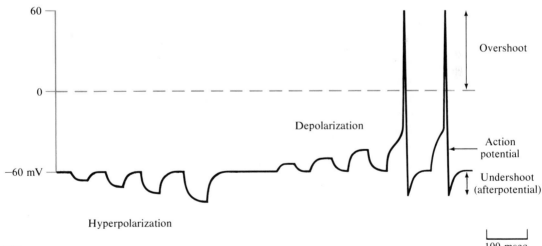

60

Overshoot

0

Depolarization

Action
potential

−60 mV

Undershoot
(afterpotential)

Hyperpolarization

100 msec

B–2

FIGURE 5-3.
Intracellular stimulation. **A.** In addition to an intracellular recording microelectrode, a second microelectrode is inserted into the cell body. This electrode is connected to a constant-current generator and used to pass current across the membrane, between the inside of the cell and an outside electrode, which may be connected to ground. **B.** The current injected into the cell is measured on a current monitor and diplayed on one trace of an oscilloscope (*B-1*). Inward-current pulses (−) hyperpolarize and inhibit the cell; outward-current pulses (+) depolarize and excite the cell (*B-2*). If depolarizing pulses are sufficiently large ($+8$ to $+10 \times 10^{-10}$ A), they will trigger an action potential. The action potential is all-or-none, that is, it does not increase in size as more depolarizing current is injected (compare the responses to $+8$ and $+10 \times 10^{-10}$ A).

(1942) discovered the surprising fact that the action potential not only abolishes the resting potential (reducing it from -60 mV to 0), but actually reverses the membrane potential for a moment, moving it all the way to $+55$ mV. This reversal of potential from 0 to $+55$ mV is called the *overshoot*. Upon reaching this peak, the potential rapidly returns to the resting level. In many cells the descending limb of the action potential does not simply return to -60 mV, but goes beyond it to -65 or even -70 mV. This is called the *hyperpolarizing afterpotential,* or *undershoot* (Figure 5-3B).

Unlike an electrotonic potential, the action potential in a neuron is conducted without any loss in the amplitude of the signal to the terminal region of the axon. There it initiates a process called *synaptic transmission.*

Synaptic Transmission

Synaptic transmission is central to an understanding of how the nervous system works because it is the process by which nerve cells communicate with one another and with effector organs. Synaptic transmission does not occur between any two nerve cells. It occurs only where the membranes of two neurons have a specific morphological and functional relationship. This relationship consists of a presynaptic fiber, a zone of apposition, and a postsynaptic fiber.

How synaptic transmission occurs in the nervous system was once a lively topic of dispute. Before development of microelectrodes (and the direct study of synaptic transmission), most electrophysiologists thought that synaptic transmission was *electrically mediated* by the direct flow of current from the presynaptic to postsynaptic cell (for interesting early exceptions to this theory see Eccles, Katz, and Kuffler, 1941; and Kuffler, 1948). A minority opinion, held mostly by neuropharmacologists, was that synaptic transmission was chemically mediated. Stimulated by T. R. Elliot's suggestion that transmission in the autonomic nervous system was chemical, and by Loewi's important discovery in 1921 that synaptic (vagal) inhibition of the heart is *chemically mediated,* pharmacologists, particularly John Langley and Henry Dale, argued that transmission in the peripheral and central nervous systems was likely also to be chemical. This opinion was rapidly confirmed once microelectrodes were developed. Between 1950 and 1953 Fatt and Katz, and Eccles and his colleagues found that synaptic excitation and inhibition at several peripheral and central synapses did indeed involve chemically mediated transmission. As a result, most physiologists rapidly accepted the notion that *all* synaptic transmission was chemical (for reviews see Eccles, 1953, 1957). But in 1957 Furshpan and Potter discovered an excitatory synapse in the crayfish central nervous system that operated by electrical means, and in 1963 Furukawa and Furshpan found a form of synaptic inhibition that was electrically mediated. Also in 1963, Martin and Pilar found an example of synaptic excitation that was both electrical and chemical, indicating that both types of process could occur together at the same synapse.

Much subsequent research has revealed that both chemical and electrical transmission occur in all nervous systems examined, although chemical transmission is generally much more abundant (for review see Bennett, 1973). Electrical and chemical synapses differ in their mediating agent, morphology, and (often) in their directionality. Each mode has a variety of subtypes. At a chemical synapse an action potential in the presynaptic neuron leads to the release of a chemical transmitter substance that diffuses across a small space (the *synaptic cleft*) separating the presynaptic and postsynaptic structures of the synapse, and interacts with receptor molecules on the external membrane of the postsynaptic cell. This chemical interaction leads to current flow in the postsynaptic cell that alters its membrane potential by producing either an inhibitory or an excitatory postsynaptic potential. At all chemical synapses the synaptic cleft is comparable in width to the usual extracellular space of 20 to 30 nm (1 nm = 10^{-9} m), and may even be slightly larger; the pre- and postsynaptic neurons are not electrically connected (Figure 5-4A and B). By contrast, at electrical synapses the pre- and postsynaptic membranes are generally in close apposition, 2 to 4 nm apart. In what appears to be the most common type of electrical synapse, the *gap junction,* the pre- and postsynaptic membranes are actually connected by protoplasmic channels that bridge the gap and join the cytoplasm of the two cells (Figure 5-4C). These channels are large enough to let not only ionic current (potassium, sodium, and other inorganic ions) flow between the pre- and postsynaptic cells, but even much larger organic molecules, such as the dye procion yellow (M.W. 625). Another type of synapse, the *tight junction,* was once thought by some to be involved in electrical transmission. At a tight junction the pre- and postsynaptic membranes may actually fuse at one or more points. Although the issue is not settled, most synaptic physiologists believe that gap junctions are the only, or at least the predominant, junctions mediating electrical transmission (Bennett, 1976).

As a result of the morphological features of electrical synapses, the current produced by an action potential in the terminals of a presynaptic neuron flows across the gap junction and alters the membrane potential of the post-synaptic cell (Figure 5-4C). Also, electrical junctions are often bidirectional: currents produced by action potentials in the postsynaptic cell can flow back into the presynaptic neuron (for an interesting exception, see Furshpan and

104

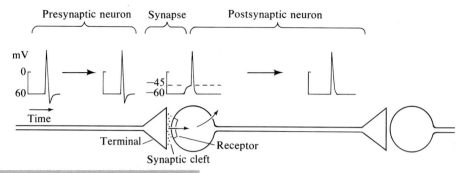

Presynaptic neuron **Synapse** **Postsynaptic neuron**

mV

Time

Terminal

Receptor

Synaptic cleft

A. Chemical Excitatory Synaptic Action

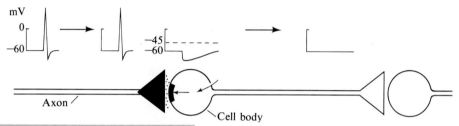

mV

Axon

Cell body

B. Chemical Inhibitory Synaptic Action

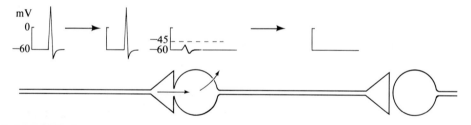

mV

C. Electrical Synaptic Action

FIGURE 5-4.

Conduction and transmission in central neurons.

In this and many subsequent figures, presynaptic terminals are shown making contact with the cell body of postsynaptic neurons. This is done for illustrative purposes only. Actually, in invertebrates the presynaptic terminals contact dendritic branches of the main axon of postsynaptic neurons.

A. Chemically mediated excitatory synaptic action. An action potential travels down the axon of the presynaptic neuron; when it invades the axon terminal, it leads to the release of a chemical transmitter substance. The transmitter diffuses across the synaptic cleft and interacts with recptor molecules at the exterior surface of the postsynaptic cell to give rise to a current flow (arrows). The current flows outward across the nonsynaptic membrane and depolarizes the postsynaptic cell past threshold, causing it to generate an action potential that is then propagated down the axon of the postsynaptic cell.

B. Chemically mediated inhibitory synaptic action. The interaction between transmitter and receptor gives rise to a current flow (arrows). The current flows inward across the nonsynaptic membrane and hyperpolarizes the cell, moving it away from its threshold level.

C. Electrically mediated synapse. Current injected into the presynaptic terminal flows directly into the postsynaptic cell producing a diphasic electrical synaptic potential. Only the outward current flow for the depolarizing component is illustrated.

Potter, 1957). At chemical synapses the extracellular space between the pre-synaptic terminal and the postsynaptic cell is not bridged and little or no current from the presynaptic action potential flows across the cytoplasmic discontinuity into the postsynaptic cell (Figure 5-4*A* and *B*). Since the agent crossing the cleft at chemical synapses is a chemical that can be released only by the presynaptic neuron terminal, these synapses are unidirectional—transmission occurs only from the presynaptic terminal to the postsynaptic cell.

Triggering of Action Potentials by Synaptic Potentials

Presynaptic neurons can generate two types of chemically mediated synaptic potentials in the postsynaptic cell: excitatory and inhibitory. An *excitatory postsynaptic potential* (EPSP) leads to an outward flow of current across the nonsynaptic membrane, which depolarizes the membrane of the trigger zone of the neuron and tends to bring it to threshold (Figure 5-4*A*). An *inhibitory postsynaptic potential* (IPSP) leads to an inward flow of current across the non-synaptic membrane, which hyperpolarizes the membrane of the trigger zone and keeps it from reaching threshold (Figure 5-4*B*). If the net depolarization produced by the activated excitatory synapses on the postsynaptic neuron exceeds by a certain critical minimum the net inhibition (produced by the activated inhibitory synapses) threshold is exceeded and an action potential is triggered. The action potential then propagates along the axon of this neuron to its terminals, where the neuron establishes contact with other nerve cells—or with muscular or glandular effector organs, giving rise to behavior.

   In later sections of this and the subsequent chapter I shall consider the mechanisms for propagation of neural information within and between cells. But the few facts already discussed are sufficient to allow us to consider, on a simple descriptive level, how neurons work and how interconnected groups of nerve cells produce behavior.

NEURONS AND REFLEX BEHAVIOR

The cell membrane of each of the four morphological regions of the neuron has distinct properties and signaling functions. The various cellular regions and the electrical signals generated by each can be illustrated by considering a simple behavior: the withdrawal reflex of the cat.

   The skin of the cat's leg contains sensory receptor organs that respond to mechanical stimulation, such as scratching. These sense organs contain the terminals of a class of sensory axons that run from the skin to the central nervous system. In the central nervous system the axons form chemical synaptic connections with several, as yet not fully specified, classes of excitatory interneurons in the spinal cord. These interneurons ultimately connect (by chemical synapses) to appropriate flexor motor neurons, which in turn connect to the flexor muscle fibers of the leg. The axons of the sensory neurons also excite another group of interneurons that inhibit the antagonistic extensor motor neurons of the leg (Figure 5-5).

The behavior mediated by this neural circuit works as follows. When the skin is scratched the sensory neurons are excited and excite in sequence several sets of interneurons, some of which excite the motor neurons, which in turn

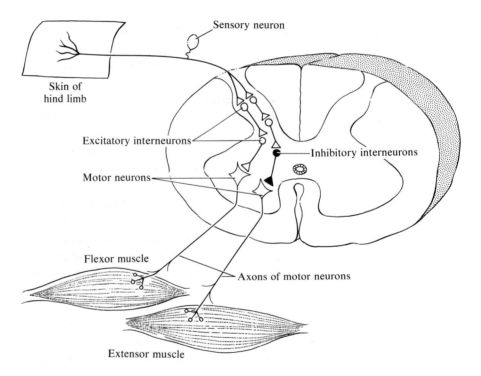

FIGURE 5-5.
Simplified circuit diagram of the flexion withdrawal reflex in the cat. A tactile stimulus activates a population of sensory neurons (one of which is illustrated) in the skin of the hind limb. The sensory neurons excite populations of several classes of interneurons, which in turn excite a population of flexor motor cells that cause the flexor muscles to contract. The sensory neurons also activate excitatory interneurons, and these engage inhibitory cells that inhibit the extensor motor neuron population and prevent it from firing.

excite the flexor muscles, causing them to contract. Other interneurons inhibit the motor neurons to the extensor muscles, preventing them from becoming active. Thus, electrical signals serve to transmit (1) sensory information from the skin to the central nervous system; (2) reciprocal innervation to motor neurons (excitation of synergistic motor cells and inhibition of antagonistic ones); and (3) motor commands from the central nervous system to the muscle.

During the reflex each nerve cell performs four distinct activities in rapid sequence, each involving somewhat different electrical processes at different sites: (1) a transducer function at the sense organs or the dendrites (the input component of the neuron); (2) an integrative function at the trigger zone, usually located in the initial portion (or segment) of the axon (the integrative component); (3) a conductile function in the axon (the conductile component); and (4) a transmitting function at the presynaptic terminal (the output component).

At the sensory-receptor membrane, in the fine nerve endings that innervate the skin, the mechanical energy of scratching the skin is transduced into electrical energy. The *generator potential* produced in these endings is smoothly graded and mirrors the configuration of the mechanical stimulus to the skin. Within a certain range, the more intense and longer-lasting the stimulus, the larger and longer-lasting will be the resulting generator potential. The membrane of the fine nerve endings that produce the generator potential usually cannot produce an action potential. The generator potential therefore propagates passively down the axon, by electrotonic means to be discussed below. The membrane of the portion of the axon adjacent to the fine nerve endings can produce action potentials. If the generator potential is sufficiently large, it will therefore depolarize this membrane to the point of threshold and trigger an action potential. Because the initial portion of the axon determines whether or not a given sensory stimulus to that neuron will generate an action potential, this zone is called the *integrative component* of the neuron.

Once initiated in the integrative component of the sensory neuron, the action potential propagates rapidly and without loss of signal strength along the axon and into the central reaches of the nervous system. When the action potential arrives at the terminal region of the axon, the output component of the neuron, the action potential sets into motion the process of synaptic transmission. This involves secretion of a small amount of a chemical transmitter. The transmitter diffuses across the synaptic cleft that separates the terminals of the sensory neuron from the postsynaptic site of the interneuron. The transmitter interacts with receptor molecules on the input component of the interneuron (usually located in the dendrites) to initiate the current flow that produces the synaptic potential in the interneuron.

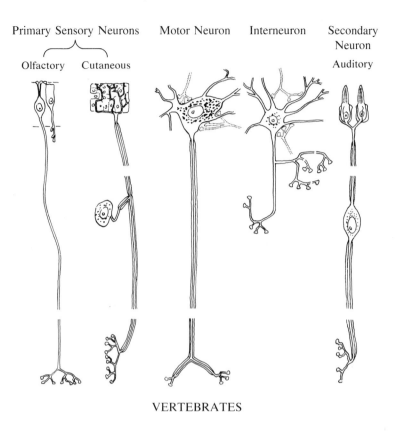

Primary Sensory Neurons   Motor Neuron   Interneuron   Secondary Neuron

Olfactory   Cutaneous   Auditory

VERTEBRATES

The synaptic potential in the interneuron and the generator potential in the receptor portion of the sensory neuron resemble each other. Both are input processes having transducing functions. Whereas the mechanoreceptor portion of the sensory neuron converts mechanical energy to electrical energy, the synaptic receptor sites in the membrane of the interneuron convert chemical energy to electrical energy. As with the generator potential, the synaptic potential is graded (within limits) and the synaptic inputs from different presynaptic neurons are additive. Both the generator potential and the synaptic potential propagate electrotonically with decrement (see below).

The trigger zone of the interneuron is similar to that of the sensory cell. It adds up all excitation (and inhibition) and determines whether the cell will fire. If the threshold of the trigger zone of the interneuron is reached, an action potential will propagate down the axon and invade the presynaptic terminals, leading to release of transmitter substance. The first interneuron activated in this reflex engages several other classes of interneurons. Finally, activity in these interneurons activates the motor neuron; an action potential propagates down the axon of the motor neuron to release transmitter that interacts with

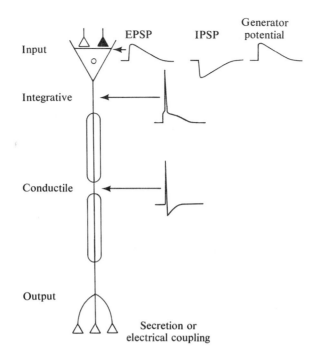

NVERTEBRATES   IDEALIZED NEURON

FIGURE 5-6.
The formal structure of various neurons. The model neuron includes the functional components
common to all neurons (the traces on the right illustrate responses of each of the components).
[After Grundfest, 1957, and Bodian, 1967.]

the receptors of the muscle membrane. This interaction in turn initiates an
action potential that produces an element of behavior, the contraction of
the muscle.

THE MODEL NEURON

A comparison of the sensory neuron, the interneuron, and the motor neuron
reveals that the three types of cells share common characteristics. For this
reason it is possible to develop an idealized or *model neuron*. The model describes the features common to all neurons regardless of shape, size, location,
or function (Figure 5-6); for example, the action potential varies only slightly
in different classes of nerve cells. As Adrian discovered, one cannot tell by
looking at the configuration or amplitude of an action potential in an axon
whether a cell is carrying sensory information into the nervous system or motor
commands out. Adrian, Hartline, and others also found that neural coding
for the intensity of a stimulus or response is not accomplished by alteration of

the shape or amplitude of the action potential of a given fiber. In most cases the intensity of a sensation or speed of motor movement is controlled by varying the frequency of action potentials in a given fiber or by varying the number of active fibers (Figure 3-3, p. 53). Finally, the type of sensation or movement does not depend upon changes in the shape of individual action potentials but on the type of nerve cells activated.

Because the action and synaptic potentials in various types of nerve cells of widely differing animals have features in common, great effort has centered on analyzing the mechanisms for generating action and synaptic potentials. These studies have shown that there is a fundamental difference in the propagation of action and synaptic potentials. Action potentials are propagated actively in the neuron, whereas synaptic potentials are propagated passively. These two means of conduction are due to the *active* and *passive properties* of the nerve-cell membrane.

PASSIVE PROPERTIES OF THE MEMBRANE

The passive propagation of electrical signals within a neuron is due to the neuron's *passive properties,* also called *cable properties* since they are similar to those of submarine cables. Like the copper wire in the core of the telegraph cable and the ocean in which the cable rests, the cytoplasm and extracellular fluid are highly conductive. Like the telegraph cable, the nerve cell has an insulating sheath, the nerve cell membrane. The similarity between nerve cells and submarine cables is so great that the equations developed by Lord Kelvin to describe the propagation of a potential change along the trans-Atlantic cable apply equally well to the propagation of small (subthreshold) electrical signals along nerve cells.

Kelvin found that the behavior of any small segment of the cable could be described by a model consisting of four *passive* electrical components: (1) a resistance of low value, representing the conducting pathways in the central core of the cable; (2) another, similar resistance, representing the conducting pathway in the ocean on the outside of the cable; and (3) a high resistance in parallel with (4) a capacitance, representing the insulating membrane. The behavior of the nerve-cell membrane in the subthreshold range can also be accurately described and simulated by a set of four passive electrical elements: the membrane resistance, the membrane capacitance, the resistance of the external medium, and the intracellular resistance of the cytoplasm. In the nerve cell, as in the submarine cable, these electrical components and their responses are called passive because they do not change their properties as a function of the voltage across them.

In contrast to the passive responses of the nerve cells, the *active responses* (rectification, action potentials, and synaptic potentials) are due to changes in membrane resistance (ionic conductance). We shall consider the mechanisms for generating active responses in the next chapter. Before examining the active properties it will be important to understand the passive properties of neurons and their role in neuronal integration.

In most nerve cells, the part of the membrane between the trigger zone and a synaptic (or generator potential) site has a very high threshold. At the same time, synaptic sites may be separated by up to a millimeter. The passive properties of the membrane therefore determine whether and to what extent synaptic potentials propagate to the trigger zone and influence its membrane potential. From knowledge of the four passive components one can calculate the *time constant* and *space constant*, which determine the time course of the synaptic potential and its spatial propagation from the site of generation to the trigger zone.

I shall first consider the resistance and capacitance of nerve cells and then describe how the time and space constants influence the integrative capabilities of the neuron.

Resistance

The passive electrical properties of neurons can be demonstrated by injecting small depolarizing or hyperpolarizing currents across the membrane of the cell (see Figure 5-3). One such property is the ratio of the change in membrane potential (ΔV_m) to the current (I) flowing across the membrane (Figure 5-7). This ratio (or proportionality factor) is called the *resistance* (R) of the membrane and is measured in ohms. Resistance describes the ability of the membrane to oppose the flow of current. Since the current flows through the cytoplasm as well as the membrane, and the geometry of the neuron is sometimes not known, the total resistance across which the current flows is called the *input resistance* (R_{input}). Ohm's law, which describes the relationship between current, voltage, and resistance in electrical circuits, also applies to biological membranes (over a restricted range of membrane potentials). Thus,

$$R_{input} = \frac{\Delta V_m}{I} \qquad (5\text{-}1)$$

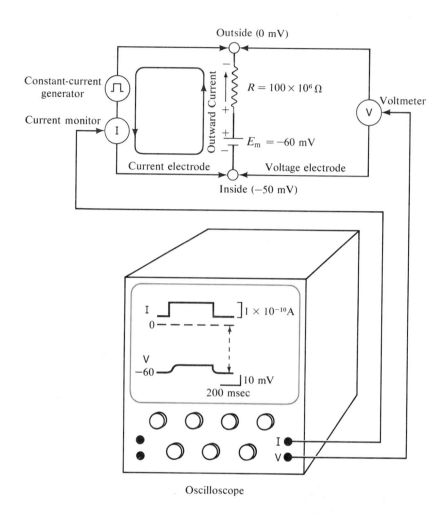

FIGURE 5-7.
Partial electrical equivalent circuit of a patch of membrane, showing the battery for the membrane potential (E_m) in series with resistance (R). Current is injected across the membrane by means of a constant-current generator ($\sqcap$) connected to the electrodes on the left. The current is monitored with an ammeter (I) and displayed as the upper trace on a dual-trace oscilloscope. The membrane potential is monitored by a voltmeter (V) connected to the electrodes on the right, and is displayed as the second trace on the oscilloscope. An outward-current pulse of 1×10^{-10} A produces a 10 mV depolarizing electrotonic potential (ΔV_m). Therefore, calculating from Ohm's law, the input resistance of the cell is 100×10^6 Ω. In an electrical circuit current is carried by electrons (1A = 6.24×10^{18} electrons per sec). In animal tissue current is carried by ions (1A = 6.24×10^{18} univalent ions per sec).

The reciprocal of resistance $(1/R)$ describes the ability of a membrane to conduct current. This ratio is called the *conductance (G)* of the membrane and is measured in mhos:

$$G_{\text{input}} = \frac{I}{\Delta V_{\text{m}}} \tag{5-2}$$

Thus, a membrane with a resistance of 10^6 ohms has a conductance of 10^{-6} mhos. A decrease in the resistance of the membrane is equivalent to an increase in conductance.

Let us assume we have a highly idealized cell body with no axonal or dendritic extensions (see Figure 5-3A). If we were to pass 0.1 nanoamps (1 nA $=$ 10^{-9} A) of current outward across the cell membrane with an intracellular electrode, the current would distribute uniformly across the total area of the cell body membrane. As a result, a 10 mV (1 mV $= 10^{-3}$ V) depolarizing change in potential registered at a nearby point within the cell would indicate that at that point the cell (membrane and axoplasm) has an input resistance of 100 megohm (1 megohm $= 10^6$ Ω):

$$R_{\text{input}} = \frac{10 \times 10^{-3} \text{ V}}{1 \times 10^{-10} \text{ A}} = 100 \times 10^6 \; \Omega$$

This value is typical for molluscan neurons. Depending on the geometry of the cell and the properties of the membrane, the input resistance of nerve cells ranges from 10^5 to 10^8 Ω; the smaller the cell, the higher the input resistance (for cells with the same type of membrane). The resistance of the membrane can be depicted as being in series with the *battery for the membrane potential,* E_{m} (Figure 5-7).[3] For most spherical cells the input resistance is a measure of the resistance of the membrane only. The *specific (intracellular) resistance of the cytoplasm* (R_{i}) is defined as the resistance of 1 cm of cytoplasm with a cross-sectional area of 1 cm^2; R_{i} is so low (50 Ωcm) that it can be ignored.

Because the measurement of R_{input} does not take into account the size and shape of the neuron, it is not useful for comparing the membrane properties of cells of different sizes and shapes. For this reason a standardized term is used: the *membrane resistance* (R_{m}), which describes the resistance across 1 cm^2

---

[3]The term E_{m} refers to the membrane potential at rest only, in the absence of extrinsic current or excitatory or inhibitory input. The term V_{m} refers to any membrane potential, resting or active; ΔV_{m} refers to any *change* in the membrane potential.

of membrane.[4] The resistance of a given cell membrane thus depends on two input factors: (1) the *membrane resistance* (R_m), which is a measure of the conducting properties of the membrane (its leakiness to ions), and (2) the *total surface area,* which determines how many parallel paths for current there are (the greater the surface area, the lower the resistance, because more ion paths are available to carry the current).

For a spherical cell without dendrites,[5] with a radius r and a surface area $4\pi r^2$, the membrane resistance R_m is given by

$$R_m = R_{input} \times 4\pi r^2 \qquad (5\text{-}3)$$

For a cell having an R_{input} of 100 megohms and a radius of 100 μm (a surface area of 1.26×10^{-3} or approximately 1×10^{-3} cm^2)

$$R_m = 1 \times 10^{-3} \text{ cm}^2 \times 10^8 \; \Omega$$
$$= 1 \times 10^5 \; \Omega\text{cm}^2$$
$$= 100{,}000 \; \Omega\text{cm}^2$$

This value is also typical for the cell-body membrane of molluscan neurons. In most other animals the neuronal cell-body membranes are leakier to ions and have considerably lower values of membrane resistance, ranging from an extreme of 20 Ωcm^2 (the node of Ranvier in the peripheral axons of the frog) to 16,000 Ωcm^2 (the Mauthner cell body in the lamprey).

Current–Voltage Relationship

By passing graded pulses of depolarizing and hyperpolarizing current across the membrane, one can determine the *current-voltage relationship* of a cell,

---

[4]In addition to R_m, the membrane resistance (in Ωcm^2), and R_i, the specific resistance of the cytoplasm or axoplasm (in Ωcm), three related terms are commonly used:

(1) r_m (in Ωcm), the *membrane resistance per unit length of a membrane cylinder,* is given by $R_m/\pi d$, where d is the diameter of the fiber. In the cell with an R_m of 100,000 Ωcm^2 and an axonal diameter of 100 μm, r_m would be about 3×10^6 Ωcm.

(2) r_i (in Ω/cm), the *intracellular (axoplasmic) resistance to current flow along the axon per unit length,* is given by $R_i/\pi r^2$. For a cell with an R_i of 50 Ωcm and an axon with a radius of 50 μm, r_i is 6.3×10^5 Ω/cm.

(3) r_o (in Ω/cm), the *extracellular resistance to current flow along the axon per unit length.*

[5]For the sake of simplicity I am ignoring the complexity of the neuron's geometry. Current injected into the cell body flows not only across the soma membrane, but also across the proximal axon and dendritic tree. To obtain an accurate estimate of specific membrane resistance, one needs to include the proximal dendritic tree in calculating the surface area (see Rall, 1959, 1976).

and examine whether the input resistance of the cell is completely passive (linear) or varies with the membrane potential (Figure 5-8). In some nerve cells, and most inexcitable cells, the ratio of $\Delta V_m/\Delta I$ is constant for small depolarizing or hyperpolarizing current steps below threshold (Figure 5-8A). In these cells R_{input} (given by the slope of the current-voltage relationship) is constant throughout this range and not dependent on membrane potential.

In most nerve cells, however, the input resistance does not behave linearly, but offers less resistance to current flow in one direction than in another. This asymmetrical property of membrane resistance represents an active response and is called *rectification*. In the steady state it can generally take two forms. One form, called *delayed rectification*, occurs when the membrane potential is depolarized from the resting level (Figure 5-8B). This can be seen when a membrane is depolarized by a constant-current pulse; the change in membrane potential declines as a result of a delayed (in relation to the onset of the current pulse) decrease in resistance.[6] Here the membrane offers less resistance to outward (depolarizing) current than to inward (hyperpolarizing) current. Delayed rectification becomes more prominent with greater depolarizations. Most nerve cells show delayed rectification close to (and beyond) the threshold for the action potential, but some cells show it even with small depolarizations. A second form of nonlinearity is called *anomalous rectification* (Figure 5-8C) because it occurs in the opposite direction of delayed rectification. It is usually seen when the membrane potential is hyperpolarized from the resting level, and is evident as less resistance to inward (hyperpolarizing) current flow than to outward (depolarizing) current flow.

Capacitance

By injecting small current pulses into cells we can learn about still another membrane property. Although the current pulse (I_m) that produces a change in membrane potential (ΔV_m) can be made to rise and decline very rapidly, the change in membrane potential rises and declines only slowly (Figures 5-8A and 5-9A1). This is due to another physical property of the membrane, its *capacitance*. A *capacitor* is an element in an electrical circuit that stores charge across its surface; it will therefore oppose any change in potential. Current flows into and out of a capacitor only when the voltage across the capacitor

---

[6]Actually, as we shall see in Chapter 6 (Figure 6-2B, p. 148) in most cells this decrease in potential is due to two factors: a decrease in resistance and the existence of an ionic battery for potassium ions (E_K) that is in series with the resistance and has a value (-75 mV) more negative than the resting potential (-60 mV).

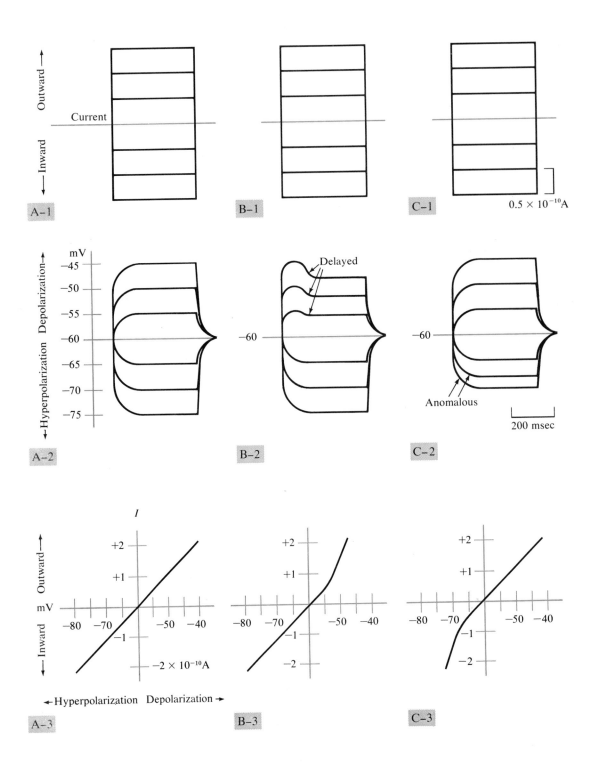

FIGURE 5-8.
Current–voltage relationships. By passing subthreshold, graded, inward- and outward-current pulses into the cell one can determine the current–voltage (*I–V*) relationship of the cell. In all three cases illustrated the current (top row) is the same.

A. Linear current–voltage relationship. Graded increases in outward- or inward-current pulses produce proportionally graded and symmetrical increases in the electrotonic potential (*A-2*). The resulting *I–V* curve (*A-3*), obtained by plotting the steady-state voltage against the injected current, is linear.

B. Delayed rectification. Graded inward-current pulses produce proportional increases in the hyperpolarizing electrotonic potential. In contrast to the linear *I–V* curve in part *A*, an outward-current pulse here produces a depolarization that declines to a new potential level after a brief delay (*B-2*). This is the result of a delayed increase in conductance to the depolarizing current (delayed rectification), which gives the steady-state *I–V* curve a characteristic upward slope in the depolarizing range of membrane potential.

C. Anomalous rectification. Graded outward-current pulses produce proportional increases in the depolarizing electrotonic potential. But inward-current pulses produce progressively smaller increments of hyperpolarizing potential change. This reflects an increase in conductance to hyperpolarizing current pulses (anomalous rectification) and gives the *I–V* curve a downward slope in the hyperpolarizing range of membrane potential.

is changing. Once the capacitor is charged to the voltage of the driving potential, no further capacitive current will flow. Thus a capacitor does not prevent voltage change, it only slows its onset and offset.

A capacitor consists of two conductors (or plates) separated by an insulator. The membrane acts as a capacitor because the intracellular space (the cytoplasm) and extracellular space of nerve cells are fluid-electrolyte compartments and therefore are good conductors, whereas the high-resistance lipid-protein membrane is a good insulator. As a result of its capacitive properties, the nerve membrane can store (or separate) charge. Thus, if a constant-current pulse produces a change in membrane potential (ΔV_m), a charge proportional to ΔV_m will build up on its inner and outer surfaces. The input capacitance of the membrane (C_{input}) is therefore defined as the ratio between the charge (q, in coulombs) separated by the membrane and the change in membrane potential:

$$C_{input} = \frac{q}{\Delta V_m} \qquad (5\text{-}4)$$

The capacitance of a cell is directly proportional to the surface area of the membrane (a larger area of inner and outer fluid conductors will allow the membrane to hold more charge) and inversely proportional to the thickness of the membrane (increased thickness decreases the effect that a charge on one conducting surface exerts on the other). Since the thickness of all cell membranes is approximately the same, about 7.5 nm, the membrane capacitance (C_m) depends primarily on surface area and is standardized to 1 cm² of membrane. The membrane potential of a cell is therefore the charge on the membrane divided by the membrane capacitance ($V_m = q/C_m$). Capacitance is measured in Farads; 1 F = 1 coulomb (6.24×10^{18} electrons) per volt.

If a square pulse of current flowed only through the resistance of the membrane it would give rise to a square pulse of voltage. However, since the membrane also acts as a capacitor and a capacitor holds charge, current flowing through the membrane takes time to produce a change in the potential across the membrane. Because the membrane capacitance is parallel with the membrane resistance, the voltage across the capacitance and resistance will be the same, so that at any given time the membrane capacitance will have a charge that is proportional to the membrane potential. Before a current pulse can fully change the membrane potential to a new value (determined by the magnitude of the current and the membrane resistance), the current flow must first alter the existing charge on the membrane capacitance and bring it to the level of charge required for the new membrane voltage. Therefore, current flowing into a cell will initially flow into and out of the capacitor, altering its charge. As the membrane potential gradually approaches its new value, less current flows across the capacitance and more current will flow across the resistance. When the capacitor is fully charged, all the current will flow across the resistance.[7]

Therefore, when a current pulse is applied to a membrane the current takes two pathways. Initially, current flows across the membrane capacitance, changing the charge across it. This component of the current is called the *capacitive current* (I_C). As the membrane capacitance becomes charged, current begins to flow through the resistance of the membrane. This second component of the current is called the *resistive current* or, in body tissues (where electricity is carried only by ions and not by electrons), *ionic current* (I_R). These two currents are schematically depicted in Figure 5-9A.

Thus, if we apply current across the membrane, the flow of the current through R_m is described by Ohm's law (Equation 5-1):

$$I_R = \frac{\Delta V_m}{R_m}$$

where ΔV_m is the change in membrane potential produced by the current I_R.

[7]Again, when the current pulse is removed, the membrane potential will return only slowly to the initial level because the initial charge on the membrane capacitance must be reconstituted. The phrase "current flowing across the capacitor" is meant to indicate current into and out of the two plates. Actually, current cannot flow *across* a capacitor because it does not cross the insulator separating the plates.

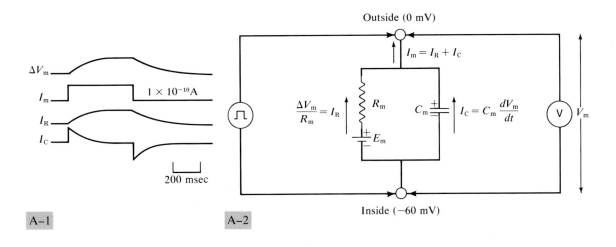

Outside (0 mV)

$I_m = I_R + I_C$

$\frac{\Delta V_m}{R_m} = I_R$

R_m

C_m

$I_C = C_m \frac{dV_m}{dt}$

E_m

V

V_m

Inside (−60 mV)

ΔV_m

I_m

I_R

I_C

1×10^{-10}A

200 msec

A-1

A-2

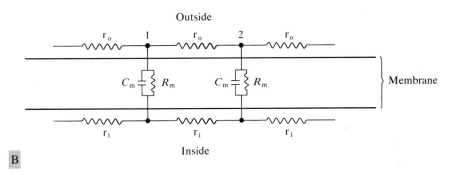

Outside

r_o 1 r_o 2 r_o

C_m R_m C_m R_m

Membrane

r_i r_i r_i

Inside

B

FIGURE 5-9.

The electrical equivalent circuit of a patch of membrane.

A. Part *A-1* compares the amplitudes and time courses of the resistive (I_R) and capacitive (I_C) components of a square outward-current pulse (I_m) that produces the electrotonic potential (ΔV_m). Part *A-2* shows the equivalent circuit for the membrane potential (V_m). The circuit consists of the battery for the resting potential (E_m) in series with the resistive element (R_m) and in parallel with the capacitive element (C_m). The recording and stimulating arrangement is the same as in Figures 5-2 and 5-3. The diagram shows the passage of an outward current (I_m) across a small patch of membrane. Part of the membrane current passes across the resistive element and is called the *ionic current* (I_R). The other component of membrane current flows into and out of the capacitive element and is called the *capacitive current* (I_C).

B. Electrical equivalent circuit of the cell membrane. For convenience, R_m and C_m are represented as occurring in discrete elements; actually, they are uniformly distributed along the cell surface. Similarly, internal and external resistances (r_i, r_o) along the lateral extent of the axon are represented as discrete elements, although these resistances are distributed throughout intra- and extracellular spaces. The new resistances along the extent of the axon, added in part *B*, are responsible for cable properties discussed in the text. (The series membrane batteries do not affect the passive properties of the membrane and have been eliminated to simplify the circuit.) [From Brinley, 1974.]

The flow of the capacitive current through C_m can be derived as follows. The capacitive current equals the rate of change of charge, or $I_C = dq/dt$. Since $q = C_m V_m$, the flow of the capacitive current is given by the size of C_m and the rate of change of voltage (dV_m/dt):

$$I_C = C_m \frac{dV_m}{dt} \qquad (5\text{-}5)$$

The total current across the membrane will thus be

$$I_m = \frac{\Delta V_m}{R_m} + C_m \frac{dV_m}{dt} \qquad (5\text{-}6)$$

The time course of I_R can be obtained from the record of the membrane potential because $\Delta V_m = I_R \times R_m$. Since the sum $I_R + I_C = I_m$, the current across the capacitance (I_C) can be obtained by subtracting I_R from I_m (Figure 5-9A).

Time Constant and Space Constant

Figure 5-9A2 illustrates an *electrical equivalent circuit* of a short segment of the membrane. When appropriately connected by resistors representing cytoplasm and extracellular space, the equivalent circuit of adjacent short segments constitutes an equivalent circuit for the entire cell body (or axon) membrane (Figure 5-9B). In electrical equivalent circuits the capacitor represents the insulating lipid-protein membrane separating the cytoplasm and the extracellular fluid. The resistances (or conductances) represent aqueous conducting channels that are interspersed throughout the membrane.

An electrical equivalent circuit is useful for understanding current flow and changes in membrane potential. It has the same response properties as the membrane to applied currents or to changes in resistance or voltage, yet the equivalent circuit represents actual physical components that can be wired together, analyzed, and quantified. For example, the passive properties of the equivalent circuit (and those of the membrane) can be fully characterized by two parameters, the time constant and the space constant. These parameters are central for understanding the contribution of the passive properties of the neuron to neuronal integration. The *time constant* characterizes the time course of a change in membrane potential, i.e., the rate with which the membrane potential changes when it is moved from one level to another. The *space constant* characterizes the rate with which a particular voltage change decays with distance. Let us first consider the time constant.

The *membrane time constant* (τ_m) is the time taken for a constant-current pulse to charge the membrane capacitance (of a spherical cell) to 63 percent of its final value (or to put it more precisely, to $1 - 1/e$ of its final value).[8] The membrane time constant can be measured directly from the voltage trace (Figure 5-10A). In our example of a molluscan neuron with an $R_m = 100,000$ Ωcm^2, a typical τ_m would be 100 msec. In different cells values for τ_m range from one to several hundred msec. In the simplest example of a spherical cell without an axon or dendrites the time constant is defined as

$$\tau_m = R_m \times C_m \tag{5-7}$$

Thus, a measure of τ_m is also useful for estimating C_m. Assuming τ_m to be 100 msec and R_m to be 100,000 Ωcm^2, C_m can be calculated to be

$$C_m = \frac{100 \times 10^{-3} \text{ sec}}{1 \times 10^5 \text{ } \Omega cm^2}$$

$$= 100 \times 10^{-8} \text{ F/cm}^2$$

$$= 1 \text{ } \mu F/cm^2$$

The *membrane space constant* or *length constant* (λ) is the distance along the axon over which a voltage applied at one point in the neuron (say the cell body) has fallen 63 percent ($1 - 1/e$) of its initial value in the axon. We can then measure the speed of the potential change along the axon produced by a current applied in the cell body by simply inserting recording electrodes (V_1, V_2, V_3, V_4) at various distances along the axon (Figure 5-10B). To simplify matters, let us assume that the axon has a large diameter, one that is comparable to that of the cell body.

The membrane space constant (in cm) is given by

$$\lambda = \frac{1}{2}\sqrt{\frac{dR_m}{R_i}} \tag{5-8}$$

where R_m is the membrane resistance, R_i the specific resistance of the axoplasm, and d the diameter of the axon or dendrite.

[8]The equation that governs the time course of the rising portion of potential during the pulse is $V_m = IR_m (1 - e^{-t/R_mC_m})$, where I is the magnitude of the current step imposed across the membrane. IR_m is the final value of the membrane potential due to the current pulse. When $t = \tau_m = R_mC_m$, $V_m = IR_m (1 - e^{-1}) = IR_m (1 - 1/e) = 0.63 \text{ } IR_m$.

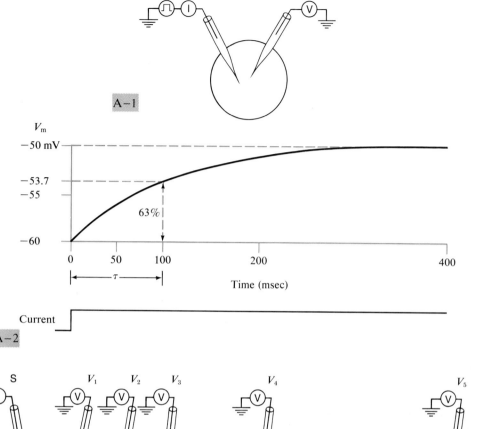

A-1

A-2

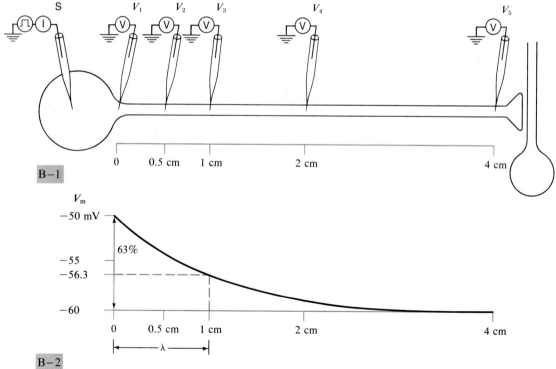

B-1

B-2

FIGURE 5-10.
Determination of the time constant and space constant.

A. Time constant. *A-1.* Experimental set up. A prolonged current step is applied through a current-passing electrode and the change in membrane potential is measured with an intracellular electrode. *A-2.* Graph of the change in membrane potential as a function of time after injection of a constant-current pulse. The current produces a 10 mV depolarization, and changes the membrane potential from -60 to -50 mV. The time that it takes to reach 63.2 percent of that is defined as the *time constant* of the cell-body membrane. For the cell in this example it is 100 msec.

B. Space constant. *B-1.* Experimental set up. A prolonged current step is applied through a current-passing electrode in the cell body and the change in membrane potential is measured with an intracellular electrode at the axonal edge of the cell body (V_1) and at various points along the axon (V_2, V_3, V_4 and V_5). *B-2.* Graph of the change in membrane potential as a function of distance. The current produces a 10 mV depolarization in the cell body (V_1). At 1 cm (V_3), this potential change has fallen by 63 percent to reach 37 percent of its initial value (3.7 mV). This distance (1 cm) is defined as the *space constant* of the axonal membrane.

The membrane space constant is a measure of the distance over which a subthreshold signal will propagate electrotonically along the axon (or dendrite). For the neuron in our example, with $R_m = 100,000\ \Omega\text{cm}^2$ and $R_i = 50\ \Omega\text{cm}$, the space constant of an axon with a diameter of 20 μm would be 1 cm. But in axons with small diameters (*d* in Equation 5-8) the length constant may be considerably less than 0.1 cm.

ACTIVE PROPERTIES OF THE MEMBRANE

By inserting a second recording electrode (V_2) into the cell body of the neuron, and still others at different points along its axon (V_3, V_4, V_5), we can also compare the passive propagation of the electrotonic potential produced in the cell body by a current pulse with the *active* propagation of an action potential (Figure 5-11). The stimulating electrode is placed in the cell body. At time t_1 a small hyperpolarizing current pulse is injected; at a later time (t_2) a symmetrical depolarizing current is injected, and at a still later time (t_3) a larger depolarizing pulse that triggers an action potential. These three events at the five positions are then recorded with five electrodes. The second electrode in the cell body will record the same potential change as the first. This tells us that the cytoplasm of the cell is an excellent conductor, so there is relatively little resistance to current flow between the two electrodes (see Note 4, p. 114). This is true any place within the cell body because its volume is large and the specific intracellular resistance of the cytoplasm is low (a hundred million times less that of the membrane). As a result, the cell body is at all times electrically uniform or *isopotential.*

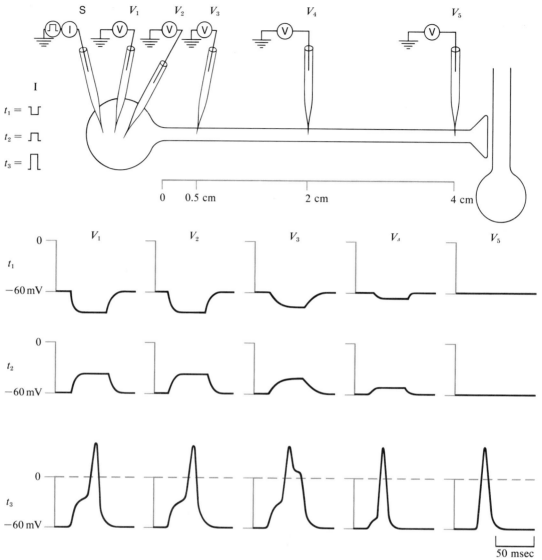

FIGURE 5-11.

Comparison of the propagation of electrotonic potential and action potential. In this hypothetical experiment the consequences of injecting various current pulses into the cell body with a stimulating electrode (S) are examined with intracellular electrodes at several points along the neuron. Electrodes V_1 and V_2 are in the cell body; V_1 is closer than V_2 to the current-passing electrodes. Electrode V_3 is in the initial segment of the axon (0.5 cm from the cell body), where the action potential is initiated. Electrode V_4 is in the axon, 2 cm from the cell body, and V_5 is further down the axon, 4 cm from the cell body. At time t_1 a small inward current is injected into the cell; at t_2 a small subthreshold outward current is injected; and at t_3 a suprathreshold outward current that triggers an action potential is injected. The inward and subthreshold outward currents produce hyper-polarizing and depolarizing electrotonic potentials, respectively, that propagate decrementally with distance and are too small to be detected beyond 2 cm. By contrast, the suprathreshold outward current triggers an action potential at the initial segment of the axon (V_3), and the action potential propagates without decrement in two directions: orthodromically, down the length of the axon to the right, and antidromically, toward the cell body on the left. The notched action potential V_3 is due to the fact that the impulse propagates in both directions from the site of initiation. This feature of impulse initiation is considered further in Chapter 7 (Figures 7-17 and 7-18). The electrotonic potential that triggers the spike also propagates down the axon, but it decays and is not recordable at V_5.

However, as the change in potential propagates electrotonically down the axon, it undergoes characteristic changes with distance: it becomes progressively smaller and rises and decays more slowly. Thus, although subthreshold potential changes initiated in the cell body or the initial part of the axon propagate along the axon for some distance (sometimes for up to 2 or 3 cm), the potential decreases progressively with distance, usually undergoing a 63 percent $(1 - 1/e)$ decrease every 1 cm (in a neuron with a long space constant). The attenuation occurs because (1) as the distance becomes greater the specific intracellular resistance of the axoplasm (R_i) becomes a significant factor in limiting current flow inside the neuron; and (2) some current escapes through the axon membrane along its length into the low-resistance extracellular space.

The slowing in the rate of rise (and decay) of the change in membrane potential with distance is due to the time that it takes to charge the membrane capacitance. At first all the current is used to charge the capacitance of the cell-body membrane near the current-injecting electrode. Later, a lesser amount of current has to charge the much more extensive distal area of the axonal membrane, so that the rate of charging and therefore the membrane potential change will be less. Moreover, the current does not begin to flow at a distant point along the axon until the capacitance of the cell body and proximal axon membrane have been partially charged. The further an axon site is from the site of current injection, the longer it will take before significant current is channeled to that site (by the progressive charging of nearby regions) and the slower will be the rate of changing the membrane potential at that site.

As we have seen, the passive properties of neurons resemble those of submarine telegraph cables. Because of attenuation of the signal along their length, core-conductor cables are not good long-range signaling devices. In telegraph cables this problem is handled by interposing booster devices at given distances. The problem has been solved in nerve cells in a similar way. Neurons use a second signaling device, the action potential, along most (often all) of their length. When the membrane is depolarized to threshold, an action potential is triggered.

The active propagation of the action potential along the axon is very different from the passive propagation of the electrotonic potential (Figure 5-11). Whereas the amplitude of the electrotonic potential decreases and its rate of rise and decline decreases with distance, the amplitude and shape of the action potential do not change. The action potential is constant because the energy used in its propagation does not come from the stimulus, as with the electrotonic potential, but is released all along the length of the axon.[9] In this respect nervous conduction is often compared to the burning of a fuse: once a critical temperature is reached at one end of the fuse it ignites the next segment of the fuse, the heat produced by that segment ignites the next, and so on down the line. Another feature distinguishes the action potential from the electrotonic potential: the action potential is all-or-none, i.e., its amplitude is not dependent on that of the initiating stimulus. An action potential is triggered once a stimulus is above threshold; a larger stimulus will not produce a larger action potential.

Active propagation offers several advantages over passive propagation for long-distance signaling. First, it assures a favorable signal-to-noise ratio. Membranes produce a small amount of random low-amplitude noise, which distorts small signals more than it does large ones. Because the all-or-none impulse is large (in relation to the noise), there is minimal distortion in the signal along the length of the axon. In addition, a large signal amplitude is guaranteed. In most cells transmitter release has a definite threshold; the terminal must be depolarized by about 25 mV (see Figure 6-15, p. 182); in the range of 25 to 75 mV greater release of transmitter requires greater depolarization. An action potential of 80 to 120 mV thus operates in an optimal range of transmitter release. Finally, in active propagation distortion in the signal's high-frequency components due to capacitive losses is minimal.

PASSIVE PROPERTIES AND NEURONAL INTEGRATION

Neuronal integration is the process by which a neuron adds the various subthreshold signals generated by its membrane and determines whether or not to discharge an action potential. As a result, integration is critically dependent

---

[9]This applies to invertebrate axons, almost all of which are not myelinated, and to the small nonmyelinated axons of vertebrates. Nerve fibers in vertebrates are covered with a thick myelin sheath, which is an excellent insulator. This sheath is interrupted periodically by nodes (the nodes of Ranvier) at which changes in ionic conductance occur. In these fibers conduction is not continuous but *saltatory*, i.e., the electrical activity jumps from node to node. Between the nodes the action potential is propagated electrotonically with little decrement. This is because the myelin sheath increases the membrane resistance and thereby the space constant of the axon. At the nodes, where ionic current can flow, the electrotonic potential triggers an action potential and the process is repeated. Saltatory conduction therefore reduces ionic exchange. The velocity of saltatory conduction in a myelinated fiber is much greater than that of an unmyelinated fiber of the same diameter. Conduction velocity depends upon the rate with which the membrane capacity in the region ahead of the action action potential is discharged to threshold. The myelin sheath decreases the membrane capacity per unit length and thereby allows rapid discharging by the action potential.

upon the passive membrane properties of the neuron. For example, in most cases the excitatory postsynaptic potential produced by the activity of a single presynaptic neuron is insufficient to depolarize the initial segment of the axon by the 15 mV or so necessary to fire the cell (Cell B, Figure 5-12). Furthermore, concurrent inhibitory postsynaptic potentials produced by other cells act to prevent discharge of the postsynaptic cell (Cell A, Figure 5-12) even though an otherwise adequate number of excitatory endings are simultaneously active. The trigger zone of the postsynaptic cell adds the various excitatory and inhibitory synaptic inputs and determines whether or not an impulse is discharged (Cell C, Figure 5-12).[10]

The passive properties of neurons contribute to neural integration in two ways. First, the time constant determines the configuration of the synaptic potential and thereby affects the temporal aspects of integration. The synaptic current produced by a synaptic action is usually brief compared to the time constant of the postsynaptic cell. The capacitive properties of the postsynaptic cell slow the voltage changes so that a synaptic potential will rise (slightly) slower and will decay considerably more slowly than the synaptic current (2 to 10 times, depending on the time constant). As a result, a second impulse in the same presynaptic neuron, following closely upon the first, is able to add its depolarization to that remaining from the previous synaptic potential in the postsynaptic cell. The process by which consecutive synaptic actions produced at the same site (by the same neurons) add in the postsynaptic cell is called *temporal summation*. The membrane time constant is the critical determinant of temporal summation. Neurons with a long time constant have a considerably greater capability for temporal summation than neurons with a short time constant (Figure 5-13*A*).

The space constant is important for the *spatial* aspects of integration. The generator potentials of sensory neurons, the synaptic potential, and the pacemaker potentials usually have to propagate over some distance from their site of origin to the trigger zone. The degree to which a depolarization of a given size at its source is decreased by propagation to the trigger zone is determined by the space constant of the cell. Cells with long space constants are more effective than cells with short space constants in allowing distant signals to propagate to the trigger zone with minimal decrement. In addition, a given depolarization produced at one synaptic locus (by one presynaptic neuron) is

---

[10]Many invertebrate and some vertebrate neurons have multiple trigger zones. In such neurons each zone sums the total excitation and inhibition produced by synaptic inputs near it. The discharge of the local trigger zone then leads to the discharge of the final common trigger zone in the initial segment of the axon.

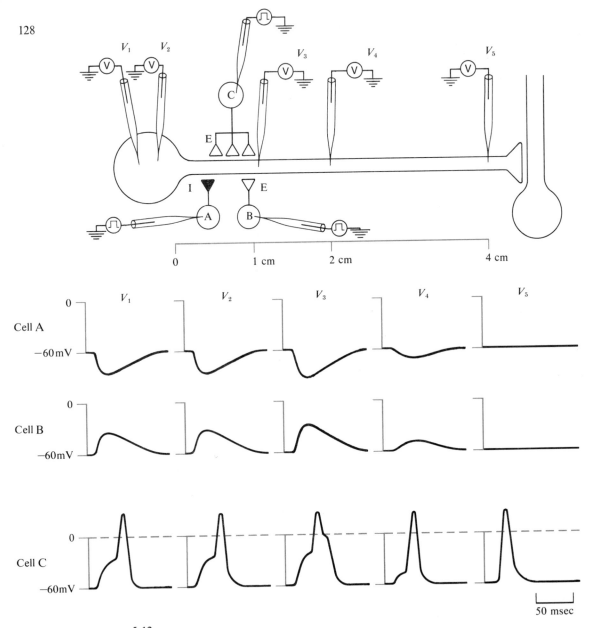

FIGURE 5-12.

Comparison of electrotonic (passive) propagation of the synaptic potential with active conduction of the action potential. In this hypothetical experiment, as in the experiment in Figure 5-11, recordings are obtained from five sites within a neuron. Two electrodes, V_1 and V_2, are in the cell body; V_3 is in the initial segment; V_4 and V_5 are further out along the axon, 2 cm and 4 cm, respectively, from the cell body.

Cell A: Stimulating a cell that mediates inhibition produces an IPSP near the initial segment of the axon. The IPSP propagates electrotonically with decrement in both directions (toward the cell body and down the axon).

Cell B: Stimulating a cell that mediates subthreshold excitation produces an EPSP near the initial segment of the axon. The EPSP also propagates electrotonically with decrement in both directions.

Cell C: Stimulating a cell that produces suprathreshold excitation initiates an action potential that propagates without decrement in both directions. The EPSP that triggers the action potential also propagates in both directions, but it decays down the axon and is not recorded at V_5.

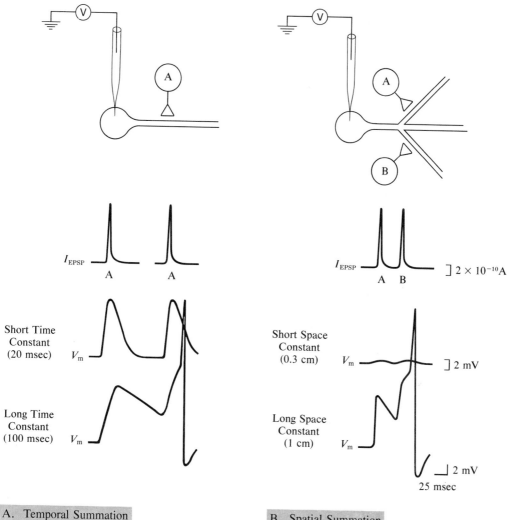

A. Temporal Summation

B. Spatial Summation

FIGURE 5-13.

Neuronal integration: temporal and spatial summation.

A. Temporal summation of two EPSPs produced by a single presynaptic neuron. Whether the cell has a short time constant (20 msec) or a long time constant (100 msec), the EPSPs will be longer than the synaptic current (I_{EPSP}). But in the cell with the long time constant the EPSPs are more likely to summate and lead to an action potential. In the cell with the short time constant the EPSPs will fail to trigger an action potential.

B. Spatial summation. Whether the cell has a short or long space constant, the synaptic current (I_{EPSP}) is assumed to be the same. In each case two presynaptic terminals end on remote portions of the dendrites and produce EPSPs that are 20 mV at their sites of origin. In the case of the short space constant (0.3 cm) the cell body is three space constants away from the source of the EPSPs, and the EPSPs will be barely detectable in the cell body. In the case of the long space constant (1 cm) the cell body is only one space constant away, and the EPSPs will decrease to about 37 percent in the cell body. Since the initial segment of the main axon is closer to the source of the EPSPs, the EPSPs will summate and trigger an action potential there.

usually not sufficient to trigger an action potential at the trigger zone. To reach threshold, the integrative component may require the additive actions of several simultaneously active synaptic inputs. The process by which several spatially discrete synaptic inputs add their action is called *spatial summation.* As one can see from Figure 5-13*B,* the space constant is the critical determinant of spatial summation.

In the final analysis, the integrative control of a neuron rests with the spatial and temporal factors that control the membrane potential of the trigger zone (or zones). The propagation of synaptic excitation and inhibition is only important for the signaling function of the neuron insofar as it permits or prevents the membrane potential of the *trigger zone* from reaching threshold. A neuron will fire an action potential only if excitation at the trigger zone exceeds inhibition by a critical amount (usually about 15 mV) sufficient to allow the membrane potential of the trigger zone to reach threshold (see Cell C, Figure 5-12).

PASSIVE PROPERTIES AND SYNAPTIC TRANSMISSION

We have seen how the passive properties of the membrane and electrical equivalent circuits explain the propagation of subthreshold signals *within* the neuron. Passive properties and equivalent circuits are also useful for exploring current flow *between* neurons. The flow of current between neurons can be examined by injecting current pulses into the presynaptic neuron and recording the potential change produced in the postsynaptic cell.

At electrical synapses some of the current injected into a presynaptic neuron flows through the bridging channels of the gap junction into the postsynaptic cell and across its membrane. This occurs because the resistance to current flow (per unit area) is lower across the junction than the resistance across the remainder of the external membrane (Figure 5-14*A*). In the electrical equivalent circuit the bridging channel can be depicted as a coupling resistance (r_c) that connects the presynaptic and postsynaptic cells (Figure 5-14*B*). The membrane of the bridging channels usually does not generate an action potential. The potential change produced in the presynaptic neuron propagates electrotonically across the gap junction into the postsynaptic cell where it produces a potential change of reduced amplitude. Both depolarizing and hyperpolarizing potential changes will propagate across the synapse. Therefore, an action potential in the presynaptic neuron that has a large hyperpolarizing afterpotential (as is the case for many molluscan neurons) produces a diphasic (depolarizing-

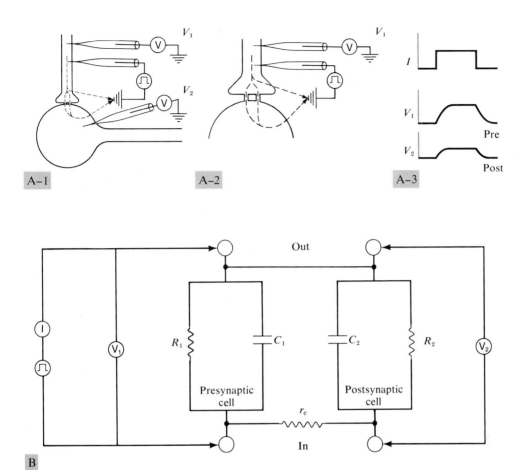

A-1   A-2   A-3

B

FIGURE 5-14.
Current flow at an electrical synapse following injection of an outward-current pulse into the presynaptic terminals.

A. *A-1.* Experimental arrangement. *A-2.* Enlarged view of the synapse. *A-3.* The injected current pulse (I) and the pre- and postsynaptic recordings (V_1 and V_2, respectively). Current injected into the presynaptic terminal will flow outward across the terminal membrane and produce an electrotonic potential recorded at V_1. In addition, current will spread across the gap junction into the postsynaptic cell and flow outward there, producing a smaller electrotonic potential in the postsynaptic cell.

B. Electrical equivalent circuit for an electrical synapse. In this circuit r_c indicates the coupling resistance due to the gap junction; R_1 and C_1, and R_2 and C_2 are the resistance and capacitance of the presynaptic and postsynaptic cells, respectively. V_1 and V_2 are voltage recording sites for the pre- and postsynaptic cells, respectively. The bar connecting the outside terminals represents the low-resistance extracellular fluid.

hyperpolarizing) electrotonic potential change in the postsynaptic cell (see Figure 5-4C). As a result of these properties electrical synapses are often called *electrotonic synapses* (Bennett, 1973).

From the equivalent circuit (Figure 5-14B) one can define the *coupling ratio* (or coupling coefficient) for voltage propagation from the presynaptic to the postsynaptic cell and estimate the decrease in amplitude of the signal. For synaptic transmission in the forward direction the coupling ratio is

$$\frac{V_2}{V_1} = \frac{R_2}{(R_2 + r_c)} \qquad (5\text{-}9)$$

In the case of synaptic transmission in the reverse direction, the coupling ratio is

$$\frac{V_1}{V_2} = \frac{R_1}{(R_1 + r_c)} \qquad (5\text{-}10)$$

where V_1 and R_1 and V_2 and R_2 are the potential changes and input resistances of cells 1 and 2, respectively, and r_c is the coupling resistance between them.[11]

At chemical synapses the extracellular space is always normal and may actually be enlarged. This is indicated in the equivalent circuit by an absence of a coupling resistance between the pre- and postsynaptic cells (Figure 5-15). Because the resistance of membranes is high and that of the extracellular space low, most of the current injected into the presynaptic neuron flows across the membrane into the low-resistance extracellular space and little or no current flows into the postsynaptic cell (Figure 5-15A). Transmission occurs at chemical synapses because a depolarizing signal in the presynaptic cell leads to transmitter release.

SUMMARY AND PERSPECTIVE

Intracellular studies of a variety of neurons in different animals reveal that the excitable surface membrane of the neuron is not homogeneous, but has at least four signaling components with distinct functions: (1) a transducer or input component, located at the sensory receptor membrane in sensory neurons and at the synaptic membrane of the dendrites (and to a limited extent

---

[11]An important feature of these relationships is that the resistance of the presynaptic neuron does not appear in the coupling coefficient. If R_2 is much larger than r_c, then V_2/V_1 will be close to one. But if R_1 is small compared to r_c, V_1/V_2 will be small. Thus, transmission in one direction can be much greater than that in the other and the synapse can be unidirectional.

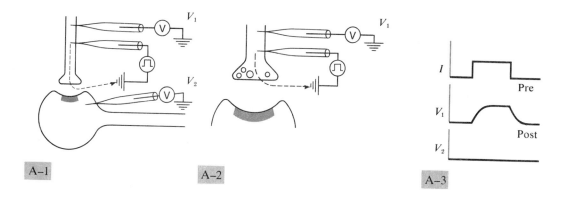

A–1   A–2   A–3

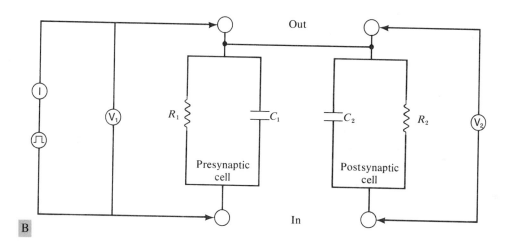

FIGURE 5-15.

Current flow at a chemical synapse following injection of an outward-current pulse into the presynaptic terminals.

A. The stimulating and recording arrangements are illustrated in parts *A-1* and *A-2*. The injected current pulse (*I*) and the pre- and postsynaptic recordings (V_1 and V_2, respectively) are illustrated in part *A-3*. As with electrical synapses, current injected into the presynaptic cell flows outward across the terminal and produces a depolarizing electrotonic potential at V_1. But because of the separation between the pre- and postsynaptic elements, the cells are effectively uncoupled electrically. As a result, current flows back to the return (ground) electrode through the low-resistance path of the synaptic cleft (part *A-2*) and no current flows across the postsynaptic membrane so that no potential change is recorded at V_2 (part *A-3*).

B. Electrical equivalent circuit for a chemical synapse. Note the lack of coupling resistance for the pre- and postsynaptic cells.

the cell body) of central neurons; (2) a decision-making or integrative component, located at the initial segment of the axon in most neurons; (3) a conductile component, located along the axon; and (4) a transmissional or output component, located at the presynaptic terminals of the neuron.

At the receptor membrane of a sensory neuron a stimulus is transformed into electrical energy as a generator potential. The generator potential then propagates passively (electrotonically) down the axon. If it is sufficiently large, it will be adequate, despite some decrement, to depolarize to threshold the membrane potential of the initial portion of the axon and so trigger an all-or-none action potential. Once initiated, the action potential propagates rapidly without decrement along the axon into the central reaches of the nervous system. When the action potential arrives at the terminal of the axon it sets in motion the process of synaptic transmission, which conveys information to the next cell. At electrical synapses the current produced by the action potential in the terminal region spreads directly into the next cell across a gap junction, which establishes electrical continuity between the pre- and postsynaptic cells. At chemical synapses there is no electrical continuity between the pre- and postsynaptic cells. Instead, the action potential in the terminals leads to the secretion of a small amount of a chemical substance, which then interacts with synaptic receptor molecules on the external surface of the next cell to initiate the ionic current flow that produces the excitatory (or inhibitory) postsynaptic potential. The various excitatory (or inhibitory) postsynaptic potentials bring the membrane potential of the trigger zone toward (or away from) the critical firing threshold. The postsynaptic neuron will fire an action potential only if excitation at the trigger zone exceeds inhibition by a critical minimum. Neuronal integration, the process by which a cell adds temporally and spatially discrete input from excitatory and inhibitory sources, is in turn critically determined by the passive properties of the cell membrane: its resistance, capacitance, and time and space constant.

Regardless of shape, size, location, and function most nerve cells of vertebrates and invertebrates, can be described in terms of an idealized neuron containing the four signaling components. Moreover, as we shall see in the next chapter, the functioning of each of these components is well understood: The resting potential, the action potential, the excitatory and inhibitory synaptic potentials, and the generator potential of sensory cells can all be explained by means of a unified theory, the ionic hypothesis, which applies to all nerve cells.

SELECTED READING

Eccles, J. C. 1973. *The Understanding of the Brain.* McGraw-Hill. Chapters 1 through 3 are an up-to-date overview of cellular neurobiology.

Stevens, C. F. 1966. *Neurophysiology: A Primer.* Wiley. A simple introductory text.

CHAPTER 6

Cellular Mechanisms of Neuronal Function

That various nerve cells can be described in terms of a common model has been immensely reassuring for students of the neurobiology of behavior. It suggests that the analysis of any nervous system (or a component thereof) involves primarily an understanding of three types of principles: (1) those underlying the signaling capabilities of individual nerve cells; (2) those that describe how individual cells interconnect and interact; and (3) those that relate patterns of interconnected cells to behavior.

In this chapter I will consider further the mechanisms for electrical signaling by individual cells, an area in which a great deal of research has been focused. Out of this research has come the *ionic hypothesis,* according to which the resting, action, and synaptic potentials all involve the same basic mechanism: the passive movement of inorganic ions down their electrochemical gradients. Various types of potential changes differ primarily in the ion species involved, the sequence in which the ionic movements are activated or inactivated, and the dependence of the ionic movement on membrane potential.

IONIC MECHANISMS OF MEMBRANE AND ACTION POTENTIALS

As we have seen, a nerve cell at rest maintains a potential difference of about 60 mV across its surface membrane, the inside being negative in relation to the outside. Working on the giant axon of the squid, Curtis and Cole (1942) and Hodgkin and Katz (1949) demonstrated that the battery that generates this voltage derives its potential energy from an unequal distribution of potassium ions across the nerve cell membrane (Table 6-1). The long-term distribution of potassium and sodium ions is maintained by a metabolic process, the Na^+–K^+ transport mechanism or pump, which keeps the intracellular Na^+ concentration low and the intracellular K^+ concentration high by actively transporting Na^+ from the inside of the cell to the outside, and K^+ from outside the cell in. As a result of the activity of this Na^+–K^+ exchange, the K^+ concentration inside a nerve cell is about 20 times higher than that of the extracellular space, and the Na^+ concentration is approximately 10 times lower inside than outside the cell. In the squid giant axon the intracellular concentration of K^+ is 400 mM and the extracellular concentration is 20 mM; the intracellular concentration of Na^+ is 50 mM and the extracellular concentration is 440 mM.

In solution, ions move randomly as a result of collisions with the solvent. The average kinetic energy of ions, attributable to their random thermal motion, is such that the rate at which an ion species moves out of a small volume is proportional to its concentration gradient. If the concentration of an ion varies from one region to another, there will be a net movement of the ion from the region of high concentration to the region of low concentration. Thus, ions move down a concentration gradient just as water flows down a stream in response to a potential energy gradient. In living systems the movement of ions into or out of a cell is restricted by the lipid-protein membrane separating the two compartments. It is because it resists or opposes the flow of ions that a cell membrane can be said to have resistance.[1] Conversely, the membrane has a certain conductance to ions because it is to some degree leaky. Membranes oppose the flow of ions in solution (more than the flow of water molecules) because they are relatively impermeant barriers and permit ions to flow only through specific and sparsely distributed pores called *channels*. Ions vary in size, three dimensional shape, and in the layers of water molecules (hydration shells) that attach to them (Table 6-1). The limitation to the movement of a particular ion species is thought to be determined by two factors: the number of available channels and the type of channels (size, shape, chemical

---

[1]The resistance of natural membranes is also a function of the concentration of ions in the membrane.

TABLE 6-1.

Comparison of ion concentrations in the squid giant axon, squid blood,
and seawater (usually used as external bathing solution).

	Concentration (mM)			Ionic crystal radius (nM)	Calculated hydration number
	Axoplasm	Blood	Seawater		
K+	400	20	10	.133	2.9
Na+	50	440	460	.095	4.5
Cl−	40–150	560	540	.181	2.9
Ca++	0.4	10	10	.099	7.0
Mg++	10	54	53	.066	10.0

SOURCE: Hodgkin, 1964; Katz, 1966.

properties, and fixed charges). Each type of channel can therefore discriminate between ions on the basis of size, shape, and energetic factors (how readily their water layers can be stripped off). As a result, a given type of channel permits permeation of only a few closely related ion species.

The membrane potential exists because K+ is not distributed equally across the membrane and because the membrane has a relatively high permeability to K+ in its resting state.[2] The exact mechanism for the selective permeability of the resting membrane to K+ (as opposed to Na+) is not known. The number and specificity of the channels that are open in the resting membrane are probably key factors (Hille, 1976). Channels that permit K+ penetration also permit some other cations (thallium, rubidium and ammonium) to penetrate,

---

[2]"Ionic permeability" indicates the ease with which an ion *can* cross the membrane in response to a chemical or electrical gradient. "Ionic conductance" indicates the ease with which ions *actually* carry current across the membrane in response to an electrical gradient. Because changes in conductance (as during the action and synaptic potentials) are paralleled by changes in permeability, I will use these terms interchangeably. The permeability for an ion species, e.g., K+, is given by

$$P_K = \frac{u \beta R T}{a F}$$

where u is the mobility of K+, β is the partition coefficient between the membrane and solution, and a is the thickness of the membrane, F is the Faraday constant, T the absolute temperature, and R the gas constant. The relation of permeability to conductance of K+ is given by

$$g_K = \frac{P_K F^3 V_m K_o K_i}{R^2 T^2 (1 - e^{-V_m F/RT})(K_i - K_o)}$$

where K_o and K_i are the external and internal K+ concentrations, respectively. Thus, g_K depends not only on P_K but also on the quantity and distribution of K+ on each side of the membrane (Katz, 1966).

but largely exclude Na$^+$. In their hydrated form, the penetrating cations are all small compared to Na$^+$ (Hille, 1976). The K$^+$ channels in the resting membrane are therefore thought to select among cations in part on the basis of size. The resting membrane is about 100 times more permeable to K$^+$ than to Na$^+$. The Na$^+$ channels, in turn, permit lithium penetration but exclude K$^+$.

Membrane Potential

The membrane potential is generated in the following way. Because K$^+$ ions are present in high concentration inside the cell (as a result of the Na$^+$–K$^+$ transport mechanism), they tend to diffuse out of the cell, down their concentration gradient. But the inside of the cell tends to maintain an overall (bulk) electrical neutrality, an equal balance of positive and negative charges. To achieve neutrality, the outward movement of positively charged K$^+$ ions must be neutralized either by Na$^+$ moving into the cell or by anions moving out. The anions inside the cell are large, being mostly negatively charged proteins or amino acids. Because the membrane is much less permeable to these large organic anions and Na$^+$ than to K$^+$, a small amount of K$^+$ will leave the cell driven by the force of the concentration gradient and penetrate alone through the membrane, unaccompanied by anions. Consequently, the outside of the membrane accumulates a positive charge (due to the excess of K$^+$) and the inside a negative charge (because of the deficit of K$^+$ and the resulting preponderance of anions).[3] Since opposite charges attract each other, the positive charge on the outside and the negative charge on the inside collect locally on either side of the membrane in the form of a cloud (Figure 6-1). This cloud of charge reflects the electrical capacitance property of the membrane—its ability to separate charges—and must be dispersed for any decrease in membrane potential to occur.

Despite the electrical attraction between the K$^+$ on the outer surface of the membrane and the anions on the inner surface, no net movement of K$^+$ into the cell occurs because of the continued force exerted by the concentration gradient for K$^+$. But the separation of charge gives rise to a potential difference: positive outside, negative inside. The more K$^+$ continues to flow, the

---

[3]The law of electrical neutrality stipulates that there are no *macroscopic* differences in charge in the bulk phase of any solution. The separation of charge that generates the membrane potential is *microscopic* and highly localized, and does not violate bulk electrical neutrality. As we shall see below (p. 144), the amount of charge separated is very small and the distance of separation (the thickness of the membrane) is microscopic.

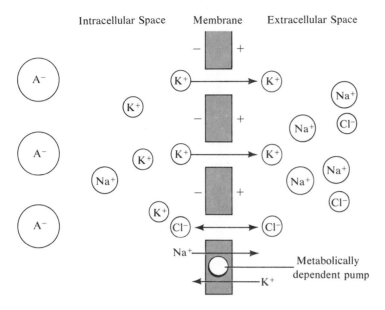

Intracellular Space   Membrane   Extracellular Space

FIGURE 6-1.

Schematic illustration of how potassium movement out of the cell establishes a resting potential that is negative inside in relation to outside. In response to the force of its concentration gradient, K⁺ leaves the cell and traverses the membrane without being accompanied by negatively charged organic molecules (A⁻), which are too large to penetrate the membrane. Potassium accumulates as a cloud on the outside surface of the membrane, attracted toward the membrane by the excess negative charge remaining in the cell. The distribution of ions that leads to the outward movement of K⁺ is established and maintained by a metabolically dependent active-transport process that pumps K⁺ into the cell and Na⁺ out against the concentration gradient. Chloride is assumed to distribute freely (according to its concentration gradient and the electrical gradient) on both sides of the cell.

more charge will be separated and the greater the potential difference will become. The electrical force due to the buildup of this potential difference opposes the force of the concentration gradient and impedes the further outward movement of the K⁺ ions. At a certain potential the electrical force due to the charge on the membrane becomes equal to the oppositely directed chemical force due to the concentration gradient, and no further net movement of K⁺ occurs. This potential is called the K⁺ *equilibrium potential* (E_K). It can be calculated from an equation developed by the physical chemist W. Nernst:

$$E_K = \frac{RT}{F} \ln \frac{[K^+]_o}{[K^+]_i} \qquad (6\text{-}1)$$

where E_K is the equilibrium potential of K^+ (defined as the inside potential minus the external potential; see Note 2, p. 98). R is the gas constant, T the absolute temperature, and F the Faraday constant; $[K^+]_o$ and $[K^+]_i$ are the concentrations of K^+ outside and inside the cell. In squid blood $[K^+]_o$ is 20 mM and $[K^+]_i$ is 400 mM (see Table 6-1). Natural logarithms are converted to logarithms at base 10 by multiplying by 2.3. Thus, multiplying by 2.3 gives a more useful expression of the Nernst equation:

$$E_K = 2.3 \frac{RT}{F} \log \frac{[K^+]_o}{[K^+]_i} \text{ mV} \tag{6-2}$$

At 18°C $RT/F = 25$, therefore

$$E_K = 58 \log \frac{[K^+]_o}{[K^+]_i} \text{ mV}$$

$$= 58 \log \frac{20}{400} \text{ mV}$$

$$= -75 \text{ mV}$$

The value of the resting potential, usually -60 (but sometimes as low as -40 mV or as high as -70 mV), is close to E_K. Small deviations from E_K are usually due to the slight permeability of the membrane to ions other than K^+, particularly Na^+ (see below). These deviations have been taken into account in a more elaborate equation devised by Goldman (1943). This equation considers (1) the concentration of all three major small ion species, and (2) their relative permeabilities (P_K, P_{Na}, P_{Cl}). Based upon the simple and reasonable assumption that the electric field (the potential gradient) *within* the membrane is constant, the Goldman equation gives the membrane potential V_m, as follows:

$$V_m = \frac{RT}{F} \ln \frac{P_K[K^+]_o + P_{Na}[Na^+]_o + P_{Cl}[Cl^+]_i}{P_K[K^+]_i + P_{Na}[Na^+]_i + P_{Cl}[Cl^+]_o} \tag{6-3}$$

Hodgkin and Katz (1949) used this equation to test the ionic hypothesis and estimated the relative permeabilities of the membrane of the squid giant axon to be $P_K:P_{Na}:P_{Cl} = 1.0:0.04:0.45$.

As these estimates illustrate, the membrane is relatively permeable to *both* K^+ and Cl^-. To what degree does Cl^- contribute to the membrane potential? The concentration ratio for Cl^-, $[Cl^-]_i/[Cl^-]_o = 40/560 = 1/16$, is roughly the reciprocal of the K^+ concentration ratio, $[K^+]_i/[K^+]_o = 400/20 = 20$. In the Nernst equation the sign is reversed for anions. The reciprocally distributed

Cl$^-$ therefore generates the same potential across the membrane as does K$^+$! In other words, the equilibrium potential for Cl$^-$ (E_{Cl}) is usually close to E_K, or about -75 mV. The reciprocal distribution of K$^+$ and Cl$^-$ arises because there are specific restrictions on the K$^+$ concentration but there are no restrictions on the Cl$^-$ concentration. This allows Cl$^-$ to adjust itself (by movement into and out of the cell) to the membrane potential determined by K$^+$. Unlike Cl$^-$, the K$^+$ concentration cannot change readily. Potassium ions must provide the positive charge needed to neutralize the negatively charged amino acids and proteins inside the cell. These intracellular anions are large and their concentration is fixed because they cannot penetrate the membrane and leave the cell. Sodium ions could of course balance this excess negative charge but they cannot enter the cell. As a result, the intracellular K$^+$ concentration becomes relatively fixed.

Thus, Cl$^-$ does not (usually) contribute to the membrane potential because its concentration gradient merely adjusts itself to be reciprocal to that of K$^+$. As a result, the Nernst potential for Cl$^-$ is equal in sign and magnitude to that of K$^+$. Deviations of V_m from E_K are therefore largely due to the small leakage of Na$^+$ into the cell.

Action Potential

Most cells generate a membrane potential across their membranes. What distinguishes nerve, muscle, and gland cells from the nonexcitable cells of the body is that their resting membrane potential can be changed for signaling. When the membrane potential of a nerve cell is made less negative by a certain critical amount, usually about 15 mV (from -60 to -45 mV), an action potential is initiated. In 1939 Cole and Curtis provided the first evidence that the generation of an action potential involves a dramatic increase in the ionic conductance of the membrane. They found that during the action potential the membrane resistance of the squid giant axon dropped fortyfold from 1000 Ωcm^2 to 25 Ωcm^2. By contrast, the membrane capacitance changed less than 2 percent.

In 1949 Hodgkin and Katz discovered that the initial increase in the ionic conductance during the action potential is due to a sudden reversal in the permeability characteristics of the membrane: it becomes temporarily more permeable to Na$^+$ than to K$^+$, presumably due to the sudden availability of previously occluded Na$^+$ channels. The resulting potential change is regenerative. Depolarization increases Na$^+$ permeability, producing an influx of a very small amount of Na$^+$ and causing a further depolarization, which increases

Na$^+$ permeability even more. This explosive event not only abolishes the resting potential (reducing it to zero), but actually reverses the membrane potential, driving it to about +55 mV, the equilibrium potential of Na$^+$ (E_{Na}).

For a brief moment, the inside of the cell is positive in relation to the outside. The value of E_{Na} can also be calculated from the Nernst equation (Equation 6-1). When [Na$^+$]$_i$ is 50 mM and [Na$^+$]$_o$ is 440 mM, then at 18°C:

$$E_{Na} = \frac{RT}{F} \ln \frac{[Na^+]_o}{[Na^+]_i} \, mV \tag{6-4}$$

$$= 58 \log \frac{440}{50} \, mV$$

$$= +55 \, mV$$

The sudden reversal of the membrane potential is transient. The progressive depolarization produced by the action potential shuts off the enhanced Na$^+$ permeability. This process is called *sodium inactivation.* Accompanying Na$^+$ inactivation is a further increase in the already high permeability to K$^+$ caused by depolarization. This process is called *delayed rectification* because it accounts for the nonlinearity of the current-voltage curve with depolarization that we considered earlier (see Figure 5-8B, p. 116). Sodium inactivation and the delayed increase in K$^+$ permeability combine to restore the membrane potential to its resting value.

The net movement of ions across the membrane during each action potential is actually very small and does not alter the overall ionic concentration gradients. This is so because the movement of a relatively small number of ions is enough to alter the charge on the membrane capacitance needed to depolarize the membrane potential by 130 mV and return it to the resting level.

The amount of charge that needs to cross the membrane each way can be calculated from Equation 5-4:

$$C = q/V$$

$$q = CV \tag{6-5}$$

$$q = 1 \times 10^{-6} \, F/cm^2 \times 0.13 \, V$$

$$= 1.3 \times 10^{-7} \, coulombs/cm^2$$

Since one mole of a univalent ion, such as Na$^+$ or K$^+$, has a charge of 10^5 coulombs, the charge $q = \dfrac{1.3 \times 10^{-7} \, coulombs/cm^2}{10^5 \, coulombs/mole} = 1.3 \times 10^{-12} \, moles/cm^2$. Actually the ionic movement is slightly greater than this. In the squid giant axon 3 to 4×10^{-12} moles of Na^+ move into the cell per impulse per cm^2 of membrane; an equal amount of K^+ leaves. This is equivalent to one K^+ ion leaving the cell out of 10,000,000 inside.

The Sodium–Potassium Pump

The voltages of the ionic batteries are determined by the concentration gradients of the ions, which are assumed to be constant. Yet, with each action potential a small amount of Na^+ leaks into the cell and some K^+ leaks out. In large cells the quantities are so small that they can be disregarded for the short term (see above). A squid giant fiber can discharge half a million action potentials without having to recharge its battery. However, activity in small cells reduces their batteries more rapidly; even the squid giant axon fails after several hours of activity when the processes that restore the gradient are inhibited. Work with metabolic inhibitors has revealed that the membrane actively transports Na^+ against a concentration gradient from inside to out, and K^+ from outside to in. This metabolically dependent (energy consuming) process is responsible for the long-term maintenance of the ionic concentration gradients for Na^+ and K^+. A component of the pump is an enzyme (called ATPase) that requires both Na^+ and K^+ for its activity. In vitro this enzyme hydrolyzes (splits) the energy-rich nucleotide adesonine triphosphate (ATP); in the intact membrane it is thought to utilize the energy from the hydrolysis of ATP in the transport process.

In the squid giant axon neither the membrane potential nor the action potentials are directly dependent on the Na^+-K^+ pump in the short term. Signaling can be independent of immediate energy generation because there is an excess of chemical energy in the neuron, in the form of charged ionic batteries (the concentration gradient). The Na^+-K^+ pump does, however, importantly contribute to the membrane and action potential in an indirect way, by maintaining the ionic concentration gradient for K^+. But in addition, as we shall see in Chapter 7, in some cells the Na^+-K^+ pump is capable of generating potentials in its own right.

Testing the Ionic Hypothesis

One of the assumptions of the ionic hypothesis is that the batteries for the membrane potential and action potential are largely determined by the concentration gradient of K^+ and Na^+, respectively. This assumption has been tested in the squid giant axon by changing the K^+ and Na^+ concentrations. Changing the K^+ concentration alters the membrane potential but has only a small effect on the action potential, whereas changing the Na^+ concentration alters the overshoot of the action potential but has only a small effect on the membrane potential. In each case the observed results are consistent with the Goldman equation (Curtis and Cole, 1942; Hodgkin and Katz, 1949).

In the squid giant axon internal and external ions can easily be changed. Baker, Hodgkin, and Shaw (1962) squeezed the cytoplasm out of the axon, like toothpaste out of a tube, and reinflated the fiber by perfusing it with artificial solutions. By this means they provided a dramatic test of the ionic hypothesis. For an axon containing 600 mM KCl and seawater as the external medium (10 mM K^+ and 470 mM Na^+), the inside of the fiber was -60 mV in relation to the outside, as predicted by the Goldman equation. As they replaced the K^+ inside with Na^+, the membrane potential fell to zero, again as predicted. Baker, Hodgkin, and Shaw then raised the external K^+ concentration to 600 mM, thereby reversing the normal ionic concentrations; they found, as predicted, that the inside of the axon was $+60$ mV in relation to the outside of the axon. The normal membrane potential had been completely reversed. Similarly, increasing the Na^+ inside the axon (and thereby decreasing the Na^+ concentration gradient) reduced the overshoot of the spike and ultimately led to failure of spike generation. These experiments parallel those demonstrating reduction of the overshoot by reduction in the external Na^+. Both sets of experiments illustrate that the difference in Na^+ concentration on the two sides of the membrane provides the battery for the action potential.

The Ionic Hypothesis and the Equivalent Circuit

How can the ionic hypothesis be related to the electrical equivalent circuit of parallel resistances and capacitances? One of Hodgkin, Huxley, and Katz's great insights was the realization that measurements of membrane resistance at rest primarily test the ability of the membrane to permit passage of K^+ (since at rest membrane resistance to other ions is much higher). But during the action potential Na^+ channels temporarily open up and the flow of ionic current through these channels exceeds, by 10 to 100 times, the current flowing at rest through the K^+ channels.

Hodgkin and Katz (1949) and Hodgkin and Huxley (1952b) proposed that the equivalent circuit for each patch of membrane be viewed as a set of channels for each major ion species (Na^+, K^+, and other ions, primarily Cl^-). The flow through each of these channels is determined by the membrane resistance to that ion. To facilitate their calculations, Hodgkin and Huxley called these ionic channels *conductance channels* (remember, $R = 1/G$). They proposed that each specific ionic conductance (g_K, g_{Na} and g_{Cl}) is in series with a battery (E_{Na}, E_K, E_{Cl}), the value of which is related to the difference across the membrane of the ionic concentrations (Figure 6-2). The value of the battery attributable to each ion is given by the Nernst equation. These batteries are relatively fixed; they change only if and when the concentration of the ion changes, and for the moment we can ignore these special circumstances. The ionic conductances are not fixed; Hodgkin, Huxley, and Katz (1952) discovered that they vary systematically with voltage and with time. If each type of ion pathway is pictured as a group of specifically shaped channels in a patch of membrane, variation in ionic conductance in that patch can be pictured as a change in the number of open or available channels, a change controlled by voltage. Whenever the conductance of a pathway is very large relative to the other ion pathways, its battery determines the potential of the membrane. At rest, membrane conductance to K^+ is much higher than that to Na^+ (and Cl^- distributes itself passively according to the K^+ concentration), so that the membrane potential is largely determined by the K^+ battery (E_K). During the action potential the conductance to Na increases temporarily, and for the moment Na^+ conductance is higher than K^+ conductance. At that moment the membrane potential moves toward and is largely determined by E_{Na}.

Some quantitative aspects of current flows and potential changes across the membrane during both rest and activity can be readily grasped by considering the equivalent circuit of the membrane (Figure 6-2A). For simplicity, membrane capacitance as well as E_{Cl} and g_{Cl} can be ignored (because in the resting state the membrane capacitance is charged and no capacitive current is flowing and because E_{Cl} is similar to and dependent upon E_K). The equivalent circuit is reduced to two parallel ionic pathways (Figure 6-2B): the Na^+ pathway (E_{Na} in series with g_{Na}) and the K^+ pathway (E_K in series with g_K). At rest, the membrane potential is determined largely by E_K because g_{Na} is very low. To simplify the computation, let us consider specific ionic conductances (g) as resistances (r), or $1/g$, and let us further assume that at rest g_{Na} is 10 times lower than g_K, e.g., r_{Na} is $10^9 \, \Omega$ and r_K is $10^8 \, \Omega$. In a loop circuit of the sort illustrated in Figure 6-2B (where I is the current flowing in the loop) the sum of the voltage drops around the loop must be zero. Starting with E_{Na} and working counterclockwise around the loop, we first encounter a negative E_{Na} battery, then a negative E_K battery, the voltage drop across $1/g_K$ (given by I_{r_K}), and the voltage drop across $1/g_{Na}$ ($I_{r_{Na}}$). Thus

$$-E_{Na} - E_K + I_{r_K} + I_{r_{Na}} = 0 \qquad (6\text{-}6)$$

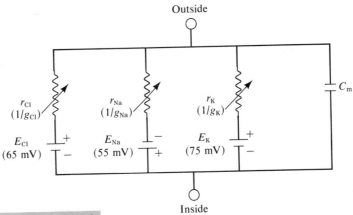

A. Equivalent Circuit

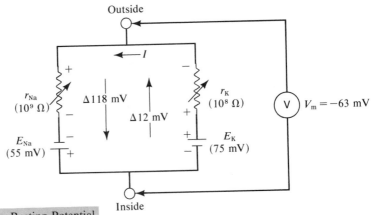

B. Resting Potential

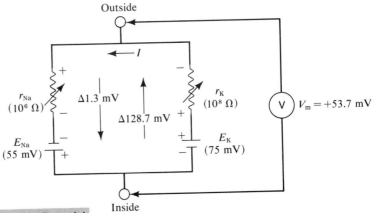

C. Action Potential

FIGURE 6-2.
Electrical equivalent circuit and current flow for the resting and active cell.

A. Equivalent circuit illustrating ionic batteries for Cl⁻, Na⁺, and K⁺ (E_{Cl}, E_{Na}, E_K) in series with the resistance for each ion species (r_{Cl}, r_{Na}, r_K) and in parallel with the membrane capacitance C_m. Resistance can be expressed as the reciprocal of ionic conductance: $r_{Cl} = 1/g_{Cl}$; $r_{Na} = 1/g_{Na}$; $r_K = 1/g_K$. Although under most circumstances a battery has a fixed value, the transmembrane potential can be changed because the conductance is variable. [Modified from Hodgkin and Huxley, 1952d.]

B. The resting potential is negative because of the relatively low resistance of the membrane to K⁺ (10^8 Ω) compared to Na⁺ (10^9 Ω). This gives rise to an outward current through the K⁺ channels (producing a voltage drop of 12 mV across r_K) and an equal inward current through the Na⁺ channels, thereby producing a voltage drop of 118 mV across r_{Na}. This current flow brings the membrane potential to -63 mV. For simplicity, the Cl⁻ channel and the membrane capacitance have been omitted.

C. Current flow during the rising phase of the action potential is produced by a sudden decrease in resistance of the Na⁺ channel from 10^9 Ω to 10^6 Ω. This causes Na⁺ to flow inward down its concentration gradient through the Na⁺ channel (causing a voltage drop of 1.3 mV across r_{Na}) and current to flow outward through the K⁺ channel (causing a voltage-drop of 128.7 mV across r_K). This depolarizes the membrane to $+53.7$ mV. Again, Cl⁻ and capacitative pathways have been ignored.

By rearranging terms, we have

$$I = \frac{E_{Na} + E_K}{r_K + r_{Na}} \tag{6-7}$$

Substituting the values for the batteries and resistances,[4]

$$I = \frac{+55(10^{-3}\ V) + 75(10^{-3}\ V)}{10^8\ \Omega + 10^9\ \Omega}$$

$$= \frac{+130(10^{-3}\ V)}{1.1(10^9\ \Omega)}$$

$$= 118(10^{-12}\ A)$$

This current corresponds to an outward flow of positive ions through the K⁺ pathway and an equal net inward current through the parallel Na⁺ pathway. The inward current through the Na⁺ pathway and the outward current through the K⁺ pathway add algebraically to E_{Na} and E_K to produce the membrane potential of the cell. Given the current, the hyperpolarizing voltage $V_{r_{Na}}$ produced by the inward current across the Na⁺ pathway can be calculated from

---

[4]Here the absolute values of the K⁺ and Na⁺ batteries are substituted, since their polarity, with respect to the current loop (Figure 6-2B and Equation 6-6), has already been accounted for.

Ohm's law (Equation 5-1, p. 111):

$$V_{r_{Na}} = 118 \times 10^{-12} \text{ A} \times 10^9 \text{ }\Omega$$

$$= 118 \text{ mV}$$

This voltage drop (due to the steady inward current across r_{Na}) will oppose E_{Na} (+55 mV) and produce a potential of -118 mV + 55 mV, or -63 mV. The same magnitude outward current flowing simultaneously in the K$^+$ pathway causes a depolarizing voltage drop (V_{r_K}) across the K$^+$ resistance (equivalent to the voltage drop across the internal resistance of a battery):

$$V_{r_K} = 118 \times 10^{-12} \text{ A} \times 10^8 \text{ }\Omega$$

$$= 11.8 \text{ mV}$$

Since the voltage drop opposes the potential of E_K (-75 mV), the potential across this branch (calculated as -75 mV + 12 mV) will also be -63 mV. This is to be expected since both branches of a parallel circuit must have the same potential. These estimates illustrate why the membrane potential is largely dominated by E_K (-75 mV). The leakage to Na$^+$ reduces the membrane potential from E_K by only 12 mV (and Cl$^-$ distributes itself according to the membrane potential).

During the rising phase of the action potential there is a sudden change in magnitude of the current flow. With depolarization, g_{Na} increases rapidly to a maximum (g_K increases more slowly and will be ignored here for simplicity). This is equivalent to changing the variable Na$^+$ resistance from a value of 10^9 at rest (Figure 6-2B) to a value of 10^6 during activity (Figure 6-2C). From Equation 6-7 the value of the resulting current flow will be:

$$I = \frac{E_{Na} + E_K}{r_K + r_{Na}}$$

$$= \frac{+55 \times 10^{-3} \text{ V} + 75 \times 10^{-3} \text{ V}}{10^8 \text{ }\Omega + 10^6 \text{ }\Omega}$$

$$= \frac{+130 \text{ mV}}{1.01 \times 10^8 \text{ }\Omega}$$

$$= 1.287 \times 10^{-9} \text{ A}$$

This current corresponds to an inward flow of positive ions through the Na$^+$ pathway and an equal net outward current flow through the inactive K$^+$ pathway. The outward flow through the inactive pathways of the membrane is carried by K^+ ions and produces a depolarization across the terminals of the equivalent circuit, corresponding to the depolarization of the cell. The depolarization is again given by Ohm's law:

$$V_{r_K} = 1.287 \times 10^{-9} \text{ A} \times 10^8 \text{ }\Omega$$

$$= 128.7 \text{ mV}$$

This voltage drop due to the outward current opposes the battery caused by E_K (-75 mV) and effects a new potential across this branch:

$$V_{r_K} = (128.7 - 75) \text{ mV}$$

$$= 53.7 \text{ mV}$$

The simultaneous inward current in the Na^+ pathway causes a hyperpolarizing voltage drop across that Na^+ resistance:

$$V_{r_{Na}} = 1.287 \times 10^{-9} \times 10^6 \text{ }\Omega$$

$$= 1.3 \text{ mV}$$

The voltage drop across the Na^+ resistance also brings this branch of the circuit to 53.7 mV ($+55$ mV $-$ 1.3 mV). During the peak of the spike the membrane potential will therefore be transiently dominated by E_{Na}

Hodgkin and Huxley also used the ionic hypothesis and a more complete form of the equivalent circuit to explain the propagation of the action potential. Propagation is dependent on an outward current across the membrane in the region in front of the action potential. This outward current produces a depolarizing electrotonic potential; when this exceeds threshold an action potential is triggered. Consider the patch of membrane at the front of the depolarizing wave (Figure 6-3A, point 1); this patch is not yet active, so the potential across it is still negative on the inside. At the active region (Figure 6-3A, point 2) the membrane is highly permeable to Na^+, which flows inward through the Na^+ channels, giving rise to an outward current (primarily K^+) that flows through inactive pathways in front of the action potential making that the active region of the membrane. This potential difference draws charge from various points along the membrane capacitance. At points on the outside, positive ions, e.g., Na^+, are drawn toward the active (more negative) region. On the inside, positive ions, e.g., K^+, are driven away from the active area (repelled by the Na^+ influx) and placed on the membrane capacitance at points in front of the active site. As a result, the potential at the membrane patch in front of the wave becomes less negative and, once it reaches a critical point,

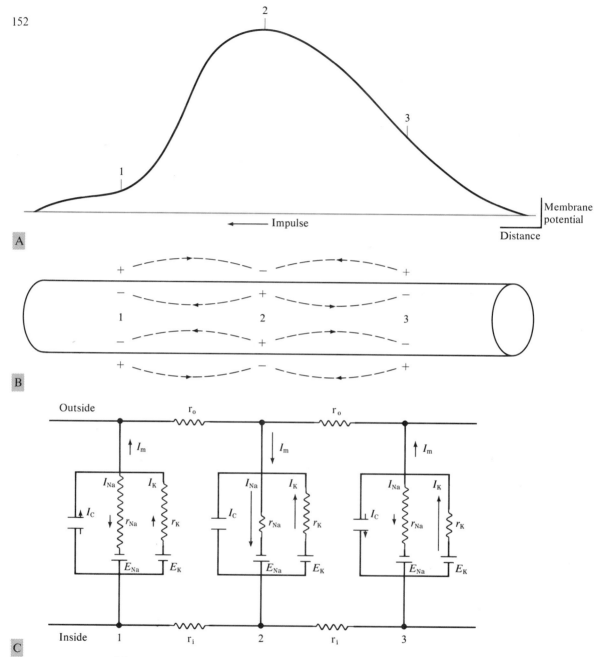

A

B

C

FIGURE 6-3.

The propagation of the action potential.

A. Diagram of the action potential propagating from right to left.

B. Diagram of local-circuit current flow during propagated action potential. Current flow is illustrated as occurring only at the three regions indicated in Part A. These correspond to (1) the beginning of the ascending limb of the spike; (2) the peak of the spike; and (3) the descending limb of the spike.

C. Circuit diagram illustrating the direction and relative magnitude of the Na⁺ and K⁺ ionic currents and the capacitive currents at the three points. In reality, membrane currents in nonmyelinated nerves do not flow only at discrete points, but vary smoothly and continuously along the entire length of the nerve traveled by the nerve impulse. [Modified from Brinley, 1974.]

g_{Na} will increase regeneratively and initiate an action potential. At the tail of the wavefront (Figure 6-3A, point 3) the spike has passed, and the membrane repolarizes due to the delayed increase in g_K. At this patch the membrane is refractory and cannot be reinvaded immediately because g_{Na} is low (as a result of Na^+ inactivation) and because g_K is very high (due to delayed rectification). Both factors prevent further spike generation.

As can be seen by referring to the circuit diagram in Figure 6-3, the ionic hypothesis for the resting and action potentials and for impulse propagation assumes that only ionic conductances change as a function of membrane voltage. The parallel capacitance is assumed not to change. Support for these critical assumptions is provided by the experiments of Cole and Curtis (1939) discussed above (p. 143).

Quantifying the Ionic Hypothesis

The ionic hypothesis was further extended and quantified by applying space- and voltage-clamp techniques to the squid giant axon (for reviews see Hodgkin, 1964; Katz, 1966). A space clamp is a device for establishing spatial uniformity over the membrane potential of a cell by forcing a large region of axon to undergo the same potential change simultaneously. This is accomplished by putting one long axial electrode, for stimulating, inside the giant axon and placing a long axial electrode (or better still, a concentric ring electrode) outside. Another electrode is placed in the cell for recording. As an impulse propagates along the axon, the membrane potential changes with time and with distance (Figure 6-3); current flows from one region of the axon to neighboring ones. In the space-clamp condition the different regions of the inside of the axon are electrically connected, by means of the low-resistance metal stimulating electrode that short-circuits the inside of the axon, making all regions isopotential. All regions of the axon then experience the action potential simultaneously, so that the complicating effects of the cable properties of the axon (the electrotonic flow of current along the axoplasm and across the membrane on either side of the active zones) are eliminated. Instead of behaving like a leaky cable, the area of nerve that is clamped behaves like an electrically isolated patch of membrane.

Voltage-clamp experiments combine a space clamp with a negative-feedback electronic device that permits changing the membrane potential rapidly to any desired value and maintaining (clamping) it there (Figure 6-4). This is accomplished by connecting a voltage amplifier to two recording electrodes, which sense the membrane potential, one electrode being inside the cell and the other

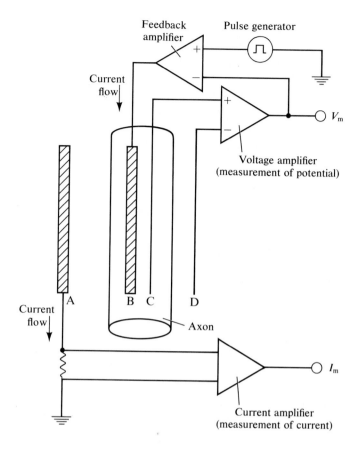

FIGURE 6-4.
Diagram of voltage-clamp technique, as applied to squid axons. The membrane potential is recorded between electrodes C and D by a voltage amplifier. The membrane potential can be held (clamped) at any arbitrary level by passing appropriate currents, between electrodes A and B, by means of a feedback amplifier. The clamping current is recorded by a current amplifier.

outside. The output of the voltage amplifier, which represents the actual membrane potential, is then connected to a feedback amplifier that compares the membrane potential with the desired potential (set by a pulse generator) and produces an error signal. The feedback amplifier then automatically delivers sufficient current through the current electrode inside the cell (electrode B in Figure 6-4) to minimize the error voltage and maintain the membrane at the desired value. This current can be measured by means of another amplifier connected to an extracellular electrode (electrode A in Figure 6-4).

The experimental advantage of the voltage clamp is twofold. First, it permits measurement with good time resolution of the current flowing through the membrane. By fully and rapidly charging the membrane capacitance, it allows the resistive current to be measured in isolation. We saw previously that in response to an impressed voltage the current through a passive membrane flows across both a capacitive and a resistive (conductance) pathway. Thus, from Equation 5-6 (p. 120):

$$I_m = \frac{\Delta V_m}{R_m} + C_m \frac{dV_m}{dt}$$

$$= G_m \Delta V_m + C_m \frac{dV_m}{dt}$$

When V_m is held constant, capacitive current will flow only briefly, at the very moment the membrane potential is stepped to a new value. Thereafter no capacitive current will flow because dV_m/dt (the rate of membrane voltage change) is zero. The membrane current then reduces to a simple function of membrane conductance (G_m) and voltage (V_m):

$$I_m = G_m \Delta V_m$$

The current that the feedback amplifier must supply to maintain the membrane potential at a given value is exactly equal to the total membrane current flowing across the clamped patch of membrane at that membrane potential. These currents provide an estimate of changes in *total ionic conductance* (and thus specific ionic conductances) produced by the change in membrane potential.

Second, in the unclamped condition a shift in membrane potential normally produces ionic conductance changes that in turn cause secondary changes in membrane potential (for example, delayed rectification). Voltage clamping prevents these secondary changes. For example, a 60 mV command voltage step that depolarizes the membrane from its resting value of -60 mV to 0 mV would normally trigger a regenerative action potential that would overshoot 0 mV in the unclamped axon. In the clamped axon the membrane can be held at 0 mV (Figure 6-5A). The 60 mV voltage step produces a brief capacitive transient current followed by a diphasic ionic current (Figure 6-5B): an early inward current followed, at 2 msec, by a delayed outward current. With more depolarizing voltage steps the early inward current becomes smaller; with a voltage step of 110 to 120 mV (bringing the membrane to between $+50$ and $+60$ mV) the inward current is nullified. Beyond that value (the *reversal potential*), the early current reverses and flows outward. By contrast, these depolarizing steps only further enhance the delayed outward current, which grows larger as the membrane becomes increasingly depolarized. The delayed current becomes smaller only as the membrane potential is stepped in the hyperpolar-

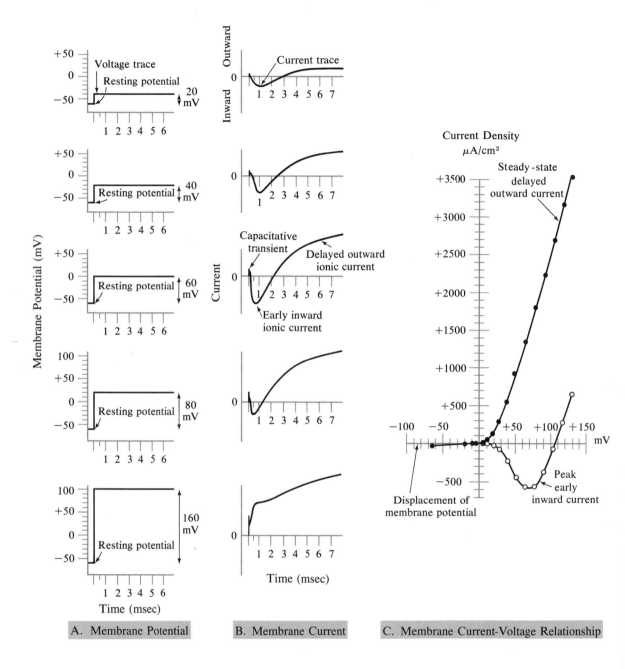

A. Membrane Potential

B. Membrane Current

C. Membrane Current-Voltage Relationship

izing direction (after first having been briefly depolarized by a step of 84 mV). The reversal potential for the delayed current is about -80 mV. One can easily visualize the variations in early and delayed currents as functions of membrane potential using graphs of the data devised by Hodgkin and Huxley. In these graphs the peaks of the early inward current and the steady-state, delayed outward current are plotted on the same scale as functions of membrane potential (Figure 6-5C). As the graphs illustrate, depolarizing steps turn on both the inward and the outward current. With increasing depolarization, both currents increase in amplitude. The inward current then declines in magnitude and reverses, with depolarizing steps greater than 110 mV.

The inward current is most readily interpreted as Na^+ rushing into the cell as a result of the increase in g_{Na} caused by the depolarization. As would be predicted from an increase in g_{Na}, this current is nullified at $+55$ mV, the Nernst Na^+ equilibrium potential. At E_{Na} the force due to electrical potential is equal to and opposite to the force due to the concentration gradient, so that no net Na^+ current flows. At still more depolarized values the force due to the electrical gradient exceeds that due to the concentration gradient and Na^+ flows from inside the cell to outside, thereby reversing the early current from inward to outward. Similarly, the delayed outward current has properties consistent with those one would expect if K^+ were its carrier. As the membrane potential is depolarized, the force of the electrical potential, which tends to keep K^+ in the cell, is reduced, allowing greater dominance of the outwardly directed force of the K^+ concentration gradient.

FIGURE 6-5.
Voltage clamp of squid giant axon.

A–B. Records from an hypothetical voltage-clamp experiment very similar to those performed by Hodgkin, Huxley, and Katz (1952). Membrane potentials are illustrated in part A as a function of time; the simultaneously obtained membrane currents are illustrated in part B. Only depolarizing steps from a resting level of -60 mV are illustrated (step values, indicating displacement of membrane potentials, are shown at the right of each graph in part A). [Modified from Stevens, 1966; after Hodgkin, Huxley, and Katz, 1952.]

C. Relation between membrane current and membrane potential based on an actual voltage-clamp experiment by Hodgkin, Huxley, and Katz. Displacement of the membrane potential from its resting value is plotted on the abscissa. Membrane current intensity 0.63 msec after the beginning of the voltage step (early current) and in the steady state (delayed current) is plotted on the ordinate. Inward current is plotted as negative. Zero on the abcissa corresponds to resting potential; hyperpolarizing displacements of the membrane potential are plotted as negative values and depolarizing displacements as positive values. In addition to the two major currents, the inward current (due to Na^+) and the delayed outward current (due to K^+), there is a third current called *leakage current*. This current does not have the dependence on time or on specific ions that is typical of the other two currents. It is thought to be produced by the nonspecific leakage of ions (particularly Cl^-) across the membrane. This current is not important for excitation. It appears in the clamp records in part B as a progressively greater displacement from the baseline with increasing depolarizing commands (compare the onsets of current following displacements of 60 mV, 80 mV, and 160 mV). [Slightly modified from Hodgkin, Huxley, and Katz, 1952.]

Hodgkin, Huxley, and Katz (1952) tested these ideas by selectively reducing the Na^+ current and using voltage-clamping to quantify the specific ionic currents, i.e., the contribution of each ion to the total membrane current at given membrane potentials (Figure 6-6). They eliminated the inward Na^+ current by substituting an impermeant cation, choline, in the external solution and found that under this condition the early inward current to a depolarizing step of 60 mV was eliminated. What remained was only the outward current, presumably due to K^+. By subtracting the ionic current in Na^+-free solution from the total current in normal seawater, the independent Na^+ and K^+ currents were obtained (Figure 6-6).

So far in discussing the voltage-clamp experiments I have only considered the ionic currents flowing across the cell membrane. Of more fundamental

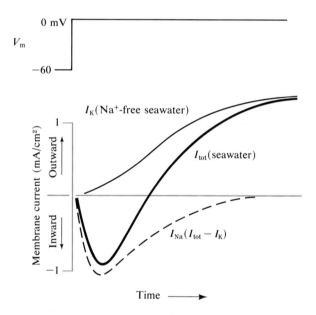

FIGURE 6-6.
The total membrane current (I_{tot}) produced by a 60 mV depolarizing potential step in seawater can be separated into two components, Na^+ and K^+. With the axon bathed in Na^+-free (choline) seawater, the current generated by the same depolarizing step is attributable to I_K. The difference between the total ionic current in seawater and that in Na^+-free seawater (I_K) is attributable to I_{Na}. In the absence of external Na^+ a 60 mV depolarization actually causes intracellular Na^+ to move out, giving rise to a small outward current. [Modified from Hodgkin, 1964.]

importance, however, are the conductance changes that underlie these currents. Once Hodgkin and Huxley were able to separate the individual currents, they were able to describe quantitatively how the Na⁺ and K⁺ conductances vary with membrane potential and with time during the action potential. They determined that the current flow in the Na⁺ and K⁺ channels of the squid giant axon simply obeyed Ohm's law (Equation 5-2, p. 113). Thus the current carried by a given ion is

$$I_{ion} = GE$$

The battery or driving force for ionic current (E) is the difference between the membrane potential (V_m) and the battery or equilibrium potential for the ion (E_{ion}). Now we can rewrite Ohm's law:

$$I_{ion} = g_{ion}(V_m - E_{ion}) \qquad (6\text{-}8)$$

The individual ionic current for Na⁺ and K⁺ can then be described as

$$I_{Na} = g_{Na}(V_m - E_{Na})$$
$$I_K = g_K(V_m - E_K)$$

This is simply a restatement of the circuit diagram in Figure 6-2. The ionic pathways have an ionic battery, E_K or E_{Na}, in series with a conductance, g_K or g_{Na}. Using these equations, Hodgkin and Huxley derived the changes in g_{Na} and g_K in response to graded changes in voltage (Figure 6-7). Like the ionic currents, the conductances are variable, depending on V_m and time. A brief depolarization (Figure 6-7A) produces a gradual increase in both g_{Na} and g_K. Sodium conductance is increased first, while the increase in g_K (and the inactivation of the increased g_{Na}) is delayed. In the unclamped axon, if the depolarization is sufficiently large (and rapid) the Na⁺ current will increase, exceed the K⁺ (and leakage) currents, and initiate a regenerative process leading to an action potential.

Since g_{Na} increases smoothly with no sudden discontinuities as the membrane is depolarized, it may seem paradoxical that there should be a sharp threshold (at 15 mV depolarization) and an all-or-none action potential in the unclamped cell. This is so because V_m and g_{Na} are linked in a regenerative way. An increased g_{Na} will cause Na⁺ to move into the cell, producing greater depolarization, which increases g_{Na} further, leading to more depolarization and pushing V_m to E_{Na}. A threshold exists because there is a relatively precise potential at which the inward Na⁺ current slightly exceeds the outward current.

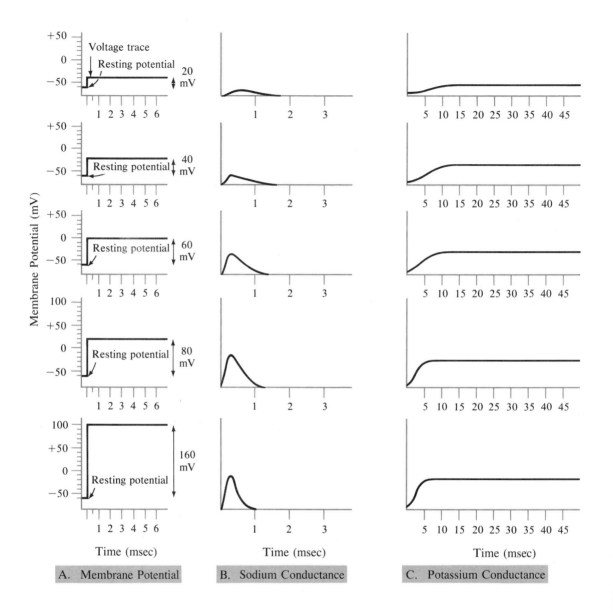

Membrane Potential (mV)

| A. Membrane Potential | B. Sodium Conductance | C. Potassium Conductance |

FIGURE 6-7.

Conductance measurements from the hypothetical voltage-clamp experiment illustrated in Figure 6-5A, B.

A. Clamping voltages for conductances in parts B and C. The magnitudes of the displacements of the membrane potential (all depolarizing voltage steps) are indicated at the right of each trace.

B. Sodium conductance for voltage steps illustrated in part A.

C. Potassium conductance for voltage steps illustrated in part A. The time scale on these curves is more compressed than on the sodium curves. Comparison of the graphs in parts B and C shows that the g_{Na} inactivates with time, whereas g_K does not. [From Stevens, 1966; after Hodgkin and Huxley, 1952d.]

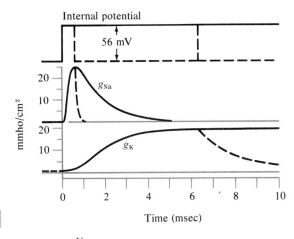

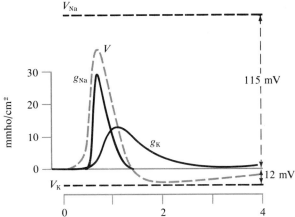

A

B

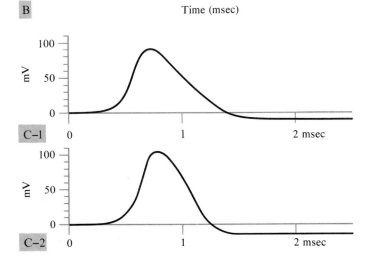

C-1

C-2

FIGURE 6-8.

Sodium and potassium conductances and the theoretical reconstruction of the action potential.

A. The time course of g_{Na} and g_K associated with a depolarization of 56 mV. The solid curves are for a maintained depolarization; the dashed curves show the effect of repolarization after 0.6 and 6.3 msec. [From Hodgkin, 1964.]

B. Calculated variations of g_{Na} and g_K during the action potential (V). [From Hodgkin and Huxley, 1952d.]

C. Comparison of the calculated action potential (*C-1*) with the actual action potential observed in the squid giant axon (*C-2*). The calculated velocity was 18.8 msec, the experimental velocity 21.2 msec. [From Hodgkin and Huxley, 1952d.]

For subthreshold depolarization the outward K^+ current exceeds the inward Na^+ current and prevents the membrane from reaching threshold. As the inward Na^+ current exceeds the outward K^+ current, even slightly, the membrane potential is forced to move in a regenerative manner toward E_{Na}. But now the delayed g_K is turned on fully and begins to return the membrane potential toward (and often slightly beyond) the resting value. In addition, another independent process becomes operative: depolarization of the membrane also turns on Na^+ inactivation (with a delay), which shuts off g_{Na}, thereby helping to bring the membrane potential back to its resting value. Thus, depolarization in the squid giant axon turns on three processes: (1) a quickly rising g_{Na}; (2) a delayed g_K that does not become inactivated (but is turned off by the repolarization of the membrane); and (3) a delayed Na^+ inactivation (Figure 6-8*A, B*).

A final step in establishing the ionic hypothesis was to determine whether the conductances measured in voltage-clamp experiments account quantitatively for the action potential, the afterpotential, impulse propagation, and other aspects of excitability. This was accomplished by means of a kinetic model based on (1) a series of empirically derived equations that describe how the Na^+ and K^+ conductances vary with membrane potential and time, and (2) the equations that describe the passive properties of the neurons (p. 110). This model is important because the normal, unclamped nerve fiber undergoes a series of complicated voltage- and time-dependent electrical changes, and it was not intuitively obvious that simple changes in g_{Na} and g_K could fully account for all of these features. The equations were based upon the measured value of g_{Na} and g_K activation and g_{Na} inactivation at various values of membrane potential and upon the values of the rate constants of these processes, estimated from the shape of the measured conductance curves (Figure 6-7). Using these equations, Hodgkin and Huxley (1952d) successfully calculated, among other things, the conducted action potential following electrical stimulation in an unclamped axon. The calculated values accurately fit the observed action potential (Figure 6-8*C*).

The ion-substitution experiments and the resulting equations are based on the assumption that movement of ions in one type of channel is independent of the presence or absence of the other ion species. This independence principle has now been confirmed in several ways. Essentially similar curves are obtained when highly selective blocking agents are used that do not require changing the ionic constituents. For example, when tetrodotoxin (TTX), a poison produced by the Japanese puffer fish, is applied on the outside surface of the cell, it specifically interferes with the transient rise in g_{Na} that occurs when the membrane is depolarized. This action is amazingly selective. Tetrodotoxin does not affect K^+ conductance (Figure 6-9A, C). Indeed, TTX does not even interfere significantly with the Na^+ channels that are operative in the resting membrane (or those that are activated by transmitter action—see below). TTX blocks only the increased g_{Na} that occurs with the depolarization. Another substance, tetraethylammonium (TEA), interferes with the increased K^+ permeability produced by depolarization only when it is applied to the inside of the axon (Figure 6-9B). TEA does not affect g_{Na}. The selectivity of pharmacological agents has allowed other distinctions to be made. For example, pronase, an enzyme that splits proteins, selectively destroys Na^+ inactivation when applied to inside of the axon. After treatment with pronase, the Na^+ current turns on in the usual way but no longer turns off; the Na^+ current remains high throughout the applied depolarizing pulse. Thus, Na^+ activation (the turning on of g_{Na}) and inactivation (the turning off of g_{Na}) are independent processes.

Recent work has been directed toward analyzing the mechanisms by which changes in membrane potential increase g_{Na} and g_K. Hodgkin and Huxley (1952d) suggested that charged molecules on the surface of the membrane may serve as "gates" at the open end of the channel (perhaps three such macromolecules per Na^+ channel), which would prevent ions from transversing the membrane. In response to a change in membrane potential, the molecules would change conformation and the gates would swing open, allowing ion movement. One prediction from this hypothesis is that the change in position of the molecules produces a capacitive current, a *gating current*, that precedes the opening of the Na^+ channel. Bezanilla and Armstrong (1974) and Keynes and Rojas (1974) found a small current with these properties. Studies of the gating current may make it possible to analyze the molecular properties of the gating structures (for review see Armstrong, 1975).

In addition to explaining the membrane potential, the action potential, and impulse propagation, the ionic hypothesis explains the excitatory and inhibitory postsynaptic potentials produced by chemical transmission.

POSTSYNAPTIC MECHANISMS OF CHEMICAL TRANSMISSION

The ionic hypothesis was applied to synaptic transmission in a classic series of studies by Fatt and Katz (1951, 1953) on the nerve-muscle synapses of the frog and crab. Fatt and Katz found that the currents for excitatory and in-

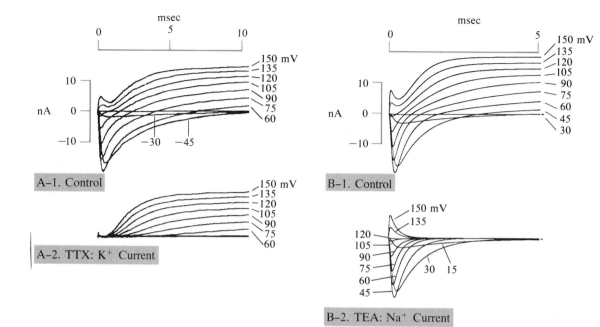

msec

150 mV
135
120
105
90
75
60

nA 0

−30 −45

A–1. Control

150 mV
135
120
105
90
75
60

A–2. TTX: K⁺ Current

msec

150 mV
135
120
105
90
75
60
45
30

nA 0

B–1. Control

150 mV
135
120
105
90
75
60
45

30   15

B–2. TEA: Na⁺ Current

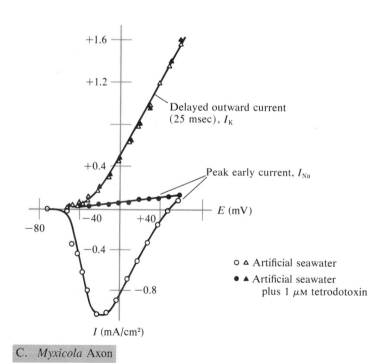

Delayed outward current
(25 msec), I_K

Peak early current, I_{Na}

E (mV)

−80   −40   +40

+1.6
+1.2
+0.4
−0.4
−0.8

○ △ Artificial seawater
● ▲ Artificial seawater
plus 1 μM tetrodotoxin

I (mA/cm²)

C. *Myxicola* Axon

FIGURE 6-9.
Selective blockage of Na$^+$ and K$^+$ channels with tetrodotoxin (TTX) and tetraethylammonium (TEA). Nine to 10 voltage-clamp records (from 30 to 150 mV) from a single node of Ranvier in a frog sciatic nerve have been superimposed. The records show 15 mV increments in the displacement of the membrane potential from a resting level of about -75 mV. Inward current is downward. Because the area of the clamped nodal membrane cannot be determined with certainty, current flow is expressed as nanoamperes (nA) rather than as current density.

A. Time course of current flow in normal Ringer's solution (*A-1*) and in the presence of TTX (*A-2*). TTX blocks that portion of total membrane current carried by Na$^+$ ions but not that carried by K$^+$. [From Hille, 1976.]

B. Time course of current flow in normal Ringer's solution (*B-1*) and in the presence of TEA (*B-2*). Effects of TEA on voltage-clamp current indicate that this agent blocks K$^+$ but not Na$^+$ current. [Modified from Hille, 1976.]

C. Current-voltage relations in *Myxicola* giant axon under voltage clamp, showing that 1 μM TTX blocks I_{Na} but not I_K. Points are ionic currents during voltage steps from rest to the indicated voltage, measured on families of currents like those in Figure 6-5*B*. Temperature 1 to 3°C. [From Binstock and Goldman, 1969.]

hibitory synaptic actions are generated in the postsynaptic cell by changes in membrane conductance. Subsequently, Eccles and his colleagues Fatt and Coombs established the validity of these principles for central neurons (Coombs, Eccles, and Fatt, 1955; Eccles, 1957).

In muscle and nerve cells receptors on the external surface of the postsynaptic membrane control ionic channels. Activation of different receptors leads to the generation of excitatory and inhibitory postsynaptic potentials. Most of the synaptic actions in the nervous system or at peripheral junctions that have been analyzed are due to increases in ionic conductance. However, certain synaptic actions have been shown to be due to decreases in conductance.

Synaptic Potentials Due to Increased Ionic Conductance

EXCITATORY POSTSYNAPTIC POTENTIALS

Excitatory synaptic actions due to an increase in ionic conductance depolarize and excite because of a sudden increase in g_{Na}, which is often accompanied by an increase in g_K. Superficially, this mechanism appears similar to that of the action potential. Actually, it differs in several ways. One difference is that the increase in g_{Na} and g_K is simultaneous and not sequential as in the action potential. In addition, the increase in g_{Na} is not regenerative. The depolarization produced by synaptic action does not lead to further increases in synaptic conductances because (unlike the action potential) the conductance channels activated by the transmitter are not dependent on voltage. Finally, the Na$^+$ and K$^+$ channels opened in response to transmitter action are pharmacologically different from those opened by the action potential. The Na$^+$ channel is not blocked by tetrodotoxin and the K$^+$ channel is usually not blocked by TEA. Because g_{Na} and g_K increase simultaneously, the *reversal potential for excitation* (E_{EPSP}) is a compound potential, and usually located close to -10 mV, about midway between E_{Na} ($+55$ mV) and E_K (-75 mV). As a result, excitatory synaptic actions tend to drive the membrane potential from its resting value (-60 mV) past threshold (-45 mV) in the direction of the E_{EPSP}, -10 mV.

Ignoring C_m, the ionic current for the excitatory synaptic potential I_{EPSP} can be calculated using Ohm's law (Equation 6-8, p. 159). To simplify the analysis, the separate membrane batteries E_{Na} and E_K and their series resistances can be combined into a single battery E_m (equal to -60 mV) representing the membrane potential of the nonsynaptic membrane in series with the nonsynaptic membrane resistance R_m (equal to 1×10^8 Ω). This membrane is in parallel with the membrane of the excitatory synapse. The synaptic membrane consists of two parallel ionic batteries E_{Na} ($+55$ mV) and E_K (-75 mV) in series with their respective resistances, which for the purpose of this discussion we will give as $r_{Na} = 3 \times 10^8$ Ω and $r_K = 3 \times 10^8$ Ω (Figure 6-10A).

FIGURE 6-10.
Chemical excitatory synaptic actions due to increases in g_{Na} and g_K.

A. Equivalent circuit of an excitatory synapse. The synaptic membrane consists of a Na$^+$ and K$^+$ battery (E_K and E_{Na}) each in series with a resistance (r_K and r_{Na}). The synaptic membrane is in parallel with the nonsynaptic membrane, consisting of the battery for the resting membrane potential (E_m) in series with the resting membrane resistance R_m.

B. Simplified equivalent circuit of a chemical excitatory synapse and current flow during synaptic action. At rest, the synaptic channel is an open circuit and no current flows through it. The synaptic action is equivalent to throwing the switch **S**, thereby closing the circuit by placing the synaptic resistance pathway, r_{EPSP}, in parallel with the nonsynaptic membrane resistance, R_m. As a result, current flows inward through the synaptic channel and outward through the nonsynaptic membrane, depolarizing it by 20 mV from a resting level of -60 mV. (This circuit diagram and those in Figures 6-11, 6-12, and 6-13 describe the synaptic current that flows with the membrane potential at rest only. In part C of these figures, where V_m is changed, one can consider the effect of a constant-current source as equivalent to a change in the value of E_m.)

C. Reversal potential. In this hypothetical experiment two electrodes are placed in the presynaptic cell and two in the postsynaptic cell. One electrode is for passing current—in the presynaptic cell to produce an action potential, and in the postsynaptic cell to systematically alter the membrane potential. The other electrode is for recording. When the membrane potential is at its resting value (-60 mV) a presynaptic spike produces a depolarizing EPSP, which increases when the membrane potential is increased (hyperpolarized) to -90 mV. But as the membrane potential is depolarized (made less negative) to -30 mV, the EPSP becomes smaller, and as the membrane potential reaches -10 mV the EPSP becomes nullified. This null potential is called the *reversal potential* of the EPSP. Further depolarization to $+20$ mV inverts the PSP, making it a hyperpolarizing potential change. On either side of the reversal potential the synaptic action drives the membrane potential toward the reversal potential.

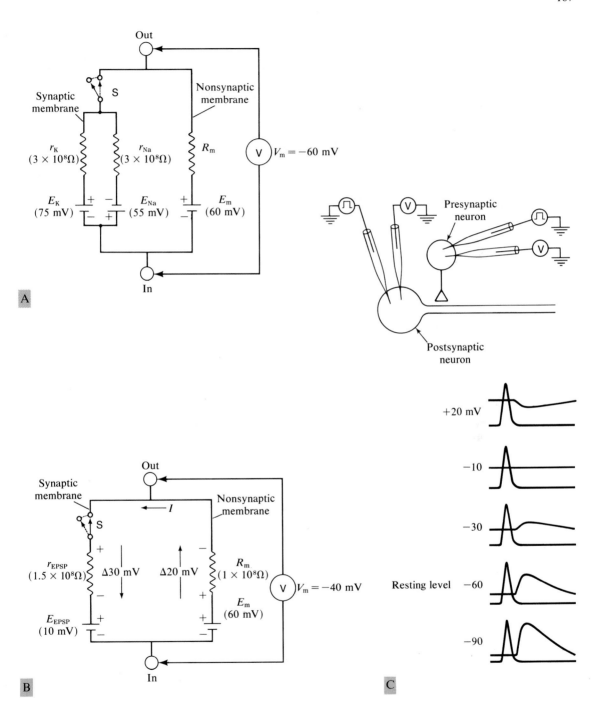

These parallel pathways can be combined into a single synaptic pathway consisting of a single synaptic battery E_{EPSP} in series with a single synaptic resistance r_{EPSP}.

Following Equation 6-8, $I_{ion} = g_{ion}(V_m - E_{ion})$, the sum of the individual currents through the two synaptic pathways of the postsynaptic cell in Figure 6-10*A* will be

$$I_{EPSP} = g_{Na}(V_m - E_{Na}) + g_K(V_m - E_k)$$

Rearranging and factoring out the term $(g_{Na} + g_K)$ yields

$$I_{EPSP} = (g_{Na} + g_K)\left(V_m - \frac{g_{Na}E_{Na} + g_K E_K}{g_{Na} + g_K}\right) \qquad (6\text{-}9A)$$

Since g_{ion} can be called g_{EPSP}, and the voltage term (E_{ion}) can be called E_{EPSP}, substituting these terms in Equation 6-8 gives us

$$I_{EPSP} = g_{EPSP}(V_m - E_{EPSP}) \qquad (6\text{-}9B)$$

We can therefore replace the two synaptic pathways with a single synaptic battery E_{EPSP} in series with synaptic conductance g_{EPSP} (Figure 6-10*B*). A comparison of Equations 6-9A and 6-9B shows that $g_{EPSP} = g_{Na} + g_k$ and that the value of E_{EPSP} is given by

$$E_{EPSP} = \frac{g_{Na}E_{Na} + g_K E_K}{g_{Na} + g_K} \qquad (6\text{-}10)$$

The EPSP is thus determined by the ratio of the two synaptic conductances and its amplitude is governed by the ion with the larger conductance. To facilitate calculating the magnitude of the synaptic battery, Equation 6-10 can be rewritten in terms of the resistances:

$$E_{EPSP} = \frac{\dfrac{E_K}{r_K} + \dfrac{E_{Na}}{r_{Na}}}{\dfrac{1}{r_K} + \dfrac{1}{r_{Na}}} \qquad (6\text{-}11)$$

Substituting -75 mV for E_K, $+55$ mV for E_{Na}, and 3×10^8 Ω for both r_{Na} and r_K gives

$$E_{EPSP} = \frac{\dfrac{-75 \times 10^{-3} \text{ V}}{3 \times 10^8 \text{ } \Omega} + \dfrac{+55 \times 10^{-3} \text{ V}}{3 \times 10^8 \text{ } \Omega}}{\dfrac{1}{3 \times 10^8 \text{ } \Omega} + \dfrac{1}{3 \times 10^8 \text{ } \Omega}} = -10 \text{ mV}$$

The equivalent resistance of the synaptic pathway can be similarly calculated:

$$
\begin{aligned}
r_{\text{EPSP}} &= \frac{1}{g_{\text{EPSP}}} \\
&= \frac{1}{\dfrac{1}{r_{\text{K}}} + \dfrac{1}{r_{\text{Na}}}} \\
&= \frac{r_{\text{K}} r_{\text{Na}}}{r_{\text{K}} + r_{\text{Na}}} \\
&= \frac{(3 \times 10^8)(3 \times 10^8)}{(3 \times 10^8) + (3 \times 10^8)} \\
&= 1.5 \times 10^8 \ \Omega
\end{aligned}
$$

These values can now be applied to the simplified equivalent circuit in Figure 6-10B to illustrate the relationship of the synaptic to the nonsynaptic membrane.

From Equation 6-9 it follows that the current flowing during the synaptic action can be described by the same relationship (Equation 6-7) that applies to the currents that determine the resting and action potentials:

$$
I_{\text{EPSP}} = \frac{E_{\text{m}} - E_{\text{EPSP}}}{R_{\text{m}} + r_{\text{EPSP}}} \tag{6-12}
$$

The amplitude of the synaptic potential (the change in membrane potential, ΔV_{m}) is given by Ohm's law:

$$
\Delta V_{\text{m}} = R_{\text{m}} \times I_{\text{EPSP}}
$$

Substituting for I_{EPSP} (Equation 6-12) we obtain:

$$
\Delta V_{\text{m}} = R_{\text{m}} \frac{E_{\text{m}} - E_{\text{EPSP}}}{R_{\text{m}} + r_{\text{EPSP}}}
$$

The amplitude of the EPSP can therefore be calculated by referring to the circuit of the excitatory synapse (Figure 6-10B). The synaptic action can be simulated by closing the switch, thereby placing the synaptic pathway in parallel with the nonsynaptic or ordinary resting membrane. This will cause current to flow inward through the synaptic pathway and outward through the nonsynaptic pathway, depolarizing the membrane. The exact value of the current can be calculated from Equation 6-12:

$$I_{\text{EPSP}} = \frac{E_{\text{m}} - E_{\text{EPSP}}}{R_{\text{m}} + r_{\text{EPSP}}}$$

$$= \frac{60 \times 10^{-3} \text{ V} - 10 \times 10^{-3} \text{ V}}{(1.0 \times 10^8 \ \Omega + 1.5 \times 10^8 \ \Omega)}$$

$$= \frac{50 \times 10^{-3} \text{ V}}{2.5 \times 10^8 \ \Omega}$$

$$= 20 \times 10^{-11} \text{ A}.$$

The change in voltage across the nonsynaptic membrane due to the current flowing outward will be

$$V_{\text{EPSP}} = \Delta V_{\text{m}} = R_{\text{m}} \times I_{\text{EPSP}}$$

$$= 10^8 \times 20 \times 10^{-11} \text{ A}$$

$$= 20 \text{ mV}$$

The resulting EPSP of 20 mV opposes the resting battery and moves the potential of the nonsynaptic branch from -60 to -40 mV, which is beyond threshold (-45 mV) for spike generation.

An inward current of the same magnitude flowing simultaneously through the synaptic pathway produces a voltage drop across the synaptic resistance, i.e., $V_{\text{s}} = 1.5 \times 10^8 \ \Omega \times 20 \times 10^{-11} \text{ A} = 30$ mV. Since the two branches of a parallel circuit must be at the same potential, the inward current flow across the synaptic membrane produces a voltage drop that adds to the synaptic battery (E_{EPSP}) of -10 mV, thereby also bringing this side of the circuit to -40 mV.

If additional parallel synaptic pathways were activated by the transmitter, the resistance of the synaptic branch would decrease further and more current would flow, producing a larger EPSP. With a larger conductance change, e.g., with an r_{EPSP} of $10^6 \ \Omega$, the EPSP amplitude would grow to almost 50 mV and the membrane potential would be transiently dominated by E_{EPSP}.

The current that flows during excitatory synaptic action can best be measured in experiments with voltage clamping. But an intuitive appreciation of the current flow and of the ions that generate I_{EPSP} can be determined simply by systematically changing the membrane potential and observing the effects on the amplitude and sign of the EPSP (Figure 6-10C). Thus, as the membrane potential is hyperpolarized from -60 mV to -90 mV, the EPSP increases, because more inward current flows through the synaptic channels (g_{EPSP}) and more outward (depolarizing) current flows through the nonsynaptic membrane. This can be seen from Equation 6-9: increasing V_m increases the driving force for I_{EPSP}. But as the membrane potential is progressively depolarized, the EPSP gets smaller, until it is abolished (Figure 6-10C).[5] The EPSP is nullified at the reversal potential (E_{EPSP}), usually -10 mV. At this potential the inward current due to Na$^+$ entering the cell is balanced by the outward current due to K$^+$ leaving the cell (-10 mV, Figure 6-10C). Further depolarization inverts the EPSP, making it a hyperpolarizing potential change ($+20$ mV, Figure 6-10C). Now the outward current is greater than the inward. The driving force on K$^+$ is greater than that on Na$^+$ because the membrane potential is closer to E_{Na} than to E_K. As a result, the amount of K$^+$ leaving the cell through the synaptic pathways is greater than the amount of Na$^+$ entering.

INHIBITORY POSTSYNAPTIC POTENTIALS

The reaction of the transmitter with the receptor at inhibitory synapses that involve an increase in ionic conductance leads to a sudden increase in the permeability of the postsynaptic cell membrane to K$^+$ or Cl$^-$. The equilibrium potential for Cl$^-$ is usually -65 mV, that for K$^+$ -75 mV. In most cells the equilibrium potential for either of these ions is more negative (hyperpolarizing) than the resting potential (-60 mV). For example, the amplitude of an IPSP due to an increased g_{Cl} can be calculated by referring to the equivalent circuit of the membrane (Figure 6-11A), which is similar to that used for the excitatory synapse (Figure 6-10B). The battery for the inhibitory synapse can be represented as E_{Cl} in series with a resistance r_{Cl}. This inhibitory synaptic pathway is in parallel with the pathway of the ordinary or nonsynaptic membrane, which includes a battery E_m in series with a resistance R_m. When the switch is thrown, negative ions flow inward through the inhibitory branch. This action is equivalent to positive current flowing outward through the synaptic membrane and drives positive current inward through the nonsynaptic membrane, thereby hyperpolarizing the cell.

The current flowing outward across the nonsynaptic pathway when the

[5]In these experiments the cell is first maintained at each level of membrane potential for one or more minutes until the membrane accommodates to the depolarization and stops generating action potentials. Only then is the analysis carried out.

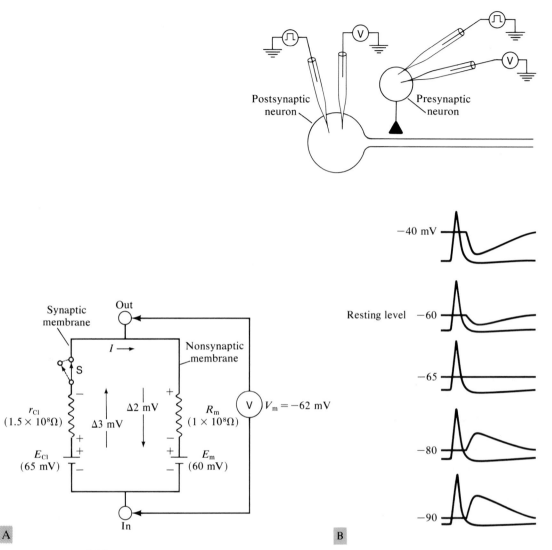

FIGURE 6-11.

Chemical inhibitory synaptic action due to increases in g_{Cl}.

A. Equivalent circuit of an inhibitory synapse and current flow during synaptic action. At rest, the synaptic channel is an open circuit and no current flows through it. The synaptic action is equivalent to closing the switch S, thereby placing the synaptic resistance pathway $r_{Cl} = 1/g_{Cl}$ in parallel with the nonsynaptic resistance R_m. Chloride current will now flow out of the cell through the synaptic channel and an equal current, mainly K^+, will flow into the cell through the nonsynaptic membrane, hyperpolarizing it by 2 mV from its resting level of -60 mV.

B. Reversal potential. Experimental arrangement as in Figure 6-10C except that the presynaptic neuron mediates inhibition. At the resting value (-60 mV) a presynaptic spike produces a hyperpolarizing IPSP, which increases in amplitude as the membrane is depolarized to -40 mV. But as the membrane potential is hyperpolarized, the PSP first becomes nullified (at -65 mV, the reversal potential) and then inverted to a depolarizing synaptic potential (-80 mV and -90 mV). Even this depolarizing action is inhibitory, however, because it only brings the membrane potential toward and not beyond the reversal level, which is considerably below the firing level (-45 mV).

switch is closed will be

$$I_{\text{IPSP}} = \frac{E_{\text{Cl}} - E_{\text{m}}}{R_{\text{m}} + r_{\text{Cl}}}$$

$$= \frac{65 \times 10^{-3} \text{ V} - 60 \times 10^{-3} \text{ V}}{1.5 \times 10^{8} \ \Omega + 1 \times 10^{8} \ \Omega} \qquad (6\text{-}13)$$

$$= 2 \times 10^{-11} \text{ A}$$

The voltage across the nonsynaptic membrane produced by this outward current will be

$$\Delta V_{\text{m}} = R_{\text{m}} \times I_{\text{EPSP}}$$

$$= 10^{8} \ \Omega \times 2 \times 10^{-11} \text{ A}$$

$$= 2 \text{ mV}$$

This voltage change is in the same direction as the resting battery, thus the potential of the nonsynaptic membrane is increased to $(-60 - 2)$ mV, or -62 mV. The same current flowing outward across the synaptic resistance produces a voltage drop of 3 mV ($1.5 \times 10^{8} \ \Omega \times 2 \times 10^{-11}$ A), which opposes the effective voltage produced by the inhibitory battery and brings this branch from -65 mV to -62 mV (Figure 6-11A).

In more effective synaptic actions additional parallel synaptic pathways would be opened up and more current would flow, so that at its limit the membrane potential would be dominated by E_{IPSP} (see below).

Even when an IPSP does not hyperpolarize the membrane, because the membrane potential is already at -65 mV (E_{Cl}), the inhibitory synaptic action still inhibits the cell from firing. This is because the IPSP tends to clamp the membrane potential at E_{Cl}. By increasing the conductance to Cl^{-}, synaptic inhibition increases the overall conductance of the membrane (G_{m}). Since the potential change produced by the EPSP equals $I_{\text{EPSP}}/G_{\text{m}}$, the synaptic current produced by excitatory synapses produces a smaller EPSP across the higher-conductance nonsynaptic membrane than across the lower-conductance membrane that exists in the absence of inhibitory synaptic actions. Thus, IPSPs due to an increased conductance inhibit in two ways. First, they invariably increase the membrane conductance (this is called the *shunting* or *short-circuiting action* of synaptic potentials due to increased conductance). Second, inhibitory synaptic actions due to increased conductance usually hyperpolarize the membrane and move it further away from threshold (except in cells that have a high resting potential where V_{m} is equal to or more hyperpolarized than E_{Cl}).

The method used to examine the *reversal potential for inhibition* is similar to that used for excitation. Let us assume the simple example of an IPSP due only to an increase of g_{Cl}. As the membrane potential is depolarized the IPSP gets larger because V_m is further away from E_{Cl}, as is evident from Equation 6-8:

$$I_{IPSP} = g_{IPSP} (V_m - E_{IPSP})$$

The force due to the concentration gradient moving Cl⁻ into the cell remains the same, but the force due to the electrical gradient moving Cl⁻ out of the cell is reduced. If the membrane potential is now hyperpolarized to −65 mV, the IPSP becomes nullified (Figure 6-11*B*). This null point is defined as the reversal potential for inhibition (E_{IPSP}). In this example it will be E_{Cl}. At the null point E_{Cl} will equal V_m and the electrical driving force acting on Cl⁻ will be exactly equal to the force due to the concentration gradient. If the membrane potential is further hyperpolarized, the electrical driving force will exceed the driving force of the concentration gradient and Cl⁻ will begin to move out of the cell, causing a depolarization. Synaptic inhibition therefore drives the membrane potential toward E_{Cl} (−65 mV) and away from threshold (−45 mV).

Excitatory and inhibitory synaptic potentials due to conductance increases exert their actions because of the relation between their respective reversal potentials and the threshold potential of the trigger zone (typically −45 mV). Excitatory synaptic actions tend to drive the membrane potential in a depolarizing direction beyond threshold (−45 mV) toward a reversal potential that is close to −10 mV. Inhibitory synaptic actions invariably tend to drive (or at least clamp) the membrane potential to a value that is below threshold.

Synaptic Potentials Due to Decreased Ionic Conductances

Work on the frog sympathetic ganglion and on *Aplysia* neurons has revealed that synaptic inhibition can be produced in cells with a high resting g_{Na} by decreasing g_{Na}; synaptic excitation can be produced in cells with a high resting g_K by decreasing g_K.[6] In these cells the synaptic receptor controls an ionic conductance channel that can be turned off, causing the membrane potential to move toward equilibrium potential of the other ions. These types of synaptic actions are slow—because the membrane resistance, and therefore the

---

[6]Gerschenfeld and Paupardin-Tritsch, 1974a, b; Tomita, 1970; Weight and Votava, 1970; Weight, 1974; and Kehoe, 1975.

time constant, increases—and they probably are not as common as those involving conductance increases.

EXCITATORY POSTSYNAPTIC POTENTIALS

Some cells have an E_K and high g_K in parallel with the nonsynaptic membrane that can be controlled by synaptic action. At rest, the membrane potential of the cell is increased because the synaptic E_K causes inward current to flow through the nonsynaptic membrane (Figure 6-12A). The transmitter depolarizes the cell by turning off this g_K.

At rest, the membrane battery (E_m) and resistance (R_m) are shunted by a parallel synaptic pathway with a battery E_K (-75 mV) in series with a resistance r_K (2.5×10^8 Ω). As a result, outward current flows through the synaptic channel r_K and inward current flows across the nonsynaptic pathway, thereby hyperpolarizing and increasing the membrane potential.

The outward current through the synaptic pathway (following Eqn. 6-12) is

$$
\begin{aligned}
I_K &= \frac{E_K - E_m}{R_m + r_K} \\
&= \frac{75 \times 10^{-3} \text{ V} - 60 \times 10^{-3} \text{ V}}{1 \times 10^8 \text{ Ω} + 2.5 \times 10^8 \text{ Ω}} \\
&= 4.3 \times 10^{-11} \text{ A}
\end{aligned}
\tag{6-14}
$$

This current then flows inward through the nonsynaptic membrane and produces a steady hyperpolarization. The voltage change is thus 4.3×10^{-11} A $\times 10^8$ Ω, or 4.3 mV. Since the voltage change is in the same direction as the nonsynaptic battery (E_m), it increases the potential of the nonsynaptic membrane from -60 to -64.3 mV. The same current flowing outward through the synaptic pathway produces a V_{r_K} of 10.7 mV (4.3×10^{-11} A $\times 2.5 \times 10^8$ Ω). This opposes E_K in the leakage pathway, changing it also to -64.3 mV (-75 mV $+ 10.7$ mV). The transmitter action is equivalent to opening the switch in the synaptic pathway, thereby eliminating the contribution of this resting g_K. This causes the membrane potential to move 4.3 mV in the depolarizing direction to E_m (Figure 6-12B).

If the membrane potential is now altered, an EPSP due to decreased conductance will behave very differently from one due to increased conductance. Depolarizing the membrane potential from -60 to -50 mV (Figure 6-12B) will increase the EPSP. The current flowing outward through the K$^+$ channel and inward through the nonsynaptic pathways will be larger at -50 mV than at -60 mV (Equation 6-8). Turning off g_K thus results in a larger synaptic

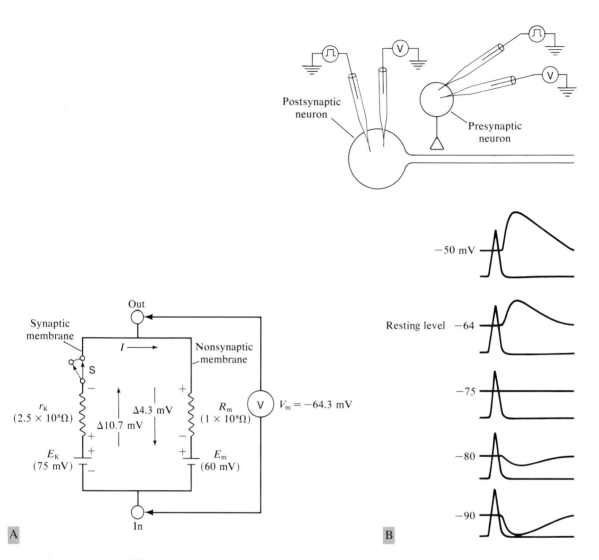

FIGURE 6-12.

Chemically mediated excitatory synaptic actions due to decreases in g_K.

A. Equivalent circuit of excitatory synapse and current flow during rest. At rest, the batteries E_m and E_K are in parallel and current flows outward through their series resistance r_K and inward through R_m. The synaptic action is equivalent to opening the switch S, thus taking r_K and E_K out of the circuit. This eliminates the outward current flowing through r_K and the inward current generated across R_m, thereby depolarizing the membrane potential from -64.3 mV to -60 mV.

B. Reversal potential. Experimental arrangement as in Figure 6-10C. At the resting level (-64 mV) a presynaptic spike produces a depolarizing EPSP, which increases in amplitude as the membrane potential is depolarized to -50 mV. But when the membrane potential is hyperpolarized to -75 mV the EPSP becomes nullified and at -80 mV it is reversed. Further hyperpolarization (-90 mV) increases the amplitude of the reversed EPSP.

current being turned off, causing a greater depolarization than at normal rest-ing potential. But if the membrane potential is hyperpolarized from the resting level to -75 mV, which is E_K, the EPSP will be nullified (Figure 6-12B) be-cause the electrical force of the membrane will cancel the chemical force due to the concentration gradient and no further net K$^+$ current will flow. At more hyperpolarized levels the EPSPs become inverted because the inward force on K$^+$ due to the membrane potential overrides the outward driving force due to the chemical gradient. Therefore, hyperpolarizing the membrane potential beyond E_K (to -80 mV or -90 mV) changes the net electrochemical driving force so that the K$^+$ current flow at rest will tend to push the membrane in a depolarizing direction. Turning off g_K hyperpolarizes the cell.

INHIBITORY POSTSYNAPTIC POTENTIALS

In cells that have a high resting g_{Na} controlled by synaptic action, the trans-mitter can turn off that conductance and produce a hyperpolarization that will drive the membrane potential toward the E_m of the nonsynaptic membrane (Figure 6-13).

At rest the magnitude of the synaptically controlled current is

$$I = \frac{E_m + E_{Na}}{R_m + r_{Na}}$$

$$= \frac{60 \times 10^{-3} \text{ V} + 55 \times 10^{-3} \text{ V}}{1 \times 10^8 \ \Omega + 1 \times 10^9 \ \Omega} \qquad (6\text{-}15)$$

$$= 10.45 \times 10^{-11} \text{ A}$$

This current flows outward through the nonsynaptic membrane and depolar-izes it (Figure 6-13A). The voltage change is 10.45×10^{-11}A $\times$ 1 $\times$ 10$^8\Omega$. or 10 mV. Since the voltage change opposes that of the battery of the nonsyn-aptic membrane E_m, it reduces the negativity of the nonsynaptic membrane from 60 to 50 mV. The current flowing inward through the synaptic pathway will produce a $V_{r_{Na}}$ of 105 mV (10.45×10^{-11} A $\times$ 1 $\times$ 10^9 Ω). This voltage drop opposes the Na$^+$ battery, also bringing it to 55 mV $-$ 105 mV, or 50 mV. Thus, E_m is significantly reduced by Na$^+$ leakage. The synaptic action is equiv-alent to opening the switch in Figure 6-13A, thus removing E_{Na} and r_{Na} from the membrane circuit. As a result, the negativity of the membrane potential is increased by 10 mV, moving from 50 to 60 mV, which is E_m.

Thus, when the membrane potential is hyperpolarized, the IPSP due to decreased conductance becomes larger (Figure 6-13B). At hyperpolarized levels the driving force on Na$^+$ is increased and more inward current flows

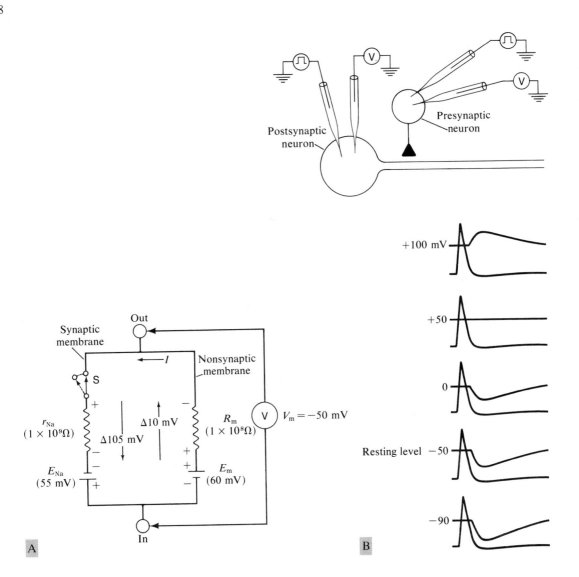

FIGURE 6-13.
Chemically mediated inhibitory synaptic actions due to decreases in g_{Na}.

A. Equivalent circuit of inhibitory synapse and current flow during rest. At rest, current flows outward through R_m and inward through R_{Na}. The synaptic action is equivalent to opening switch S; this turns off the slight Na$^+$ leak that normally keeps V_m away from E_m. As a result, the membrane hyperpolarizes and moves to E_m.

B. Reversal potential. Experimental arrangement as in Figure 6-10C. At the resting level (-50 mV) the presynaptic spike produces a hyperpolarizing IPSP, which increases in amplitude as the membrane potential is hyperpolarized (-90 mV). But as the membrane potential is decreased to 0, the IPSP becomes smaller. When the membrane potential is reversed and brought to the equilibrium potential for Na$^+$, the IPSP becomes nullified. Further depolarization (to $+100$ mV) reverses the IPSP to a depolarizing potential.

through the Na^+ channel, driving current outward through the adjacent K^+ channel (Equation 6-8). This depolarizing current counteracts the effects of the hyperpolarizing current. Turning off g_{Na} decreases the inward current and hyperpolarizes the membrane potential. But when the membrane potential is depolarized to E_{Na} (Figure 6-13B), the synaptic potential becomes nullified. If it is depolarized beyond E_{Na} (to $+100$ mV), the IPSP becomes inverted to a depolarizing synaptic potential. At this level the synaptic conductance of Na^+ generates an outward current that hyperpolarizes the cell; turning that conductance off causes a depolarization.

The effectiveness of a synaptic action due to conductance increase is determined by two changes it produces: changes in conductance and voltage. The inhibitory action of an IPSP due to decreased conductance is strictly related to its ability to change the membrane potential. As a result, such an IPSP is less effective than that due to increased conductance because the latter can also short-circuit nearby excitatory synaptic actions as well as conductance channels generating action potentials. On the other hand, an EPSP due to decreased conductance enhances the effectiveness of concomitant and nearby excitatory synaptic actions, and therefore can be more effective than an EPSP due to increased conductance (see Figure 9-37C, p. 402).

PRESYNAPTIC MECHANISMS OF CHEMICAL TRANSMISSION

At chemical synapses a presynaptic neuron releases a chemical transmitter substance following the invasion of its terminals by an action potential. How do the electrical events of the action potential initiate the secretion of the transmitter?

The action potential in the presynaptic terminals is associated with the usual sequence of conductance changes. Depolarization leads first to an increase of g_{Na}, resulting in an intense influx of Na^+, followed by a rapid suppression of g_{Na} and an increase in g_K. Are these ionic permeability changes necessary for transmitter release? This question has been examined by Katz and Miledi (1967c; see also Kusano *et al.*, 1967, and Bloedel *et al.*, 1966) using the giant synapse of squid (Figure 6-14A). They applied tetrodotoxin to block the increase in g_{Na} due to the spike and examined its effect on the transmitter release that is evoked by electrical stimulation of the presynaptic fiber (Figure 6-14B). Tetrodotoxin produced a gradual decline in the amplitude of the evoked presynaptic action potential and a concomitant decline in the postsynaptic potential. The amplitude of the postsynaptic potential, and therefore the amount of transmitter release, was directly related to the amplitude of the presynaptic

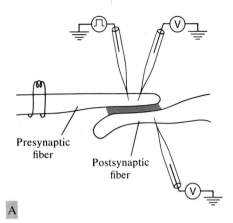

Presynaptic
fiber

Postsynaptic
fiber

A

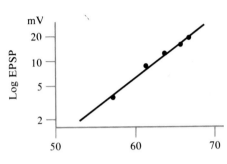

mV

20

Log EPSP

10

5

2

50   60   70

Presynaptic Spike Amplitude (mV)

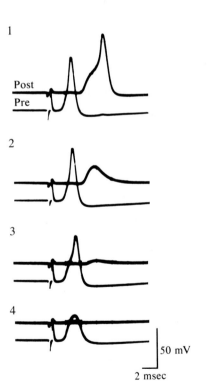

1

Post
Pre

2

3

4

50 mV

2 msec

B

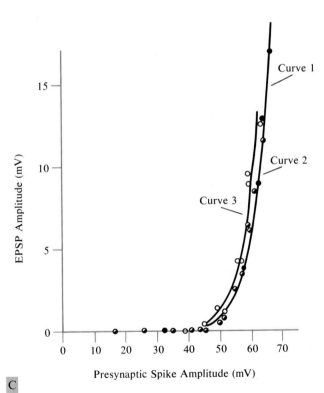

15

EPSP Amplitude (mV)

10

5

0

Curve 1

Curve 2

Curve 3

0   10   20   30   40   50   60   70

Presynaptic Spike Amplitude (mV)

C

FIGURE 6-14.

Mechanism of transmitter release. Synaptic transmission at the squid giant (stellate ganglion) synapse. [From Katz and Miledi, 1967c.]

A. Experimental arrangement with recording electrodes in both the pre- and postsynaptic fibers of the giant synapse. A current-passing electrode has also been inserted in the presynaptic terminal.

B. Transmitting effectiveness of the nerve impulse during the development of tetrodotoxin (TTX) paralysis. In part *B-1*, 7 min after adding TTX the presynaptic spike still produces a suprathreshold EPSP that triggers an action potential. In *B-2* and *B-3*, 14 to 15 min after adding TTX the presynaptic spike gradually becomes smaller and in so doing produces smaller EPSPs. In *B-4*, when the presynaptic spike is reduced to 40 mV it fails to produce a synaptic action.

C. Input–output curve for different-size action potentials in TTX. The EPSP has a definite threshold; in this experiment the presynaptic spike had to be 40 mV to produce an EPSP. Beyond this threshold the relationship of presynaptic spike height to EPSP is logarithmic; usually a 10 mV increase in the presynaptic spike produces a tenfold increase in the EPSP. A semilog plot of this relationship is shown above the curve. Presynaptic spike potentials are shown by solid circles (curve 1). After complete TTX paralysis,and without altering the position of the electrodes, electrotonic potentials were produced by passing current pulses of 1 msec (half dark circles) and 2 msec (open circles) in duration to the presynaptic fiber (curves 2 and 3, respectively).

spike. The relationship is logarithmic: usually a 10 mV increase in the presynaptic spike produced a tenfold increase in the postsynaptic potential (Figure 6-14C). These experiments indicated that the action potential is essential for the release of transmitter and that Na^+ permeability may be the mediating agent.

With the preparation treated with TTX, Katz and Miledi inserted a second intracellular current-passing electrode into the presynaptic fiber (Figure 6-15A). They then applied a current pulse directly to the terminal and found that electrotonic depolarization of the presynaptic terminal (without a regenerative g_{Na}) also led to a postsynaptic potential. Increasing the strength of the stimulating current produced greater (*presynaptic*) electronic potentials in the presynaptic terminals that resulted in synaptic potentials of progressively greater amplitude in the postsynaptic cell (Figure 6-15B). The postsynaptic potential had a clear threshold. It first appeared when the presynaptic terminal was depolarized by 25 mV or more. As the presynaptic depolarizing potential increased beyond this threshold value, the EPSP grew rapidly in amplitude until the presynaptic depolarization reached 60 mV, at which point the depolarization failed to increase the size of the EPSP. The steep input-output relationship under these circumstances is very similar to that seen with the normal spike (compare the curves in Figure 6-14C). Thus, the regenerative Na^+ current is not necessary for the release of transmitter substance. An electrotonic potential in the terminal, matched approximately in size and shape to the depolariz-

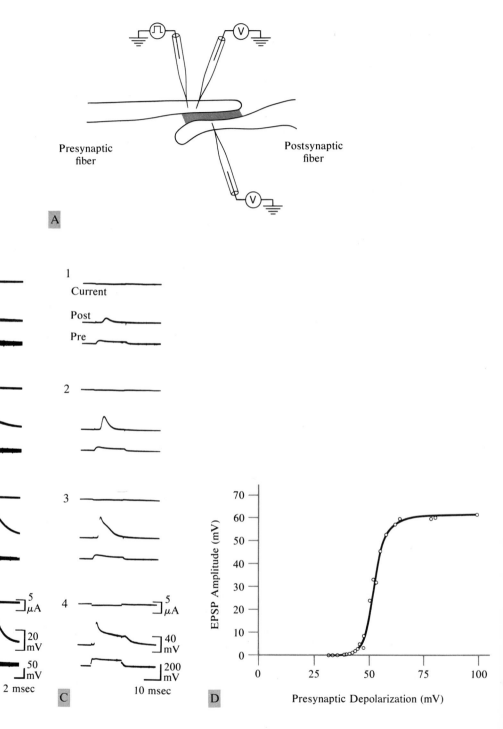

FIGURE 6-15.

Release of transmitter substance at the squid giant synapse without nerve impulses. [From Katz and Miledi, 1967c.]

A. Recording arrangement as in Figure 6-14.

B. The three traces in parts *B-1* to *B-4* and *C-1* to *C-4* represent (from top to bottom) the current injected into the presynaptic cell, the PSP generated in the postsynaptic cell, and the electrotonic potential produced by the current pulse in the presynaptic neuron. Progressively larger depolarizations of the presynaptic terminal generated by intracellular current pulses produce a progressive increase in the response of the postsynaptic cell (*B-1* to *B-4*). The action potentials have been blocked with TTX. There is a characteristic synaptic delay between the onset of the depolarization in the presynaptic fiber and the EPSP in the postsynaptic fiber.

C. Progressive increase in the EPSP in response to long current pulses after iontophoretic loading of the terminal with TEA in the presence of TTX. (The intensity of the depolarizing current pulse increases from parts *C-1* to *C-4*).

D. Input-output curve based on the experiment illustrated in part *C;* the initial level of the presynaptic membrane potential was −69 mV. In this experiment the threshold for transmitter release (as indicated by the appearance of an EPSP in the postsynaptic cell) was about 40 mV. In other experiments the threshold was lower, about 25 mV.

ing action of the normal action potential, is as effective in triggering the release of transmitter and producing a postsynaptic response.

Katz and Miledi next injected TEA into the presynaptic terminal to block K^+ permeability and determine its contribution to release (Figure 6-15*C*). They found the input-output curve to be exactly the same in the presence of both TEA and TTX as it was with TTX alone (compare Figures 6-14*C* and 6-15*D*). Thus, changes in g_{Na} or g_K are not directly responsible for transmitter release.

One ion known to be essential for transmitter release is calcium, which is normally higher in concentration outside the cell than inside. At the vertebrate nerve-muscle synapse Katz and Miledi (1967a) found that the inward movement of Ca^{++} is an important link between depolarization of the nerve terminal and the release of acetylcholine. When Ca^{++} is removed from the bathing solution transmission stops. The nerve impulse still arrives in the presynaptic terminal, but it no longer elicits a postsynaptic response.

Increasing the amount of Ca^{++} facilitates synaptic transmission even in the presence of TTX. Increasing the magnesium ion concentration depresses release (Katz and Miledi, 1968, 1970). For example, raising the external Ca^{++} concentration fivefold (11 to 55 mM) changes the slope of the input-output curve so that only a 7.5 mV (instead of the usual 10 mV) depolarization is necessary to produce a tenfold increase in the postsynaptic potential. By contrast, when the Mg^{++} concentration is raised about threefold (54 to 184 mM),

18.5 mV (instead of 10) are necessary to produce a tenfold change. The suggestion that presynaptic depolarization causes an inward movement of Ca^{++} is also supported by evidence in the squid giant synapse that Ca^{++} moves into the cell during the action potential. Unlike the rise of g_{Na}, this increase in g_{Ca} is not eliminated by TTX. Moreover, Katz and Miledi (1969a, b) found that in the presence of TTX and TEA depolarization of the presynaptic terminals actually generates a small regenerative inward Ca^{++} current that is normally masked by the much larger Na^+ and K^+ currents. This Ca^{++} response is restricted to the presynaptic terminals. Additional evidence of Ca^{++} inflow is provided by experiments with aequorin, a protein that fluoresces in the presence of Ca^{++}. When aequorin is injected into the presynaptic terminals of the squid synapse, presynaptic stimulation leads to increased fluorescence (Llinás and Nicholson, 1975).

To find out at what stage of transmitter release external Ca^{++} is involved, Katz and Miledi used a nerve-muscle preparation bathed in TTX and calcium-free Ringer solution. Besides the normal microelectrode for recording inside the muscle fiber, they also used two external electrodes. One electrode, filled with NaCl, served as a stimulating electrode that depolarized the terminals, and the other electrode, filled with $CaCl_2$, was used to raise the local Ca^{++} concentration at a critical moment before or after the depolarizing pulse (Figure 6-16). By this means Katz and Miledi found that Ca^{++} must be present during the depolarization. When the Ca^{++} pulse is delayed until the end of the depolarization, no release occurs. This suggested that the depolarization of the action potential releases transmitter by changing the properties of the terminal axon membrane in such a way that Ca^{++} channels in the membrane are opened. As a consequence of this increased g_{Ca}, calcium ions can move down a very steep concentration gradient toward the inside of the membrane and reach the critical sites involved in transmitter release. The critical role of Ca^{++} for release has now been directly confirmed by Miledi (1973), who observed an increase in the release of a transmitter following injection of Ca^{++} into the presynaptic terminal of the squid giant synapse.

Before we examine further how Ca^{++} leads to release, it is important to consider some quantitative features of the release process.

Quantal Release of Transmitter

At all chemical synapses so far examined the release of synaptic transmitter substance has been shown to be quantized. The synaptic potential is made up of small elementary units (*quanta*) of fixed size. The reason the input-output

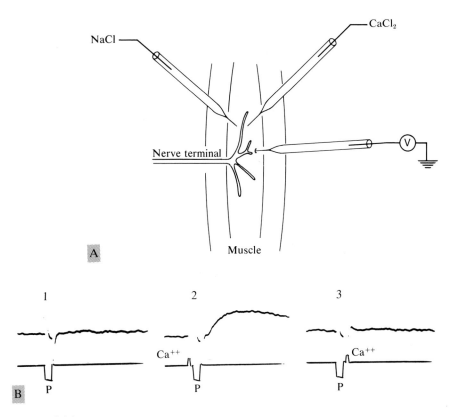

FIGURE 6-16.
The timing of Ca^{++} action in transmitter release. [From Katz and Miledi, 1967a.]

A. Experimental set up at a frog nerve-muscle synapse. Nerve impulses have been eliminated with TTX, and brief depolarizing pulses are applied to the presynaptic nerve terminals by means of an extracellular NaCl electrode. The preparation is bathed in a low-Ca^{++} Ringer's solution, which normally blocks transmitter release, and current pulses of Ca^{++} are delivered to the nerve terminal by means of a second extracellular electrode filled with $CaCl_2$.

B. Timing of Ca^{++} influx. A brief depolarizing pulse (P) is presented alone (*B-1*), just after a pulse of Ca^{++} (*B-2*), and just before (*B-3*). Only when the Ca^{++} pulse precedes the depolarizing pulse (*B-2*) by a critical interval is transmitter released. Thus, the utilization of Ca^{++} must occur during, or perhaps *immediately* after, the depolarization.

curve and the synaptic potential appear graded is that the number of quantal units released is graded. The first clue for quantal transmission came when Fatt and Katz (1951) obtained recordings from the frog nerve-muscle synapse in the absence of presynaptic stimulation and observed small, spontaneously occurring potential changes of about 0.5 to 1.0 mV in amplitude (Figure 6-17). Since the synaptic potentials at vertebrate nerve-muscle synapses are called

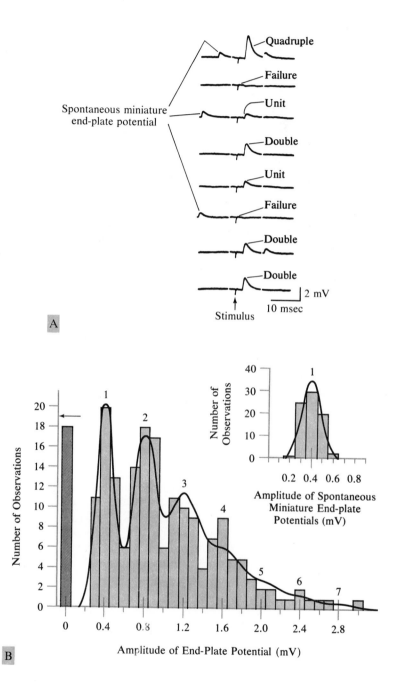

A

B

end-plate potentials (e.p.p.), because the postsynaptic region has a specialization called the *end plate region,* Fatt and Katz called these small spontaneously occurring potentials *miniature end-plate potentials.* The time course of the miniature end-plate potentials and their response to various drugs are almost indistinguishable from those of the end-plate potential evoked by nerve stimulation. This synapse uses acetylcholine (ACh) as its transmitter and, like the end-plate potential, the miniature end-plate potentials are enhanced and prolonged by prostigmine, a compound that inhibits the splitting (hydrolysis) of ACh by acetylcholinesterase. The miniature end-plate potentials are also abolished by agents that block the ACh receptor, such as *d*-tubocurarine (*d*-tbc). In the absence of stimulation the miniature end-plate potentials occur at random. The frequency of their occurrence is increased by depolarizing the presynaptic terminal; they disappear following degeneration of the presynaptic nerve and reappear with reinnervation, indicating that they are caused by the random release of ACh from the presynaptic nerve terminal.

One possible explanation for the fixed size of the spontaneous synaptic potentials is that they represent the responses of the receptor to single ACh molecules, and that the receptor response is quantized. This hypothesis was tested by ACh iontophoresis: a positive potential source is applied to an electrode containing acetylcholine chloride and the positively charged ACh is passed out of the pipette in controlled amounts as a function of current. By applying a small amount of ACh to the receptor membrane in this manner, del Castillo

FIGURE 6-17.

Quantal nature of synaptic transmission.

A. Quantal components of end-plate potentials (e.p.p.) were first demonstrated in the frog by del Castillo and Katz, but similar results are obtained in mammalian muscle. These intracellular recordings are from a rat nerve-muscle synapse, showing a few spontaneous miniature e.p.p.'s and the responses to a single impulse in the phrenic nerve in a Ca^{++}-deficient and Mg^{++}-rich medium. The e.p.p.'s show considerable fluctuations; two impulses produced failures and two (3rd and 5th traces) produced a unit potential; others produced responses that are two or four times the amplitude of the unit potential. Note the stimulus artifact produced by current flowing between the stimulating and recording electrodes in the bathing solution. [From Liley, 1956b.]

B. Histogram illustrating the distribution of amplitudes of evoked e.p.p.'s at a mammalian end plate. Synaptic transmission has been reduced with a high-Mg^{++} solution. The distribution of the spontaneous miniature e.p.p.'s is shown in the inset. Peaks of the amplitude histogram of the evoked e.p.p.'s occur at 1, 2, 3, and 4 times the mean amplitude of the *spontaneous* potentials (0.4 mV). A Gaussian curve is fitted to the spontaneous miniature e.p.p.'s and used to calculate the theoretical (Poisson) distribution of the evoked e.p.p. amplitudes, represented by the continuous line. The dark bar (0 mV) indicates the actual number of failures; the arrow on the ordinate indicates the expected number of failures. [From Boyd and Martin, 1956.]

and Katz (1957) did not find quantized receptor responses resembling minia-ture end-plate potentials. The potential change produced by the interaction of the postsynaptic receptor and a small number of ACh molecules is much smaller than a 0.5 mV potential change produced by the miniature end-plate potential.

Subsequently, Katz and Miledi (1971, 1972) successfully estimated what appears to be the elementary ionic conductance event—the interaction of one ACh molecule (or at most a very few) with a single ACh receptor. They did this by analyzing the fluctuations in membrane potential (*ACh noise*) caused by applying small amounts of ACh to the receptor membrane. Based on a model in which the electrical noise of the membrane produced by ACh repre-sents the sum of many randomly occurring elementary (shot) events, Katz and Miledi estimated that the elementary ACh potential produced by the opening of a single conductance channel is only about 0.3 μV. A miniature end-plate potential of 0.5 mV therefore represents the summation of about 10^3 elemen-tary depolarizations. This is only a lower limit. The collisions necessary to produce an elementary conductance change probably involve one or two ACh molecules. Some of the released ACh does not interact with receptor molecules but is lost by diffusion in the synaptic cleft or by hydrolysis. Thus, it is likely that at least 10^4 molecules are necessary to produce a miniature end-plate potential. This number is similar to that estimated on the basis of actual ACh released, which suggests that a miniature synaptic potential is produced by a packet containing up to 5×10^4 ACh molecules.

The question now arises whether during *normal* transmission ACh is released in quanta or released in continuously graded amounts.

In a classic series of studies at the frog nerve-muscle synapse del Castillo and Katz (1954a, b, c, d) found that the size of the end-plate potential was markedly reduced by bathing the preparation in high-Mg^{++} or low-Ca^{++} solutions (which interfere with the release of chemical transmitters); the end-plate potential exhibited a small amplitude (0.5 to 2.5 mV) and often failed. The minimum response above zero (the *unit synaptic potential*) was identical in size and shape to the spontaneously occurring miniature end-plate potential. The larger end-plate potentials were integral multiples of the unit potential. Thus, an amplitude histogram of the responses to a large number of stimuli revealed a peak of failures (zero responses) followed by a multimodal response distribution. The first response peak occurs at the voltage of the unit potential; this voltage is also identical to the amplitude of the spontaneous miniature end-plate potential. The voltage of each subsequent peak is an integral mul-tiple of the value of the first response peak: that of the second response peak is twice that of the first, and that of the third peak is three times that of the first (Figure 6-17). The series of peaks is broad rather than sharply defined because of the statistical variability in the size of individual miniature end-plate potentials (inset, Figure 6-17B).

In view of the dramatic step-wise fluctuations in the amplitude of the end-plate potentials at low levels of release and the finding that the unit potential has the same mean amplitude as the spontaneously released miniature end-plate potentials, del Castillo and Katz suggested that the normal end-plate potential (40 to 50 mV) is caused by the release of more than 300 quanta and that the exact number of quanta released varies from stimulus to stimulus. Based on these observations, del Castillo and Katz proposed the *quantal hypothesis for synaptic transmission,* according to which the amplitude of the end-plate potential fluctuates in response to consecutive stimuli because varying numbers of quanta are released.

Further investigations of the release process indicated that it operates in a random manner predictable from a binomial distribution. The fate of each quantum in response to an action potential in the presynaptic terminals resembles a binomial or Bernoulli trial (similar to tossing a single coin in the air to determine whether it comes up heads or tails) with only two possible outcomes, success or failure—the quantum is or is not released. For a population of releasable quanta, therefore, each action potential represents a series of independent binomial trials (comparable to tossing a handful of coins to see how many come up heads). In a binomial distribution p stands for the average probability of success and q (or $1 - p$) stands for the mean probability of failure. The probability of a quantum being released by an action potential is independent of the release of other quanta by that action potential. Both the average probability (p) that individual quanta are released and the store (n) from which the quanta are released are assumed to be constant. Any reduction in the store is assumed to be quickly replenished after each stimulus. Once n and p are known, the binomial probability law allows one to estimate the mean number of quanta (m, called the *quantal content* or *quantal output*) that are released to make up the end-plate potential in a series of stimuli, where

$$m = np \qquad (6\text{-}16)$$

The application of the binomial statistic to transmitter release can be illustrated with the following example. A terminal has a releasable store of five quanta ($n = 5$); assume that $p = 0.1$, then q (the probability that a quantum is not released from the terminals) is $1 - p$, or 0.9. We can now determine for any given number of stimuli, say 100, how many times a stimulus will release no quanta (failure), a single quantum, two quanta, three quanta, or any number of quanta (up to n). Thus, the probability that none of the five available quanta will be released by a given stimulus is the product of the individual probabilities that each quantum will not be released: $q^5 = (0.9)^5$, or 0.59. We would thus expect to see 59 failures in a hundred stimuli. The probabilities of observing zero, one, two, three, four, or five quanta are represented by the successive terms of the binomial expansion

$$(q + p)^5 = q^5 \text{ (failures)} + 5 q^4p \text{ (1 quantum)} + 10 q^3p^2 \text{ (2 quanta)} +$$

$$10 q^2p^3 \text{ (3 quanta)} + 5 qp^4 \text{ (4 quanta)} + p^5 \text{ (5 quanta)}$$

Thus, in 100 stimuli the binomial expansion would predict 33 unit responses, 7 double responses, one triple response, and zero quadruple and quintuple responses.

If instead of 5 quanta there are n quanta, the probability of occurrence of multiunit responses is given by the expansion

$$(q + p)^n = \binom{n}{0} q^n + \binom{n}{1} pq^{n-1} + \binom{n}{2} p^2q^{n-2} + \binom{n}{3} p^3q^{n-3} + \ldots$$

$$\ldots + \binom{n}{x} p^xq^{n-x} \text{ (general term)} \ldots \binom{n}{n} p^n \text{ (last term)}$$

where $\binom{n}{0}, \binom{n}{1}, \binom{n}{2}, \binom{n}{x}$ denote the number of possible selections of 0, 1, 2, or x quanta from a pool of n quanta taken 0, 1, 2 or x at a time. Since

$$\binom{n}{x} = \frac{n!}{(n - x)!x!}$$

the probability that x quanta (0, 1, 2, 3, 4 . . .) will be released by a given stimulus will therefore be given by the general term of the binomial expansion, which can be expressed as

$$P_x = (q + p)^n = \binom{n}{x} p^xq^{n-x} = \frac{n!}{(n - x)!x!} p^xq^{n-x} \qquad (6\text{-}17)$$

With this general formulation one can predict the probability of seeing a failure, a unit, a double, triple, quadruple, or quintuple response to a *single stimulus* as well as the expected number of each of these responses in a *series of stimuli*. For a series of stimuli the expected number of stimuli that release x quanta is given by $N_x = N \times P_x$, where N is the total number of stimuli. But the binomial is a *two*-parameter distribution. These predictions require that two of the three parameters n, p, or m be known. For many synapses it is difficult to determine two parameters. Although m can often be reliably estimated directly, or at least indirectly, estimates of n and p are always indirect and in some cases not highly reliable.[7]

However, at low levels of release (in high-Mg^{++}, low-Ca^{++} solutions) p is often very low compared to n, so that p approaches zero and n becomes very large. Under these conditions the binomial distribution can be approximated by the Poisson distribution. The great advantage of the Poisson distribution is that it is a *one*-parameter distribution. Knowing m, the mean quantal output, one can describe the entire distribution. Moreover, as we will see below, the Poisson statistics provide several easy ways to estimate m.[8]

The relationship of the binomial distribution (implied by the quantal hypothesis) and the Poisson distribution (often used to evaluate the quantal

---

[7]A common but indirect way of estimating p derives from the finding at various nerve-muscle synapses that the second of two consecutive stimuli (separated by say 100 msec) produces a depressed synaptic response (Liley and North, 1953). If the amplitude of the second response is examined as a function of separation between the two stimuli, and the curve is extrapolated to a zero-time separation to avoid errors introduced by the recovery process, one can obtain a value for zero-time depression. Assuming that the depression at zero-time is due completely to depletion of the available quanta (n), then p can be estimated. For example, if the extrapolated zero-time amplitude of the second response is 10 percent less than the first, one can assume that the releasable store of the transmitter in the terminal (n) has also been reduced by 10 percent as a result of the first stimulus, and that p is therefore 0.1. However, this estimate is based on the assumption (not fully reliable) that p is not changed by the first stimulus.

At some synapses m can be measured directly. This can be done in two ways: (1) by recording spontaneous miniature synaptic potentials and estimating m from the ratio of the average PSP to the unit PSP ($\overline{v}/\overline{v}_1$, Equation 6-25); and (2) by working at low temperature and counting individual quanta emitted by each stimulus (Katz and Miledi, 1965b). If m is known, the value of p can be calculated either from the amplitude variations of the evoked PSPs or from the number of failures. In the binomial distribution the variance (var) is $n \times p\,(1 - p)$ and the number of failures (N_0) in a series of N stimuli is $N_0 = N\,(1 - p)^n$. The variance (var) can be estimated from the variance of the quantal content distribution over a series of evoked EPSPs. Since $m = np$ and m is known, one can solve for n or p and then substitute n or p into either of the equations to obtain the other parameter.

[8]Although the Poisson distribution is easier to apply than the binomial—because only m needs to be estimated to know the distribution—the basic assumptions of a Poisson distribution are more stringent than those of the binomial distribution. In addition to satisfying the binomial assumptions (that the release of quanta is discrete and independent), the Poisson distribution demands (1) that the store of releasable quanta (n) be relatively large, and (2) the probability (p) of any given quantum being released is relatively low. If these assumptions are not met at a particular synapse (for example, if p is greater than 0.3), the Poisson distribution will not accurately approximate the binomial distribution (Wernig, 1975). However, since p and n are often not known, the Poisson distribution remains a useful statistical estimate. Its applicability to a particular synapse can be estimated by determining statistically how well the actual distribution of failures and responses fits the theoretical distribution predicted by the Poisson equation (Equation 6-24 and Figure 6-17*B*).

hypothesis) can be illustrated by considering the first term of the binomial expansion, which predicts the probability of failure. Substituting zero for x in Equation 6-17, we have

$$P_0 = q^n \tag{6-18}$$

Since $q = 1 - p$ and $p = m/n$, we have

$$P_0 = q^n = (1 - p)^n = (1 - m/n)^n \tag{6-19}$$

As n becomes very large, the limiting value of this expression becomes $P_0 = e^{-m}$. This expression can be developed in the following way:

$$\text{let} \quad t = -m/n, \tag{6-20}$$

$$\text{therefore} \quad n = -m/t \tag{6-21}$$

From Equations 6-19, 6-20, and 6-21, we have

$$P_0 = q^n = (1 - m/n)^n = (1 + t)^{-m/t} = [(1 + t)^{1/t}]^{-m} \tag{6-22}$$

But as n approaches infinity (Equation 6-20), t approaches zero, and $(1 + t)^{1/t}$ approaches e by definition. Thus we have

$$P_0 = q^n = (1 - m/n)^n = e^{-m} \tag{6-23}$$

The prediction for a single unit response can be obtained by substituting 1 for x in Equation 6-17:

$$P_1 = npq^{n-1} = mq^{n-1}$$

Since q^n has a limiting value of e^{-m} as n becomes very large, then

$$P_1 = me^{-m}$$

The prediction for double and triple responses will be

$$P_2 = \frac{m^2 e^{-m}}{2}$$

$$P_3 = \frac{m^3 e^{-m}}{6}$$

The general expression for the probability of any number (x) of unit PSPs is given by the Poisson distribution:

$$P_x = \frac{m^x e^{-m}}{x!} \tag{6-24}$$

Experimentally, m can be obtained in four independent ways. A direct estimate (usually referred to as m_1) can be obtained by dividing the average amplitude of the synaptic potential in a given series ($\bar{v}$) by the average amplitude of the spontaneously occurring miniature synaptic (or end-plate) potential $\bar{v}_1$ or its equivalent, the unit synaptic potential:

$$m_1 = \bar{v}/\bar{v}_1 \tag{6-25}$$

A second method of direct determination is to count the number of quanta released by each stimulus of a series of stimuli and divide by the number of stimuli. This can be done only if m is small and release is dispersed in time (by cooling to a low temperature), or if release consists only of unit responses and failures.

A third (less direct) method for determining m is derived from the Poisson distribution (Equation 6-24). Given $x = 0$, both $x!$ and m^x become 1, and the probability (P_0) of producing a failure from any single stimulus is given by

$$P_0 = e^{-m}$$

But this probability P_0 is also equal to the number of failures in a series of stimuli (N_0) divided by the total number of stimuli (N) in that series:

$$N_0/N = e^{-m}$$

Taking the natural log on both sides gives

$$\ln \frac{N_0}{N} = \ln e^{-m}$$

which can be rewritten as

$$\ln \frac{N_0}{N} = -m \ln e$$

Since $\ln e = 1$, dividing by -1 and rewriting results in

$$m_0 = \ln \frac{N}{N_0} \tag{6-26}$$

Thus m_0 can be obtained from the ratio of the number of failures following presynaptic stimulation to the total number of stimuli.

A fourth (and least direct) means of estimating m derives from the property of the Poisson distribution that the variance is equal to the mean. Since the mean of the Poisson distribution is equal to m, the variance equals m, and the standard deviation (the square root of the variance) is equal to $\sqrt{m}$. Thus, the coefficient of variation (the ratio of standard deviation over the mean) can be expressed as

$$\text{CV} = \frac{\sqrt{m}}{m} \tag{6-27}$$

By squaring both sides of the equation and solving for m we have

$$m_{\text{cv}} = \frac{1}{(\text{CV})^2} \tag{6-28}$$

In practice, the coefficient of variation of the quantal content can be estimated from the coefficient of variation of the synaptic potential amplitude for a series of stimuli. Due to fluctuations in the unit PSP size, however, it may be necessary to take certain corrections into consideration (Katz, 1966; Martin, 1966).

These four tests for m were first carried out by Katz and his colleagues at the frog nerve-muscle synapse (del Castillo and Katz, 1954a, b, c, d; and Katz and Miledi, 1965a). Quantal analyses with essentially similar results have also been obtained at other nerve-muscle junctions in vertebrates and invertebrates, at certain peripheral synapses in sympathetic ganglia, and in the central synapses of the spinal cord of the frog and the cat, and in the squid and *Aplysia* (see for example Katz and Miledi, 1963; Kuno, 1964a, b; Kuno and Miyahara, 1968, 1969; Miledi, 1967; Castellucci and Kandel, 1974). The analysis has been found applicable to several noncholinergic as well as cholinergic synapses. In most cases there has been fairly good agreement among different determinations of m, particularly when m is small.

Values for m vary, from about 100 to 300 at the vertebrate nerve-muscle synapse, the squid giant synapse, and *Aplysia* central synapses, to 1 to 3 in the synapses of the sympathetic ganglion and spinal cord of vertebrates. Estimates of p range from 0.15 to 0.5, and those for n from 1000 (at the vertebrate nerve-muscle synapse) to 3 (at single terminals of the crayfish).

The parameters n and p are statistical terms; the physical processes represented by them are not known. Although the parameter n is usually referred to as the *readily releasable* (or *readily available*) store of quanta, it may actually represent the number of release sites in the presynaptic terminals that are loaded with vesicles. The number of release sites is thought to be fixed, but the fraction loaded with vesicles is thought to be variable (Wernig, 1972; Zucker, 1973). The loading process is called *mobilization;* the emptying process is called *discharge* or *release.* The parameter p probably represents a compound probability depending on at least two functions: p_1, the probability of mobilizing a vesicle into a release site (reloading) after an impulse; and p_2, the probability that an action potential discharges a quantum from an active site (Zucker, 1973).

Additional insights into the release process and perhaps into the functional meaning of n and p will come from morphological studies. These have already shown that the presynaptic terminals of cholinergic neurons contain small, pale, electron-lucent organelles (20 to 40 nm in size) called *synaptic vesicles* (see Figure 6-18B). It seems attractive to think that these vesicles store ACh and that the all-or-none release of their content into the synaptic cleft explains the quantal nature of release. When cholinergic terminals are examined biochemically by subcellular fractionation, much of the ACh is found to be associated with the vesicle fraction (Whittaker and Zimmerman, 1974). A similar relationship applies in adrenergic neurons between norepinephrine and dense-core vesicles (Geffen and Jarrott, 1976).

Morphological studies have also provided a scheme of how these organelles extrude their contents during synaptic transmission. The membrane of the synaptic vesicle appears to fuse with the external membrane of the presynaptic terminal. The fused membranes then open into the synaptic cleft, causing the vesicle to empty its total contents by a process called *exocytosis* (Figure 6-18A; see Katz, 1966). The vesicle membrane is then thought to be incorporated into the surface membrane and recycled for reuse (Figure 6-18B; see Heuser and Reese, 1974; Hurlbut and Ceccarelli, 1974).

As is evident from these arguments, quantal analysis of synaptic transmission serves as a bridge between cellular electrophysiology, morphology, and molecular mechanisms by providing a means for assaying, electrophysiologically, the release of multimolecular packets of chemical transmitter substance. For example, the quantal hypothesis often allows one to distinguish between presynaptic and postsynaptic factors that might underlie changes in the effectiveness of synaptic transmission. The average amplitude of a synaptic potential (the synaptic efficacy, $\bar{E}$) produced by a presynaptic impulse is given by

$$\bar{E} = m \times \bar{q} \tag{6-29}$$

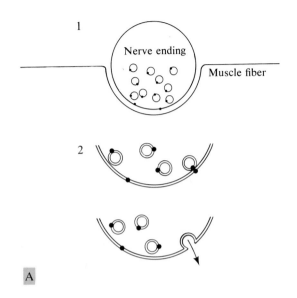

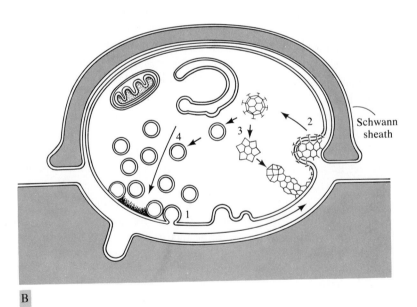

FIGURE 6-18.

Models for quantal release and recycling of the synaptic vesicles.

A. Illustration of the del Castillo and Katz hypothesis of quantal release. The model postulates that critical collisions between the synaptic vesicles and the axon membrane lead to temporary membrane fusion and an all-or-none discharge of the vesicular contents into the synaptic cleft. *A-1.* Black dots on the vesicles and the axon membrane represent reactive sites. *A-2.* Release of transmitter occurs only when the reactive sites collide. Although the collision rate is assumed to be high, critical collisions (between reactive sites) are statistically rare. The probability of occurrence varies directly with the number of release sites in the axon membrane. [From del Castillo and Katz, 1957.]

B. Illustration of the Heuser and Reese hypothesis for recycling of the synaptic-vesicle membrane at the frog nerve-muscle synapse. As in the del Castillo and Katz hypothesis (part *A*), the synaptic vesicles are thought to discharge their content of transmitter as they coalesce with the plasma membrane at specific regions adjacent to the muscle (1). Following release of the transmitter, the vesicle membrane is retrieved by being pinched off at regions of the plasma membrane adjacent to the Schwann (glial) sheath (2). To facilitate retrieval, the pinched-off vesicles are provided with a special coat. The coated vesicles lose their coats and the coat fragments return to the membrane (3). The synaptic vesicles are then reused (4). [From Heuser and Reese, 1974.]

where $\bar{q}$ is the average amplitude of the unit synaptic potential (the synaptic potential produced when only one quantum is released) and m the average quantal content.

The quantal content m is a measure of the amount of transmitter released and is therefore a reflection of certain presynaptic properties: (1) the size of the presynaptic terminals; (2) the number of terminal branches of a single presynaptic fiber; and (3) possible alterations of quantal release associated with changes in synaptic efficacy. On the other hand, given that each packet of transmitter contains the normal number of transmitter molecules,[9] the average quantal size ($\bar{q}$) indicates the response of the postsynaptic membrane to a single quantum of transmitter. Quantal size therefore depends on properties of the postsynaptic cell, such as the input resistance (which can be independently estimated) and the sensitivity of the postsynaptic receptor to the transmitter substance.

Calcium Influx and Quantal Release

Katz and Miledi (1969a) propose a simple unifying hypothesis relating Ca^{++} influx to quantal release. They postulate that Ca^{++} increases the probability of release by facilitating the mobilization of vesicles into release sites. Spe-

[9]It is commonly thought that barring pharmacological interference, extreme use, or disease states, vesicles fill optimally. However, this assumption has been tested at only a few synapses.

cifically, depolarization of the terminal produced by the action potential opens up channels to Ca^{++} ions, which then move down their concentration gradient into the cell. Inside the cell Ca^{++} moves to critical release sites where it causes or facilitates a fusion of the vesicular membrane with the membrane of the axon terminal, leading to the increased release of quantal packets of transmitter. Katz and Miledi propose that Ca^{++} changes the properties of the terminal membrane so that it presents many more reactive discharge sites to the colliding vesicles and thereby raises the statistical chances for the release of transmitter quanta. According to this notion, an action potential can release a quantum from a given active site only if a critical number of Ca^{++} ions (one to four) has combined with a loaded release site. Magnesium antagonizes the Ca^{++} action (Rahamimoff, 1974).

Thus the membrane potential controls transmitter release by controlling Ca^{++} conductance. Because g_{Ca} is highly voltage-dependent, slight changes in membrane potential can have powerful control over transmitter release, and thus over synaptic effectiveness. At most synapses a 25 mV depolarization is necessary to increase intracellular Ca^{++} levels sufficiently to initiate release. Once this threshold is reached, the synaptic input-output relationship is quite steep because further depolarizations produce large additional increases in g_{Ca}. In some synapses even a small maintained depolarization of the terminals may partially inactivate g_{Ca}, whereas slight hyperpolarization may remove inactivation. Therefore, one way to alter synaptic effectiveness is to change the membrane potential of the presynaptic terminals (Figure 6-19). In this way a synaptic connection that was previously weak might become highly effective, and one that was effective might be made to stop transmitting.

Factors Controlling the Membrane Potential of the Presynaptic Terminals

The factors that control the membrane potential of presynaptic terminals and their importance for transmitter release have not yet been well analyzed. Both

FIGURE 6-19.

Hypothetical experiment summarizing some possible consequences of changes in the membrane potential of the presynaptic terminal on the amplitude of the postsynaptic potential.

A. Experimental arrangement with intracellular electrodes for recording in both the pre- and postsynaptic elements, as in Figure 6-15. A current-passing microelectrode is inserted into the presynaptic terminal to alter its membrane potential. Action potentials are initiated by stimulating the presynaptic fiber with extracellular electrodes. An electrode is inserted in the postsynaptic cell to record the EPSPs.

B-C. At the normal resting potential of the presynaptic terminal an action potential produces a PSP of about 20 mV. When the membrane potential of the presynaptic terminal is hyperpolarized by 5 mV the presynaptic spike produces a PSP of 40 mV, which triggers an action potential in the postsynaptic cell. When the presynaptic terminal is returned to the resting level and then depolarized by 5 mV, the resulting action potential produces a PSP of about 10 mV. Similar effects might be produced by synaptic actions that alter the conductance of the presynaptic terminals.

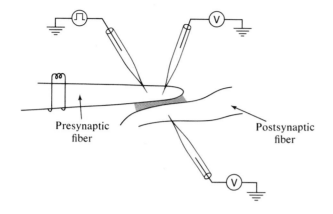

A

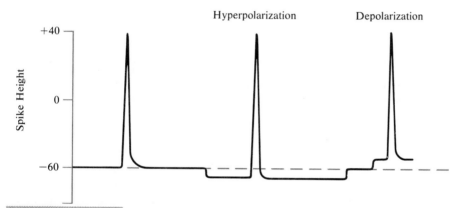

Hyperpolarization Depolarization

+40

Spike Height

0

−60

B. Presynaptic Fiber

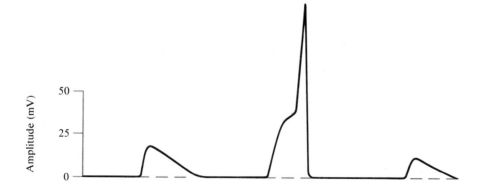

Amplitude (mV)

50

25

0

C. Postsynaptic Fiber

intrinsic (homosynaptic) and extrinsic (heterosynaptic) processes are thought to be important.[10]

Intrinsic processes include the summated afterpotentials of a train of spikes; these can make synaptic efficacy a function of recent activity in the presynaptic neuron itself (see p. 438). Extrinsic factors include neuronal activity from axons of other cells that synapse directly on the presynaptic terminals. The most compelling evidence for synapses on presynaptic terminals comes from studies at the lobster nerve-muscle synapse. Here an inhibitory axon makes two types of synapses: one on the muscle, producing conventional postsynaptic inhibition, and one on the presynaptic terminals of the excitatory axon. The presynaptic synapse on the terminals of the excitatory axon acts to depress transmitter release and thereby produce presynaptic inhibition. *Presynaptic inhibition* reduces the number of transmitter quanta released by the excitatory axon without affecting receptor sensitivity (Figure 6-20; see Dudel and Kuffler, 1961). At both pre- and post-synaptic sites the inhibitory axon releases the

[10]The terms "homosynaptic" and "heterosynaptic" are contractions of the older terms "homonymous synaptic" and "heteronomous synaptic" (Eccles, 1953). The terms reflect whether the changes in efficacy at a given synapse result from activity in *that* synaptic pathway (homosynaptic) or in a different, usually neighboring, synaptic pathway (heterosynaptic) that impinges upon the test synaptic pathway.

FIGURE 6-20.
Presynaptic inhibition. [From Dudel and Kuffler, 1961.]

A. Experimental arrangement with intracellular electrodes for recording from lobster abductor muscle. Individual excitatory and inhibitory axons are dissected and stimulated separately by means of extracellular electrodes.

B. Part *B-1* shows a single excitatory junctional potential (e.j.p.) of 2 mV, produced by stimulating the excitatory axon. Part *B-2* shows a single depolarizing inhibitory junctional potential (i.j.p.) of 0.2 mV, produced by stimulating the inhibitory axon. The i.j.p. is small and depolarizing because the membrane potential (86 mV) is slightly more negative than the reversal level for the i.j.p. (-80 mV). *B-3.* The inhibitory axon is stimulated 1.5 msec after the excitatory axon; as a result, the i.j.p. coincides with the peak of the e.j.p. and does not alter its amplitude. *B-4.* The inhibitory axon is stimulated 3 msec before the excitatory axon, with the result that the i.j.p. occurs before the e.j.p. Now the e.j.p. is greatly reduced due to presynaptic inhibition of the terminals of the excitatory axon by the inhibitory axon.

C. Same experiment as in part *B.* Graph of peak amplitude of e.j.p. and i.j.p. (i.j.p. is again very small) as a function of the interval separating stimulation of the excitatory and inhibitory axons. Control (at right) is the size of the e.j.p. without inhibitory stimulation. When the i.j.p. follows the e.j.p., it does not alter the e.j.p. amplitude; when the i.j.p. precedes the e.j.p. by 6 msec or less, it depresses the e.j.p.

D. Amplitude histograms of extracellularly recorded e.j.p.'s from a single junctional area (stimulus 5/sec). In *D-1* the junction is not inhibited. In *D-2* inhibitory impulses precede the excitatory ones by 2 msec; there is now an increase in the number of failures (from 72 to 570) and a decrease in quantal output (m) (from 2.4 to 0.56). The dashed line is the calculated theoretical (Poisson) distribution. Multiples of the unit potential are indicated by numbers above the peaks; the large arrow indicates the average size of the spontaneous miniature end-plate potentials. The position of the first peak, the unit potential, remains unaffected during the inhibitory action, while the quantal output decreases, indicating that the decrease in size of the e.j.p. (in parts *B-4* and *C*) is due to presynaptic rather than postsynaptic inhibition.

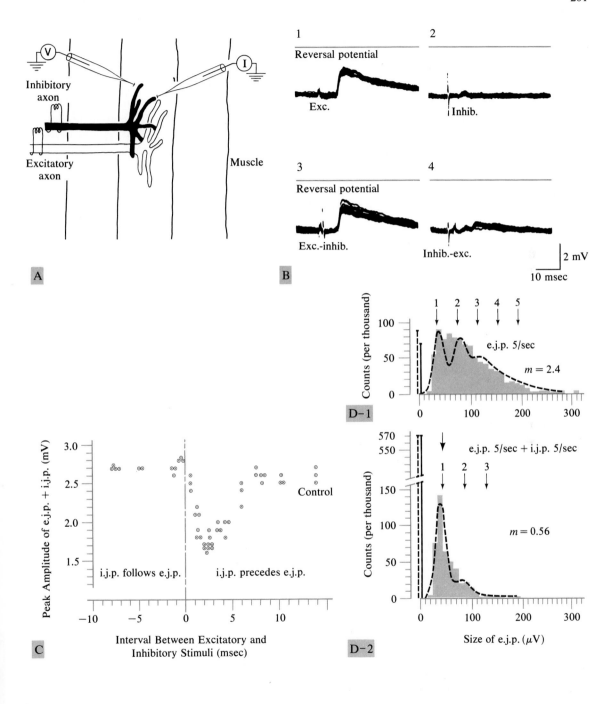

A — Inhibitory axon / Excitatory axon / Muscle

B
1
Reversal potential
Exc.

2
Inhib.

3
Reversal potential
Exc.-inhib.

4
Inhib.-exc.

2 mV
10 msec

C
Peak Amplitude of e.j.p. + i.j.p. (mV)

3.0
2.5
Control
2.0
1.5
i.j.p. follows e.j.p.   i.j.p. precedes e.j.p.

−10   −5   0   5   10

Interval Between Excitatory and
Inhibitory Stimuli (msec)

D-1
Counts (per thousand)
1 2 3 4 5
100
50
0
e.j.p. 5/sec
$m = 2.4$
0   100   200   300

D-2
Counts (per thousand)
570
550
e.j.p. 5/sec + i.j.p. 5/sec
1   2   3
150
100
$m = 0.56$
50
0
0   100   200   300

Size of e.j.p. (μV)

same transmitter substance, gamma-aminobutyric acid (GABA), which leads to an increased permeability to Cl$^-$ (Takeuchi and Takeuchi, 1966). In the terminals of the excitatory axon the increased g_{Cl} reduces the amplitude of the presynaptic spike; the concomitant inflow of Cl$^-$ neutralizes some of the positive charge carried in by Na$^+$ and thereby leads to a slightly smaller depolarization and, presumably, a smaller inflow of Ca^{++}. Presynaptic inhibition is also found in the early relays of the mammalian central nervous system, where it is associated with a depolarization of the presynaptic terminal that reduces the height of the action potential (Eccles, 1964). There is also evidence in *Aplysia* for *presynaptic facilitation*, an opposite type of extrinsic regulation by which activity in one pathway enhances the activity of another pathway (see p. 575).

Presynaptic interactions have the effect of making synaptic efficacy a function of recent (heterosynaptic) activity in anatomically related pathways (Figure 6-21).

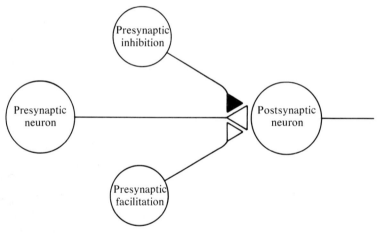

FIGURE 6-21.
Axo-axonal synapses that can act as governors mediating transmitter release.

CHEMICAL NATURE OF TRANSMITTERS AND RECEPTORS

Although the mechanisms of chemical transmission are remarkably similar in all neurons, nerve cells synthesize and release a variety of transmitter substances. Cells that synthesize one transmitter are thought not to release another. The transmitter substance that a neuron synthesizes, packages, and releases is thus likely to be a specific, differentiated property of the cell. It is therefore somewhat surprising, given the large number of nerve cells, that perhaps as few as a dozen substances may serve as transmitters. In invertebrates the transmitter roles of acetylcholine (ACh), dopamine, 5-hydroxytryptamine (5-HT) or serotonin, octopamine, GABA, and glutamate have been well established. In vertebrates there is good evidence for two additional substances, norepinephrine and glycine. These substances are simple compounds; some of them (glycine, GABA, glutamate) are amino acids, others are closely related substances (Table 6-2).

TABLE 6.2.
Structure and metabolism of common transmitter candidates.

Compound	Structure	Synthesis	Inactivation
Acetylcholine (ACh)	$H_3C\ \overset{O}{\overset{\|}{C}}-O-(CH_2)_2\ \overset{+}{N}\ (CH_3)_3$	Choline acetylation	Hydrolysis by AChE
Dopamine	HO— (ring) —CH_2CH_2—NH_2, HO—	Tyrosine hydroxylation, dopa decarboxylation	Reuptake, MAO, COMT
Norepinephrine	HO— (ring) —CH(OH)—CH_2—NH_2, HO—	Dopamine-β-hydroxylation	Reuptake, oxidative deamination by MAO, 3-O-methylation by COMT
5-Hydroxytryptamine (5-HT), or serotonin	HO— (indole ring) —CH_2CH_2—NH_2	Tryptophan hydroxylation, 5-OH-tryptophan, decarboxylation	Reuptake, MAO
Histamine	(imidazole ring)—CH_2CH_2—NH_2	Histidine decarboxylation	N-methylation by histamine-N-methyl transferase; oxidative deamination (MAO ?)
Excitatory amino acids, e.g., glutamate, aspartate	$CH_2\ (CH_2)_n\ CH\ NH_2$ $\|$ $\|$ COOH COOH $n = 0\text{-}1$	————	Reuptake, decarboxylation, NH_3 fixation
Inhibitory amino acids, e.g., GABA, glycine	$CH_2\ (CH_2)_n\ NH_2$ $\|$ COOH $n = 0\text{-}4$	GABA by glutamate decarboxylation	GABA—reuptake transamination and oxidation to succinate

SOURCE: Mountcastle and Baldessarini, 1974.

Less is known about the chemical properties of receptor molecules (Figure 6-22). The best information available is on the ACh receptor, where experiments have been greatly aided by the discovery that a venom (α-bungarotoxin) released by elapid snakes selectively binds to the receptor. With the help of this toxin (and other techniques) the ACh receptor of various nerve-muscle synapses has been purified and shown to be a protein with a molecular weight of about 200,000 daltons. The protein has several subunits; the one that contains the presumed transmitter binding site has a molecular weight of about 40,000 daltons. There are thought to be about 12,000 to 30,000 ACh binding sites per μm^2 in regions of the postsynaptic cell that have the highest transmitter sensitivity (Klett, Fulpius, Cooper, and Reich, 1974; Karlin, 1974; Michaelson and Raftery, 1974).

A key question in analyzing the structure and function of receptors is how the receptor molecule's recognition and binding of transmitter relates to its control of ionic permeability. These two functions are often separated conceptually: the recognition and binding of transmitter is ascribed to a *receptor site* and ionic permeability to an *ionophore* (Figure 6-22B). Whether one component controls both functions or whether the two functions represent different components is generally not known. Preliminary studies suggest that a purified macromolecule containing the receptor site for ACh may also contain the ionic channels necessary for the conductance change (Michaelson and Raftery, 1974).

SUMMARY AND PERSPECTIVE

The membrane potential, the action potential, the excitatory and inhibitory synaptic potential, and the generator potential of sensory cells can all be explained by means of a unified theory, the ionic hypothesis, which applies to all nerve cells. According to this hypothesis, the membrane, action, and synaptic potentials all involve the passive movement of inorganic ions down their concentration gradient.

All nerve cells maintain across their surface membrane a potential of about -40 to -60 mV, called the resting or membrane potential. This potential results from unequal distribution across the membrane of K^+ ions. An energy-dependent Na^+-K^+ exchange keeps the K^+ concentration inside a nerve cell about 20 times higher than outside, and Na^+ concentration inside the cell approximately 10 times lower than outside. The value of the membrane potential is usually close to the equilibrium potential for K^+ (E_K), which is typically -75 mV. Small deviations from E_K are usually due to a slight leakage of the membrane to other ions, particularly to Na^+.

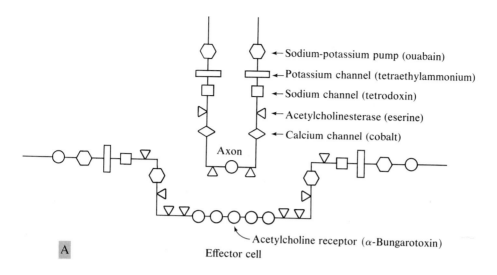

Sodium-potassium pump (ouabain)
Potassium channel (tetraethylammonium)
Sodium channel (tetrodoxin)
Acetylcholinesterase (eserine)
Calcium channel (cobalt)

Axon

Acetylcholine receptor (α-Bungarotoxin)

A

Effector cell

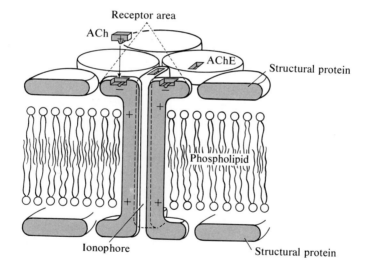

Receptor area

ACh

AChE

Structural protein

Phospholipid

B

Ionophore

Structural protein

FIGURE 6-22.

A. Specific macromolecules that can be identified in the surface membranes of the presynaptic terminal and postsynaptic cell at a cholinergic synapse. Each of these molecules can be bound by a specific block-ing agent, indicated in parentheses. [From A. Karlin, unpublished.]

B. Macromolecular organization of the surface membrane of the postsynaptic receptor area. The basic structure of the nonsynaptic membrane is a biomolecular leaflet of phospholipid molecules that is stabilized by structural proteins. Enzymes, receptor sites, and other protein molecules are applied to either the inner or outer surfaces. Other protein molecules, such as the channels for Na^+, K^+, and Ca^{++}, penetrate the entire lipid plane of the membrane. The synaptic channel is controlled by receptor sites for ACh. [Modified from de Robertis, 1971. Copyright by the American Association for the Advancement of Science.]

When the membrane potential is made less negative by a critical amount, usually by about 15 mV (from -60 to -45 mV), an action potential is initiated. During the action potential there is a sudden change in the relative permeability characteristics of the membrane—it becomes relatively more permeable to Na^+ than to K^+, presumably as a result of sudden increased availability of previously closed Na^+ channels. The resulting potential change is regenerative. Depolarization increases Na^+ permeability, resulting in an inrush of a small quantity of Na^+, which in turn produces a further depolarization that increases Na^+ permeability even further. This explosive event reverses the membrane potential, driving it toward $+55$ mV, the Na^+ equilibrium potential (E_{Na}). The sudden reversal of the membrane potential is transient and self-limiting. The progressive depolarization of the action potential ultimately leads to a shutting off or inactivation of the enhanced Na^+ permeability of the membrane. The decrease in Na^+ permeability is accompanied by an increase in the membrane's already high permeability to K^+. These two processes combine to bring the membrane potential back to the resting value.

The ionic hypothesis for the membrane and action potentials was quantified and extended by measuring the ionic currents that flow through the membrane in the squid giant axon. These studies used the voltage clamp, which permits changing the membrane potential rapidly to any desired level and maintaining it there. The current that a feedback amplifier must supply to maintain the membrane potential at a given level equals the total ionic current that flows at that membrane potential. It is possible to separate the total current into component currents carried by Na^+ or K^+, by selectively eliminating the current carried by the other ion species. This can be done by ion substitution, or by using pharmacological agents (tetrodotoxin and tetraethylammonium) that selectively block one or the other ion channels. Knowing the currents, Hodgkin and Huxley were able to describe how the Na^+ and K^+ conductances (g_{Na} and g_K) vary with membrane potential and with time during the action potential. Using a kinetic model, they demonstrated that the conductances measured with the voltage clamp quantitatively account for the action potential, the afterpotential, and impulse propagation.

Another remarkable feature of the ionic hypothesis is its ability to account for chemically mediated excitatory and inhibitory synaptic actions. The external surface of the postsynaptic membrane of nerve and muscle cells contains receptor molecules that control the ionic channels for excitatory and inhibitory synaptic actions. Synaptic potentials can be produced either by increasing or decreasing ionic conductances. Most synaptic actions involve mechanisms that increase ionic conductance. Thus, most excitatory postsynaptic potentials (EPSPs) depolarize and excite because the reaction of the transmitter with the receptors leads to a sudden increase in g_{Na} and, in many instances, a simultaneous increase in g_K. Because g_{Na} and g_K increase simultaneously, the reversal potential for excitation (E_{EPSP}) is usually located about midway between E_{Na} (+55 mV) and E_K (−75 mV), typically at −10 mV. As a result, excitatory synaptic actions tend to drive the membrane potential of the trigger zone from its resting value (−60 mV), past threshold (−45 mV), in the direction of the E_{EPSP}, −10 mV. In some cells the nonsynaptic membrane is in parallel with a high g_K, which tends to keep the membrane potential hyperpolarized. In these cells EPSPs can be generated by a reduction of g_K.

Inhibitory postsynaptic potentials (IPSPs) inhibit by keeping the membrane potential of the trigger zone from reaching the threshold for spike generation. They usually do this by both increasing membrane conductance and hyperpolarizing the membrane potential. The reaction of the transmitter with the receptor leads to a sudden increase in the membrane permeability to K^+ or Cl^-. The equilibrium potential for Cl^- is usually −65 mV, that for K^+ −75 mV. In most cells the equilibrium potential for either of these ions is more negative than the resting potential (−60 mV). In some cells E_m is partially shunted by a high g_{Na}; IPSPs can be generated in these cells by turning off the g_{Na}.

The release of the synaptic transmitter substance is quantized. The synaptic potential is made up of small, elementary (quantal) units, each containing several thousand transmitter molecules. The presynaptic terminals of neurons contain small packets called synaptic vesicles. These organelles are thought to store the transmitter substance. Morphological studies suggest that during transmitter release the membrane of the synaptic vesicle fuses with the external membrane of the presynaptic terminal. The fused membranes open into the synaptic cleft, causing the vesicles to discharge their contents in a quantal, all-or-none fashion by a process called exocytosis. Quantal analysis of synaptic transmission serves as a bridge between electrophysiological, morphological and molecular mechanisms by providing a means for assaying, electrophysiologically, the release of multimolecular packets of chemical transmitter substance.

The process by which transmitter release occurs may be outlined as follows. Depolarization of the terminal by the action potential increases g_{Ca} so that Ca^{++} moves down its concentration gradient into the cell. Inside the cell Ca^{++} is thought to move to release sites where it presumably facilitates fusion of the vesicular membrane with the membrane of the axon terminal, leading to exocytosis and the release of the transmitter quanta. According to this notion, an action potential can release a quantum from an active site only if a critical number of Ca^{++} ions have combined with a loaded release site.

Changes in membrane potential can affect transmitter release, presumably by controlling Ca^{++} permeability. Because Ca^{++} permeability is highly voltage-dependent, slight changes in the membrane potential can exercise powerful control over transmitter release, and hence over synaptic effectiveness. Two types of processes are thought to govern the membrane potential of the terminals: (1) intrinsic or homosynaptic factors, such as the summated afterpotentials of a train of spikes, which makes synaptic efficacy a function of recent activity of the neuron; and (2) extrinsic or heterosynaptic factors, such as the axon terminals from other cells that synapse directly on the presynaptic terminals. These synapses on the presynaptic terminals can control membrane potential and permeability of the terminals and thereby depress or facilitate transmitter release, producing presynaptic inhibition or facilitation. Presynaptic interactions have the effect of making synaptic efficacy a function of recent activity in anatomically related pathways.

The detailed mechanisms of release are still poorly understood. Definitive evidence that transmitter is released from vesicles is lacking, although it would be surprising if vesicles were not the releasing agents. There is much more uncertainty, however, about how the vesicle moves to the discharge sites and combines with the membrane for release. Most models of release contain two compartments a readily releasable pool of transmitter and a storage pool, but little is known about their respective morphological sites. The best guess is that the readily releasable pool consists of vesicles that are already in contact with the presynaptic terminal membrane, perhaps those vesicles that are already located in release sites. The storage pool may then be any remaining vesicles in the terminal that can be moved into release sites. Finally, nothing is known about how patterned activity affects these pools over a long term. Techniques are needed that will permit distinct anatomical structures to be identified with the three physiological concepts—readily releasable pool, storage pool, and mobilization—so that they can be studied as a function of different patterns of presynaptic activity.

SELECTED READING

Eccles, J. C. 1964. *The Physiology of Synapses,* Springer. A detailed review of early intracellular studies of synaptic transmission.

Ginsborg, B. L. 1967. Ion movements in junctional transmission. *Pharmacological Reviews,* 19:289–316. A description of ionic mechanisms, illustrating the types of current flow produced by various increased-conductance synaptic actions.

Hodgkin, A. L. 1964. *The Conduction of the Nervous Impulse,* Liverpool University Press. A brief and simple account of the ionic hypothesis and electrical conduction in the axon. A must for students.

Katz, B. 1966. *Nerve, Muscle, and Synapse,* New York: McGraw-Hill. A clear account of axonal conduction and synaptic transmission. Another must.

Katz, B. 1969. *The Release of Neural Transmitter Substances,* Springfield: Thomas. A review of transmitter release.

Martin, A. R. 1976. Junctional transmission, II: Presynaptic mechanisms. In "Cellular Biology of Neurons" (Vol. 1, Sec. 1, *Handbook of Physiology, The Nervous System*), edited by E. R. Kandel, Baltimore: Williams and Wilkins. An up-to-date survey of presynaptic factors in release.

CHAPTER 7

Identified Neurons

The first major question confronted in the cellular study of invertebrate central nervous systems was whether the nerve cells in one region of the brain differ from one another. This question, which is fundamental to an understanding of how behavior and learning are mediated in the nervous system, precipitated a lively and persistent dispute.

When computer technology was first applied to biology, in the years following World War II, the similarities between the computer and the brain caught the imagination of both biologists and mathematicians. Like the brain, the computer can discriminate, learn, and remember, yet these remarkable performances require only a small variety of components. Flexibility is achieved by wiring the same components together in a great many ways; different results can even be achieved simply by combining varying numbers of the same type of component. Some mathematically oriented neurophysiologists thought that the brain might function in an analogous manner.[1] They argued that the nerve cells of the brain might have sufficiently similar properties to be thought of as identical elements, having interconnections of roughly equal value.

On the surface this idea seemed consistent with the model neuron emerging

[1]Pitts and McCollough, 1947; Beurle, 1956; Rosenblatt, 1958; Culbertson, 1963; Winograd and Cowan, 1963.

from initial intracellular studies of peripheral axons and central neurons (see Figure 5-6, p. 109). One implication of this model was that despite striking differences in size, shape, location, and function, all neurons shared important common properties. By neglecting small differences between neurons, large parts of the nervous system, particularly the cerebral cortex and some subcortical structures, could be treated as networks of identical elements. Their function could then be described by mathematical laws which emphasized the probabilistic features of neuronal functioning and conceived of behavior as the result of changes in the statistical distribution of activity in networks of similar cells.[2]

Such a bulk or probabilistic approach was also consistent with the *aggregate-field* view of brain function developed by the physiological psychologist Karl Lashley (1929), whose work dominated thinking on the relation of brain and behavior in the period between 1930 and 1955. The thrust of Lashley's work was a challenge to *cellular connectionism,* the idea, derived from Ramón y Cajal, that the neuron is the basic signaling unit of the brain and that behavior is mediated by specific and localizable synaptic connections between sensory and motor structures. Lashley examined maze learning in rats following cortical ablations and found that the removal of any specific cortical area did not have any specific effect on the animals' relearning of the maze. Animals did as well with one part of the cortex removed as with another. What seemed to matter was not which *area* but how *much* of the cortex was removed. Based on these findings, Lashley postulated two laws: *mass action* and *equipotentiality.* The law of mass action stated that learning is dependent on the amount of functioning cortex rather than on the integrity of specific cortical areas. The law of equipotentiality stated that (with some exceptions, such as pattern discrimination) one part of the cortex is equal to another in its importance for maze learning. Lashley therefore emphasized the whole as more than the sum of its parts, and minimized the significance of individual neurons and of specific neuronal connections. What was important in Lashley's aggregate-field view was brain *mass,* not specific neuronal architecture.

Lashley's results represented a serious challenge to the idea that the cortex, like the spinal cord, mediated behavior on the basis of specific interconnections. The results even raised slight doubts about the organization of the spinal cord. E. G. Boring summarized this quandary in 1950 in his influential *History of Experimental Psychology* (pp. 687–688):

[2]Ashby, 1952; von Neumann, 1956; Beurle, 1956; Winograd and Cowan, 1963; Rosenblatt, 1958; Culbertson, 1963; John, 1967.

> It is plain that connectionism—explanations in terms of synaptic connections of fibers—works, in the peripheral nervous system, largely in the spinal cord, and probably at the lower brain levels. In the cortex we do not know . . . . The truth about how the brain functions may eventually yield to a technique that comes from some new field remote from physiology or psychology.

Coming as it did both from a remote field, mathematical biology, and in the wake of Lashley's challenge to cellular connectionism, the probabilistic view of brain function had a great and persistent impact on psychology (for current restatements see John, 1972; Adey, 1972). Even in the recent edition of their thoughtful book *Theories of Learning* (1975, pp. 548–49) Hilgard and Bower write:

> . . . Appropriate networks of "model neurons" can be easily designed that display rather amazing discriminative and performance capabilities. . . . Orderly discriminative performance can be produced (through reinforcement) by a large network that begins with completely random interconnections.
> . . . From the viewpoint of many neurophysiologists, the nerve-net theories are held to ignore too much of what is known about the neuron and the structure of the nervous system. Be that as it may, it seems obvious that if neurophysiology is to have an explicit theory capable of making contact with behavioral data, the form will be little different from the kind of scheme presently being designed by those working on nerve-net automata.

In recent years, however, Lashley's findings have been reinterpreted and the probabilistic conception of the brain has been shown to be inadequate. A variety of studies has demonstrated that maze-learning, used by Lashley in his tests, is not a suitable task for localizing brain function because it involves several motor and sensory capacities. Deprived of one, an animal can still learn the maze with the others. In addition, Lashley's experiments were carried out before the precise topography of the sensory and motor areas of the neocortex had been mapped, so that his lesions were not restricted to a single functional region. Finally, Lashley lesioned only the neocortex. As more specific tests were developed and callosal connections and subcortical areas were explored, specific functions could often be precisely localized (Myers and Sperry, 1958; McCleary, 1961; Segaar, 1962; Thompson, 1969).

Another powerful rebuttal of Lashley's challenge to the importance of cellular connections came from developmental studies of the nervous system, and specifically from the work of Roger Sperry (1951). Sperry examined the formation of synaptic connections in the brains of lower vertebrates during regeneration following lesions. He found that visual perception and motor coordination are reestablished because the nerve fibers regenerate and apparently reform their original connections in a highly specific fashion. The cells invariably connect only with certain cells and never with others. A similar picture emerges from studies of the neocortex of the cat and monkey. Vernon Mountcastle, studying the somatosensory system, and David Hubel and Torsten Wiesel, in their researches on the visual system, have found that the sensory cortexes of mammals are also organized in highly structured and precise ways, with different cell types having different response properties (Mountcastle, 1957; Hubel and Wiesel, 1965).

Moreover, as a greater variety of nerve cell bodies was studied, subtle but important differences between neurons were found. Although the differences were usually not present in the axon, they could not be neglected in any realistic discussion of neuronal circuitry. The cell bodies of neurons were found to vary in their intrinsic properties—membrane potential, action potential, firing pattern, morphology, and transmitter biochemistry. As a result, it now appears that many, perhaps most, neurons may prove sufficiently distinct from one another to be unique individuals (Sperry, 1963, 1965b).

Evidence for the uniqueness of neurons has come from studies of both vertebrate and invertebrate nervous systems. The strongest single finding in support of the concept of any brain consisting of unique cells was the discovery of *identifiable neurons* in invertebrates. These cells are sufficiently distinctive that each can be reliably recognized in every individual of a species. The fact that some neurons in invertebrates are unique and identifiable does not, of course, guarantee that each neuron in the cerebral cortex of mammals will prove a distinct individual. But it does prove that some brains are constructed at least in part from unique nerve cells. In addition to providing strong support for the idea that the nervous system is organized on the basis of specific cells with distinct properties, identified cells allow a precise study of the relation of single nerve cells to behavior.

In this chapter I shall first relate the history of the identification of nerve cells and then describe the diversity that exists among neurons of a single ganglion. Finally, I shall consider the mechanisms that account for neuronal diversity. The next chapter will describe studies of the patterns of interconnections between identified cells based on the specificity of their synaptic interrelationships.

IDENTIFICATION OF NEURONS

In 1836 the German physiologist Christian Gottfried Ehrenberg described the giant fibers ("colossal fibers") of crustacea and gave the first histological account of what later became recognized as an identified nerve-cell process: the lateral and medial giant fibers of the crab. The idea that certain nerve cells have an invariant location in the nervous system and can be repeatedly identified in all members of the same species came from studies of nematode and annelid worms. The idea probably originated with F. Leydig, who in 1864 described the giant fiber system ("huge dark-rimmed fibers") of the nerve cord of the earthworm *Lumbricus* and the paired giant nerve cells ("ganglion balls") in the segmental ganglia of the leech (Leydig, 1864, 1886). He noted the distinctive structure and invariant position of these nerve cells in the leech, one on each side of each segmental ganglion. These findings were confirmed by Gustaf Retzius (1888, 1891), who also described an even larger pair of cells, the "colossal cells," later named after him by Kuffler and Potter (1964; see Figure 7-1). Invariant cells were soon discovered in other annelid worms (Kükenthal, 1887; Friedländer, 1888; Bürger, 1890).

Invariant cells were also discovered among vertebrates. In 1859 L. Mauthner described giant axons in the spinal cords of teleost fish. In 1888 W. Goronowitsch discovered the remarkably shaped giant cell bodies of these axons in the medulla. This completed the description of the symmetrical neurons we now call Mauthner's cells, the most prominent and best-studied vertebrate identified neurons (Diamond, 1971).

The finding that at least some cells are identifiable soon raised a different and more radical possibility. In 1897 Istvan Apáthy suggested that in some invertebrates the entire nervous system was invariant; regardless of the size of the animal there was always the same number of nerve cells. Apáthy examined the abdominal ganglia of both young and fully grown leeches and found that the number of cells ranged narrowly between 350 and 400. Apáthy also confirmed Leydig's and Retzius's descriptions of the remarkable constancy of the cell bodies and processes of individual ganglion cells.

The first rigorous demonstration that a large part of a nervous system can be described in terms of an invariant number of uniquely identifiable cells was accomplished by Richard Goldschmidt in a study of the rostral ganglia of the nematode worm *Ascaris* (Figure 7-2). In his famous lecture to the German

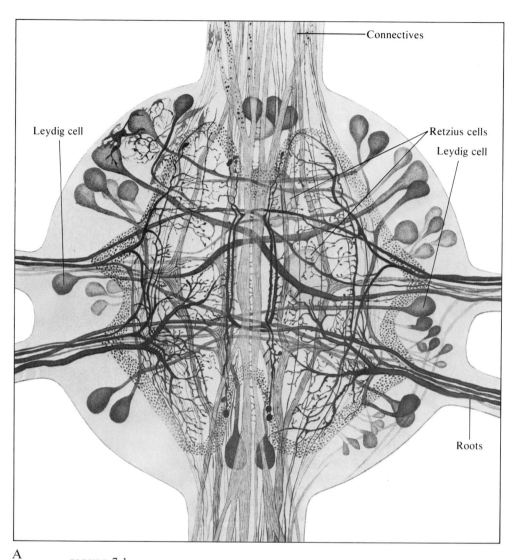

Connectives

Leydig cell

Retzius cells

Leydig cell

Leydig cell

Roots

A

FIGURE 7-1.
Retzius's drawings of the identified cells in a segmental ganglion of an annelid worm, the leech, based on a methylene blue preparation of the fourth abdominal ganglion. **A.** Dorsal surface illustrating the position of the two giant ("colossal") cells identified by Retzius (and later named after him by Kuffler and Potter) and the cells identified by and named after Leydig. **B.** Detailed drawing of the two Retzius cells. [From Retzius, 1891.]

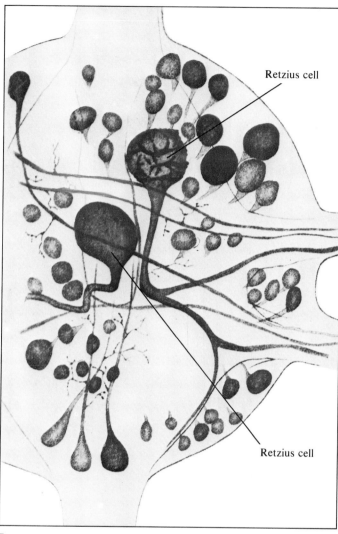

Retzius cell

Retzius cell

B

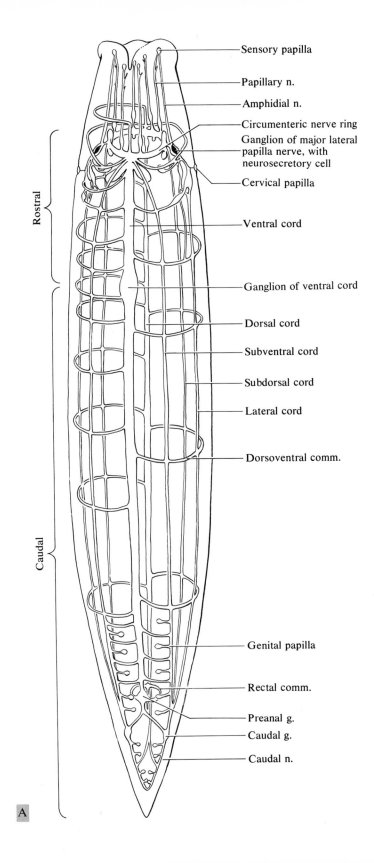

Sensory papilla

Papillary n.

Amphidial n.

Circumenteric nerve ring

Ganglion of major lateral papilla nerve, with neurosecretory cell

Cervical papilla

Ventral cord

Ganglion of ventral cord

Dorsal cord

Subventral cord

Subdorsal cord

Lateral cord

Dorsoventral comm.

Genital papilla

Rectal comm.

Preanal g.

Caudal g.

Caudal n.

Rostral

Caudal

A

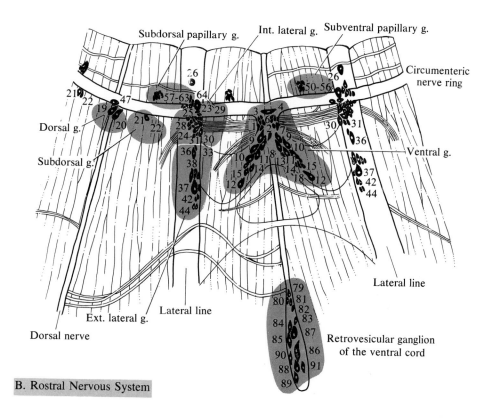

B. Rostral Nervous System

FIGURE 7-2.

The central nervous system of the nematode worm *Ascaris lumbricoides*, illustrating the neurons identified by Goldschmidt (1908). [Modified from Bullock and Horridge, 1965.]

A. Plan of the nervous system, illustrating the five longitudinal cords (ventral, subventral, dorsal, subdorsal, and lateral) and the horizontal commissures. The nervous system can be roughly divided into a rostral portion, containing 162 cells, and a caudal portion, containing about 40 cells.

B. Rostral portion of the nervous system, illustrating the identified cells. The esophagus and lips have been removed and the cylindrical body wall has been incised and unrolled. Most of the identified cells are marked with their identifying number.

Zoological Association in 1907 Goldschmidt spelled out his extraordinary findings and their implications (pp. 130–31):

> The resolution of the many questions that have come up through the last years in relation to the more detailed study of the nervous system has been made especially difficult because of the extremely complicated construction of the nervous system. Despite many single findings of research, this complexity has not allowed us—even in a single example—an insight into its function. Such an insight can only be achieved if one could take a simple organism and study its many components completely; that is, all the ganglion cells, nerve fibers and connections which make up its nervous system. This goal is not as impossible as it seems; I contend that I have almost succeeded. The preparation I have used is the nervous system of *Ascaris lumbricoides,* which presents perhaps the only possibility for this type of research due to the small number of elements of which it is composed, the isolated condition of its ganglia, and the known peculiarity of the nematodes, whereby the muscle goes toward the nerve in order to get its innervation. The results of this research will be published soon, with the illustrations that are necessary to understand it properly. Here in my talk today I will only discuss a few of the points of the coming paper.
>
> One point is the almost startling constancy of the elements of the nervous system. *There are 162 ganglion cells in the center, never one more nor less.* From each of these cells emerge very distinctive processes which connect in a very typical way with others. The constancy extends itself to the relative size of cells, to details such as the angles at which the axons leave the cell body, and to the position of the nucleus in the cell body. It must, without a doubt, have a meaning for the law of the specific ganglion cell function.
>
> A further interesting point is the symmetry of the cells in the nervous system. Every cell on the left half of the body is repeated in the same place on the right side. An exception is some cells located exactly in the midline and also the two cells of the abdominal ganglion which exist only on the right side of the body.

During the time that Goldschmidt described his findings on the invariance of cells in the nervous system (1906–1912), Martini (1908) published a series of studies on nematodes and related worms. He found that in a given species the cells of most larval organs—epidermis, muscles, esophagus, gut, rectum, sexual apparatus—were invariant as to number, grouping, and shape. In some instances this invariance persisted into adulthood. Martini recognized this finding as the "crowning of the determined development" that makes it possible "to speak of a homology of cells and of an exact cellular comparative anatomy."

It soon became clear, however, that if a nervous system had some invariant cells this did not mean that the entire system was invariant. Invariant neurons are found in a large number of vertebrates and invertebrates, but only in some worms is the total number of nerve cells invariant. In the ganglia of most invertebrates the number of cells increases during postembryonic development. Occasionally, the number of cells may decrease prior to death. The increase in number of cells often involves only certain types and certain regions of the brain. Other cell types remain fixed in number throughout postembryonic life. The stimulus for cell division and the factors that permit the division of some cell types but not others (or new differentiation from the stem cells) are not known, but there is a correlation between an increase in cell number and increased functional demands. Some cell types, often large cells, may never vary in number because they never experience demands for functional elaboration, or if they do they respond by enlarging and undergoing DNA replication but not cellular replication. Other, usually smaller, cell types respond to increased demand by replicating, or their stem cells are stimulated to differentiate; these cells may not be invariant. (For a review of studies of changes in the number of nerve cells during development see Kandel and Kupfermann, 1970.)

IDENTIFIED NEURONS IN GASTROPODS

In 1877 Ihering described giant neurons in gastropods, and in 1894 de Nabias described as invariant the cell bodies of the symmetrical dorsal and ventral metacerebral giant cells of gastropod molluscs. De Nabias found the *metacerebral cells* in various species of *Helix* and in other pulmonates and took them as evidence of a perfect symmetry in the neuronal architecture of gastropods. Although Mazzarelli (1893) and Bottazzi (1899) described the large size of the cell bodies of the neurons of *Aplysia* at the beginning of the twentieth century, it was not until 1942 that Angelique Arvanitaki and her collaborators described unique cells in this genus.

Arvanitaki and Tchou (1942) described seven cells on the basis of the position of their cell bodies (Figure 7-3*A*). Later, in 1958, Arvanitaki and

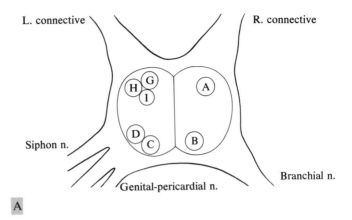

A

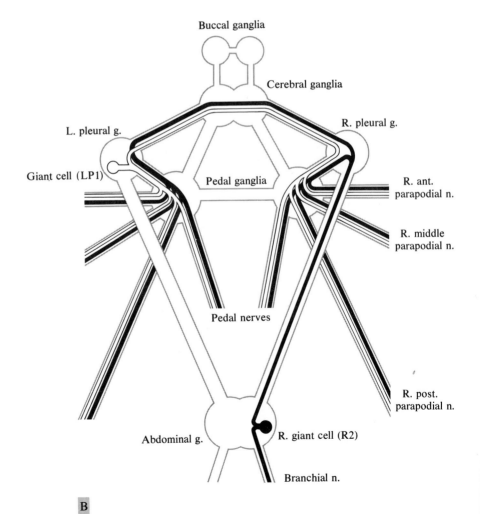

B

Chalazonitis examined the firing patterns of three of these cells and found that, in addition to the location of the cells, the firing pattern of each was also unique. This discovery had two important results. First, it added a functional criterion for identifying cells, which proved useful because position is by itself often insufficient for recognizing a cell repeatedly. Second, the fact that the cells were found to have intrinsically different firing patterns indicated that a surprising degree of functional diversity may exist in the neurons of a single ganglion.

In 1963 Hughes and Tauc mapped the pathways of the efferent axons of two identified cells by stimulating the connectives and peripheral nerves and found that these pathways were invariant (Figure 7-3B). Kandel and Tauc (1966a) developed still other criteria for invariance in studies of the giant metacerebral cells of *Helix* (Figure 7-4). In addition to position, efferent pathway, and firing pattern, they described the biophysical properties of the neurons and the physiology and pharmacology of the synaptic connections made on these cells by spontaneously firing interneurons and by afferent inputs activated by stimulating nerves and connectives. Finally, they found that the anatomical symmetry, noted by de Nabias (1894), was functionally preserved. The processes of the two cells were symmetrically distributed to peripheral nerves and connectives (Figure 7-4B) and the synaptic convergence on the cells was parallel. In each case the laterality of the input was signaled by latency and by synaptic efficacy: the ipsilateral input produced EPSPs that were consistently more effective and of shorter latency than the contralateral ones (Figure 7-4B, C). The two symmetrical cells also had a similar and invariant biophysical property (anomalous rectification). Both cells responded identically to acetylcholine and some of the synaptic input appeared to be cholinergic.

FIGURE 7-3.
Identified cells in *Aplysia*. **A.** First map of identified cells in the abdominal ganglion of *Aplysia* (*A. fasciata*). [Modified from Arvanitaki and Tchou, 1942.] **B.** Schematic drawing of the major central ganglia of *Aplysia* indicating the invariant axon pathway of the two identified giant cells: R2, in the abdominal ganglion, and LP1, in the left pleural ganglion. The axon pathway of R2 is indicated by the dark lines, the axon pathway of LP1 by the light lines. Despite the fact that the two cells occur in different ganglia, similarities in the axon pathways and physiological properties suggest that the cells may have been symmetrical at an earlier phase of development. [Modified from Hughes and Tauc, 1961.]

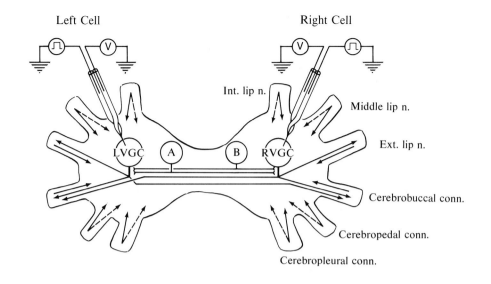

Left Cell

Right Cell

Int. lip n.

Middle lip n.

Ext. lip n.

LVGC   A   B   RVGC

Cerebrobuccal conn.

Cerebropedal conn.

Cerebropleural conn.

A

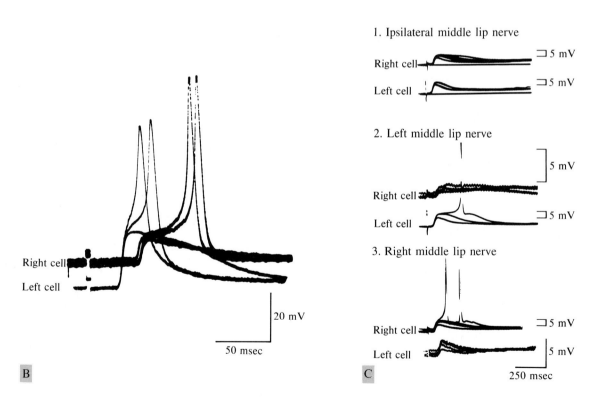

Right cell

Left cell

20 mV

50 msec

B

1. Ipsilateral middle lip nerve

Right cell 5 mV

Left cell 5 mV

2. Left middle lip nerve

Right cell 5 mV

Left cell 5 mV

3. Right middle lip nerve

Right cell 5 mV

Left cell 5 mV

250 msec

C

FIGURE 7-4.

Symmetrical identified cells in the cerebral ganglia of the pulmonate snail *Helix*.

A. A schematic drawing of the main features of the input-output organization of the left and right (ventral) metacerebral giant cells (LVGC and RVGC) of *Helix* and the experimental arrangement for studying them. A and B are two unidentified metacerebral cells. Outgoing arrows indicate axon pathways of the metacerebral cells; incoming arrows indicate afferent synaptic input to the metacerebral cells. Solid arrows indicate ipsilateral input, dashed arrows contralateral input.

B. Symmetry of the axon pathways of the two metacerebral cells. Simultaneous recordings from the left and right metacerebral cells showing backward (antidromic) invasion of both cells following stimulation of their axons in the left cerebrobuccal connective. The two cells respond with different latencies and action-potential configuration (see Figure 7-23, p. 252, and related discussion in the text about factors determining the configuration of the antidromic action potential). The ipsilateral (left) cell responds with a shorter latency than does the contralateral (right) cell. Several oscilloscope sweeps have been superimposed to illustrate the inflection in the ascending limb of the spike and the two components in the antidromic action potential.

C. Symmetry of the afferent input. Recordings from left and right cells illustrate the types of synaptic input produced in each cell following stimulation of either the ipsilateral or contralateral middle lip nerve. *C-1* illustrates the similarity in each metacerebral cell of the latency and amplitude of the input from each cell's ipsilateral middle lip nerve (separately recorded). In *C-2* and *C-3* simultaneously recorded responses of the two cells to stimulation of a middle lip nerve are compared. In *C-2* the left middle lip nerve is stimulated; this nerve is ipsilateral to the left cell and contralateral to the right cell. In *C-3* the right middle lip nerve is stimulated; it is ipsilateral to the right cell and contralateral to the left. In each case the ipsilateral nerve produces EPSPs of shorter latency and greater amplitude than does the contralateral nerve. [From Kandel and Tauc, 1966a.]

IDENTIFIED NEURONS IN *APLYSIA*

Abdominal Ganglion

Because occasional cells in the ganglia of gastropod molluscs could be identified as unique on the basis of only a few criteria, it became apparent that by using several criteria one might be able to identify many more cells in a ganglion. Using eight criteria—position, spontaneous firing pattern, efferent axon pathway, pattern of spontaneous postsynaptic potentials, synaptic response to stimulation of nerves and connectives, response to iontophoretically applied acetylcholine, fine structure, and direct synaptic connections with other identified cells—Frazier, Kandel, Kupfermann, Waziri, and Coggeshall (1967) identified almost all of the very large cells (30 cells) and eight prominent cell *clusters* visible on the dorsal and vental surfaces of the abdominal ganglion of *Aplysia* (Figure 7-5). Clusters are groups of undetermined numbers of cells that have similar properties and cannot be immediately distinguished from one another (Figure 7-6). But by introducing an additional criterion, the behavioral function of the cell, it becomes possible to further distinguish, as

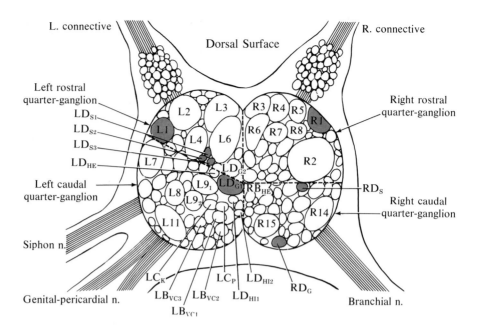

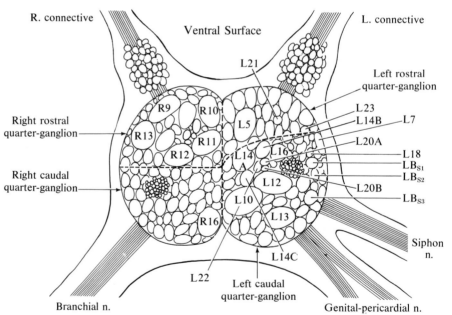

FIGURE 7-5.

Map of identified cells in the abdominal ganglion of *Aplysia californica* indicating the most common positions of the identified cells. The identified cells are labeled L or R (left or right hemiganglion) and assigned a number. The hemiganglia are arbitrarily subdivided into quarter-ganglia. Cells that share similar properties generally have similar labels, e.g., $L9_{G1}$ and $L9_{G2}$, and L14A, L14B, and L14C. Cells that are members of clusters are identified by the cluster name and a subscript identifying the behavioral function of the cell, e.g., LD_{HI1} and LD_{HI2}, two heart inhibitors belonging to the LD cluster. [From Frazier *et al.*, 1967; and Koester and Kandel, unpubl.]

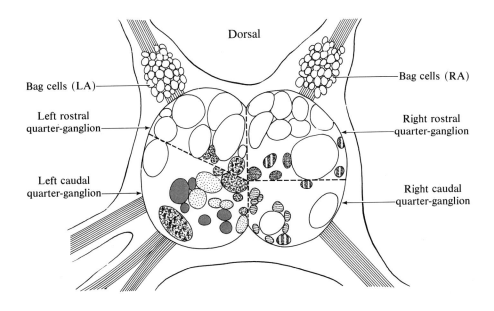

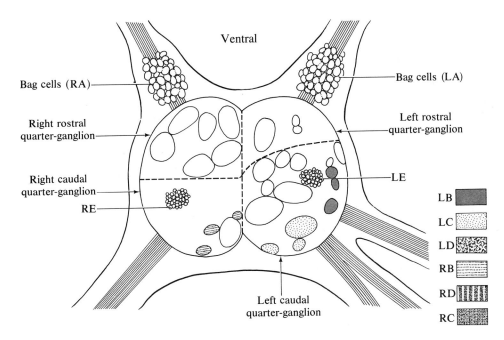

FIGURE 7-6.
Map of identified cell clusters in the abdominal ganglion of *Aplysia*. Each cluster consists of an homogenous group of cells, which at first could not be distinguished from each other. Each cluster is labeled with a letter (beginning with A) and the prefix L or R (left or right hemiganglion). Using functional criteria it is now possible to distinguish between some cells in these clusters (see Figure 7-5). [From Frazier *et al.*, 1967; and Koester and Kandel, unpubl.]

unique, neurons belonging to identified clusters (Koester and Kandel, unpub.). With this criterion, the map by Frazier and his colleagues was extended to include 25 additional identified cells (bringing the total to 55) and two more cell clusters (bringing the total to 10).[3] Thus, in a typical 100 gm animal, in which the abdominal ganglion contains about 1,500 cells, 55 cells can be identified as individuals and about 1,000 cells can be identified as members of specific cell clusters.

In attempting to develop criteria for identifying cells, it was useful to ignore the occasional aberrant cases and proceed as though the identified cells were relatively constant in location and morphological and functional characteristics. This strategy worked—many cells could be identified easily. As experience with a cell or cell group increased, the cell could be located in different animals with increasing ease and precision. Although the location as well as some functional characteristics of certain cells may vary (at times considerably) from ganglion to ganglion, the variations are neither so great nor so common as to hamper identification. To what degree structural and physiological deviations are accounted for by genetic, developmental, or experiential differences is not known, but could in principle now be tested in inbred strains of laboratory-cultured animals brought up under identical environmental conditions.

Frazier and his colleagues established an open-ended system of cell nomenclature for *Aplysia,* one that easily includes newly identified cells. Cells are first labeled R or L (right or left) to indicate the hemiganglion in which they are found. Cells of the same hemiganglion are assigned successive numbers, beginning with 1. (Cells in the right and left hemiganglia with the same number, e.g., L3 and R3, do not necessarily bear any special relationship to one another.) For descriptive purposes, Frazier and his colleagues subdivide the left and right hemiganglia into rostral and caudal quarter-ganglia (Figure 7-5). Clusters are labeled with the letter R or L, indicating the hemiganglion in which they are found, and a letter, beginning with A in each hemiganglion. As cells in identified clusters become distinguishable on the basis of functional criteria they are labeled with the name of the cluster and a subscript indicating the effector organ innervated by the cell. For example, LD_G is a gill motor cell of the LD cluster, and LB_S is a siphon motor cell of the LB cluster (Figure 7-5).

Maps essentially similar to those obtained for *Aplysia californica* have now

---

[3]Most instructive in the use of behavioral criteria to identify members of identified clusters has been the successful characterization of eight cells in the LD cluster. These are two gill motor neurons (LD_{G1} and LD_{G2}), three siphon motor cells (LD_{S1}, LD_{S2}, and LD_{S3}), two heart inhibitors (LD_{HI1} and LD_{HI2}), and one heart excitor (LD_{HE}).

been obtained for two other species of *Aplysia: A. brasiliana* (Blankenship and Coggeshall, 1973) and *A. dactylomela* (Kandel and Castellucci, unpublished). The identification of homologous cells in different species makes it possible to compare the neural circuitry of homologous behavior in close relatives.[4]

IS THE ENTIRE STRUCTURE INVARIANT?

The entire nervous system of *Ascaris* is invariant morphologically. Is there cell-for-cell invariance as well in the *Aplysia* abdominal ganglion? Ganglia from different-sized animals, presumed to be at different stages of maturation, were examined by Frazier and co-workers (1967) and by Margaret Skyles (1975, unpublished). They found that all the large identified cells are present in the ganglia of small juveniles. However, the total number of cells is not invariant—the number increases as the animal matures. The increase occurs primarily among the small cells, e.g., the bag-cell cluster (LA and RA) and small cells surrounding the central neuropil region. These findings suggest that the cells so far identified may be early and invariant elements in the organization of the ganglion. Since each identified cell cluster appears to relate to a nearby identifiable cell, it is possible that the ganglion may be made up of a constant number of functional units consisting of a fixed but small number of large cells and a variable number of small cells. What causes the increase in the number of smaller cells is not known but, as suggested above, it could be a response to increased functional demands. For example, the bag cells, a group of neuroendocrine cells that release the egg-laying hormone, undergo an increase in functional demand as the animal matures sexually. In response to this demand each cell increases the number of processes it sends into the connective tissue sheath. In addition, the cells increase in number, going from 50 or less cells per cluster when the animal is a young juvenile up to 400 per cluster when the animal becomes sexually mature.

FUNCTIONAL ORGANIZATION

Identified cells have been used to examine the anatomical and functional architecture of the ganglion and, consequently, the rules by which cells are grouped together. Are there subgroupings? What is accomplished by the subgrouping? What relationship to one another do symmetrical cells have?

[4]See Kandel, *The Behavioral Biology of Aplysia,* forthcoming.

Examination of these relationships has revealed that slight variations in the positions of the cell bodies of identified nerve cells are often found. As a result of what could be mechanical forces, a given cell body is often found in different positions in different animals. Moreover, the cell bodies in a cluster are not necessarily next to one another; often cells of another type are interspersed among member cells of a cluster.

Because of positional variability, one might expect little functional correlation between cells in a cluster. However, that is not always the case (Figure 7-5). All the mechanoreceptor sensory neurons for the siphon skin (LE cluster) are grouped together, as are those for the mantle shelf (the RE cluster). The bag cells are also collected together (LA and RA clusters). The remaining clusters are less homogenous functionally. This is because the population of motor cells for any given effector organ, except that of the LB vasoconstrictor cells and the L14 ink-gland motor cells, is distributed throughout several cell clusters. As a result, most of the identified clusters (with the exception of the LE, RE, LA and RA clusters) contain a variety of motor cells. Despite this functional differentiation, however, member cells are related. They invariably share with each other, and sometimes with an identified cell outside the cluster, some common innervation that allows them to participate in aspects of the same behavior. In addition, members of a cluster tend to share the same transmitter biochemistry. For example, all the LD cells are excited by the firing of a common command unit (unidentified Interneuron II). With one exception (LD_{S1}), all the LD cells so far examined are cholinergic. Moreover, many clusters can be divided into more homogenous subgroups, e.g., the heart inhibitors, LD_{HI1} and LD_{HI2}; the gill motor cells, LD_{G1} and LD_{G2}; and the vasoconstrictors, LB_{VC1}, LB_{VC2}, and LB_{VC3}.

Finally, in symmetrical ganglia every cell in each ganglion has a functional partner of the same size, position, and synaptic connection on the opposite side. Although the *Aplysia* abdominal ganglion is overtly asymmetrical, it is actually slightly symmetrical (Figure 7-5). Most obviously symmetrical are the LA and RA clusters of neurosecretory cells (the bag cells) and the two large identified cells L1 and R1 (Figure 7-7A). As is the case for the symmetrical metacerebral cells (see Figure 7-4), the efferent processes of L1 and R1 have a symmetrical distribution; each cell sends a major axon into the ipsilateral connective (Figure 7-7A). Finally, the two mechanoreceptor cell clusters LE and RE are symmetrical (see Figure 7-5; Byrne, Castellucci, and Kandel, 1974b). Nevertheless, the symmetry is not perfect even for these elements. For example, cell L1 receives some synaptic input from interneurons that R1 does not receive. The two sensory clusters do not innervate symmetrical regions of the body; the LE cluster innervates the siphon, the RE cluster the mantle shelf.

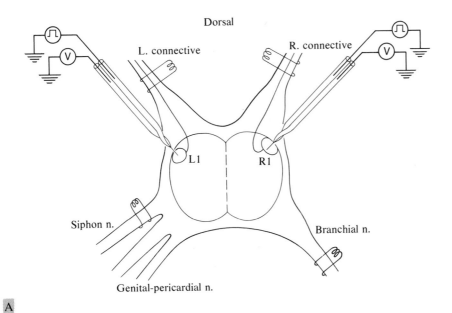

A

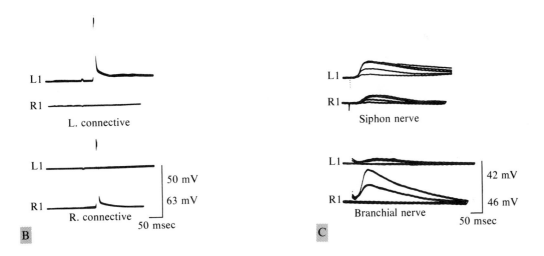

B

C

FIGURE 7-7.
The two symmetrical cells in the abdominal ganglion. **A.** Experimental arrangement for studying cells L1 and R1. **B.** Comparison of pathways of efferent axons of L1 and R1, showing responses to stimulation of the left and right connectives. **C.** Pathways of afferent synaptic input. Responses of L1 and R1 to stimulation of the siphon nerve (located in the left hemiganglion) and the branchial nerve (located in the right hemiganglion). As with the symmetrical metacerebral cells (Figure 7-4c), the ipsilateral input is of shorter latency and greater amplitude than the contralateral. [From Frazier et al., 1967.]

Also, right and left hemiganglia cells with similar properties (LD and RD cells) are often not symmetrical in location.

That cells of a cluster share transmitter biochemistry, reflex function, and symmetry was first shown in vertebrates for spinal motor neurons (Eccles, 1964) and in invertebrates for the third abdominal ganglion of the lobster (Otsuka, Kravitz, and Potter, 1967; see also Figure 7-11). Similar findings have since been described for ganglia in other animals.

Buccal and Pleural Ganglia

Maps of identified nerve cells in *Aplysia* are also now available for two symmetrical pairs of ganglia, the buccal and pleural. Using multiple criteria, Gardner (1971; Gardner and Kandel, 1972) identified 11 large cells in each buccal ganglion (Figure 7-8). Each cell has a symmetrical counterpart in the other ganglion. The symmetrical cells have similar properties, including com-

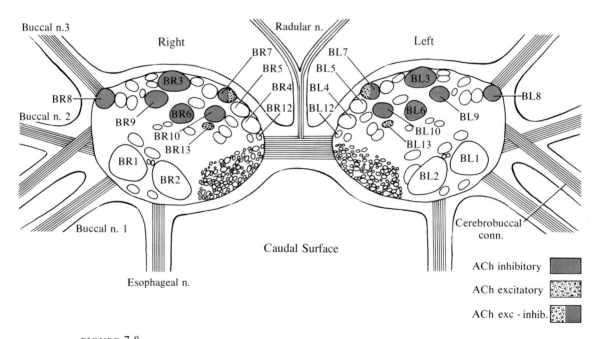

FIGURE 7-8.
Map of some of the identified cells in the symmetrical buccal ganglia of *Aplysia* and their sensitivity to ACh. Note the symmetrical positions and response properties of cells in the two ganglia. [After Gardner, 1971. Copyright by the American Association for the Advancement of Science.]

mon synaptic input, similar firing patterns, and response to ACh. Eight of the 11 identified cells on each side are innervated by two identified, presumably cholinergic, interneurons (see Chapter 8).

The two pleural ganglia vary in that the left ganglion contains an identifiable giant cell LP1 not found in the right ganglion (Hughes and Tauc, 1961). This cell is thought to be related to R2, in the right abdominal hemiganglion: both are similar in appearance and electrophysiological properties, both are cholinergic, and both have symmetrically distributed efferent axons (see Figure 7-3*B*). Kehoe (1972b) has identified three symmetrical clusters of cells in the pleural ganglia: the anterior, medial, and posterior pleural neurons (Figure 7-9).

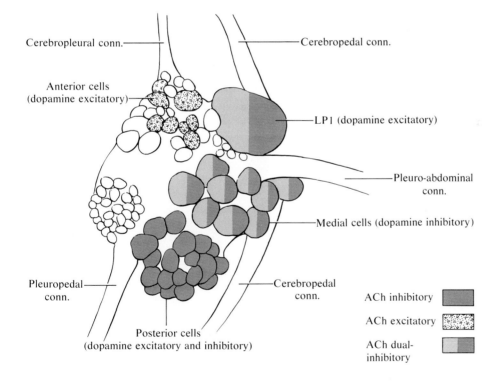

FIGURE 7-9.
Map of the identified cell clusters in the pleural ganglia of *Aplysia* and their sensitivity to ACh. The response to dopamine is also indicated. [From Kehoe, 1972b]

IDENTIFIED NEURONS IN OTHER OPISTHOBRANCHS

Levitan, Tauc, and Segundo (1970; and Wollacott, 1974) have identified cells of the buccal ganglia of the cephalaspid *Navanax inermis,* and Dorsett (1967) and Willows (1968) have identified many cells in the periesophageal ganglia of the nudibranch *Tritonia* (Figure 7-10). These maps form the bases of impor-

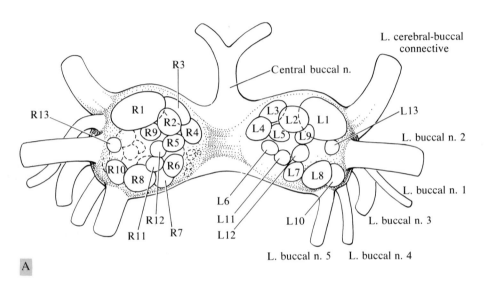

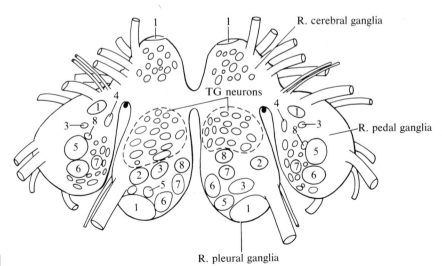

tant behavioral studies that we will consider in later chapters. Partial maps are also available for the brains of the pulmonates *Helix aspersa* (Kerkut, French, and Walker 1970; Parmentier and Case, 1972), *Otala lactea* (Gainer, 1972), *Helisoma trivolis* (Kater and Fraser Rowell, 1973), and *Onichidium verrucu-latum* (Katayama, 1970, 1973).

IDENTIFIED NEURONS IN OTHER INVERTEBRATES

Maps of identified cells have been obtained for the fourth abdominal ganglion (Otsuka, Kravitz and Potter, 1967) and stomatogastric ganglia (Maynard, 1972; Mulloney and Selverston, 1974a) of the lobster, the abdominal ganglion of the crayfish (Kennedy, Selverston, and Remler, 1969), as well as the segmental ganglia of the leech (Nicholls and Baylor, 1968; Stuart, 1970; Ort, Kristan, and Stent, 1974), the locust (Burrows and Hoyle, 1973), and the cockroach (Cohen and Jacklet, 1967). Because of their conceptual and historic interest, I will consider here only the studies of the lobster and leech.

Otsuka, Kravitz, and Potter (1967) identified 21 pairs of neurons in the third abdominal ganglion of the lobster that innervate muscles. They used four criteria: (1) position of the cell in the ganglion; (2) which muscle the cell innervated; (3) whether the innervation was inhibitory or excitatory; and (4) whether the cell contained a high concentration of either gamma-aminobutyric acid (GABA), the inhibitory transmitter in this ganglion, or glutamate, the putative excitatory transmitter. They found that the cell bodies occurred in clusters that were relatively constant in location and number and that the constituent cells of these clusters showed one or more common features (Figure 7-11). The most important characteristic of a cluster appeared to be the type of transmitter synthesized by the cells. Thus, three cells that contained high concentrations of GABA were regarded as members of one cluster even though the cells mediate inhibition to three unrelated muscles. Two other cells containing high concentrations of GABA but of unknown function are also located near this cluster. Given common transmitter synthesis, cells were next grouped according to reflex function. Fourteen cells were found to be excitatory motor

FIGURE 7-10.

A. Map of some of the identified cells in the symmetrical buccal ganglia of the opisthobranch *Navanax inermis*. [From Woolacott, 1974.] **B.** Map of the identified cells in the nervous system of the opisthobranch *Tritonia diomedia*. [From Willows, 1968.]

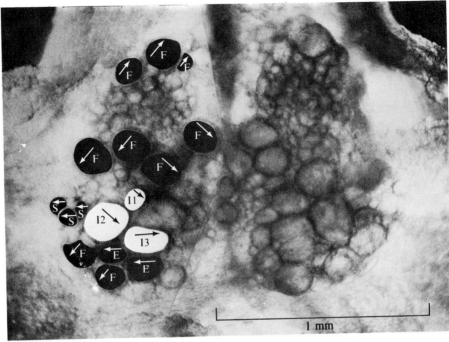

FIGURE 7-11.
Map of some of the identified cells in the third abdominal ganglion of the lobster. The cells marked in black are excitatory motor neurons controlling extensors (E), flexors (F), and swimmerets (S). The three cells marked in white (I1, I2, and I3) are inhibitory motor neurons containing high concentrations of γ-aminobutyric acid. The arrows indicate whether the axons are crossed or uncrossed; a diagonal arrow indicates that the axon leaves through the third root. [From Otsuka, Kravitz, and Potter, 1967.]

neurons to muscle; adjacent motor cells were found to innervate the same or synergist muscles. These cells all contain a low concentration of GABA. Finally, since these ganglia are symmetrical, each cell also has a functional partner of the same size and position on the opposite side of the ganglion.

The developmental factors determining clustering of cell bodies are unknown, but Otsuka and co-workers (1967) have advanced a number of reasonable hypotheses. They suggest that the cluster of inhibitory neurons containing GABA derives from a common embryological origin and that the clustering of cells according to motor function makes it easier to form common connections needed for common reflex functions. The finding that clustering according to common transmitter is more fundamental than clustering according to common reflex connections suggests that differentiation of transmitter biosynthesis occurs before the development of appropriate connections.

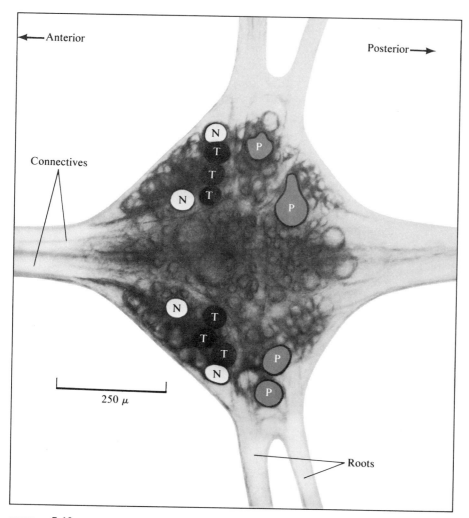

FIGURE 7-12.
Map of some of the identified sensory neurons in the ganglion of the leech. Three types of mechanoreceptor neurons have been identified: three touch cells (T) on each side, two pressure cells (P) on each side, and two cells on each side responding to noxious stimulation (N). [From Nicholls and Baylor, 1968.]

Studies of the leech by Nicholls and Baylor (1968) established a behavioral function for 14 of the cells described by Retzius 80 years earlier (see Figure 7-1). Nicholls and Baylor described seven symmetrical pairs of identified sensory neurons in each segmental ganglion: three cells on each side respond to touch, two to pressure, and two to noxious stimuli (Figure 7-12). Each cell

of a group has a characteristic constellation of functional properties. A similar specificity has been noted by Stuart (1970) for 17 pairs of motor neurons. (For a more complete map of the segmental ganglia of the leech, see Ort *et al.*, 1974.)

DIVERSITY AMONG IDENTIFIABLE NEURONS

The ability to identify nerve cells reliably is based not merely on the invariance of their location, but also on differences in morphological and functional properties. Studies of the functional properties of identified cells are thus useful for three reasons. One, the degree to which the cells of a ganglion differ can be rigorously specified. Two, the biophysical and biochemical mechanisms that account for the diversity of nerve cells can be determined. Three, these studies are a prerequisite for attempts to relate specific biophysical and biochemical mechanisms of neuronal function to behavior. Here I will consider primarily the diversity of cells in the abdominal ganglion of *Aplysia*, but I will also refer to studies of diversity in other invertebrate ganglia in order to illustrate the generality of the findings.

Firing Patterns

The first feature that impresses one in examining the cell bodies of the neurons of the abdominal ganglion of *Aplysia* and other invertebrate ganglia is that all neurons do not have the same firing pattern. Some cells have a stable resting potential and are *silent* in the absence of synaptic input; others have an unstable resting potential and fire spontaneously (Tauc, 1960; Arvanitaki and Chalazonitis, 1958; Frazier *et al.*, 1967). Cells that are spontaneously active can be distinguished further. Some discharge *single* action potentials in a highly regular manner; these are called *beaters* or *pacers*. Others fire regularly recurrent groups or *bursts* of action potentials; these are called *bursters*. Other cells show irregular bursting rhythms (Figure 7-13). Although there are important exceptions, there is a tendency for cells in the abdominal ganglion with similar firing patterns to be clustered together (Figure 7-14).

The finding that cells vary in their degree of rest and activity, and even in the type of firing pattern, suggests that the membranes of different neuronal cell bodies have a greater variety of electrogenic mechanisms than that encountered in voltage-clamp studies of isolated axons (Chapter 6). This suggestion

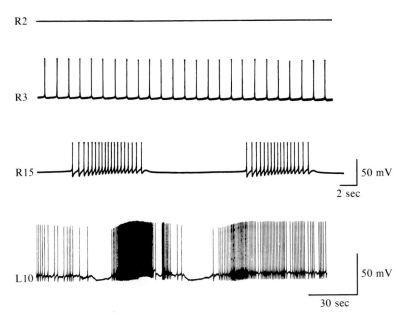

FIGURE 7-13.
Four types of firing patterns in the abdominal ganglion: R2, silent; R3, regular
beating rhythm; R15, regular bursting rhythm; L10, irregular bursting rhythm.
(Time and voltage calibrations for L10 are different from those for the other cells.)

has been amply confirmed. Studies have shown that the cell bodies of different
identified cells vary in the ionic mechanisms of their membrane potential,
action potential, and afterpotential, and in the location of the site for impulse
initiation. In addition, identified cells also vary in their biochemical characteristics and in their morphology. These differences account for the remarkable
diversity that exists among cells.

Mechanisms for Generating Membrane Potentials

The resting membrane of the squid giant axon behaves as a fairly good Nernst
potassium electrode (see p. 138). The membrane is primarily permeable to K^+
and this permeability accounts for the resting potential. Deviations of the
membrane potential from the ideal behavior of a K^+ electrode can be accounted for by a slight permeability of the membrane to other ions, particularly

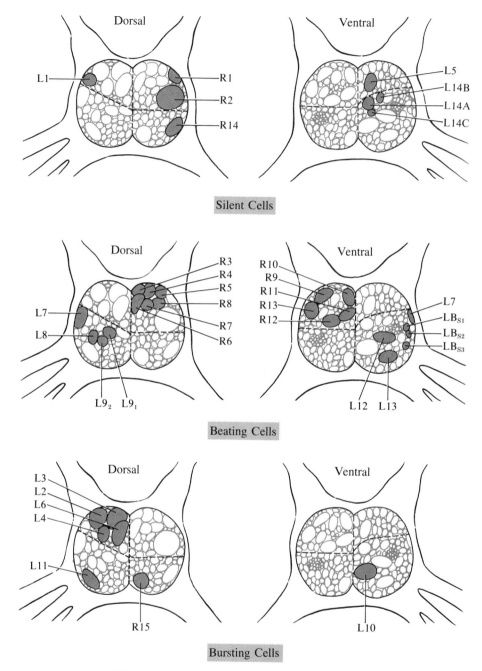

FIGURE 7-14.
Distribution of cells in the abdominal ganglion according to firing pattern.

Na$^+$. The Na$^+$–K$^+$ transport mechanism does not contribute importantly to the membrane potential (see p. 145). In the cell bodies of many *Aplysia* neurons, however, as much as 30 percent (15 to 20 mV) of the membrane potential is attributable to an active Na$^+$–K$^+$ exchange mechanism (Carpenter and Alving, 1968). This occurs because the Na$^+$–K$^+$ transport mechanism is not electrically neutral; it pumps more Na$^+$ out than K$^+$ in. The outward flow of positive ions causes an inward flow in adjacent membrane patches, which hyperpolarizes the cell membrane. Transport systems that can generate a potential are called *electrogenic*.

The electrogenic Na$^+$–K$^+$ transport mechanism can be activated by increasing Na$^+$ on the inside surface of the membrane and by increasing K$^+$ on the outside surface. Its effectiveness can therefore be reduced and even blocked by reducing intracellular Na$^+$ or extracellular K$^+$. The action of the transport system is dependent on the energy metabolism of the cell. It is thought to be mediated by a Na$^+$- and K$^+$-dependent enzyme that derives energy for the active transport of these ions by splitting the energy-rich molecule adenosine-triphosphate. The active transport process is selectively blocked by cardiac glycosides, such as ouabain. Because it is metabolically dependent, the transport system is also sharply dependent on temperature and in *Aplysia* is suppressed by cooling below 10°C.

Kerkut and Thomas (1964) first demonstrated an *electrogenic Na$^+$ pump* in an identifiable cell in the pulmonate gastropod *Helix*. In this cell the pump does not normally contribute to the membrane potential. But when the Na$^+$ concentration of the cell is raised artificially by intracellular injection, a large hyperpolarization of the membrane potential is produced. This potential change is eliminated by procedures that block the Na$^+$–K$^+$ transport. To measure the current produced by the Na$^+$ pump, Thomas (1969) examined the cell under voltage clamp (Figure 7-15). Using a Na$^+$-sensitive glass microelectrode, he was also able to follow the concomitant changes in intracellular Na$^+$ concentration. He found that, above a baseline level, the current produced by the electrogenic pump and the rate of Na$^+$ extrusion are both proportional to the concentration of intracellular Na$^+$ (Figure 7-16).

Similar results were obtained by Cooke, Leblanc, and Tauc (1974) in cell R2 of the abdominal ganglion of *Aplysia*. In addition to measuring membrane potential with an intracellular electrode, transmembrane current with a voltage clamp, and intracellular Na$^+$ activity with a Na$^+$-sensitive intracellular electrode, Cooke and co-workers also injected radiolabeled ^{24}Na$^+$ into the cell and monitored the outward movement (efflux) of ^{24}Na$^+$. They found that following injection the ^{24}Na$^+$ efflux paralleled the hyperpolarization of the

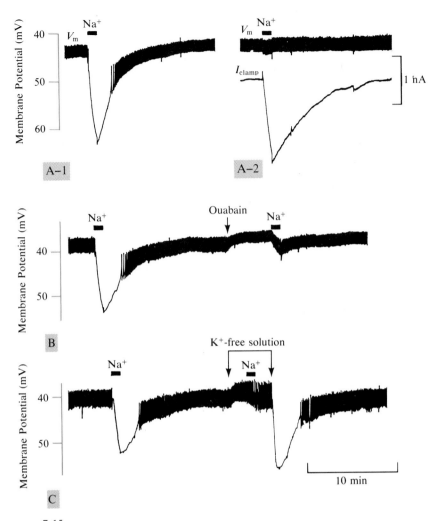

FIGURE 7-15.

An electrogenic Na$^+$ pump in snail neurons. Sodium was injected into the cell during the period indicated by the bar. [From Thomas, 1969.]

A. Response of the cell to Na$^+$ injection with and without voltage clamping. *A-1.* The cell was not clamped; the increase in intracellular Na$^+$ stimulated the active-transport process, leading to an electrogenic Na$^+$ extrusion that hyperpolarized the membrane. *A-2.* The experiment was repeated, but the membrane potential was clamped at the resting level and not allowed to undergo a hyperpolarizing change. The outward current passed by the feedback amplifier (lower trace) is a measure of Na$^+$ current flowing outward through the membrane as a result of the injection. Injection current was 38 nA.

B. Response of the membrane potential to Na$^+$ injection before and after the application of ouabain (2×10^{-5} weight/volume), which blocks active Na$^+$–K$^+$ transport. Injection current was 35 nA.

C. Response of the membrane potential to Na$^+$ injection in normal Ringer's solution and in a K$^+$-free Ringer's solution. In the K$^+$-free solution the Na$^+$–K$^+$ pump is blocked. As soon as K$^+$ is restored, the Na$^+$ current, produced by the pump, is turned on. Injection current was 34 nA.

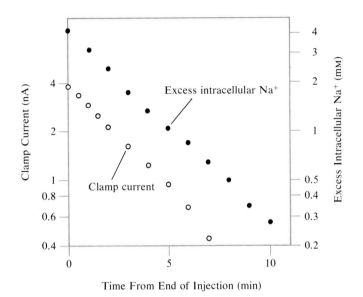

FIGURE 7-16.
Relation between clamp current and excess Na+, both plotted on logarithmic scales against the time from the end of injection of sodium
acetate. [From Thomas, 1969.]

membrane and the outward current. Half of the actively extruded Na+ was
electrogenic, suggesting that for every K+ ion pumped in two Na+ ions are
pumped out of the cell.

   In many *Aplysia* neurons, such as the silent cell R2, the pump is active even
at normal intracellular Na+ concentrations and contributes substantially to the
membrane potential (Carpenter and Alving, 1968; and see Figure 7-17). A
similar dependence of the membrane potential on electrogenic Na+ transport
was found by Gorman and Marmor (1970) in an identified neuron of the
opisthobranch mollusc *Anisodoris nobilis*. As in *Aplysia*, the membrane potential of the *Anisodoris* cell can be experimentally dissociated into two components. One component is a K+ conductance that resembles the mechanism
that controls the membrane potential of the squid axon. It is dependent upon
ionic gradients and permeability (primarily to K+, but also to Na+) in accor-

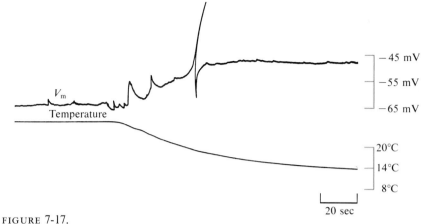

FIGURE 7-17.

Electrogenic Na+ pump contributes to the normal resting potential in *Aplysia*. Sodium–potassium active transport is strongly temperature dependent; reducing the temperature from 20°C to 8°C reduces the resting potential of cell R2 from 60 mV to about 48 mV. This experiment indicates that the electrogenic Na+ pump accounts for about one-third of the resting potential. [From Carpenter and Alving, 1968.]

dance with the Goldman equation (p. 142). The other component depends on electrogenic Na+ transport (Figure 7-18).[5]

The exact magnitude of the contribution of the electrogenic Na+ pump to the membrane potentials of various neurons in the abdominal ganglion varies. But irrespective of firing pattern, all identified cells examined have some contribution from this source (Carpenter and Alving, 1968; Pinsker and Kandel, 1969; Carpenter, 1973b; Cooke *et al.*, 1974). Thus, a component of the membrane potential in most cells is immediately dependent on metabolism and therefore highly influenced by temperature. Such cells may function very differently at warm temperature (22 to 27°C) than at cold (12 to 17°C). Differences in the functional properties of the cells at the two temperatures are likely to be important for behavior because *Aplysia californica* is exposed to a temperature range that extends from 27°C to 12°C.

[5]The difference between the squid giant axon and the *Aplysia* and *Anisodoris* cell bodies may not reflect a difference in the active transport mechanisms. Even in squid the pump is not completely neutral but is thought to carry three Na+ ions out for every two K+ carried in (Hodgkin and Keynes, 1955) and contributes a very small component (1.3 mV) to the resting potential (de Weer and Geduldig, 1973). The difference between the two types of membrane seems to reside in their specific resistance. In the squid it is about 10^3 ohms/cm² (Cole and Hodgkin, 1939), whereas in the R2 cell body of *Aplysia* it is 10^5 ohms/cm² (Carpenter, 1973a), and in the gastroesophageal neuron of *Anisodoris* it is 10^6 ohms/cm² (Gorman and Mirolli, 1972). As a result, Na+ current that produces only a small potential change in the squid axon would produce a sizeable potential change in these cell bodies. What accounts for the difference between *Aplysia* and *Anisodoris* neurons and the cells in *Helix* is not clear. Indeed Christoffersen (1972) (in contradiction to Kerkut and Thomas, 1964) describes a small (7 mV) electrogenic-pump contribution to the resting potential in *Helix* cells. Christoffersen attributes the small size of the pump contribution to the fact that resistance of the cell membrane in *Helix* is only one-tenth that of *Aplysia*. Carpenter (1970) attributes the high specific resistance of the membrane in *Aplysia* to a low permeability to K+.

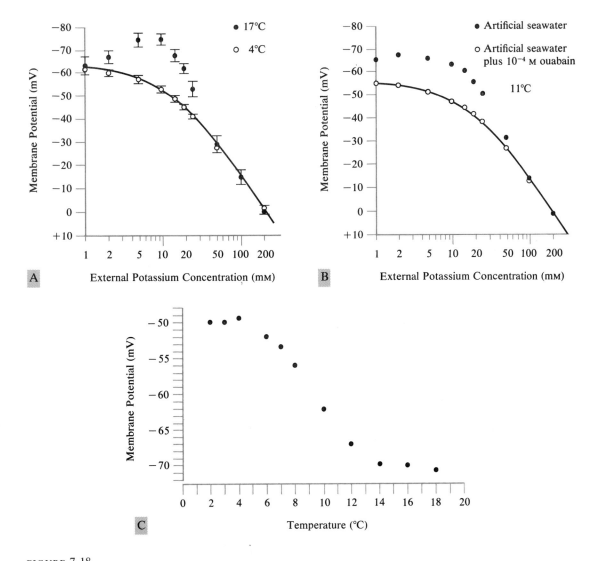

FIGURE 7-18.
Separation of the resting potential in a molluscan neuron into two components: contributions of K^+ conductance and an electrogenic Na^+ pump. [From Marmor and Gorman, 1970; and Gorman and Marmor, 1970.]

A. Dependence of the membrane potential on external K^+ and temperature. Each point is an average from five experiments (SE shown). The curve drawn here is the theoretical curve, based on a modified form of the Goldman equation (p. 142). Only at 4°C, when most of the active Na^+-K^+ transport is blocked, can the membrane potential be accurately predicted by the Goldman equation.

B. Dependence of the membrane potential on external K^+ before and after blockade of the Na^+ pump. The temperature was 11°C. The curve drawn here is the theoretical curve, based on the Goldman equation. Only in ouabain, which blocks active Na^+-K^+ transport, can the membrane potential be accurately predicted by the Goldman equation.

C. Relationship between temperature and membrane potential.

Mechanisms for Generating Action Potentials

The upstroke of the action potential in squid giant axon and in vertebrate axons is due to an increase in Na permeability (see Chapter 6). For many years the spikes of all nerve-cell bodies were thought to use a similar mechanism. However, Geduldig and Junge (1968; and Geduldig and Gruener, 1970) have found that in the cell body of R2, Ca^{++} contributes to the action potential. When cell R2 is exposed to either Na^+-free or Ca^{++}-free solutions an action potential is recorded with only partial reduction of the overshoot; elimination of both ions abolishes the spike (Figure 7-19). In a Ca^{++}-free medium the overshoot behaves like a Nernst Na^+ electrode in response to changes in Na^+ concentrations; in a Na^+-free medium it behaves like a Nernst Ca^{++} electrode. The two ionic components of the spike can be distinguished pharmacologically and are thought to move through different membrane channels (Figure 7-20). The Na^+ component is selectively blocked by TTX, which blocks Na^+ channels in the squid axon (see Chapter 6; and Nakamura, Nakajima, and Grundfest, 1965). The Ca^{++} component is selectively blocked by cobalt ion, which blocks Ca^{++} channels in barnacle muscle cells (Hagiwara and Takahashi, 1967). Voltage-clamp experiments by Geduldig and Gruener (1970) indicate that during the normal action potential in the cell body Na^+ carries most of the inward current (Figure 7-21) and causes the rapid depolarization accompanying the action potential. In normal seawater the early inward current is relatively unaffected when Ca^{++} is lowered from its normal value of 11 mM to 0 mM. However, when the cell is fired repeatedly the Ca^{++} contribution to the action potential appears to be increased. Using aequorin, a light-emitting photo-protein activated by Ca^{++}, Stinnakre and Tauc (1973) found that every action potential in cell R2 is accompanied by a transient influx of Ca^{++}. With a train of action potentials, the successive light responses (indicating Ca^{++} influx) increase, as does the height of the action potential. When cobalt ion is added to the bathing medium the Ca^{++} influx is blocked, and the spike height remains constant throughout the train of action potentials (Figure 7-22).

Since the cell body of R2 has a Ca^{++}-dependent action potential, is the action potential propagated in the axon also partially dependent on Ca^{++}? Kado (1973) and Junge and Miller (1974) have examined this question and found that the spike in the axon is due only to Na^+. Thus, spike-generation in the cell-body and axonal membranes can use different ionic mechanisms (see also Wald, 1972). Cells vary in their Na^+ requirements. The spike-generating mechanisms of cells R15, L7, L10, and L11 do not require Ca^{++} and are largely

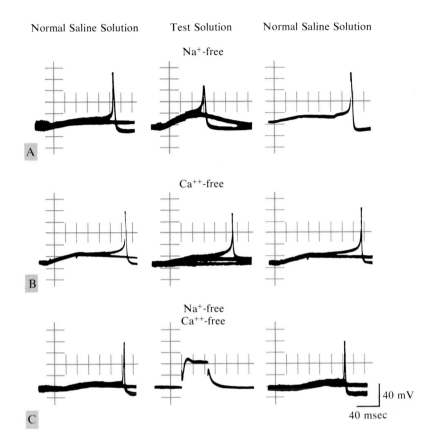

Normal Saline Solution   Test Solution   Normal Saline Solution

FIGURE 7-19.

Sodium and Ca^{++} components of inward current in the action potential of cell R2. The left and right records of all tests (A,B,C) show an action potential in normal saline (seawater) solution. In some records two sweeps are superimposed; subthreshold and suprathreshold (spike-generating) responses are shown. Zero membrane potential is indicated by the horizontal line. The data illustrate that the cell body of R2 is capable of firing in the absence of either Na^+ or Ca^{++}, but not in the absence of both ions. [From Geduldig and Junge, 1968.]

A. The role of Na^+ is examined by replacing Na^+ with tris (hydroxymethyl) aminomethane, an impermeant organic cation (TRIS). The action potential is reduced in amplitude.

B. The role of Ca^{++} is examined by replacing Ca^{++} with TRIS. The amplitude of the action potential is almost the same as that seen in normal saline solution.

C. The combined roles of Na^+ and Ca^{++} are examined by simultaneous replacement of Na^+ and Ca^{++} with TRIS. Depolarization of the membrane does not give rise to an action potential.

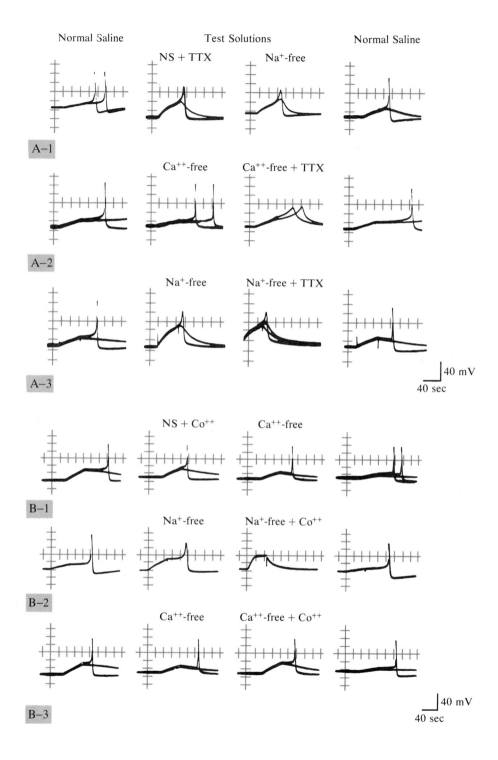

Normal Saline   Test Solutions   Normal Saline

NS + TTX   Na⁺-free

A–1

Ca⁺⁺-free   Ca⁺⁺-free + TTX

A–2

Na⁺-free   Na⁺-free + TTX

A–3

40 mV
40 sec

NS + Co⁺⁺   Ca⁺⁺-free

B–1

Na⁺-free   Na⁺-free + Co⁺⁺

B–2

Ca⁺⁺-free   Ca⁺⁺-free + Co⁺⁺

B–3

40 mV
40 sec

FIGURE 7-20 (*Facing page*).
Pharmacological separation of Na+ and Ca++ channels in cell R2. TTX = the addition of 30 μM tetrodotoxin into the bathing solution; Co++ = the addition of 30 μM cobalt chloride (CoCl₂) into the bathing solution. [From Geduldig and Junge, 1968.]

A. Blockade of the Na+ component of the spike by TTX. *A-1*. Effects of TTX compared to effects of a Na+-free medium. Both treatments lower the amplitude of the spike but do not eliminate the spike. *A-2*. Effect of TTX in a Ca++-free medium. TTX blocks the Na+ component of the spike and the Ca++-free medium eliminates the Ca++ contribution. When both components are eliminated no spike occurs. *A-3* Effects of TTX in a Na+-free medium. In a Na+-free medium the action potential is due to Ca++ exclusively. Since TTX specifically blocks Na+ channels, no effects are observed when TTX is added to a Na+-free solution.

B. Blockade of the Ca++ component of the spike by cobalt. *B-1*. Effects of Co++ compared to effects of a Ca++-free medium. Neither treatment eliminates the spike, since the Na+ component of the spike is not affected. *B-2*. Effect of Co++ in a Na+-free medium. Cobalt blocks the Ca++ component of the spike. In a Na+-free medium no spike can occur in the presence of Co++ since both Na+ and Ca++ components are eliminated. *B-3*. Effects of Co++ in a Ca++-free medium. In a Ca++-free medium the action potential is due to Na+ exclusively. Since Co++ specifically blocks Ca++ channels, no effects are observed when Co++ is added to a Ca++-free solution.

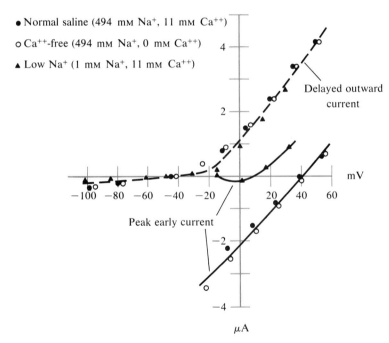

FIGURE 7-21.
Relative contributions of Na+ and Ca++ to the peak early inward and delayed outward currents of the action potential in cell R2, as assessed by voltage clamp of R2 in low-Na+ and Ca++-free solutions. In each case Na+ and Ca++ have been replaced by impermeant ions. The current-voltage relationships of both the peak early inward current (solid lines) and the delayed outward current (dashed line) are plotted. In the squid giant axon (see p. 164) the early inward current is mediated by Na+ ions. In cell R2 in *Aplysia* neither zero Ca++ nor low Na+ affect the delayed outward current, indicating that this current is mediated by K+, as it is in the squid giant axon. Low Na+ affects the peak early current, indicating that a large part of this current is mediated by Na+; zero Ca++ does not greatly affect the peak early current. These findings indicate that the inward current of the action potential is largely mediated by Na+, and that even though Ca++ can substitute for Na+ in its absence, Ca++ does not under normal circumstances (in the presence of Na+) contribute importantly to the rising phase of a single action potential. [From Geduldig and Gruener, 1970.]

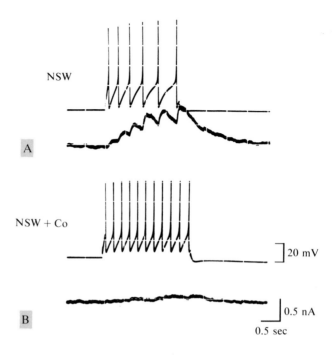

FIGURE 7-22.

Relationship of Ca++ influx and the height of action potentials in a train. The Ca++ influx is reflected in the levels of light emission from an intracellulary injected photo-protein, aequorin. **A.** In normal seawater (NSW) increased Ca++ influx (lower trace) is correlated with increases in later spikes in a train in cell R2 (upper trace). **B.** In seawater with cobalt (NSW + Co++) the Ca++ influx is suppressed and the spike height remains constant throughout the train. This indicates that even though Ca++ does not contribute importantly to the first action potential of a train (see Figure 7-21), it does contribute to later action potentials. [From Stinnakre and Tauc, 1973.]

or exclusively dependent on Na+ (Strumwasser, 1967b and 1974; Carpenter, 1973b; Bryant and Weinreich, 1975). The functional significance of the combined Na+–Ca++ mechanism is not clear, but in some cells Ca++ is used for intracellular signaling and may serve to activate one or more cellular processes. For example, in a variety of neurons one component of the delayed increase in g_K (accompanying the depolarization produced by the action potential) is turned on by the influx of Ca++ during the ascending climb of the action potential (Meech and Standen, 1975).

Mechanisms for Generating Afterpotentials

One of two mechanisms can contribute to the hyperpolarizing afterpotentials of neurons. The brief hyperpolarizing afterpotential that follows a single spike is usually due to a persistence of the increase in the K^+ conductance turned on by the action potential (see p. 161). But when neurons fire repetitively they often produce a more prolonged hyperpolarization called *posttetanic* hyperpolarization. In cells R2 and R15 this posttetanic hyperpolarization is also due to an increased K^+ conductance (Brodwick and Junge, 1972; Junge and Stephens, 1973; for *Helix* neurons see Gola, 1974). However, in some cells the train of action potentials also increases the intracellular Na^+ concentration and causes a hyperpolarization due to a Na^+ pump. In crayfish sensory cells this second mechanism completely accounts for postsynaptic hyperpolarizing afterpotentials (Nakajima and Takahashi, 1966). In the sensory neurons of the leech a change in K^+ conductance and an electrogenic Na^+-pump mechanism operate simultaneously (Jansen and Nicholls, 1973).

Hyperpolarizing afterpotentials serve as an important rate-limiting mechanism for repetitive firing in neurons. When the cell fires repetitively, both the amplitude and rate of decay of the brief afterpotential following each spike determines the *interspike interval* and thereby limits the frequency of firing. The posttetanic afterpotential determines the *interburst interval* and thereby determines how soon a cell will fire again after a period of repetitive firing.

Sites for Impulse Initiation

Cells also vary from one another in the site within the neuron at which the action potential is initiated. In an elegant study Tauc (1962a, b) examined the site of initiation of the action potential in cell R2 and found that it occurred in the axon about 1 to 1.5 mm from the cell body. By recording simultaneously in the cell body and the axon of R2, he found that independent of how the cell is fired—whether naturally, through synaptic stimulation; autoactively; or antidromically, by stimulating the axon in the connective—the impulse is invariably initiated in the axon and only subsequently invades the cell body (Figure 7-23). The axon serves as a trigger zone because it has a much lower threshold than the cell-body membrane (15 mV compared to 45 mV). The

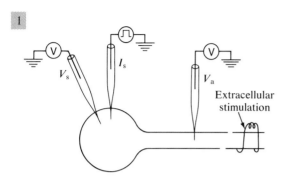

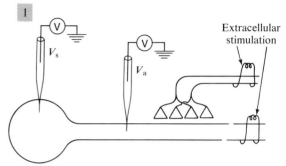

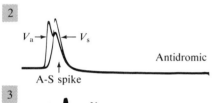

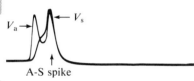

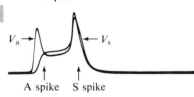

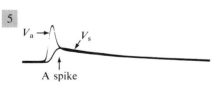

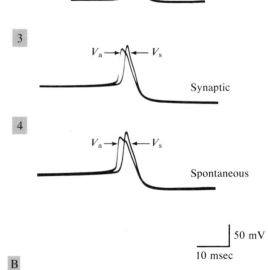

A

B

FIGURE 7-23.

Site of origin of the action potential in cell R2. Simultaneous intracellular recordings in the cell body or soma (V_s) and axon (V_a). [From Tauc, 1962a.]

A. Antidromic activation. *A-1.* Experimental arrangement, consisting of intracellular recording microelectrodes in the cell body (V_s) and initial segment of the axon (V_a), and an intracellular stimulating electrode in the cell body (I_s). Extracellular stimulating electrodes on the distant portion of the axon are used to produce antidromic activation. (Parts *A-2* to *A-5* are superimposed records.) *A-2.* At the resting membrane potential the recording electrode in the axon (V_a) registers a notched double action potential. The first component is due to the action potential in the axon near the recording microelectrode. The second component coincides with the action potential in the cell body and is an axonal reflection of the action potential in the cell body. The action potential in the cell body recorded at V_s also has two components (the A-S spike). One component (the A spike) is a 45 mV electrotonic potential at the foot of the action potential in the cell body due to impulse activity in the axon propagating decrementally toward the cell body. The second component (the S spike) is the action potential in the cell body. At the resting membrane potential, invasion of the cell body by the axon action potential rapidly triggers an action potential in the cell body and the two components appear as one in the cell body (the A-S spike). *A-3* and *A-4.* As the membrane of the cell body is progressively hyperpolarized, the ability of the axon action potential to trigger an action potential in the cell body is impeded because the membrane potential of the cell body is moved further from threshold. As a result, the separation between the A and S components of the spikes becomes more prominent, as does the notch in the axon spike. *A-5.* With further hyperpolarization, the axon action potential fails to trigger an action potential in the cell body: the S spike, the second component of the axon spike, is not generated. As a result, the electrotonic potential produced in the cell body by the axon spike (the A spike) is completely unmasked.

B. Comparison of antidromic activation with synaptic activation and spontaneous activity. *B-1.* Experimental arrangement, consisting of intracellular recording electrodes in the cell body and axon and extracellular stimulating electrodes on an afferent pathway (for synaptic activation) and on the distant portion of the axon (for antidromic activation). The action potential occurred spontaneously on occasion. *B-2.* Antidromic activation. *B-3.* Synaptic activation. *B-4.* Spontaneous firing. With all three modes of activation, initiation of the action potential begins at a trigger zone in the axon, as is evident by the invariably shorter latency of the axon spike compared to the spike in the cell body (the A-S spike). The depolarizing synaptic potential (*B-3*) and the pacemaker potential during spontaneous activity (*B-4*) facilitate the discharge of the cell body by the axon spike. This is evident by the decrease in the delay between the onset of the spike in the axon and the onset of the spike in the cell body and by the increased amplitude (reduced short-circuiting) of the second component of the axon spike (the S spike) in *B-3* and *B-4* as compared to *B-2.*

excitability of the membrane between the trigger zone and the cell body is transitional and varies from 15 mV near the trigger zone to 45 mV near the cell body (Figure 7-24). Even though the cell body of R2 has a high threshold, it is excitable and is usually invaded by a full action potential. Thus, in this cell, as in many vertebrate nerve cells (see Figure 5-6, p. 108), normal synaptic activation leads to initiation of an impulse in the axon. The impulse then propagates in two directions: away from the cell body and out of the ganglion, and toward the cell body.

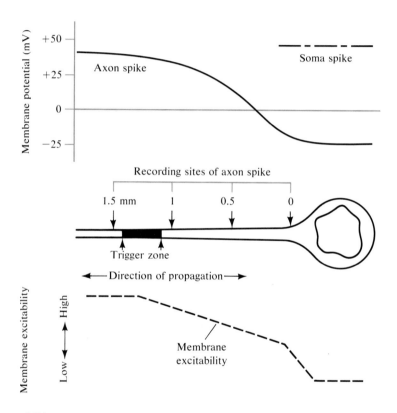

FIGURE 7-24.
Schematic illustration of the maximum height of the axon spike recorded at different distances from the soma. As the axon spike approaches the cell body it decreases in amplitude; in the cell body it is recorded as a subthreshold electrotonic potential. The broken line (above) represents the maximum amplitude of the cell-body (soma) spike recorded in the cell body. The resting membrane potential is the same throughout the cell body and proximal axon (about −50mV). The dashed line (below) represents the hypothetical change in membrane excitability (threshold) in different regions of the neuron. [Modified from Tauc, 1962a.]

A different situation may apply for certain autoactive neurons. Barbara Alving (1968) tied off the cell bodies of some autoactive cells so as to insulate them electrically from their axons. She found that the isolated cell bodies of regular-bursting cells continued to show characteristic rhythms, as did regularly firing (beating) cell bodies (Figure 7-25). Moreover, isolated cell bodies of silent cells remained silent (see also Chen, von Baumgarten, and Takeda, 1971). Alving suggests that whereas impulse initiation occurs in the axon in silent cells, in autoactive neurons it takes place in the cell body.[6]

[6]To be certain of this point, it will be necessary to record from the axons of the tied-off and intact cells to determine whether they have pacemaker potentialities. In the lobster cardiac ganglion the axons of autoactive cells continue to be active in the absence of the cell body (Mayeri, 1973).

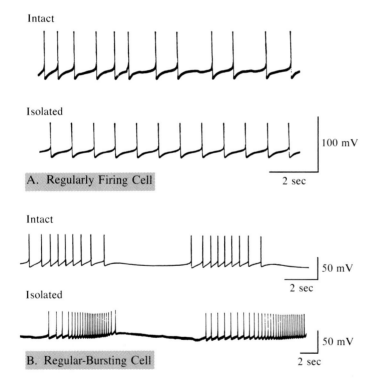

Intact

Isolated

100 mV

A. Regularly Firing Cell

2 sec

Intact

50 mV

2 sec

Isolated

50 mV

B. Regular-Bursting Cell

2 sec

FIGURE 7-25.
Autoactive beating and bursting pacemaker activity in the isolated cell body. **A.** Firing patterns of a cell with a regular beating rhythm before and after the cell body had been isolated (by ligating the axon hillock). **B.** Firing patterns of a cell with a regular bursting rhythm before and after the cell body had been isolated. [From Alving, 1968.]

Signaling Function of Autoactive Rhythms

Spontaneous neural activity has long been observed in the isolated nervous system of vertebrates (see p. 50). Until recently there was no way of deciding whether this activity was synaptically driven—due perhaps to a circulation of impulses in cells interconnected with each other in loop-like arrangements— or whether it reflected a truly endogenous autoactive property of the cell, like the pacemaker mechanisms of certain heart cells. Several lines of evidence now indicate that the activity of many identified autoactive beating and bursting cells in the abdominal ganglion of *Aplysia* is truly endogenous and reflects an unstable membrane potential.

The prevalence of autoactivity in the abdominal ganglion of *Aplysia* (in about 50 percent of the identified cells) indicates that a major method for information transformation in this ganglion involves the modulation of auto-active rhythm by synaptic input. Autoactive cells may also have a special role in behavior. By sustaining ongoing activity in the nervous system, they could provide a cellular basis for internal drive, or motivational state. In addition, the release of an autoactive rhythm (by disinhibition) could provide a means for generating the central programs of fixed-action patterns. Finally, variations in the firing pattern of autoactive cells could control bodily functions that have circadian (daily) rhythms (Strumwasser, 1967a).

In view of the importance of autoactivity for neural integration and its potential importance for behavior, a considerable effort has been spent on analyzing the biophysical mechanisms responsible for autoactivity.

Mechanisms of Various Autoactive Rhythms

BEATING CELLS

The neurosecretory white cells (R3 to R13) fire in a highly regular fashion. The beating rhythm of these cells is intrinsic to the cell membrane; there are no accompanying synaptic potentials to account for the rhythm. In fact, the cells have very little synaptic input even when nerves to the ganglion are electrically stimulated (Frazier *et al.,* 1967). The rhythm can be changed by injecting depolarizing current pulses that trigger action potentials in the cell. The rhythm of an autoactive cell can be reset by interrupting its oscillatory behavior whereas the rhythm of a cell driven by the synaptic activity of another cell cannot. Moreover, Alving (1968), and Chen, von Baumgarten, and Takeda

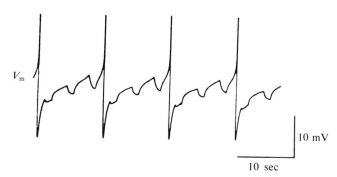

FIGURE 7-26.
Variation in the apparent input conductance of the cell body during pacemaker discharge in a regularly firing cell (R14), illustrating anomalous rectification (a decrease in input conductance with depolarization). The upper trace records intracellular current applied through a stimulating electrode. The lower trace is an intracellular recording of the hyperpolarizing electrotonic potentials produced by the current pulses. [From Carpenter, 1973b.]

(1971) found that the cell body isolated from the synaptic region in the axon still shows autoactive capabilities (Figure 7-25A). Little is known, however, about the detailed mechanisms of the beating rhythm of autoactive cells except that some have a high resting g_{Na}, a novel g_K (in addition to the usual g_K), and anomalous rectification (Figure 7-26).

The high resting Na+ current tends to drive the membrane in the depolarizing direction, toward threshold for spike intiation (Carpenter and Gunn, 1970). The novel type of g_K was first discovered by Connor and Stevens (1971a, b) in the opisthobranchs *Anisodoris* and *Archidoris,* and by Neher (1971) in *Helix.* These cells are not spontaneously active, but would beat in response to long depolarizing current pulses. Using voltage-clamp techniques, Connor and Stevens and Neher found that, as expected, the action potentials were associated with an inward current (presumably due to Na+ or Ca++) and a delayed outward current due to K+ (see p. 164). In addition, a third current was found, an *early (fast) outward K+ current* with novel properties (see also Hagiwara, Kusano, and Saito, 1961). The early K+ current becomes rapidly activated near the resting potential, when the membrane is depolarized beyond

−55 mV, and for this reason is also called the fast K⁺ current. But once activated, it also inactivates rapidly. Removal of inactivation requires a conditioning hyperpolarization. During normal beating activity the spike afterpotential produced by the delayed K⁺ current provides this hyperpolarization. The depolarization between spikes (called the *pacemaker potential,* and due to the steady g_{Na} or the application of a depolarizing current and the decay of the delayed K⁺ current) then activates the fast K⁺ current. Because this current is turned on more rapidly and at more negative (hyperpolarized) values of the membrane potential than either the inward Na⁺ current or the delayed K⁺ current, the subthreshold behavior of these neurons is dominated by the early K⁺ current, which slows the depolarizing drift of the pacemaker potential (Figure 7-27). As the early K⁺ current inactivates rapidly it allows the membrane potential to drift more rapidly toward threshold.

The early K⁺ current has some features that could account for anomalous rectification (Stevens, 1969; Gola, 1974; Barrett and Crill, 1972). Because the early K⁺ current is inactivated at the resting level, the input resistance in the range of the resting potential is anomalously rectifying. It is higher at the resting level (and at more depolarized levels) than at hyperpolarized levels, where early K⁺ currents are only partially inactivated. But the exact relationship of anomalous rectification to the early K⁺ current is still not known.

FIGURE 7-27.
Early K⁺ current and its role in beating neurons.
A. Voltage range for the early K⁺ current. The first column is a series of voltage-clamp currents in response to depolarizing command pulses. Resting membrane potential 48 mV; temperature 14.5°C. From this potential, all the depolarizing command pulses initiate only an inward current (Na⁺ and/or Ca⁺⁺) and a delayed outward K⁺ current. In the second column a 45 mV hyperpolarizing prepulse of 500 msec duration preceded each of the depolarizing command pulses; a current artifact indicates the end of each prepulse. The depolarizing command pulses are identical to those of the first column. Following the prepulse the depolarizing command pulses activate a third current, an early outward K⁺ current, which is not evident in the first column because it is partially or fully inactivated at the resting level. As a result of the activation of the early outward K⁺ current, the inward (downward) current is completely masked. The third column is a computer simulation, and illustrates the presumed time course of the three currents in the middle column: the early outward (K⁺) current, the inward (Na⁺ or Ca⁺⁺) current, and the delayed outward (K⁺) current. [From Neher, 1971.]

B. Connor and Stevens's model of the subthreshold (pacemaker) behavior of a silent cell in *Anisodoris* during repetitive firing in response to depolarizing current. *B-1.* Comparison of computed and recorded action potentials during repetitive firing. *B-2.* Computed membrane currents from which the overall behavior of the cell was reconstructed in *B-1*. Currents flowing during the action potentials are not shown because their large magnitude obscures subthreshold behavior. [From Connor and Stevens, 1971b.]

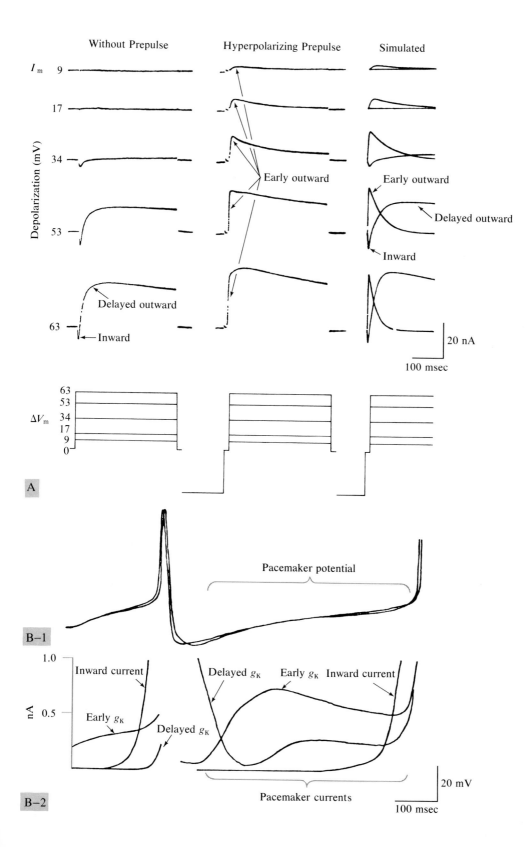

Without Prepulse

Hyperpolarizing Prepulse

Simulated

I_m

9

17

34

53

63

Depolarization (mV)

Early outward

Delayed outward

Inward

Early outward

Delayed outward

Inward

20 nA

100 msec

ΔV_m

63
53
34
17
9
0

A

Pacemaker potential

B–1

1.0

0.5

nA

Inward current

Early g_K

Delayed g_K

Delayed g_K

Early g_K

Inward current

Pacemaker currents

20 mV

100 msec

B–2

BURSTING CELLS

Seven cells in the abdominal ganglion (L2, L3, L4, L6, R15, L10, and L11) generate recurrent bursts of spikes. Five of these (L2, L3, L4, L6, and R15) produce regularly occurring bursts (see Figure 7-14).

When a bursting cell is depolarized it beats; when it is hyperpolarized it is silent. In the intermediate range of membrane potential the interval between the onset of each burst is a continuous function of membrane potential or polarizing current (Figure 7-28). This feature alone nicely illustrates the endogenous nature of the autoactive bursting rhythm and its control by membrane potential. Other evidence for the endogeneous nature of the rhythm is that direct hyperpolarization of the cell fails to reveal synaptic potentials that could account for the rhythm. In addition, intracellularly applied depolarizing pulses that generate spikes reset the rhythm (Figure 7-29A), as do hyperpolarizing pulses (Figure 7-30) that subtract spikes from a burst (Strumwasser, 1967b; Waziri, Frazier and Kandel, 1965). Finally, and most critical, the cell body shows bursting capability even when isolated from its synaptic region in the axon (Figure 7-25B; Alving, 1968).

The membrane of the bursting cells generates the six currents found in pacing cells: an electrogenic Na$^+$ pump, a g_K for generating the membrane potentials, an inward Na$^+$ (or Ca^{++}) current, a late K$^+$ current (delayed rectification), an early K$^+$ current (Figure 7-31B; Faber and Klee, 1972; Gola, 1974), and a relatively high (voltage-independent) resting g_{Na} that pulls the membrane potential in the depolarizing direction (Faber and Klee, 1972; Carpenter, 1973b; Smith, Barker, and Gainer, 1975; Gola, 1974; Junge and Stephens, 1973). In addition, there are two more currents: a slow Na$^+$ current and a slow K$^+$ current. These two slow currents introduce a new feature, a self-sustained membrane oscillation that underlies (and normally triggers) the recurrent

FIGURE 7-28.
Burst-onset interval in a burst-generating cell (L3) in the abdominal ganglion of *Aplysia*, in response to changes in membrane potential. The upper trace is the intracellular record and the lower trace indicates the current through an independent stimulating electrode.

A. With no polarizing current (*A-1*) the cell fires regularly without bursts. When the cell is hyperpolarized (*A-2*) regular bursts are generated. With increasing hyperpolarization (*A-3* to *A-5*) the burst-onset interval is progressively prolonged. With further hyperpolarization (not indicated) the bursting can be completely suppressed. [From Frazier *et al.*, 1967.]

B. Graph of burst-onset intervals as a function of polarizing current. [From Frazier, *et al.*, 1967; and unpublished observations.]

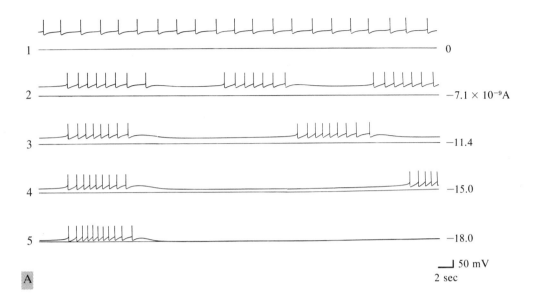

A

1 — 0

2 — -7.1×10^{-9} A

3 — -11.4

4 — -15.0

5 — -18.0

50 mV
2 sec

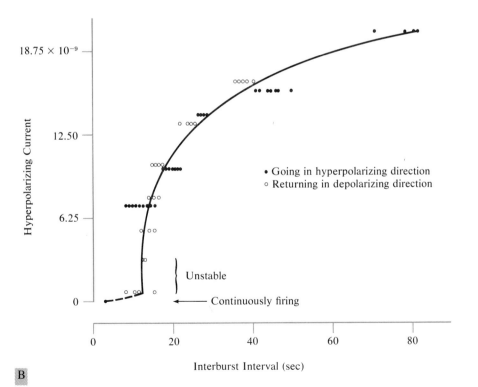

B

18.75×10^{-9}

Hyperpolarizing Current

12.50

6.25

0

• Going in hyperpolarizing direction
○ Returning in depolarizing direction

Unstable

Continuously firing

0   20   40   60   80

Interburst Interval (sec)

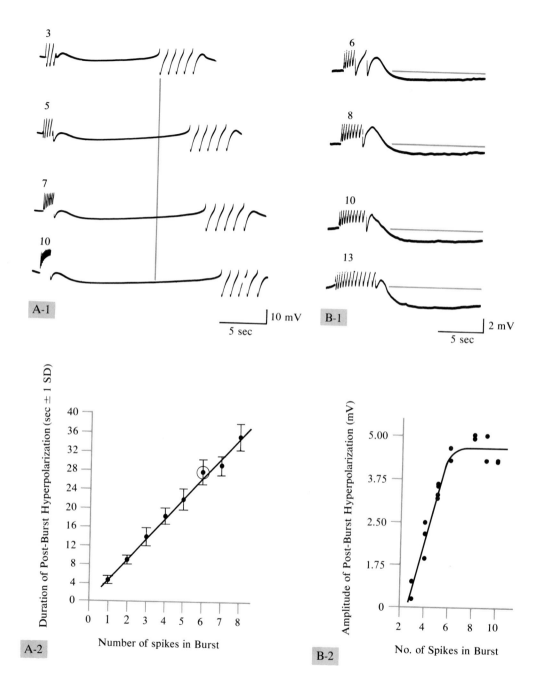

A-1

10 mV
5 sec

B-1

2 mV
5 sec

A-2

Number of spikes in Burst

B-2

No. of Spikes in Burst

FIGURE 7-29 (*Facing page*).

Pacemaker potential in burst-generating cell L3 has properties of a posttetanic afterpotential, duration and amplitude of which are functions of the number of spikes in the burst. [From Frazier *et al.*, 1967; Pinsker and Kandel, unpubl.]

A. Duration of the hyperpolarizing afterpotential is a function of the number of spikes in the burst. *A-1*. Increasing the number of spikes from 3 to 10 increases the duration of the hyperpolarization. *A-2*. Graph of duration of the mean postburst hyperpolarization as a function of the number of spikes in the burst. The cell was firing in bursts of 6 action potentials per burst (circled point on the graph). Action potentials were added or subtracted by means of brief depolarizing and hyperpolarizing current pulses. Bars indicate S.E.M.

B. *B-1*. Increasing the number of spikes per burst from 6 to 13 increases the amplitude of the hyperpolarization. *B-2*. Plot of the amplitude of the postburst hyperpolarization as a function of the number of spikes in the burst.

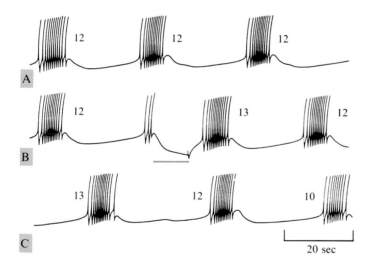

FIGURE 7-30.

Resetting of the spontaneous bursting rhythm of cell R15 by removing spikes. (The record is a continuous trace; the rows have been aligned so that the onset of the first burst in row B is coincident with the onset of the first burst in row A.) The number of spikes in the second burst in row B is reduced by means of a direct hyperpolarizing pulse (bar). This causes the next burst to be initiated much earlier. [From Strumwasser, 1967b.]

264

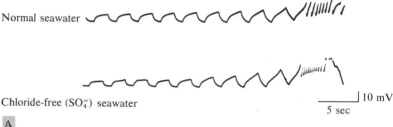

Normal seawater

Chloride-free (SO$_4^=$) seawater

$\boxed{\text{A}}$

| 10 mV

5 sec

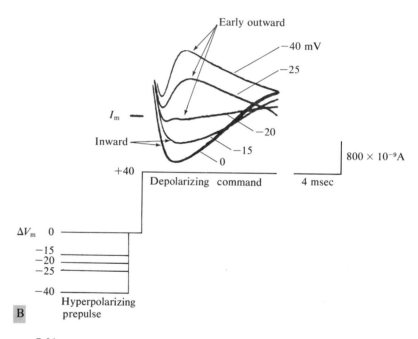

Early outward

−40 mV

−25

I_m −

Inward

−20

−15

0

800 × 10^{-9}A

+40

Depolarizing command

4 msec

ΔV_m 0

−15
−20
−25

−40

Hyperpolarizing
prepulse

$\boxed{\text{B}}$

FIGURE 7-31.
Electrophysiological properties of burst generating cell L3.

A. Changes in the amplitude of the electrotonic potential (apparent input conductance) during pacemaker potential in normal seawater and Cl$^-$-free seawater. The decreased conductance (anomalous rectification) during the course of the pacemaker potential is not dependent on external Cl$^-$. [From Waziri, Frazier, and Kandel, 1965; Pinsker and Kandel, unpubl.]

B. Voltage-clamp records illustrating early K$^+$ current in response to a 40 mV depolarizing command pulse presented after 400 msec hyperpolarization prepulses (see Figure 7-27A) varying from 0 mV to 40 mV. In this cell the inward and early outward K$^+$ currents had similar time courses. When the hyperpolarization prepulse is −20 mV, the inactivation of the early (K$^+$) current is removed, as is indicated by the decrease in the peak early inward (Na$^+$) current. With hyperpolarizing prepulses of −25 and −40 mV, the early K$^+$ current is clearly manifested as an outward current. [From Faber and Klee, 1972.]

burst discharge (Figure 7-32; Strumwasser, 1968; Mathieu and Roberge, 1971; Junge and Stephens, 1973). When the ganglion is perfused with TTX (to block the g_{Na} channel of the spike) and a low-Ca^{++} solution (to block the g_{Ca} channel), the bursting cells show large (30 mV) recurrent membrane oscillations with a periodicity similar to that of the burst cycle (Figure 7-32). Voltage-clamp experiments by Gola (1974), Wilson and Wachtel (1974), and Smith, Barker, and Gainer (1975) indicate that the depolarizing phase of the TTX-resistant oscillations represents an unusual g_{Na} that differs from the resting g_{Na} because it is voltage dependent. It also differs from the g_{Na} of the action potential because its threshold of activation is below (more negative than) the threshold for the action potential. This Na$^+$ current does not inactivate

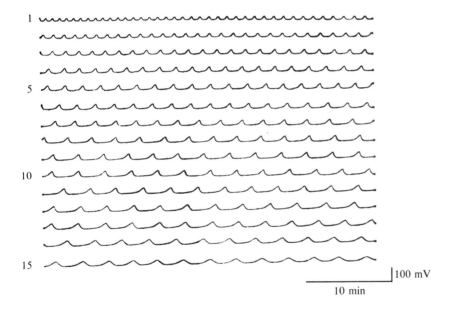

100 mV

10 min

FIGURE 7-32.
Membrane oscillations underlying the autoactive burst pattern of cell R15 in the absence of action potentials. Continuous intracellular recording from R15 in the presence of Ca^{++}-free artificial seawater containing TTX (25 μg/ml), which blocks the action potentials. Lines 1–15 represent a continuous record, reading from top to bottom; each line is 40 min long. Because actions potentials are blocked, the oscillations in g_{Na} and g_K that normally underlie the bursts in this cell can be seen. [Modified from Strumwasser, 1971.]

and is highly temperature dependent.[7] The hyperpolarizing phase of the oscillation is due to a g_K that is activated by depolarization and which turns on and off with a very slow time course of several seconds.

Superimposed on the self-sustained membrane oscillations produced by the slow g_{Na} and g_K are recurrent bursts of action potentials triggered by the depolarizing phase of each membrane oscillation. The bursts of action potentials reinforce the slow oscillations and help to sustain and amplify the bursting rhythm. Each burst of action potentials is followed by a posttetanic hyperpolarizing afterpotential that decays with time. Its duration and amplitude are functions of the number of spikes in the burst (Figure 7-29; Waziri, Frazier, and Kandel, 1965; Pinsker and Kandel, unpub). The decay of the afterpotential, the high resting g_{Na}, and the slow g_{Na} are thought to interact to produce a depolarizing interburst pacemaker potential that brings the membrane potential to threshold and triggers the next burst. In turn, the progressive activation of the slow g_K is thought to shut off the burst. In the range of the interburst pacemaker potential the membrane characteristics of the bursting cells are nonlinear and show anomalous rectification (Figure 7-31A), presumably in part due to the inactivation of the early g_K.

As in the beating cells, the postburst hyperpolarization provides the conditioning hyperpolarization for the early K[+] current, and the depolarization produced by the interburst pacemaker potential activates the early K[+] current. This current in turn slows the gradual upswing of the pacemaker potential. A progressive inactivation of the early K[+] current during the late phase of the interburst pacemaker potential combined with the decay of the slow g_K could produce the apparent conductance decrease associated with anomalous rectification.[8] The further inactivation of the early g_K during the onset of the burst

---

[7]A similar current attributable to g_{Ca} has been found in some bursting cells of *Helix* (Eckert and Lux, 1976).

[8]Electrotonic potentials produced by constant-current pulses provide a good measure of the passive properties of neurons and ionic conductance changes that are *independent* of membrane potential, such as those produced by synaptic actions. However, in examining ionic conductances that are *dependent* on membrane potential, the electrotonic potentials themselves may affect the measurement of these conductances by the changes they produce in membrane potential. In this case a change in amplitude of the electrotonic potential cannot be attributed unambiguously. For example, in the region of membrane potential traversed by the pacemaker potential, a decrease in electrotonic potential could be caused either by an increase in g_K (as appears to be the case for the early phase of the pacemaker potential) or by an increase in g_{Na} (as appears to occur near threshold). In the latter case a constant-current hyperpolarizing pulse also produces an increased electrotonic potential (see Figure 7-31A). Moving the membrane potential in the hyperpolarizing direction will turn off some of the increase in g_{Na} and force the membrane potential further in the hyperpolarizing direction, producing a larger electrotonic potential than would occur in the absence of the increased g_{Na}.

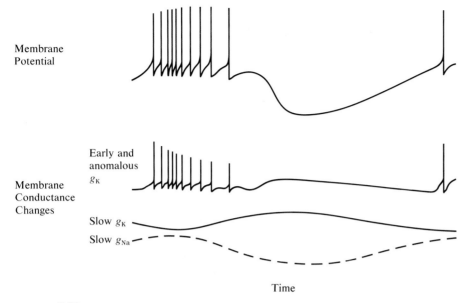

Membrane
Potential

Membrane
Conductance
Changes

Early and
anomalous
g_K

Slow g_K

Slow g_{Na}

Time

FIGURE 7-33.

A model for burst generation. The model consists of four currents in addition to those responsible for the action potential: (1) a steady, time- and voltage-independent g_{Na} that tends to pull the membrane potential in the hyperpolarizing direction; (2) an early g_K; (3) a slow g_{Na} (in *Helix* the function of this current is served by g_{Ca}; and (4) a slow g_K. The burst is postulated to be turned off by two processes: the buildup of the slow g_K and the decay of the slow g_{Na}. The two factors drive the membrane potential in the hyperpolarized direction into the range where the early (anomalously rectifying) K⁺ conductance is reactivated. As the slow g_K decays with time, the constant g_{Na} again becomes effective in moving the membrane potential in the depolarizing direction. The slow depolarization produces a decrease in the early (anomalously rectifying) g_K, which tends to keep the membrane going in the depolarizing direction, thereby turning off the g_K even more. As the g_K decreases, the Na⁺ current becomes even more effective in depolarizing the membrane. The positive interaction of a decreasing g_K and a consequently more effective Na⁺ current have a regenerative depolarizing tendency. This produces the pacemaker depolarization that turns on the slow g_{Na} and then terminates in the burst. As the early g_K inactivates further, it leads to an increase in spike height and rate of firing. [Based on Waziri, Frazier, and Kandel, 1965; Faber and Klee, 1972; Gola, 1974; Smith, Barker, and Gainer, 1975.]

probably accounts for the progressive increase in spike height and the shortening of the interspike interval after the second spike.

From these data one can build a model for bursting cells that might work as follows (Figure 7-33). The burst is turned off by a progressive activation of the slow g_K (and the posttetanic hyperpolarizing afterpotential), which drives the membrane potential in the hyperpolarized direction into the range where the early g_K is activated. As the slow g_K decays with time, the resting g_{Na} again becomes effective in moving the membrane potential in the depolarizing direction and activating the early g_K, which in turn slows the depolarization. But as the depolarization progresses, it produces a gradual decrease in the early g_K, which tends to keep the membrane moving in the depolarizing direction, thereby turning off the early g_K even more. The positive interaction of the decrease in the two K$^+$ conductances and the consequently more effective Na$^+$ current have a regenerative depolarizing tendency. This produces the interburst pacemaker depolarization, which turns on the slow g_{Na} that triggers the burst. As the early g_K becomes further inactivated during the repetitive firing, there is an increase in spike height and rate of firing until the slow g_K becomes sufficiently activated to turn off the burst.

It is interesting to note that one could obtain a burst-generating mechanism with fewer currents. The depolarizing driving force of the Na$^+$ leak alone could trigger the burst and the progressive build up of Na$^+$-inactivation or the activation of the delayed K$^+$ current could be sufficient to turn off the burst. The presence of additional current (the early K$^+$ and the slow K$^+$ and Na$^+$ currents) illustrates the overdetermined nature of the bursting mechanism. The multiple reinforcing currents are presumably designed to insure that the cell functions optimally under a wide range of conditions.

Protein Synthesis

Neurons synthesize many species of proteins and other large molecules (macromolecules). The functions of most of these are still unknown. Ultimately, it should be possible to explain each cell's unique properties in terms of the specific proteins and other macromolecules synthesized by it. A particularly fascinating aspect of this problem is whether autoactive cells synthesize proteins that are different from those in silent cells and that are necessary for their characteristic firing pattern.

Using SDS-polyacryl-amide gel electrophoresis, a technique that predominately separates proteins according to size, D. Wilson (1971) found that neurons with different autoactive firing patterns do indeed synthesize different proteins. The regular-bursting cell R15, the irregular-bursting cell L11, and the beating cell R14 synthesize a large amount of a small protein or polypeptide with a low molecular weight (about 12,000 daltons). By contrast, the silent cell R2, synthesizes primarily protein of higher molecular weight (Figure 7-34). Gainer

FIGURE 7-34.
Patterns of protein synthesis in identified silent cell R2, cell R14, which sometimes has a regular beating rhythm, and cells R15 and L11, which have a bursting rhythm. Isolated ganglia were incubated in a medium containing the labeled amino acid ^{3}H-leucine (duration of incubation is indicated for each cell). Single cells were dissected and extracts were electrophoresed on SDS-polyacrylamide gels, which sieve proteins according to molecular weight. Calibrations of molecular weights (as determined by migration of standards) is indicated at the top of each histogram. Each gel slice is 1¼ mm in length. In the range of lower molecular weights (about 12,000 daltons) there is a peak of radioactivity in the autoactive cells R14, R15, and L11 that is lacking in the silent cell R2. (In these gels 8.5×10^6 cpm $= 1$ μg leucine.) [From D. Wilson, 1971.]

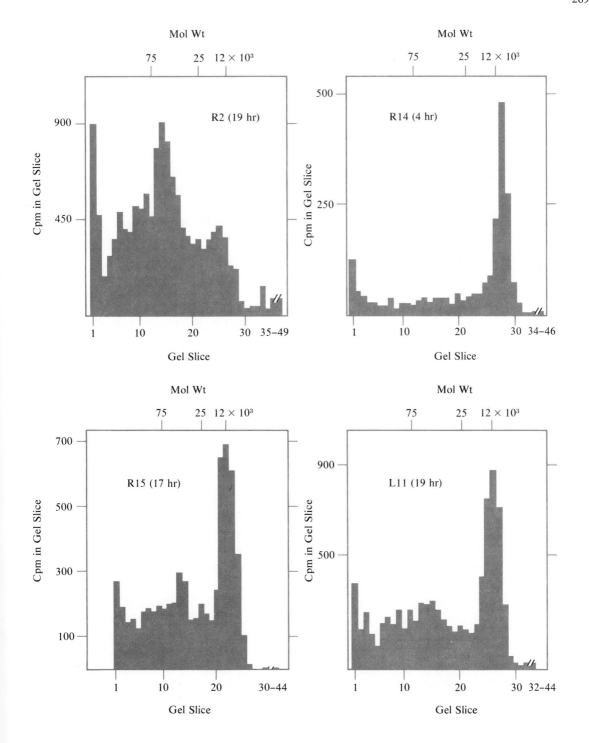

(1972) has suggested, however, that the material of low molecular weight is not responsible for autoactivity but represents one or more neurosecretory products. Gainer (1972; and Gainer and Wollberg, 1974) found polypeptides of low molecular weight (about 5,000 daltons)—similar to those found by Wilson in L11 and R15—in two groups of neurosecretory cells, the bag cells and the white cells.

The idea that this material represents neuroendocrine products rather than a burst-generating polypeptide is supported by the work of Kupfermann and Weiss (1976). They found that the bursting cell R15 has a neuroendocrine function. When extracts of R15 are injected into an animal, they cause weight gain and increased uptake of water. This hormonal action is mediated by material of low molecular weight, in the 2,000 to 20,000 dalton range.

Transmitter Biochemistry and Pharmacology

Among the proteins that most distinguish nerve cells are the enzymes involved in the synthesis of transmitter substances. Chemical synapses in the brains of animals of widely differing phyla all seem to use the same class of chemical substances as transmitter agents (see Table 6-2, p. 203). Typically, these are small molecules: amino acids or closely related compounds.

The study of the biochemistry of synaptic transmission in *Aplysia* began with the work of Tauc and Gerschenfeld (1961, 1962). They found that various unidentified cells in the abdominal ganglion may have different receptors to the same transmitter substance. When ACh was perfused onto the abdominal ganglion some cells became hyperpolarized and inhibited, whereas other cells became depolarized and excited. Similar effects were observed when ACh was applied iontophoretically to the nerve cell bodies, indicating that ACh can act directly on the cell-body membrane and not by way of interposed interneurons. These findings also indicated that even though cell bodies of *Aplysia* neurons are free of synapses (see p. 80) they contain receptors to ACh. Thus, the cell bodies of molluscan neurons can serve as models for studying the pharmacology of synaptic transmission.

Tauc and Gerschenfeld (1961, 1962; and Gerschenfeld, 1973) were able to divide the cells of the abdominal ganglion into two groups on the basis of their responses to ACh: cells showing a hyperpolarizing response, and cells showing a depolarizing response (Figure 7-35). Additional and more complex responses to ACh have since been described, and these are considered in Chapter 8.

Tauc and Gerschenfeld also presented pharmacological evidence that ACh is the inhibitory transmitter for cells with hyperpolarizing responses to ACh

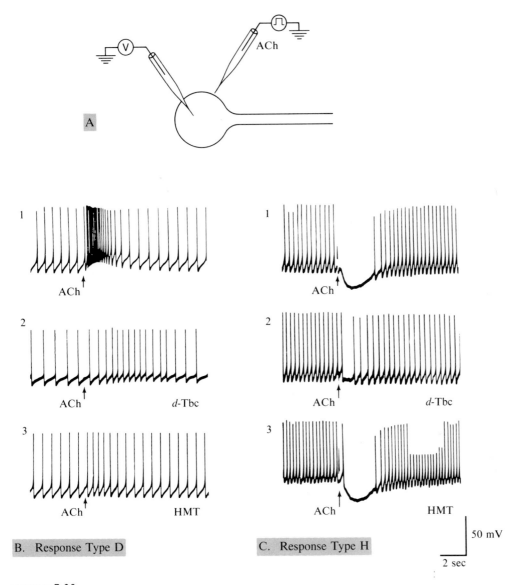

A

1

ACh

2

ACh   *d*-Tbc

3

ACh   HMT

B. Response Type D

1

ACh

2

ACh   *d*-Tbc

3

ACh   HMT

C. Response Type H

50 mV

2 sec

FIGURE 7-35.

Pharmacological properties of neurons in the abdominal ganglion of *Aplysia.* [From Tauc and Gerschenfeld, 1961.]

A. Experimental arrangement for applying brief iontophoretic pulses of ACh to the membrane of the cell body. Acetylcholine is a positively charged ion and can be released onto the outer surface of the cell membrane by filling an extracellular micropipette with acetylcholine chloride and electrically controlling the rate of diffusion out of the pipette. The responses to the extracellular pulses of ACh are recorded with intracellular electrodes. **B** and **C.** Application of ACh to various cells reveals two response types: D (depolarizing) and H (hyperpolarizing). *Trace 1.* Acetylcholine depolarizes one type of cell (the D cell) and accelerates its spontaneous firing, and hyperpolarizes another type (the H cell) and inhibits its firing. *Trace 2.* The drug *d*-tubocurarine (*d*-tbc), is placed in the bathing solution (in a concentration of 10^{-4} g/ml) and the same quantity of ACh as in trace 1 is applied to the cell. The action of ACh on both D and H cell membranes is blocked. *Trace 3.* Hexamethonium (HMT) bromide (in a concentration of 10^{-4} g/ml) blocks the action of ACh on the D cell membrane, but not on the H cell membrane.

and the excitatory transmitter for cells with depolarizing responses. They found that both depolarizing and hyperpolarizing responses could be blocked by atropine and *d*-tubocurarine (curare), two pharmacological agents known to block ACh transmission in vertebrates by binding to ACh receptors (Figure 7-35). The two types of responses to ACh could be further distinguished. Another blocking agent, hexamethonium, blocked only the receptors for the depolarizing response and not those for the hyperpolarizing response (Figure 7-35). Curare also blocked the polysynaptic inputs to the cells produced by stimulating nerves or connectives (Figure 7-36).

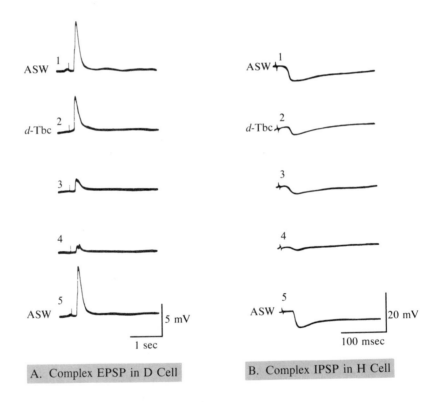

A. Complex EPSP in D Cell

B. Complex IPSP in H Cell

FIGURE 7-36.
Pharmacological properties of complex (polysynaptic) excitatory and inhibitory potentials produced in unidentified cells of the abdominal ganglion by stimulation of nerves or connectives. **A.** Complex EPSP due to activation of many cells producing both monosynaptic and polysynaptic actions in a D cell. **B.** Complex IPSP in an H cell; notches in the action potentials indicate the complex nature of synaptic input. In both parts *A* and *B*, trace 1 shows a normal complex PSP in artificial seawater (ASW). Traces 2 to 4 show progressive blockade of the PSP in *d*-tbc. Trace 5 shows that the blockade is reversed by washing out the *d*-tbc with artificial seawater. [From Tauc and Gerschenfeld, 1961.]

In the course of identifying cells in the abdominal ganglion, some cells were found to be presynaptic to others, which permitted a pharmacological and electrophysiological analysis of synaptic actions based on known direct (mono-synaptic) connections. Using the monosynaptic connections between one identified presynaptic cell (L10) and its follower neurons, Kandel, Frazier, Waziri, and Coggeshall (1967), and later Blankenship, Wachtel, and Kandel (1971) and Kehoe (1972b, c), found that the follower cells of L10 responded to ACh and that the direct synaptic connections between L10 and its follower cells were blocked by cholinergic blocking agents. The ionic mechanisms for both the synaptic potential and the ACh responses were found to be the same (see Chapter 8). The evidence for the cholinergic nature of transmission at these synapses was further strengthened by the biochemical studies of Giller and Schwartz (1968, 1971).

Taking advantage of the fact that the cell bodies of neurons in this ganglion are large and free of synapses, Giller and Schwartz performed biochemical studies on individually dissected cells. They found that cell L10 contains the enzyme choline acetyltransferase, which catalyzes the synthesis of ACh from choline and acetate. The enzyme is active in the cell body. When choline labeled with radioactivity (^{3}H-choline) was injected into L10, 85 percent of the labeled material was converted to ACh within one hour (Koike, Kandel, and Schwartz, 1974). The labeled material is released by impulse activity after an initial latency, presumably reflecting the amount of time necessary to transport ACh from the cell body to the terminals (Koike *et al.*, 1974).[9] As is the general case for chemical synaptic transmission, stimulated release is sensitive to the Ca^{++}/Mg^{++} ratio of the bathing solution (see p. 183). Increasing the Mg^{++} concentration decreased the stimulated release; increasing the Ca^{++} concentration increased the stimulated release (Figure 7-37). Comparable injections of choline into the noncholinergic cells did not lead either to conversion of choline to ACh or to an increase of firing release. Similarly, injection of ^{3}H-choline into cell R2, a cholinergic neuron that does not have synapses in the abdominal ganglion, led to rapid conversion to ACh but did not give rise to an increase of firing release.

Thus, the evidence for cholinergic synaptic transmission in cell L10 is almost complete (see Chapter 8). Acetylcholine simulates the sign of the natural transmitter and produces the same ionic conductance changes when applied iontophoretically to the follower cells. The cell body of the presynaptic neuron

---

[9]The material released was found to be choline rather than ACh. But there is good reason to believe that this is due to breakdown of the secreted ACh by the degradative enzyme acetylcholinesterase outside the cell (Koike *et al.*, 1974).

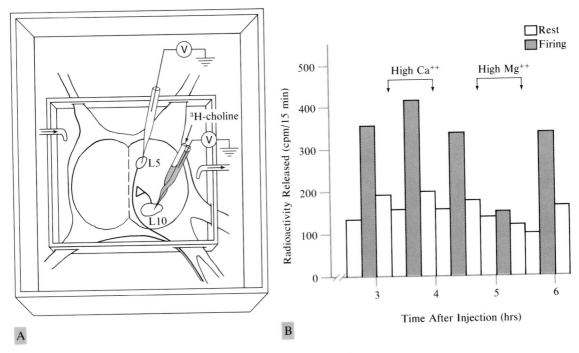

FIGURE 7-37.
Release of labeled material from cell L10. [From Koike, Kandel, and Schwartz, 1974.] **A.** Experimental arrange-ment. A labeled precursor of ACh, ^{3}H-choline, is injected into L10; intracellular recordings are taken from cell L5 to identify L10. **B.** The graph compares release of the labeled material in response to stimulation (2 impulses/sec) and while the cell is at rest. The ganglion was alternatively bathed in normal seawater and high-Ca^{++} and high-Mg^{++} solutions. The high-Ca^{++} solution produced a slight increase in stimulated release while the high-Mg^{++} solution blocked stimulated release.

contains choline acetyltransferase and synthesizes ACh. Acetylcholine is trans-ported from the cell body to the terminals and appears to be released by nerve impulses. There is also good but less complete evidence for the cholinergic nature of cells R2, L11, LD$_{G1}$, LD$_{G2}$, LD$_{HI1}$, LD$_{HI2}$, LB$_{VC1}$, LB$_{VC2}$, and LB$_{VC3}$. Again, as with other differentiating characteristics of neurons, many of the cells having the ability to synthesize ACh tend to be grouped together in clus-ters distributed throughout the ganglion (Figure 7-38).

Serotonin (5-hydroxytryptamine, or 5-HT) serves as a transmitter for other cells of the abdominal ganglion (Figure 7-38; Eisenstadt *et al.*, 1973). The first step in the biosynthesis of serotonin is the conversion of the amino acid tryptophan (by hydroxylation) to 5-hydroxytryptophan by the enzyme trypto-phan hydroxylase. The substance 5-hydroxytryptophan is then converted (by

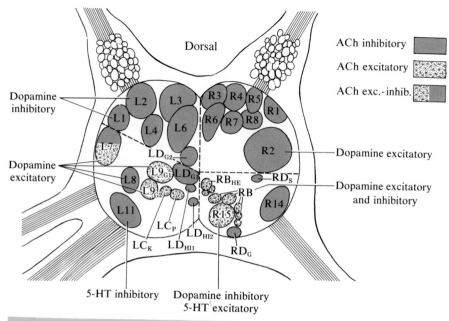

A. Acetylcholine- and Dopamine-Receptive Neurons

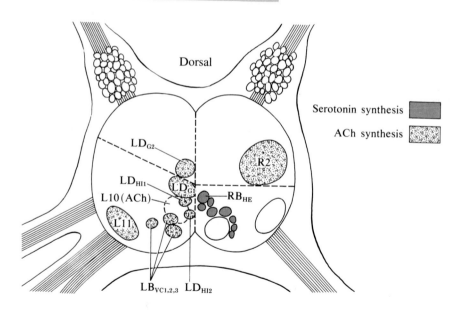

B. Cholinergic and Serotonergic Neurons

FIGURE 7-38.
Map of cells in the abdominal ganglion of *Aplysia* that synthesize or respond to certain types of transmitter. **A.** Cells that respond to ACh (see key). For some of these cells the response to dopamine and serotonin (5-HT) is also indicated, but the maps for these two transmitters are less complete. [Based on Kandel *et al.*, 1967; and Ascher, 1972.] **B.** Cells that synthesize ACh and serotonin. [Based on Giller and Schwartz, 1971; and Eisenstadt *et al.*, 1973.]

decarboxylation) to serotonin (5-hydroxytryptamine) by the enzyme aromatic amino acid decarboxylase. When the precursor amino acid L-³H-tryptophan is injected into a cell belonging to the RB cluster, it is converted to serotonin (Eisenstadt et al., 1973).[10]

Cells seem to be restricted in their capability for synthesizing (or packaging) transmitter substance. The cholinergic cells L10 and R2 cannot convert L-³H-tryptophan to serotonin because they lack the enzyme tryptophan hydroxylase, and serotonergic RB cells cannot convert choline to ACh because they lack the enzyme choline acetyltransferase (Eisenstadt et al., 1973). Thus, at least one component of the synthetic apparatus for each of the transmitter substances is a unique property of each neuron. Not all enzymes in the synthetic pathway are specific, however. The cholinergic cell R2 can convert 5-hydroxytryptophan to serotonin (5-hydroxytryptamine) presumably because all identified cells in the ganglion have the enzyme aromatic amino acid decarboxylase (Weinrich, Dewhurst, and McCaman, 1972; Eisenstadt et al., 1973).

Less is known about other transmitter candidates: dopamine, gamma-aminobutyric acid, and glutamate. Of these, dopamine has been most studied. Receptors for dopamine, determined by iontophoresis, have been found on several identified cells (Ascher, 1972). A comparison of that map with those available for ACh and serotonin receptors (Kandel, Frazier, Waziri, and Coggeshall, 1967; Faber and Klee, 1974) shows that although the cell may synthesize and release only one transmitter it can synthesize receptors to more than one transmitter (Figure 7-38).

Morphology

Three recently developed intracellular marking techniques permit study of the light-microscopic morphology of individual nerve cells and their processes: (1) injection of one of several fluorescent Procion dyes (Stretton and Kravitz, 1968, 1973); (2) injection of cobalt chloride followed by incubation of the ganglion in ammonium sulfate, which leads to precipitation of the cobalt chloride (Pitman, Tweedle, and Cohen, 1972); and (3) injection of the enzyme horseradish peroxidase followed by incubation of the ganglion in diaminobenzidine, which leads to a reaction product that is electron dense (Muller and

[10]The natural synaptic actions of the RB_HE cells on the heart are also simulated by serotonin and blocked by serotonin antagonists (Liebeswar, Koester, and Goldman, 1973). For data on the serotonergic nature of the metacerebral cells of *Aplysia* see Weinreich, McCaman, McCaman, and Vaughn, 1973; Eisenstadt et al., 1973; and Gerschenfeld and Paupardin-Tritsch, 1974b.

McMahan, 1975). These markers are retained in the cell and can be used to study the three-dimensional geometry of individual cells.[11]

Morphological studies of identified invertebrate neurons based upon intracellular injection of markers have revealed that the three-dimensional structure of specific identified neurons is fairly distinctive and roughly the same in all animals of a species. In crayfish and lobster the three-dimensional structure can actually serve as a fingerprint of the cell (Stretton and Kravitz, 1968, 1973). In *Aplysia,* too, different identified cells have distinctive morphologies. Most impressive are the differences in the dendritic complexity of identified cells, which varies with the complexity of the synaptic input of the cells. The white cells, which on the basis of physiological data are known to receive few connections, have a very modest dendritic field. By contrast, cell L7, which receives many connections, has a rich and extensive dendritic tree. Cells with moderate synaptic input (such as L3) have a moderately complex dendritic tree (Figure 7-39). But in *Aplysia,* unlike crayfish or lobster, there is variability in the pat-

[11]Under some circumstances, however, the stains cross electrical synapses (see Furshpan and Potter, 1968; Bennett, 1973).

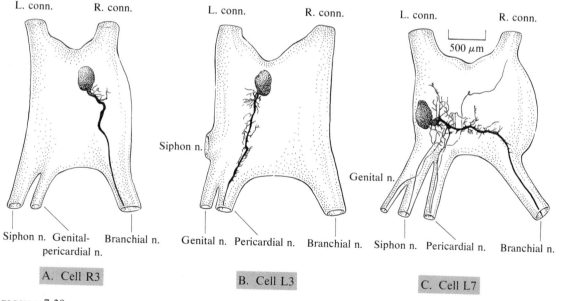

FIGURE 7-39.
Intracellular injection of Co++ into various cell types illustrates the relation between the extent of dendritic arborization and extent of synaptic input. **A.** A neurosecretory cell (R3) has a small dendritic field and little input **B.** Cell L3 has a moderate dendritic field and moderate input. **C.** Cell L7 has an extensive dendritic field and extensive input. [From Winlow and Kandel, 1976.]

tern (and sometimes in the existence) of the secondary and tertiary branches, although the path of the primary axon is fairly invariant (Winlow and Kandel, 1976).

SUMMARY AND PERSPECTIVE

The existence of identified nerve cells in invertebrates has been known since the second half of the nineteenth century and was well established for several nematode and annelid worms by the beginning of this century. The most remarkable of these examples was the nematode worm *Ascaris;* all 162 cells in the nervous system are invariant. Some invariant cells have now been mapped in *Aplysia,* in other opisthobranch and pulmonate molluscs, and in the ganglia of the crayfish, lobster, leech, cockroach, and cricket. The discovery of unique nerve cells, distinguishable in every member of the species, is one of the most telling arguments against the notion that the nervous system is made up of stereotyped units with similar properties.

Although the axons of most nerve cells seem remarkably similar, a number of electrophysiological differences between neurons are evident in the cell bodies. For example, in *Aplysia* the cell bodies of different identified cells have different autoactive firing properties. Some cells are silent, some fire regularly (beating cells), some in regular bursts (bursting cells), and some in irregular bursts. The cell bodies also vary in the mechanism that controls the membrane potential, the action potential, and the afterpotentials.

What accounts for this diversity? Biophysical studies of identified cells have shown that the membrane of the axon usually produces only three currents: a high resting K^+ conductance that generates the membrane potential, and the sequential increases in Na^+ and K^+ conductance that account for the action potential. But the cell bodies of central neurons often have additional current generators that control or contribute to the cell's activity. Various combinations of membrane ionic mechanisms thus produce the diversity of resting and firing patterns.

The membrane potential of cell bodies of many neurons in the abdominal ganglion is controlled by an electrogenic Na^+ pump that operates in parallel with the high K^+ conductance. The relative contribution of these two mechanisms to the membrane potential varies among cells. In some cells the Ca^{++} as well as Na^+ contributes to the inward current of action potential. Afterpotentials can be caused by an increase in K^+ conductance, by the activity of an electrogenic Na^+ pump, or (in the leech) by both. Even greater accretions of conductances are found in neurons with autoactive capabilities. Beating cells have the usual K$^+$ conductance and electrogenic Na$^+$ pump, for generating the membrane potential, and the usual inward Na$^+$ (and/or Ca^{++}) current and a delayed outward K$^+$ current, for generating the action potential. These cells also have a high resting Na$^+$ current and an early K$^+$ conductance, both of which contribute to the depolarization between spikes that make up the pacemaker potential. In addition to these six currents, the cell bodies of burst-generating neurons have two other currents: a slow cyclical Na$^+$ conductance that triggers the burst, and a slow K$^+$ conductance that aids in turning it off. The diversity of currents that can be generated by cell bodies of neurons seem truly remarkable, particularly since only three species of ions (Na$^+$, K$^+$, and Ca^{++}) are often involved.

Biophysical studies of the ionic currents of the membrane allow new insights into the functioning of the cell bodies of neurons. It has long been appreciated that the cell body of a neuron contains the biochemical synthetic machinery necessary for cellular function. It has also been known that the cell body imposes integrative restraints on neuronal function because of its location in relation to synaptic input. But prior to the detailed biophysical studies of identified cells it was not realized that the excitable membrane of the cell body differs markedly from the axon membrane in that the cell body has a much greater variety of current generators than does the axon. The diversity of current generators in the cell bodies of different identified cells is perhaps the most distinctive electrophysiological characteristic of cell types.

Identified cells also differ in the types of macromolecules they synthesize and in the biochemistry of their transmitter substances. For example, some cells synthesize acetylcholine, others synthesize serotonin; cells that synthesize one of these transmitter substances cannot synthesize or release the other. On the other hand, cells can synthesize and utilize receptors to more than one species of transmitter compounds. In addition to serotonin and acetylcholine, some cells have receptors to dopamine, gamma-aminobutyric acid, or glutamate. There is also a surprising degree of diversity in the morphology of neurons. As additional criteria are applied, such as behavioral function, previously indistinguishable cells often can be distinguished. Indeed, most cells may ultimately prove unique.

However, much more work in analyzing the diversity among neurons is needed. For example, it is not known to what degree the several ionic currents quantitatively account for a particular autoactive pattern, such as bursting. And little is known about transmitters other than acetylcholine and serotonin. As a result, the nature of the transmitter of most identified cells (in most invertebrate ganglia) is not known. There is even less knowledge about other biochemical aspects of cellular function. Except for transmitters, neurohormones, and receptors, few functional properties of a cell have yet been related to a specific biochemical process. There is a great need for experiments that relate the synthesis of specific macromolecules with ionic conductance patterns in cells.

SELECTED READING

Kennedy, D., A. I. Selverston, and M. P. Remler. 1969. Analysis of restricted neural networks. *Science,* 164:1488–1496. Reviews identified cell mapping in the crayfish.

Nicholls, J. G., and D. A. Baylor. 1968. Specific modalities and receptive fields of sensory neurons in CNS of the leech. *Journal of Neurophysiology,* 31:740–756. Relates Retzius's study on identified cells to their sensory function.

Otsuka, M., E. A. Kravitz, and D. D. Potter. 1967. Physiological and chemical architecture of a lobster ganglion with particular reference to gamma-aminobutyrate and glutamate. *Journal of Neurophysiology,* 30:725–752. The first study to examine the transmitter biochemistry of individual cells and relate it to their motor and topographical properties.

Tauc, L. 1967. Transmission in invertebrate and vertebrate ganglia. *Physiological Reviews,* 47:521–593. Compares the cellular physiology and pharmacology of vertebrate and invertebrate ganglia.

CHAPTER 8

Patterns of Synaptic Connection

A major problem in the study of the nervous system is the analysis of the interconnection of nerve cells. This problem has three aspects. First, how invariant are the connections? As we have seen, many nerve cells in invertebrate ganglia are unique and present in all members of the species. Does this invariance also apply to the synaptic connections between cells? Does a given identified cell always connect to the same follower cells? Second, what are the rules that determine the physiological types of connections between neurons? And third, how do different patterns of interconnection relate to different classes of behavior?

In this chapter I shall consider two major patterns of connection between identified neurons, the invariance of these patterns, and the rules that determine their physiological properties. The first is the *elementary convergent aggregate*: convergent connections made by a population of neurons on a single follower cell. The second is the *elementary divergent aggregate*: divergent connections made by a single neuron on the population of follower cells. Compared to convergent aggregates, divergent aggregates have been studied more extensively, and the major part of this chapter will therefore be concerned with this pattern. Later, in Chapters 9 and 10, I shall consider how different patterns of interconnection relate to different classes of behavior.

ELEMENTARY DIVERGENT AGGREGATES

One of the most general and elementary patterns of interconnection is that made by a single neuron with a population of follower cells (Figure 8-1). This

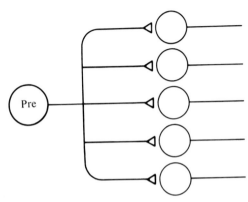

FIGURE 8-1.
Schematic drawing of an elementary divergent aggregate, consisting of a presynaptic neuron and a population of five follower neurons. *For purposes of simplification, the connections in this and subsequent illustrations are shown terminating on the* cell body. *Actually, all synapses in invertebrate ganglia end on the* axon and dendrites *of the postsynpatic cell.*

pattern is found in all nervous systems; I shall refer to it as the *elementary divergent aggregate.* Although simple, this pattern of interconnection is instructive and can provide answers to the following questions: (1) Does a given neuron always synapse on the same cells in the same way? (2) Are presynaptic neurons limited to one type of synaptic action (inhibition or excitation), or can different synaptic branches act differently on different neurons? (3) Does a neuron release the same transmitter substance from all of its terminals, or can it release different substances from different terminals? (4) Is the sign of the synaptic potential (and the underlying ionic conductance mechanism) specified by the chemical structure of the transmitter substance released by the presynaptic cell or by the configuration of the receptor molecule(s) of the postsynaptic cell? (5) Do all the follower cells of a presynaptic neuron have only one type of receptor for a given transmitter released by the presynaptic neuron, or can a single postsynaptic cell have more than one type of receptor to the same transmitter?

The search for rules to describe the patterns of interconnection among cells of divergent aggregates has had a long and interesting history. Ramon y Cajal first described the elementary divergent aggregate at the beginning of the century. He thought that cells connect to each other in reliable ways, that is, a given cell always connects to certain types of cells and not to others. But Cajal also thought that neurons mediate only excitatory actions; his diagrams show no inhibitory synaptic actions (for review see Cajal, 1911). Sherrington, on the other hand, appreciated the existence of inhibitory action early in his work and suggested that inhibition was a separate process, of opposite sign to synaptic excitation. He formulated the principle of *reciprocal innervation:* excitation of synergist motor neurons is accompanied by inhibition of antagonist motor neurons. Reciprocal innervation suggested to Sherrington that neurons are basically divalent in action and that they mediate both excitation and inhibition (Figure 8-2*A*). "The single afferent nerve-fiber would therefore in regard to one set of its terminal branches be *specifically excitor,* and in regard to another set of its central endings be *specifically inhibitory.* It would, in this respect, be duplex centrally."[1]

It was not until 1941, however, that this hypothesis was tested. D. P. C. Lloyd found that the 1A afferent fibers (fibers from stretch receptors) from a given muscle mediated direct excitation to the motor neurons of that muscle and direct inhibition to the motor neurons of the antagonist muscle. Although not identical, the time course and latency (see below) of the excitatory and inhibitory synaptic actions proved remarkably similar, and there was good evidence that the excitatory connections were monosynaptic. Lloyd therefore argued that the inhibitory connections were also monosynaptic.

Whether these synaptic actions in the central nervous system were chemically or electrically mediated was not known at that time. However, Sir Henry Dale and his colleagues had already provided evidence that transmission was chemical at peripheral synapses, at the neuromuscular junction, and in autonomic ganglia. Dale eleborated these findings in a series of reviews (for example, see Dale, 1935). He suggested that neurons are biochemically specific and have a metabolic unity that extends to all their branches, that is, that neurons were likely to release the same chemical transmitter substance from all of their terminals.

> . . . In the cases for which direct evidence is already available, the phenomena of regeneration appear to indicate that the nature of the chemical function, whether cholinergic or adrenergic, is characteristic for each particular neurone

[1]Sherrington, 1906, pp. 104–107.

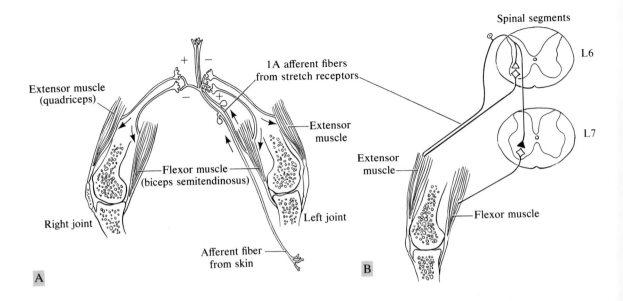

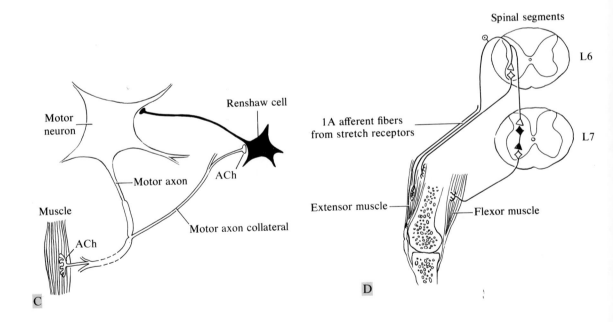

and unchangeable. When we are dealing with two different endings of the same sensory neurone the one peripheral . . . and the other central, can we suppose that the discovery and identification of a chemical transmitter . . . [at the peripheral synapse] would furnish a hint as to the nature of this process at a central synapse. The possibility has at least some value as a stimulus to further experiment.[2]

In 1952 Brock, Coombs, and Eccles examined direct excitation and direct inhibition in the spinal cord of the cat, using intracellular microelectrodes, and provided evidence that these central synaptic actions were also chemically mediated. Brock and co-workers assumed that both direct excitation and direct inhibition were monosynaptically mediated and, following Dale's suggestion, they concluded that the different branches of 1A afferent fibers probably produced their opposite synaptic actions by utilizing the same transmitter (Figure 8-2B; see Brock, Coombs, and Eccles, 1952; Eccles, 1953).

Two years later, while studying the effects of antidromic (backward) stimulation of the axons of the motor neurons, Eccles, Fatt, and Koketsu (1954) discovered recurrent (feedback) inhibition, that is, activity in some motor cells feeds back on other motor cells and inhibits their activity. They immediately appreciated that recurrent inhibition must be mediated by a pathway, described previously by Renshaw (1946), which consists of recurrent collateral

[2]Dale, 1935, p. 329

FIGURE 8-2.
Various views of neuron specialization.

A. Sherrington's diagram of dual-action neurons mediating reciprocal innervation: excitation to synergist motor neurons and inhibition to antagonist motor neurons. The flexor (biceps semitendinosus) and extensor (quadriceps) muscles of the left and right knee joint are illustrated. A sensory neuron mediates monosynaptic excitation by way of an axon to the left flexor motor neurons and monosynaptic inhibition by way of an axon to the left extensor motor neurons. The neuron also produces reciprocal actions in the contralateral side. Excitation (+), inhibition (−). [Modified from Sherrington, 1906.]

B. Brock, Coombs, and Eccles' diagram of dual-action sensory neurons mediating reciprocal innervation to the extensor and flexor muscles of the knee joint. As in part A, one sensory neuron (a stretch receptor) and its 1A afferent fiber is shown for each muscle; it divides in the spinal cord into two branches. One branch mediates direct excitation to the synergist motor neurons, the other branch (black terminal) mediates direct inhibition to antagonist motor neurons. [Modified from Eccles, 1953.]

C. Diagram based upon work of Eccles, Fatt, and Koketsu illustrating a neuron specialized for inhibition (Renshaw cell) interposed in the recurrent pathway between the motor axon collateral and the motor neuron. [Based on Eccles, Fatt, and Koketsu, 1954.]

D. Diagram based upon work of Eccles, Fatt, Landgren, and Winsbury illustrating reciprocal innervation. As in parts A and B, a stretch-receptor sensory neuron is shown dividing in the spinal cord into two branches. One branch produces direct excitation to the synergist motor neurons. The other excites an inhibitory interneuron that inhibits the antagonist motor neuron. [Modified from Eccles, Fatt, Landgren, and Winsbury, 1954.]

branches of motor-neuron axons synapsing on a small interneuron in the central nervous system (the Renshaw cell), which in turn synapses back on the motor neuron. Because the motor neuron releases acetylcholine at its endings in the muscle fiber, Eccles and his co-workers again followed Dale's suggestion and predicted that the collateral branches of the motor-neuron axons would also release ACh at their central synapse on Renshaw cells. This prediction was supported, and ACh was the first transmitter identified in the central nervous system (Figure 8-2C). The findings that the action of ACh in the Renshaw cell was excitatory, as was its peripheral action on muscle, and that the recurrent inhibitory pathway seemed to require an interneuron specialized for inhibition, caused Eccles, Fatt, and Landgren (1956) to reexamine the direct inhibitory pathway of the 1A afferent fibers to the motor neuron of the antagonist muscles. They found that the latency of the IPSP in the motor neuron was 0.8 msec longer than that of the EPSP. This difference could be accounted for if there were an inhibitory element interposed in the direct inhibitory pathway. In fact, Eccles, Fatt, and Landgren (1956) found a group of cells in the spinal cord whose properties seemed to conform to those required for inhibitory interneurons in this pathway (Figure 8-2D; see Hultborn, Jankowska, and Lindström, 1968).

The finding that two different inhibitory pathways to motor neurons each contained an interposed neuron suggested to Eccles an important extension of Dale's principle, the notion of *single function.*

> Conceivably the inhibition could be produced by the same transmitter substance that is responsible for excitatory synaptic transmission. . . . It appears that this solution was not practicable in the morphogenesis of the central nervous system, where instead a change in the transmitter substance was secured by inserting a special interneurone even on the direct inhibitory pathway with all the consequential complications and the delay of an additional synaptic mechanism. It is therefore postulated that any one transmitter substance always has the same synaptic action, i.e., excitatory or inhibitory, at all synapses on nerve cells. *According to this principle, any one class of nerve cells will function exclusively either in an excitatory or in an inhibitory capacity at all of its synaptic endings, i.e., there are functionally just two types of nerve cells, excitatory and inhibitory.*[3]

[3]Eccles, 1957, p. 213; italics mine.

Eccles provided further support of this hypothesis in studies of crossed inhibition in the spinal cord as well as in studies of pathways in the higher centers of the vertebrate brain. Encouraged by the experimental success of his principle of single function, Eccles advanced a second extension of Dale's principle, the principle of *single ionic mechanism:*

> The second principle is that, at all of the synaptic terminals of a nerve cell, the transmitter substance opens just one type of ionic gate, that characterizing either excitatory or inhibitory synapses. . . . There is no simple explanation for the rigid operation of the second principle. For example, it cannot be simply derived from Dale's principle, because a transmitter substance such as acetylcholine acts as an excitatory type of transmitter substance at some synapses (sympathetic ganglia, neuromuscular junctions, and the synapses of motor axon collaterals on Renshaw cells), and as an inhibitory type of transmitter at others (vagus on heart, synapses on H- and D-cells in *Mollusca*). *It seems that there must be some principle of neurogenesis whereby the outgrowing axonal branches of a neurone can make effective synaptic contacts only of an excitatory or of an inhibitory type.*[4]

Dale's principle and the derived notions of single function and single ionic mechanism have been supported by consistent and quite compelling indirect evidence from vertebrates (see Eccles, 1969). Direct examination of Dale's principle has generally not been possible in vertebrates. Individual cells must first be identified and the actions mediated by various branches of a presynaptic neuron then traced to individual postsynaptic cells. In vertebrates, and particularly in mammals, it has usually not been possible to map connections between individual neurons, only between populations of cells. By contrast, invertebrate ganglia can be isolated, the ionic concentrations of the bathing solutions changed, and the ionic conductance changes produced in a single neuron by a transmitter specified. With this procedure, direct and common connections can easily be mapped electrophysiologically, and thus the physiological and pharmacological properties of single synaptic processes can be examined (Figure 8-3).

[4]Eccles, 1969, pp. 112–113; italics mine. H- and D-cells are cells that respond to acetylcholine with hyperpolarization and depolarization, respectively (see pp. 270–272 here).

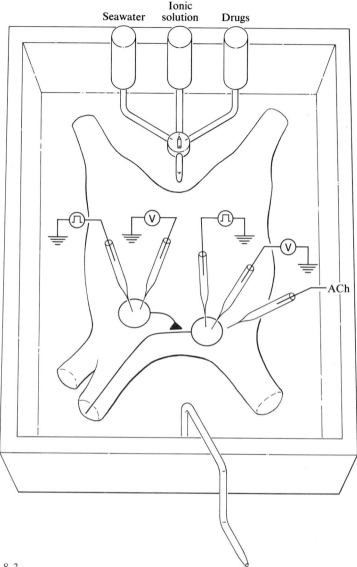

FIGURE 8-3.

Techniques for studying synaptic connections between cells in *Aplysia*. Cell bodies of identifiable neurons are directly visible on the surface of ganglia. Intracellular recordings can be obtained from both presynaptic neurons and postsynaptic follower cells, and independent stimulating electrodes can be inserted into both cells for changing the membrane potential. Although the cell body is free of synapses (these occur on the axon in the neuropil), it contains receptors to the transmitter substances of the neurons that synapse on the cell. One can examine the properties of these receptors by iontophoresing a transmitter substance, e.g., ACh, on the cell body and recording the response produced with the intracellular electrode. A ganglion can be isolated for up to a day in a chamber perfused with seawater. In a culture medium it can be carried for several days. The ionic concentration of the bathing solution can be altered and various drugs can be introduced.

ELECTROPHYSIOLOGICAL CRITERIA FOR DIRECT AND COMMON CONNECTIONS

Chemically Mediated Direct Connections

A direct (monosynaptic) connection between two cells is inferred when every action potential in the presynaptic neuron produces an elementary PSP of brief and constant latency in the follower neuron. "Elementary" refers to the fact that the synaptic potential is unitary; that is, it is produced by a single neuron. All action potentials in a presynaptic cell produce synaptic potentials of identical shape and (except under certain circumstances) similar size in a given postsynaptic cell.

Latency is the time between the onset of the action potential in the presynaptic neuron and the onset of the synaptic potential in the postsynaptic cell. It includes conduction time and synaptic delay. *Conduction time* is the time that it takes for an action potential to propagate the distance from the cell body of the presynaptic neuron to the terminals on the postsynaptic cell being sampled. In intraganglionic connections in invertebrates this distance is short, a millimeter or less. *Synaptic delay* refers to the time required for the action potential to release transmitter, the transmitter to diffuse across the synaptic cleft, and the transmitter to interact with the receptor to give rise to the current flow that generates the PSP; most of the synaptic delay is in the release step. Finding a brief (5 to 15 msec) and constant latency speaks for a short and direct pathway, without the variable and time-consuming steps introduced by interneurons.

Postsynaptic potentials with brief and constant latency are strong electrophysiological evidence for a direct connection between neurons (Figure 8-4*A*). Nevertheless, such evidence does not completely rule out the possibility of an interneuron. A number of additional tests can help to distinguish further between direct and indirect (polysynaptic) connections. One procedure is to raise the firing threshold of cells. This can be done by increasing the concentration of divalent cations (Ca^{++} and Mg^{++}) in the bathing solution (Figure 8-4*B*). Calcium and magnesium have opposite effects on synaptic release (see p. 183), but this can be counterbalanced by raising the concentrations of both ions proportionally, keeping the ratio between them constant. Yet even with its threshold increased the presynaptic neuron can be made to discharge an action potential simply by supplying more outward current. Nevertheless, since the amplitude of the synaptic potential produced in the postsynaptic cell will not be significantly changed, it will be much less likely to discharge an action potential in a postsynaptic cell whose threshold has also gone up considerably.

1

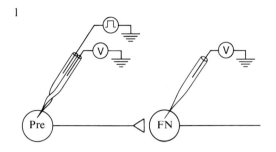

1

TEA

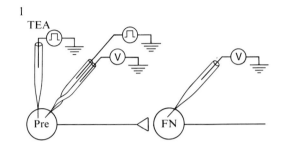

2. Excitatory connection

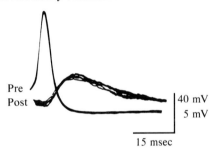

Pre
Post

40 mV

5 mV

15 msec

2. High Ca^{++}, Mg^{++}

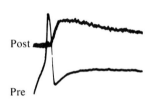

Post

Pre

3. Inhibitory connection

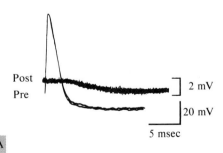

Post
Pre

2 mV

20 mV

5 msec

A

3. TEA

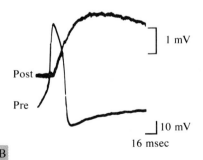

1 mV

Post

Pre

10 mV

16 msec

B

Under these conditions, an indirect connection to the follower cell that is mediated through an interneuron is likely to fail. An elegant further means of demonstrating direct connections is to inject tetraethylammonium chloride (TEA) into the presynaptic neuron (Kehoe, 1969a). This prolongs the action potential by blocking the delayed K^+ current. As TEA diffuses through the cell to the terminals, the action potential in the terminals is prolonged, as is the duration of transmitter release. In the case of a direct connection this would appear as a smooth increase in amplitude and duration of the postsynaptic potential without a change in latency (Figure 8-4*B*). By contrast, in an indirect (polysynaptic) pathway a step increase in latency would be seen (Figure 8-5).[5]

In the final analysis, however, an argument for monosynaptic connections requires morphological in addition to physiological evidence. For a few invertebrate neurons, some light-microscopic evidence is available to support inferences made from electrophysiological studies. In no case, however, is there yet any electron-microscopic verification.

[5]TEA spreads across electrotonic junctions and therefore cannot rule out an interposed electrotonically coupled interneuron (Deschênes and Bennett, 1974).

FIGURE 8-4.

Chemically mediated direct connections. Criteria for monosynaptic connections include a short and constant latency between the ascending limb of the presynaptic action potential and the postsynaptic potential. Because most of the latency is due to conduction time and is independent of the amount of transmitter released by the presynaptic neuron, the latency also remains unaltered in solutions with high concentrations of divalent cations, which increase the threshold of neurons. [From Kandel *et al.*, 1967; and Castellucci and Kandel, 1974.]

A. Test for short and constant latency. *A-1.* Experimental arrangement for simultaneous intracellular recordings from presynaptic and follower neurons. The presynaptic neuron is depolarized and caused to fire repeatedly. To examine the constancy of the latency in successive PSPs, several oscilloscope sweeps are superimposed. *A-2.* Excitatory connection. *A-3.* Inhibitory connection.

B. Test of whether latency is independent of the level of transmitter released, even in a high-Ca^{++}, high-Mg^{++} solution. *B-1.* Experimental arrangement for firing the presynaptic neuron and increasing its transmitter release by injecting tetraethylammonium (TEA) into the cell body of a ganglion bathed in a high Ca^{++}, high-Mg^{++} solution. *B-2.* Control prior to TEA injection. *B-3.* Following TEA injection, the action potential is broadened (because the increase in g_K normally produced by the presynaptic spike is partially blocked) and produces a larger EPSP but with no change in latency.

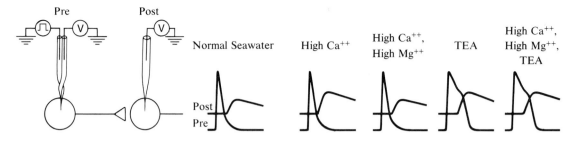

Direct (Monosynaptic) Connection

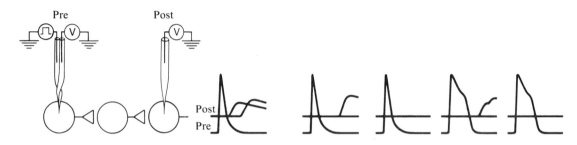

Indirect (Polysynaptic) Connection

FIGURE 8-5.
Schematic drawings based on a hypothetical experiment to distinguish direct (monosynaptic) and indirect (polysynaptic) connections. This is done by bathing the ganglion in ionic solutions that increase the threshold and increasing transmitter release by intracellular injection of TEA into the presynaptic neuron. (1) In normal seawater the latency of a direct connection is constant whereas that of an indirect connection is usually variable. (2) A high-Ca^{++} solution increases the size of the EPSP but increases the firing threshold even more. As a result, an interposed interneuron would be less likely to fire in response to an action potential in the presynaptic neuron. (3) A high-Ca^{++}, high-Mg^{++} solution increases the threshold of neurons even more than the high-Ca^{++} solution, and does so without enhancing synaptic transmission because Mg^{++} counteracts Ca^{++} action at the synapse but both act to raise the threshold. This solution is therefore even more likely than a high Ca^{++} solution to block the firing of an interposed interneuron. (4) TEA injected into the presynaptic neuron prolongs the action potential and enhances transmitter release. An interneuron in a polysynaptic pathway is likely to fire repetitively as a result of the enhanced PSP, but the latency of some of the summed PSPs that the interneuron produces on the follower neuron are likely to be prolonged, and the PSP in the follower cell is likely to have notches or steps—one for each spike in the interneuron. (5) The combination of a high-Mg^{++}, high-Ca^{++} solution and intracellular TEA has the advantages of both methods. It raises the threshold of interposed interneurons and increases transmitter release.

Chemically Mediated Common Connections

One can also identify common presynaptic inputs to two or more neurons by recording from them simultaneously and observing whether a PSP in one cell is invariably associated with a PSP in the other cell. Synchronization of the PSPs would suggest that the two neurons share a common presynaptic neuron (Figure 8-6).

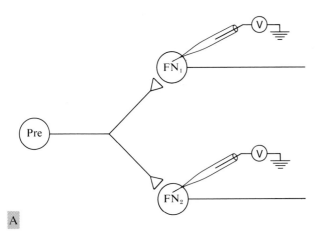

A

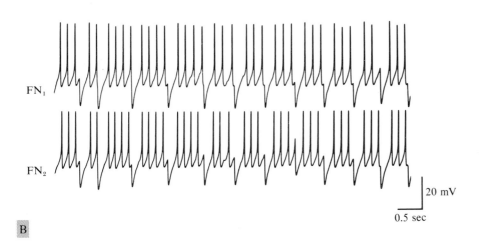

FN$_1$

FN$_2$

20 mV

0.5 sec

B

FIGURE 8-6.
Common connections. **A.** Schematic drawing of the experimental arrangement. Intracellular recordings are made from two follower neurons connected to a common presynaptic neuron. **B.** Simultaneous recordings from two follower neurons. The IPSPs in the two cells are invariably synchronous, suggesting innervation from a common presynaptic neuron. [From Mayeri, Koester, Kupfermann, Liebeswar, and Kandel, 1974.]

Electrically Mediated Connections

Some cells are interconnected by electrically transmitting connections. In such cases an action potential in the presynaptic neuron produces a virtually synchronous electrical (or electrotonic) synaptic potential in the postsynaptic cell. In *Aplysia* these are usually diphasic. Moreover, passing hyperpolarizing or depolarizing current pulses into the presynaptic neuron produces an electrotonic potential in it and an attenuated electrotonic potential of the same sign in the follower cells. Electrical synapses between neurons are usually bidirectional. In most cases, therefore, passing current into the postsynaptic cell also produces an attenuated electrotonic potential in the presynaptic cell if the recording is close to the synapse (Figure 8-7).

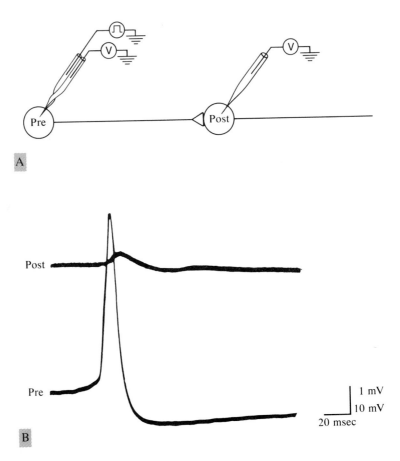

A MULTIACTION CHOLINERGIC NEURON IN
THE ABDOMINAL GANGLION OF *APLYSIA*

Opposite Actions Mediated to Different Follower Neurons

I will first consider a group of identified neurons in the abdominal ganglion of *Aplysia*: cells L2, L3, L4, L6, and R15. These are the only identified neurons in the abdominal ganglion that show regular bursting rhythm, and they all send their main axons into the pericardial branch of the genital-pericardial nerve (Frazier *et al.*, 1967; Winlow and Kandel, 1976).

The bursting rhythm in all five neurons is autoactive and not dependent on afferent synaptic input (see p. 260). However, these neurons receive an intermittent synaptic bombardment from interneurons that modulate their endogenous rhythms. The main modulating synaptic input in each of the four bursting neurons in the left rostral quarter-ganglion (cells L2, L3, L4, L6) is a unitary IPSP. The IPSPs in these four cells are synchronous, suggesting that each cell receives an inhibitory connection from a common presynaptic neuron (Figure 8-8*A*). One of the several modulating synaptic inputs on the bursting cell in the right caudal quarter-ganglion (cell R15) is an elementary EPSP. The EPSP in the neuron of the right caudal quarter-ganglion and the IPSPs in the neurons of the left rostral quarter-ganglion are also synchronous and seem to represent a common connection mediated by different branches of the same neuron (Figure 8-8*B*).

Synchronous PSPs of *opposite* sign can be seen among certain neurons in the ganglion. The first such case was encountered by Strumwasser (1962), who suggested that these opposite synaptic actions were mediated by different branches of a single neuron. To establish that these opposite synaptic actions were actually mediated by a single cell it was necessary to identify such a cell and to examine the actions of its different branches both electrophysiologically and anatomically. Kandel, Frazier, Waziri, and Coggeshall (1967) searched the ganglion and identified a neuron, L10, that produced IPSPs in cells L2, L3, L4, and L6 and synchronous EPSPs in R15 (Figure 8-8*C*). The PSPs in each of the follower neurons were always present and followed the action potential of L10 with constant and short latency and in an all-or-none manner, suggesting that the connections between the interneuron and the follower cells were direct (Figure 8-9*A*). The latencies of the PSPs remained unchanged when the

FIGURE 8-7.
Electrically mediated connections. **A.** Schematic drawing of recording conditions. **B.** The action potential in the presynaptic neuron produces a diphasic synaptic potential with a short latency in the postsynaptic cell. [From Waziri, 1969.]

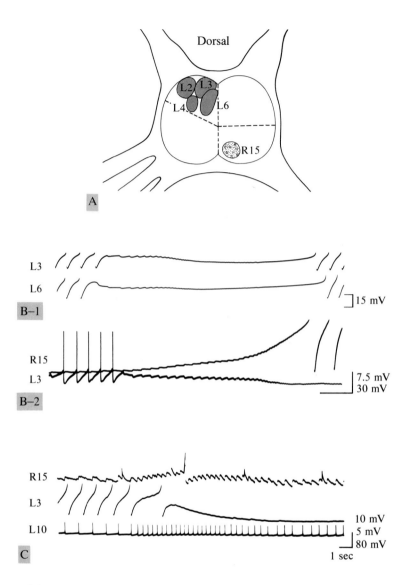

FIGURE 8-8.

A single cell (L10) produces PSPs of the opposite sign in different follower cells. [From Kandel, Frazier, Waziri, and Coggeshall, 1967.]

A. Schematic drawing of the dorsal surface of the abdominal ganglion indicating the position of the five identified bursting cells.

B. Common connections between bursting cells. *B-1.* Simultaneous intracellular recordings from L3 and L6, two bursting cells of the left rostral quarter-ganglion. These cells show a synchronous IPSP. *B-2.* Simultaneous intracellular recordings from a left rostral bursting cell (L3) and the right caudal bursting cell (R15). These cells show a synchronous EPSP.

C. Simultaneous intracellular recordings from L10 and follower cells R15 and L3. The firing of action potentials in L10 produces depolarization and spike generation in R15 and hyperpolarization and inhibition in L3.

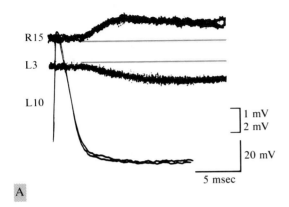

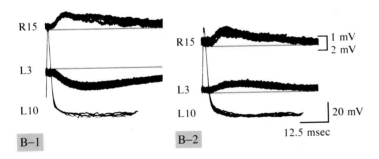

FIGURE 8-9.

Simultaneous intracellular recordings from presynaptic neuron L10 and follower neurons R15 and L3. Cell L10 was stimulated intracellularly and its action potential was used to trigger a sweep of the oscilloscope. Several sweeps were then superimposed, to compare the latencies of successive responses. [From Kandel, Frazier, Waziri, and Coggeshall, 1967.]

A. The action potentials in L10 produce synchronous EPSPs and IPSPs in R15 and L3, respectively. The PSPs in each follower neuron are of constant latency and the (superimposed) successive responses are of constant shape.

B. In Figure *B-1* the membrane potential of the follower cells is at the resting level. In Figure *B-2* the membrane potential of the follower cells was hyperpolarized, causing the hyperpolarizing PSP in L3 to be inverted to a depolarizing PSP. Under both conditions the IPSP rises more slowly than the EPSP and has a longer time course.

ganglion was bathed in solutions with high cation concentrations designed to increase the neuronal threshold (Koester and Kandel, unpub.), or when TEA was injected into L10, prolonging its action potential and increasing the amplitude of the PSPs (Kehoe, 1972c; Bryant and Weinreich 1975).

The two types of PSPs produced by L10 in its different follower cells are opposite in sign but they are not mirror images. The EPSP has a faster rise time, an earlier peak, and a shorter duration than does the IPSP (Figure 8-9B).

To test further whether the connections between interneuron L10 and its follower cells are monosynaptic, the dendritic processes of the left rostral and right caudal follower cells of L10 were traced in the light microscope and found to come in close contact with the processes of L10 (Kandel *et al.*, 1967). Similar results were obtained when L10 and the follower cells were injected with an intracellular marker, cobalt chloride, and their axonal process traced in cleared whole-mount preparations of the ganglion (Winlow and Kandel, 1976). Although electron-microscopic studies are necessary to verify the existence of a direct contact between L10 and the follower neurons L2, L3, L4, L6 and R15, the anatomical findings are consistent with a monosynaptic connection (Figure 8-10). The physiological and anatomical findings in this case provide good evidence that a neuron need not be restricted in its synaptic actions and that it can excite some of its follower neurons while inhibiting others. The connec-

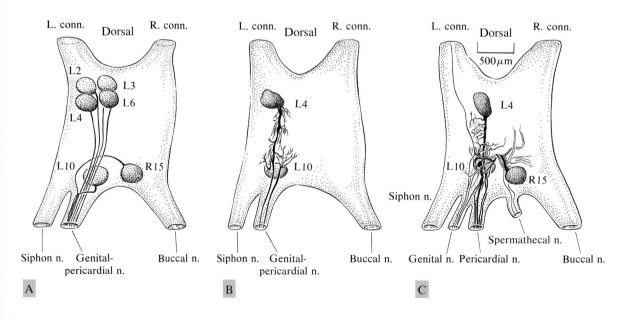

tions are invariant in the species—L10 always excites R15 and always inhibits L2, L3, L4, and L6. A homologous cell having an identical pattern of interconnection has been found in the abdominal ganglion of the species *Aplysia brasiliana* (Blankenship and Coggeshall, 1973).

Are the EPSPs and IPSPs mediated by L10 to its different follower cells produced by different transmitters? Or is the same transmitter released from all terminals but capable of eliciting both excitatory and inhibitory actions on different postsynaptic cells? Tauc and Gerschenfeld's survey of cells in the abdominal ganglion of *Aplysia* (see Figure 7-35, p. 271) revealed that many cells in this ganglion respond to acetylcholine with either a depolarizing or a hyperpolarizing response (D- or H-type cells). Based on this finding, Tauc and Gerschenfeld (1961) predicted that some cholinergic neurons might produce opposite synaptic actions from different terminals.

The possibility that all the branches of L10 release ACh was investigated by Kandel and co-workers (1967; and see Blankenship *et al.*, 1971). They applied ACh iontophoretically to the surface of the cell body of each of the follower cells and found that it caused a hyperpolarization and inhibition of cells L2, L3, L4, and L6 and a depolarization and excitation of R15 (Figure 8-11). In both cases the effect of the ACh simulated that of the interneuron.

To establish that a chemical substance serves as the transmitter at a given synapse one needs to demonstrate that it produces ionic conductance changes identical to those occurring naturally. This can be shown, as a first approximation, by altering the membrane potential of the postsynaptic cell and determining whether the reversal potential of the action produced by the presumed (experimental) transmitter substance is similar to that of the naturally occurring synaptic potential (see Figures 6-10 to 6-13). As a next step, the concentrations of critical ions in the extracellular medium are changed to determine

FIGURE 8-10.

Light-microscopic examination of the axonal pathways of L10 and its follower cells and their intertwining processes.

A. Schematic diagram (dorsal view) of *en passant* contacts made by the process of L10 as its axon wraps around the processes of five of its follower cells. The diagram is a reconstruction from serial paraffin sections observed under the light microscope. [From Kandel, Frazier, Waziri, and Coggeshall, 1967.]

B and **C.** The pathways of the main axons of L4 and L15 and of these two cells and L10 as revealed by intracellular injection of cobalt ions precipitated by ammonium sulfate. [From Winlow and Kandel, 1976.]

From reconstructions in both parts *A* and *C*, the axons of the follower neurons are seen to come in close contact with the axon of L10. These findings support the idea that connections between L10 and its followers are monosynaptic.

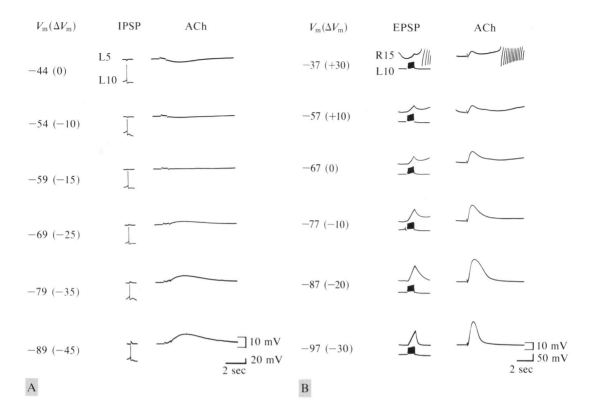

FIGURE 8-11.

Comparison of the reversal potentials of naturally occurring PSPs with those of ACh-induced potentials in the same cells. [From Blankenship, Wachtel, and Kandel, 1971.]

A. Effects of membrane potential changes on the IPSPs produced by L10 and ACh-induced potentials in an inhibitory follower of L10, cell L5. *Left column:* Intracellular recordings from L5 at different membrane potentials (V_m) show how the IPSPs produced by a single spike in L10 are affected by changes in the membrane potential (ΔV_m). *Right column:* Potentials produced in cell L5 by iontophoretic application of ACh (at the same membrane potentials as for the naturally occurring IPSPs in the left column). The resting potential is -44 mV ($\Delta V_m = 0$ mV); negative values for ΔV_m represent hyperpolarization from the resting level. Hyperpolarizing the cell first decreases and then nullifies (at a reversal potential of -59 mV) both the IPSP and the ACh-induced responses. Further hyperpolarization reverses both responses.

B. Effects of membrane potential changes on the EPSPs produced by L10 and the ACh-induced potentials in an excitatory follower of L10, cell R15. *Left column:* Intracellular recordings from R15 at different membrane potentials show how EPSPs produced by trains of impulses in L10 (16-18 spikes per train) are affected by changes in the membrane potential. *Right column:* Potentials produced in R15 by iontophoretic application of ACh (at the same membrane potentials as for the naturally occurring EPSPs). The resting potential is -67 mV ($\Delta V_m = 0$ mV); positive values for ΔV_m represent depolarization from the resting level, negative ones represent hyperpolarization. When ΔV_m is -30 mV, anomalous rectification reduces input resistance so that excitatory responses appear smaller. Beyond that, it is difficult to depolarize the cell successfully. Hyperpolarizing and depolarizing the membrane potential produces parallel changes in the EPSP and ACh-induced responses.

which ion or combination of ions provides the battery (or driving force) for the synaptic current. Changing the driving force for the synaptic actions of the natural transmitter should also alter the driving force for the presumed transmitter (see p. 168). In addition, pharmacological agents that block the receptors for the natural transmitter should also block the presumed transmitter substance. It must also be shown that the presynaptic neuron contains the enzymes necessary to synthesize the transmitter, and the enzymes must be active in the neuron. Finally, impulse activity in the presynaptic neuron must be capable of releasing the transmitter substance (for discussion of the criteria for establishing the transmitter substance at a synapse see Katz, 1966; Gerschenfeld, 1966; Werman, 1966; 1969). All of these tests have been carried out on cell L10.

Simultaneous comparison of the reversal levels of the naturally occurring IPSPs and the ACh-induced potentials in inhibitory follower cells of L10 revealed identical mean reversal potentials: -57 mV, or 17 mV more hyperpolarized than the resting potential (Figure 8-11A). The reversal potentials for excitatory synaptic actions could not be obtained directly but had to be extrapolated from graphs of the response amplitude at higher levels of membrane potential.[6] Nonetheless, the mean value for the extrapolated reversal level for both the naturally occurring EPSPs and ACh-induced potentials was -10 mV, or 30 mV depolarized from the resting membrane potential (Figure 8-11B). Thus, for both the excitatory and inhibitory actions mediated by L10 there is good agreement between the natural and ACh-induced responses (Blankenship *et al.*, 1971).

Both ACh and the natural synaptic transmitter also produced similar increases in ionic conductances. The naturally occurring and ACh-induced inhibition increase the conductance to Cl$^-$ ions. When Cl$^-$ was replaced in the external medium by propionate, an impermeable organic anion, both the naturally occurring IPSP and the concomitantly recorded ACh-induced potential of the inhibitory follower cells rapidly inverted to depolarizing potentials at the resting membrane potential, because the concentration gradient for Cl$^-$ was reversed and, in response to ACh, Cl$^-$ moved out of the cell (Figure 8-12A). Whereas in normal seawater ACh hyperpolarized the cell, in Cl$^-$-free solutions it induced firing. By contrast, substituting impermeant cations for Na$^+$ has no effect on the ACh response. Certain other types of cholinergic inhibition are also mediated by Cl$^-$ in snail neurons (Kerkut and Meech, 1966; Chiarandini, Stefani, and Gerschenfeld, 1967) and in the pleural ganglion in *Aplysia* (Kehoe, 1967, 1972a).

---

[6]The excitatory follower cells could not be depolarized beyond -20 mV because the delayed rectification in these cells greatly reduces the resistance of the membrane at depolarized levels.

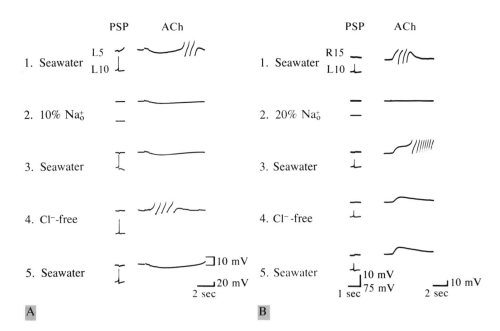

A

B

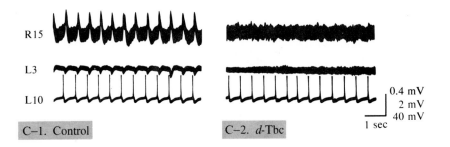

C–1. Control   C–2. *d*-Tbc

A very different ionic mechanism occurs in cells that receive an excitatory action from L10 (Figure 8-12 B). Replacement of Cl$^-$ in the seawater solution produced no significant change in the EPSP or in the ACh responses. By contrast, reductions in external Na$^+$ concentration produced a marked diminution in the amplitude of the ACh response. These data indicate that cholinergic excitation in this cell increases g_{Na}. Thus, cell L10 mediates opposite synaptic actions to different follower cells by acting on receptors that control different ionic conductances.

The cholinergic receptor-blocking agent *d*-tubocurarine blocked both the EPSP and IPSP (Figure 8-12 C), providing further support for the idea that ACh is the transmitter released from both the excitatory and inhibitory terminals of L10. However, the two receptors can be pharmacologically distinguished (Kandel *et al.*, 1967). As Tauc and Gerschenfeld had found, although the excitatory receptor to ACh is blocked by hexamethonium, the inhibitory receptor is not (see p. 271). Independent biochemical evidence also supports the

FIGURE 8-12.
Ionic mechanisms of the synaptic actions produced by L10.

A. Effects of different ionic solutions on naturally occurring IPSPs and ACh-induced potentials in cell L5, an inhibitory follower of L10. All records were taken at the same membrane potential. *A-1.* Response in normal seawater. *A-2.* Responses in seawater containing only 10 percent of the normal amount of Na$^+$. Cell L10 does not fire in this solution because its action potential depends on Na$^+$, but the ACh-induced potential is not affected. The smaller amplitude of the ACh-induced potential (compared to that in normal seawater) is due to decreased input resistance. *A-3.* Action potentials and IPSPs return when normal seawater is returned. *A-4.* Responses in Cl$^-$-free seawater. The cell's action potential is unaffected, but both the naturally occurring IPSP and the ACh-induced potential are inverted to depolarizing responses. *A-5.* Return to normal seawater. [From Blankenship, Wachtel, and Kandel, 1971.]

B. Effects of different ionic solutions on naturally occurring EPSPs and ACh-induced potentials in cell R15, an excitatory follower of L10. All recordings were taken at the same membrane potential. *B-1.* Responses in normal seawater. *B-2.* Responses in seawater containing only 20 percent of the normal amount of Na$^+$. Interneuron L10 does not fire and no EPSP is produced. The ACh-induced potential is barely detectable. *B-3.* The action potentials and EPSPs as well as the ACh-induced potentials recover when normal seawater is returned. *B-4.* In Cl$^-$-free seawater the EPSPs as well as the ACh-induced potentials are unaffected. *B-5.* Return to normal seawater. The differences in amplitude of the three ACh-induced responses in normal seawater are due to the phase of the interburst interval at the time that ACh is applied. [From Blankenship *et al.*, 1971.]

C. Blockade of both excitatory and inhibitory synaptic actions produced by L10 with *d*-tubocurarine. *C-1.* Simultaneous recordings from L10 and an excitatory (R15) and inhibitory (L3) follower cell in control solution (seawater). Each action potential in L10 produced an EPSP in R15 and an IPSP in L3. *C-2.* After the addition of *d*-tbc (10^{-4} g/ml) both synaptic actions are blocked. [From Kandel *et al.*, 1967.]

idea that L10 is cholinergic. Giller and Schwartz (1968, 1971) and McCaman and Dewhurst (1970) found that L10 contains choline acetyltransferase, the enzyme that catalyzes the last step in the synthesis of ACh. This enzyme is active in the cell body; 85 percent of the ^{3}H-choline injected into the cell body of L10 was transformed to ACh within one hour (Koike, Kandel and Schwartz, 1974). Moreover, labeled material could be released by intracellular stimulation. (Figure 7-37*B*, p. 274).

Taken together these results indicate that cell L10 always mediates synaptic connections to the same follower cells. These actions are remarkably selective. Different branches of L10 produce opposite actions on different follower cells. The morphological evidence supports the hypothesis that the connections between L10 and its follower cells are direct; electrophysiological evidence demonstrates opposite synaptic actions on different identified follower cells; and the combined physiological, pharmacological, and biochemical evidence suggests that each branch of L10 releases the same transmitter, ACh, which produces different synaptic effects by acting on different receptors, controlling different ionic permeabilities in different follower cells. Thus, the principles of single function and single ionic mechanisms do *not* hold here. In this case the sign of the synaptic action is not determined by the chemical structure of the transmitter substance released by the presynaptic neuron, but by the chemical configuration of the receptor molecules on the postsynaptic cell and by the ionic channels controlled by the receptors.

Dual Actions Mediated to a Single Follower Neuron

The finding that a presynaptic neuron using one transmitter can mediate opposite actions through different terminals by opening up different ionic channels in the postsynaptic cells redirected attention away from the chemical specificity of the transmitter to the properties of the postsynaptic receptor. If the same transmitter can open up two different ionic channels in different follower cells, and the two ionic mechanisms are independent, then a single postsynaptic neuron could employ them both. For such a postsynaptic neuron, one chemical substance could serve as both excitatory and inhibitory transmitter.

Wachtel and Kandel (1967, 1971) found support for this idea. They found one identified left-caudal follower cell (L7) of L10 that receives both excitation and inhibition from L10. Other examples of dual action have also been described (Kehoe, 1967, 1969a; Kehoe and Ascher, 1970; Pinsker and Kandel, 1969; Gardner and Kandel, 1972).

EXCITATORY–INHIBITORY ACTIONS

The dual synaptic action of L10 on L7 is frequency dependent. When the presynaptic neuron produces action potentials at low frequency it excites the follower cell, but when the presynaptic neuron fires at high frequency it inhibits the follower cell. When L10 fires slowly every spike produces an EPSP in L7, but when it fires rapidly the EPSPs diminish in size and invert to IPSPs. If the firing of L10 is again slowed, the IPSPs persist for a number of spikes before the initial EPSP state is restored (Figures 8-13 and 8-14).

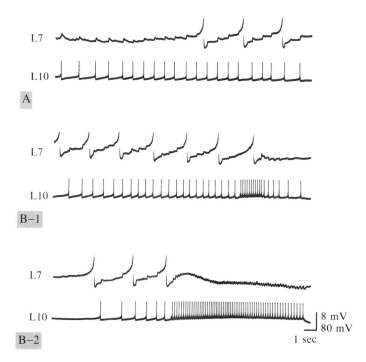

FIGURE 8-13.
Simultaneous intracellular recordings from neuron L10 and its follower cell L7 showing a dual-action connection between them. [From Wachtel and Kandel, 1967. Copyright by the American Association for the Advancement of Science.]

A. Cell L10 is firing at a slow rate (about 1/sec) and every action potential produces an EPSP in L7.

B. *B-1* and *B-2* are continuous records. *B-1.* Cell L10 was initially fired at a slow rate, producing EPSPs in L7. When it was briefly speeded up, the EPSPs grew smaller and soon inverted to IPSPs. *B-2.* For a short time after L10 was returned to the slow rate the IPSPs persisted, but gradually the initial EPSP state was restored. When L10 was again speeded up, the EPSPs at first summated to depolarize L7, but then inverted to IPSPs, which in turn summated to hyperpolarize L7. The IPSP amplitude remained stable as long as high-frequency firing of L10 was maintained.

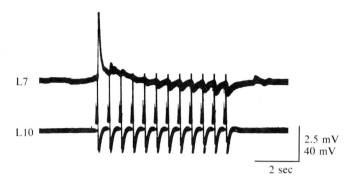

FIGURE 8-14.
Simultaneous high-gain intracellular recordings from L10 and L7 showing a very rapid conversion of an EPSP to an IPSP. [From Wachtel and Kandel, 1971.]

Each action potential in L10 invariably produces PSPs in L7. The first several action potentials of a train produce EPSPs, the later ones IPSPs. These data suggest that the dual action of L10 may be mediated monosynaptically, without the interposition of interneurons. This inference was supported by a comparison of the latency between the ascending limb of the action potential in L7 and the two types of synaptic potentials (see Figure 8-15). In a given train of action potentials the latency of the EPSP (before the conversion) and that of the IPSP (after the conversion) were equal and were comparable to the longest latencies of the monosynaptic PSPs mediated by L10 to its purely inhibitory and purely excitatory follower cells (Figure 8-15). A similar conclusion was reached by Bryant and Weinreich (1975) after injection of TEA into L10. As was evident in the comparison of the purely excitatory and purely inhibitory action of L10 (see Figure 8-9), the two actions that L10 mediates to L7 are not exactly mirror images. The excitatory action again peaks earlier and decays more rapidly than does the inhibitory action.

Both the excitatory and inhibitory PSPs in L7 are blocked by curare (Wachtel and Kandel, 1967) but hexamethonium again selectively blocks only the EPSP (Koester, Dieringer, and Kandel, unpubl.). When ACh is applied iontophoretically to the cell body of L7, a small amount elicits only a depolarizing response, whereas a larger amount elicits a diphasic response, suggesting the existence of two independent (perhaps spatially distributed) postsynaptic receptors to ACh, one excitatory and one inhibitory (Figure 8-16*A*).

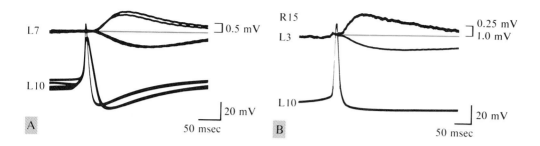

FIGURE 8-15.
Comparison of the two types of PSPs produced by L10 in cell L7 with those produced in R3 and R15. [From Wachtel and Kandel, 1971.]

A. Simultaneous high-speed recordings from L10 and its follower cell L7. Several traces have been superimposed by aligning the ascending limbs of the action potentials to permit a comparison of the EPSP produced when L10 fires slowly and the IPSP produced when L10 fires rapidly. The IPSP in L7 is associated with the shorter and broader action potentials in L10. The change in configuration of the action potential in L10 results from firing it at the high frequency necessary to produce the inversion of the IPSP in L7.

B. Simultaneous high-speed recordings from L10 and two of its follower cells, R15, and L3. Several sweeps have been superimposed to permit a comparison of the EPSP in R15 with that in L3. A comparison of the traces in parts *A* and *B* shows that the EPSPs in L7 and R15 are quite similar to each other, as are the IPSPs in L7 and L3. In each case the IPSP peaks later and is more prolonged than the EPSP.

The conversion from synaptic excitation to inhibition could be accomplished by one of two mechanisms. The EPSP and IPSP might be mediated by separate presynaptic branches of L10 ending on two species of postsynaptic receptors of L7, one leading to excitation and the other to inhibition. The PSP conversion could then be explained by a presynaptic mechanism, such as the blocking of the excitatory branch at high rates of stimulation, or the facilitation of the inhibitory branch. Alternatively, the conversion could occur postsynaptically as a result of a change in the responsiveness of one or both species of receptor. Either the excitatory receptors could undergo *desensitization* (decreased responsiveness to a constant amount of ACh) or the inhibitory receptors could undergo *sensitization* (increased responsiveness to a constant amount of ACh). These alternatives were examined by Wachtel and Kandel (1971) by applying constant amounts of ACh to L7 iontophoretically, thereby controlling the presynaptic factors. The tests showed that the excitatory component of the ACh-induced potential nevertheless decreased rapidly (Figure 8-16*B*).

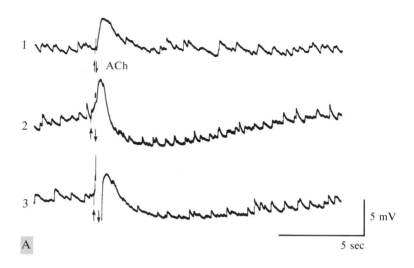

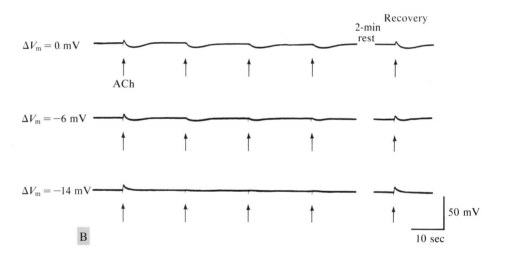

FIGURE 8-16.

Responses of cell L7 to iontophoretic application of ACh. [From Wachtel and Kandel, 1967 and 1971.]

A. Two-component responses. *A-1.* A very brief pulse of ACh (arrows) results in a depolarization. *A-2.* A longer pulse of ACh produces a diphasic response consisting of an early, brief depolarization followed by a longer hyperpolarization. *A-3.* With a slightly stronger pulse the early depolarization is effective in producing an action potential.

B. Desensitization of the excitatory component of the diphasic response of cell L7 with iontophoretic pulses of ACh presented approximately every 18 sec. Control responses in the column at the right show recovery following two min of rest. Changes in ACh response were examined at the resting level and with the membrane potential hyperpolarized by 6 and 14 mV.

This finding is not consistent with blockade of an excitatory branch and is best explained as the result of receptor desensitization (presumably due to a reduced ability of the receptors to open up ionic channels).[7] In addition, the inhibitory receptor sometimes showed a concomitant increase in amplitude (sensitization). Similar conclusions were reached by Bennett (1971) and by Koester, Dieringer, and Kandel (unpub.) in an analysis of the conductance changes accompanying the conversion of the ACh-induced response.

The two receptors control independent ionic conductances (Figure 8-17).

[7]For an earlier description of desensitization of *Aplysia* neurons to ACh, see Tauc and Bruner, 1963.

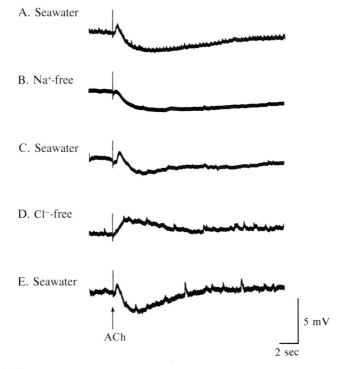

FIGURE 8-17.
Ionic mechanisms of the diphasic response of cell L7 to ACh. **A.** Control in normal seawater. **B.** In Na$^+$-free seawater ACh produces a purely hyperpolarizing response, indicating that the depolarizing component in normal seawater involves an increase in g_{Na}. **C.** Control in normal seawater following Na$^+$-free solution. **D.** In Cl$^-$-free seawater ACh produces a purely depolarizing response, indicating that the hyperpolarizing component in normal seawater involves an increase in g_{Cl}. **E.** Control in normal seawater following Cl$^-$-free solution. [From Blankenship, Wachtel, and Kandel, 1971.]

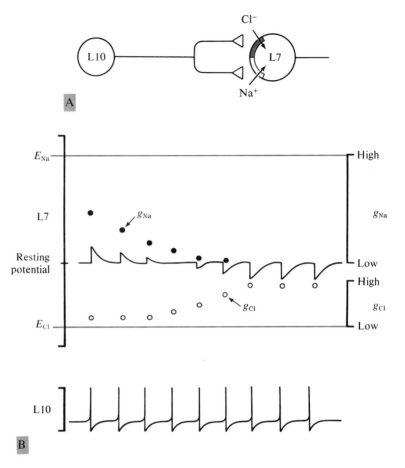

FIGURE 8-18.

Model of a two-component (inhibitory–excitatory) synapse. [From Wachtel and Kandel, 1971.]

A. Schematic drawing of the dual-action synapse. A single cell, L10, releasing a single transmitter, synapses on both inhibitory and excitatory receptors in a single cell, L7. The connection from L10 to L7 could be mediated either by a single branch that ends on both type of receptors, or, more likely, by separate branches (as shown) that end on the separate receptors. *In this and subsequent diagrams of synaptic connections in Chapter 8 the type of synaptic action is indicated in the* post-synaptic cell. *A grey receptor "window" means inhibition, a white window excitation. Elsewhere in the book, the type of synaptic action is indicated in the* presynaptic *terminal (see for example Figure 9-24, p. 383).*

B. The presumed mechanism for converting synaptic excitation to inhibition. When L10 fires, the ACh it releases initially causes only an EPSP in L7 because the increase in g_{NA} produced by ACh is larger than the increase in g_{Cl}. Sustained firing of L10 causes desensitization of the excitatory receptor (decreasing g_{Na}) and possible facilitation (sensitization) of the inhibitory receptor (increasing g_{Cl}). These two effects summate, first to reduce the magnitude of the EPSP and then to convert it to an IPSP. A steady state is eventually reached, in which IPSPs of constant size are produced.

The initial depolarizing component of the diphasic ACh-induced response involves an increase in g_{Na}, whereas the hyperpolarizing component of the ACh-induced response involves an increase in g_{Cl} (Blankenship *et al.,* 1971). These results support the hypothesis that the ionic mechanisms of the dual response to ACh, and by analogy the dual action of L10 on L7, result from a combination of the sodium mechanism employed by the excitatory follower cells and the chloride mechanisms employed by the inhibitory follower cells (Figure 8-18).

DUAL INHIBITORY ACTIONS

Once it was established that the synaptic action of L10 on L7 is sensitive to the firing frequency of L10, the possibility was raised that other synapses of L10 might also be dependent on the firing frequency of L10. Pinsker and Kandel (1969) therefore examined both the excitatory and inhibitory follower cells of L10 for their frequency sensitivity. They found that changing the frequency of firing did not lead to an additional synaptic action in the excitatory follower cells, but it did unmask a second inhibitory action in the inhibitory follower cells of the left rostral quarter-ganglion. When neuron L10 fires a single action potential it mediates the previously described IPSP of about 500 msec in duration by means of an increase in g_{Cl} controlled by a receptor that is blocked by *d*-tubocurarine (the early IPSP). But when the neuron is fired repetitively the duration of the hyperpolarization becomes progressively greater and sometimes reaches 10 to 15 seconds (Figure 8-19*A*). This prolongation (the late IPSP) is due to a second class of receptors with quite different properties (Pinsker and Kandel, 1969; Kehoe and Ascher, 1970). At times even a single spike from L10 can trigger a two-component IPSP, especially in cell L2 (Figure 8-19*B*). A similar two-component IPSP was first analyzed in the pleural ganglion of *Aplysia* by Kehoe (1967, 1969a, 1972a, b).

The distinction between the early and late IPSPs is especially evident when the membrane potential of the follower cell is hyperpolarized beyond the Cl⁻ equilibrium potential (about -65 mV). The early IPSP inverts to a depolarizing postsynaptic potential (Figure 8-19*A*) but the late IPSP fails to invert at this level of membrane potential. Similar results for the synaptic current of these IPSPs have been obtained by Wilson (1971) in voltage-clamp studies of these synapses.

The early IPSP can be selectively blocked by *d*-tubocurarine or atropine; the second component can be selectively blocked by cooling the solution bathing the ganglion to 10°C (Pinsker and Kandel, 1969) or by adding methyl-

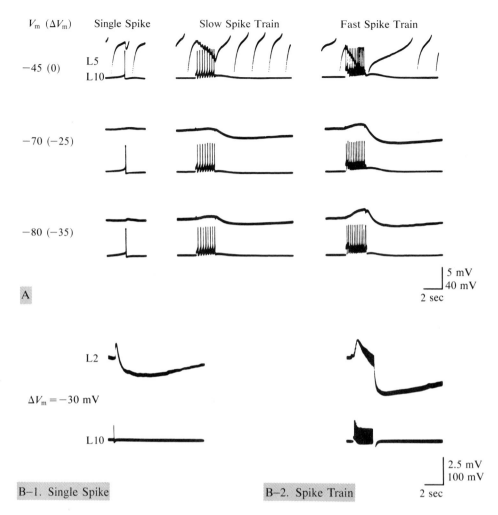

V_m (ΔV_m) Single Spike Slow Spike Train Fast Spike Train

−45 (0) L5
L10

−70 (−25)

−80 (−35)

5 mV
40 mV
2 sec

A

L2

$\Delta V_m = -30$ mV

L10

2.5 mV
100 mV
B−1. Single Spike B−2. Spike Train 2 sec

FIGURE 8-19.

A two-component (inhibitory–inhibitory) synapse. [From Pinsker and Kandel, 1969. Copyright by the American Association for the Advancement of Science.]

A. The amplitudes of the early and late IPSPs in cell L5 are a function of membrane potential and the frequency of presynaptic spikes produced by cell L10. (The tops of the spikes of L5 in the top row have been cut off.) V_m is the membrane potential of cell L5; ΔV_m indicates hyperpolarizing changes in L5 from the resting potential. A single spike in L10 produces only an elementary (early) IPSP in L5, which is nullified at a reversal potential $\Delta V_m = -20$ mV (not shown) and becomes progressively more depolarizing at $\Delta V_m = -25$ mV and −35 mV (see also Figure 8-11A, p. 300). When slow or fast trains of spikes are fired in L10, by passing an intracellular current pulse, a second component (late IPSP) becomes evident. At $\Delta V_m = 0$ mV (top row) this second component is seen only as a slowed return of the early IPSP and a delayed firing of L5. At $\Delta V_m = -25$ mV and −35 mV the early IPSP is inverted to a depolarizing response but the late IPSP is still clearly evident as a hyperpolarization.

B. Early and late IPSP in follower cell L2 following a single spike and a train of spikes in interneuron L10. The membrane potential of L2 was hyperpolarized by 30 mV to invert the early IPSP. The late IPSP is particularly well developed and even a single presynaptic spike can trigger it.

xylocholine (Kehoe and Ascher, 1970). Measurements of membrane conductance in the cell body also show differences for the two components. The first component is associated with a conductance increase detectable in the cell body, the second is not. Iontophoretic application of ACh to the synapse-free cell body or the addition of ACh to the perfusing solution also produces a two-component IPSP. The early component is selectively blocked by curare, the late component by cooling (Figure 8-20).

The fact that the late IPSP can be selectively blocked by cooling, the difficulty of reversing it by hyperpolarizing the membrane potential, the failure to detect a conductance change in the cell body, and the finding that the late IPSP is selectively depressed by ouabain led Pinsker and Kandel (1969) to suggest that the late IPSP is dependent on energy metabolism and due to an electrogenic Na^+ pump (see also Ascher, 1968). However, Kehoe and Ascher (1970) have since shown that the late IPSP is probably due to a remote synapse involving a g_K (see also Kunze and Brown, 1971). When the early IPSP was blocked by d-tubocurarine, part of the late IPSP was inverted at 85 to 90 mV, which is close to E_K. However, the temperature sensitivity and time course of the late IPSP are not typical of a PSP due to increased conductance, suggesting again that at least one step in the generation of this conductance may be dependent on energy metabolism.

Electrical Synaptic Actions

In addition to mediating four types of chemical synaptic actions—excitation, inhibition, excitation–inhibition, and dual–inhibition—L10 also makes electrical connections with at least two cells, L20 and L21. In both cells L10 produces small and brief (20 msec) diphasic potentials with very brief latency compared to the chemical synaptic actions produced in L3. Hyperpolarizing or depolarizing current pulses injected into L10 produce attenuated membrane potential changes in the two cells (Figure 8-21; Waziri, 1969, 1971).[8]

The five known synaptic actions mediated by L10 to its follower cells are summarized in Figure 8-22. The two combinations of synaptic actions so far demonstrated for the followers of L10 (Na^+–Cl^- and Cl^-–K^+) are by no means the only ones possible for these cells. Considering only the three ionic mechanisms for *increased* conductance found in *Aplysia*, there are six logically possible combinations (Table 8-1). If we include ionic mechanisms for *decreased* conductance (p. 174), the number of possible combinations would be even greater.

---

[8]Kehoe (1972c) has also on occasion found what appear to be weak electrical connections to R15 and some of the bursting cells in the left upper quadrant.

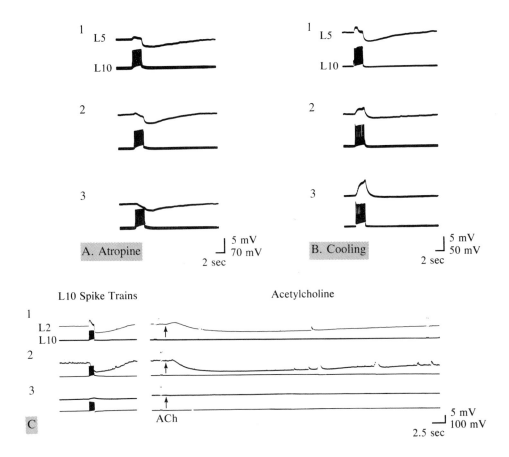

FIGURE 8-20.

Selective blockade of the early and late IPSPs produced in follower cell L5 by spike trains in L10. In all records the membrane potential of cell L5 was kept at a constant, hyperpolarized value beyond the reversal potential for the early IPSP, thus making the early IPSP depolarizing. [From Pinsker and Kandel, 1967; and unpublished observations.]

A. Atropine (5×10^{-4} w/v) selectively blocks the early IPSP and thereby unmasks the early phase of the late IPSP. *A-1.* Before atropine. *A-2.* Five minutes after atropine the early response is partially blocked. *A-3.* Twenty minutes after atropine the early response is totally blocked.

B. Reducing the temperature from 19° to 10°C produces a progressive and selective decrease in the late IPSP. *B-1.* Before cooling (temperature 19°). *B-2.* At 14.2°C the late IPSP is partially blocked. *B-3.* At 10°C the late IPSP is completely blocked. The early IPSP has increased in size because removal of the late IPSP unmasks the true size and time course of the early IPSP.

C. Simultaneously recorded responses of follower cell L5 to spike trains in interneuron L10 and to iontophoretic application of ACh. *C-1.* Two-component responses (early and late phases) to both L10 and ACh. *C-2.* Application of *d*-tubocurarine (5×10^{-4} g/ml) blocks the early IPSP and the early phase of the ACh-induced response. *C-3.* With *d*-tbc, cooling the ganglion to 10°C abolishes both the late IPSP and the late phase of the ACh-induced response. The residual depolarization of cell L5 during the train of spikes in L10 is due to a resistive coupling artifact between the stimulating and recording electrodes.

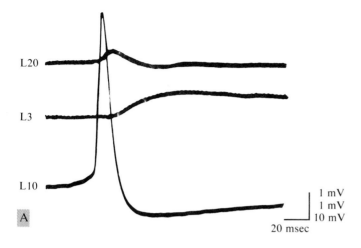

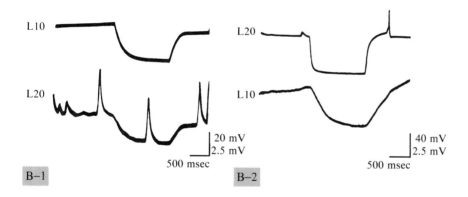

FIGURE 8-21.
Electrical synaptic action mediated by cell L10 to cell L20. [From Waziri, 1969.]

A. Simultaneous recordings from L10 and two follower cells, L3 and L20. The action potential
in L10 produces a diphasic electrical PSP in L20 and a chemical IPSP in L3. For comparison
purposes, the IPSP in L3 was inverted by hyperpolarizing the membrane potential beyond the
inhibitory reversal level. Note that the electrical PSP in L20 has a very short latency and begins
with the onset of the action potential, whereas the latency of the chemical PSP in L3 is about
15 msec. The duration of the depolarizing phase of the PSP in L20 is 20 msec; that in L3 is about
800 msec.

B. *B-1.* The electrical synapse between L10 and L20 is bidirectional. A hyperpolarizing current
pulse in L10 produces a large electrotonic potential in L10 and a small hyperpolarizing electrotonic
potential in L20. *B-2.* A large hyperpolarizing pulse in L20 produces a small hyperpolarizing
electrotonic potential in L10. (Note the differences in voltage calibrations.)

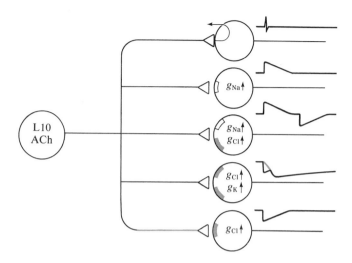

FIGURE 8-22.

Schematic illustration of the five types of synaptic actions mediated by different branches of interneuron L10 to follower cells. These actions include diphasic electrical coupling, chemical excitation (mediated primarily by Na⁺), two-component excitation-inhibition chemical action (Na⁺ and Cl⁻), two-component inhibition-inhibition chemical action (Cl⁻ and K⁺), and chemical inhibition (Cl⁻).

TABLE 8-1

Logically possible combinations in Aplysia *of synaptic actions due to increased conductance.*

	I_K	I_{Cl}	E_{Na}
I_K		$I_K I_{Cl}$	$I_K E_{Na}$
I_{Cl}	$I_{Cl} I_K$ Demonstrated		$I_{Cl} E_{Na}$
E_{Na}	$E_{Na} I_K$	$E_{Na} I_{Cl}$ Demonstrated	

MULTIACTION CHOLINERGIC NEURONS
IN THE BUCCAL GANGLIA

To examine the generality of the findings based on multiaction cell L10 in the abdominal ganglion, Gardner (1971; and Gardner and Kandel, 1972) examined the interconnections between cells in the two symmetrical buccal ganglia of *Aplysia*. Eleven symmetrical cells can be identified in each of the two ganglia (Figure 7-8, p. 232). Eight of the 11 identified cells on each side are innervated by two identified interneurons (BL4 and BL5 in the left ganglion and BR4 and BR5 in the right ganglion). Each interneuron mediates hyperpolarizing PSPs to six identified follower cells and depolarizing PSPs to two other identified cells, all of which are on the same side. These connections are invariant and appear to be monosynaptic; their latency is short, constant, and unaffected by solutions with high concentrations of divalent cations.

The hyperpolarizing PSPs in all six cells receiving inhibitory connections (BL3 and BR3, BL6 and BR6, BL8 to BL11, and BR8 to BR11) are similar and behave as single-component IPSPs due to increased g_{Cl}. The depolarizing PSPs in one type of excitatory follower cell (BL13 and BR13) also behave as single-component PSPs, due to increased g_{Na}. In the other type of excitatory follower cell (BL7 and BR7) the PSP is actually diphasic. Changing the membrane potential in this type cell revealed that the PSP consists of two components, distinguishable from one another by their reversal potentials, pharmacological properties, and ionic mechanisms (Figure 8-23).

Near the resting level, the PSP in this type follower cell often resembles an elementary, one-component EPSP (Figure 8-23C). However, as the membrane potential is depolarized, a late inhibitory component, which is small or absent at the resting level, becomes evident. The reversal potential for the first component (-10 mV) resembles the elementary EPSP in the excitatory followers (BL13 and BR13), whereas the reversal potential of the second component (-63 mV) resembles the elementary IPSP in the inhibitory followers (e.g., BL3). Since the firing threshold of cells BL7 and BR7 is about -40 mV, the first component is excitatory and drives the membrane potential beyond threshold, whereas the second component is inhibitory. Both types of synaptic actions result from a direct connection.

Hexamethonium blocks the excitatory component without significantly affecting the inhibitory component, whereas relatively weak concentrations of curare (5×10^{-5} g/ml) selectively reduce the hyperpolarizing component, often leaving the depolarizing component relatively unaffected. Higher concentrations of curare (2.5×10^{-4} g/ml) block both components in the dual-action followers as well as the elementary IPSP and EPSP in the purely

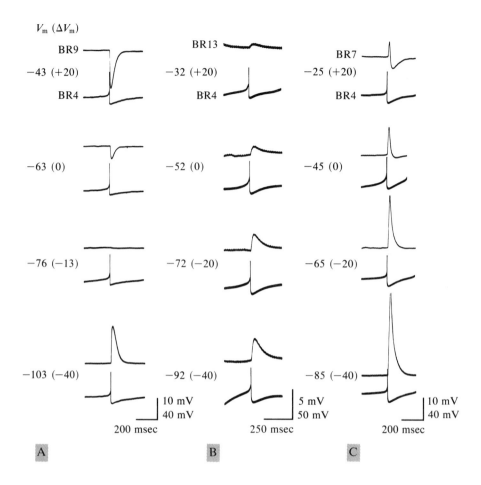

$V_m (\Delta V_m)$

BR9 —

-43 (+20)

BR4 —

BR13 —

-32 (+20)

BR4 —

BR7 —

-25 (+20)

BR4 —

-63 (0)

-52 (0)

-45 (0)

-76 (-13)

-72 (-20)

-65 (-20)

-103 (-40)

-92 (-40)

-85 (-40)

10 mV
40 mV

200 msec

5 mV
50 mV

250 msec

10 mV
40 mV

200 msec

A

B

C

FIGURE 8-23.

Multiple synaptic actions mediated by a single neuron, BR4, in the buccal ganglion. [From Gardner and Kandel, 1972. Copyright by the American Association for the Advancement of Science.]

A. Single-component IPSP produced in follower cell BR9 by action potentials in BR4. Changes in membrane potential are indicated by ΔV_m; positive values are depolarizations from the resting level, negative values are hyperpolarizations. The absolute value of the membrane potential is indicated by V_m (resting value of V_m is -63 mV).

B. Single-component EPSPs in cell BR13 produced by an action potential in interneuron BR4 (resting value of V_m is -52 mV).

C. Two-component PSPs produced in cell BR7 by an action potential in interneuron BR4. As the membrane potential is progressively depolarized by 20 mV from the resting level (-45 mV), the presence of a second hyperpolarizing component becomes evident. This synaptic component is usually small or masked at the resting level. The ganglion was bathed in a high-Ca^{++} solution to increase threshold and block any possible interposed interneuronal activity.

inhibitory or excitatory followers. Selective blockade of either of the two components reveals that the isolated EPSP peaks earlier and decays faster than the isolated IPSP. This difference in time course of the two components accounts for the diphasic configuration of the dual synaptic potential (Figure 8-24).

Cells BL7 and BR7 respond to iontophoretic applications of ACh with diphasic (depolarizing–hyperpolarizing or, more frequently, hyperpolarizing–depolarizing) responses that resemble the PSPs produced in these cells by presynaptic action potentials. In some experiments both depolarizing–hyperpolarizing and hyperpolarizing–depolarizing sequences have been obtained by moving the ACh electrode from one position on the cell body to another (Figure 8-25A). The variability in the sequence of the components may be due to a difference in the spatial distribution of the two types of receptor on the cell-body membrane (Stefani and Gerschenfeld, 1969; Levitan and Tauc, 1972; and see Figure 8-28). The components of the ACh-induced response resemble the corresponding components of the PSP in their reversal potential, their sensitivity to cholinergic blocking agents, and their response to changes in ionic concentrations.

The two components of the dual synapse also differ in their response to repeated presynaptic stimulation. Both components decrease with repeated stimulation, but the decrease is greater for the hyperpolarizing component, presumably because of a preferential desensitization of the inhibitory receptor. A similar decrease in the inhibitory component occurs with consecutive ACh pulses (Figure 8-25B and C). These changes in response amplitude are not associated with changes in membrane potential. Moreover, the rate of depression for the hyperpolarizing component is the same whether or not the depolarizing component of the PSP or the ACh-induced response is blocked with hexamethonium. Thus, here, as with the dual synapse in the abdominal ganglion, receptor desensitization may serve a physiological role in regulating synaptic function.

It is interesting to compare the multiaction cells in the buccal and abdominal ganglia. In both ganglia a single presynaptic neuron can act on inhibitory and excitatory receptors in different follower cells. In addition, some follower cells in each ganglion have both types of receptors to ACh so that the presynaptic cell can mediate both excitatory and inhibitory synaptic actions to a single follower cell. In the dual-action follower cell of the abdominal ganglion the two types of receptors have different kinetic properties, so that an action potential in the presynaptic cell produces depolarizing PSPs at low rates of firing and hyperpolarizing PSPs at high rates of firing. The action of the excitatory receptors at first masks the action of the inhibitory receptors. But the excitatory receptors

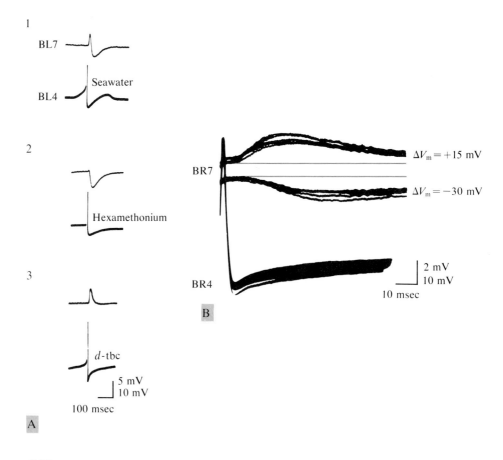

FIGURE 8-24.

Pharmacological properties and time course of the excitatory and inhibitory components of dual synaptic action in the buccal ganglia. [From Gardner and Kandel, 1972. Copyright by the American Association for the Advancement of Science.]

A. Differential sensitivity of the two components of the PSP to cholinergic blocking agents. In all three parts of this figure the membrane potential of follower cell BL7 was maintained at a depolarization of 20 mV. *A-1.* In normal seawater an action potential in BL4 produces a diphasic PSP in BL7. *A-2.* With the ganglia bathed in seawater containing 10^{-4} g/ml hexamethonium, the depolarizing component of the dual response is reversibly blocked and an action potential in BL4 produces only a hyperpolarizing response in BL7 and reveals the time course of this component. *A-3.* With the ganglia bathed in seawater containing 7.5×10^{-5} g/ml *d*-tubocurarine, the hyperpolarizing synaptic component is reversibly blocked and an action potential in BL4 produces a depolarizing PSP in BL7 and reveals the time course of this component, which is more rapid than that of the hyperpolarizing component (A-2).

B. A second estimate of the time course of the two PSPs of the dual synaptic action in BR7 mediated by BR4. The ganglia were bathed in seawater with a high Ca^{++} concentration. Multiple oscilloscope sweeps have been superimposed. The upper trace of the two traces for BR7 was obtained with the follower cell hyperpolarized ($\Delta V_m = + 15$ mV) to the reversal potential of the hyperpolarizing component of the dual response. At this level of membrane potential the time course of the first component is somewhat better illustrated. The lower trace for BR7 was obtained with the follower cell depolarized by 30 mV. This depolarization brings the membrane potential closer to the reversal potential for the first component and allows a better estimate of the time course of the second component alone. Here again the excitatory response peaks earlier and decays faster than the inhibitory response.

PSP ACh Responses

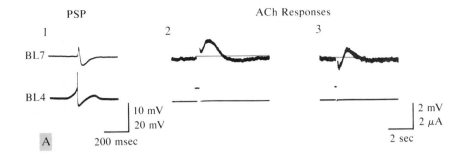

1   2   3

BL7

BL4

10 mV
20 mV

2 mV
2 μA

A   200 msec   2 sec

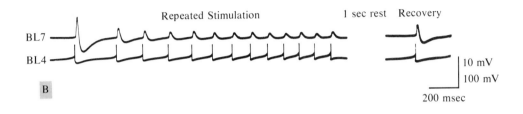

Repeated Stimulation   1 sec rest   Recovery

BL7

BL4

10 mV
100 mV

B   200 msec

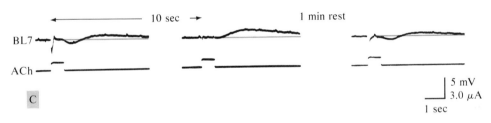

10 sec →   1 min rest

BL7

ACh

5 mV
3.0 μA

C   1 sec

FIGURE 8-25.
Comparison of diphasic synaptic potentials and ACh-induced potentials. [From Gardner and Kandel, 1972. Copyright by the American Association for the Advancement of Science.]

A. Comparison of diphasic PSPs and ACh-induced responses in cell BL7; the cell was depolarized by 20 mV. Ganglia were bathed in solutions containing a high concentration (60 mM) of Ca^{++}. *A-1.* Simultaneous intracellular recordings showing a depolarizing–hyperpolarizing PSP produced in BL7 by an action potential in BL4. *A-2.* Diphasic depolarizing–hyperpolarizing response produced in the same cell by iontophoretic application of ACh. *A-3.* By altering the position and pulse parameters of the ACh electrode, the response changed from depolarizing–hyperpolarizing (shown in *A-2*) to hyperpolarizing–depolarizing.

B. Frequency-sensitive decrease of the diphasic PSPs and ACh-induced responses in cell BL7. Simultaneous recordings from interneuron BL4 and follower cell BL7. The membrane potential of BL7 was depolarized 20 mV to show both components of the dual response. Repetitive firing of BL4 depresses the amplitude of both components of the dual response, but the hyperpolarizing component decreases more than the depolarizing component. After a 1-sec rest there is partial recovery of both components.

C. Desensitization of the dual response in follower cell BL7 to ACh. The membrane potential was depolarized by 20 mV to show both components of the response. Three pulses of ACh were applied iontophoretically onto the cell body of BL7. The first two pulses were spaced 10 sec apart; the second and third pulses were separated by 1 min. The first pulse yields only a depolarization. After a 1-min rest the diphasic response is restored.

are readily desensitized at high rates of stimulation. The inhibitory receptors are also activated by the first action potential but their action only becomes unmasked at high rates of stimulation. By contrast, in the buccal ganglion both the excitatory and the inhibitory receptors are equally activated by ACh, so that a single presynaptic action potential produces a diphasic PSP. At high rates of firing both components decrease, but the inhibiting component is more affected, apparently because the inhibitory receptors desensitize more rapidly than the excitatory ones. It appears possible, therefore, that a nervous system may employ otherwise similar receptor components in different ways by varying the sequence of their activation and their kinetics for desensitization.

A MULTIACTION CHOLINERGIC NEURON IN THE LEFT PLEURAL GANGLION

A multiaction neuron has also been discovered in the left pleural ganglion of *Aplysia* by JacSue Kehoe (1972a, b). This cell mediates a purely excitatory PSP (due to increased g_{Na}) to cells in the anterior region, a purely inhibitory PSP (due to increased g_{Cl}) to cells in the anteromedial region, and a two-component IPSP to cells of the medial region. The connections again are invariant and monosynaptic. Kehoe injected tetraethylammonium (TEA) into the presynaptic neuron and found that prolongation of the presynaptic action potential prolonged all the synaptic actions including both components of the two-component IPSP. This provides good evidence that both components are mediated directly without an intervening interneuron (Figure 8-26; Kehoe, 1969a). These two postsynaptic components are due to increased g_{Cl} and g_K. The Cl^- component is blocked by curare and the K^+ component by TEA or methylxylocholine (Figure 8-27A). The purely inhibitory, purely excitatory, and two-component inhibitory actions are simulated by ACh (Figure 8-27B).

FIGURE 8-27.
Selective blockade of the two components of a dual-inhibitory PSP in the left pleural ganglion. [From Tauc, 1969; based on Kehoe, 1969a.]

A. Records from the pre- and postsynaptic neurons showing dual inhibition. *A-1.* Extracellular application of *d*-tubocurarine (10^{-3} g/ml) selectively blocks the early IPSP. *A-2.* Extracellular application of TEA (10^{-4} M) selectively blocks the late IPSP.

B. Effects of iontophoretic application of ACh on a follower cell as a function of different levels of membrane potential (V_m). Application of *d*-tbc (10^{-4} M) selectively blocks the early IPSP.

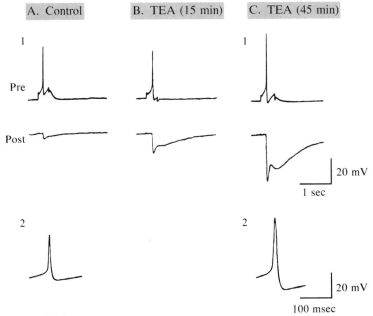

A. Control   B. TEA (15 min)   C. TEA (45 min)

FIGURE 8-26.
Two-component IPSPs in the left pleural ganglion. TEA was injected intracellularly in the presynaptic neuron to increase transmitter release. As the size and duration of the presynaptic spike increased progressively following the injection, the amplitude of both phases of the IPSP also increased. **(A)** Before injection; **(B)** 15 min after injection; **(C)** 45 min after injection. *A-2* and *C-2* are fast-sweep records of the action potentials illustrated in *A-1* and *C-2*. [From Kehoe, 1969a.]

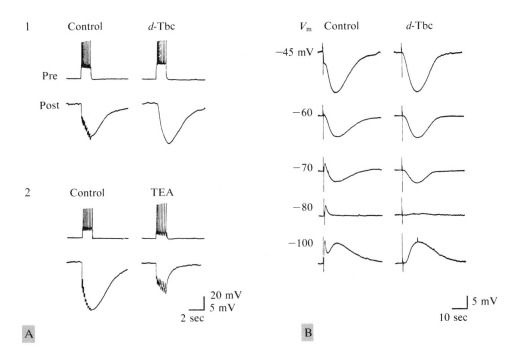

PHARMACOLOGICAL PROPERTIES OF THE
THREE TYPES OF ACh RECEPTORS

The ACh receptors found in the vertebrate peripheral nervous system have been classified as nicotinic or muscarinic based on (1) their responses to the two ACh agonists, nicotine and muscarine, and (2) the selective blockade of the receptors by different pharmacological agents. Kehoe (1972c) analyzed the properties of the three types of ACh receptors innervated by multiaction cholinergic neurons in the pleural and abdominal ganglia. She found that the receptors that control the IPSPs due to Cl^- conductance and those that control the EPSPs due to the Na^+ conductance have properties somewhat similar to the nicotinic receptors of vertebrates. These receptors are stimulated by nicotine and nicotine-like substances and are not affected by muscarine. They are blocked by curare and dihydro-β-erythroidine. The excitatory receptor resembles the nicotinic receptors of the autonomic ganglia (called type 2) because it is blocked by hexamethonium but not α-bungarotoxin. The receptor for the early IPSP due to Cl^- conductance somewhat resembles the type 1 nicotinic receptors of skeletal muscle because it is not blocked by hexamethonium but is blocked by curare and α-bungarotoxin (Kehoe, Sealock, and Bon, unpub.). However, both receptors differ from vertebrate nicotinic receptors in that they are blocked by atropine.

The third receptor, which mediates the slow IPSP due to K^+ conductance has novel properties. It is unaffected by either nicotine or muscarine and its analogs and is not blocked by atropine, curare, or α-bungarotoxin. It is stimulated by arecoline and blocked by methylxylocholine, tetraethylammonium and phenytrimethylammonium (Kehoe, 1972c). Also, whereas conductance changes produced by the other two receptors are not temperature sensitive, the action of the third receptor is blocked by cooling (Pinsker and Kandel, 1969; Kehoe, 1972c). These data are summarized in Table 8-2.

MULTIACTION CHOLINERGIC NEURONS
IN OTHER INVERTEBRATES

Levitan and Tauc (1972) have described the ACh sensitivity of the neurons of the buccal ganglia of the cephalaspid opisthobranch *Navanax* and found some cells that give depolarizing–hyperpolarizing responses. When ACh is applied to the cell-body surface facing the neuropil, it causes a simple mono-

TABLE 8-2
*Pharmacological properties of three types of cholinergic
synaptic actions in* Aplysia.

	EPSP	Fast IPSP	Slow (late) IPSP
Atropine	Blockade	Partial blockade	No effect
Hexamethonium		No effect	
d-Tubocurarine Dihydro-*β*-erythroidine Strychnine Brucine		Blockade	
α-bungarotoxin	No effect		
Methylxylocholine TEA Cooling		No effect	Blockade
Comparable to vertebrate ACh receptors	?Nicotinic (type 2)	?Nictotinic (type 1)	None

SOURCE: Slightly modified from Kehoe, 1972b; and Ascher and Kehoe, 1975.

phasic hyperpolarization due to increased g_{Cl}. When ACh is applied to the axon in the neuropil region, it produces a monophasic depolarizing response due to increased g_{Na}. With large iontophoretic currents both components can be obtained (Figure 8-28). The receptor for the depolarizing component seems to have the properties of the excitatory ACh receptor in *Aplysia;* it is blocked by tubocurarine, hexamethonium, and atropine. But the receptor for the hyperpolarizing component is unusual; it is not blocked by any of the typical blocking agents. The presynaptic neuron that mediates these actions—if there is a single one—has not yet been found. Possible cholinergic dual-action cells have also been described in the stomatogastric ganglion of the lobster (Maynard and Atwood, 1969; Maynard, 1972; Mulloney and Selverston, 1974b).

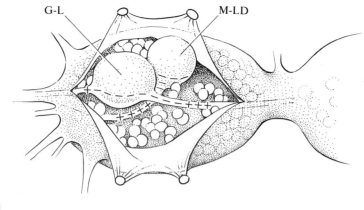

A

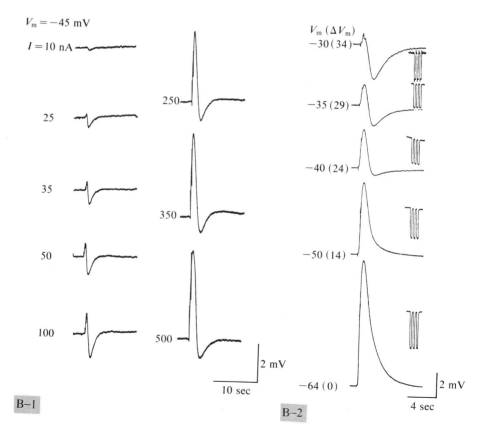

$V_m = -45$ mV

$I = 10$ nA

25

35

50

100

250

350

500

2 mV

10 sec

B-1

$V_m (\Delta V_m)$
$-30 (34)$

$-35 (29)$

$-40 (24)$

$-50 (14)$

$-64 (0)$

2 mV

4 sec

B-2

FIGURE 8-28.
Diphasic ACh-induced responses in the buccal ganglion of *Navanax*. [From Levitan and Tauc, 1972.]

A. Schematic diagram of the dorsal face of the left half of the buccal ganglion of *Navanax*, illustrating regions of ACh sensitivity. The connective tissue that normally envelops the ganglion has been cut and reflected with pins, exposing a giant cell, G-L (cell L1), and a medium-size left dorsal cell, M-LD. The part of the somatic membrane facing inward responds to ACh with a hyperpolarizing potential having a reversal level near -50 mV. This region of the neuron is marked $(-)$. Application of ACh to the axonal processes in the neuropil produces depolarizing potentials with reversal potentials of -30 mV and 0 mV. Areas of the neurons showing the depolarizing responses are marked $(+)$.

B. The ACh-induced response as a function of the intensity of iontophoretic current and membrane potential. *B-1.* Changing current intensity. With the ACh electrode fixed in the neuropil and the membrane potential held constant at -45 mV, small iontophoretic currents (10 to 50 nA) produce a response in which the hyperpolarizing phase is predominant. Larger currents (100 to 500 nA) produce a progressive increase in a depolarizing component. The recordings were made in a high-Mg^{++} medium (to block synaptic activity due to other cells). *B-2.* Changing membrane potential. Diphasic response to a 200-msec application of ACh (40 nA). At the resting level of membrane potential (-64 mV) a monophasic depolarization is seen. At -40 mV a depolarizing phase is followed by a hyperpolarizing phase. The reversal potential for the hyperpolarizing phase is about -50 mV.

MULTIACTION SEROTONERGIC AND DOPAMINERGIC NEURONS

Multiple synaptic actions are not limited to cholinergic cells. Gerschenfeld and Paupardin-Tritsch (1974b) have found that the metacerebral cells of *Aplysia,* which are serotonergic,[9] excite nine cells and inhibit four other cells in the buccal ganglia. The excitatory follower cells have two types of receptors conjointly producing two different types of synaptic potentials. A slow synaptic action (mediated by the A receptor) is blocked by curare, 7-methyltryptamine, and LSD 25; a fast synaptic action (A′ receptor) is blocked by bufotenin, which also blocks the slow synaptic action. Both receptors produce increased g_{Na} and probably also increased g_K.

The inhibitory connections are of two types. Three cells produce an IPSP due to increased g_K; this IPSP decreases with hyperpolarization and is reversed at -80 mV. One cell, however, produces an atypical IPSP that increases in amplitude with hyperpolarization. This would not be expected from a synaptic action produced by increased g_{Cl} or g_K because depolarization would move the membrane potential further away from E_{Cl} or E_K and would increase the amplitude of an IPSP due to increased g_{Cl} or g_K (see Figure 6-11). Studies based on iontophoretic application of serotonin to these cells indicate that the IPSP is due to decreased ionic conductance (Figure 8-29). As we have seen

[9]Eisenstadt *et al.,* 1973; Weinreich *et al.,* 1973.

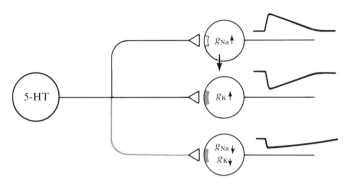

FIGURE 8-29.
Schematic drawing showing three types of synaptic actions mediated by metacerebral cells. The metacerebral cells synthesize 5-HT and release it through branches terminating on various buccal neurons. Some buccal follower cells are excited by an increase in g_{Na}; some cells are inhibited by an increase in g_K. A third class of buccal neurons are thought to be inhibited by a decrease in both g_{Na} and g_K. [From Gerschenfeld and Paupardin-Tritsch, 1974b.]

in Chapter 6 (p. 177), synaptic transmitters can produce hyperpolarizations by turning off the resting g_{Na}, causing the membrane potential to move away from E_{Na} toward E_K (and E_{Cl}). In the case of the connections of the metacerebral cells, both g_{Na} and g_K are turned off by serotonin, causing the membrane to move away from a compound equilibrium level for the two ions (presumably close to 0 mV) toward E_{Cl}. Hyperpolarizing the postsynaptic cell more moves it further from this compound equilibrium potential and thereby increases the synaptic potential.

Berry and Cottrell (1975) found that a dopaminergic cell in the pulmonate snail *Planorbis* mediates inhibition to some cells, excitation to others, and excitation–inhibition to a third group of cells.

MULTIACTION NEURONS IN VERTEBRATES

Multiaction neurons are not restricted to invertebrates. One type of multiaction cell has been found in the vertebrate nervous system by Roper (1976). As was first shown by Otto Loewi (1921) the vagus nerve produces inhibition of the heart through the release of ACh. The preganglionic cholinergic fibers of the vagus nerve synapse on and excite postganglionic (principal) cells located in a small parasympathetic ganglion in the auricle of the heart. In the mud puppy

(Roper, 1976) the principal cells not only inhibit heart muscle fibers but also excite one another. Both actions are chemically mediated and are blocked by cholinergic blocking agents.

In addition, Yasargil and Diamond (1968) have found that Mauthner's neurons, identified symmetrical neurons in the brain of goldfish (see p. 215), excite ipsilateral motor neurons and inhibit contralateral ones. Although the inhibition is mediated by an axon that crosses the midline, its latency is equal to that for the EPSP, suggesting that excitation and inhibition may be mediated directly by different branches of the Mauthner's cell. Although this has not yet been established, Mauthner's neurons provide an interesting possibility of a dual-action neuron located within the vertebrate central nervous system (Figure 8-30).

Although direct evidence that a single *mammalian* neuron can mediate different synaptic actions is still lacking, there is good evidence that postsyn-

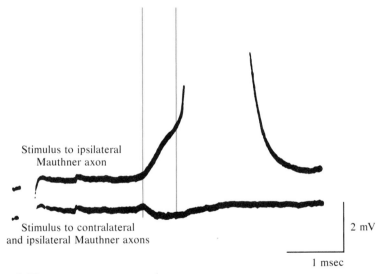

Stimulus to ipsilateral
Mauthner axon

Stimulus to contralateral
and ipsilateral Mauthner axons

2 mV

1 msec

FIGURE 8-30.
A possible example of a dual-action neuron in a vertebrate brain: intracellular recordings from a motor neuron in the goldfish. In the upper trace the ipsilateral Mauthner's axon is stimulated (onset is indicated by the shock artifact on the left), leading to an EPSP and a spike, which has been cut off in this trace. The lower trace shows the response in the same motor neuron when both Mauthner's axons are fired synchronously. An IPSP due to activity in the contralateral Mauthner's axon inhibits excitation and prevents spiking. (The PSPs mediated by the Mauthner's cells are between the vertical grey lines). Even though the (contralateral) pathway for the inhibitory action is slightly longer than that of the excitatory action, the time of onset of the IPSP (left vertical line) is similar to that of the EPSP, suggesting that Mauthner's cells may mediate chemical excitation to the ipsilateral motor neurons and chemical inhibition to the contralateral motor neurons. [From Yargasil and Diamond, 1968.]

aptic cells may have two species of receptors to a single transmitter. In the cat, the Renshaw cell in the spinal cord and cells in the sympathetic ganglion have two pharmacologically different excitatory receptors to ACh: a nicotinic receptor, mediating a brief early excitatory response that is blocked by curare, and a muscarinic receptor, mediating a late slow excitatory response that is insensitive to curare (Curtis and Ryall, 1966; Koketsu, 1969; Libet, 1970). Weight and Votava (1970) found that the two receptors control different ionic conductance mechanisms. The nicotinic receptor leads to an increase in conductance primarily to Na^+, whereas the muscarinic receptor leads to a decrease in conductance to K^+. These two cases exemplify vertebrate synapses that have been examined in detail because their transmitter is known. But multiple receptors may well exist among less well-studied synaptic actions.

ELEMENTARY CONVERGENT AGGREGATES

The elementary convergent aggregate consists of a population of neurons synapsing on a single cell. Shimahara and Tauc (1975a) analyzed the synaptic action on cell R2 in the abdominal ganglion by the group of neurons that innervate it. They found some cells that produce elementary IPSPs and some that produce elementary EPSPs. They also found neurons that mediate each of three types of dual PSPs: excitation–excitation, excitation–inhibition, and inhibition–inhibition. In addition, some cells mediate dual electrical and chemical connections to R2. Thus, many types of connections can converge on a single cell (Figure 8-31). It will be of obvious interest to examine the pharmacological properties of these connections in the future. Gerschenfeld, Ascher, and Tauc (1967) found that a single cell (R15) receives two types of excitatory synaptic input via two different transmitters. One EPSP is apparently cholinergic, the other apparently serotonergic. Studies by Ascher (1972) indicate that this cell also has receptors to dopamine (Figure 7-38). Morphological studies also support the idea of multiple types of transmitter actions on a single cell. Most cells that have been examined electron-microscopically receive input from various classes of synaptic terminals with each class containing different types of synaptic vesicles (see for example Frazier *et al.,* 1967).

HIGHER-ORDER PATTERNS OF INTERCONNECTION

Divergent and convergent aggregates interconnect to produce higher-order patterns of interaction. Three patterns have been examined in studies of the multiaction neurons in the abdominal and buccal ganglia: (1) higher-order interneurons that converge onto interneurons; (2) feedback connections by

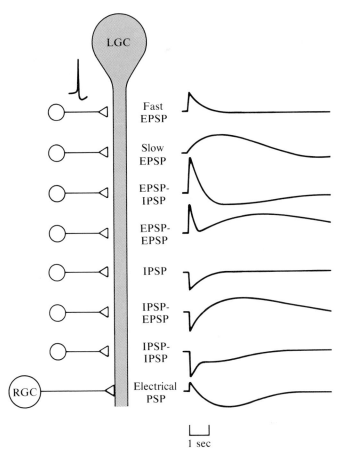

FIGURE 8-31.
Different types of unitary postsynaptic potentials produced by direct stimulation of different identified presynaptic neurons synapsing on the giant cell in the left pleural ganglion. [From Shimahara and Tauc, 1975a.]

follower cells, including lateral (direct) connections between these cells; and (3) feed-forward connections to the follower cells of an interneuron from higher-order interneurons that converge on the interneuron.

Convergent Connections

By recording the synaptic input to L10 one can identify at least four types of PSPs, all inhibitory, presumably reflecting the activity of four as yet unidentified interneurons (Figure 8-32). For convenience in describing their inter-

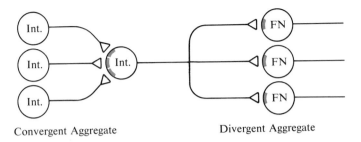

Convergent Aggregate   Divergent Aggregate

FIGURE 8-32.
Schematic illustration of the connection between the convergent and divergent aggregates of L10 (only three follower cells of L10 are shown).

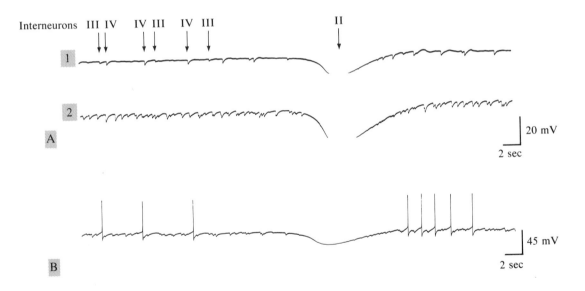

FIGURE 8-33.
Interneurons converging onto L10. [From Kandel, Frazier, and Wachtel, 1969.]

A. The three unidentified interneurons (II, III, and IV) that converge on L10 (Interneuron I) produce IPSPs in L10. A fourth unidentified interneuron (V) cannot be recognized in these records (see Figure 8-36B). *A-1*. The three types of IPSPs occur infrequently and can be distinguished from one another. *A-2*. The IPSPs occur frequently and those attributed to Interneurons III and IV cannot be individually discerned, although they can still be distinguished from the prolonged IPSP produced by Interneuron II.

B. In the isolated ganglion the autoactive rhythm of L10 is primarily modulated by the IPSPs attributed to Interneurons II, III, and IV.

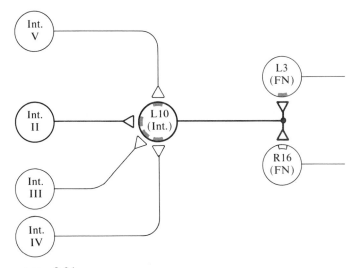

FIGURE 8-34.
Schematic illustration of the inhibitory actions mediated by the four unidentified interneurons (II, III, IV, and V) ending on L10. (Only two follower
cells of L10 are shown.) [Modified from Kandel *et al.,* 1967.]

action, L10 has been called Interneuron I, and the four that converge on it,
Interneurons II, III, IV, and V (Kandel *et al.,* 1967). Since these cells have
not been identified, one cannot be certain whether each inferred interneuron
represents a single cell. The activity attributed to Interneurons III, IV, and V
is likely to represent single cells but activity attributed to Interneuron II seems
to represent several closely coupled cells. The actions of Interneurons II, III,
and IV are illustrated in Figure 8-33. When these interneurons are active, L10
is tonically inhibited. When the interneurons become inactive the autoactive
pacemaker drive of L10 is reestablished. Figure 8-34 is a schematic illustration
of the convergent aggregate on L10, consisting of four interneurons making
inhibitory connections.

Feedback Connections by Follower Neurons

The follower cells of L10 were examined two at a time to determine whether
any (lateral) connections existed between them. No such connections were
found (Kandel *et al.,* 1967; Waziri and Kandel, 1969). Also, none of the follower cells feed back onto L10.

Feed-Forward Connections

FEED-FORWARD SUBSTITUTION

Each of the four interneurons that converge on L10 makes a feed-forward connection with at least one, and up to 13 identified follower cells of L10. Moreover, the four interneurons seem themselves to be dual-action neurons. Each of the higher-order interneurons of the convergent aggregate can therefore inhibit L10 and substitute its action in a follower cell for the action of L10. This action at two sites is called *feed-forward substitution.* The substitution can be equivalent, e.g., excitation for excitation or nonequivalent, e.g., excitation for inhibition.[10]

Figure 8-35 shows three types of interconnection that mediate equivalent substitution between Interneurons II, III, IV, V via cell L10 onto two of its follower cells, L3 and R16. Interneurons II, III, and IV inhibit L10 and excite R16. The inhibition of L10 results in disexcitation of R16 (Figure 8-36*A*). Activity in these interneurons produces an equivalent excitatory-excitatory substitution; this has the effect of transferring the control of R16 from one excitatory input to another. Interneuron II inhibits L3 as well as disinhibiting it by inhibiting L10, thereby producing an equivalent inhibitory-inhibitiory

[10]"Equivalent" refers to equivalence of sign, not effectiveness of synaptic action.

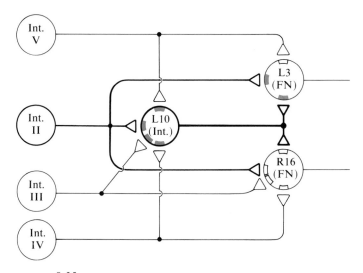

FIGURE 8-35.
Composite diagram illustrating feed-forward substitutions mediated by un-identified Interneurons II, III, IV, and V on two follower cells of L10. [From Kandel and Wachtel, 1968.]

substitution. This has the effect of transferring control of L3 from one inhibitory input to another. Finally, Interneuron V excites L3 and inhibits L10, which in turn disinhibits L3. Activity in Interneuron V produces a nonequivalent excitatory-inhibitory substitution; this has the effect of increasing the excitatory drive upon L3, while simultaneously suppressing the inhibitory drive from L10 (Figure 8-36*B*).

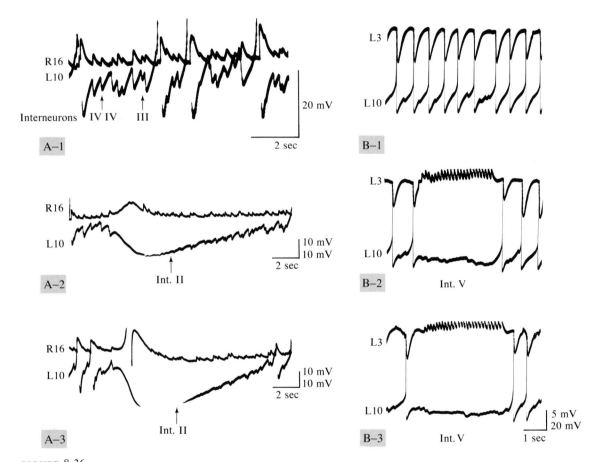

FIGURE 8-36.
Equivalent and nonequivalent feed-forward substitution. [From Kandel *et al.*, 1967.]

A. Equivalent substitution. *A-1.* Cell R16 receives an EPSP from L10 and two smaller, fast EPSPs synchronous with IPSPs in L10 and attributable to unidentified interneurons (III and IV). *A-2* and *A-3.* The traces are at lower gain and illustrate an additional slow EPSP in R16 synchronous with a slow IPSP in L10 and attributable to unidentified Interneuron II.

B. Nonequivalent substitution. The traces in B-1 show the inhibitory connection of L10 to L3 in the absence of activity in Interneuron V. A burst of PSPs attributable to unidentified Interneuron V (parts 2 and 3) produces inhibition in L10 and excitation in L3. Thus, Interneuron V produces nonequivalent substitution, substituting excitation for inhibition.

In addition to inhibiting L10, Interneuron II also receives inhibition back from L10. Thus, whereas the interaction between an interneuron and its follower cells involves simple direct connections without feedback, the interaction between these two interneurons involves both feed-forward and feedback loops. The introduction of loop connections adds new and complex dimensions to the properties of neural aggregates.

Figure 8-37 is a more elaborate diagram of a part of the complex convergent-divergent aggregate (Figure 8-35) and includes (1) the feedback inhibition

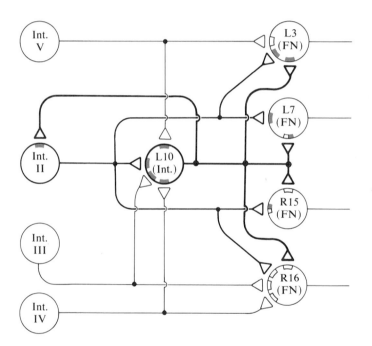

FIGURE 8-37.

Composite diagram illustrating feed-forward substitution mediated by unidentified Interneurons II, III, IV, and V to some follower cells of L10 (for convenience only four follower cells are shown). Each of the four unidentified interneurons inhibits Interneuron L10 and connects with particular follower cells of L10. An interneuron can disexcite or disinhibit followers of L10 indirectly by means of its action on L10. In addition, each interneuron can also substitute its own actions for those of L10 in some followers. Interneuron II inhibits L10 and is itself inhibited by L10. This reciprocal inhibition leads to a switching of control between the two cells. [From Gardner and Kandel, 1972. Copyright by the American Association for the Advancement of Science.]

from L10 to Interneuron II; (2) two more follower cells (L7 and R15); and (3) connections from Interneuron II to L7 and R15. This diagram illustrates two additional types of substitutions mediated by Interneuron II, dual excitation-inhibition for excitation on R15 and inhibition for dual excitation-inhibition on L7. The range of actions attributed to Interneuron II is illustrated in Figure 8-38.

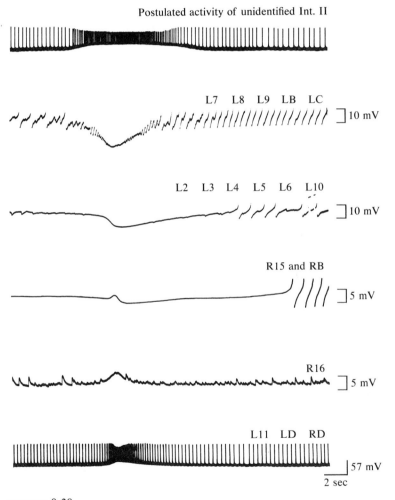

FIGURE 8-38.
Multiple actions attributed to unidentified Interneuron II on different identified follower cells of L10. A synchronous PSP burst characteristic of Interneuron II was recorded in different follower cells, two or three at a time. The responses of the cells were then lined up together for display purposes. [From Waziri and Kandel, 1969.]

FEED-FORWARD SUMMATION

In the abdominal ganglion the two main multiaction cells (L10 and Interneuron II) make mutually inhibitory connections, which leads to feed-forward substitution. In each buccal ganglion the two multiaction cells make mutually excitatory connections, so that the actions of one interneuron can be added to similar actions of the other in common follower cells. This action at two sites is called *feed-forward summation* (Figure 8-39). Each pair of multiaction neurons in the buccal ganglia—BL4 and BL5, and BR4 and BR5—is electrically coupled. Also each pair of interneurons receives a common excitatory input (Figure 8-40A). This combination of electrical coupling and common pre-

FIGURE 8-39.
Similarity of feed-forward summation and substitution. **A.** In feed-forward substitution one interneuron inhibits another and interposes its own action on a common follower neuron. **B.** In feed-forward summation one interneuron excites another (electrically) and the actions of the two are combined at a common follower neuron.

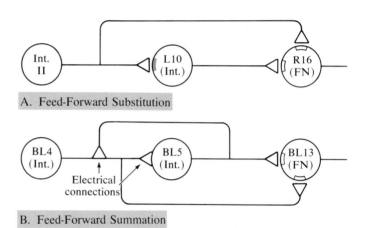

FIGURE 8-40.
Feed-forward summation in the buccal ganglia of *Aplysia*. [From Gardner, 1971. Copyright by the American Association for the Advancement of Science.]

A. Schematic illustration of feed-forward summation in the left buccal ganglion (an identical, symmetrical network is also found in the right ganglion). Neurons BL4 and BL5 receive both common and separate synaptic inputs from unidentified interneurons (unlabeled cells). The two cells are electrically coupled, as indicated by the close-contact axon collaterals, and each mediates chemical inhibition, chemical excitation, and chemical excitation–inhibition to a common population of followers. (Only one of each type of follower cells is shown.)

B. Simultaneous recordings from two ipsilateral neurons (BL4 and BL5) and an inhibitory follower cell (BL3). *B-1.* With BL4 hyperpolarized to prevent firing, action potentials in BL5 produce IPSPs in BL3. *B-2.* With BL5 hyperpolarized, action potentials in BL4 produce IPSPs in BL3. *B-3.* Traces from *B-1* and *B-2* are superimposed to show the similarity of the PSPs produced in a follower cell by two ipsilateral neurons.

C. Recordings from two neurons, BL4 and BL5, and a common follower, BL3 (ganglion was bathed in seawater with a high-Ca^{++} concentration). An action potential in either neuron is capable of producing an IPSP in BL3 and a small electrical coupling potential in the other neuron. Closely spaced action potentials in the two neurons produce notched two-component IPSPs. Synchronous action potentials in the two neurons produce PSPs that summate and sometimes give the appearance of a large unitary IPSP.

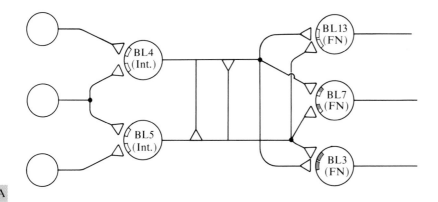

A

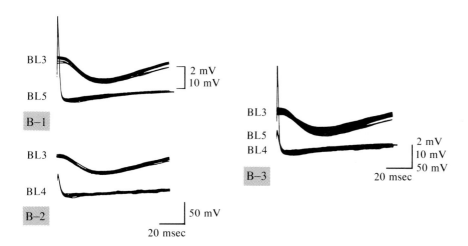

B–1

B–2

B–3

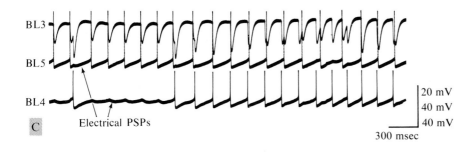

C

Electrical PSPs

synaptic input tends to cause each pair of cells to fire synchronously, thereby producing large, summated EPSPs and IPSPs in their respective follower cells (Figure 8-40B). The shared inputs to ipsilateral pairs of interneurons are of the same sign, and produce a cascading of activity that results in amplified synaptic output from the interneuron pair (Gardner, 1971). Since the two interneurons also have separate inputs (Figure 8-40A), they can also provide distinct channels for transmitting neural information from separate higher-order neurons to the follower-cell population.

SUMMARY AND PERSPECTIVE

To bridge the gap between studies of single cells and behavior it is necessary to examine the pattern of interconnections between cells. Fortunately there are now good techniques for mapping electroanatomically the direct and common connections between individual cells. This makes it possible to determine not only the existence of a connection, but also its sign and effectiveness. In combination with biochemical and pharmacological techniques, one can (in favorable cases) also specify the nature of the chemical transmitter substance involved.

Two elementary patterns of interconnections have been analyzed in *Aplysia:* (1) the elementary divergent aggregate, the synaptic connections made by a single neuron on a population of follower cells, and (2) the elementary convergent aggregate, the connections made by a population of cells that converge on a single neuron. Studies of these neural aggregates show that the unique properties of identified cells extend to their synaptic connections. The connections made between the two types of aggregates are invariant: presynaptic neurons always make connections to the same follower cells and not to others.

Another conclusion that emerges from the study of these neural aggregates is that a great deal of functional complexity arises from three main principles of interconnection: (1) the sign of the synaptic action is not determined by the transmitter but by the properties of the receptors on the postsynaptic cell; (2) the receptors in the follower cells of a single presynaptic neuron can be pharmacologically distinct and can control different ionic channels; and (3) a single follower cell may have more than one kind of receptor for a given transmitter, with each receptor controlling a different ionic conductance mechanism. As a result of these three features, cells can mediate opposite synaptic actions to different follower cells or to a single follower cell.

One multiaction cell can innervate another and both can converge on a common follower-cell population. Two types of connections between multiaction cells have been found: chemical inhibition and electrical excitation. Chemical inhibition of one multiaction cell by another gives rise to feedforward substitution. Thus, in the abdominal ganglion, four multiaction neurons converge upon one cell, L10. Each of these neurons can inhibit L10 and substitute its own synaptic action (of the same or of opposite sign) for the action of L10. The four neurons that synapse onto L10 differ from one another in the number of feed-forward connections and type of substitutions they make. By inhibiting L10, they all exert a common effect: the release of the follower population from the action of L10. Through their varying feedforward connections, the individual neurons also exert additional and more specific effects on selected members of the follower population.

In the buccal ganglion two multiaction neurons that make identical connections to a common follower cell population are electrically coupled. Activity in one neuron elicits activity in the other, leading to a feed-forward summation of their actions on common follower cells.

The versatile functions of multiaction neurons permit them to mediate complex integrative actions. One of the major questions to be considered in Chapter 10 is the function of multiaction neurons in behavior. But even at this point it is clear that multiaction neurons could serve to integrate various combinations of effector responses.

In all instances so far examined the actions of neurons seem to be mediated by a single transmitter substance. Multiple transmitters in a single cell are possible because some synaptic terminals contain two types of vesicular profile. However, it seems unlikely that *different* branches would release different transmitter substances. It would be highly uneconomical for a cell to localize and segregate the macromolecular machinery necessary for synthesizing, storing, and transporting different transmitter compounds in different terminals.

Because there are several receptor types to a single transmitter and several possible combinations of receptor species in a single cell, it would, in principle, be feasible to construct a whole nervous system using only one transmitter substance, simply by varying the types, combinations, and locations of the receptors in the postsynaptic cells as well as their sequence of activation. That more than one transmitter exists in the nervous system poses a challenge to neurobiological theory. Different transmitters are clearly not needed to trigger different types of conductance. But the question arises, are different transmitters perhaps necessary for other purposes, such as cellular recognition, maintenance of synaptic contacts, and the activation of second messengers for prolonging synaptic action or modulating neuronal function? To answer this question we must examine the biochemical and behavioral consequences of various types of transmitter substance.

Adequate morphological data about the different types of synaptic connections as well as biochemical insight into the structure of the different classes of receptors is lacking for a full understanding of synaptic transmission. Three types of receptors to acetylcholine have been characterized pharmacologically, each controlling a specific ionic conductance increase (Na^+, K^+, Cl^-). Five receptors to serotonin have been described: in addition to the three types of conductance increase (Na^+, K^+, Cl^-) there are two types of conductance decrease (K^+ and Na^+). It would obviously be of interest to relate the pharmacological properties of these various receptor types to their biochemical structure. Such studies might ultimately clarify how the ionic channel relates to the structure of the receptor molecule.

SELECTED READING

Ascher, P., and J. S. Kehoe. 1975. Amine and amino receptors in gastropods. in *Handbook of Psychopharmacology,* Vol. 14, edited by L. Iverson, S. Iverson, and S. Snyder. New York: Plenum, pp. 265–310.

Dale, H. H. 1935. Pharmacology and nerve endings. *Proceedings of the Royal Society of Medicine,* 28:319–332. A classic review of the early biochemical and pharmacological work on cholinergic and adrenergic neurons.

Eccles, J. C. 1957. *Physiology of Nerve Cells.* Johns Hopkins Press. Summarizes the model of the nerve cell based on studies of the spinal motor neuron in which each transmitter is postulated to mediate only one type of synaptic action.

Gerschenfeld, H. M. 1973. Chemical transmission in invertebrate central nervous systems and neuromuscular junctions. *Physiological Reviews,* 53:1–119. A review of transmitter biochemistry and pharmacology of invertebrates.

PART III

NERVE CELLS
AND BEHAVIOR

Because invertebrates have relatively few nerve cells, specific identified cells can be causally related to a given behavior. Using identified cells, one can investigate how the physiological, morphological, and biochemical properties of a cell are expressed in behavior. This part of the book illustrates the mutuality of behavioral and neurobiological studies. The study of behavior can enlighten our understanding of neuronal functioning, just as the study of nerve cells can enlighten our understanding of behavior.

Behavioral responses can be classified into two broad overlapping categories: *reflex patterns* and *fixed-action patterns.* Operational criteria for distinguishing these categories more effectively, based on stimulus-response relationships, are discussed in Chapter 9. Cellular studies of elementary forms of the two types of behavior—reflex acts and fixed acts—are then described. These studies suggest that differences in neuronal architecture account for the differences between the two classes of behavior. The fixed acts described here are generated by cells whose electrical interconnection, or burst-generating tendency, produces a central program that provides the sequence and timing of behavior. This central machinery is lacking in the neurons that produce reflex acts. Chapter 10 describes cellular studies of the more complex central programs found in fixed-action patterns, behavior that involves several motor systems or a sequence of different behavioral responses.

Both reflex acts and fixed acts can be modified by learning, but in different ways. Learning tends to alter the *form* of reflex responses and the *frequency* of occurrence of fixed acts. This distinction is developed further in Chapter 11, which considers possible sites and mechanisms of neuronal plasticity, the process whereby neurons change their functional properties. A review of cellular studies shows that the chemical connections between nerve cells, which account for the graded properties of reflex acts, can contribute to the plasticity of response. By contrast, the electrical connections between cells, which account for the properties of fixed acts, can contribute to the stereotypy of responses.

Chapters 12 and 13 describe the cellular mechanisms underlying three simple forms of behavioral modification (habituation, dishabituation, and sensitization) and the relationships between short- and long-term retention (memory). These simple behavioral modifications produce changes in the effectiveness of previously existing chemical synaptic connections. Chapter 14 compares the findings of studies of *Aplysia* with those for other invertebrates, illustrating that the principles involved in the generation of behavior and learning are common to different species.

The final chapter speculates on the relevance of studies of normal behavioral mechanisms to the study of abnormal behavior. The cellular study of behavioral abnormalities promises to be a particularly exciting and fruitful avenue for future investigation.

CHAPTER 9

The Neuronal Organization
of Elementary Behavior

The discovery in invertebrate ganglia of identifiable nerve cells that make specific connections with one another and with receptor and effector structures permits an exact study of the relation of individual nerve cells to behavior. Studying behavior in terms of individual cells can, in turn, shed light on the functional importance of certain biophysical and biochemical features of neurons. The identified neurons of an invertebrate ganglion vary in the value of their resting membrane potential, the presence or absence of spontaneous activity, their transmitter biochemistry, and the distribution and types of receptor molecules. Even the branches of a single cell can vary in their actions. What is the functional role of this remarkable diversity? To answer that question it is necessary to determine how the distinctive biophysical properties and the interconnections of the cell determine certain features of a behavior.

   In this and the next chapter, I shall consider the neuronal circuitry underlying the main types of behavior controlled by the abdominal ganglion. From these studies I attempt to extract, in a preliminary way, the principles that govern three aspects of behavior: (1) how different cellular properties and patterns of interconnection are related to the two major classes of behavior, reflex patterns and fixed-action patterns; (2) how elementary synergist components of a complex behavior are integrated; and (3) how antagonist behavioral responses are coordinated so that there is only one response to a given stimulus.

THE SEARCH FOR UNITS OF BEHAVIOR

While there is agreement among psychologists that behavior is best defined in terms of overt responses (see p. 30), there is less agreement on what constitutes components or units of behavior. One approach is to break down behavior according to adaptive functions: feeding, mating, sex, defense, escape, attack, circulatory adjustments, and so on. This schema generates many particular units rather than a few general ones. It is somewhat comparable to classifying animals one by one rather than by species.

The most serious effort to define general, useful units of behavior has been the attempt by ethologists to distinguish reflex and fixed-action patterns. The boldest formulation of this distinction has been made by Lorenz (Lorenz and Tinbergen, 1938; Lorenz, 1965). According to Lorenz, reflex patterns always require an eliciting stimulus. They are smoothly graded in both amplitude (being dependent on the intensity of the stimulus) and form (being dependent on the pattern of the stimulus). In addition, the execution of any reflex movement in a sequence requires feedback from an earlier movement. Although a reflex pattern may not vary greatly in threshold, many of its features can be modified. In fact, some reflex (conditioned) responses are learned. Therefore, details of expression of the reflex may not be genetically determined.

By contrast, fixed-action patterns are not as dependent on an external stimulus, which only serves to release a stereotypic response. Once initiated, the intensity and duration of a fixed-action pattern is independent of stimulus strength; further increments of the stimulus do not produce a stronger or longer-lasting response. Fixed-action patterns are generated by a central nervous system mechanism (a *central program*) that does not require feedback from the periphery either for dictating the sequence of behavior or maintaining the response. The threshold for fixed-action patterns varies greatly, depending upon the state of the organism, and occasionally, in the absence of an overt stimulus, the pattern is released spontaneously (vacuum behavior). Every animal is born with a repertory of fixed-action patterns characteristic of the species. Additional patterns are not readily developed and even the forms of characteristic patterns are not readily modified by experience. Fixed-action patterns are therefore the closest to purely innate behavior. Individuals isolated from birth, who have had no opportunity to learn from others, express fixed-action patterns fully consistent with the behavior of the species.

Subsequent work has shown that there are a number of exceptions to Lorenz's definition of fixed-action patterns. For example, some components of fixed-action patterns are learned, as in the song patterns of certain species of birds (Marler and Hamilton, 1966). In fact, the notion that a behavior is innate because it is not learned was based on the erroneous assumption that the development of any behavior is determined only by the interaction of two processes: the genetic program and learning. However, a variety of embryonic and environmental factors unrelated to learning, e.g., nutrition, temperature, and humidity, are also crucial to the development of behavior. Since many of these factors interact with the genetic program, it is often impossible in practice to untangle them (see for example Lehrman, 1953). At best, behavior can be characterized by whether it is highly dependent or relatively independent of environmental factors (Hinde, 1970).

Another false implication of the early distinction between reflex- and fixed-action patterns is that each defines an exclusive behavioral category. This heavy claim cannot be supported. Many fixed-action patterns are variable under some circumstances. Moreover, many types of behavior have features common to both reflex and fixed-action patterns. One example is the scratch reflex. When a dog is tickled he will scratch where tickled. The response resembles a reflex in that it is oriented to the site of stimulation and the intensity of the response is proportional to the intensity of the stimulus. But the response (the movement of the leg doing the scratching) does not necessarily subside with cessation of the tickling stimulus, and often greatly outlasts it. Thus, scratching has features of a fixed-action pattern.

Although the distinction between reflex and fixed-action patterns cannot be rigidly maintained, many ethologists hold that it is still the most useful starting point for classifying behavior. "Just because the diversity of nature does not fit readily into man-made pigeonholes, many concepts in biology prove useful at a particular level of analysis but yet must not be regarded as absolute. The category of fixed-action pattern is such a one" (Hinde, 1970, p. 21).

Rather than force a rigid theoretical distinction on what is clearly a continuum of behavioral responses, it is important to define operationally the extreme points on the continuum. First, it is useful to distinguish between elementary and complex behavior. Elementary behavior consists of a single response in a single motor system; complex behavior consists of repeated responses or the recruitment of several motor systems. I refer to the elementary responses as *acts* and the complex responses as *patterns.* Thus, in the *reflex act* a stimulus elicits a single graded response from a single motor system; in the *reflex pattern* more than one motor system is involved and one motor component may trigger another. Similarly, in the *fixed act* the stimulus elicits a single all-or-none event in only one motor system; in the *fixed-action pattern* (or *super act*) the same fixed act is repeated or is accompanied by other fixed acts involving other motor systems.

Perhaps the most reliable operational distinction between reflex and fixed behavior can be made on the basis of the *stimulus-response relationship* of each, that is, how the intensity and pattern of the stimulus controls the amplitude and pattern of the response. At one end of the behavioral continuum are the smoothly graded stimulus–response relationships of reflex responses, such as the flexion reflex of the cat or the pupillary response of man. The amplitude and temporal pattern of such responses are a function of the intensity and temporal pattern of the stimulus. At the other end of the continuum are the steep stimulus–response relationships of stereotypic responses, such as sneezing, coughing, vomiting, and orgasm. The amplitude and pattern of these responses are relatively independent of the strength and temporal pattern of the stimulus, and increments in stimulus strength or duration produce only small increments in the amplitude or duration of the response.

Since elementary behavior consists of only a single episode, reflex and fixed acts can be distinguished by simply determining how the amplitude of the response varies as a function of the intensity of the stimulus. The stimulus–response curve for a reflex (graded) act has a gradual slope; that of a fixed (all-or-none) act resembles a step function (Figure 9-1). The steep slope leads one to suspect that there is a regenerative element in the stimulus–response relationship. Beyond threshold there are features in the response (the ampli-

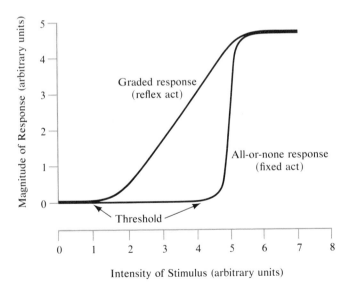

FIGURE 9-1.
Comparison of graded, reflex acts and all-or-none, fixed acts. The stimulus–response curve of the reflex act has a gradual slope, that of the fixed act has a steep slope.

tude, duration, and form of the output) that are not apparent in the input. One can consider the regenerative aspects of the output pattern as the behavioral representation of the central program. That a behavioral response is all-or-none does not mean the response is identical each time it is elicited; the amplitude of the response may vary at any given time. Nevertheless, the response is usually all-or-none in relation to the stimulus—stronger stimuli do not produce larger or longer-lasting responses. Certain internal and environmental factors may alter the parameters of the response, but the stimulus-response relationship will always be steep.

The stimulus–response relationship for complex behavior (patterns) is not so simple. Whereas in elementary behavior the amplitude of only one response varies, in complex behavior the amplitude of each of several component responses can vary. In addition, the sequence of the responses can vary. In fixed-action patterns, however, both tend to be stereotypic. Moreover, some fixed-action patterns occasionally occur spontaneously, without any obvious external stimulus, whereas all reflex patterns require an eliciting stimulus.

At this early stage in the physical analysis of behavior, cellular analyses of fixed and reflex acts can be particularly useful because they can provide an independent way of examining the distinctions between these two categories of behavior and can suggest other criteria for classifying behavior. In turn, the existence of two types of behavior with different stimulus–response characteristics focuses cellular studies on several crucial questions. What types of neural circuitry are required to generate graded behavior? What types of circuitry produce stereotypic behavior? Since fixed acts are thought to be determined by a central program, other questions emerge. Where in the neural circuitry of various fixed acts are central programs located? How can a central program be triggered without an external stimulus? Finally, because graded reflexes are thought to be more modifiable than stereotypic fixed acts, one would like to know what in the neural circuitry accounts for the easy modifiability of graded behaviors and the rigidity of the circuits of stereotypic behaviors.

In an attempt to search for rules relating cellular function to behavior, I have been guided by two assumptions. One is that these rules can initially best be discerned in a single neural structure, ideally a simple brain. Focusing on one structure permits reliable comparisons of the neural architecture of different behavioral responses. A second assumption is that the tools used in these studies must ultimately include morphological and biochemical techniques in addition to electrophysiological ones, so as to provide a broader cellular biological basis, and therefore a surer basis, for establishing relationships between interconnected groups of cells and behavior.

The brain of both vertebrates and invertebrates controls four types of effector systems: somatic-motor, neuroglandular,[1] neuroendocrine, and visceral-motor. Unlike some invertebrate segmental ganglia that control only one type of effector system, the abdominal ganglion of *Aplysia* controls elementary behavioral responses involving each of the four. The principles encountered in these elementary responses will help us to examine the neural circuitry of more complex behavior in *Aplysia* and other opisthobranchs, in particular feeding and escape swimming.

In this chapter I shall consider various examples of elementary behavior in *Aplysia:* first a pair of reflex acts (that make up a reflex pattern) and then several types of fixed acts. In the next chapter I shall describe examples of fixed-action patterns, as well as how fixed-action patterns can occur spontaneously. Also in these two chapters, and in Chapters 11 and 12, I shall take up the question of why reflex acts are more readily modifiable than fixed acts.

A SOMATIC-MOTOR REFLEX ACT:
DEFENSIVE WITHDRAWAL OF THE MANTLE ORGANS

The Behavior

Aplysia has an external respiratory organ, the gill, which is housed in the mantle cavity, a respiratory chamber covered by a fold of tissue called the mantle shelf. At its posterior end the mantle shelf forms a fleshy spout, the siphon, which normally protrudes out of the mantle cavity between the parapodia, the winglike extensions of the body wall and foot (Figure 9-2). Because the mantle organs are bodily appendages, the control of these organs resembles that of somatic-motor functions. The gill is a large respiratory organ. It receives deoxygenated blood from the body cavity through an afferent vein that courses through its inner edge; it conveys oxygenated blood to the heart by means of an efferent vein on its outer margin. The organ is composed of about 16 respiratory units, called pinnules, that connect the afferent and efferent veins.

A stimulus of weak or moderate intensity (100 to 1600 g/cm²) applied to the siphon or mantle shelf triggers a two-component reflex. Each of these

---

[1]There are two types of neuroglandular controls. In one, neural control causes contraction of muscle, leading to the emptying of a gland. In the other, neural control acts directly on the gland, causing release of the hormone. Neural control of the ink gland in *Aplysia*, which we will consider in this chapter, falls into the former category.

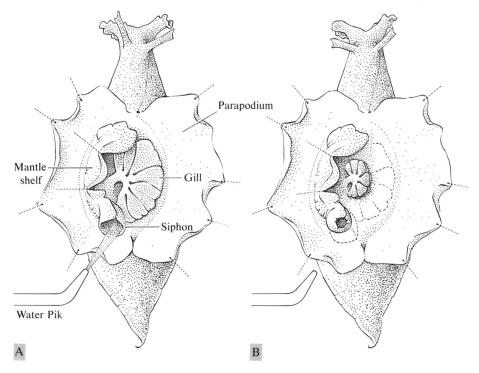

Parapodium

Mantle shelf

Gill

Siphon

Water Pik

A   B

FIGURE 9-2.
The defensive-withdrawal reflex of the siphon and gill in *Aplysia* (dorsal view of an intact animal). Since the parapodia and mantle shelf ordinarily obscure the view of the gill, they must be retracted to allow direct observation. The tactile receptive field for the gill-withdrawal reflex consists of the siphon and the edge of the mantle shelf. **A.** Relaxed position. **B.** Defensive-withdrawal reflex in response to a weak tactile stimulus to the siphon (a jet of seawater delivering an effective pressure of 250 g/cm²). The relaxed position of the gill is indicated by the dotted lines.

components is itself a reflex act, and together they form a reflex pattern. The first component, the *siphon-withdrawal reflex,* consists of contraction of the siphon and withdrawal behind the parapodia. The second, the *gill-withdrawal reflex,* consists of contraction of the gill and withdrawal into the mantle cavity (Figure 9-3). Both components have a short latency and their amplitudes are a graded function of the intensity of the tactile stimulus (Figure 9-4). By means of these acts the animal can protect the gill and siphon under the mantle shelf and parapodia in response to potentially threatening stimuli. In addition to the reflex act of the defensive withdrawal, the mantle organs also produce a fixed

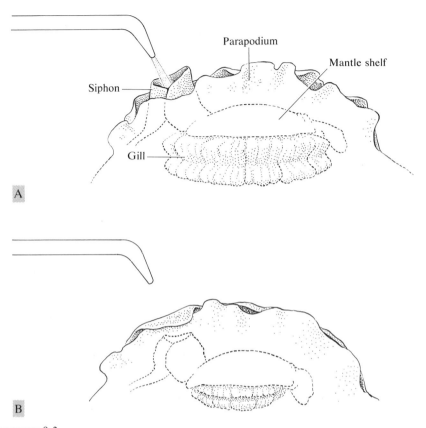

FIGURE 9-3.
The defensive-withdrawal reflex. Side view of posterior part of an intact *Aplysia*. **A.** Relaxed position, illustrating siphon and gill (seen through the parapodia). **B.** Withdrawal of the siphon between the parapodia and the concomitant withdrawal of the gill into the mantle cavity during the defensive withdrawal reflex elicited by a jet of seawater to the siphon.

act, respiratory pumping, to be considered later.[2] The mantle organs are therefore a good effector system for examining how the nervous system controls two different behavioral responses in a single system.

During the defensive withdrawal reflex the gill and siphon contract simultaneously, so that observing one motor component of the reflex provides an

[2]In addition to these two behaviors, the gill shows a third behavior. Individual gill pinnules contract when stimulated directly (the pinnule response). A stronger stimulus to the pinnules can cause the entire gill to contract. These responses are mediated by peripheral neurons (Peretz, 1970, Kupfermann *et al.*, 1971). The relationship of this behavior to defensive withdrawal of the gill is discussed in Kupfermann *et al.*, 1974, and Kandel, *The Behavioral Biology of Aplysia*.

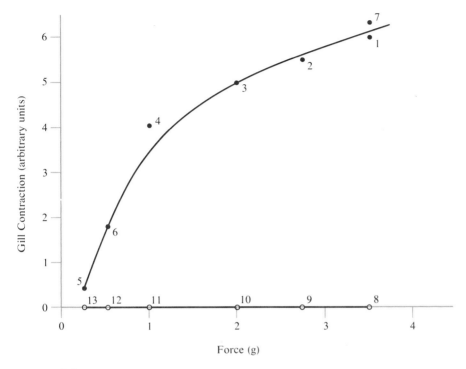

FIGURE 9-4.
Graded gill response to tactile stimuli of different intensities. In a simplified preparation (see Figure 9-5B) constant-force stimuli of different amplitudes were applied to the siphon skin every 25 min. The evoked gill contractions were recorded by means of a photocell; 6 arbitrary units (au) equals about 65 percent of a maximal gill contraction. The numbers on the data points indicate the sequence of stimuli. After stimulus 7 the abdominal ganglion was removed, abolishing the reflex (stimuli 8 to 13). In some preparations a small component (about 5 percent) of the reflex remained in response to stimulation of 4 g after the ganglion was removed. The diameter of the probe was 0.5 mm, so that 1 g produced a pressure of 400 g/cm². [From Kandel, Brunelli, Byrne, and Castellucci, 1976.]

index of the activity of the other. This is convenient because gill movement can only be measured in restrained animals (the parapodia and mantle shelf must be retracted to expose the gill). Siphon movements, however, can be observed and measured in completely unrestrained animals (Figures 9-2, 9-3).

The Neuronal Circuit

The neural control of behavior in *Aplysia* and other molluscs often involves both central ganglia and peripheral neurons. In vertebrates peripheral motor neurons are restricted to certain viscera, such as the stomach and intestine. In *Aplysia* and other molluscs they are more extensive and innervate somatic (locomotor and appendageal) muscles. These peripheral pathways can mediate some responses in the absence of central ganglia.[3] Analysis of a behavior in *Aplysia* therefore requires an understanding of the relative contributions of peripheral and central neurons.

The early neuronal analysis of behavior in molluscs was confounded by the inability to distinguish between central and peripheral pathways (for a review of early research see Kandel and Spencer, 1968). However, the development of appropriate controls has now made it possible to distinguish these components in several dually innervated effector systems (Kupfermann *et al.*, 1971, 1974). In general, weak stimuli applied at some distance from a target effector structure activate only central pathways. By contrast, strong stimuli at a distance or direct stimulation also activate peripheral pathways.

The neural circuit for the gill-withdrawal reflex elicited by weak stimuli to the siphon is located in the abdominal ganglion. Removal of the ganglion abolishes this reflex act (Figure 9-4). At stronger stimuli, additional (as yet unanalyzed) motor pathways that run outside of the ganglion can be activated (Kupfermann *et al.*, 1971, 1974; Peretz *et al.*, 1976). Here I will consider only the centrally mediated response to stimuli of weak or moderate intensity. The siphon-withdrawal reflex elicited by weak or moderate stimuli to the siphon is mediated in part (55 percent) by the abdominal ganglion and in part (45 percent) by peripheral motor neurons. Both parts of the circuit are activated simultaneously, and both have been analyzed to some degree (Perlman, 1975; Bailey *et al.*, 1975).

MOTOR NEURONS

The neural circuit of the gill component has been studied using two preparations. In one the whole animal is restrained in a chamber with aerated circulating seawater and the abdominal ganglion is externalized through a small slit in the animal's neck (Figure 9-5*A*). Tactile stimuli are delivered to the siphon skin by means of measured jets of seawater. The gill response is monitored with a photocell. In the second, simpler preparation the siphon and gill and

---

[3]Bullock and Horridge, 1965; Kandel and Spencer, 1968; Peretz, 1970; Peretz *et al.*, 1976; Newby, 1973; Kupfermann *et al.*, 1971, 1974; Carew *et al.*, 1974; Lukowiak and Jacklet, 1972, 1974; Perlman, 1975; Bailey *et al.*, 1975.

their connections to the abdominal ganglion are isolated from the rest of the animal and pinned in the chamber (Figure 9-5B). This allows measured tactile stimuli to be delivered to precise points on the skin by means of an electro-mechanical stimulator. Gill contractions are again monitored with a photocell. In both preparations one or more nerve cells can be impaled with microelectrodes and fired with intracellular current pulses while the movements of the external organs of the mantle cavity are monitored with a photocell or movie camera. In the simplified preparation intracellular recordings can also be obtained from gill muscles (Kupfermann and Kandel, 1969; Carew, Pinsker, Rubinson and Kandel, 1974; Kupfermann et al., 1974; Peretz, 1969).

Using these preparations, Kupfermann and his colleagues (1974) found six cells that, when stimulated, produce movements of the gill (Figure 9-6). Five of these are in the left caudal quarter-ganglion (L7, LD_{G1}, LD_{G2}, $L9_{G1}$ and $L9_{G2}$); one is in the right caudal quarter-ganglion (RD_G). Stimulation of four of these cells (L7, LD_{G1}, LD_{G2} and RD_G) produces particularly large and brisk gill contractions. The two other cells ($L9_{G1}$ and $L9_{G2}$) produce smaller contractions. Some cells cause characteristic combinations of movements. Based on studies of motion pictures, the gill movements have been classified as follows (Figure 9-7):

1. Movements of the entire gill:
   A. *Rotation.* The entire gill can rotate forward as if pivoting on its anterior insertion.
   B. *Contraction.* The entire gill can contract due to a reduction in size of each of the individual pinnules.

2. Movements of the pinnule halves:
   A. *Flaring.* The dorsal and ventral sets of pinnules can move further apart; the flaring movements expose the efferent vein.
   B. *Antiflaring.* The dorsal and ventral sets of pinnules can move closer together; the antiflaring movements cover the efferent vein.

3. Movements of the pinnules in relation to one another:
   A. *Bunching.* Individual pinnules can move closer together in a fan-like movement in which the boundaries between the pinnules are obscured.
   B. *Spreading at the base.* Individual pinnules can also move closer together with the boundary between pinnules accentuated because of constriction at the base of individual pinnules.

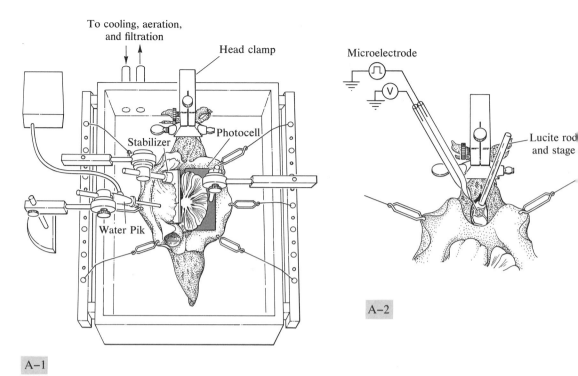

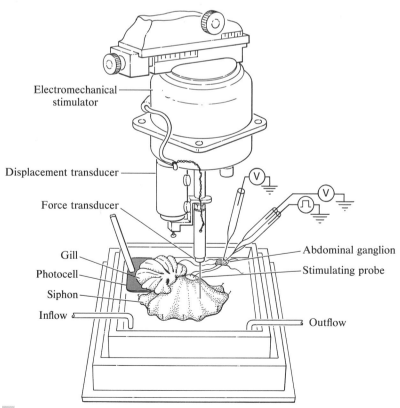

FIGURE 9-5. (*Facing page*).
Three preparations used to study gill and siphon withdrawal.

A. Experimental setup for studying simultaneously the cellular and behavioral aspects of gill withdrawal in the restrained animal. *A-1.* Overhead, dorsal view of an intact animal with parapodia and mantle shelf retracted. The animal is immobilized in a small aquarium containing cooled, filtered, and aerated circulating seawater. The behavioral responses are then monitored with either a 16 mm movie camera or a photocell placed under the gill. A light is directed on the photocell transducer. When the gill contracts, a voltage change proportional to the gill contraction is recorded. The edge of the mantle shelf is pinned to a substage and a constant, measured tactile stimulus is delivered to the siphon by brief jets of seawater from a Water Pik. In contrast to the probe stimulus used on the simplified preparations illustrated in part *B* (and in Figure 9-4), which has a diameter of 0.5 mm and activates only an area of about 0.25 mm², the Water Pik is designed to stimulate a larger area on the siphon skin (about 300 mm²). Because many more sensory neurons are activated, less pressure is required to initiate a maximal reflex. Thus, the reflex produced by a stimulus of 250 g/cm² from the Water Pik is comparable to the reflex produced by a stimulus of 1600 g/cm² from the probe. [Modified from Pinsker *et al.*, 1970. Copyright by the American Association for the Advancement of Science.] *A-2.* Detailed view. A small slit is made in the neck and the intact abdominal ganglion and its peripheral nerves and connectives are externalized and pinned to a lucite stage. Identified motor cells are impaled with double-barreled microelectrodes for stimulation and recording. Gill movements are recorded with a photocell or movie camera simultaneously with the intracellular recordings. [From Kupfermann *et al.*, 1970. Copyright by the American Association for the Advancement of Science.]

B. Simplified (isolated reflex) test system used for studying the gill-withdrawal reflex. The abdominal ganglion remains connected to the gill by means of the branchial and pericardial-genital nerves, and to the siphon by means of the siphon nerve. Contraction of the gill and siphon are monitored while intracellular recordings are obtained from motor neurons (usually L7 or LD_{G1}). Highly controlled tactile stimuli (ranging from 20 to 2000 g/cm²) are applied to the siphon skin by means of a constant-force electromechanical stimulator. [After Byrne *et al.*, 1974a.]

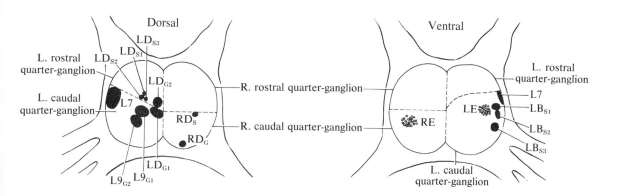

FIGURE 9-6.
Identified central cells and cell clusters of the abdominal ganglion involved in gill and siphon withdrawal.

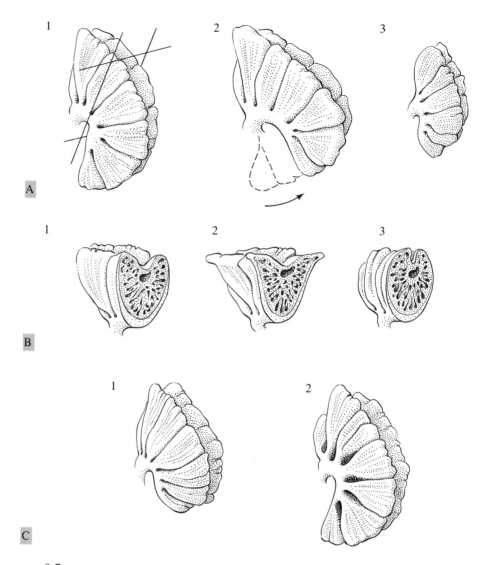

FIGURE 9-7.

Types of gill movement. Although contraction of the gill usually involves a combination of two or more movements, each type of movement is illustrated separately. [From Kupfermann *et al.*, 1974.]

A. Movement of the whole gill. *A-1.* The relaxed gill; the lines indicate the plane of section in Part *B.* *A-2.* Forward rotation of the gill. *A-3.* Contraction due to reduction in volume.

B. Movement of the pinnule halves; the gill is shown in cross section in order to better illustrate the movement. *B-1.* Relaxed gill. *B-2.* Flaring movement of pinnule halves. *B-3.* Antiflaring movement of pinnule halves.

C. Movement of individual pinnules in relation to one another. *C-1.* Bunching of individual pinnules. *C-2.* Spreading at base of pinnules.

The type of action produced by a motor cell is related to the cell's pattern of activity. All cells cause a reduction in gill volume. The three autoactive cells, L7, $L9_{G1}$, and $L9_{G2}$, also produce spreading of the pinnules at their bases and closure of the two gill halves (antiflaring). Cell L7 produces large movements, whereas $L9_{G1}$ and $L9_{G2}$ produce weaker movements. In contrast to these autoactive cells, cell LD_{G1} is often silent, and LD_{G2} and RD_G are usually silent (Figure 9-8). Cells LD_{G1}, LD_{G2} and RD_G produce a bunching of the pinnules and rostral rotation of the gill. Cells LD_{G1} and RD_G also produce antiflaring of the gill halves. Cell LD_{G2} causes flaring of the two gill halves, exposing the efferent vein.

The reduction in gill volume produced by each cell can be conveniently recorded on a photocell, which provides a measure of the relative effectiveness of each cell. Three cells, L7, LD_{G1}, LD_{G2}, are remarkably effective; a few spikes elicit significant contractions. In favorable instances single spikes in these cells produce one-for-one twitches (Figure 9-9A). The other three cells are less effective.

All six motor cells send their axons out to peripheral nerves in the gill. The neuromuscular connections of three cells, L7, LD_{G1}, and LD_{G2}, have been studied in detail (Carew et al., 1974). All three make chemically mediated direct excitatory connections to gill muscle. Pharmacological and biochemical studies indicate that LD_{G1} and LD_{G2} are cholinergic, whereas L7 is not. Although L7 and LD_{G1} use different transmitters, both cells seem to make direct connections with some muscle fibers (Figure 9-9B).

Seven siphon motor neurons have also been found in the abdominal ganglion: LD_{S1}, LD_{S2}, LD_{S3}, RD_S, LB_{S1}, LB_{S2}, and LB_{S3} (Kupfermann et al., 1974; Perlman, 1975). These control different longitudinal and circumferential movements (contraction and constriction). The central motor component of the reflex pattern thus has at least 13 central cells. Five cells produce movements of the gill, seven cells produce movements of the siphon, and one cell (L7) produces movements of the gill, the mantle shelf and, indirectly, the siphon (Figure 9-10). In addition to these central motor neurons the siphon is innervated by about 30 peripheral motor neurons that resemble the LB_S central neurons. The peripheral motor cells produce small contractions of different parts of the siphon (Bailey et al., 1975). The motor control of the defensive reflex thus consists of cells with specific yet overlapping innervation, with some cells having more extensive fields of innervation than others (Figure 9-10). This type of hierarchical organization is common in the motor systems of invertebrates (Kennedy and Davis, 1976).

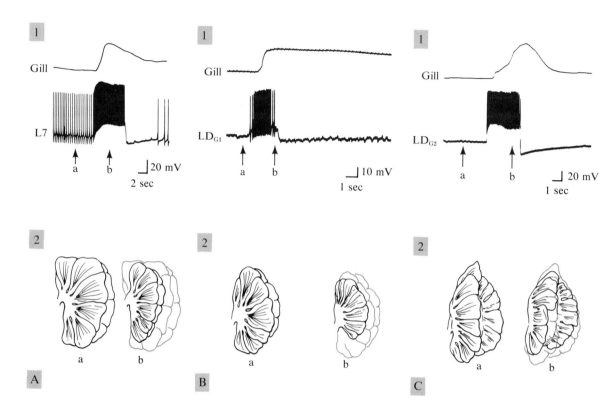

FIGURE 9-8.
Gill contractions produced by each of three gill motor cells in an intact preparation. [From Kupfermann *et al.*, 1974.]

A. Gill contraction produced by intracellular stimulation of motor cell L7. *A-1.* Simultaneous photocell record of gill contraction and intracellular recording from cell L7. Cell L7 fires spontaneously, and this activity contributes to the tonus of the gill in its resting state. A high-frequency burst of action potentials, produced by a depolarizing current pulse, was superimposed upon the steady firing and produced a brisk contraction. *A-2.* Tracings from single cinematographic frames exposed at the points indicated in the intracellular recording shown in part *A-1:* (a) tone of the gill in the relaxed state; (b) gill contraction elicited by high-frequency firing of L7.

B. Gill contraction produced by intracellular stimulation of motor cell LD_{G1}. *B-1.* Simultaneous photocell record of gill contraction and intracellular recording from LD_{G1}. *B-2.* Tracings from single cinematographic frames exposed at the points indicated in the intracellular recording shown in part *B-1:* (a) relaxed gill; (b) gill contraction following a high-frequency train of action potentials in LD_{G1}.

C. Gill contraction produced by intracellular stimulation of motor cell LD_{G2}. *C-1.* Simultaneous photocell record of gill contraction and intracellular recording from cell LD_{G2}. *C-2.* Tracings from single cinematographic frames exposed at the points indicated in the intracellular recording shown in part *C-1:* (a) relaxed gill; (b) gill contraction following high-frequency firing of LD_{G2}.

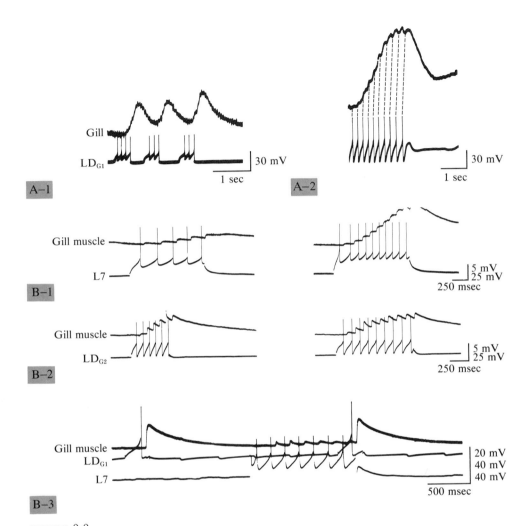

FIGURE 9-9.
Responses of the gill and individual gill muscle fibers to stimulation of individual motor cells.

A. Gill contraction (photocell record) produced by intracellular stimulation of motor neuron LD_{G1}. *A-1.* Smooth contractions produced by 3 to 4 spikes in cell LD_{G1}. *A-2.* Individual twitches produced by LD_{G1} when it fires at high frequencies. [From Kupfermann and Kandel, 1969.]

B. Intracellular recordings of excitatory junction potentials (e.j.p.'s) in gill muscle fibers due to intracellular stimulation of L7, LD_{G1}, or LD_{G2}. *B-1.* The e.j.p.'s in the gill followed L7 spikes one-for-one with a fixed latency at both lower frequency (5/sec) and higher frequency (10/sec). At the higher frequency the e.j.p.'s potentiate. Different frequencies of activity in L7 were elicited by depolarizing current pulses of different intensities. *B-2.* Response of a gill muscle fiber to intracellular stimulation of LD_{G2} (short and long current pulses). Potentiation resulted in both cases. *B-3.* Intracellular recordings of e.j.p.'s in a gill muscle fiber produced by action potential from LD_{G1} and L7. The presence of e.j.p.'s from both motor cells indicates that more than one motor neuron innervates a given muscle fiber. Alternatively, it is possible, though less likely, that the two motor neurons innervate different muscles that are electrically coupled. [From Carew, Pinsker, Rubinson, and Kandel, 1974].

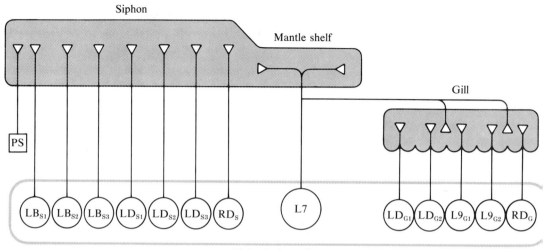

FIGURE 9-10.

The motor component of the defensive-withdrawal reflex. There are 13 identified central motor neurons in the abdominal ganglion innervating the gill and siphon, and an identified cluster of about 30 peripheral siphon motor neurons (PS). Of the central motor neurons seven innervate the siphon; five innervate the gill; and one cell, L7, innervates the gill, mantle shelf, and siphon.

QUANTITATIVE CONTRIBUTIONS OF INDIVIDUAL MOTOR NEURONS

The most effective area of the body for eliciting a gill-withdrawal response to a weak stimulus, i.e., the receptive field with the lowest threshold, is the edge of the siphon. Weak tactile stimulation of this structure produces large EPSPs in all gill (and siphon) motor neurons, causing them to discharge a brief burst of 5 to 15 action potentials in relative synchrony (Figure 9-11A). The resulting gill movements produce a great reduction in volume, slight or no closure of the two gill halves, and a spreading of the pinnules at the base. The gill movements thus contain elements of activity by all three major motor cells, with that of cell L7 dominant.

The contributions of individual motor cells to the gill-withdrawal reflex can be determined by hyperpolarizing single motor neurons by 40 to 50 mV. This prevents the cell from reaching threshold for firing and temporarily removes it from the reflex pathway. Each cell is examined first at its resting potential and firing normally to a siphon stimulus, and then hyperpolarized and removed from the reflex, while changes in the amplitude of the gill con-

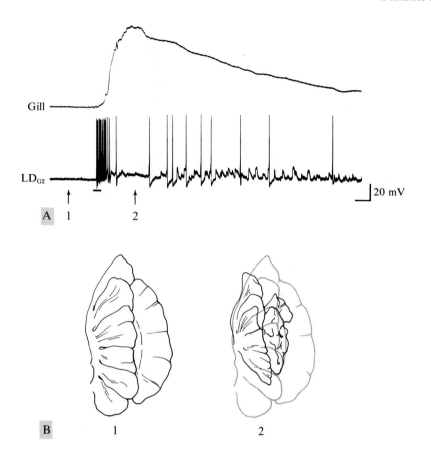

Gill

LD$_{G2}$

$\rfloor$20 mV

A   1   2

B   1   2

FIGURE 9-11.
Gill-withdrawal reflex and motor-neuron firing elicited by a tactile stimulus to the siphon of an
intact animal. **A.** Gill contraction (photocell response) and intracellular recording from cell LD$_{G2}$.
B. Tracings from single cinematographic frames obtained simultaneously at the points indicated
in the recordings in part *A:* (1) relaxed gill; (2) gill during maximal contraction. (Compare this
figure with Figure 9-8.)

traction are recorded simultaneously (Kupfermann *et al.*, 1971, 1974). Reversible removal of a cell from a functional circuit is comparable in effect to surgical
removal, and is essential for demonstrating a causal relationship between a
given cell and a behavior. Only in this way can one show that removal of a
cell's functional contribution from the neural circuit removes a component
of that behavior.

By this means the average contribution of motor neurons L7 and LD$_{G1}$ to
the gill-withdrawal reflex was found to be about 35 percent each. The contrac-

tions mediated by the two cells are additive; together they seem to account for about 70 percent of the entire reflex response. The other motor neurons (LD_{G2}, RD_G, $L9_{G1}$, and $L9_{G2}$) each contribute less, although their exact shares have not yet been determined. Altogether, the gill motor neurons in the abdominal ganglion seem to account for the withdrawal response of the gill to weak- or moderate-intensity stimuli applied to the siphon. Removing the ganglion surgically (see Figure 9-4) or blocking its chemical synaptic transmission with seawater solutions containing a high-Mg^{++} concentration abolishes the reflex response to these stimuli (Figure 9-12).[4]

When the contributions of individual motor cells to siphon withdrawal are examined, the abdominal ganglion is found to contribute about 55 percent

[4]With moderate stimuli a small component (about 5 percent) remains after surgical removal of the ganglion (Kupfermann *et al.*, 1971, 1974; Carew *et al.*, 1976). This component may be mediated by peripheral pathways.

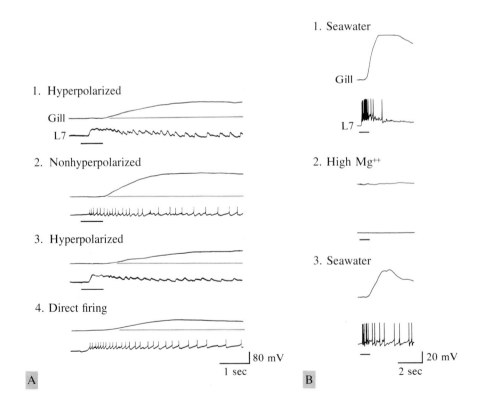

of the total response (Perlman, 1975). The remaining 45 percent is at least in part mediated by a cluster of peripheral motor neurons located along the course of the siphon nerve (Bailey *et al.*, 1975). Like the gill motor cells, the siphon motor cells vary in their effectiveness. Three cells (RD_{S1}, LD_{S1}, LB_{S1}) account for about 30 percent of the total response, or more than 60 percent of the contribution of the abdominal ganglion. The remaining cells make smaller contributions (Perlman, 1975).

SENSORY NEURONS

The synaptic input produced in the gill and siphon motor neurons by stimulation of the siphon is both direct and indirect. The direct (monosynaptic) input is mediated by approximately 24 sensory neurons located in an identified cluster (the LE cluster) near the motor neurons. These neurons are silent but have a low threshold to mechanical stimuli (Figure 9-13). The sensory neurons respond to only one modality, touch-pressure; they do not respond to light, heat, or cold. Each sensory neuron innervates a particular receptive field on the skin and is discharged by stimulation of that area. Although it is possible that each sensory neuron innervating the siphon is unique and always innervates the same patch of skin, this can be said with certainty only for five of the 24 cells. Like the fields of innervation of the motor neurons, the receptive fields of the sensory neurons vary in size from small (20 mm² in area) to large (400 mm²), covering the entire siphon. Several small fields in the siphon are arranged in a shingle-like manner. These are overlapped by medium-sized fields, and both types are overlapped by a single giant field that covers the entire siphon (Figure 9-14).

FIGURE 9-12.
Central mediation of the gill-withdrawal reflex. [From Kupfermann *et al.*, 1974.]

A. Effects on the gill-withdrawal reflex of functional removal of cell L7 from the neural circuit by hyperpolarization. The gill-withdrawal reflex was elicited by a jet of seawater (indicated by the short bar) applied to the siphon every 5 min. On alternate trials cell L7 was hyperpolarized so that the excitatory input could not discharge it. In the last set of traces L7 was directly fired by a long depolarizing pulse that was adjusted to fire L7 in a pattern comparable to that seen during gill contraction. The size of the gill contraction due to direct firing of L7 was approximately equal to the reduction of gill contraction caused by removing L7 from the reflex.

B. Effects on the gill-withdrawal reflex of functional removal of the abdominal ganglion by blockade of synaptic transmission. In each part of the experiment the gill reflex was elicited by a jet of seawater (indicated by the short bar) applied to the siphon. In the first pair of traces the ganglion was bathed in normal seawater. In the second pair it was bathed in a high-Mg^{++} solution, which blocks synaptic transmission in the ganglion. The third pair of traces shows the gill response and the synaptic output to cell L7 after the ganglion was returned to normal seawater.

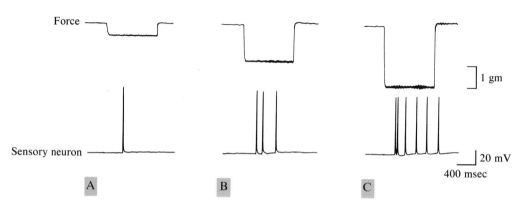

FIGURE 9-13.

Responses of a sensory neuron to tactile stimulation of the siphon. Recordings were made in the cell body of a sensory neuron innervating the siphon skin; 800-msec constant-force stimuli were applied to the siphon skin at three intensities: **(A)** weak (0.5 g); **(B)** moderate (1.5 g); and **(C)** strong (2.5 g). A force of 0.5 g is equivalent to a pressure of 200 g/cm². [From Byrne, unpubl.]

FIGURE 9-14 (*Facing page*).

Postulated organization of the receptive fields of 24 sensory neurons innervating the tip of the siphon skin. This model is based on a summary of the fields observed in different experiments. Fields 1 through 8 roughly cover the left edge of the siphon, fields 9 to 16 cover the middle third, and fields 17 to 24 cover the right edge of the siphon. Each point on the skin is covered by about eight cells. Small overlapping fields cover the entire siphon skin. Large fields completely overlap several small fields and also partially overlap each other. Field 24 is a giant field that covers the entire siphon and overlaps all of the other fields. [From Byrne, Castellucci, and Kandel, 1974b.]

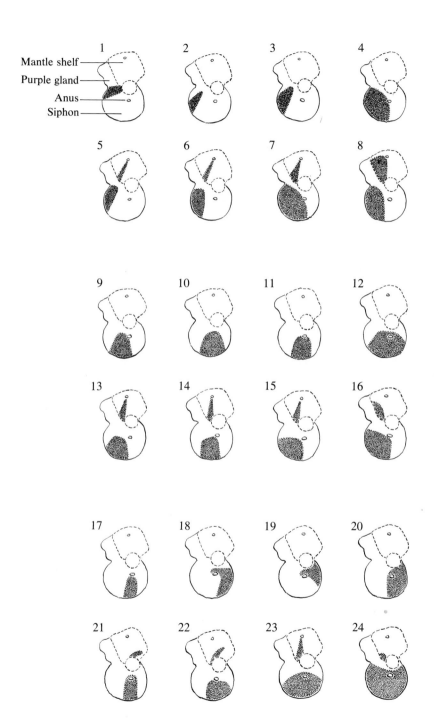

Because the receptive fields overlap, a stimulus to a point on the siphon activates about eight sensory neurons, each with a receptive field of different size (small, medium and large). Why each point on the skin is represented in the nervous system by several mechanoreceptor neurons is not known. One attractive idea is that some sensory neurons (perhaps those with small receptive fields) convey specific information to motor neurons about the site and intensity of the mechanical stimulus, whereas others (perhaps those with large fields) help to set the general excitability or arousal level of the animal.[5] Alternatively, sensory neurons with receptive fields of various sizes may connect to different motor cells. In fact, motor cell L7, which has the largest area of motor control (gill, mantle shelf, and siphon), also receives the most extensive connections from the sensory neurons of the siphon skin. Cells that produce more restricted movements, e.g., L9$_{G1}$ and L9$_{G2}$, seem to have a more restricted input.

In addition to the cluster of sensory cells innervating the siphon skin, a symmetrical cluster in the right hemiganglion (the RE cluster) innervates the mantle shelf. This cluster has been less well studied but the properties of the cells are similar to those of the siphon sensory cells. The areas innervated by the two clusters (siphon and mantle shelf) make up the entire receptive field of the defensive-withdrawal response to tactile stimuli of weak or moderate intensity. [6]

The LE sensory neurons make direct connections to the gill motor neurons and the central siphon motor neurons (Figure 9-15). They also make direct connections to the peripheral siphon motor neurons by means of collateral axons. In addition, they make indirect connections with the motor neurons via at least three central interneurons. Two of these are excitatory to some motor neurons, including L7; one (L16) inhibits all LE sensory neurons as well as the two excitatory interneurons (Figure 9-16). These connections have been examined using several techniques—latency measurements, tests in solutions with high concentrations of divalent cations, and following TEA injection into the presynaptic neuron— and appear to be monosynaptic by all criteria.

Based on these connections, a neural circuit for the reflex response can be constructed. The sensory neurons make direct excitatory connections with the gill and siphon motor neurons and indirect connections via excitatory interneurons. The sensory neurons also connect to an inhibitory interneuron that feeds back to the sensory neurons as well as forward to the excitatory interneurons (Figure 9-17).

[5]For discussion of the possible dual function of sensory systems, see Cohen, 1965; Wiersma, 1976.

[6]Very strong stimuli applied to other sites, such as the head, neck, tail, and gill can also elicit gill withdrawal.

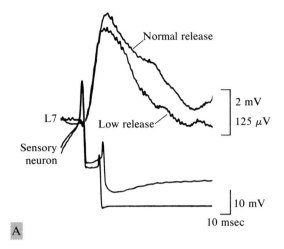

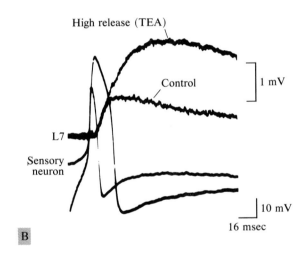

FIGURE 9-15.
A direct (monosynaptic) connection between an LE sensory neuron and a motor cell (L7). The latency is independent of the level of synaptic transmitter release. [From Castellucci and Kandel, 1974.]

A. Comparison of synaptic latency with normal and low release. Superimposed records from the sensory and motor neurons show normal release in normal artificial seawater (55 mM Mg^{++}, 11 mM Ca^{++}) and low release in a solution that depresses release (165 mM Mg^{++}, 8 mM Ca^{++}). Note that the voltage calibrations for the EPSPs are different.

B. Comparison of synaptic latency with normal and increased release. Superimposed records from a pair of neurons before (control) and 10 min after (high release) intracellular injection of TEA into the sensory neuron. The amplitude and duration of the presynaptic spike of the sensory neuron were increased by TEA, resulting in an increase in the EPSP in L7.

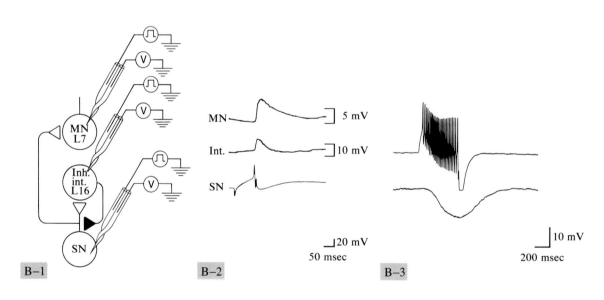

FIGURE 9-16.
Interconnection of the excitatory and inhibitory interneurons participating in the gill-withdrawal reflex. [From Castellucci, Byrne, and Kandel, unpubl.]

A. Excitatory interneuron. *A-1.* Schematic drawing of synaptic interrelationships. *A-2.* A sensory neuron (SN) produces a monosynaptic EPSP in the excitor interneuron (L22) and motor neuron (L7). *A-3.* The excitor interneuron in turn produces a monosynaptic EPSP in L7.

B. Inhibitory interneuron. *B-1.* Schematic drawing. *B-2.* A sensory neuron produces a monosynaptic EPSP in the inhibitory interneuron (L16). *B-3.* When L16 is fired intracellularly at a high frequency it produces a slow hyperpolarization in the sensory cell.

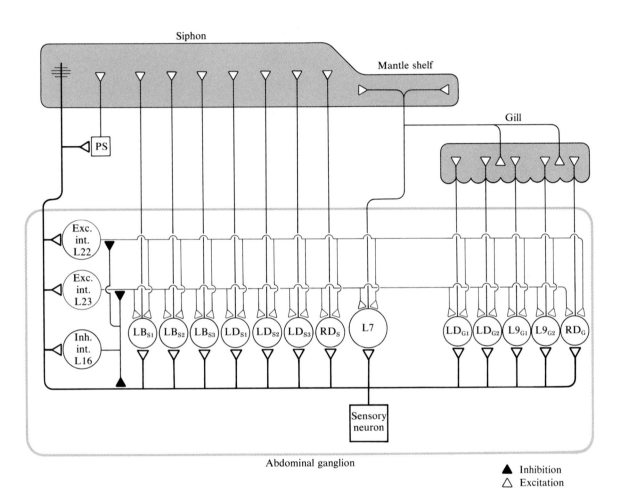

FIGURE 9-17.
Schematic neural circuit of the defensive-withdrawal reflex, indicating the sensory, interneuronal, and motor-neuronal components of the total reflex. The population of sensory neurons innervating the siphon skin consists of about 24 cells. These neurons have direct connections to the motor neurons and indirect connections via two excitatory and one inhibitory interneuron. The three interneurons and the 13 central motor cells in the abdominal ganglion are all unique, identified cells. The peripheral siphon motor cells (PS) are not uniquely identified. The same population of sensory neurons activates both the siphon and gill motor neurons. This accounts for the fact that the two acts of the reflex pattern occur simultaneously. [From Kupfermann *et al.*, 1974.]

QUANTITATIVE CONTRIBUTIONS OF INDIVIDUAL SENSORY NEURONS

When the siphon is stimulated, a complex EPSP is produced in the motor cells, causing them to discharge. This EPSP presumably consists of direct contributions from sensory neurons and the excitatory interneurons. Techniques analogous to those used in evaluating the contributions of individual motor cells to behavior can be used to estimate the contributions of individual sensory cells to the amplitude and duration of the complex EPSPs in the motor cells (Byrne, Castellucci, and Kandel, 1974a).

Since the sensory neurons fire first in response to a siphon stimulus, one might predict that they would contribute only to the early part of the complex EPSP and that the late part would be generated primarily by the interneurons. This has proved not to be the case, however. A comparison of the discharge pattern of a population of mechanoreceptor neurons with the simultaneous complex EPSP in motor neuron L7 reveals that the sensory cells discharge throughout the complex EPSP (Figure 9-18). This is due to the fact that the axons of the various sensory neurons have different conduction velocities and each sensory neuron may fire more than once in response to a single stimulus. Thus the mechanoreceptor discharge may account for a significant portion of the total duration of the complex EPSP and not simply the early phase. This would suggest that much of the complex EPSP is actually the result of the summation of monosynaptic EPSPs of the sensory neurons.

To pursue this possibility further, the ganglion and siphon skin were placed in separate chambers. The complex EPSP elicited in the motor neurons by a tactile stimulus to the siphon was examined with the ganglion first in normal seawater and then in a solution with a high concentration of divalent cations, which block most polysynaptic pathways. The similarity of the EPSPs in the two solutions was surprisingly good (although not perfect), indicating that 50 to 80 percent of the complex EPSP in the motor neurons can be accounted for by the monosynaptic contributions of the sensory neurons (Figure 9-18 B).

Because of the overlapping distribution of the receptive fields in the siphon skin, a weak stimulus applied to a 0.5 mm point activates about eight sensory cells, causing each cell to fire one or two action potentials. One might expect, therefore, that individual action potentials in a single neuron would contribute significantly to the amplitude of the complex EPSP in the motor neuron. To test this idea, a weak tactile stimulus that consistently fires a sensory cell and produces a complex EPSP in the motor cell was delivered to the siphon skin. The sensory cell was then fired with an intracellular current pulse, causing a single action potential and a short-latency monosynaptic EPSP in the motor neuron. The size of the EPSP produced by a single action potential was usually

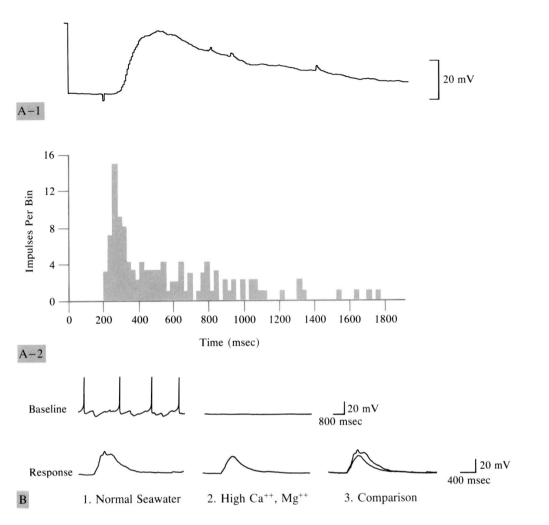

FIGURE 9-18.

Contribution of sensory neurons to the duration of the complex EPSP in a motor cell following stimulation of the siphon skin. [From Byrne *et al.*, unpubl.]

A. Comparison of the duration of a complex EPSP in a motor neuron with a histogram of the sensory neuron discharge. Data were averaged from 26 experiments using different preparations. *A-1.* Computer-averaged complex EPSP from motor neuron L7. *A-2.* Averaged histogram of the mechanoreceptor discharge. The discharge takes place during almost the entire complex EPSP in the motor neuron, indicating that the activity of the sensory cell contributes to the entire duration of the complex EPSP.

B. Comparison of complex EPSPs in seawater and in a high divalent-cation solution. *B-1.* Upper trace is background activity in L7; lower trace is the response of L7 to an 800-msec tactile stimulus to the skin. The cell was hyperpolarized to prevent firing so that synaptic potentials could be seen more clearly. *B-2.* Upper trace is background activity in L7 when the ganglion was bathed in a high divalent-cation solution, which blocks the polysynaptic input; lower trace is the response to the same stimulus as in part *B-1* but with the cell in a high divalent-cation solution. *B-3.* The two EPSPs from *B-1* and *B-2* are superimposed. The magnitude of the response is only slightly reduced by the high divalent-cation solution, indicating that a large part of the response is monosynaptic. The difference between the two EPSPs indicates how much of the response is due to polysynaptic input. In different experiments the response was reduced by 20 to 45 percent.

about one-eighth as large as the complex EPSP produced by the tactile stimulus. This finding is consistent with the idea that about eight sensory neurons contribute to the EPSP at this particular stimulus strength. Next, a very weak stimulus barely threshold for the sensory neuron was examined. When the stimulus did not cause the cell to fire (because of random fluctuations in the threshold of the sensory neuron) the complex EPSP was reduced, in extreme cases by as much as 70 percent! A comparison of the sizes of the complex EPSPs with and without the contribution of one sensory neuron indicates that at some locations as few as three sensory neurons contribute to the reflex (Figure 9-19).

These estimates are consistent with the distribution of the receptive field on the skin (see Figure 9-14). A complex EPSP seems to be composed of the contributions of at least three and usually eight sensory neurons, the exact number depending on the site of the stimulation, the intensity of the stimulus, and the number of sensory cells connecting to the motor cell under investigation.

If these experiments are quantitatively correct, one should be able to account for the reflex in terms of its known neural components. To test this, a force was applied to the siphon skin and the gill contraction measured while intracellular recordings were obtained from both a sensory neuron and a major motor cell, either L7 or LD_{G1} (Byrne et al., 1974b; Byrne, Castelluci, and Kandel, 1976). A tactile stimulus to the skin, presumably activating eight sensory cells, caused a single spike in the sensory neuron; in addition it caused discharge of the motor neuron and contraction of the gill. The sensory neuron was then fired repetitively (about eight times) in order to substitute a temporal pattern for the spatial pattern of the natural stimulus (activation of eight sensory cells). Repetitive firing of the single sensory neuron produced a discharge in the motor neuron and led to a reflex contraction of the gill! The fact that a single sensory neuron can elicit the behavior indicates that the reflex may be accounted for by its known neural components (Figure 9-20).

Quantitative Analysis of a Behavior

Can a behavior be *fully* explained in terms of the sum of its neural parts? Some argue that it cannot, that behavior has defining properties that cannot be reduced to the cellular level and therefore cannot be accounted for by examining the isolated interaction of a few neural components. According to this view, the interaction of all the components with each other, as well as with

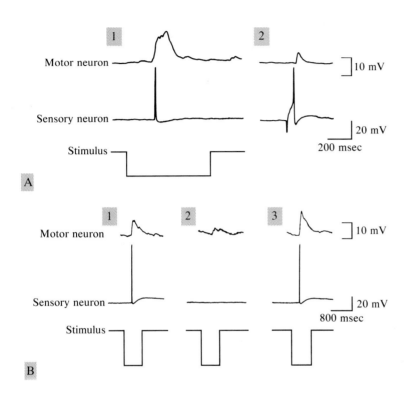

FIGURE 9-19.
The contribution of an elementary EPSP produced by a single spike in an individual sensory cell to the complex EPSP in a motor cell. [From Byrne, Castellucci, and Kandel, unpub.]

A. The contribution of an individual sensory neuron to the amplitude of the complex EPSP in response to a weak stimulus can be estimated by comparing the EPSP produced by intracellular firing of the sensory cell and the overall complex EPSP. *A-1.* A weak stimulus delivered to the siphon skin fired an action potential in the mechanoreceptor being monitored and produced a complex EPSP in the motor neuron. *A-2.* A single spike in the mechanoreceptor triggered with an intracellular pulse produced a short-latency EPSP in the motor cell.

B. A very weak (barely threshold) stimulus to the siphon skin occasionally fails to discharge the sensory cells. As a result, the complex EPSPs during firing and nonfiring can be compared. *B-1.* The stimulus is of sufficient intensity to fire the mechanoreceptor being monitored and to produce a complex EPSP in the motor neuron. *B-2.* The mechanoreceptor does not fire in response to the same intensity stimulus and the amplitude of the complex EPSP is reduced. *B-3.* The next stimulus fires the mechanoreceptor and the EPSP returns to its previous amplitude.

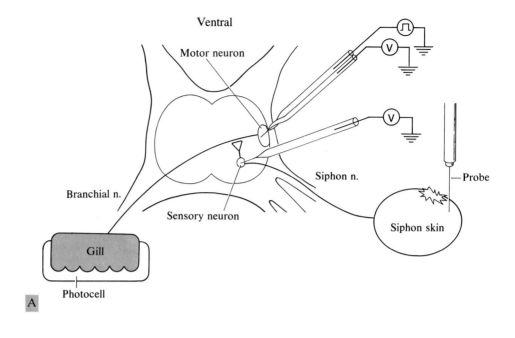

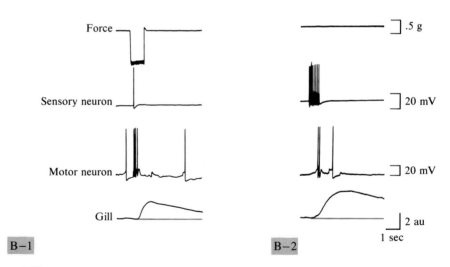

FIGURE 9-20.

Reconstitution of the gill-withdrawal reflex from its component parts. [From Kandel, Brunelli, Byrne, and Castellucci, 1976.]

A. Experimental arrangement for simultaneous recordings of force to the skin, sensory neuron discharge, motor neuron response and gill contraction.

B. Reconstruction. *B-1.* Stimulation of a point on the skin produces a discharge of the sensory cell, repetitive firing of the motor cell, and a gill contraction (simultaneous recordings). *B-2.* Each point on the skin activates about eight sensory cells, each of which fires one to two spikes. In an attempt to simulate the natural behavior of the total population of sensory cells, a single cell was fired repetitively, producing nine spikes. This procedure mimics the natural input of the entire population of sensory cells to the motor cell, causing it to fire and to produce gill contractions similar to those produced with natural stimulation.

receptor and effector organs, reveals principles that cannot be inferred by examining the components in isolation. Even if few would quarrel with the idea that an elementary behavior in an animal like *Aplysia* might be accounted for by its known neural component parts, the question whether a reductive analysis is possible for the complex behavior of mammals, particularly man, is highly controversial. Quantitative analyses of elementary behavior in invertebrates are interesting precisely because they afford a simple test of the reductionist hypothesis. If an elementary behavior cannot be reductively analyzed, then complex behavior certainly cannot. Also, quantitative analyses of elementary behavior in invertebrates allow one to develop the necessary analytic and mathematical techniques needed to apply reductionist strategies to more complex behavior, in invertebrates as well as vertebrates.

Gill withdrawal has not yet been sufficiently analyzed quantitatively to permit us to answer confidently whether the sum of the known neural parts of this behavior accounts for the entire behavior. Nevertheless, some preliminary clues have emerged. The receptive fields of the LE and RE sensory neurons are coextensive with the receptive field of the behavior. The threshold of the sensory neurons (and that of the complex EPSP in the motor cells) is similar to that of the reflex (.25 g). The sensory neuron discharge, the complex EPSP, and the amplitude of the gill contraction are all linearly related to the intensity of the tactile stimulus (Byrne, unpub.). The sensory neurons make direct connections to the motor cells, and these connections contribute significantly to the amplitude and duration of the complex EPSPs that discharge in the motor neurons to mediate the reflex. Each point on the skin can activate about eight sensory cells, and various cells contribute about one-eighth of the entire complex EPSP. Thus, firing a sensory neuron repetitively (eight times, to simulate the activity of eight cells) generates the behavior. These data indicate that the sensory neurons probably account for most of the tactile input from the siphon skin. Moreover, the six motor cells account quantitatively (in terms of amplitude) and qualitatively (in terms of type of movement) for most, and perhaps all, of the gill movement.

Thus, to a surprising degree the withdrawal reflex can be explained in terms of the properties of its neural components examined in isolation, for the following reasons: (1) the mechanoreceptor discharge is proportional to stimulus intensity; (2) there is fairly good linear summation of the chemical synaptic actions produced by the sensory neurons onto motor cells; (3) the spike frequency is a linear function of the amplitude of the complex EPSP; and (4) the muscle contraction is linearly related to motor neuron discharge (Byrne *et al.*,

1976). These several features also account for the graded properties of this reflex act.

I would emphasize that although cellular analysis has accounted for almost all the neural components of the gill-withdrawal reflex elicited by stimulation of the siphon, this reflex is only a limited aspect of the full behavioral repertory of the gill. Gill withdrawal activated by stronger stimuli or from other sites involves additional pathways, including peripheral ones, that have not yet been analyzed. However, as studies of the siphon-withdrawal reflex have shown, even peripheral pathways can now be systematically analyzed. In principle, then, even the more extensive neural circuit that controls gill withdrawal under a wide range of stimulus conditions could be worked out. Moreover, as we shall see in Chapters 12 and 13, much can be learned about behavior and its modification from analyzing it under restricted conditions of activation.

A VISCERAL-MOTOR FIXED ACT: RESPIRATORY PUMPING

The Behavior

The mantle organs also participate in a fixed act, a respiratory pumping movement. Although the defensive-withdrawal reflex of the gill and siphon is smoothly graded, the pumping movements occur in discrete steps and appear to be all-or-none. In addition to being elicited by external stimuli, the pumping movements can be emitted spontaneously in response to a central command, sometimes regularly at several-minute intervals. The rate of pumping is increased by anoxia or feeding, conditions that increase the animal's respiratory load. Like the defensive reflex of the mantle organs, respiratory pumping involves the gill, siphon, and parapodia, but the movements in the two behaviors differ considerably (Kupfermann and Kandel, 1969; Kupfermann et al., 1974).

Respiratory pumping consists of a contraction and rostral rotation of the gill, flaring of the gill halves so as to expose the efferent vein, and bunching of the individual pinnules, accompanied by a contraction of the siphon and parapodia that usually begins as a wave at the rostral tip of the parapodia and ends posteriorly at the siphon (Figure 9-21). These movements circulate seawater through the mantle cavity, aerate the gill, and cleanse the mantle cavity of detritus. This fixed act is a component of a larger fixed-action pattern that also involves changes in heart rate and blood pressure (see p. 436 and Fig. 4-13).

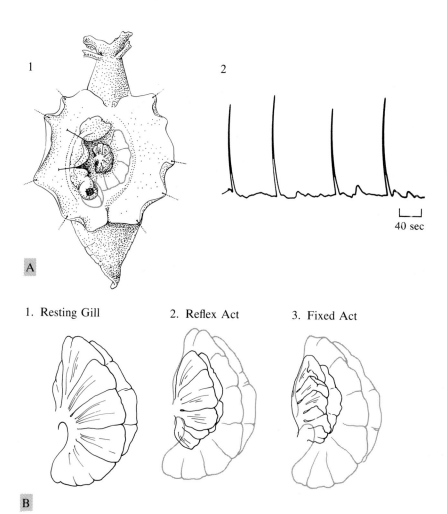

1. Resting Gill   2. Reflex Act   3. Fixed Act

FIGURE 9-21.

Respiratory pumping movements. [From Kupfermann *et al.*, 1974.]

A. The respiratory pumping movement is a stereotypic fixed act that is often generated in an all-or-none manner. *A-1.* Top view of an *Aplysia* showing centrally commanded contraction of the gill and the siphon (dotted lines indicate organs in the resting state) and the parapodia. *A-2.* Photocell recording of several successive spontaneous gill contractions showing similarity in the amplitude and duration of several consecutive fixed acts.

B. Comparison of reflexive and fixed responses. Tracings from single cinematographic frames of the gill in a restrained, intact preparation. The anterior part of the gill is toward the top of the figure. *B-1.* Relaxed gill. *B-2.* Reflexive gill contraction elicited by stimulating the siphon with a jet of seawater. *B-3.* Centrally commanded fixed act. This contraction occurred without a recognizable external stimulus.

The Neuronal Circuit

Whereas the contractions produced by intracellular stimulation of L7 are similar to those that occur in reflexive gill withdrawal, the contractions produced by stimulation of LD_{G1} and LD_{G2} are similar to those that occur during the respiratory pumping movements. Thus the two types of gill movement can be reconstructed from the actions of the three major motor cells. Reduction in gill volume with no flaring or antiflaring of the gill halves and spreading of the pinnules, characteristic of the reflex contraction, is based upon the combined actions of all three cells, with the action of L7 predominant. Reduction in gill volume with rostral rotation, bunching of the pinnules, and flaring of the two gill halves, characteristic of the respiratory pumping movement, is based primarily upon the action of LD_{G1} and LD_{G2}, with L7 not participating in the early phase. Intracellular recordings from these cells during the generation of the reflex and fixed acts are consistent with this inference. During reflexive withdrawal all the motor cells fire synchronously, but during centrally commanded withdrawal the motor neurons do not fire synchronously. Cells LD_{G1} and LD_{G2} fire first while cell L7 is inhibited; L7 fires only later as a rebound effect called *rebound excitation* (Figure 9-22). This is thought to be due to removing Na^+ inactivation by hyperpolarization, thereby increasing (transiently) the excitability of the cells.

The central command consists of bursts of synaptic potentials that occur

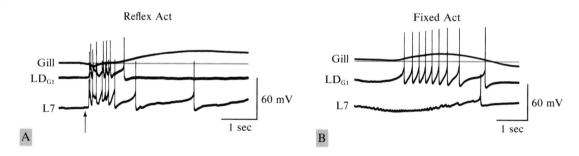

FIGURE 9-22.

Comparison of the firing patterns of motor cells LD_{G1} and L7 during reflexive and centrally commanded gill contractions.

A. Photocell record of gill contraction made simultaneously with intracellular recordings from two gill motor neurons during a reflex act elicited by tactile stimulation of the siphon.

B. Photocell record of gill contraction made simultaneously with intracellular recordings of the two gill motor neurons in part *A* during a centrally commanded fixed act. Motor neuron LD_{G1} is excited and L7 is inhibited during the fixed act. Both motor neurons are excited during reflexive contraction.

synchronously in the various motor neurons. These PSPs are ascribed to Interneuron II, which is made up of two or more tightly coupled unidentified cells in the ganglion (see p. 333). Interneuron II makes synaptic connections to all gill and siphon motor cells, but produces different actions on different cells. In triggering the pumping movements, Interneuron II first produces a large EPSP burst and high-frequency discharge in gill motor cells LD_{G2} and RD_{G1}, and then, slightly later, a small EPSP and a few spikes in LD_{G1}. In addition, Interneuron II inhibits L7, $L9_{G1}$, and $L9_{G2}$. Since L7, $L9_{G1}$, and $L9_{G2}$ are spontaneously active, inhibition of these cells is followed by rebound excitation, which adds to the late stage of the contraction (Figure 9-23). The synaptic excitation and inhibition produced by the central command controls the firing sequence of the motor neurons so that the discharge of LD_{G2} and RD_G precedes that of LD_{G1}, which in turn precedes that of L7, $L9_{G1}$, and $L9_{G2}$. This sequence

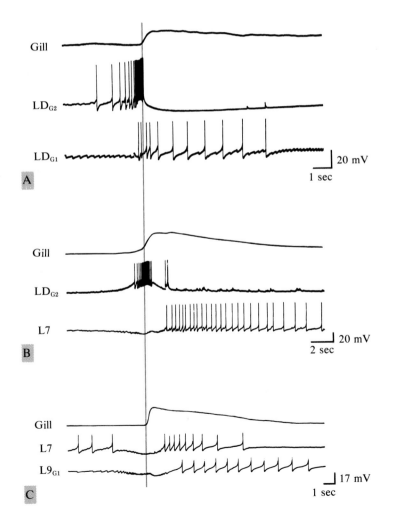

FIGURE 9-23.
Phasing of the action potentials in the gill motor neurons during centrally commanded respiratory pumping by the burst of PSPs produced by Interneuron II. (Gill responses shown were recorded with a photocell.) Records in parts *A, B,* and *C* are from different experiments and are aligned to illustrate the temporal patterning of the activity of the motor neurons. Each motor cell consistently fires at a fixed phase of the movement. Cell LD_{G2} and RD_G (not shown) always fire first; LD_{G1} fires next; L7, $L9_{G1}$, and $L9_{G2}$ are inhibited during the early phase and fire only later, as a rebound excitation following inhibition, and L7 always fires before the L9 cells. [From Kupfermann *et al.*, 1974.]

establishes the pumping motion that circulates blood through the gill, while the concomitant contractions of the parapodia and mantle shelf move fresh seawater through the mantle cavity (Kupfermann *et al.,* 1974).

Is this fixed act truly fixed? When carefully analyzed, fixed-action patterns sometimes vary in intensity, in characteristic amplitude steps, but tend not to vary in form. The relation between successive movements is quite stereotypic (see for example Barlow, 1968). This is true for successive respiratory pumping actions, which may vary in intensity in a step-wise manner, while the movements themselves tend to be stereotypic. The relative stereotypy results from the fixed firing sequence of the neurons. Whether the spontaneous contraction is large or small, LD_{G2}, RD_G, and LD_{G1} fire first, while L7 and L9 cells are inhibited. During the rebound excitation following inhibition, L7 fires before the L9 cells. Interneuron II also excites some siphon motor cells (LD_{S1}, LD_{S2}, LD_{S3}, and RD_S) and inhibits others (L7, LB_{S1}, LB_{S2}, and LB_{S3}). Here also the firing sequence is fixed; what varies is the intensity of the command—the size and frequency of the PSPs produced—presumably reflecting characteristic variations in the firing frequency of Interneuron II (see Figure 10-19).

The central command for respiratory pumping does not require sensory feedback from the periphery for its generation. It occurs, as well, in the completely isolated abdominal ganglion (Waziri and Kandel, 1969).

Figure 9-24*A* illustrates the neural circuit of this fixed act. The central command can be triggered by excitatory input from the siphon and gill. The siphon-triggering is perhaps mediated by the siphon sensory neurons, and the gill-triggering is probably mediated by peripheral stretch receptor neurons. In addition, the central command can be released both spontaneously and in response to internal (metabolic) variables, such as variations in oxygen tension of the blood (Kupfermann *et al.,* 1974).

FIGURE 9-24 (*Facing page*).
Different neuronal controls of a single effector system.

A. Neural circuit for the centrally commanded respiratory pumping movements mediated by Interneuron II. Since the cells that make up Interneuron II are not yet identified, the synaptic connections of Interneuron II are indicated with grey lines. Individual peripheral motor neurons (PS) innervating the siphon have not been identified.

B. The converging inputs into identified motor neurons involved in reflexive gill contraction (defensive withdrawal reflex) and in the centrally commanded respiratory pumping movements. All neurons receive excitatory synaptic input from sensory neurons (SN) and interneurons (not shown) mediating reflex withdrawal. By contrast, some motor neurons are excited and some are inhibited by Interneuron II, which mediates the centrally commanded movements. By means of these connections the same motor population can be switched from a reflex to a fixed response. Interneuron II can be triggered in several ways: by strong stimulation of the siphon, by strong mechanical stretching (contraction) of the gill, or in response to central commands. The centrally commanded contractions are spontaneous and increase in frequency by stimuli that increase respiratory demand, such as feeding, anoxia, and ink or debris in the mantle cavity. (SR = stretch receptor; SN = sensory neurons).

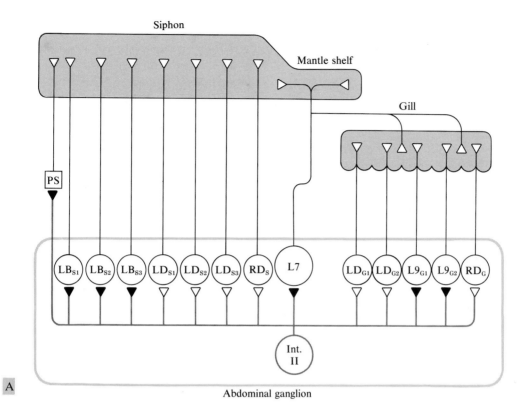

A

B

MULTIPLE CONTROLS OF A SINGLE EFFECTOR SYSTEM

The fact that the mantle organs are involved in both reflex and fixed acts raises the question: How does the nervous system generate two types of responses in a single effector system? Two main findings have emerged from the analysis of behavioral responses of the mantle system. First, the various motor neurons of the mantle organs are independent, parallel elements that are not connected with each other. Second, both the motor and sensory neurons are arranged in a hierarchical manner. Some motor neurons move only the gill, some motor neurons move only the siphon, and one motor cell moves the siphon together with the gill and mantle shelf. The sensory neurons also seem to differ in their central connections. As a result, reflex acts or central commands can play on different combinations of a common pool of motor neurons and thus activate either different effector movements or different sequences of the same set of movements. This arrangement permits an effector organ or an entire effector system to be controlled in different ways for very different purposes (Figure 9-24B). In addition, as we shall see in Chapter 10, the entire mantle system or components of it can be controlled in varying combinations with other effector systems, such as the heart and kidney.

The motor neurons innervating the external organs of the mantle can be activated in any of the three modes: (1) a reflex mode; (2) a fixed-act mode triggered by an external stimulus; and (3) a fixed-act mode without an obvious external stimulus. How does the nervous system switch from one to another of the three states? A comparison of the cellular responses during the three modes shows that the choice of response is determined by the activity of one or the other of two sets of neurons that provide synaptic input to the common pool of motor neurons (Figure 9-24B). The stimulus that elicits the reflex activates sensory neurons (and interneurons) that excite the cells of the motor-neuron pool; the interneuron(s) that emit the central command for the (elicited or spontaneous) fixed act excite some cells of the motor-neuron pool and inhibit others. Although the efficacy and sign of the synaptic input varies with different motor neurons, every gill and siphon motor neuron receives some synaptic input during both the reflex and fixed responses (Figure 9-24B). This organization is an example of Sherrington's principle that appropriate integration and execution of even simple behavior requires the action of a surprisingly large part of the nervous system (Sherrington, 1906).

Although the reflex and fixed acts involve different sequences of activation of the motor-neuron pool, the elicited and spontaneous fixed acts involve identical activation of the motor neurons—only the source of activation of the central command is different. For the elicited response the interneurons mediating the central command are excited by sensory neurons from the siphon and from stretch receptors in the gill. For the spontaneous response the interneurons are activated either by an unidentified central source (perhaps a periodic burst generated by the command cell in response to a metabolic change) or by removal of a tonic inhibitory input.

Sherrington also developed the concept of a *final common motor neuron* to explain how the central nervous system commands different responses from the same effector organs. Different combinations of excitatory and inhibitory inputs converge on final common motor neurons to produce various reflex behaviors. Whereas Sherrington's concept was developed for reflex actions, studies in *Aplysia* show that both reflex and fixed acts can share the same pathway. The central controls for each do not utilize separate pools of motor neurons.[7]

The neuronal organization of reflexive withdrawal in *Aplysia* resembles that of defensive reflexes in vertebrates. There is one main difference. In mammals defensive (flexion) reflexes are polysynaptic (Lloyd, 1949). In the gill and siphon withdrawal reflexes of *Aplysia* the sensory input to the motor cells has a large monosynaptic component.

Reflex acts in vertebrates are moderately well understood but there is no comparable understanding of fixed acts. In particular, there is no knowledge about how the central program of fixed acts is generated. To examine the nature of central programs, I will first consider two very simple examples of fixed acts in *Aplysia:* inking and egg-laying.

A GLANDULAR FIXED ACT: INKING

The siphon, the same site at which stimulation elicits a graded reflex response to weak stimuli, can at a higher stimulus strength give rise to an all-or-none action, inking. By comparing the cellular properties and interconnections of a fixed act with those of a graded reflex, the rules that relate different types of neuronal circuits to different classes of behavior can be examined.

---

[7]However, movements of the gill can be produced from still a third source: stimulation of the gill pinnules directly (Peretz, 1970; Newby, 1973; Kupfermann *et al.*, 1971). These movements are mediated by a separate neuronal machinery that does not seem to involve the final common motor pathway for defensive withdrawal and respiratory pumping. Thus, in addition to the central motor neurons that are necessary for coordinating the reflex and fixed acts of the gill with other effector responses (the siphon, the parapodia, and the heart), the gill contains peripheral neurons that mediate local responses (to direct stimulation) of individual gill pinnules, alone or in combination with other pinnules.

The Behavior

Many *Aplysia* species, including *A. californica*, have a purple gland at the edge of their mantle shelf. A strong, noxious stimulus to the siphon (or to the foot or head) will elicit ink from the acinar vesicles of the purple gland; the ink is expelled from the mantle cavity as a dark purple cloud by the respiratory pumping movements of the mantle organs (Figure 9-21). In a limited volume of water the ink may serve as camouflage and completely hide the animal from view (Figure 9-25). The ink may also be noxious to some predators. Inking does not normally occur spontaneously; it has a high threshold to tactile stimulation, compared with gill withdrawal or other defensive responses. Once that high threshold has been reached, the stimulus-response relationship is very steep and inking usually occurs in an all-or-none fashion (Figure 9-26). Inking behavior is a complex fixed-action pattern consisting of recurrent inking episodes and respiratory pumping movements. I will focus on only one component, the isolated inking episode.

FIGURE 9-25.
Inking response in *Aplysia*.

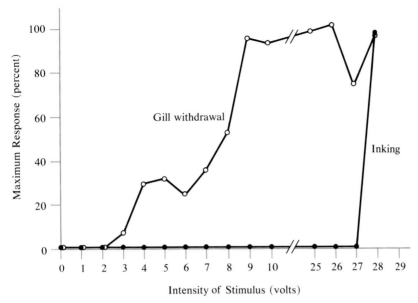

FIGURE 9-26.
Comparison of the gradual input-output curve of the graded gill-withdrawal reflex and the steep input-output curve of inking. Both responses were simultaneously recorded and elicited by an electrical stimulus to the siphon. [From Carew and Kandel, unpubl.]

The Neuronal Circuit

The cellular mechanisms of inking can be studied in an isolated abdominal ganglion and purple gland. Using this preparation, Carew and Kandel (1973, unpub.) found three identified cells with similar properties—L14A, L14B, and L14C—the firing of which leads to the release of ink from the purple gland (Figure 9-27). Each cell releases ink from a different portion of the gland. Cell L14A releases ink from the anterior three-quarters of the gland, L14B from the posterior half of the gland, and L14C from a small transitional region in the middle. As in the gill and siphon motor system, the three ink-gland cells innervate overlapping areas of the purple gland. The most extensive overlap occurs in the middle portion (Figure 9-28).

Each of these three cells has a high threshold; a 30–40 mV depolarization is required to initiate an action potential. Once threshold is surpassed, the release of ink by action potentials in an ink-gland cell is graded. Increasing the number of spikes in a cell produces progressively greater release of ink from the

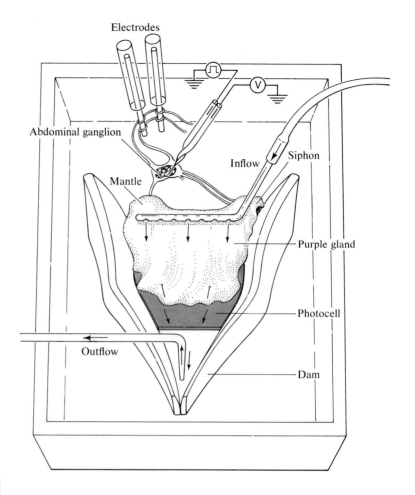

Electrodes

Abdominal ganglion

Mantle

Inflow

Siphon

Purple gland

Photocell

Outflow

Dam

A

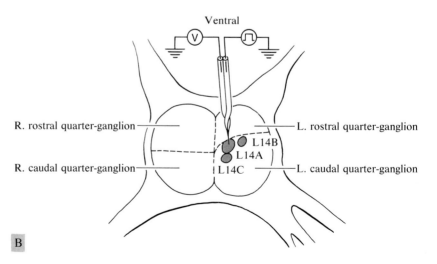

Ventral

R. rostral quarter-ganglion

L. rostral quarter-ganglion

L14B

L14A

R. caudal quarter-ganglion

L14C

L. caudal quarter-ganglion

B

FIGURE 9-27 (*Facing page*).

Experimental arrangement for studying inking. [From Carew and Kandel, unpubl.]

A. The purple gland is first excised but left attached to the abdominal ganglion by means of the peripheral nerves. The gland and the ganglion are then pinned to the floor of the chamber. Cooled and aerated seawater is perfused across the gland, and the water is channeled across a photocell transducer (imbedded in the floor of the chamber) and collected by a vacuum pipette. A light is directed on the photocell transducer so that when ink is released from the gland and flows across the photocell transducer, a voltage change proportional to the amount of ink released will be produced. One or more cells in the abdominal ganglion are simultaneously monitored, and the connectives, which carry the sensory input from the head and neck regions to the ganglion, can be electrically stimulated.

B. Experimental arrangement of the preparation, showing the position on the ventral surface of the ganglion of the three L14 motor cells that effect inking.

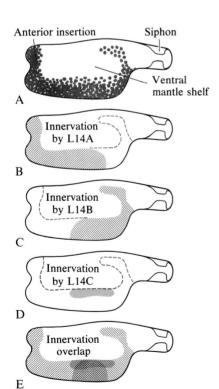

FIGURE 9-28.

Areas of the purple gland innervated by each of three neurons (L14A, L14B, L14C) that mediate inking. **A.** The distribution of ink-containing vesicles in the purple gland. **B-D.** The fields of motor innervation of each of the three cells. **E.** Overlap of the three fields. [From Carew and Kandel, unpubl.]

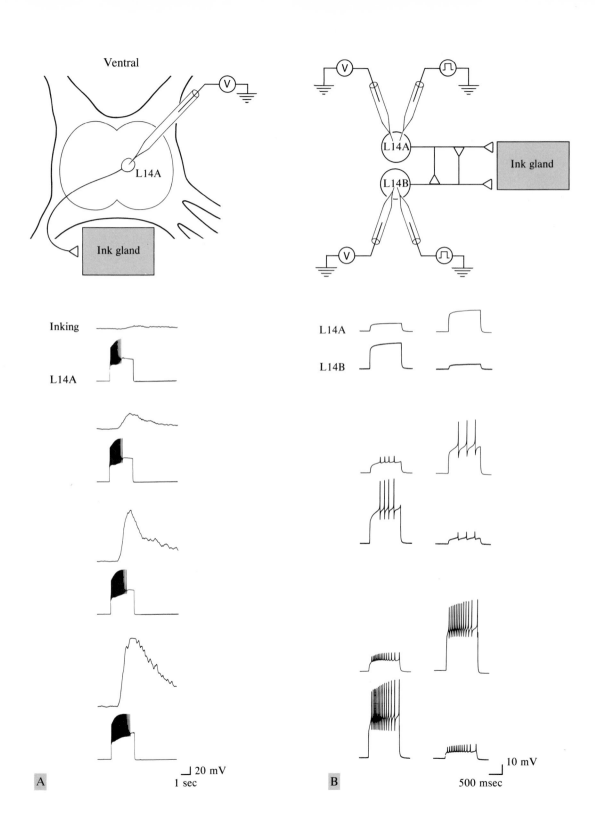

Ventral

Ink gland

Inking

L14A

20 mV
1 sec

A

L14A

L14B

Ink gland

10 mV

500 msec

B

gland (Figure 9-29). Thus the regenerative (all-or-none) feature evident in the behavioral response does not seem to result from connections between the motor cells and the ink gland.

The three ink-gland motor cells are electrically coupled to one another. Injecting current pulses into one cell produces electrotonic potentials in the other two. The coupling ratio between L14A and L14B (the ratio of the electrotonic potential produced in the coupled cell to that produced in the injected cell) is particularly high, ranging from 0.9 to 0.25 (Figure 9-29B). The coupling ratio between either of these cells and L14C is somewhat lower (0.4 to 0.1). Because the cells are tightly coupled (the coupling ratio is high) they usually act as a single motor unit. In addition, all three cells receive almost identical and synchronous synaptic input from the connectives that carry input from the siphon, or the head and neck, to the abdominal ganglion. For example, a train of strong stimuli (6/sec for 2.5 sec) to the connectives produces synchronous synaptic potentials and spike discharge in all three motor cells, accompanied by massive release of ink from the purple gland (Figure 9-30).

To determine how much of the inking response is mediated by the three L14 motor cells, all three were reversibly removed from the neural circuit by hyperpolarizing the membrane potential (by 40 to 50 mV) to prevent the cells from reaching threshold and discharging action potentials. Under these circumstances, a train of strong stimuli failed to release ink. With the cells at their normal resting potential, identical stimulation produced a repetitive spike discharge in the three cells and massive inking. Thus, most if not all of the inking response is mediated by the three L14 motor cells (Figure 9-31).

FIGURE 9-29.
The ink-gland (L14) motor cells. [From Carew and Kandel, unpubl.]

A. Stimulation of cell L14A with a depolarizing current pulse releases ink from the purple gland. The upper trace in each pair of records is a photocell record of ink release. Increasing the depolarizing pulses increases the number of action potentials generated by L14A and produces progressively greater release of ink. The cell has a high threshold and requires 30 to 40 mV depolarization to initiate action potentials.

B. Electrical synaptic connections between two of the three motor cells. Increasing the depolarizing current pulses to L14B produces increasing potential changes in L14A (first column); increasing current pulses to L14A produces increasing potential changes in L14B (second column).

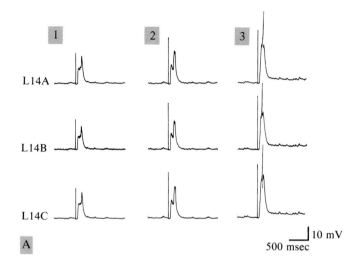

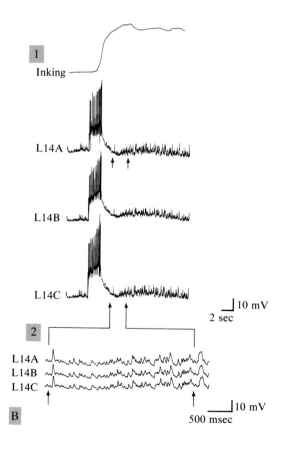

FIGURE 9-30.
Common connections of the ink-gland (L-14) cells.
[From Carew and Kandel, unpubl.]

A. Synchronous EPSPs in all three motor cells are elicited by stimuli of increasing strength (columns 1, 2, 3) applied to one of the connections.

B. In the intact animal inking is elicited by stimulating the head. Inking can be elicited in an isolated preparation by delivering a strong train of stimuli to the connectives that carry fibers from the head ganglia. *B-1.* Synchronous firing of the three cells in response to connective stimulation elicits a full inking response. *B-2.* A fast-sweep record of the interval marked by the arrows in *B-1,* illustrating synchrony of the evoked shower of synaptic potentials.

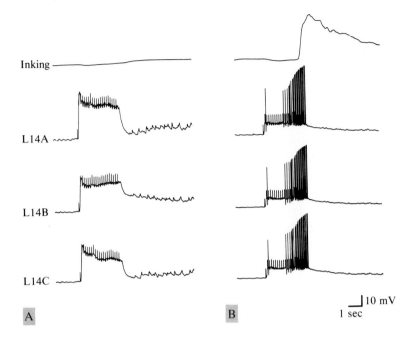

FIGURE 9-31.
The quantitative contribution of the three L14 cells to the inking response. [From Carew and Kandel, unpubl.]

A. With the L14 cells hyperpolarized, a train of stimuli to the connectives produces EPSPs in the L14 cells but no release of ink. Upward-going deflections riding on the depolarization in the L14 cells are shock artifacts.

B. With the L14 cells at their normal resting potential, an identical train of stimuli to the connectives produces a high-frequency discharge of action potentials in the L14 cells and massive inking. This indicates that inking is probably entirély controlled by these three cells.

PROPERTIES OF THE NEURONAL CIRCUIT AND PROPERTIES OF THE BEHAVIOR

Since the cells are similar and fire synchronously, there should be a good correspondence between their cellular properties and the characteristics of inking behavior. Moreover, differences in response properties between the ink-gland motor cells and the gill motor cells might reveal differences in the mechanisms for generating fixed acts, such as inking, and reflex acts, such as gill withdrawal.

Inking does not occur spontaneously, whereas the gill has a resting level of contraction (a resting tone) and often undergoes slight changes in tone spontaneously. Inking is a high-threshold behavior and limited to one behavioral purpose. Gill movements are involved in different behavioral responses. One type of gill movement occurs as part of a low-threshold reflex; another type occurs as part of a centrally commanded fixed act. Finally, inking is an all-or-none behavior, whereas the reflex withdrawal of the gill is graded. The differences between inking and gill movements in spontaneity, threshold, stereotypy, and gradedness are reflected in the cellular properties and synaptic connections of the ink-gland and gill motor cells, respectively.

The differences in behavioral threshold and spontaneity are related to the membrane potential of the motor cells (Carew and Kandel, unpubl.). Some of the gill motor neurons, such as L7, have a low membrane potential (-35 mV) and are spontaneously active (thereby contributing to the gill's resting tone), whereas all the ink-gland motor cells are silent and have a very high membrane potential (-65 mV). Because of its low membrane potential, cell L7 has a low threshold and great sensitivity for slight increases in excitatory synaptic activation. By contrast, the ink-gland motor cells have an unusually high threshold of firing (30 mV) because of their high membrane potential.[8]

The fact that the gill participates in several behavioral responses while the ink gland is limited to one behavior is related to differences in the connections between the respective motor cells: the gill motor cells are not interconnected, whereas the ink-gland motor cells are electrically interconnected. As a result, the gill motor cells can respond individually, whereas the ink-gland motor cells tend to respond together to synaptic input. Moreover, the gill motor neurons receive synaptic input from several command cells, e.g., Interneuron II and L10, that have different effects on different cells (Figure 9-32). The ink-

---

[8]When the input to them is not completely synchronous, the electrical coupling may also contribute to the high threshold. Cells that are electrically coupled have a lower input resistance than uncoupled cells because the electrical synapses produce additional parallel current pathways for the synaptic input to each cell. A given synaptic current will produce a PSP whose amplitude is determined by the input resistance of the cell. The lower the input resistance the smaller the PSP. As a result, a synaptic input that generates sufficient synaptic current to produce a large suprathreshold EPSP in an uncoupled cell with high input resistance will produce a small (subthreshold) PSP in cells that are electrically coupled and therefore have a low input resistance (Bennett, 1968; Willows and Hoyle, 1969; Kandel and Kupfermann, 1970; Getting and Willows, 1974). This factor may not be critical for inking, however, because the input to the ink-gland motor cells is most often synchronous. Each cell will therefore be at the same potential as the other two so that no current can flow between them. Thus, synchronous input effectively uncouples electrotonically coupled cells (Getting and Willows, 1974).

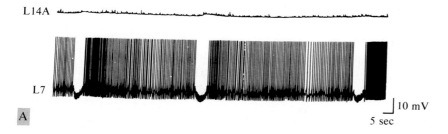

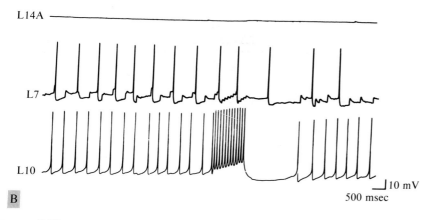

FIGURE 9-32.

Comparison of the firing patterns and synaptic input from autoactive interneurons in an ink-gland motor neuron (L14A) and gill motor neuron (L7). **A.** Cell L14A is silent but L7 is spontaneously active. In addition, L7 receives large bursts of IPSPs from autoactive Interneuron II, whereas L14A receives only a small depolarization. **B.** At a low rate of firing, L10 produces EPSPs in L7 but not in L14A. [From Carew and Kandel, unpubl.]

gland motor cells receive little spontaneous input from central command cells and most inputs in the various motor cells tend to be synchronous.

The graded nature of the gill-withdrawal reflex and the all-or-none nature of inking reflect the stimulus–response relationships of their respective motor components. Tactile or electrical stimulation of the mantle organs, the head, or the connectives (coming from the head ganglia) evokes graded complex EPSPs in both the gill and ink-gland motor cells. The gill motor cells transform the graded input into a graded increase in the frequency of action potentials; the ink-gland motor cells transform the graded input into an all-or-none burst of action potentials (Figure 9-33). Thus, stimuli of weak to moderate intensity produce suprathreshold EPSPs and a brisk discharge of action potentials in the gill motor cells. But the same stimuli produce only subthreshold EPSPs in the ink-gland motor cells because the resting potential of these cells is high. As the stimulus becomes stronger, the gill motor cells increase their frequency of firing in a graded fashion until they reach a maximum. The sudden generation

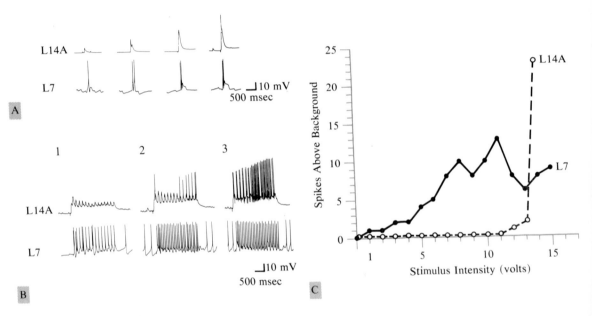

FIGURE 9-33.
Comparison of the synaptic input-output relationships in ink-gland and gill motor cells. [From Carew and Kandel, unpubl.]

A. Comparison of responses to progressively stronger stimuli applied to the connectives. In L7 these stimuli readily reach threshold and produce spiking because of the cell's low membrane potential. Only a strong stimulus discharges L14A, however, because of the cell's high resting potential.

B. Comparison of responses to trains of stimuli. A weak- or moderate-intensity train of stimuli to the connectives produces a progressively more effective discharge in L7, whereas only small EPSPs or a few action potentials are produced in L14A (*B-1* and *B-2*). However, as the stimulus strength is further increased (*B-3*), a critical point is reached at which L14A suddenly fires much more effectively and discharges in an accelerating burst; the discharge frequency of L14A is now higher than that of L7. Maximal firing of L14A therefore seems to be produced by certain patterns of prolonged, phasic input.

C. Comparison of spike discharges in L7 and L14A in response to stimuli of increasing strength. The stimulus–response curve for the ink-gland motor neurons is steeper than that for the gill motor neurons (compare the stimulus-response curves for the inking and gill-withdrawal reflexes, Figure 9-26).

of a burst of action potentials in the ink-gland motor cell seems to account for the all-or-none properties of the behavior. It has therefore been studied in some detail in an attempt to understand the central program for inking.

THE CENTRAL PROGRAM FOR INKING

How is the sudden acceleration of the discharge of the ink-gland motor cells achieved? It might be due to a sudden increase in the excitatory synaptic current when either the intensity or duration of the stimulus is increased. Neither of these increases occurs, however. In fact, as the stimulus intensity is increased, the synaptic potentials increase in a graded fashion. Moreover, the later synaptic potentials of a train have the same amplitude as the early ones, or decrease slightly (Figure 9-34), indicating that the synaptic current does not increase

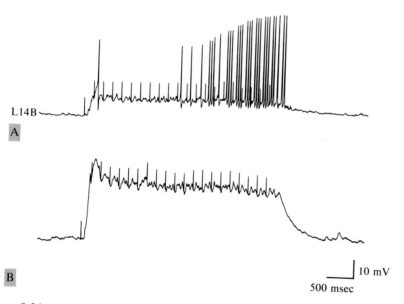

FIGURE 9-34.
The role of EPSPs in the production of the accelerating burst in L14 cells. [From Carew and Kandel, unpubl.]

A. An accelerating burst in cell L14B at the resting potential.

B. The cell was hyperpolarized, allowing one to see the synaptic potentials occurring during the burst. The initial large EPSP undoubtedly accounts for the initial burst of spikes in Trace A, but the subsequent acceleration in L14 probably cannot be accounted for by EPSP facilitation because the EPSPs do not get larger.

abruptly. The acceleration seems to involve two factors: (1) the inactivation of an early K⁺ conductance channel in the nonsynaptic membrane of the ink-gland motor cells (Figure 7-27), and (2) more effective transmission across the electrical synapses between the motor cells (Carew and Kandel, unpub.).

Carew and Kandel (unpub.) analyzed the role of the K⁺ conductance mech-anism in the nonsynaptic membrane by using a train of constant depolarizing current pulses to simulate (albeit imperfectly) the potential changes produced by a train of synaptic potentials. Although the current pulses were constant, the later pulses in the train produced larger electrotonic potentials than did the early pulses. If the early electrotonic potentials were close to threshold, the later ones triggered action potentials. This increase in electrotonic potential with repeated pulses indicates that later synaptic current in a train will produce more effective membrane depolarizations than will the early transients (Figure 9-35).

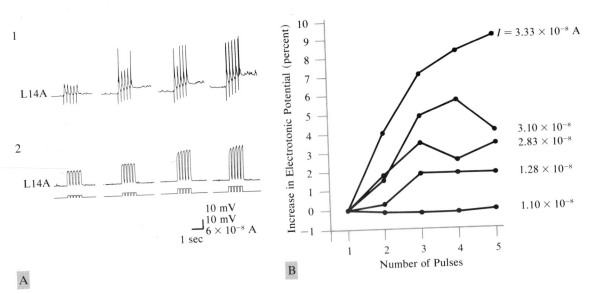

FIGURE 9-35.
Analysis of the central program for inking. [From Carew and Kandel, unpubl.]

A. Comparison of changes in PSPs and electrotonic potential in response to increasing stimulus strength. *A-1.* There is a tendency for PSPs later in the burst to be of larger amplitude than PSPs earlier in the burst. *A-2.* The electrotonic potential (apparent input resistance) increases with each successive electrotonic pulse in the train (bottom trace is the current monitor).

B. Amplitude of the electrotonic potential as a function of the number of pulses in the train at different levels of current strength (indicated by numbers on the right). As the amplitude of the current pulse is increased, the electrotonic potential increases with successive pulses of the train.

Ion-change and voltage-clamp experiments suggest that the increases in electrotonic potential are due to inactivation of an early K^+ conductance (Figure 9-36; Carew and Kandel, unpub.; Byrne, Dieringer, and Koester, unpub.). This conductance is normally activated at membrane potentials more depolarized than -50 mV. Once activated, the early K^+ current inactivates rapidly. The inactivation is abolished by returning the membrane potential to more hyperpolarized levels. Thus the increase in electrotonic potential with repeated pulses occurs only in a limited range of membrane potential on either side of threshold. Hyperpolarization of the membrane potential, which removes it from the voltage range in which activation of the early K^+ current occurs, first diminishes and finally abolishes the increase in electrotonic potential (Figure 9-36). Since the ink-gland motor cells have an unusually high resting potential (-65 mV), inactivation of the early K^+ conductance is normally abolished at the resting level. The initial depolarizations produced by the early EPSPs in a train therefore turn on the early K^+ current, which reduces (short circuits) the effectiveness of the early EPSPs. But the continued depolarization produced by the subsequent EPSPs of the train inactivates the early K^+ current, resulting in a progressive increase in input resistance. This factor can account for at least part of the accelerating response of the motor cells to constant synaptic input.

A second factor that may contribute to the accelerated firing is the fact that the ink-gland motor cells are electrically coupled. An action potential in one motor cell produces electrically mediated EPSPs in the other two motor cells. If these EPSPs cause the two cells to fire, they in turn will produce electrical EPSPs in the first motor cell. During repetitive firing both the height and duration of the action potentials increase (due in part to the progressive inactivation of the early K^+ conductance). These changes in the configuration of the action potential enhance the positive feedback between the electrically coupled cells. Larger and slower action potentials are transmitted more effectively than smaller and faster spikes through electrical connections, making the electrical EPSPs larger. Enhancement of the electrical EPSPs during repetitive firing would further enhance the steepness of the stimulus–response curve.

TWO WAYS OF TRIGGERING THE CENTRAL PROGRAM FOR INKING

Inking can be triggered in two ways: (1) by a strong (suprathreshold) stimulus to one site, and (2) by weaker (subthreshold) stimuli to two different sites. The two ways of triggering involve different synaptic mechanisms. Inking in response to a strong stimulus at one site (of the sort we have considered above) involves synaptic actions due to an increase in membrane conductance (presumably to Na^+ and perhaps to K^+). Inking in response to two closely timed

400

1. Resting potential   **2. −10 mV**   **3. −15 mV**

L14A
Current

5 mV
4.3 × 10⁻⁸ A
500 msec

A

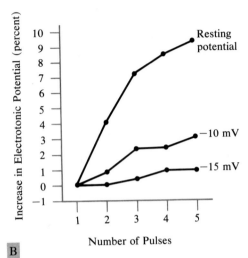

B

1. Normal Seawater

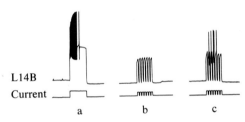
L14B
Current

      a   b   c

2. Na⁺-free Seawater

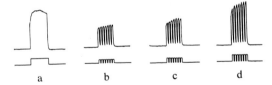

    a   b   c   d

3. Normal Seawater

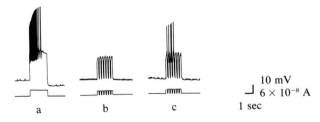

    a   b   c

10 mV
6 × 10⁻⁸ A
1 sec

C

FIGURE 9-36.

The increase in input resistance of ink-gland motor cells is dependent on the value of the membrane potential and not dependent on the presence of Na^+ in the external medium. [From Carew and Kandel, unpubl.]

A. *A-1.* With an ink-gland motor cell (L14B) at resting potential, repeated constant depolarizing current pulses produce a progressive increase in the size of the electrotonic potentials and consequently the cell's apparent input resistance. *A-2.* When the cell is hyperpolarized 10 mV below the resting potential, the progressive increase the input resistance of this particular specimen of cell L14A decreased when the cell was hyperpolarized. to 15 mV below resting potential, the increase in apparent input resistance is almost completely abolished. The first electrotonic pulse seen when the cell is hyperpolarized has a slightly higher amplitude than the rest because the input resistance of this particular specimen of cell L14A decreased when the cell was hyperpolarized.

B. Graph of the data in part *A.* Data are expressed as a percentage of the amplitude of the *first pulse* in the train. The membrane potential at which each curve was generated is indicated on the right.

C. Input resistance increases in Na^+-free seawater. *C-1.* In normal seawater a long (2.5 sec) strong depolarizing intracellular current pulse produces a burst of action potentials in L14B (trace *a*). Small and brief repeated depolarizing pulses (3/sec for 2.5 sec) produce a progressive increase in the input resistance of the cell, as shown by the increase in the amplitude of the electrotonic potential (trace *b*). Increasing the amplitude of the intracellular current pulses produces a further resistance increase, giving rise to spikes that begin to accelerate (trace *c*). *C-2.* The ganglion and purple gland have been perfused for 45 min with Na^+-free seawater. A prolonged pulse identical to the one in *C-1a* produces no action potentials, indicating that Na^+ has effectively been replaced in the extracellular space (trace *a*). Current pulses identical to *C-1b* and *C-1c* still produce a progressive increase in input resistance in *b* and *c,* respectively. A further increase in the amplitude of the intracellular pulses (which now cannot produce action potentials as they did in normal seawater) produces a more pronounced resistance increase (trace *d*). These data indicate that the increase in amplitude of the electrotonic pulses is not due to partial activation of a Na^+ current. *C-3.* Washing the ganglion with normal seawater for 30 min restores action potentials in L14B. The input resistance increase and impulse acceleration from repeated pulses is still evident.

subthreshold stimuli involves a synaptically mediated decrease in membrane conductance (presumably to K^+), which amplifies the action of an otherwise ineffective sensory input (Carew and Kandel, 1976).

A single brief electrical stimulus to the connectives or peripheral nerves produces two types of EPSP: a fast EPSP (lasting approximately 0.1 to 0.2 sec) followed by a slow EPSP (lasting approximately 10 sec to 3 min) (Figure 9-37*A*). The two types of EPSPs are affected differently by changes in membrane potential: depolarization reduces the amplitude of the fast EPSP and hyperpolarization increases its amplitude. These results are consistent with EPSPs due to increased conductance, which typically have their reversal potential close to 0 mV. By contrast, depolarization increases the amplitude of the slow EPSP and hyperpolarization decreases its amplitude. The slow EPSP inverts at approximately -70 mV (a hyperpolarization of about 5 mV from the resting potential), which is in the range of the Nernst K^+ potentials in *Aplysia* neurons (Figure 9-37*A*). These results are consistent with an EPSP due to decreased conductance of K^+. These inferences are supported by measurement of input conductance (Figure 9-37*B*).

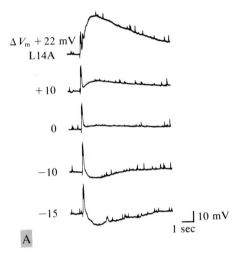

ΔV_{m} + 22 mV
L14A

+ 10

0

−10

−15

10 mV
1 sec

A

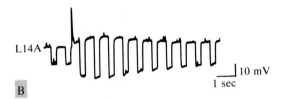

L14A

10 mV
1 sec

B

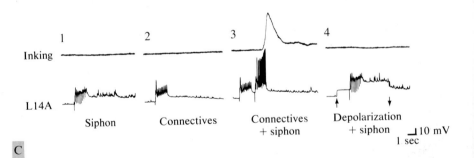

1	2	3	4

Inking

L14A

Siphon   Connectives   Connectives   Depolarization
                          + siphon   + siphon

10 mV
1 sec

C

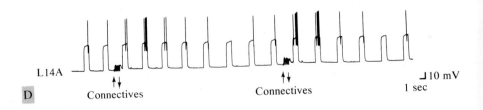

L14A

Connectives   Connectives

10 mV
1 sec

D

FIGURE 9-37.

Two types of EPSPs in ink-gland motor cells. [From Carew and Kandel, 1976.]

A. Effects of membrane potential on fast and slow EPSP. In parts *A* and *B*, EPSPs were produced by a single (1.5 msec) electrical pulse to the connectives. Numbers on the left indicate displacement of the membrane potential from the resting level (0 mV). Depolarization reduces amplitude of the fast EPSP (as would be expected of an EPSP due to increased conductance) and increases the slow EPSP (as would be expected of an EPSP due to decreased conductance). Hyperpolarization increases the fast EPSP and reduces the slow EPSP, inverting it at approximately -70 mV.

B. Decreased input conductance during slow EPSP. Repeated hyperpolarizing current pulses are injected from one intracellular electrode; the membrane potential is recorded with a second electrode. The electrotonic potential increases during the slow EPSP.

C. Synaptic amplification of sensory input. Upper trace is a photocell record of ink response; lower trace is an intracellular record from L14A. *C-1.* The siphon skin is electrically stimulated (1.5 msec pulses, 6/sec for 2 sec) and produces a subthreshold complex EPSP in L14A. *C-2.* Connectives are stimulated (1.5 msec pulses, 6/sec for 2 sec) and produce a slow EPSP. *C-3.* Following identical connective stimulation, a siphon stimulus is timed to coincide with a slow EPSP, which now triggers a burst of action potentials in L14A leading to the release of ink from the ink gland. *C-4.* Depolarization produced by passing current through an independent intracellular electrode does not produce amplification when superimposed on siphon input.

D. Synaptic amplification of electrotonic pulses. Constant-current depolarizing pulses barely threshold for spike initiation are repeatedly delivered. The connective is stimulated (6/sec for 1 sec, first set of arrows) to produce a slow (decreased-conductance) EPSP. Following stimulation, the pulses are amplified during the time course of the slow EPSP because of an increase in the input resistance. Stronger connective stimulation (second set of arrows) produces greater amplification.

As is the case for other EPSPs due to increased conductance, those generated in the ink-gland motor cells excite because they move the membrane potential toward threshold. Because they are produced by an increase in the conductance of the membrane, these EPSPs also decrease the amplitude of other EPSPs. The same synaptic current produces a smaller synaptic potential across a membrane with decreased conductance. Thus, in the absence of presynaptic interactions, the depolarization produced by two separate EPSPs due to increased conductance is at best equal to and usually less than the algebraic sum of the individual EPSPs. EPSPs due to decreased conductance also excite by depolarizing the membrane. However, because they are produced by a decrease in the conductance of the membrane, these EPSPs can amplify concomitant excitatory synaptic actions. Thus, when a train of stimuli to the siphon skin (which by themselves are considerably below threshold for inking) is timed to coincide with the small depolarization of a slow EPSP (produced by connective stimulation), the previously subthreshold siphon stimulus will trigger the characteristic accelerating burst of action potentials in the motor cells and the concomitant release of ink (Figure 9-37C). This amplification of input from the siphon skin is not due simply to the summation of the depolarization produced by the slow and fast EPSPs. Comparable depolarization produced by intracellular current or by another fast EPSP does not amplify the fast EPSP produced by the same siphon input. The amplification is due to the decreased conductance produced by the slow EPSP. Depolarizing electrotonic potentials that are barely threshold are also amplified by the slow EPSP (Figure 9-37C and D). Electrotonic pulses are amplified even when the ink-gland cells are hyperpolarized to the reversal level of the slow EPSP so that no potential change is produced. Thus the slow EPSP provides a post-synaptic mechanism for the amplification of one input by another.

The two types of EPSPs also have different modulating effects on the electrical coupling between the ink-gland motor cells. During the fast EPSP the coupling ratio between the cells decreases (a finding first described by Spira and Bennett, 1972, for IPSPs due to conductance increases). The significance of this finding for inking behavior is unclear; it may be irrelevant. However, during the slow EPSP the coupling ratio between the motor cells increases (Figure 9-38).[9] This may increase positive feedback among the motor cells, with the result that synaptic potentials and action potentials in one cell depolarize the others more effectively. Thus the slow EPSP can enhance the fast EPSP from another pathway in two ways: by decreasing input conductance,

[9]The electrotonic coupling between ink-gland cells is not uniform: the coupling ratio between L14A and L14B is greater than that between either of these two cells and L14C. The degree of modulation varies inversely with the effectiveness of coupling. Both the decreased coupling ratio during the fast EPSP and the increased coupling ratio during the slow EPSP are proportionately greater for the weak connections L14A and L14B make with L14C than for the strong connections L14A and L14B make with one another. This follows from the equivalent circuit for electrically coupled cells (Figure 5-14). As we have seen, the coupling ratio between two cells (1 and 2) is given by the equation $V_2/V_1 = R_2/(R_2 + r_c)$, where V_1 and V_2 are the potential changes in cell 1 and cell 2, R_2 is the input resistance of cell 2, and r_c is the coupling resistance between cells 1 and 2. The greater r_c is in relation to V_2 (the weaker the coupling), the greater will be the enhancement of coupling due to a decreased-conductance EPSP. This assumes (as is likely to be the case) that the coupling resistance does not change during these EPSPs, and changes in coupling occur because the two types of EPSPs alter the nonsynaptic input resistance of the postsynaptic cells.

FIGURE 9-38.

Comparison of the effects of slow and fast EPSPs on the coupling ratio of the ink-gland motor neurons. [From Carew and Kandel, 1976.]

A. Increase in electrotonic coupling during the slow EPSP. A-1. Current is injected in L14A and membrane potential changes in all three cells are recorded (diagrammed at right). A-2. The gain in L14C and L14B was increased to match the pulse amplitude with that of L14A (column a). During a slow EPSP the electrotonic potential in each cell increases (column b), due to increased input resistance. The increase is relatively greater in L14B and (especially) L14C than in L14A, reflecting an increase in coupling.

B. Decrease and increase in electrical coupling during fast and slow EPSPs. Electrical coupling between the ink-gland motor cells is monitored with hyperpolarizing pulses (1/sec) as in part A. The connective was stimulated (1.5 msec pulses, 6/sec for 2 sec, indicated by arrows). Stimulation produced a fast EPSP during the train and a slow EPSP following the train. The percent change in coupling between L14A and L14C is plotted for the fast and slow EPSP as a function of time. Control (100 percent) is the coupling ratio prior to stimulation. The coupling ratio decreases during fast EPSP and increases during slow EPSP.

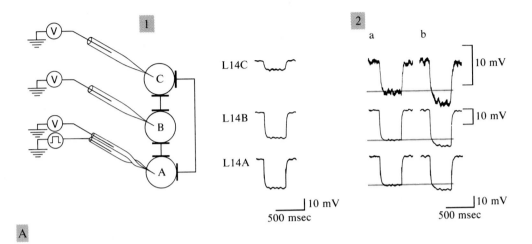

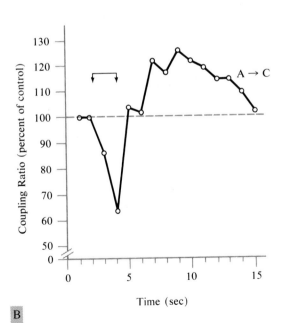

which causes the synaptic current from another input to produce a larger synaptic potential, and by increasing the coupling ratio, which enhances positive feedback among the motor cells.

Inking elicited by a brief noxious stimulus to the head or siphon is predominately triggered by the fast EPSP. The slow EPSP provides a second means of activating the inking system. If a single strong stimulus is not sufficient to produce inking, the slow EPSP assures that inking will be triggered even by a considerably weaker stimulus if it follows the first stimulus within a few seconds.

A NEUROENDOCRINE FIXED ACT: EGG-LAYING

Another all-or-none behavior controlled by the abdominal ganglion is egg-laying. *Aplysia* is a prolific egg producer. Large animals have been known to lay a total of 4.78×10^8 eggs in 27 separate episodes during the five months of their life in which they are capable of reproducing (MacGinitie, 1934). The eggs are laid in strings. The total length of all strings in one laying can reach almost one-third of a mile (60,565 cm). The largest single egg string found was 17,520 cm long and was laid at a rate of 41,000 eggs per minute.

During egg-laying the animal grasps the egg string in the fold of the upper lip and covers it with sticky mucus. Moving its head back and forth, it then lays the emerging string in an irregular pile (Figure 9-39). The total sequence is a fairly complex fixed-action pattern. Here I shall focus on only one component: the neural mechanism for the all-or-none release of eggs from the ovotestis. Egg-laying is not mediated by a few cells but by two large identified clusters of neuroendocrine cells. But like the motor circuit for inking, each cluster functions as a single cell.

FIGURE 9-39.
Egg-laying in *Aplysia*. Animals usually deposit a large bolus of eggs in an all-or-none manner. Prior to depositing, the egg cases are coated with a gelatinous material that binds them together into long strands.

The Behavior

Under natural conditions *Aplysia* lays large masses of eggs at one site within a relatively brief period. Egg-laying is triggered by the secretory product of the bag cells (Kupfermann, 1967, 1970; Strumwasser, Jacklet, and Alvarez, 1969). Two clusters of these neuroendocrine cells, each consisting of 250 to 400 cells, are located at the junction of the left and right connectives with the abdominal ganglion (Figure 9-40). Each cell sends one axonal process into the sheath that surrounds the ganglion and connectives. Before ending in the sheath, the axonal processes form a ring or cuff around the axons that run in the connectives (Figure 9-40). The processes of the bag cells do not synapse on other neurons or effector cells.

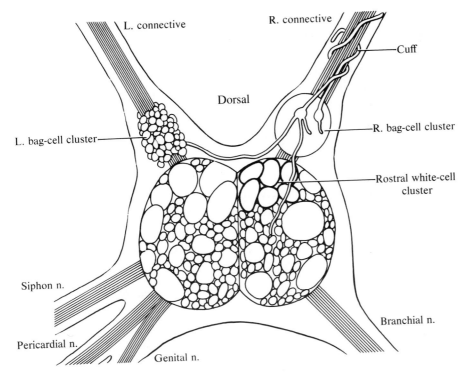

FIGURE 9-40.
Diagram of the abdominal ganglion showing right and left bag-cell clusters. The two cells shown in the right cluster illustrate the axonal processes of the bag-cells. Some processes end in the connective tissue sheath that overlies the ganglion; other processes form a cuff around the connective. This cuff is thought to be the site of electrical coupling of cells within a cluster. [Based on R. E. Coggeshall, 1967.]

The cell bodies produce large (150 to 250 nm), round, moderately dense-core granules that presumably contain the egg-laying hormone. These granules are transported into the processes lying in the connective tissue sheath (Figure 9-41). Since the sheath is highly vascularized (a neurohaemal organ), it is thought that the bag cells release their secretion into the sheath, from which it can enter the bloodstream (Figure 4-6). Injections into an animal of as little as 20 percent of the homogenates of all the bag cells results in eggs being laid within one hour. The active hormone is thought to be a small protein or polypeptide weighing 1,500 to 6,000 daltons (Toevs and Brackenbury, 1969; Arch, 1972a, b; Loh, Sarne, and Gainer, 1975).[10]

The Neuronal Circuit

All the bag cells have identical electrophysiological properties (Kupfermann and Kandel, 1970). As with the ink-gland motor cells, the bag cells are silent and do not connect with central neurons that fire spontaneously. They also do not respond to stimulation from the peripheral nerves. But a strong electrical stimulus to the connectives causes a prolonged discharge that lasts from several minutes to half an hour (Figure 9-42). This repetitive response resembles an all-or-none event: once threshold has been achieved a self-sustaining response is generated, so that a single stimulus can trigger a discharge of the bag cells that outlasts the stimulus by many minutes.

Once having responded to a triggering stimulus, the bag cells become relatively refractory for some time to further input (Figure 9-42). During the refractory period, stimuli that were effective in starting the burst sometimes evoke single spikes, but for up to an hour after the discharge has stopped, they are ineffective in eliciting the responses seen in rested preparations. Activity can sometimes be elicited by using a greater number of shocks per time, but if

[10]During the egg-laying season this protein makes up almost 50 percent of the cell's total protein. The hormone is synthesized throughout the year, even during the period when animals are not laying eggs. When the bag cells are depolarized with a high-K$^+$ solution, the polypeptide is released into the perfusion solution. Release is dependent upon the presence of normal Ca^{++} concentration (Arch, 1972a, b), suggesting that the mechanism of release is similar to those of other neurohormones and transmitter substances. Presumably this polypeptide is the egg-laying hormone. Arch (1927b) and Loh, Sarne, and Gainer (1975) have found that the cells first synthesize a large (25,000 dalton) precursor molecule, which then breaks down into smaller units (1,500 to 6,000 daltons). It is these smaller molecules that presumably serve as the active hormone. The bag-cell hormone thus seems to resemble some other polypeptide hormones, such as insulin, which is synthesized as a large, inactive, covalently bonded precursor molecule and then split into an active hormone and an inactive molecule (Steiner et al., 1971).

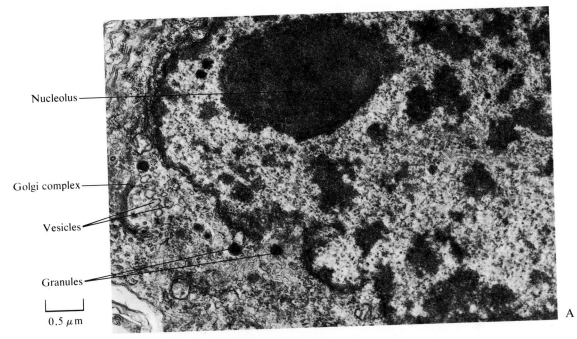

Nucleolus

Golgi complex

Vesicles

Granules

0.5 μm

A

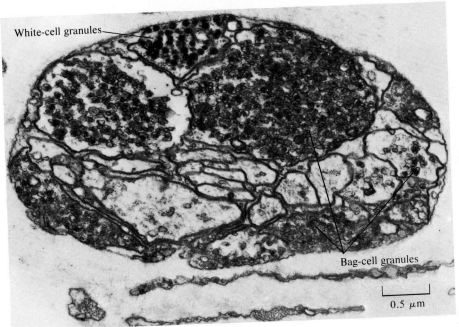

White-cell granules

Bag-cell granules

0.5 μm

B

FIGURE 9-41.
Morphology of the bag-cells. [From Frazier *et al.*, 1967.]

A. The nucleus and perinuclear cytoplasm of the soma of a bag cell. The nucleolus in the bag-cell nucleus is remarkable in that many of the particles presumed to be ribosomes are aligned in rows. The cytoplasm contains a number of large granules that seem to be formed in the Golgi complex and are thought to contain the hormone released by the bag cells. These granules travel into the bag-cell processes and are thought to be released in the sheath that surrounds the ganglion.

B. Nerves in the sheath covering the ganglion contain the bag-cell processes and processes of another group of neurosecretory cells, the white cells.

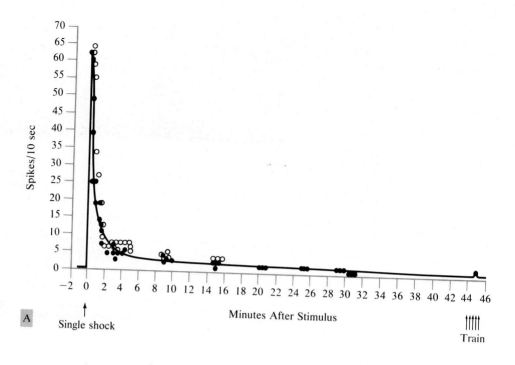

A

Single shock

Train

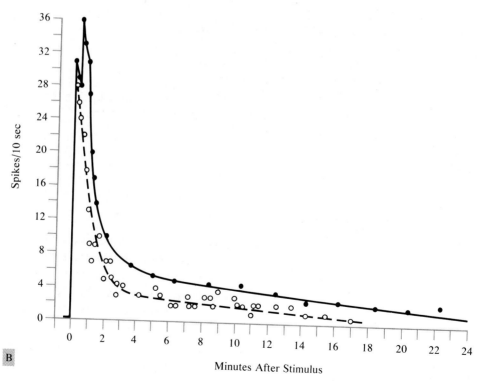

B

FIGURE 9-42.

The sustained discharge in bag cells. [From Kandel and Kupfermann, 1970.]

A. Time courses in two different preparations. Discharges were elicited by a single shock applied to a connective. In the first experiment (solid circles) single shocks applied to the connectives 15 min after cessation of spike activity (44 min following initial stimulus) were ineffective in triggering further activity. A train of stimuli (presented immediately after the single shocks) resulted in only a very weak discharge. Data are plotted over 10-sec intervals and therefore do not reveal the typical increase in firing frequency during the first few seconds.

B. Time courses elicited in a single preparation by repetitive stimulation of a connective. Solid circles indicate the time course of the first discharge; open circles indicate the course of a second discharge elicited two hours following termination of the first. Several attempts to elicit a sustained discharge at different times during the two-hour period between discharges were unsuccessful.

insufficient recovery time is allowed, the duration of the response is considerably shorter and the maximum frequency of the discharge is much less than the initial response.

A remarkable aspect of the activity of bag cells is that throughout the discharge, from onset to termination, all the cells in a cluster invariably fire synchronously; moreover, they all show the same variations in size of even graded potentials (Figure 9-43). Following stimulation of one connective, cells in the contralateral bag-cell cluster are also triggered into activity, but with a much longer latency than the ipsilateral cells. After both clusters become active, the spikes of all cells in the two clusters sometimes become synchronous. However, unlike cells within a cluster, the two clusters always become independent as the spike frequency gradually decreases with time (Figure 9-44). When the two clusters fire out of synchrony, the discharge of cells in one cluster is often synchronous with a small potential (presumably an electrical EPSP) in the other. Because the discharge of one cluster always precedes that of the other, bag cells on one side appear capable of acting as a pacemaker population for the opposite cluster. During repetitive activity the pacemaker roles can switch spontaneously, and when this occurs the amplitude of the spikes is reduced (in Figure 9-44 such altered spikes are indicated by dots).[11]

[11]Recordings from cells in the opposite bag-cell clusters therefore suggest a reverberation of activity between these two groups of cells. But this reverberation is not essential for sustaining activity within a bag-cell cluster. Bisected ganglia in which the two bag-cell clusters are physically separated from each other can sustain activity, although the duration of the discharge is reduced. This decrease in duration could be secondary to physical injury produced by cutting the crossing bag-cell process. Alternatively, the contralateral bag-cell cluster might actually serve as a secondary or facultative pacemaker population that contributes to sustaining the discharge. Except for the possible contribution of the contralateral bag-cell cluster, the mechanism of the sustained repetitive firing is not understood (Kupfermann and Kandel, 1970).

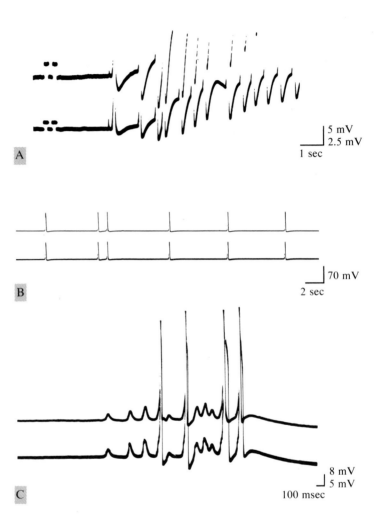

A

5 mV
2.5 mV
1 sec

B

70 mV
2 sec

C

8 mV
5 mV
100 msec

FIGURE 9-43.
Simultaneous recordings from bag cells in the same cluster. **A.** High-gain recordings illustrate the initiation of discharge following a single stimulus. **B.** Low-gain recordings of the same cells several minutes later show they are still firing in perfect synchrony. **C.** High-gain recordings from a pair of bag cells in another experiment illustrate the close parallelism of the spikes and graded potentials of cells in the same cluster. [From Kupfermann and Kandel, 1970.]

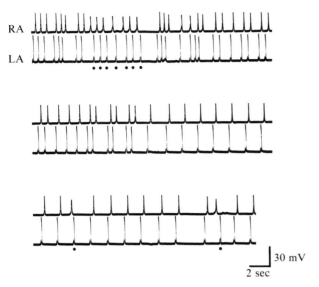

FIGURE 9-44.
Simultaneous recordings from two bag cells in opposite clusters several minutes after stimulation (records are continuous). Action potentials in the left bag-cell cluster (LA) generally preceded those of the right cluster (RA), but on occasion the pacemaker role was reversed and discharges in cells of the right cluster preceded those in the left. Dots indicate when the pacemaker role is reversed. Note that in each case of reversal there is a characteristic change in the spike configuration of the two cells. When the right cell becomes the pacemaker it has a smaller spike than it does as follower. When the left cell becomes the follower it has a more pronounced and rapidly rising prepotential at the foot of its spike than it does as pacemaker. [From Kupfermann and Kandel, 1970.]

Neither the synchronization nor the persistent activity of the bag cells requires connections with other cells of the ganglion. The synchronization is thought to be accomplished by electrical connections of the bag cells' remote axonal processes (Kupfermann and Kandel, 1970), perhaps at the cuff the processes form around the connective (Figure 9-45). Although experiments designed to test this hypothesis have not been conclusive, the extraordinary degree of synchrony is most consistent with the notion that the cells are electrically coupled to one another. The synchronized and prolonged firing of all 250 to 400 bag cells within a cluster and the excitatory coupling between the two clusters on opposite sides may account for the all-or-none deposition of large egg masses. The prolonged firing presumably guarantees an adequate amount of hormone for the laying of a large mass of eggs. The prolonged refractory period sets an upper limit on the amount of hormone released in a

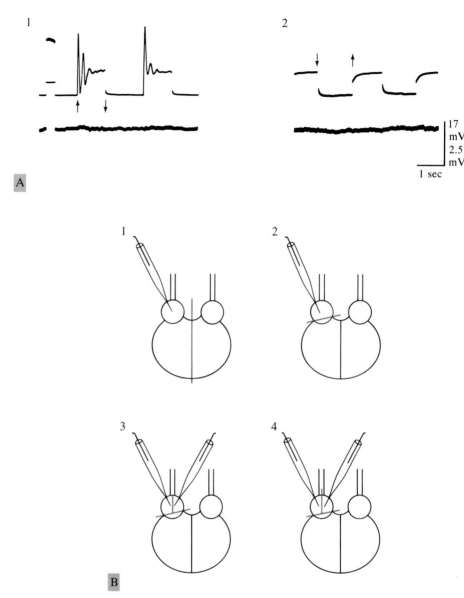

FIGURE 9-45.
Tests to examine electrical connections between bag cells in the same cluster. [From Kupfermann and Kandel, 1970.]

A. Failure to demonstrate electrical coupling between two bag cells. Depolarizing currents (*A-1*) and hyperpolarizing currents (*A-2*) injected into one cell failed to produce a membrane potential change that can be recorded in the cell body of the other.

B. Four types of incisions through the abdominal ganglion and the two bag-cell clusters. *B-1.* Separation of the two hemiganglia so as to isolate the two cell clusters. *B-2.* Separation of one cluster from the remaining part of the abdominal ganglion. *B-3.* Same as in *B-2* except that the bag-cell cluster is also bisected. The bisecting cut is made just up to the point at which the connective joins the bag-cell cluster. *B-4.* Same as in *B-3* except that the cut is extended a short distance into the connective. Only the cut in *B-4* eliminates the electrical coupling, suggesting that coupling occurs where the bag-cell processes form a cuff around the connective (see Figure 9-40). The remoteness of the site may also explain why an electrotonic potential produced in one cell cannot be recorded in another (see part *A*).

given period of time. Presumably this permits the cells adequate time for restoring the hormone.

The stimulus–response relationship of the bag cells, like that of the ink-gland motor cells, is very steep. The response of the cells is not linearly related to the stimulus. Once the threshold for activation is exceeded, the stimulus triggers a central program lasting many minutes. This response does not require cues from other cells or peripheral feedback for its timing or maintenance.

COMMON FEATURES OF FIXED ACTS: INKING AND EGG-LAYING

Inking and egg-laying are all-or-none responses. Both have high thresholds and occur infrequently; once generated, they soon become refractory. The neural circuits of the two responses and their central programs share common features. In each case the motor cells are silent, have a high resting potential, a high threshold for spike generation, few or no connections from spontaneously active neurons, and are (or appear to be) electrically coupled. Although these features alone assure a high-threshold response, other factors contribute. In particular, the motor cells discharge in response to specific sensory input. The ink-gland motor cells respond only to strong stimuli. The bag cells respond only to electrical stimulation of the connectives from the head ganglia. They do not fire in response to electrical stimulation of any of the four peripheral nerves innervating the mantle and tail region. The natural stimulus presumably reaches them by means of receptors in the head or neck region.

The all-or-none characteristic of these two behavioral responses implies an amplifying or regenerative capability. This capability seems to result from the ability of the ink-gland motor cells and bag cells to act on input in an explosive way. A threshold stimulus triggers an accelerating train of spikes in the L14 cells and a prolonged after-discharge in the bag cells. In both cases the motor cells and their electrical interconnection can act to amplify the input signal. In each case at least part of the central program is not dependent on chemical synaptic connections. The presence of electrical rather than chemical interconnecting synapses seems designed to assure stereotypic responses.

A COMPARISON OF REFLEX AND FIXED ACTS

A comparison of gill withdrawal with inking and egg-laying provides some preliminary insights into at least one way by which a nervous system can generate both reflex and fixed acts. The graded properties of gill withdrawal stem from the relatively linear input-output relationships that exist at several points in its neural circuit—increasing the strength of the stimulus causes each sensory neuron to fire more often. The temporal and spatial summation of the chemical synaptic action produced by the sensory cells seems to add linearly on the motor cells (at least over a limited range). The frequency of the action potential in the gill motor cells also seems to be linearly related to the synaptic input from the sensory neurons, and the amplitude of the gill contraction is linearly related to the firing frequency of the motor cells (Byrne, unpub.). Although the analysis of the inking and egg-laying circuits is less complete than that for gill withdrawal (because only the motor cells have so far been identified) the ink-gland motor cells and the bag cells convert a graded synaptic input into an all-or-none output in the form of a train of action potentials.

As a result, inking and egg-laying are not as continuously controlled by the stimulus as the gill-withdrawal reflex is. Rather, a central program for inking and egg-laying exists in the form of the burst capabilities (or motor tape) of the motor cells. Thus, whereas the output of the motor neurons of the reflex response is dependent on the amplitude and pattern of the synaptic input, the output of the motor cells of these fixed acts has elements of a pattern not present in the input.

SUMMARY AND PERSPECTIVE

Although there is no fully satisfactory classification system for behavior, the behavioral responses of animals may be grouped on the basis of stimulus–response relationships into two broad and overlapping classes: *reflex* and *fixed-action* responses. In a reflex response the amplitude and pattern of the output are graded functions of the intensity and pattern of the input. In a fixed-action response the input triggers an all-or-none response as the result of a central program; the response is then played out independently of the input. This centrally programmed response is sometimes released spontaneously. Reflex and fixed-action responses can be subdivided according to complexity. *Elementary* responses involve only a single episode in one organ system and are called either reflex or fixed *acts*. *Complex* responses involve several organ systems or recurrent episodes and are called reflex or fixed-action *patterns*.

The abdominal ganglion generates both reflex and fixed acts as well as fixed-action patterns. Gill and siphon withdrawal are simple, low-threshold reflex acts. The same effector organs are also involved in a fixed act. The alternation of reflex and fixed acts can take place because the motor cells are not connected to one another and so can be switched from one input to another— from sensory neurons mediating the reflex act to central command cells mediating the fixed act. Each motor cell produces a relatively specific movement. During the reflex act all the motor neurons fire in synchrony; during the fixed act they fire in sequence (the central command excites some motor neurons and inhibits others, which fire later by rebound excitation). As a result, the motor movements of the mantle system differ during the two behavioral responses. The low threshold and graded properties of the gill-withdrawal reflex result from certain characteristics of the neural circuit. Several motor cells are autoactive and are therefore responsive even to small changes in excitatory drive. In addition, the sensory neurons have a low threshold; stronger stimuli cause each cell to fire more often. Each spike in each sensory cell makes a direct excitatory connection with some motor cells. The excitary synaptic potentials produced by the same cell summate temporally and those produced by different sensory cells summate spatially, producing greater discharge of the motor cells and thus greater gill contractions.

Egg-laying and inking are fixed acts. Both behavioral responses have a high threshold. This is due both to the high membrane potential and high threshold of the participating motor neurons. Both responses also have nonlinear input-output characteristics. In each case this seems to be related in part to the electrical coupling between motor cells, which produces positive feedback as more spikes are generated. In addition, the motor cells have repetitive firing tendencies. Electrically interconnected motor cells thus provide one way to achieve a stereotypic central program.

Electrophysiological studies of behavior need to be complemented by studies of both the morphology of nerve cells and transmitter biochemistry if we are to realize a broader neurobiological view of the characteristics of the neural circuitries of different classes of behavior. The study of egg-laying is an unfortunately rare example of an integrated neurobiological analysis of behavior. In addition to the electrophysiological studies of egg-laying, which have provided data on the locus and mechanism of the central program, morphological and biochemical studies have also been conducted. These reveal that the bag cells involved in egg-laying are neurosecretory and that the synchronous discharge of these cells causes the release of a polypeptide neurohormone into the neurohemal sheath. This hormone causes muscle fibers in the ovotestis to contract, thereby triggering the first step in the release of eggs. In addition, the hormone seems to act on other nerve cells to produce some of the other behavioral concomitants of egg-laying (see p. 436).

SELECTED READING

Hinde, R. A. 1970. *Animal Behaviour. A Synthesis of Ethology and Comparative Psychology,* second edition. New York: McGraw-Hill. A scholarly overview of ethology; Chapter 3 contains a good critique of the concept of fixed-action patterns.

Marler, P., and W. J. Hamilton III. 1966. *Mechanisms of Animal Behavior.* New York: Wiley. A particularly well-written advanced textbook of ethology with good discussions of invertebrate behavior.

CHAPTER 10

The Neuronal Organization
of Complex Behavior

Fixed acts and fixed-action patterns are generated by central programs. In this chapter I shall consider further the mechanisms that generate and trigger these central programs. In particular, I shall examine how the program for a fixed-action pattern can be released spontaneously in the absence of an external stimulus and what factors in the central program determine the stereotypy of a fixed-action pattern irrespective of its stimulus for release. Then I shall examine the way in which several behavioral responses of different effector systems are controlled so that at any one time only one is expressed. Finally, I shall discuss the relationship of the neural circuitry of fixed acts to that of the more complex fixed-action patterns. Studies of that relationship seem to indicate that the neural circuit of a complex behavior is built up, by accretion, from the simpler circuits used in elementary behavior.

CIRCULATORY FIXED-ACTION PATTERNS

Behavior

In *Aplysia,* as in vertebrates, the nervous system controls the heart beat, the force of contraction, and vasomotor tone. These homeostatic functions are in turn modulated by various reflex actions as well as by fixed-action patterns

(Pinsker, Feinstein and Gooden, 1974; Koester, Mayeri, Liebeswar and Kandel, Kandel, 1974; Dieringer and Koester, unpub.). For example, mechanical stimuli to the head produce defensive head withdrawal and concomitant slowing of heart rate. Stimulation of the tail leads to locomotion and speeding up of the heart rate. A powerful escape response in *Aplysia* is elicited by a starfish, *Astrometis,* or by salt crystals applied to the siphon or tail. These stimuli trigger pedal (locomotor) waves and a marked increase in heart rate (Dieringer and Koester, unpub.). Animals also show reflex changes following submersion in seawater and exposure to air that are analogous to the diving reflex of vertebrates. Submersion causes an increase in heart rate while exposure causes a slowing of heart rate (Pinsker, Feinstein, and Gooden, 1974). Finally, animals experience a prolonged and profound increase in heart rate following eating.

Cardiovascular function is also modulated by fixed-action patterns. One stereotyped cardiovascular adjustment is an increase in cardiac output, due to a brief acceleration of heart rate (and vasodilation) (Figure 10-1*A*). This phasic increase often occurs in all-or-none or stepwise fashion, and its occurrence is increased during egg-laying (Mayeri, 1975). These increases can be elicited by the egg-laying hormone, but detailed analyses of the stimulus–response relationships have not as yet been carried out. Because the behavior can occur spontaneously and often in quantized form, I shall regard it here as a simple fixed-action pattern. In addition, there is a profound and stereotypic slowing of heart rate that accompanies the respiratory pumping movements (Figure 10-1*B*; and see p. 378 and Figure 4-13, p. 88). The simultaneous occurrence of stereotypic fixed acts in two separate organ systems will permit us to explore how synergist fixed acts are coordinated into a common fixed-action pattern.

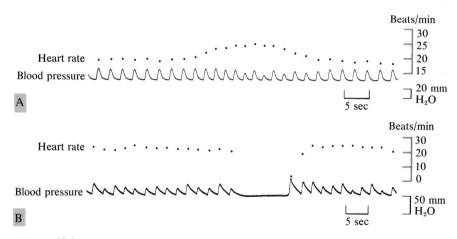

FIGURE 10-1.
Two cardiovascular homeostatic fixed-action patterns. **A.** Phasic increase in heart rate. **B.** Phasic decrease in heart rate. [From Koester *et al.,* 1974.]

Circulatory System of *Aplysia*

The cardiovascular system of *Aplysia* (Figure 10-2) is a good invertebrate system for studies of the neural control of circulation. Unlike the hearts of insects and crustaceans that have a neurogenic pacemaker, the molluscan heart resembles those of vertebrates in having a myogenic pacemaker. The rate and force of heart beat is modulated by exictatory and inhibitory neural control. Although the disposition of the musculature is different in vertebrates, the blood vessels of *Aplysia* are also under neural control. The vascular muscle is innervated and consists of a circular band along the proximal half of the gastroesophageal artery of the main trunk of the abdominal aorta.

The circulatory system of *Aplysia* differs from that of vertebrates in being half open. The arteries that carry blood from the heart and the venous sinuses that return blood to the heart are not connected by capillaries. Blood from the arterial system collects in the haemocoel, and is returned to the heart by two parallel venous sinus pathways, one through the kidney, and the other through the gill.

The major pumping action of the heart is accomplished by the ventricle; the contraction of the auricle only serves to fill the ventricle. A pair of muscular semilunar valves (atrioventricular valves) located at the base of the ventricle prevent regurgitation of blood into the auricle during ventricular contraction. There is no readily discernible pacemaker muscle in the heart. However, contraction of the ventricle often begins at its base, near one of the insertions of the atrioventricular valve flaps into the wall of the ventricle. A single semilunar valve between the ventricle and arteries prevents back flow to the ventricle. The ventricles and auricles are innervated by separate branches of the pericardial nerve (Figure 10-2).

Neuronal Circuits

MOTOR NEURONS

The motor neurons controlling circulation have been studied by Mayeri, Koester, Kupfermann, Liebeswar, and Kandel (1974) in a partially intact preparation in which the abdominal ganglion is exposed for intracellular recording and the heart is perfused to maintain its beating. The motor-neuron population can be largely accounted for by two heart excitors (RB_{HE} and LD_{HE}), two heart inhibitors, (LD_{HI1} and LD_{HI2}), and three vasoconstrictors (LB_{VC}) (Mayeri *et al.*, 1974). Six of these cells are located in the caudal quadrant of the left hemiganglion. One cell (RB_{HE}) lies in the caudal quadrant of the right hemiganglion (Figure 10-3).

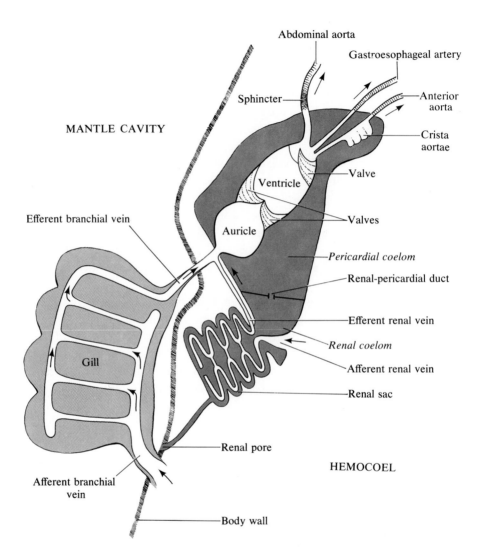

FIGURE 10-2.
Schematic diagram of the circulatory system of *Aplysia*. The heart is myogenic and has two chambers. The outflow from the heart is by means of three main arteries: the abdominal aorta, the gastroesophageal artery, and the anterior aorta. The circulation is open; there are no capillaries. The arterial blood is distributed to the organs of the body and then collects in the body cavity (the hemocoel). From there blood flows to the heart by way of two large venous sinuses: the afferent branchial vein, which empties into the gill, and the afferent renal vein, which supplies the kidney. The true coelomic cavity of the body is restricted to the pericardial and renal coelom. [From Mayeri *et al.,* 1974.]

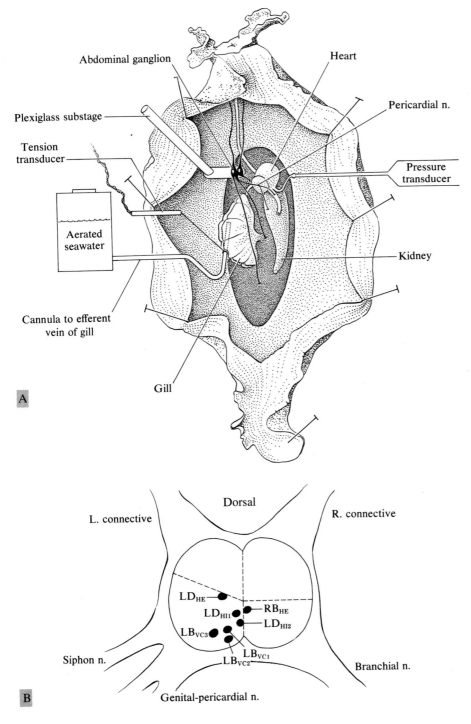

Abdominal ganglion

Heart

Pericardial n.

Plexiglass substage

Tension
transducer

Pressure
transducer

Aerated
seawater

Kidney

Cannula to efferent
vein of gill

Gill

A

Dorsal

L. connective

R. connective

LD$_{HE}$

LD$_{HI1}$ — RB$_{HE}$

LD$_{HI2}$

LB$_{VC3}$

LB$_{VC1}$

Siphon n.

LB$_{VC2}$

Branchial n.

B

Genital-pericardial n.

FIGURE 10-3.
Preparation for studying neural control of the heart. **A.** Ventral view of partially dissected animal used for studying the neural control of circulation. The gill, an external organ, has been pulled through a slit in the dorsal body wall. **B.** Schematic diagram of the dorsal surface of the abdominal ganglion showing the most common positions of the cardiovascular motor neurons. [From Mayeri *et al.,* 1974.]

The two heart excitors have quite different actions. Cell LD_{HE} has a phasic accelerating effect that usually lasts only a few seconds; a brief burst of spikes produces only one extra beat. This action is mediated to the heart muscles by discrete, fast-rising excitatory junction potentials. Cell LD_{HE} is usually silent; it fires only occasionally in response to its synaptic input and has never been observed to fire sufficiently to have an effect on heart rate. The function of this cell is not known (Figure 10-4A). The other heart excitor, RB_{HE} (or R17), has a long-lasting accelerating action. A brief (2–3 seconds) burst of spikes produces a maximal increase in heart rate of 50 percent, which declines only slowly over the next two or more minutes. Unlike LD_{HE}, which produces discrete excitatory junction potentials, RB_{HE} produces a slow, smoothly graded depolarization of heart muscle fibers. Also unlike LD_{HE}, RB_{HE} fires at a steady rate of 0.5 to 2 spikes per second. Much of this ongoing activity is due to excitatory input from several interneurons. When RB_{HE} is prevented from firing by hyperpolarization, the heart rate decreases by about 20 percent. Thus, in addition to its phasic action, RB_{HE} contributes an excitatory tone to the heart, which supplements the myogenic pacemaker activity (Figure 10-4B).

Unlike the two heart excitors, which are quite different in their properties, the two heart inhibitors are similar. Cells LD_{HI1} and LD_{HI2} have identical synaptic input, motor effect, and firing patterns. When stimulated intracellularly they decrease the amplitude or frequency of the heart beat; often they alter both. They are typically silent and fire only occasionally in brief bursts of activity driven by excitatory input. The inhibitory action is brief, and the cells do not exert a tonic inhibitory effect on the heart (Figure 10-5A).

The vasoconstrictor population consists of three nearly identical cells: LB_{VC1}, LB_{VC2}, and LB_{VC3}. The cells have identical synaptic input and firing patterns, but they do not have identical innervation patterns. Cells LB_{VC1} and LB_{VC2} innervate the vascular muscle of the abdominal aorta and the proximal third of the gastroesophageal artery; cell LB_{VC3} innervates only the abdominal artery. The abdominal and gastroesophageal arteries provide the main blood supply for the hepatopancreas and the gastrointestinal tract. Activity in LB_{VC1} and LB_{VC2} reduces blood flow to these organs and increases blood pressure. The vasoconstriction caused by all three cells is mediated by discrete, fast-rising, depolarizing excitatory junction potentials in arterial muscle fibers (Figure 10-5B).

The fact that the biochemical properties of the heart motor neurons in *Aplysia* are similar to those of vertebrates illustrates that cellular controls underlying behavior may be quite general. In both *Aplysia* and vertebrates the inhibitory action of the heart motor neurons is brief and mediated by ACh; excitation is long-lasting and mediated by a monoamine, norepinephrine in

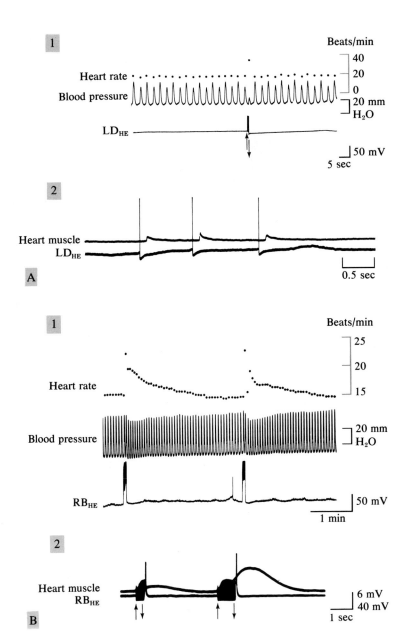

FIGURE 10-4.
Heart excitatory motor neurons. [From Mayeri *et al.*, 1974.]

A. Heart excitor LD$_{HE}$. *A-1*. Phasic increase in heart rate produced by a brief burst of spikes elicited by intracellular depolarizing current pulses in the motor cell (arrows). *A-2*. Intracellular record of e.j.p.'s produced in a heart muscle fiber by action potentials in LD$_{HE}$.

B. Heart excitor RB$_{HE}$. *B-1*. Long lasting increase in heart rate caused by a 3-sec burst of action potentials elicited in RB$_{HE}$ by intracellular depolarization. The membrane potential of RB$_{HE}$ was hyperpolarized. The large, tonically occurring potentials in the RB$_{HE}$ record are synaptic potentials enhanced by the hyperpolarization. *B-2*. Intracellular recording of slow, smoothly graded depolarizations of a heart muscle fiber caused by bursts of spikes in RB$_{HE}$ elicited by intracellular depolarization.

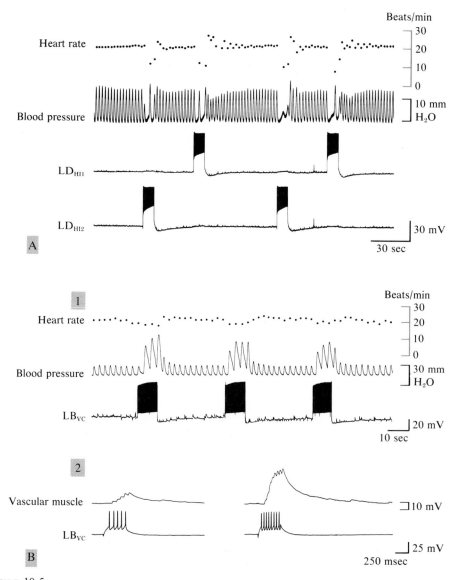

FIGURE 10-5.

Heart inhibitors and vasoconstrictors. [From Mayeri *et al.*, 1974.]

A. The two LD heart inhibitors were recorded simultaneously. Each cell was fired twice by 10-sec pulses of depolarizing current. Both cells have similar effects on heart rate and blood pressure.

B. Motor effects and excitatory junction potentials of the vasoconstrictor motor neuron. *B-1.* An LB$_{VC}$ vasoconstrictor cell was fired three times by a 10-sec depolarizing current pulse. These bursts caused constriction of the abdominal aorta and gastroesophageal artery. The proximal end of the gastroesophageal artery was cut and connected by a cannula to a transducer for measuring blood-pressure so that constriction of the abdominal aorta caused an increase in the recorded blood pressure. The slight decrease in heart rate that normally accompanies vasoconstriction was observed only occasionally; presumably it is an indirect effect caused by the decreased rate of emptying of the ventricle. *B-2.* Simultaneous intracellular recordings from a vasoconstrictor muscle cell at the base of the abdominal aorta and an LB$_{VC}$ motor neuron. The vasoconstrictor motor neuron was activated by intracellular depolarizing current pulses. Each action potential in the motor neuron is associated with an elementary e.j.p. in the muscle cell. The e.j.p.'s show both summation and facilitation.

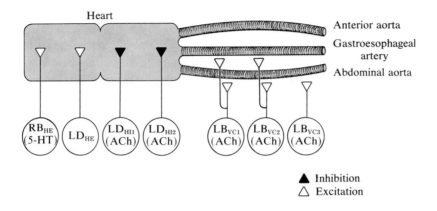

FIGURE 10-6.
Motor neurons to the heart and the vasculature. Positions of the terminals do not necessarily correspond to specific innervation sites in the heart or aorta. [From Mayeri *et al.*, 1974.]

vertebrates and serotonin in *Aplysia*. In vertebrates excitation is mediated by an intracellular, second messenger, cyclic AMP. Perhaps a similar second messenger is involved in *Aplysia*. (For a discussion of second messengers see p. 588.)

Figure 10-6 summarizes the data on the heart motor-neuron population and illustrates two features already encountered in the gill and siphon motor system. First, the motor output to the circulatory system is mediated by a few cells, each of which exerts a powerful motor effect. Second, the seven motor pathways to the heart and aortae are arranged in parallel: There are no direct or indirect synaptic connections between any of the motor cells.

The transmitter biochemistry of six of the seven circulatory motor neurons is known. Cell RB_{HE} is serotonergic; the LD_{HI} and LB_{VC} cells are cholingergic; the transmitter of LD_{HE} is unknown. The heart of *Aplysia* is accelerated by serotonin and inhibited by low concentrations of ACh (Liebeswar, Koester, and Goldman, 1973; Liebeswar, Goldman, Koester, and Mayeri, 1975). Acetylcholine also causes vasoconstriction. Acetylcholine-blocking agents block the inhibitory actions of ACh and the LD_{HI} motor neurons; they also block the vasoconstrictor actions of ACh and the LB_{VC} motor neurons. Serotonin-blocking agents block the accelerating actions of serotonin and the RB_{HE} motor neurons on heart rate. Biochemical assays for the synthesizing enzymes, based on intracellular injection of transmitter precursors (see p. 274) are also consistent with the pharmicological data: RB_{HE} can synthesize serotonin and the LD_{HI} and LB_{VC} cells can synthesize ACh.

INTERNEURONS

Even though the motor neurons are not directly interconnected, their firing patterns are often correlated (either in or out of phase) with each other and with the motor neurons of the respiratory complex. The coordinated activities of the cells are manifested in the two fixed-action patterns: phasic increase and decrease of heart rate and cardiac output. The phasic increase is associated with excitation of the heart excitor (RB_{HE}) and inhibition of both the heart inhibitors (LD_{HI}) and vasoconstrictors (LB_{VC}). The phasic decrease (which accompanies respiratory pumping) is associated with inhibition of the excitor (RB_{HE}) and the vasoconstrictor cells and excitation of the inhibitors (Figure 10-7).

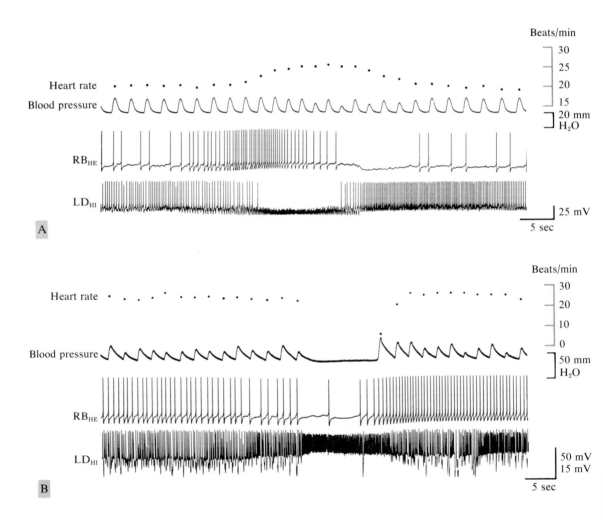

Two interneurons play a crucial role in these fixed-action patterns. High-gain recordings from the motor cells during these fixed-action patterns reveal that the coordination between them is due to common synaptic inputs. The input to the members of each set of similar motor neurons (the two inhibitors and the three vasoconstrictor cells) seems identical; however, much of the input to antagonist motor cells (RB_{HE} and LD_{HI}) is opposite in sign (Figure 10-7).

Centrally Programmed Increase of Heart Rate

The large EPSPs generated in the excitor (RB_{HE}) during the phasic increase in heart rate and the concomitant IPSPs in the two inhibitor and three vasoconstrictor cells are synchronous with each other. These common inputs are mediated by the identified cholinergic cell, L10 (Koester et al., 1974). Here, as with the other connections of L10 (see pp. 300 and 308). ACh experimentally simulates the excitatory actions of L10 on RB_{HE} and its inhibitory actions on the inhibitors and vasoconstrictor cells. The PSPs in the motor cells follow the spikes of L10 one-for-one and with constant latency, even at high frequencies, indicating that L10 makes direct connections to the heart motor and vasoconstrictor cells (Figure 10-8).

The connections made by L10 (to the excitors, inhibitors, and vasoconstrictors) suggest that L10 is capable of controlling cardiac output (Figure 10-9). This finding raises two questions. Can the activity of L10 alone account for the stereotyped increases in heart rate and cardiac output? If so, how is the central program for this simple fixed-action pattern generated? What triggers L10 into activity? What is responsible for the time course of its firing?

By stimulating L10 directly, Koester and co-workers (1974) found that L10 alone can generate the fixed-action pattern (Figure 10-10). To determine further whether this behavior is fully accounted for by the connections of L10 to the known heart motor neurons, L10 was stimulated both with and without each of the motor cells being reversibly removed from the neural circuit by hyperpolarization. These studies revealed that the action of L10 on heart rate is completely mediated by RB_{HE}. When this cell was hyperpolarized and

FIGURE 10-7.

Patterned activity in heart motor neurons during fixed-action patterns. **A.** Excitation of RB_{HE} and inhibition of RB_{HI} are accompanied by an increase in heart rate and a delayed decrease in blood pressure. **B.** Excitation of LD_{HI} and inhibition of RB_{HE} are associated with the inhibition of heart beat that accompanies respiratory pumping movements. [From Koester et al., 1974.]

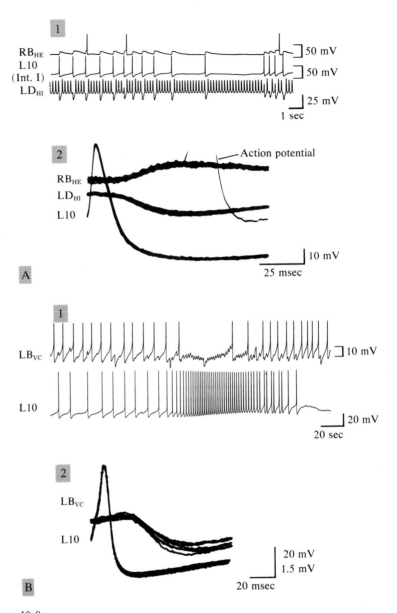

FIGURE 10-8.
Identified cell L10 (Interneuron I) produces opposite synaptic action on heart motor neurons. [From Koester *et al.*, 1974.]

A. All three cells recorded simultaneously. *A-1.* Each action potential in L10 causes an EPSP in RB_{HE} and an IPSP in LD_{HI}. *A-2.* Several superimposed sweeps illustrate brief and constant latency between action potentials in L10 and PSPs in RB_{HE} and LD_{HI}. In one sweep the EPSP in RB_{HE} was of sufficient magnitude to cause an action potential.

B. Inhibitory connections from cell L10 to the vascoconstrictor motor cells (LB_{VC}). *B-1.* A spontaneously occurring burst of spike activity in L10 causes a phasic decrease in the spontaneous firing of LB_{VC}. Action potentials in cell L10 are associated with IPSPs in cell LB_{VC}. *B-2.* Several superimposed traces show that the latency between the L10 spike and the IPSP it produces in LB_{VC} is brief and constant.

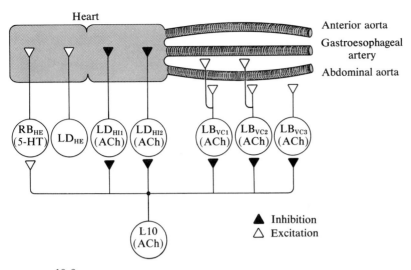

FIGURE 10-9.
Diagram of the synaptic connections made by cell L10 to the cardiovascular motor cells. Where known, the transmitter utilized by each cell is indicated. Cell LD_{HE} has not yet been examined with regard to possible input from L10. [From Koester *et al.*, 1974.]

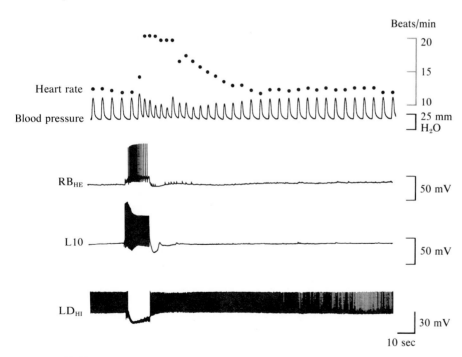

FIGURE 10-10.
The effect of intracellular stimulation of cell L10 on cardiovascular motor neurons. A 12-sec burst of spikes in L10 excites RB_{HE} and inhibits LD_{HI}. This is associated with a long-lasting increase in heart rate and a short increase, followed by a decrease, in blood pressure. [From Koester *et al.*, 1974.]

reversibly removed from the motor circuit, the acceleration of heart rate caused by a burst of L10 spikes (produced by depolarizing current pulses) was abolished (Figure 10-11).

To see how the central program is generated, the spontaneous firing of L10 was examined as a function of heart rate. A good correlation was found between the burst tendency of L10 and the phasic increases in heart rate and cardiac output. Each large phasic increase in heart rate was preceded by a high-frequency burst of spikes in L10. Thus, this simple circulatory fixed-action pattern results from the spontaneous high-frequency burst activity that L10 generates. These experiments also clarify further the variant and invariant aspects of this fixed-action pattern. The phasic increases in heart rate sometimes vary in magnitude in a stepwise manner. When this occurs, the large phasic increases in heart rate are always associated with high-frequency bursts in L10 and small phasic increases are accompanied by low-frequency bursts of spikes in L10. In each case the relationship of the neural elements to each other is still stereotypic. The variation in magnitude is due to the different burst intensities of L10. The invariance of the form is due to the invariant connections made by L10, which insure constant temporal relationships among its follower cells (Figure 10-12).

Cell L10 can undergo its characteristic firing patterns, including several types of bursts, in the isolated ganglion, that is, in the absence of peripheral feedback. The firing pattern of L10 is modulated by inhibitory synaptic input from several interneurons (II, III, IV, and V) located within the abdominal ganglion (see Figure 8-33, p. 332). Although this inhibitory activity can determine the frequency with which L10 shows each of these firing patterns, the intrinsic ability to generate these patterns does not require even these intraganglionic synaptic connections. As is the case for the autoactive burst activity of other cells (see p. 260), the burst-generating capability of L10 is an intrinsic membrane property. Bursting can be reset by injecting a train of action potentials into L10 by means of a depolarizing intracellular current pulse, or by changing the membrane potential (see Figure 7-28, p. 261). Also, when L10 is hyperpolarized to block spike activity, no synaptic activity is evident, which could account for the spontaneous firing or the bursting pattern. Because L10 makes powerful direct connections to all important circulatory motor neurons, changes in its endogenous firing patterns produce profound changes in behavior, and its periodic burst activity determines the total central program of a fixed-action pattern (Figure 10-13).[1]

---

[1]In addition to its phasic burst activity, L10 may be silent or may fire regularly. Regular firing of L10 contributes to the firing of RB_{HE}, which in turn speeds up the heart rate, adding an excitatory tone to the myogenic pacemaker.

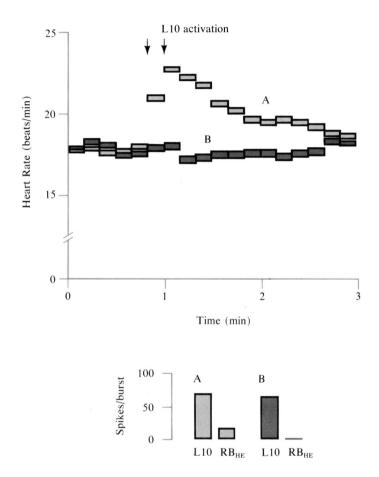

FIGURE 10-11.
Cell L10 mediates its central program by means of motor neuron RB_{HE}. Each block in the upper graph represents a 10-sec average of heart rate from two separate tests (A and B). After a one-minute control period, L10 was activated for 10 seconds (arrows). In test A the spikes evoked in L10 produced several spikes in RB_{HE} (lower graph), which resulted in a long-lasting increase in heart rate. In test B the same procedure was repeated, but with RB_{HE} hyperpolarized to prevent it from firing in response to its synaptic input. Under these conditions the same burst of spikes in L10 elicited no action potentials from RB_{HE} and the effect of L10 on heart rate was eliminated. [From Koester et al., 1974.]

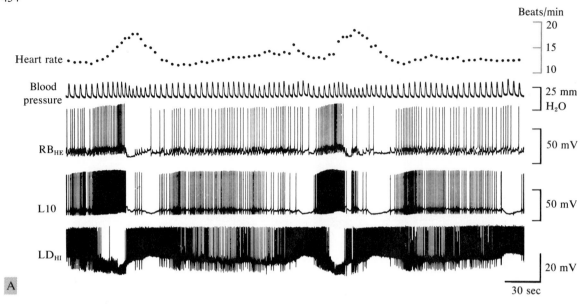

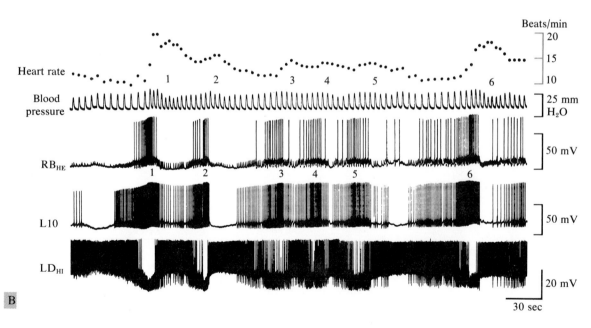

FIGURE 10-12.
Effects of various patterns of autoactivity in cell L10 on heart rate. [From Koester *et al.*, 1974.]

A. Each high-frequency burst of activity in L10 results in excitation of RB_{HE} and inhibition of LD_{HI}. These alterations in firing pattern are associated with a phasic increase in heart rate plus an initial slight increase in pulse pressure followed by a more prolonged decrease.

B. Variations in heart rate with variation in L10 burst patterns. High-frequency bursts (1 and 6) produce large increases in heart rate; lower-frequency bursts (2, 3, 4, and 5) produce lesser increases in heart rate.

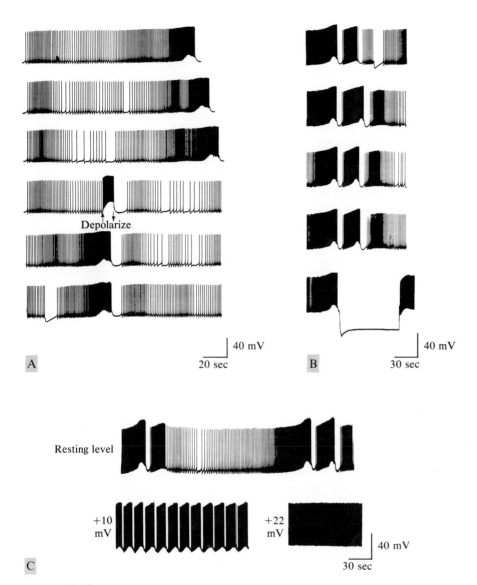

FIGURE 10-13.
Evidence for autoactive bursting in L10. [From Kandel, Carew, and Koester, 1976.]

A. Resetting of regularly occurring burst pattern by a train of action potentials produced by an intracellular depolarizing pulse. (Traces are continuous; the bursts are lined up to show the regularity.) In the fourth trace a burst was produced prematurely by depolarizing the cell with an otuward current pulse. This resets the rhythm. The next burst does not occur when it would if it were triggered by a regularly occurring EPSP. Rather, it occurs at the time of the burst elicited by the depolarizing current pulse, illustrating that the burst is locked to the previous (intrinsic) burst and not to an external trigger.

B. Failure to detect underlying EPSPs during the burst of L10. The cell was hyperpolarized (to reveal EPSPs) during an expected burst. Four consecutive bursts are aligned; the cell was hyperpolarized at the onset of the fifth burst. If the rhythm were due to rhythmic synaptic input, EPSPs would be evident.

C. Control over firing rate and bursting pattern by the membrane potential of L10. [From Koester, unpublished observations.]

Hormonal Control of Heart Rate

Attempts to analyze the factors that change the firing mode of L10 from silent to regular firing and then to various modes of bursting are just beginning. One factor that produces a striking increase in heart rate is egg-laying (Mayeri, unpub.). Mayeri has found that the bag-cell hormone that releases eggs from the ovotestis during egg-laying (see p. 408) also releases the central program for increased cardiac output. Perfusion of the isolated ganglion with seawater containing extracts of the bag cell causes L10 to go into a bursting mode.

Control by the egg-laying hormone is exerted at two sites. The hormone triggers the central program by altering the firing mode of L10. In addition, the hormone increases the effectiveness of the synaptic connections made by L10, thereby insuring the effective expression of the central program.

Centrally Programmed Decrease of Heart Rate

The respiratory pumping movements that circulate water through the mantle cavity are associated with a profound decrease of heart rate during which RB_{HE} is inhibited and LD_{HE} is excited (Figure 10-1B). This fixed-action pattern is mediated by Interneuron II, the command element for the gill and siphon components of respiratory pumping (see p. 378). Unlike L10, Interneuron II has not yet been identified; it is probably two or more tightly coupled cells that fire in partial synchrony.

The synaptic potentials attributed to Interneuron II occur in short bursts lasting 5 to 10 seconds. This characteristic command can be triggered by stimulating the organs of the mantle cavity, particularly the gill pinnules, and the frequency of the command is greatly increased following an increased respiratory load, such as feeding or anoxia. The activity of Interneuron II can be monitored by recording from one of its known follower cells (Figures 10-15, 10-16, and 10-18).

In addition to making connections with all the gill and siphon motor cells (see p. 383), Interneuron II connects to all of the cardiovascular motor neurons. Its connections here are opposite to those of L10 (Figure 10-14). Interneuron II excites the heart inhibitors and inhibits the RB_{HE} heart excitor, thus producing inhibition of heart beat (Figure 10-15). Interneuron II also inhibits the vasoconstrictors and weakly excites LD_{HE}.

To determine whether the phasic decrease in heart rate that accompanies respiratory pumping is mediated by the known cardiovascular motor cells to

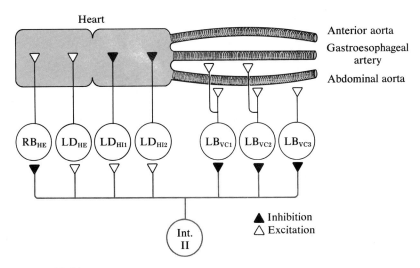

FIGURE 10-14.

Diagram of the synaptic connections made by Interneuron II to the cardiovascular motor neurons. Interneuron II and its terminals are drawn in light lines to indicate that this cell is not identified. It is believed to consist of two or more cells that are tightly coupled in their activity patterns, perhaps by means of electrical connections. [From Koester *et al.,* 1974.]

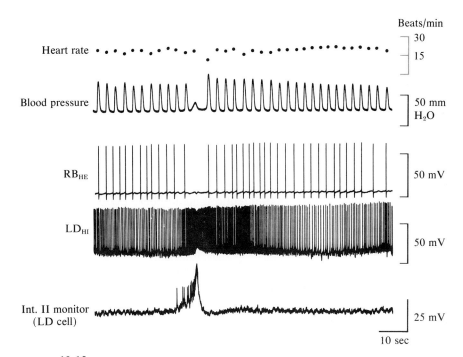

FIGURE 10-15.

Synaptic input from Interneuron II to cells RB_{HE} and LD_{HI}. A burst of activity in Interneuron II (monitored in an LD cell known to be excited by Interneuron II) excites LD_{HI}, inhibits RB_{HE}, resulting in a decrease in pulse pressure and heart rate. [From Koester *et al.,* 1974.]

which Interneuron II connects, key cells were hyperpolarized and reversibly removed from the neural circuit during the generation of Interneuron II bursts. Since Interneuron II bursts produce a high-frequency burst of spikes in the LD_{HI} cells and only brief and weak inhibition of RB_{HE}, it seems likely that most of the action potentials in Interneuron II are mediated by the two LD_{HI} cells. In fact, when the membrane potentials of both inhibitors are hyperpolarized so that they do not fire, burst activity in Interneuron II fails to produce the fixed-action patterns (Figure 10-16).

Command Elements and Central Programs

Perhaps the most dramatic finding to emerge from research in invertebrates is the discovery that single cells (command neurons) can trigger complete motor sequences. In 1938 C. A. G. Wiersma found that a single electrical stimulus applied to any one of the four giant fibers in the crayfish produced a complete escape response: the animal turned in its eyestalk, moved the antennae and legs forward, pulled the swimmerets (leglike structures) upward, and gave a strong tail flip. Later, Wiersma (1952) discovered single fibers that trigger the defensive reflex, in which the animal raises its claws and head while supporting its body on its tail and fourth and fifth thoracic legs. Subsequently, Wiersma and Ikeda (1964) and Kennedy (1971) described other command elements that produce leg and abdominal movements, abdominal postural adjustments, and rhythmic beating of the swimmerets. Since then, a variety of command cells has been analyzed (see Kennedy and Davis, 1976).

From studies of command cells in crayfish, several generalizations have emerged. (1) Command cells generally exert a widespread effect, controlling a total block or sequence of behavior. (2) More than one cell can initiate a particular behavioral sequence, and there is often overlap in the effector actions mediated by a number of different command fibers. (3) Despite the overlap in effector actions among several command cells, it is often possible to distinguish differences in their detailed field of action, suggesting that the movement produced by each element may be unique.

Because the connections and transmitter biochemistry of cell L10 in *Aplysia* are known, this cell affords an opportunity for examining how a command cell can mediate powerful control over a block of behavior and how the command functions relate to the central program of a behavior. As is true in gen-

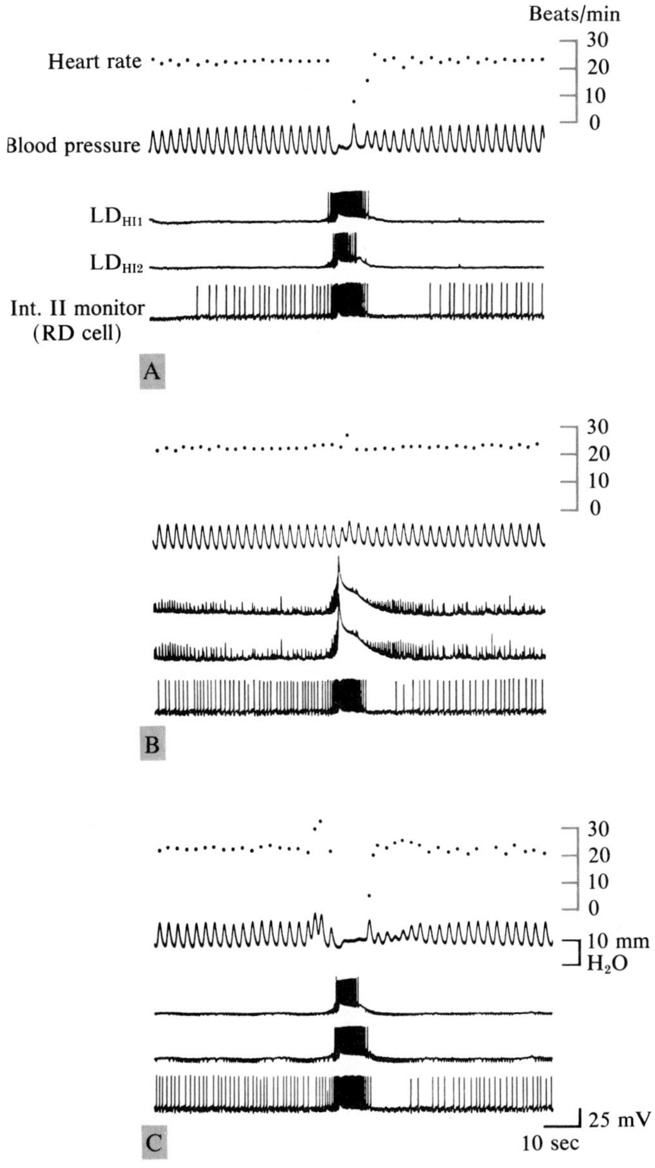

FIGURE 10-16.
Interneuron II inhibits the heart through the inhibitory motor neurons LD_{HI}. The traces in each set (A,B,C) are simultaneous records from the heart, the two heart inhibitors, and an RD cell known to be excited by Interneuron II and used as a monitor of Interneuron II activity. A burst in Interneuron II causes a burst in the two heart inhibitors. [From Koester *et al.,* 1974.]

A and **C.** A burst in Interneuron II causes the LD_{HI} cells to fire action potentials, thus inhibiting the heart beat.

B. When the two LD_{HI} cells are prevented from firing by hyperpolarizing their membrane potentials, inhibition of the heart beat by Interneuron II is eliminated.

eral for command cells, the action of L10 is not limited to one motor system; it affects the activity of several different motor systems (heart, aorta, siphon, mantle shelf, and gill). Moreover, L10 acts on these motor pathways by making direct connections with each of the motor cells. It makes excitatory connections with the excitatory motor cell of the heart (RB_{HE}), inhibitory connections with inhibitory motor cells and the vasoconstrictors, dual excitatory and inhibitory connections with a gill motor neuron (cell L7), an inhibitory connection with at least one siphon motor neuron (LB_{S1}), and an excitatory connection to R15, the neuroendocrine cell that controls water balance. Moreover, the central program for the cardiovascular fixed-action pattern commanded by L10 resides in its own autoactive firing capability.

I have considered only the behavioral actions of L10 that relate to the heart and blood pressure. The functional significance of the motor effects on the gill, siphon, and mantle shelf, on water balance, and on the bursting cells of the left upper quadrant (L2 to L6, whose behavioral role is still unknown) has not yet been worked out.

Interneuron II also activates several different motor pathways and has command functions. Since Interneuron II is probably a cluster of two or more interconnected cells, it is more appropriate to refer to it as a command *element* rather than cell.

In the somatic motor system of crustacea command cells code for specific postures or movements (see Kennedy and Davis, 1976). In visceral-motor systems the two command elements studied code for specific homeostatic functions (see also Mendelson, 1971). Thus, L10 codes for increased cardiac output; it increases heart rate (by exciting the excitors and inhibiting the inhibitors and Interneuron II) and lowers vasomotor tone (by inhibiting the vasoconstrictors). Interneuron II codes for an even more complex homeostatic function, a coordinated pumping sequence involving both the respiratory and the cardiovascular systems.

Both L10 and Interneuron II can generate their central program independently of peripheral feedback. Central programming of motor output was first described in the somatic nervous systems of insects and crayfish (Wilson, 1968; Kennedy and Davis, 1976). Thus, command cells and central programs, two features that characterize somatic motor systems, are found in the neural circuit controlling circulation and respiration in *Aplysia*. These observations suggest that the visceral-motor and somatic-motor controls may share common principles of motor organization.

Integration of Fixed Acts into a Fixed-Action Pattern

Although the respiratory and circulatory systems of *Aplysia* are anatomically separate, they are hydraulically coupled by the blood (Figure 10-2) and act together to provide for gaseous exchange in the tissues of the body. To function efficiently the activities of these two independent systems must be coordinated. A clue to how this coordination is achieved can serve as an example of how in general various fixed-acts can be integrated into a complex fixed-action pattern (Figures 10-17 and 10-18).

The respiratory and cardiovascular systems participate in stereotypic responses that are synchronous with each other: each respiratory pumping movement is linked to an inhibition of heart rate. This coordinated motor sequence

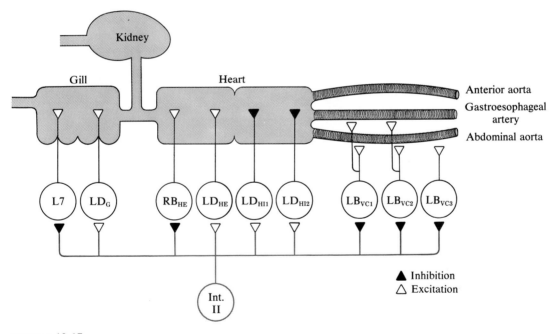

FIGURE 10-17.
Diagram of the anatomical relationships between the circulatory and respiratory systems. The diagram also indicates the synaptic connections of Interneuron II to the cardiovascular motor neurons and to two gill motor neurons, L7 and LD_G. Interneuron II and its terminals are drawn in light lines to indicate that the element is not yet identified. [From Koester *et al.*, 1974.]

442

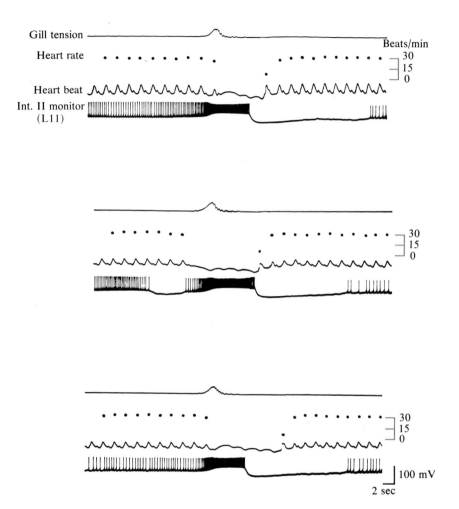

Gill tension

Heart rate

Heart beat

Int. II monitor
(L11)

Beats/min

30
15
0

30
15
0

30
15
0

100 mV

2 sec

FIGURE 10-18.

A complex fixed-action pattern consisting of gill contraction and heart inhibition produced concomitantly by bursts of activity in Interneuron II. Three sets of simultaneous recordings of gill tension, heart rate, heart beat (photocell record), and the activity of Interneuron II (monitored in cell L11) illustrate the fixed relationship of the respiratory and cardiovascular components of the fixed-action pattern. [From Koester et al., 1974.]

has two effects. First, water is pumped out of the mantle cavity and is replaced by fresh seawater, thereby increasing the flow of fresh seawater past the respiratory exchange surfaces of the gill. Second, gill contraction forces a large volume of fluid toward the auricle. Were the heart not inhibited during this gill contraction, blood from the gill might be pumped backwards into the kidney through the efferent renal vein (Figure 10-17). Thus, heart inhibition and inhibition of the vasoconstrictors function to prevent backflow of blood by reducing the vascular resistance across the heart and the aorta.

How are these separate acts integrated so that variations in one are accompanied by appropriate adjustments in the other? Cellular studies of the two acts reveal that both are triggered by a common central program attributable to activity of a common command element, Interneuron II (Koester *et al.*, 1974). The excitatory actions of Interneuron II on gill motor cells LD_{G_1}, LD_{G_2}, and RD_G and the inhibitory and rebound excitatory actions on L7 and the L9 cells produce the gill contraction and pumping movements, whereas the excitatory actions of Interneuron II on the two heart inhibitors and the inhibitory actions on the vasoconstrictors produce the concurrent inhibition of heart rate and vasodilation. The entire pattern of motor activity seems to be centrally determined. In the isolated ganglion all the motor cells receive the same burst of synaptic input from Interneuron II in the absence of peripheral feedback.

Unifying the command of the respiratory and cardiovascular components of the fixed-action pattern assures orderly relationships between movements. When one component varies, the other also varies in a characteristic way (Figure 10-19). Each variation in intensity of the fixed-action pattern seems to be related to a variation in the intensity or pattern of Interneuron II activity. The orderly recruitment of the different effector components is due to the invariant connections made by the central command elements. Like L10, Interneuron II varies in its pattern of firing (having several preferred firing modes), but it does not vary in the type of synaptic connections that it makes.

Coordination of Antagonist Fixed Acts

The neural circuit for the cardiovascular system consists of a pool of independent motor cells that receive various combinations of synaptic input from two antagonist command interneurons. How are antagonist fixed-action patterns coordinated so that at any time only one and not the other is expressed? This coordination cannot occur at the level of the motor cells because they are not connected with each other (Figure 10-6). Rather, coordination is achieved at the command level. Interneuron II makes a strong inhibitory connection to

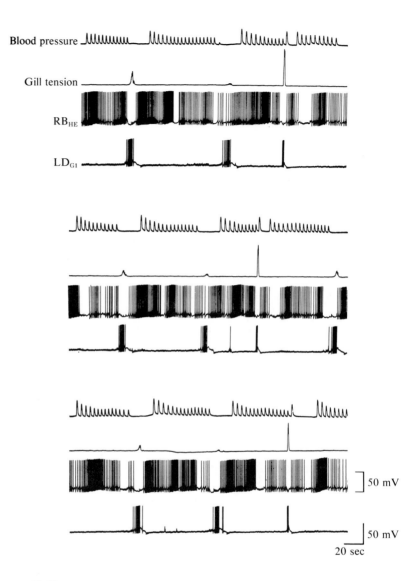

FIGURE 10-19.

Variations in the expression of a fixed-action pattern. Variations in heart inhibition and gill contraction are produced simultaneously by variations in Interneuron II activity. Ten consecutive bursts in Interneuron II produce gill contractions. Traces are continuous. Each Interneuron II burst excites gill motor neuron LD_{G1}, inhibits heart excitor RB_{HE}, and thus causes a gill contraction and heart inhibition. In each set of records the intensity of the third burst in Interneuron II is especially effective and causes a relatively large gill contraction. The large gill contraction in each row is followed by a phasic increase in blood pressure, due to the haemolymph pumped by the gill into the heart. Note that the variations in amplitude fall roughly into three groups: small, medium, and large. [From Koester *et al.,* 1974.]

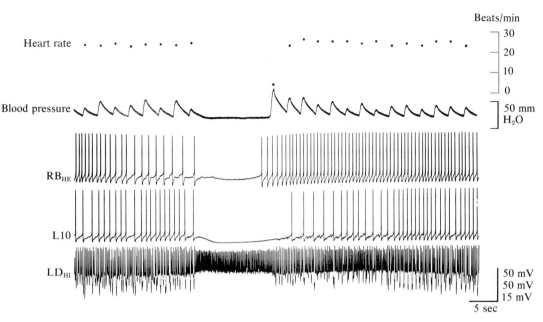

FIGURE 10-20.
Inhibitory action of Interneuron II on heart beat and on cells L10, RB$_{HE}$, and LD$_{HI}$. A burst in Interneuron II produces effects on the heart and gill by direct connections to motor neurons and inhibition of interneuron L10, which produces effects opposite to those of Interneuron II. A burst in Interneuron II is associated with inhibition of L10, excitation of LD$_{HI}$, and inhibition of RB$_{HE}$. These effects are accompanied by inhibition of heart beat. [From Koester *et al.*, 1974.]

L10 (Figure 10-20), and there is indirect evidence that L10 also makes an inhibitory connection to Interneuron II (see p. 336). The two interneurons act at two levels. When either interneuron is activated it inhibits the other interneuron and substitutes its own synaptic inputs to the motor neurons for the inputs from the inhibited interneuron. I have previously (p. 334) referred to this type of organization as feed-forward substitution. In the case of the heart inhibitors and the excitor RB$_{HE}$, each interneuron substitutes an opposite synaptic input to the motor neurons—inhibition for excitation or excitation for inhibition. In the case of the vasoconstrictor motor neurons each interneuron substitutes inhibition for inhibition. For example during the centrally commanded slowing of heart rate the spontaneous firing of L10 is interrupted by a burst of inhibitory input from Interneuron II (Figure 10-20). This results in excitation replacing inhibition in LD$_{HI}$ and inhibition replacing excitation in RB$_{HE}$. Because of the inhibitory interaction between command elements, the selection of only one of the two central programs in assured (Figure 10-21).

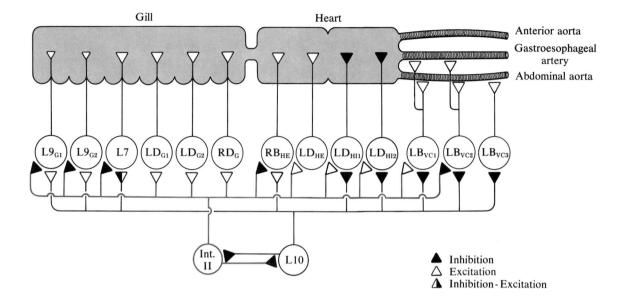

FIGURE 10-21.
Diagram of the synaptic connections made by L10 and Interneuron II on cardiovascular neurons and gill motor neurons. The interaction between the two command interneurons is also represented. The two interneurons appear to produce motor effects by acting at two levels. When either interneuron is activated it inhibits the other interneuron and substitutes its own synaptic inputs to the motor neurons for the inputs from the inhibited interneuron. This type of organization is called feed-forward substitution. In the case of LD$_{HI}$ and RB$_{HE}$ each interneuron substitutes opposite synaptic inputs to the motor neurons: inhibition for excitation or excitation for inhibition.

DEFENSIVE SWIMMING AND FEEDING

So far I have primarily considered elementary behavior (reflex and fixed acts) and simple examples of complex behavior, the fixed-action pattern involving both circulation and respiration. Although few in number, the principles already described account for a surprising degree of complexity in behavior. They can account for the integration of synergist components of behavior and coordination of antagonist ones in the same or different effector systems. Coordination of circulation and respiration, two antagonist systems, involves two command elements, each mediating its own central program and making inhibitory connections with the other. This neural circuit is schematically illustrated in Figure 10-22. A feature of this system is that under the appropriate

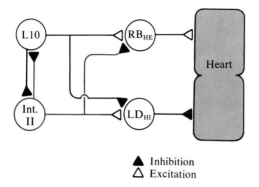

▲ Inhibition
△ Excitation

FIGURE 10-22.
Simplified diagram of reciprocal inhibition between
two spontaneously active dual-action command
cells, L10 and Interneuron II.

circumstances it will oscillate, leading to a recycling of activity between the
two autoactive command cells, Interneurons I and II (Figure 10-23). When
both interneurons are spontaneously active, a burst of spikes in Interneuron I
will inhibit the firing of Interneuron II. This inhibition is followed by disinhibition (removal of inhibition) and rebound excitation, which in turn inhibits
Interneuron I. Inhibition of Interneuron I is followed by disinhibition and
rebound excitation, leading once again to inhibition of Interneuron II and a
new cycle of activity. Thus, mutual inhibition of the two autoactive interneurons, followed by disinhibition and rebound excitation, can lead to a sustained recycling of activity. Recycling of activity is a common feature of certain
fixed-action patterns, such as swimming, feeding, locomotion, and copulation. It is attractive to think that some of these responses may utilize similar
mechanisms.

These complex behavioral responses of invertebrates are now beginning to
be analyzed. There has been substantial progress in the analysis of defensive
swimming (of the opisthobranch *Tritonia* and the leech, *Hirudo medicinalis*)
and feeding (of *Aplysia* and various gastropod molluscs: *Helisoma trivolis*,
Pleurobranchaea, and *Navanax*. As a result we can begin to draw an outline of
the possible neural circuitry underlying these behavioral responses and compare elementary and complex behavior.

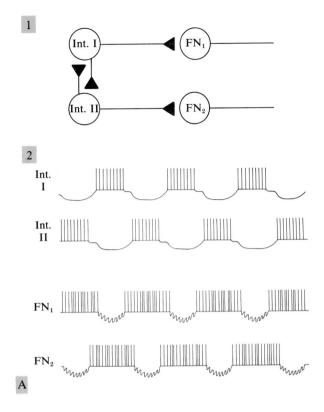

A

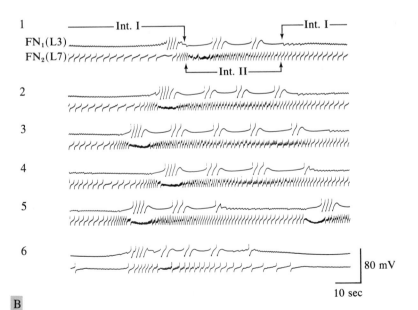

B

80 mV

10 sec

FIGURE 10-23.

Sustained oscillations due to alternating activity in two autoactive command cells, L10 (Interneuron I) and Interneuron II. [From Kandel, Frazier, and Wachtel, 1969.]

A. *A-1.* Diagram of two autoactive command cells, L10 (Interneuron I) and Interneuron II, connected by mutually inhibitory connections. For purposes of illustration, each command cell innervates only one autoactive follower cell. *A-2.* Schematic traces illustrate the consequences for the two follower cells of alternating activity in the command cells. Interneuron II bursts first and inhibits L10 and FN_2. As a result of the inhibition of L10, FN_1 is disinhibited, thereby allowing its autoactive rhythm to be expressed. Cell L10 is then disinhibited and its autoactive rhythm is established, causing FN_1 to be inhibited. When the burst in L10 terminates, FN_1 and Interneuron II are disinhibited and their rhythms are expressed, thereby initiating a new cycle of activity.

B. Actual examples of the consequences of prolonged alternation of activity in L10 and Interneuron II for two autoactive follower cells: FN_1 (L3) and FN_2 (L7). One follower cell, FN_1, receives an inhibitory connection from L10; the other, FN_2, receives an inhibitory connection from Interneuron II. The alternating IPSPs in the two follower cells reflect the alternating activity of the two interneurons.

Defensive Swimming of *Tritonia*

The most extensively studied complex behavioral response of invertebrates is the defensive swimming of the opisthobranch *Tritonia,* studied by Dennis Willows and his colleagues. Willows (1973) found that following contact with a starfish or other predators *Tritonia* shows a stereotypic swimming respons... This involves pushing off from the substrate and swimming by alternating flexion of dorsal and ventral longitudinal body wall muscles. The escape response has four components: local reflex withdrawal, preparation for swimming, swimming, and termination.

When touched, the animal withdraws and contracts its extended body. Next it prepares to swim by elongating its head and tail into two large paddle-like structures. The animal then makes violent swimming movements, consisting of a series of ventral and dorsal flexions. Although poorly coordinated, these are effective in moving the animal away from the noxious stimulus. Finally, the swimming motions end with gradually weakening dorsal flexion followed by relaxation. This highly stereotypic sequence lasts nearly 30 seconds (Figure 10-24; Willows, Dorsett, and Hoyle, 1973b).

The local withdrawal following contact is thought to be mediated by some peripheral neurons, not by the central nervous system. But if the stimulus to the skin is sufficiently strong, local withdrawal is immediately followed by centrally mediated preparation for swimming, which begins with a generalized and more prolonged withdrawal apparently driven by many nerve cells. These cells may remain active throughout this and the next stage (swimming). The withdrawal movement involves the contraction of the ventral and dorsal mus-

450

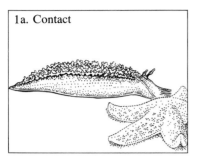

1a. Contact

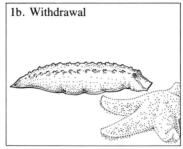

1b. Withdrawal

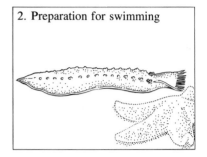

2. Preparation for swimming

3. Swimming

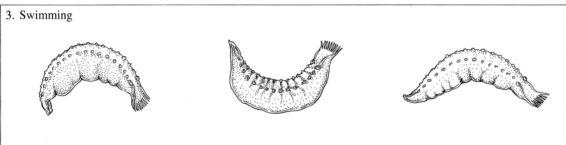

4. Termination of swimming

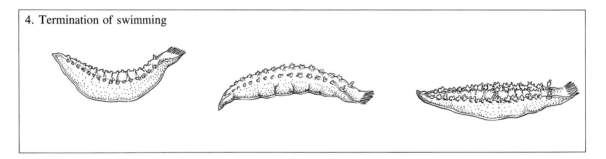

FIGURE 10-24.

Escape swimming in *Tritonia,* a fixed-action pattern consisting of four stages. [From Willows, 1971. Copyright by Scientific American, Inc. All rights reserved.]

Stage 1. The relaxed animal with branchial tufts and rhinopores extended contacts a predator. After contact with a starfish the animal withdraws reflexly and bends ventrally.

Stage 2. Preparation for swimming. The animal elongates and enlarges the oral veil while bending slightly in the dorsal direction.

Stage 3. Swimming. The animal first makes vigorous ventral flexion, then vigorous dorsal flexion. This cycle is repeated several times.

Stage 4. Termination. After a final dorsal flexion the animal returns to an unflexed position with the extremities still withdrawn, oral veil and tail enlarged. One to five weak dorsal flexions occur before the animal regains its original relaxed posture.

cle groups; the ventral muscles, driven by ventral (motor) neurons, tend to dominate over the dorsal muscles, driven by dorsal (motor) neurons.

A group of sensory neurons that respond to the eliciting stimulus is thought to connect to two groups of motor cells: (1) the dorsal-flexion neurons that initiate swimming, and (2) two symmetrical cell clusters, called trigger neurons, that are thought to be important in helping activate the dorsal-flexion neurons. The cells in each trigger cluster are coupled with one another and with the cells in the contralateral cluster by means of bidirectional electrical connections (Figure 10-25). An action potential in one cell produces diphasic electrically

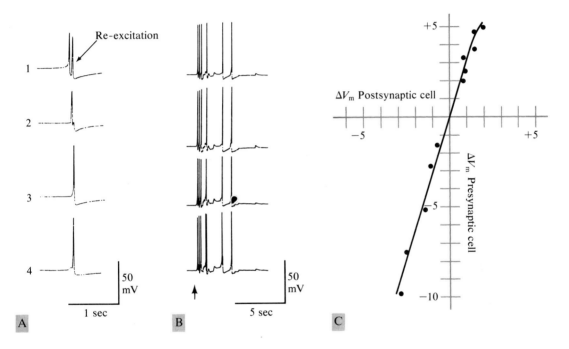

FIGURE 10-25.

Transmission and re-excitation within electrically coupled trigger group neurons in *Tritonia*. Simultaneous recordings were made from four neurons from a population of trigger cells in the left and right ganglia.

A. A spontaneously occurring impulse in one trigger cell (*A-1*) spreads to the others exciting them (*A-2, A-3, A-4*). This led to reexcitation of the first cell (*A-1*).

B. Trigger cell neurons received synchronous excitatory input. At the arrow, a small drop of salt solution (which can also trigger the fixed-action pattern) was placed on the center of the oral veil. After a short decay, a burst of spikes in all four trigger neurons occurred.

C. Electrical coupling between two trigger neurons measured over a range of applied potentials. Ordinate: changes in membrane potential produced by current in the presynaptic neuron as recorded intracellularly by means of an independent recording electrode. Abscissa: changes in membrane potential recorded in the postsynaptic neuron in response to stimulation of the first. For this pair of neurons the coupling ratio is 0.3. It is constant over a wide range of both depolarizing and hyperpolarizing currents. [From Willows *et al.*, 1973.]

mediated PSPs in the others. Contact with a starfish produces a brief (2 to 20 seconds), brisk, and nearly synchronous discharge of all the cells. The synchrony results from common input (Figure 10-25A) and from the electrical interconnection between the cells (Figures 10-25C and D). The electrical coupling also leads to recirculation of impulse activity through positive feedback. An action potential in one cell can initiate an action potential in other cells of the cluster, often with some delay, which permits the first cell to be reexcited by action potentials in other cells. By reexciting each other, cells in a cluster can maintain activity for many seconds. This regenerative activity terminates when the impulse activity becomes nearly synchronous in a majority of the cells of the cluster. Specifically, termination is due to hyperpolarization of the cells produced by the addition of post-spike hyperpolarizing afterpotentials (Getting and Willows, 1974).[2]

The connections between the sensory neurons and trigger neurons as well as those between the sensory neurons and the ventral- and dorsal-flexion motor neurons are thought to be indirect. But discharge of the trigger neurons, by a natural stimulus, is followed by a slow depolarization in the dorsal-flexion (and to some degree in the ventral-flexion) neurons. There are thought to be about 20 dorsal- and 20 ventral-flexion motor neurons. Activity in these neurons drives the swimming movements, which consist of five cycles of alternating ventral and dorsal flexion movements. During swimming the flexion and extension motor cells undergo strictly alternating burst activity (Figure 10-26), presumably due to inhibitory connections between antagonist motor cells. Within each motor group the burst of impulses is closely correlated. Correlation of firing within a group is thought to be accomplished by weak excitatory interconnections. Finally, some cells, the general excitatory interneurons, fire bursts of spikes in phase with both the ventral- and dorsal-flexion motor neurons. These cells are thought to provide positive feedback to both groups of motor cells, keeping them partially depolarized so that they can generate their alternating pattern. Swimming is presumed to be terminated by an active process (perhaps by inhibitory neurons) that shuts off the dorsal- and ventral-flexion motor neurons. The sequence of neuronal firing underlying this fixed-action pattern is present in the isolated nervous system and does not require feedback from peripheral neurons for its timing or maintenance.

Willows and co-workers (1973) and Getting (1975) have proposed the following neural circuit to account for this behavior (Figure 10-27). Sensory input

---

[2]In the absence of synchrony the spike afterpotential in any one cell is reduced (shunted) by the current drawn by the membrane of the inactive cells connected to the active cell. During synchrony the shunting action is decreased by the activity of the other cells (Getting and Willows, 1974).

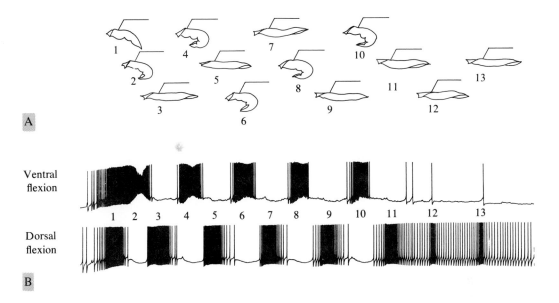

Ventral
flexion

Dorsal
flexion

A

B

FIGURE 10-26.
Central program for swimming in *Tritonia*. **A.** Swimming movements. **B.** Intracellular recordings. The top record represents a cell that drives the animal's downward (ventral) flexion; the lower record shows the cell that drives the upward (dorsal) flexion. The numbers between records correspond to the numbers in part *A* and show the types of recordings obtained during the corresponding phases of the swimming movement. [From Willows, 1971. Copyright by Scientific American, Inc. All rights reserved.]

from the oral veil excites both the trigger cells and the dorsal-flexion neurons. If the excitation of the trigger cells is sufficiently strong, many trigger cells become activated and excite other trigger cells, which in turn reexcite the trigger cells initially fired. Discharge of the trigger-cell population is thought to reexcite the dorsal- and flexion-motor neurons; these excite the general excitor neurons and inhibit the ventral-flexion motor neurons. The general excitors excite both ventral- and dorsal-flexion motor neurons and keep them depolarized so as to generate the alternating burst activity. The burst sequences are terminated by a terminator neuron.

The circuit for this complex behavior has three parts, two of which resemble the circuits found in elementary behavior. The high threshold, all-or-none trigger neurons are similar in their properties and interconnections to the bag cells that control egg-laying. But whereas the mechanism for maintaining activity in the bag cells is not known, in *Tritonia* one can clearly see how the activity in an electrically coupled group of cells can be maintained and then terminated. The antagonist motor neurons with mutually inhibitory connec-

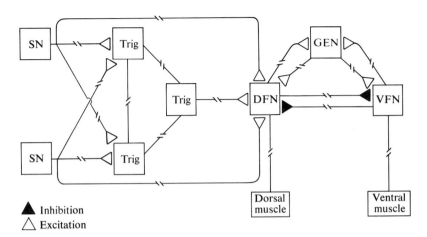

FIGURE 10-27.

Model of the neural organization controlling escape swimming, as proposed by Willows and co-workers (1973) and by Getting (1975). Excitation from sensory cells in the epithelium reaches the electrically coupled trigger-neuron system (TN) both contralaterally and ipsilaterally as well as the dorsal-flexion motor neurons (DFN). Connections are not necessarily direct (indicated by the broken lines). Other connections, not shown, may go from the sensory neurons to the ventral-flexion motor neurons (VFN), the general excitor neurons (GEN), and other neurons. These neural connections drive withdrawal and bending away from the stimulus source (stage 1, Figure 10-24). Also not shown are connections among the ventral flexors and among the dorsal flexors that tend to synchronize the activity of the synergist motor neurons. A vigorous volley from the trigger neurons excites the dorsal flexors and indirectly inhibits the ventral flexors, permitting dorsal bending (part of stage 2, Figure 10-24). When this burst decays, the ventral flexors are released from inhibition and generate a burst in response to activity in the general excitors. Inhibition of the dorsal flexors disinhibits the ventral flexors. Impulses in either flexor group excite the general excitors, which reexcite the flexors. Alternation between the antagonist flexor groups is coordinated by reciprocal inhibition. Coordination within a population of flexor neurons is insured by mutual excitation. The bursting is thought to be shut off abruptly by the activity of an unidentified group of specialized terminator neurons normally inhibited by the general excitors and the ventral and dorsal flexors. (Boxes indicate cell clusters of unspecified numbers of cells).

tions are similar to Interneurons I (L10) and II. They provide a basic oscillatory circuit for recurrent activity (see Figure 10-22). In *Tritonia* this oscillatory capacity is much better developed than in L10 and Interneuron II. In addition, the circuit in *Tritonia* controlling swimming includes a new element not encountered in elementary behavior: the general excitor, which depolarizes and excites both populations of motor neurons.

The escape swimming of *Tritonia* illustrates nicely the differences between fixed acts and fixed-action patterns. Whereas inking and egg-laying are each realized by single acts, escape swimming consists of a sequence of different acts (withdrawal, preparation for swimming, swimming) as well as recurrence of the same act (dorsal and ventral swimming, flexion movement). The neural circuitry is also more complex, involving a large number of cells. It is therefore slightly reassuring that despite this complexity, the principles of this complex behavior are similar to those of elementary behavior.

Earlier (p. 32) I discussed the idea that complex behavior may be built up from simpler behavioral elements. Analysis of the escape swimming of *Tritonia* suggests that the neural circuitry of a complex behavior may in fact be made up of a combination of the same *circuit modules* (or specific combinations of components) that control elementary forms of behavior. As in elementary behavior, each module in the circuitry of a fixed-action pattern mediates a particular fixed act. The interconnection of the modules could then determine the sequence for generating the various constituent fixed acts of the fixed-action pattern.

Swimming of the Leech

The graceful swimming of the medicinal leech, *Hirudo medicinalis,* is more complex than the escape swimming of *Tritonia.* Leeches swim by extending and flattening their tubular bodies and then undulating front to back, which propels them forward. This behavior has been studied by Stent and his colleagues Kristan, Ort, and Calabrese (for review see Kristan and Stent, 1976).

The swimming movements involve three sets of muscles. The elongation and flattening of the body is produced by tonic contraction of muscles that run from the dorsal to the ventral body wall. The alternating crests and troughs of the undulating movement are produced by contraction and relaxation of the dorsal and ventral longitudinal muscles of the body wall in each segment. Contraction of the dorsal muscles produces a body trough; contraction of the ventral muscles produces a body crest. Intersegmental coordination of the contraction cycle will progress in the posterior direction.

The leech nervous system consists of a head ganglion and a tail ganglion (each a fusion of several ganglia) and 21 symmetrical segmental ganglia. The head ganglion is not necessary for swimming. The rhythm is produced by a central program that is distributed among cells of the segmental ganglia. The segmental motor apparatus consists of about 20 bilaterally symmetrical pairs of excitor and inhibitor motor cells that innervate both dorsal and ventral longitudinal muscles (Ort, Kristan, and Stent, 1974). Activity alternates between antagonist motor cells during swimming. The firing of the dorsal excitor cells and the ventral inhibitor cells produces contraction of the dorsal body wall and distention of the ventral body wall; the firing of the ventral excitor cells and the dorsal inhibitor cells produces the opposite action. The inhibitor cells play a particularly important role in coordinating these movements: in addition to inhibiting muscle they also inhibit antagonist excitor cells. The dorsal excitors are interconnected by means of electrical synapses, as are the ventral excitors and the inhibitors. All the homologous excitors and inhibitors on the two sides of the ganglion are also interconnected electrically. The electrical connections presumably act to synchronize the activity of appropriate motor cells.

The program for swimming is located within the central nervous system (Kristan and Calabrese; cited in Kristan and Stent, 1976). Sustained alternations of ventral and dorsal excitors can be triggered in the isolated nerve cord by a brief electrical stimulus to a segmental nerve. Thus, as in the escape swimming of *Tritonia*, coordination of the swimming rhythm of the leech does not require sensory feedback. Kristan and Calabrese postulate that each segmental ganglion contains the neurons necessary for generating its own part of the central program.

The central program of each ganglion is thought to be accounted for by three groups of neurons: the dorsal and ventral inhibitors and a postulated group of excitatory oscillators. The oscillators of each segment excite both the dorsal and ventral inhibitors. Once activated, the dorsal inhibitors inhibit the ventral inhibitors and disinhibit the ventral excitors. The continued activity of the dorsal inhibitors leads to a progressive depression of their inhibitory action and to a release of the ventral inhibitors, thereby inhibiting the dorsal inhibitors and disinhibiting (removing inhibition from) the dorsal excitors. The phasic activity of the excitors that drive the swimming is therefore produced by alternating inhibition and disinhibition from the antagonist inhibitors. This model emphasizes an important feature of the role of inhibition and disinhibition in the generation of behavior. As we have seen (Figure 10-23), inhibition can do more than shut off excitation. In cells that are autoactive (or driven by a continuously excitatory source) inhibition can have a timing function: an inhibited cell's activity is expressed after an appropriate delay, by means of disinhibition.

This speculative neural model for leech swimming shares features with the model proposed for *Tritonia* swimming. In both cases antagonist motor cells are kept activated by a common neuronal system (general excitors in *Tritonia*, excitatory oscillators in the leech). However, leech swimming involves two additional features: intersegmental integration and proprioceptive feedback. Intersegmental integration is thought to be accomplished by inhibitory connections between segments: the dorsal inhibitors in one segment inhibit the ventral inhibitors of the next (posterior) segment. The proprioceptive feedback acts to compensate for changes in load, e.g., different swimming media. As the viscosity of the medium increases, the leech increases the period of its swim cycle. Kristan and Stent (1976) suggest that proprioceptive feedback operates in each segment. If a swimming movement is not properly realized anywhere along the body, any segment that senses the error sends a signal to the most anterior segment to initiate a new body wave. The presence of feedback may explain why the leech always swims gracefully even in an unfamiliar environment, whereas *Tritonia* sometimes looks somewhat uncoordinated.

Feeding

Feeding, one of the most complex and interesting behavioral responses of gastropods, has been studied in four animals: the opisthobranchs *Aplysia*, *Pleurobranchaea*, and *Navanax*, and the pulmonate snail *Helisoma*. Neural analysis of feeding is only beginning, however, and in no case has a neural circuit of this behavior been worked out in satisfactory detail. Nonetheless, preliminary analyses suggest some interesting although tentative hypotheses about how feeding might operate.

As is the case with other instinctive behavior (Chapter 1), feeding consists of two components: a labile appetitive component and a more stereotypic consummatory component (a fixed-action pattern). The appetitive component consists of an orienting and food-seeking response. In *Aplysia* the animal assumes a feeding posture by attaching its tail to the substrate and then swings its head and anterior body back and forth in search of food (see Figure 4-14, p. 89). The appetitive movements are usually triggered by food, but sometimes they occur spontaneously as vacuum behavior.[3]

Contact with food is followed by a fixed-action ingestive response, which is similar in most gastropods (Figure 10-27). It consists of back-and-forth oscillation (retraction and protraction) of the buccal mass, resulting in rasp-like movements of the feeding organ, the *odontophore*, which has a toothy surface, the *radula*. Contraction of the protractor muscles of the buccal mass propels the radula forward. In many gastropods the radula then scrapes across the seaweed and shovels the food into the buccal cavity, or mouth, by contraction of the retractor muscles of the buccal mass (Figure 10-28). The complete feed-

[3]Jordan, 1917; Frings and Frings, 1965; Lickey, 1968; Jahan-Parwar, 1970; Kupfermann and Pinsker, 1968; Kupfermann, 1974a; for further discussion see Kandel, *Behavioral Biology of Aplysia*, forthcoming.

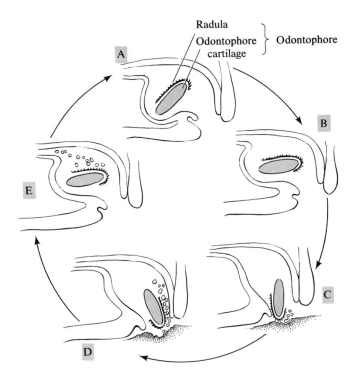

FIGURE 10-28.
Diagram of the ingestion process typical of a rasping gastropod (based on *Helisoma trivolvis*). Positions A through E make up one feeding cycle. **A.** Rest position of the buccal mass, radula, and odontophore cartilage. **B.** Buccal mass in initial protraction. **C.** Buccal mass fully protracted; independent protraction of odontophore and radula. **D.** Slight retraction of the odontophore, as well as independent radular movement (these independent movements of mass and radula are not found in *Aplysia*). **E.** Completion of retraction of the buccal mass, radula, and odontophore. [From Kater, 1974.]

ing sequence depends on repeated cycles of protraction and retraction of buccal mass and radula.

In *Aplysia* the protraction–retraction sequence is generally not used for scraping seaweed, but for biting and swallowing. The animal opens its mouth and protrudes its odontophore and the two radular halves. The radular halves open and close to grasp and bite the seaweed and pull it into the buccal cavity; they open during protraction and close during retraction (Figure 10-29). Biting is followed by swallowing, which involves the same protraction–retraction sequence as in biting except that the radula does not fully protract. The swal-

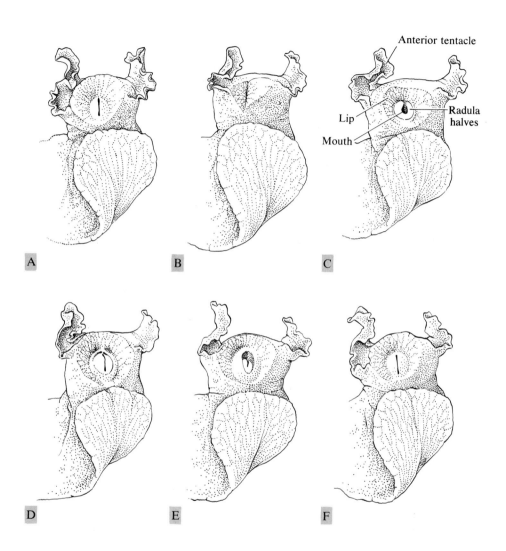

FIGURE 10-29.
Details of the biting response in *Aplysia*. Unlike some other gastropods (see Figure 10-28) *Aplysia* obtains its food by biting and tearing, not by the rasping action of the odontophore. For clarity, the food that initiated the response has been omitted. (The drawings are based on frames from motion picture film. **A.** Initial position. **B.** Lip-closing phase, during which the seaweed is grasped. **C.** Opening of the lips; first appearance of the two radula halves (clear white structure), which are widely separated. **D.** Full closure of the radula halves, which have now grasped the seaweed. The mouth has not yet started to close, since the radula with its attached odontophore has not yet started to retract. **E.** Partial retraction and backward rotation of the radula. **F.** Full retraction of the radula. [From Kupfermann, 1974a.]

lowing movements appear to be triggered by receptors on the inner surface of the mouth (Kupfermann, 1974a).

The head movements of the orienting response of *Aplysia* are mediated by the pedal ganglia and can be eliminated by cutting the cerebropedal connectives (Kupfermann, 1974b). By contrast, cutting the cerebrobuccal connectives does not eliminate the orienting response but does eliminate the animal's ability to protract and retract the odontophore. The consummatory movements of the odontophore are therefore under the control of the cerebral and buccal ganglia (Kupfermann, 1974b).

How the orienting response is mediated is not known, but Kupfermann and his colleagues have developed some preliminary ideas about the mechanisms underlying the consummatory response. The antagonist protraction (opening) and retraction (closing) movements of the radula are mediated by identified cells in the buccal ganglion (Kupfermann and Cohen, 1971; Cohen, Weiss, and Kupfermann, 1974; see Figure 7-8, p. 230). These motor cells are not interconnected, but seem to be driven by a set of protractor and retractor command elements (Kupfermann and Cohen, 1971). Although the evidence is incomplete, the protractor command cells—the symmetrical pair of identified cholinergic dual-action cells (BL4 and BL5, BR4 and BR5) of the buccal ganglion—seem to make direct connections to motor neurons effecting protraction of the radula and inhibitory connections to motor neurons effecting retraction.

The program for the repetitive protraction–retraction sequence seems to be centrally determined (Rose, 1972), but can be modified by sensory feedback from stretch receptors in the buccal mass (Kupfermann and Cohen, 1971). Muscle movements controlled by the retractor cells excite stretch receptors in the buccal mass, and these in turn excite the protractor command cells and inhibit the retractor motor cells. Another set of dual-action cells, the symmetrical metacerebral cells (see Figure 7-7, p. 231), serve as higher-order modulatory command cells. In their amplifying actions they resemble the general excitors involved in the swimming of *Tritonia* and neurons mediating dishabituation of the gill-withdrawal reflex of *Aplysia* (Figure 12-28, p. 583, and Figure 12-30, p. 590).

The metacerebral cells act centrally, on command cells and motor neurons, and peripherally, on the muscle. Centrally, they inhibit the protractor command cells and the protractor motor neurons and excite the retractor motor cells and perhaps the unidentified retractor command cells (Kupfermann and Weiss, 1974; Figure 10-30). One would think, therefore, that the metacerebral cells should retract the odontophore, but they do not. They produce in the retractor motor cells only small EPSPs that generally are not effective in firing the cells unless the cells have also been depolarized by another input. Thus, centrally, the metacerebral cells serve only as gates, facilitating other excitatory inputs to the retractor motor cells (Figure 10-30).

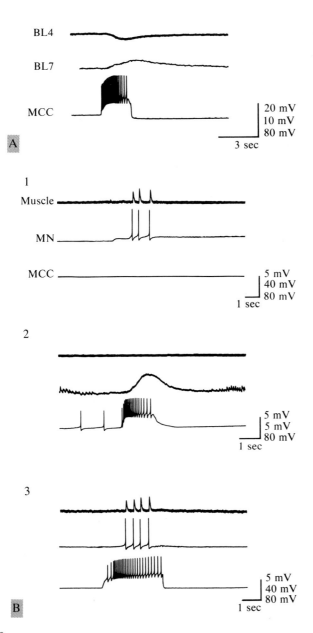

FIGURE 10-30.

A higher-order dual-action cell (the metacerebral cell) innervates another dual-action cell, command cell BL4. [From Weiss, Cohen, and Kupfermann, 1975.]

A. A train of action potentials in a metacerebral cell (MCC) produces inhibition in the protractor command cell (BL4) and excitation of a motor cell (BL7) to the strap muscle.

B. *B-1.* Action potentials in a motor neuron (MN) produce e.j.p.s in the strap muscle that closes the two radula halves. *B-2.* Firing the metacerebral cell produces a slow depolarization of the motor neuron that fails to fire it. *B-3.* If the motor neuron is depolarized close to its firing level, action potentials in the metacerebral cell will discharge action potentials in the motor neurons that produce e.j.p.'s in the muscle.

Peripherally, however, the metacerebral cells exert a powerful action. Although firing the metacerebral cells does not cause contraction of the muscles, the activity of these cells greatly enhances the contraction produced by the motor neurons (Weiss, Cohen, and Kupfermann, 1975). This enhancement is not primarily due to action on the nerve-muscle synapse, but to a direct facilitation of the contractile process of the muscles. The metacerebral cells send axons to both retractor and protractor muscles and may act as general excitors for all phases of the consummatory component of feeding. The metacerebral cells are serotonergic and, like the serotonergic heart-excitor cells, also produce a prolonged action that outlasts the stimulus by one or more minutes. It is inviting to think that here again an intracellular messenger like cAMP is involved. Figure 10-31 is the postulated neural circuit for the fixed-action pattern of feeding in *Aplysia*. Cells that have not been identified and are only postulated to exist are drawn in with grey lines.

An analysis of the neural circuitry controlling feeding in *Helisoma* has been initiated by Kater and co-workers (Kater and Fraser Rowell, 1973; Kater, 1974). As is characteristic of many gastropod molluscs, ingestion is achieved by the scraping action of the radula, which is moved forward and back by the protraction and retraction of the buccal mass. These movements are produced by seven bilateral pairs of protractor muscles and four pairs of retractor muscles (Figure 10-32) as well as by eight pairs of minor muscles. The protractor and retractor muscles are each innervated by distinct populations of identifiable motor neurons. There are five pairs of bilateral protractor motor neurons. These are electrically coupled to each other and fire in phase with each other and 180° out of phase in relation to eight bilateral retractor neurons, which are also electrically interconnected and fire in phase with each other (Figure 10-32). The antiphasing of the protractors and the retractors is accomplished

FIGURE 10-31.
Postulated neural circuit controlling feeding in *Aplysia*. Unidentified cells and their connections are drawn in light lines. [Based on data in Cohen and Kupfermann, 1971; Weiss, Cohen, and Kupfermann, 1975; and unpublished observations.]

A. The muscles that protract the radula and open its two halves and those that retract it and close the two halves are innervated by motor cells that are not interconnected. The motor cells are innervated by command cells. The two command cells for protraction are identified multiaction cells (B4 and B5). The retractor command cells are not yet identified, but are postulated to also be dual-action cells. The protractor and retractor command cells make mutually inhibitory connections. The command cells are innervated by one or more stretch receptors (SR) that excite the protractors and inhibit the retractors.

B. The dual-action protractor and retractor command cells are innervated by higher-order dual-action metacerebral cells. These excite the retractor command cells and motor cells and inhibit the protractor command cells and some of the protractor motor cells. The metacerebral cells also innervate the buccal musculature, causing a facilitation of motor neuron action, by means of direct enhancement of contraction.

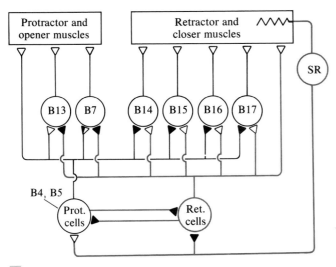

A

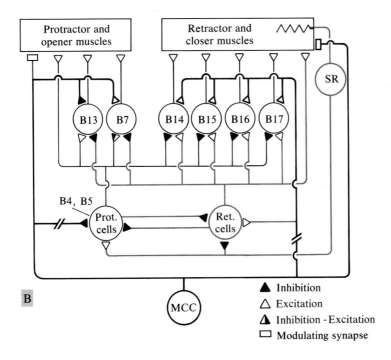

B4, B5

	Inhibition
	Excitation
	Inhibition - Excitation
	Modulating synapse

B

MCC

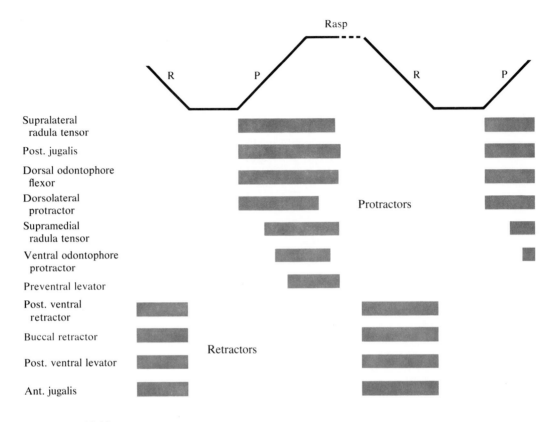

FIGURE 10-32.
Summary of the cyclical activity patterns of individual muscles of the buccal mass of *Helisoma trivolvis*. Movements are schematically represented as protraction (P) and retraction (R). The interval between protraction and retraction, during which the rasping action of the odontophore occurs, can be highly variable compared with the interval between retraction and protraction. The bars following the name of each muscle indicate periods of myogram activity. The period of the cycle depicted is about 3.5 seconds. [From Kater, 1974.]

by a population of command elements that excites the retractors and inhibits the protractors (Figure 10-33). Inhibition of the protractors is followed by rebound excitation, which drives the protractor muscles. The phase relationship is characteristic for each protractor motor neuron; some motor cells invariably rebound from inhibition before others (Figure 10-34). The cells of the command population are electrically coupled to each other and provide the central program for feeding.

How the command elements generate cyclical output is not known, but excitatory feedback between the electrically coupled command cells is thought to be involved. The ability to generate rhythmic output is therefore attributed

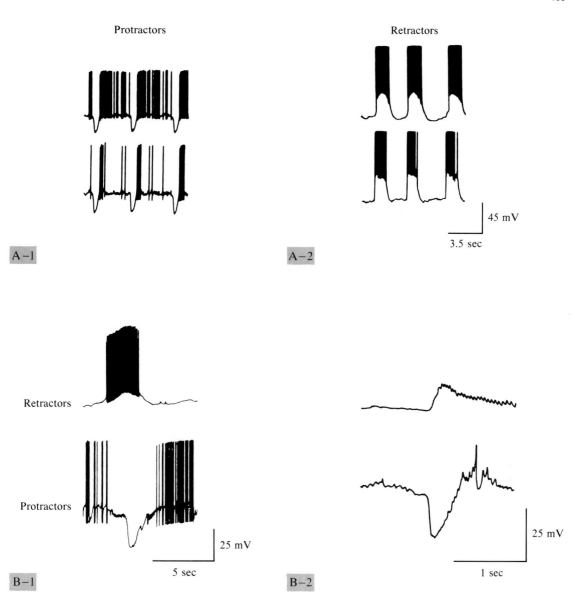

FIGURE 10-33.
Protractor and retractor motor cells involved in *Helisoma trivilvis*.

A. Cells within the same group tend to fire synchronously. [From Kater, 1974.]

B. *B-1*. Retractors and protractors fire 180° out of phase. *B-2*. Synaptic potentials of opposite sign. The retractor EPSP and the protractor IPSP (followed by a depolarization) are revealed by hyperpolarizing both cells so as to eliminate spike generation. [From Kater, 1974.]

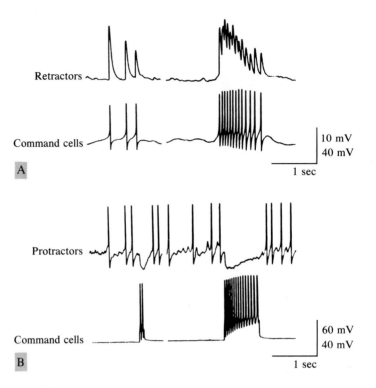

FIGURE 10-34.

Command inputs to retractor and protractor motor neurons involved in feeding in *Helisoma trivolvis*. [From Kater, 1974.]

A. Action potentials produced by intracellular injection of depolarizing current in a command cell result in EPSPs of short latency (10 msec) in a retractor motor neuron (held hyperpolarized to block action potentials).

B. Another command neuron produces IPSPs of short latency that summate to produce marked hyperpolarizations in protractor motor neurons.

to a property of a population of electrically interconnected cells and not to any single cell (Kater, 1974). This program is modulated by sensory feedback from stretch-receptor neurons in the buccal mass, feedback that regulates the firing frequency and duration of the motor neuron burst (Kater and Fraser Rowell, 1973).

The stretch receptors mediate the same sign of action to both the motor cells and the command elements, exciting the retractor motor neurons and inhibiting the protractor motor neurons. The receptors appear to be activated by contraction of the retractor muscles, and thus send a positive feedback to the retractor motor neurons causing them to fire until the buccal mass is fully retracted. The protractor motor neurons are excited by protractor command cells during the second phase of the cycle, and receive from the stretch receptors inhibitory feedback that persists until optimal retraction is assured, thereby relieving the stretch on the stretch receptors. A hypothetical model of how this sequence can be generated is illustrated in Figures 10-35 and 10-36. The proposed mechanism consists of dual-action command cells that are electrically coupled to each other. In addition, peripheral modulation of the central program occurs by means of feedback to the motor neurons from the dual-action stretch receptors. However, whether these opposite synaptic actions are mediated directly and by individual cells is not clear (Kater, 1974).

The feeding behavior of *Helisoma* and *Aplysia* is controlled by a complex central program, but again the principles can be reduced to simpler elements.

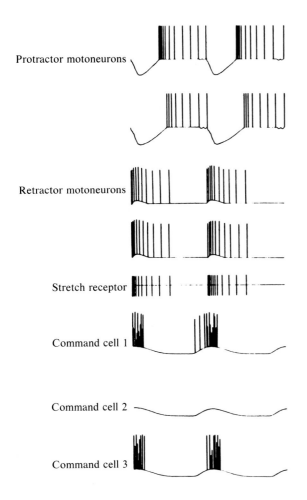

Protractor motoneurons

Retractor motoneurons

Stretch receptor

Command cell 1

Command cell 2

Command cell 3

FIGURE 10-35.
Postulated summary of firing patterns of neurons controlling feeding behavior of *Helisoma trivolvis*. Records show characteristic activity observed in each element during one cycle (about 3 sec) of feeding activity. Protractor motor neurons fire in antiphase to retractor motor neurons as a result of synaptic input from a population of command cells (of which three are shown). The protractor group consists of at least seven neurons in each symmetrical buccal ganglion, the retractor group at least 11 neurons per ganglion, and the command elements at least eight neurons per ganglion. [From Kater, 1974.]

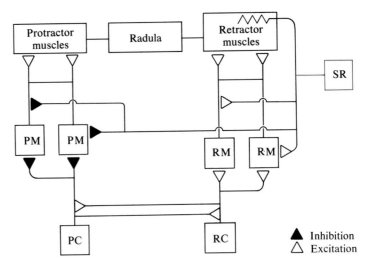

FIGURE 10-36.
Diagram of the postulated functional interconnections that give rise to feeding in *Helisoma* (PC = protractor command cells; RC = retractor command cells; SR = stretch-receptor cells, unidentified; PM, RM = protractor and retractor motor neurons). Boxes indicate populations of cells. [Modified from Kater and Fraser Rowell, 1973; Kater 1974.]

In *Aplysia,* and perhaps in *Helisoma,* dual-action command cells provide reciprocal motor controls. In *Helisoma* dual-action sensory cells also seem to provide for proprioceptive feedback. In *Helisoma* the central program may reside in the pattern of interconnection of a population of electrically coupled trigger cells. These cells resemble the bag cells in *Aplysia* that mediate egg-laying and the trigger cells in *Tritonia.*

In *Pleurobranchaea* the output to the feeding muscles is being studied by Davis and his colleagues (Davis, Siegler, and Mpitsos, 1973). The output consists of alternate bursts of impulses in antagonist motor neurons. Here the motor program appears to be even more complex because the cells generating this program are not located in one ganglion but distributed in several (the buccal, the cerebropleural, and perhaps others). How are these differently located central neurons coordinated? So far the results are preliminary, but Davis suggests that coordination is achieved by interneurons in the buccal ganglion. These interneurons coordinate the activity of cells in the several ganglia by conveying to them both an exact copy (an *efference copy*) and a rough copy (a *corollary discharge*) of the efferent pattern of motor activity (Davis *et al.,* 1973). The interneurons conveying the exact copy are thought to fire in perfect synchrony, with the burst of efferent impulses going to motor neurons in the same buccal nerve. Interneurons conveying the rough copy are thought to be active only during the same phase of the feeding cycle as the motor neurons in the buccal root. The interneurons that convey the efference copy, as well as those that convey the corollary discharge, send their axons to buccal and cerebropleural ganglia by means of the cerebrobuccal connectives. Cutting the connectives dissociates the rhythms in the buccal and cerebropleural ganglia. Figure 10-37 shows the postulated circuit.

A very different type of feeding behavior has been studied in the voracious, carnivorous cephalaspid opisthobranch *Navanax inermis* by M. V. L. Bennett and his colleagues Spira and Spray. Like other gastropods, *Navanax* shows an

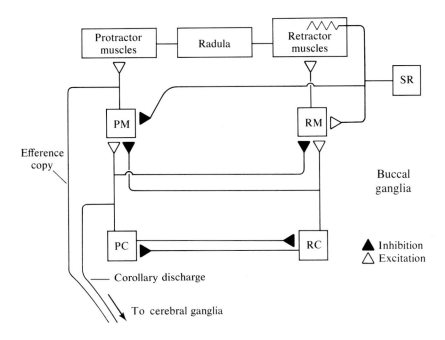

FIGURE 10-37.
Hypothetical circuit controlling feeding in *Pleurobranchaea.* The circuit illustrates the possible role of feedback copies of motor activity in coordinating the activities of the buccal and cerebral ganglia. Mutually inhibitory dual-action command cells, PC and RC, are postulated to drive protraction and retraction, respectively, by means of their reciprocal innervation of the motor cells. Branches of the motor cells are thought to produce an exact (efference) copy of the motor output from the buccal ganglia and convey it to the cerebral and pleural ganglia. Branches of the protractor command cells convey a rough copy (corollary discharge) of the motor output of the buccal ganglia and convey it to the cerebral and pleural ganglia (not shown). These feedback copies of the motor output of the buccal ganglia serve to coordinate the activity of the cerebral and pleural ganglia with that of the buccal ganglia. Boxes indicate postulated populations of cells. [Based on Davis *et al.,* 1973.]

appetitive response on chemical contact with food. *Navanax* stalks its prey by using two symmetrical head-shield organs to contact the trail. If one organ loses contact with the trail the animal corrects its course until it is again on center. If the trail is lost the animal will orient by swinging its head back and forth in search of food, much like *Aplysia*. Once the prey is contacted, however, the consummatory response of *Navanax* is remarkably different from most gastropods. *Navanax* protracts its pharynx, directing it at the prey, and then rapidly expands it (by synchronous contraction of its radial muscles and concomitant closure of the sphincter at the caudal margin of the pharynx), thereby sucking the prey in at great speed. Lacking teeth, radula, or jaws, *Navanax* swallows its prey whole and takes it directly, with a series of peristaltic movements, into its midgut (Paine, 1963; Murray, 1971).

The motor component of the consummatory response has been studied. Several motor cells in the buccal ganglion control the pharynx. The cells are electrically coupled (Levitan *et al.,* 1970) and fire synchronously to cause rapid expansion. They are triggered by a group of mechanoreceptor sensory neurons that innervate the lip and the anterior third of the pharyngeal wall (Spray and Bennett, 1975). As the ingested prey is pushed into the esophagus, the stimulus of pharyngeal inflation produces in the motor cells inhibitory postsynaptic potentials that increase the conductance of the nonsynaptic membrane and functionally uncouple the motor cells, allowing them to fire independently (Spira and Bennett, 1972). Preliminary experiments suggest that the uncoupling is produced by the activity of a second group of mechano-receptor sensory neurons (located in the buccal ganglion) that respond to tactile stimulation and to stretch of the pharynx and make inhibitory connections on the motor neurons (Spray and Bennett, 1975). Although independence of motor function following the synaptic uncoupling has not yet been demonstrated for the motor cells, in principle the independent activity of the motor cells could produce the wave of pharyngeal peristalsis that conveys the prey down to the midgut.

Behavioral Choice

Selection between antagonist responses involves some degree of behavioral choice (Davis, Mpitsos, and Pinneo, 1974a, b). For example, at any one time an animal expresses only one of two incompatible visceromotor responses: it either increases or decreases its heart rate. As we have seen, this can be accomplished by mutually inhibitory connections between command elements. But more interesting behavioral choices involve actual selection between possible alternatives: eating or copulating, fighting or fleeing. By examining the responses the animal makes to offered alternative stimuli, a heirarchy of behavioral responses can be developed.

Davis and his colleagues have begun to examine the hierarchy of behavior of *Pleurobranchaea* (Davis *et al.,* 1974a, b). According to Davis, egg-laying dominates feeding and feeding is thought to dominate all other behavior. When an animal is presented simultaneously with stimuli for feeding and fighting, the animal will feed while suppressing the neural circuit governing fighting. Davis suggests that the realtive dominance of feeding may be achieved by one of two ways: (1) the afferent chemosensory pathway that excites feeding may concomitantly inhibit other types of behavior; or (2) the command interneurons that control feeding may inhibit other behaviors. Although there is as yet no evidence to support either hypothesis, Davis raises the attractive possibility that dual-action command cells are organized hierarchically and that the interaction between higher-order command elements could determine behavioral choice.

SUMMARY AND PERSPECTIVE

As we saw in Chapter 9, electrically coupled motor cells generate part of the stereotypic central programs for the fixed acts of inking and egg-laying. In more complex neural circuits, such as those mediating the respiratory and cardiovascular adjustments in *Aplysia,* motor cells form part of a final common pathway for several independent reflex and fixed-action patterns. In these instances the motor cells are not interconnected and the program is located central to the motor neurons.

Respiratory and circulatory adjustments involve cyclic actions within a single effector system as well as the integration of actions by different effector systems. Coordination is achieved by command elements—individual nerve cells or synchronously active groups of cells—that determine motor pattern and sequence. One command cell, L10, has been studied in detail (see Chapter 8). This cell is a dual-action cell that mediates the central program of a simple fixed-action pattern—a phasic increase in cardiac output—by means of its inhibitory and excitatory synaptic actions on different circulatory motor cells. The autoactive burst capability of this single cell constitutes the central program for this fixed-action pattern, which is triggered whenever the cell fires a high-frequency burst of spikes. Although the burst capability of L10 can be modulated by inhibitory synaptic input (and by one or more hormones), it is an intrinsic membrane property and is not dependent on synaptic connections with other central neurons or the periphery.

An unidentified command element, Interneuron II, common to both the respiratory and cardiovascular systems, mediates the command for respiratory pumping and inhibition of heart rate. The presumed burst capability of this element is thought to constitute the central program for integrating the fixed-action pattern.

The bursts of L10 and Interneuron II vary in the details of their patterns in characteristic ways. This variability seems to account for the stepwise variability in the amplitude (intensity) of the fixed-action patterns. By contrast, the form of the fixed-action patterns—the sequence and relationships of the behavioral components—is relatively consistent. Constancy of form results from the invariant connections the command cells make with follower motor neurons.

Because the motor cells for cardiovascular and respiratory control are independent and not interconnected, different sensory inputs and command cells can play upon different combinations of motor cells at various times to achieve different behavioral purposes. For example, coordination of antagonist behavioral components is achieved by mutually inhibitory interactions between command neurons. In addition, when the command cells are spontaneously active, reciprocally inhibitory connections can give rise to an alternating sequence of activity—a feature that distinguishes fixed acts from the more complex fixed-action patterns. Mutually inhibitory connections may also provide a neural circuit for a simple form of behavioral choice.

Similar principles of organization are found in both elementary and complex behavior. It is therefore attractive to consider the possibility that complex behavioral responses can be generated by simple extensions or combinations of the mechanisms encountered in the neural circuits of elementary behavior. Indeed, the neural circuits of the two fixed-action patterns that have so far been examined—escape-swimming and feeding—contain the same basic circuit modules, or combinations of components, encountered in simpler behavior. But these fixed-action patterns have additional features: feedback to the central nervous system from peripheral receptors and, perhaps, feedback in the form of exact or rough copies of the pattern of motor activity. These features could be combined with the basic circuit modules to produce behavioral processes of an even higher order, such as complex behavioral choice.

In this and the preceding chapter I have overstepped the data in an attempt to draw generalizations from a few examples and to illustrate the types of hypotheses that can now be seriously entertained. But detailed analyses of complex reflex and fixed-action patterns need to be undertaken to test these hypotheses. In addition, more extensive studies of the transmitter biochemistry of neurons mediating behavior are needed. Preliminary findings on cardiovascular regulation and feeding have already demonstrated the importance of biochemical studies for the understanding of behavior. These studies have shown that the biochemical and physiological features of neural control of heart rate are similar in widely different phyla. In *Aplysia,* as in vertebrates, the inhibitory motor cells to the heart have a relatively brief synaptic action and use acetylcholine as their transmitter. At the same time, the excitor cells have a prolonged action and use a monoamine: serotinin in *Aplysia* and nor-epinephrine in vertebrates. This similarity across phyla suggests that trans-mitters are likely to be specific to types of behavioral responses. This idea receives further support from studies of feeding. Here, as in circulation, sero-tonin produces long-lasting actions. How this prolonged action is mediated is not known. But there is the possibility (see p. 341) that in addition to mediating neural information transmitters may regulate biochemical processes in the target cell—perhaps via an intracellular messenger such as cyclic AMP—and in this way prolong a synaptic action.

SELECTED READING

Kennedy, D., and W. J. Davis. 1976. The organization of invertebrate motor systems. In "Cellular Biology of Neurons" (Vol. 1. Sect. 1, *Handbook of Physiology*). Baltimore: Williams and Wilkins. A broad overview of cellular studies of invertebrate motor systems, emphasizing general principles.

Krasne, F. B. 1973. Learning in crustacea. In *Invertebrate Learning,* Vol. 2, W. C. Corning, J. A. Dyal, and A. O. D. Willows, eds. New York: Plenum Press. A good discussion of crayfish behavior and its modifiability, empha-sizing cellular analyses.

Maynard, D. M. 1972. Simpler networks. *Annals of the New York Academy of Science* 193:59–72. The pioneering study of the stomatogastric ganglion of the lobster.

Mulloney, B., and A. I. Selverston. 1974. Organization of the stomatogastric ganglion of the spiny lobster, III: Coordination of the two subsets of the gastric system. *Journal of Comparative Physiology,* 91:53–78. A detailed analysis of a fixed-action pattern.

Willows, A. O. D. 1973. Learning in gastropod mollusks. In *Invertebrate Learn-ing,* Vol. 2, W. C. Corning, J. A. Dyal, and A. O. D. Willows, eds. New York: Plenum Press. A recent review of gastropod behavior.

Wilson, D. M. 1968. The flight-control system of the locust. *Scientific American,* 218(5):83–90. A general account of the classic studies of the central control of insect flight.

CHAPTER 11

Neuronal Plasticity:
Possible Mechanisms for
Behavioral Modification

Behavior can be generated by the coordinated activity of invariant cells that are connected in precise ways. Yet behavior produced by such precisely "wired" systems can be modified by learning. How does this come about? How can behavior be modified when the cells and their connections are invariant? In this chapter I shall first sketch two alternative explanations for learning and memory. I shall then review findings in cellular physiology that indicate that even structurally invariant neurons and synapses are capable of functional changes. These *plastic changes* result in prolonged alterations of cellular and synaptic function that, in principle, could account for simple behavioral modifications. The cellular principles described in this chapter are tested in the next chapter with examples of actual behavioral modifications.

NEUROPHYSIOLOGICAL HYPOTHESES
OF BEHAVIORAL MODIFICATION

Several simple behavioral responses in invertebrates can now be largely explained in terms of the biophysical properties of identified cells and their invariant interconnection with each other and with effector organs. The same principles of neuronal organization in invertebrates seem to apply to vertebrates. For example, connections in the visual systems of both vertebrates and invertebrates seem to be formed in specific ways: a given set of cells invariably connects only to certain other cells (Sperry, 1951; 1963; 1965b; Lettvin, *et al.,* 1959; Hubel and Wiesel, 1962, 1963; Attardi and Sperry, 1963; Yoon, 1972).

These studies present us with two interesting and interrelated paradoxes: the development of connections between most neurons in the nervous system is determined by genetic and developmental processes, and yet behavior can be modified. How does a predetermined nervous system allow for the obvious changeability of behavior? Also, since learning and other behavioral modifications are characteristically of a long duration, lasting for minutes, days, and even years, how can a pattern of neural activity produce a long-term modification in the functioning of a set of precisely wired connections? Does learning require the addition of further nerve cells or connections to the neural circuitry?

A number of solutions for these dilemmas have been proposed. The two that have proven most interesting experimentally are the theories of plastic and dynamic change in neurons.

Plastic Change Hypothesis

The plastic change hypothesis states that learning involves a functional or plastic change in the properties of neurons or in their interconnections. This hypothesis was advanced at the turn of the century and derives primarily from the morphological studies of Ramón y Cajal, the major early proponent of cellular connectionism and one of the first to appreciate that the nervous system is constructed of discrete cellular units (Cajal, 1911). The first suggestion that learning is due to a change in synaptic function was made by Lugaro (1899), who proposed that learning resulted from a morphological change in the connections between neurons similar to that which occurs during formation of synapses in embryonic life.

Several years later Tanzi (1893) formulated a more specific hypothesis for learning based on a synaptic change due to usage:

> A nervous impulse which passes very frequently through a junction between neurons would cause an increased flow of nutritive substances and lead to hypertrophy just as occurs in exercised muscle. In the nervous system hypertrophy would manifest itself by a lengthening of the cellular ramifications and a reduction of the distance between the connections. The conductivity of the nervous tracks would be greater because resistance to the current is directly related to the distance between connections. Therefore (mental) exercise that reduces the gap between junctions can augment the effectiveness of neurons.[1]

The ideas of Lugaro and of Tanzi were combined and extended by Cajal in his Croonian Lecture to the Royal Society of London (1894) and again in the section on higher-order mental function that concludes his great work on neuroanatomy, *Histologie du Système Nerveux des Hommes et des Vertébrés:*

[1]Quoted in Cajal, 1911, p. 886.

The extension, the growth, and the multiplication of the appendages of the neuron do not stop at birth; they continue and nothing is more striking than the difference between the length and the number of the cellular ramifications, of the second and third order, in a newborn and an adult man.

The new cellular extensions do not grow at random; they have to orient themselves according to major neural currents or according to intracellular associations which are the object of repeated action of will. We think that formation of those new branches is followed by an increased blood flow which brings necessary nutrition. The mechanisms are probably chemo-tactile like the ones we observed during histogenesis of the spinal cord.

The ability of the neurons to grow in an adult and their power to create new associations can explain learning and the fact that man can change his ideological systems. Our hypothesis can even explain the conservation of very old memories such as memories from youth in an old man and in an amnesiac or in a mental patient, because the association pathways that have existed for a long time and have been exercised for many years are probably very powerful and were formed at the time when the plasticity of the neuron was at its maximum[2].

Modern forms of the plastic change hypothesis have been put forward by two influential theorists, the psychologists Jerzy Konorski (1948) and Donald Hebb (1949). A current version of this hypothesis states that even though neurons and their anatomical interconnections develop according to a fixed plan, the strength or effectiveness of certain anatomical connections between neurons is not entirely predetermined (see for example Eccles, 1953; Kandel, 1970). Experience can alter the functioning of neurons and the effectiveness of certain synapses. This hypothesis predicts that neurons, and in particular some classes of synapses, are capable of changing their properties as a result of altered activity.

Hebb emphasized the distinction between short-term memory, lasting minutes and hours, and long-term memory, lasting days, and suggested that plastic changes at synapses account for long-term memory. Konorski advanced a similar hypothesis to explain all memory:

> The application of a stimulus . . . leads to changes of a two-fold kind in the nervous system. . . . The first property, by virtue of which the nerve cells *react* to the incoming impulses with a certain cycle of changes, we call *excitability*, and the changes arising in the centres because of this property we shall call *changes due to excitability*. The second property, by virtue of which certain permanent functional transformations arise in particular systems of neurons as the result of appropriate stimuli or their combination we shall call *plasticity* and the corresponding changes *plastic changes*.[3]

[2]Cajal, 1911 (1955 edition), pp. 888–90.
[3]Konorski, 1948, pp. 79–80.

Konorski assumed that there is some upper limit for the number of synaptic contacts that one neuron can make on another. In some circuits this limit is reached before birth; these circuits mediate fixed acts and fixed-action patterns. In plastic circuits the maximum density is not reached until after birth; in these systems learning improves the effectiveness of synaptic transmission by means of an outgrowth of synaptic contacts.

Dynamic Change Hypothesis

The dynamic change hypothesis states that learning involves a persistence of activity in chains of interconnected interneurons. This idea was put forward by the physiologist Alexander Forbes (1922), who based it on the finding that in some spinal reflexes (such as the scratch reflex) there is a persistence of activity (an after-discharge) following stimulation. The idea of reverberating circuits was further elaborated by Lorente de Nó (1938) who found that cortical neurons are often interconnected in the form of closed chains. Neural activity could therefore be sustained by the circulation or reverberation of impulses within a closed chain of interconnected self-reexciting neurons (Figure 11-1). Such a memory mechanism does not require an anatomical change or even a functional change in the properties of neurons.

Psychologists found the idea of persistent neural activity very attractive, and Hilgard and Marquis (1940) suggested that a reverberating mechanism accounts for short-term memory. Hebb (1949) elaborated this notion into the

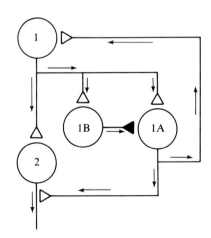

FIGURE 11-1.
A hypothesis of memory based on dynamic recirculation of activity in closed neuronal circuits. According to this hypothesis, prolonged neural activity can be maintained by the circulation of impulses in a closed neural chain. Cell 1 excites cells 2, 1B, and 1A. Cell 1A reexcites cell 1, thereby maintaining the circulation of activity in the chain. However, many neural circuits with this structure have proved to contain cells (such as 1B here) that inhibit other cells in the circuit (such as 1A) and thus prevent rather than facilitate reexcitation.

dual-trace hypothesis of memory. In Hebb's view a short-term process due to reverberatory neuronal activity leads, after a time, to the permanent neuronal changes that underlie long-term memory.

> Let us assume then that the persistence or repetition of a reverberatory activity (or "trace") tends to induce lasting cellular changes that add to its stability. The assumption can be precisely stated as follows: When an axon of cell A is near enough to excite cell B and repeatedly or persistently takes part in firing it, some growth process or metabolic change takes place in one or both cells so that A's efficiency as one of the cells firing B, is increased.[4]

The idea of reverberatory circuits also appealed to physiologists because there were many examples in the drawings of neural circuitry by Cajal (1911) and Lorente de Nó (1934) where neurons connected to each other in circular paths. Cajal, and in his earlier work Lorente de Nó, thought that all these connections were excitatory. But as studies on the physiological functions of neural networks advanced, it became clear that neurons can mediate inhibition as well as excitation (see Chapter 6). What appeared anatomically to be a self-reexciting loop often contained an inhibitory connection that could prevent reexcitation (Fig. 11-1). In addition, most memories survive convulsive seizures and other drastic events likely to disrupt neuronal activity. As a result of these findings, it seems less likely that dynamic activity might be a common neural basis of even short-term memory, although it could serve that role in special cases. On the other hand, studies of the plastic capabilities of neurons have turned out to be surprisingly rewarding. A variety of experiments have shown that built into the functioning of many and perhaps all chemically transmitting synapses is a remarkable capability for short- and long-term functional modification.

In this chapter I shall consider the evidence for neuronal plasticity independent of its relationship to behavior. I do this by examining three questions. Are all synapses capable of plastic changes? Are other components of the neuron also modifiable? And finally, what patterns of stimulation are most effective in inducing plastic modifications? In the next chapter I shall consider specifically the role of neuronal plasticity in behavioral modification.

The plastic capabilities of neurons have been studied in two ways: by examining the effects of usage produced in a neural pathway by simple repeated stimulation, and by examining the effects of more complex stimulus sequences based on those used in studies of habituation, sensitization, classical conditioning, and instrumental conditioning.

[4]Hebb, 1949, p. 62.

EFFECTS OF USAGE AT VERTEBRATE SYNAPSES

The plasticity of a central synapse following usage was first demonstrated in 1947 by Larrabee and Bronk while studying the stellate ganglion of the cat. This ganglion consists of a simple pathway: the preganglionic axons connect directly to the postganglionic neurons through a single synapse (Figure 11-2). One can therefore study, indirectly, changes in synaptic transmission without recourse to intracellular recording simply by recording extracellularly the summed activity of the axons of the postganglionic cell. Changes in the summed activity indicate a change in the number of cells that are discharged by a stimulus applied to the presynaptic axons. Larrabee and Bronk found that after a brief period of repetitive stimulation of the presynaptic fibers (15 / sec for 10 sec), the number of postsynaptic cells discharged by the stimulated monosynaptic pathway was greatly enhanced for about three minutes, while neighboring, unstimulated pathways were unaffected (Figure 11-2B). Because it persisted *after* the period of stimulation, or tetanization, this phenomenon was called *posttetanic facilitation* or potentiation. Two years later, in 1949, Lloyd described a similar facilitation in the monosynaptic reflex of the cat spinal cord (Figure 11-3), thus showing that plastic changes can occur in central neurons (Lloyd, 1949).

A few minutes is clearly not a very long time in view of the duration of most learning, but it is long compared with the millisecond events that had characterized the time courses of previously known neural actions. Indeed, some behavioral modifications are relatively short lived and a plastic mechanism such as posttetanic facilitation might underlie them. Following Lloyd's work, Spencer and Wigdor (1965) and Beswick and Conroy (1965) found that the duration of posttetanic facilitation could be extended by using longer periods of stimulation. After tetanization of 15 to 30 minutes, posttetanic facilitation can last up to two hours (Figure 11-4).

Because monosynaptic systems are anatomically simple and can be thoroughly studied with cellular techniques, some understanding has been gained of the mechanism of elementary plastic changes in these systems. Larrabee and Bronk (1947) found that the generation of action potentials produced in the postsynaptic cells by posttetanic facilitation did not alter the excitability of the cells. Thus, a similar number of action potentials produced in the postsynaptic cell by antidromic stimulation of its efferent axon did not affect the response of the postsynaptic neuron to synaptic excitation. Moreover, post-

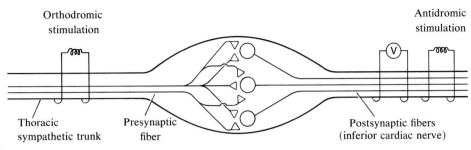

A

1. Orthodromic Stimulation

2. Antidromic Stimulation

B

FIGURE 11-2.
Posttetanic facilitation of activity in an autonomic ganglion: the stellate ganglion of the cat. [From Larrabee and Bronk, 1947.]

A. Schematic diagram of stellate ganglion illustrating monosynaptic connections between the presynaptic (preganglionic) axons in the thoracic sympathetic trunk and the postsynaptic (postganglionic) cells in the inferior cardiac nerve. Compound (summated) action potentials are recorded extracellularly from the axons of the postsynaptic cells following either orthodromic stimulation of the axons of the presynaptic cells or antidromic stimulation of the axons of the postsynaptic cells (stimulus rate is 15/sec for 10 sec).

B. Effects of orthodromic and antidromic stimulation. Orthodromic stimulation produces a facilitation of the response to stimulation of the presynaptic axons; the facilitation persists for about 120 sec. The facilitated amplitude (recorded from the postsynaptic axons) could be due to synchronous firing of a larger number of axons than had been firing prior to the tetanus; alternatively, it could be due to an increase in the spike height in the axons that had been firing previously. In order to decide between these possibilities, the postsynaptic axons were antidromically tetanized and the effects of this procedure on the amplitude of the postsynaptic response was noted. Antidromic tetanization produced no facilitation. This indicates that the facilitation produced by orthodromic stimulation is due to an increase in the efficacy of the preganglion–postganglion synapse, leading to the firing of a larger number of ganglion cells and thereby increasing the amplitude of the summated action potential recorded in the postganglionic nerve.

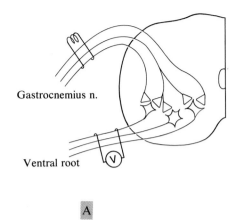

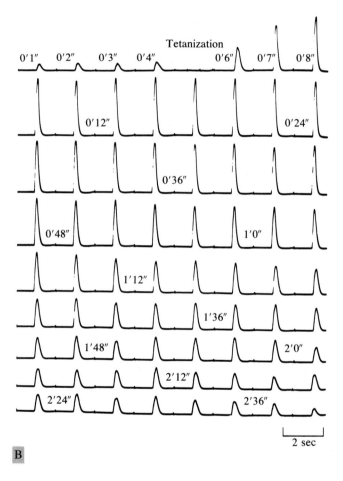

FIGURE 11-3.
Posttetanic facilitation of the monosynaptic spinal reflex in the cat. [From Lloyd, 1949.]

A. Experimental arrangement used by Lloyd to study the monosynaptic reflex discharge in the spinal cord. The ventral root contains the axons of the gastrocnemius motor neurons. The reflex is elicited by stimulating the gastrocnemius nerve that carries the afferent axons (1A fibers) from stretch receptors in the gastrocnemius muscle. As in Figure 11-2, the compound action potential recorded extracellularly on the ventral root represents the summed discharge of many axons.

B. A stimulus is applied every 2 sec. Numbers on the top line indicate consecutive responses; numbers in subsequent lines indicate time after tetanization. Four consecutive control responses of the compound action potential are recorded in the ventral root (0'1" to 0'4"). After a 12-sec tetanization at 555/sec, the reflex response is facilitated, as indicated by the increase in the compound action potential, which signifies an increase in the number of cells discharged by the reflex volley. The facilitation persisted for slightly over 2 min.

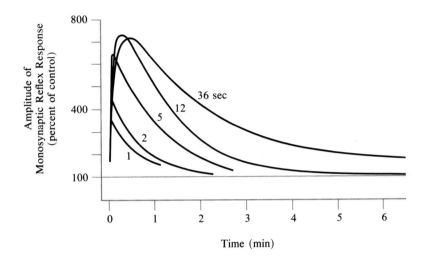

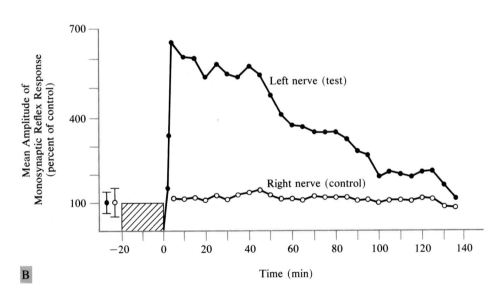

FIGURE 11-4.

Amplitude and time course of posttetanic facilitation in the monosynaptic spinal reflex as a function of duration of the tetanization.

A. Posttetanic facilitation follows tetanization of the gastrocnemius nerve at 500/sec for 1, 2, 5, 12, and 36 sec. [From Lloyd, 1949.]

B. Posttetanic facilitation following tetanization of the left tibial nerve at 500/sec for 20 min. Tetanization was not delivered to the control (right tibial) nerve. Following tetanization, test stimuli were applied to both tibial nerves at 10-sec intervals to test for potentiation. Each plotted point is an average of 10 to 20 individual responses. The hatched bar indicates the period of tetanization. Zero time represents the end of the tetanization period. Recordings were obtained from the ventral root of lumbar segment L7. [From Spencer and April, 1970.]

tetanic facilitation produced by stimulating one afferent pathway did not increase the response of the cell to synaptic activation via another afferent pathway. These several experiments suggested that posttetanic facilitation is *homosynaptic,* that it is restricted to the stimulated synaptic pathway and results from a change in the synapse itself. This suggestion was confirmed in 1952 by Brock, Coombs, and Eccles, who obtained intracellular recordings from motor neurons in the cat and found that posttetanic facilitation of the monosynaptic spinal reflex pathway involves a change in the amplitude of the excitatory postsynaptic potential (Figure 11-5).

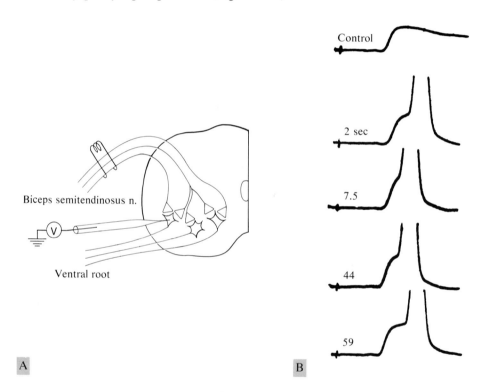

FIGURE 11-5.

Intracellular study of posttetanic facilitation in the monosynaptic reflex. [From Eccles, 1953.]

A. Experimental arrangement for recording intracellularly from a biceps semitendinosus motor neuron. The responses recorded in the neurons are generated by single volleys in the biceps semitendinosus dorsal root, which carry the afferent fibers from stretch-receptor neurons.

B. Control response before and selected responses following tetanization (450/sec for 15 sec). The EPSP was first depressed for about 1 sec and then facilitated for the next 1.5 min. (post-stimulation times are given with each record). The facilitated EPSP was large enough to trigger an action potential.

Increased effectiveness of synaptic transmission could be caused either by a presynaptic mechanism—an increase in the number of transmitter quanta released by the presynaptic terminals per unit impulse—or by a postsynaptic mechanism—an increased responsiveness of the postsynaptic receptor to a constant quantal output, or an increase in input resistance, or a decrease in threshold. Larrabee and Bronk (1947) had already done one experiment that made it seem unlikely that posttetanic facilitation was due to a change in receptor sensitivity. They found that if acetylcholine, the excitatory transmitter in the ganglion, was added to the bathing solution, the effects of tetanic stimulation were depressed rather than enhanced. In addition, Lloyd (1949) found that posttetanic facilitation was temporally correlated with a hyperpolarizing afterpotential in the presynaptic fibers, suggesting a presynaptic mechanism.

Posttetanic facilitation also occurs at the synapse between motor neurons and skeletal muscle. By examining these synaptic potentials (the end-plate potentials) using quantal analysis, Liley (1956a) was able to analyze the mechanisms of posttetanic facilitation in the rat nerve-muscle synapse. He found that the amplitude of the spontaneously released miniature end-plate potentials did not change during posttetanic facilitation, indicating that the sensitivity of the ACh receptors was not altered. Liley found that posttetanic facilitation involves an increase in the number of quanta released from the presynaptic terminal (see also Dudel and Kuffler, 1961; Kuno, 1964b; Martin and Pilar, 1964; Bittner and Kennedy, 1970; Zucker, 1973). There is also an increase in the frequency of the spontaneously released miniature end-plate potentials (Figure 11-6). Rosenthal (1969) found that posttetanic facilitation at the frog nerve-muscle synapse can be quantitatively accounted for by an increase in the probability of transmitter release and not by an increase in the available pool of transmitter (Equation 6-16, p. 189). She also found that posttetanic facilitation is reduced by lowering the external Ca^{++} concentration. Further support for a possible role for Ca^{++} was provided by Weinreich (1971). He examined posttetanic facilitation using tetrodotoxin to selectively block the regenerative Na^+ channel and thereby abolish the action potential in the presynaptic terminals. Using extracellular electrodes, he depolarized the nerve terminals electrotonically to produce transmitter release (see Figure 6-14; p. 180) and found that repeated pulses produced posttetanic facilitation comparable in amplitude to that produced by action potentials in control experiments without TTX (Figure 11-7). Posttetanic facilitation was also present in Na^+-free solutions. Both experiments exclude a role for Na^+ ions in facilitation. Moreover, the facilitation could be enhanced by factors that enhanced

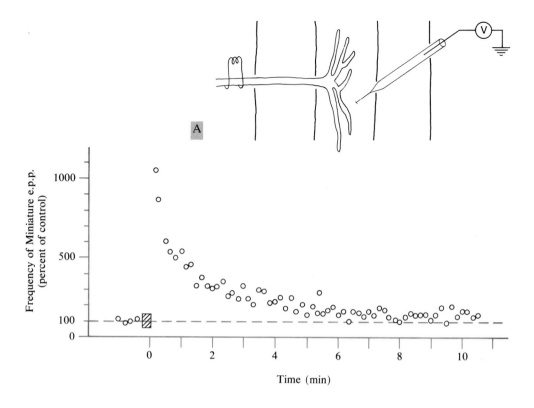

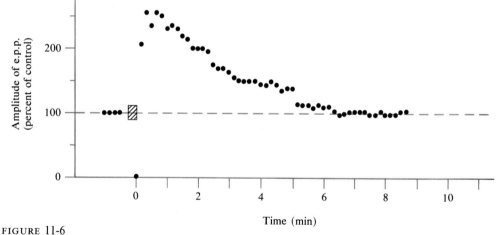

FIGURE 11-6

Parallel changes in the frequency of spontaneously occurring miniature end-plate potentials and the amplitude of the end-plate potential following tetanization. [From Liley, 1956a.]

A. Experimental setup for recording intracellularly from single muscle fibers at the end-plate region where the motor nerve synapses on the muscle fiber.

B. Frequency of miniature end-plate potentials (upper graph) and amplitude of the end-plate potential (lower graph) as percentages of control. The preparation was bathed in 10^{-6} d-tubocurarine chloride so as to reduce the amplitude of the end-plate potential. The hatched bar represents tetanization (200/sec for 15 sec); zero time represents the end of tetanization. The time course of the posttetanic facilitation of the end-plate potential is roughly paralleled by the increase of the frequency of the spontaneously released miniature end-plate potentials. The average amplitude of the miniature end-plate potentials also changed. This indicates that facilitation is likely to be due to increased release of transmitter rather than increased sensitivity to transmitter.

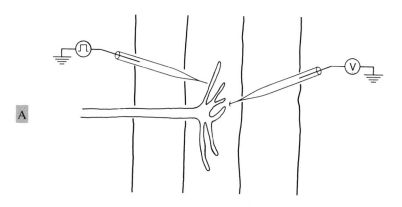

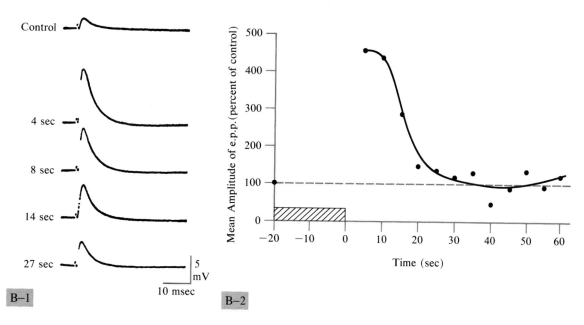

FIGURE 11-7.
Posttetanic facilitation in a nerve-muscle synapse treated with tetrodotoxin to abolish the inward Na⁺ current in the nerve terminal. [From Weinreich, 1971.]

A. Experimental setup. The end-plate potentials are recorded intracellularly from the end-plate region of a muscle fiber. The presynaptic nerve terminals of the motor axons are depolarized electrotonically by applying extracellular current pulses with a NaCl-filled microelectrode.

B. End-plate responses recorded from a single synapse before and after tetanization of the motor nerve terminal. Pulses were 1.0 msec in duration and 3.5 μA in intensity. *B-1.* Sample records of end-plate potentials before (control) and at several intervals following tetanization (50/sec for 20 sec). *B-2.* Time course of potentiation. Each point represents the mean amplitude of five posttetanic end-plate potentials; time given for each mean is that of the fifth response. The control amplitude is shown by the horizontal line and was computed by averaging about 50 pretetanic responses. Zero time indicates the end of tetanization.

Ca^{++} movement into the nerve terminals, such as an increase in the number of tetanic stimuli, increasing external Ca^{++} concentration, or increasing the intensity of depolarizing pulses. These experiments were at first explained on the basis of the residual Ca^{++} hypothesis. According to this hypothesis, each impulse leaves the terminals with a slight increase in free Ca^{++}. The facilitation is related to this summed residual Ca^{++} (Rosenthal, 1969). However, although Ca^{++} is likely to be important to posttetanic facilitation, this hypothesis does not seem to explain the quantitative features of posttetanic facilitation, and other, as yet unknown, processes must be operative (Younkin, 1974).

The finding that repeated stimulation readily produces posttetanic facilitation at different central and peripheral synapses at first suggested that there was a direct relationship between use (frequency of stimulation) and synaptic efficacy (Eccles, 1953; 1964). This idea, at least in its simple and general form, has not been supported. That the relationship between synaptic usage and efficacy is complex has been revealed by the discovery of posttetanic and low-frequency depression.

Posttetanic depression, a phenomenon opposite to posttetanic facilitation, was first encountered by Evarts and Hughes (1957) and Hughes, Evarts, and Marshall (1956) while studying the synaptic connections made by the optic nerve fibers and the principal cells of the lateral geniculate nucleus in the cat. Repetitive stimulation (500/sec) of the optic nerve fibers led to a decrease in postsynaptic efficacy lasting several hours (Figure 11-8). This synaptic depression does not develop following either very brief or very prolonged stimulation, but is most marked following tetani lasting 0.5 min. With longer periods of stimulation the posttetanic depression is replaced by posttetanic facilitation. In the monkey the depression is even more profound and dominates the facilitation, so that tetanization lasting 6 to 20 hours produces depression that lasts for days (Fentress and Doty, 1971).

One need not stimulate at these unusually high rates, however, to produce depression in synaptic transmission. The monosynaptic reflexes—which undergo posttetanic facilitation when the fibers are stimulated at high rates—show a depression, called *low-frequency depression,* at rates as low as once every 10 seconds (Figure 11-9*A*). Recovery is fairly rapid, however, generally requiring only 10 to 20 seconds (Lloyd and Wilson, 1957). Quantal analyses of low-frequency depression at this synapse and at the frog nerve–muscle synapse indicate that quantal output is reduced but quantal size (receptor sensitivity) is unaltered (Kuno, 1964a; del Castillo and Katz, 1954b).

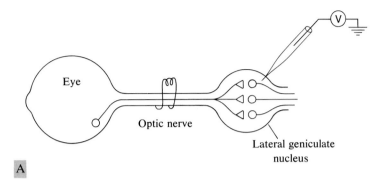

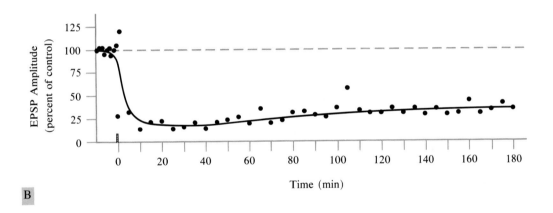

FIGURE 11-8.

Prolonged posttetanic depression following tetanization of the contralateral optic nerve in the cat.

A. Experimental setup in a cat (anesthetized with nembutal) for extracellular recording of the postsynaptic response of the axons of neurons whose cell bodies lie in the lateral geniculate nucleus. Stimuli are applied to the presynaptic axons in the optic nerve. The presynaptic axons belong to ganglion cells of the retina.

B. Graph of the relative amplitudes of the postsynaptic responses following tetanization (500/sec for 20 sec); zero time represents the end of tetanization. Subnormality was still present three hours following tetanization. [Modified from Evarts and Hughes, 1957.]

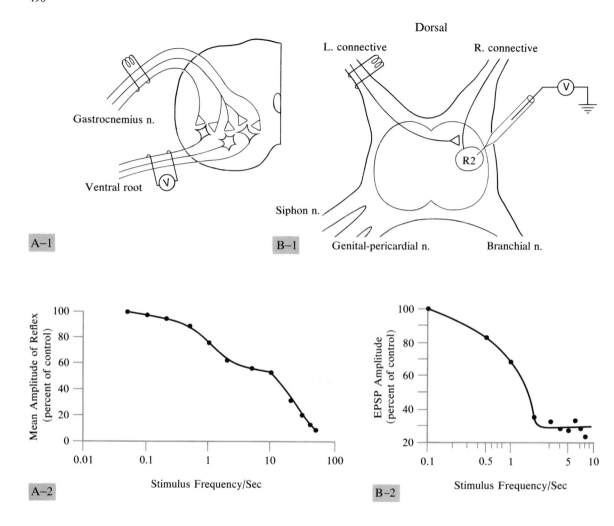

FIGURE 11-9.

Low-frequency depression.

A. Monosynaptic reflex of the cat. *A-1*. Experimental setup. Dorsal (sensory) roots are stimulated to activate the afferent (1A) fibers from the stretch receptors of the gastrocnemius muscle. The compound action potential produced by the motor neurons is recorded extracellularly from the ventral root. *A-2*. Relation between the mean amplitude of the monosynaptic responses and frequency of stimulation in the range of 0.05 per sec (3 per min) to 50 per sec. Within this range of frequencies, raising the frequency increases the depression of the synapse. [From Lloyd and Wilson, 1957.]

B. Monosynaptic EPSP in *Aplysia*. *B-1*: Experimental setup. The left connective is stimulated extracellularly, while monosynaptic responses are recorded intracellularly from cell R2. *B-2*: Relation between the mean amplitude of the monosynaptic EPSPs and frequency of stimulation in the range of 0.1 per sec (6 per min) to 10 per sec. Within this range of frequencies, raising the frequency increases the depression of the synapse. [Modified from Kandel and Tauc, 1965b.]

EFFECTS OF USAGE AT INVERTEBRATE SYNAPSES

Synaptic plasticity is found not only in the vertebrate brain; it is also well developed in the nervous systems of invertebrates. Here one can examine changes in the effectiveness of central synapses following the stimulation of a single cholinergic fiber. For example, in the abdominal ganglion of *Aplysia* a single axon in the right connective makes a monosynaptic excitatory connection with cell R15 (Gerschenfeld, Ascher, and Tauc, 1967; Schwartz, Castellucci, and Kandel, 1971; Parnas and Strumwasser, 1974). Repetitive stimulation (5/sec for 30 sec) of this axon produces a large increase in the amplitude of the monosynaptic EPSP that lasts for many minutes (Figure 11-10). As in vertebrates, some synaptic pathways in *Aplysia* show low-frequency depression when stimulated at low rates. For example, stimulation every 10 seconds of an axon in the right connective that makes a monosynaptic connection with cell R2 leads to synaptic depression (Figure 11-9*B;* and see Kandel and Tauc, 1965a, b; Bruner and Tauc, 1966). The depression resembles the low-frequency depression described for the monosynaptic spinal reflex of the cat (Figure 11-9*A*). As in the monosynaptic reflex, high-frequency stimulation of the axon that synapses on R2 produces facilitation.

Plastic changes are not limited to excitatory synapses; posttetanic facilitation also occurs at inhibitory synapses (Dudel and Kuffler, 1961; Waziri, Kandel and Frazier, 1969).

PLASTICITY OF ELECTRICAL AND CHEMICAL SYNAPSES

If alterations in synaptic efficacy resulted only from changes in the presynaptic control of the amount of chemical transmitter substance released or from changes in the sensitivity of the postsynaptic receptor, then the capability for plastic change might be limited to chemical synapses. Electrical synapses would not be expected to undergo posttetanic facilitation. This hypothesis was tested by Martin and Pilar (1964), who compared electrical and chemical synaptic transmission in the dual chemical-electrical synapse of the chick ciliary ganglion (Figure 11-11). They found that following tetanization only the chemically mediated synaptic potential was facilitated; the electrically mediated synaptic potential was not affected. A similar conclusion was reached by Bennett (1968) and Nicholls and Purves (1970) from studies of a variety of electrical synapses in both vertebrates and invertebrates. These experiments suggest that one reason for the apparent preponderance of chemical over electrical transmission in the central nervous system of vertebrates and inverte-

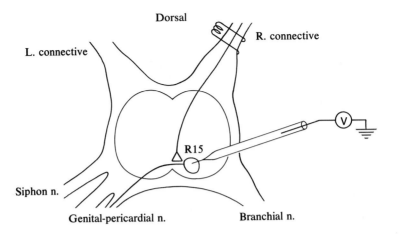

A

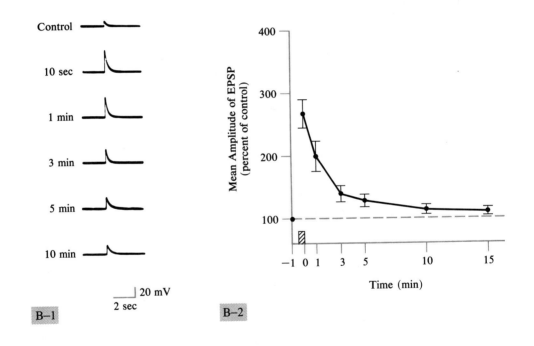

B-1

20 mV

2 sec

B-2

FIGURE 11-10.
Posttetanic facilitation of a monosynaptic EPSP in the abdominal ganglion of *Aplysia*. [From Schwartz, Castellucci, and Kandel, 1971.]

A. Experimental setup for intracellular recording from cell R15, which was hyperpolarized to prevent firing. The EPSP is elicited by stimulating the right connective.

B. *B-1*. Sample records from a single experiment. The right connective was first stimulated every 10 sec to obtain a control value of the EPSP amplitude. A brief tetanus (5/sec for 30 sec) was then applied to the right connective. Following tetanization the EPSP was again evoked at the control rate of one every 10 sec. The EPSP remained facilitated for more than 10 minutes. *B-2*. Graph based on an average of 10 experiments described in *B-1*. In each experiment the control value (100 percent in the graph) was taken from the mean of about 10 pretetanic EPSPs. The hatched bar at the bottom of the graph indicates tetanization; zero time indicates the end of tetanization.

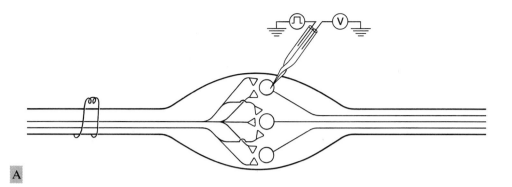

A

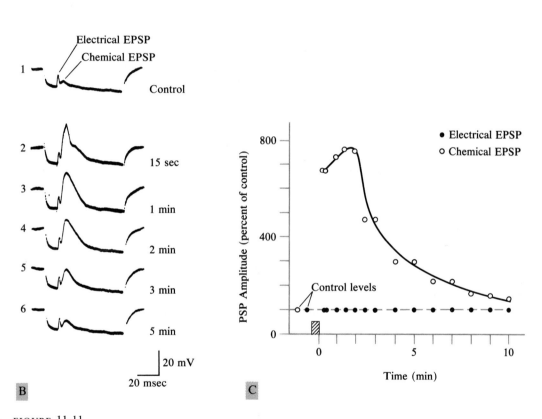

B

C

FIGURE 11-11.
Comparison of the effect of conditioning tetanus on a chemical and electrical EPSP in cells of chick ciliary ganglion. [From Martin and Pilar, 1964.]

A. Experimental setup. Intracellular recording from cell bodies in the ciliary ganglion in response to electrical stimulation of afferent fibers.

B. Sample responses from a single experiment. The chemical EPSP was reduced in amplitude by adding d-tubocurarine chloride to the bathing solution (in a concentration of 5×10^{-6} g/ml). The control response was recorded before tetanization (5/sec for 20 sec) of the presynaptic nerve; the posttetanic responses were recorded at 15 sec, 1, 3, 5, and 10 min. The chemical EPSP is potentiated, whereas the electrical EPSP remains unchanged.

C. Time course of posttetanic facilitation of intracellularly recorded chemical EPSPs in the ganglion cell. The hatched bar indicates tetanization (50/sec for 15 sec). The EPSP is potentiated by almost 800 percent; the electrical EPSP remains unchanged.

brates may be the ability of chemical synapses to undergo alterations in efficacy due to previous activity. The plastic changes themselves can be regarded as a type of information storage of the sort that could underlie memory. By contrast, the relative stability of electrical synapses seems less suited for mediating the plastic changes necessary for behavioral modification and more suited for mediating stereotypic responses (p. 415).

In tissue culture, however, a variety of non-neural cells readily form electrotonic connections on contact (Furshpan and Potter, 1968). This phenomenon may also occur between developing neurons in tissue culture (Fischbach, 1972). Moreover, electrotonic connections can be uncoupled by injury and, briefly, by synaptic activity (Bennett, 1973; and p. 404 here). For example, the two electrically coupled motor neurons that innervate the pharynx in the opisthobranch *Navanax* fire synchronously during rapid expansion. But as the ingested prey is pushed into the esophagus, the stimulus due to pharyngeal inflation produces increased-conductance IPSPs in the two motor cells. The IPSPs short-circuit the nonsynaptic membrane of the motor cells and functionally uncouple them, so that they can fire independently (Spira and Bennett, 1972, 1973). Uncoupling does not require inhibitory synaptic actions. The electrotonically coupled ink-gland motor cells in *Aplysia* are uncoupled by increased-conductance EPSPs (see Figure 9-38, and Carew and Kandel, 1976). Moreover, decreased-conductance EPSPs can enhance coupling between the ink-gland motor cells (Figure 9-38). Also, in certain instances electrotonic PSPs can facilitate because the presynaptic spike broadens at high frequencies of firing and the electrical synapse has low-pass frequency-filtering characteristics, so that it preferentially conducts slow signals (Bennett, 1972; Getting and Willows, 1974). It is conceivable, therefore, that under certain conditions even electrical synapses could play a restricted role in neural plasticity. Moreover, recirculation of impulses in a group of electrically interconnected cells can provide a dynamic means for the short-term maintenance of neuronal activity (see Figure 10-25A, p. 451).

NEURAL ANALOGS OF LEARNING

The finding that activity of a given pathway can lead to changes in the synaptic efficacy of the chemical synapses in that pathway brings one a step closer to understanding how behavior might be modified. But behavioral modification often involves more complex stimulus sequences than posttetanic facilitation and depression. These are homosynaptic processes—only activity in the repetitively stimulated pathway is affected. Yet even simple modifications, such as dishabituation and sensitization, and certainly more complex ones, such as classical and operant conditioning, involve activity in two separate pathways (see p. 7). Can plastic changes also be induced heterosynaptically? Can activity in one pathway produce prolonged alterations in the activity of another pathway?

To examine these questions, more complex stimulus sequences, of the sort used in learning experiments, have been applied to isolated preparations. The consequences for synaptic transmission have been examined in an attempt to develop neural analogs of behavioral modification paradigms.

The common learning paradigms—associative classical conditioning, associative instrumental conditioning, nonassociative classical conditioning (pseudoconditioning and sensitization), and habituation—share certain features. They are all elementary forms of learning, they are all general and found in a wide variety of animals, and they are all operationally well defined (see p. 5). This specificity of definition allows one to develop neurophysiological analogs. In the most general sense, a learning paradigm describes how a stimulus must be patterned over time or combined with other stimuli to elicit specific modes of behavior. Consequently, learning paradigms may serve as experimental designs, that is, they may be viewed as stimulus sequences that can be applied even to simplified and isolated preparations using electrical instead of natural stimuli (Table 11-1).

Electrical stimuli are applied to peripheral nerves or central axonal tracts and the resulting electrical responses of single neurons or groups of neurons

TABLE 11-1.
Learning and its neural analogs. [*From Kandel, 1967.*]

	Learning	Neural analog
Stimulus	Natural	Electrical
Response	Effector response (muscle or gland)	Electrical response (PSPs or action potential)
Preparation	Intact or slightly modified animal	Simplified animal or isolated neural structures
Learning paradigms	Associative classical and instrumental conditioning; non-associative classical conditioning (pseudo-conditioning and sensitization) and habituation	Same

are monitored. In most neural analogs the relationship of stimulated and monitored structures to specific sensory and motor pathways is not known. The neural structures only serve as convenient biological systems in which to investigate the effects of stimulus patterns that simulate the stimulus and reinforcement patterns used in studies of behavioral modifications in intact animals. The relationship of classical and instrumental conditioning and habituation to their neural analogs is illustrated in Figure 11-12. Neural analogs represent only a first step towards establishing a relationship between plastic changes in neurons and behavioral modifications; in these analogs the experimenter must look for changes in neuronal properties that resemble the changes in behavior that constitute learning.

Classical Conditioning

The essential features of classical conditioning are: an unconditioned stimulus (US) that is effective in evoking a predictable response, the unconditioned response (UR), and a conditioned stimulus (CS) that originally does not evoke a response identical to the unconditioned response. Presentation of CS followed by US at a specified time is referred to as a *reinforcement.* Any new or altered response that occurs to the CS, and that resembles the UR, is called a conditioned response (CR). After successful conditioning, repeated unpaired (unreinforced) presentations of the CS result in *extinction* of the CR. The CR often recovers spontaneously after an extinction procedure.

Classically conditioned responses are classified by two independent criteria: (1) the nature of the responses to the CS—the CR is either a new response (classical CR) or evolves from a preexisting response to the CS (alpha CR); and (2) the associative specificity of CS–US pairing required for acquisition of the CR. In associative conditioning the acquisition of the CR is dependent on precise pairing of a CS and US. If repeated unpaired presentations of the US or random pairings of CS and US produce a CR that is indistinguishable from that produced by the standard pairing procedure, then nonassociative conditioning has been demonstrated. If dissimilar CRs are elicited by such procedures, then associative conditioning has been demonstrated. Thus, four types of outcome are possible (see Table 11-2): associative classical conditioning; associative alpha conditioning; nonassociative classical conditioning (pseudoconditioning); and nonassociative alpha conditioning (sensitization).

Psychologists have traditionally regarded only associative classical conditioning as interesting. Little attention has been paid to associative alpha

CELLULAR ANALOGS OF LEARNING PARADIGMS

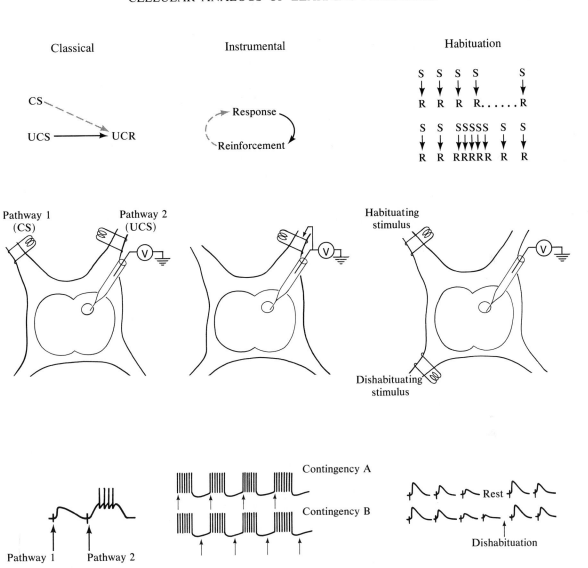

FIGURE 11-12.
Cellular analogs of classical conditioning, instrumental conditioning, and habituation. The upper part of each figure is a conventional statement of the paradigm. The middle part illustrates how these statements can be applied to an isolated ganglion using electrical instead of natural stimuli. The lower part illustrates examples of intracellularly recorded responses that have been used in cellular analogs. [From Kandel, 1967.]

TABLE 11-2.
Different types of conditioned responses generated by classical conditioning.

Specificity of CR to CS-US pairing	Response to CS	
	Establishment of a new response	Evolution of a preexisting (alpha) response
Specific	*Associative classical conditioning*	*Associative alpha conditioning*
Nonspecific	*Nonassoicative classical conditioning (pseudo-conditioning)*	*Nonassociative alpha conditioning (sensitization)*

conditioning, and the adaptive importance of nonassociative processes, such as pseudo-conditioning and sensitization, has been minimized (see p. 21). But from the larger viewpoint of biology, any successful outcome of classical conditioning is of interest. And from the viewpoint of cellular physiology, each presents a potentially independent mechanism of plastic change. Moreover, it is possible for several mechanisms of learning (such as associative classical conditioning and sensitization) to occur together in response to the same conditioning procedure and for them to interact in producing the observed behavior (Hilgard and Marquis, 1940; Kimble, 1961; Woodworth and Schlosberg, 1954). Thus, the attempt to translate these behavioral categories into neural analogs may allow insights into the relationships between the categories.

The classical conditioning paradigm has been applied to the isolated abdominal ganglion of the European species *Aplysia depilans* and the American genus *Aplysia californica* (Kandel and Tauc, 1963, 1964, 1965a, b; Jahan-Parwar and von Baumgarten, 1967; von Baumgarten, 1970). Two different afferent pathways converging on a single identified cell, R2, were used. One pathway, the test pathway (analogous to the CS pathway), was stimulated regularly at 10-second intervals, and the stimulus parameters were adjusted so that the pathway produced only a small EPSP (analogous to the alpha response to the CS). A second (priming) pathway (analogous to the US pathway) was then activated by a brief train of strong stimuli that was effective in producing a discharge of spikes in R2 (analogous to the UR). The two inputs were paired for one or more minutes with the test preceding the priming stimulus by about 300 msec. As a result of repeated presentation of the priming stimulus, the test EPSP was greatly augmented (by 100 to 800 percent). The facilitation then declined over the next 10 to 40 minutes after priming. During peak facilitation the initially ineffective test EPSP became effective in triggering an action potential. Kandel and Tauc (1964, 1965a) called this type of facilitation *heterosynaptic facilitation* (Figure 11-13). This type of facilitation has also been found in some neighboring unidentified cells, in the left hemi-ganglion (Castellucci *et al.,* 1970), and in the left pleural ganglion (Shimahara and Tauc, 1975b; see p. 575).

By introducing controls for associative conditioning analogous to those used in behavioral experiments, Kandel and Tauc found that the facilitation in R2 is not specific to associative pairing of the test and priming stimuli; both pairing and nonpairing produce facilitation (Figure 11-14). Neither is the facilitation specific to the input that is paired with the priming stimulus; an input that is not paired is facilitated as much as the paired input. The facilitation is therefore analogous to nonassociative alpha conditioning, or sensitization. It also resembles sensitization in several other ways. Like sensitization (Grether, 1938), heterosynaptic facilitation is less a function of the test than of the reinforcing stimulus. Both small and large test EPSPs are facilitated and the duration of their facilitation is comparable.[5] By contrast, weak priming (reinforcing) stimuli produce no facilitation, and there is a direct relationship between the strength of the priming stimulus and the degree of facilitation (Kandel and Tauc, 1965a).

These findings suggested that heterosynaptic facilitation could mediate behavioral sensitization. To determine whether this type of facilitation could be produced by natural stimuli, Kandel and Tauc performed experiments in a partially intact animal, while recording extracellularly from the axon of R2. Using a natural priming stimulus (stroking the siphon) instead of an electrical one, they still obtained heterosynaptic facilitation (Figure 11-15).

Heterosynaptic facilitation does not appear to result from a gross change in the input resistance of the nonsynaptic membrane of the postsynaptic cell. The input resistance membrane remains constant during facilitation. Heterosynaptic facilitation can be produced using a test pathway consisting of only a single fiber that produces an elementary EPSP in R2 (Figure 11-16). Although changes in receptor sensitivity could not be excluded, Kandel and Tauc suggested that heterosynaptic facilitation involves a presynaptic mechanism in which some fibers in the priming pathway synapse on the terminal of the test fiber. This synapse on the presynaptic terminal is thought to modulate long-term release of the chemical transmitter substance in the test pathway (Figure 11-17).

Heterosynaptic facilitation differs from posttetanic facilitation. When the same test synapse is examined under roughly comparable conditions, heterosynaptic facilitation lasts longer than posttetanic facilitation (Kandel and Tauc, 1965b; Tauc and Epstein, 1967). Moreover, posttetanic facilitation can be blocked by cooling or substituting Li^+ for Na^+, whereas these procedures do not affect heterosynaptic facilitation at the same synapse (Figure 11-18).

---

[5]The small EPSPs show a greater percentage increase than do the large test EPSPs. But both reach a comparable amplitude, suggesting a common ceiling. This ceiling may represent the common reversal level of the EPSPs.

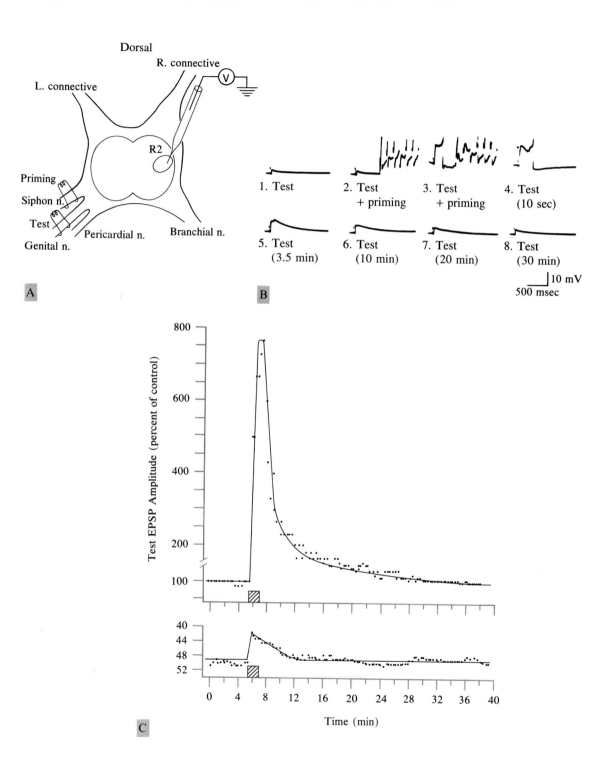

A

Dorsal

R. connective

L. connective

R2

Priming

Siphon n.

Test

Genital n.

Pericardial n.

Branchial n.

B

1. Test

2. Test
+ priming

3. Test
+ priming

4. Test
(10 sec)

5. Test
(3.5 min)

6. Test
(10 min)

7. Test
(20 min)

8. Test
(30 min)

10 mV
500 msec

C

Test EPSP Amplitude (percent of control)

800

600

400

200

100

40
44
48
52

0   4   8   12   16   20   24   28   32   36   40

Time (min)

FIGURE 11-13 (*Facing Page*).
Heterosynaptic facilitation. [From Kandel and Tauc, 1965a.]

A. Experimental arrangement for intracellular recording in cell R2. A test stimulus is applied to the genital nerve once every 10 sec throughout the experiment. A priming stimulus is applied to the siphon nerve.

B. Sample records from one experiment. *B-1*. Response to the test stimulus before pairing. *B-2*. First of nine trials pairing the test stimulus and priming stimulus (6/sec for 1 sec). *B-3*. Seventh pairing trial. Note accompanying depolarization and increase in response to test stimulus. *B-4* to *B-8*. Examples of the response to the test stimulus 10 sec, 3.5, 10, 20, and 30 min after pairing.

C. Time course of heterosynaptic facilitation (a different run of the experiment in part A). The test EPSP is plotted as a function of time and of pairing; the period of pairing is indicated by the hatched bar. The graph at the bottom indicates the average membrane potential. The time course of the depolarization that accompanies the facilitation is brief compared to facilitation.

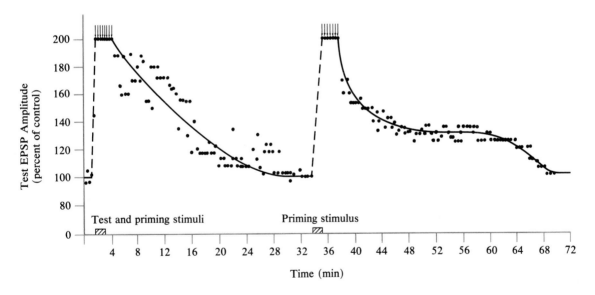

FIGURE 11-14.
Heterosynaptic facilitation of the test EPSP with and without pairing of the test and priming stimuli. The hatched bars at the bottom of the graph indicate the period during which the priming stimulus was applied. Arrows at the top of the data (during maximal facilitation) represent action potential triggered by the test EPSP. The graphs are from two consecutive runs in the same cell. Note that the graphs are quite similar although not identical. [From Kandel and Tauc, 1965a.]

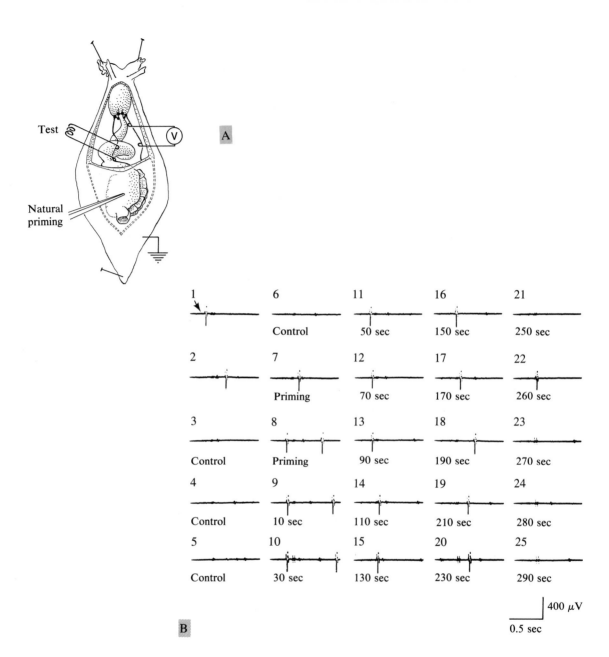

FIGURE 11-15.
Heterosynaptic facilitation elicited naturally in the intact animal. [From Kandel and Tauc, 1965a.]

A. Experimental arrangement for recording the postsynaptic activity of cell R2 in the whole animal by monitoring the action potential produced by the axon of R2 in the right connective. Electrical test stimuli are applied to the left connective. A natural mechanical stimulus, stroking the siphon and mantle with forceps, served as the priming stimulus.

B. The strength of the test stimulus is progressively decreased (trials 1-3) so that it is subthreshold for discharging R2 during the control period (trials 3-6). A priming stimulus was applied to the siphon after trial 6 and during trials 7 and 8. As a result, the previous subthreshold test stimulus became effective in triggering spikes in R2 (trials 7-22). The enhanced effectiveness declined over the following several minutes.

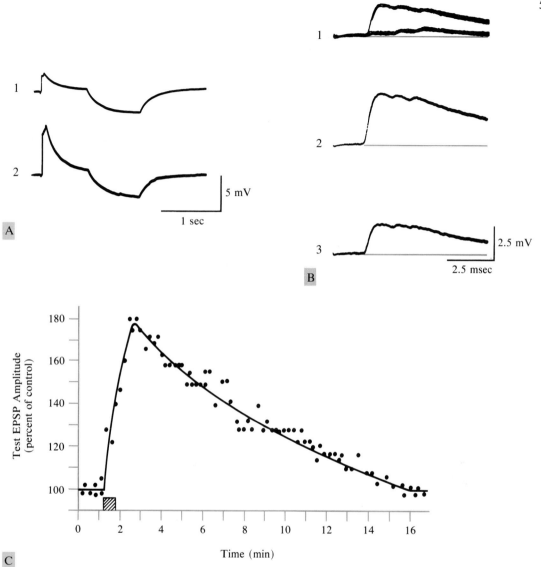

FIGURE 11-16.

Analysis of heterosynaptic facilitation in cell R2. [From Kandel and Tauc, 1965b.]

A. Simultaneous measurement of the test EPSP and input resistance measured by an electrotonic potential produced by a hyperpolarizing constant-current pulse during facilitation. *A-1.* Before facilitation. *A-2.* After facilitation. Heterosynaptic facilitation increases the EPSP but produced no change in the amplitude of the electrotonic potential.

B. Heterosynaptic facilitation of a unitary, presumably monosynaptic, EPSP in R2. The EPSP is elicited by stimulating the right connective. *B-1.* Control. Several sweeps are superimposed and stimulus strength slightly reduced to illustrate that the threshold for the presynaptic axon is reached. At threshold the EPSP is produced in an all-or-none manner. *B-2.* Peak facilitation after presentation of the priming stimulus; two late components (also present in *B-1*) have also increased. *B-3.* Return to control 14 min after priming.

C. Time course of heterosynaptic facilitation of the unitary (presumably monosynaptic) test EPSP produced by stimulating the right connective. Four suboptimal priming stimuli were applied to the siphon nerve (hatched bar).

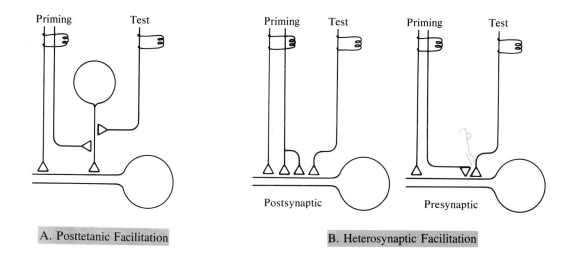

A. Posttetanic Facilitation

B. Heterosynaptic Facilitation

FIGURE 11-17.
Alternative mechanisms for facilitation of the unitary PSP following heterosynaptic stimulation. [From Kandel and Tauc, 1965b.]

A. Posttetanic facilitation. Test and priming pathways engage a common interneuron. The priming pathway discharges the interneuron repetitively, producing posttetanic facilitation at its synapse. When subsequently the test pathway activates the interneuron, each action potential in it will produce a larger PSP until the posttetanic facilitation of the interneuron subsides.

B. Heterosynaptic facilitation. *Left.* Postsynaptic facilitation. The priming pathway activates the same receptors as the test pathway and enhances their sensitivity. As a result, the same amount of transmitter produces a larger EPSP. *Right.* Presynaptic facilitation. Part of the priming pathway synapses directly (or indirectly, by means of an interneuron) on the terminal of the test pathway, enhancing its synaptic effectiveness.

FIGURE 11-18 (*Facing Page*).
Comparison of heterosynaptic and posttetanic facilitation of the same synaptic connection. [From Epstein and Tauc, 1970.]

A. Experimental setup. Intracellular recordings are made from cell R2. A test pathway (left connective) and priming pathway (branchial nerve) are stimulated extracellularly.

B. Heterosynaptic facilitation of a unitary EPSP in cell R2 when the preparation has been cooled to 7°C. At this temperature the heterosynaptic facilitation is similar to that occurring at room temperature (cf. Figure 11-16*B*).

C. Tetanization of the test pathway. At 7°C the tetanization (that produces posttetanic facilitation at room temperature) is reversed to a posttetanic depression. Differences in the temperature sensitivity of heterosynaptic and posttetanic facilitation at this synapse suggest that these processes are due to independent mechanisms.

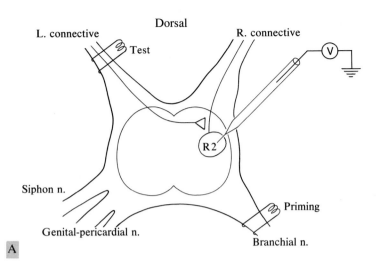

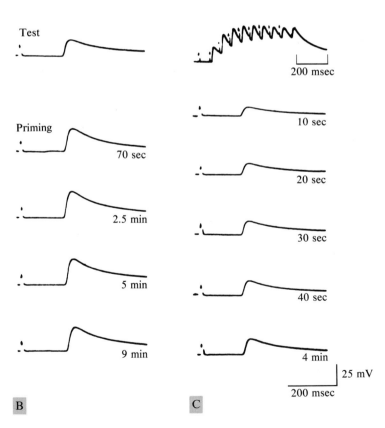

Two other ways have been found to enhance the effectiveness of one pathway by the activity of another. In the giant cell of the left pleural ganglion of *Aplysia,* polysynaptic posttetanic facilitation occurs in an interneuron common to both test and priming pathway (Figure 11-17a; Bruner and Kehoe, 1970). In the giant metacerebral cells of the snail *Helix aspersa* (Figure 7-4, p. 224) a postsynaptic mechanism accounts for the increase in synaptic effectiveness. The metacerebral cells show a pronounced anomalous rectification. The priming stimulus causes a depolarization of the postsynaptic cells which moves the membrane potential along its current-voltage relationship to a zone of increased membrane resistance (Kandel and Tauc, 1966b). As a result, the same synaptic current produced by the test stimulus now produces a larger test EPSP (Figure 11-19). Thus a formally similar end result can be produced by three quite different mechanisms.

Instrumental Conditioning

The basic feature of associative instrumental (operant) conditioning is that the frequency of occurrence of a behavioral response can be altered by presenting a reinforcing stimulus following each occurrence of the response (Figure 11-12). Initially the response (the operant) is not under experimental control, but occurs spontaneously at some typical frequency (the operant level). For example, a hungry rat will occasionally press a lever in an experimental cage. The reinforcing stimulus can be either positive (reward) or negative (punishment), depending on whether the frequency of the operant is increased or decreased by the stimulus. By selecting appropriate reinforcing stimuli, e.g., food as positive reinforcement or shock as negative reinforcement, most operant levels can be driven up or down in a predictable manner. Thus, the hungry rat will press the lever more frequently if pressing makes food pellets available; it will press less frequently if pressing brings on a painful stimulus.

In addition to producing associative (contingent) changes in the frequency of the operant, instrumental conditioning can produce nonassociative (noncontingent) changes. The two can be distinguished by appropriate controls, e.g., either random presentation of the reinforcing stimulus or changes in the schedule of reinforcement. If the effects produced by the control procedure are indistinguishable from those produced by the conditioning regimen, then the change in frequency of the behavior is considered noncontingent; if the effects are distinguishable, they are contingent.

To develop an analog of this paradigm, one needs to select as an operant a spontaneously emitted cellular response. Certain identified cells (L2, L3, L4, L6, and R15) show fairly regular autoactive firing patterns consisting of bursts

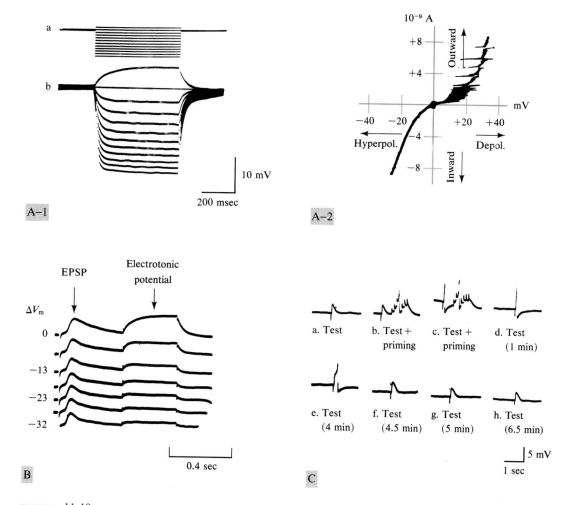

FIGURE 11-19.

A postsynaptic mechanism for producing facilitation of one pathway by another in the metacerebral cells of the snail. [From Kandel and Tauc, 1966b.]

A. Anomalous rectification. *A-1.* Current–voltage relationships: (a) monitor of current steps injected into the cell (each step is 3.3×10^{-10}A); (b) electrotonic potentials produced by injections of equal increments of current *A-2.* Dynamic current-voltage curve obtained directly on an oscilloscope. The transients on the positive-voltage axis are the lower parts of the action potentials. Zero represents the resting level (49 mV). Between -20 mV (hyperpolarization) and $+10$ mV (depolarization) the input resistance increases as the membrane is moved in the depolarizing direction (anomalous rectification). Further depolarization causes a decrease in input resistance (delayed rectification).

B. Effects of hyperpolarizing the membrane potential on the amplitude and configuration of an EPSP and electrotonic potential produced by a square-wave constant-current pulse. The amplitudes of both the EPSP and the electrotonic potential are both decreased, indicating that the decrease in the EPSP is due to a decrease in the input resistance.

C. Depolarization of the membrane potential, produced by stimulating a priming pathway, increases the EPSP produced by stimulating a test pathway. Records have been aligned to allow accurate reading of changes in membrane potential. (*a*) Test PSP before stimulation of the second pathway. (*b*) First of seven stimulus trains to the second pathway. (*c*) Fifth stimulus train to the second pathway; the membrane depolarization and the increase in amplitude of the test PSP parallel each other. (*d–h*) Decline in amplitude and depolarization of the test PSP in the 6.5 min following stimulus to the second pathway. The increase in the size of the PSP is due to anomalous rectification in the cell; thus, when the membrane potential declines to its original value, the PSP decreases to its control size.

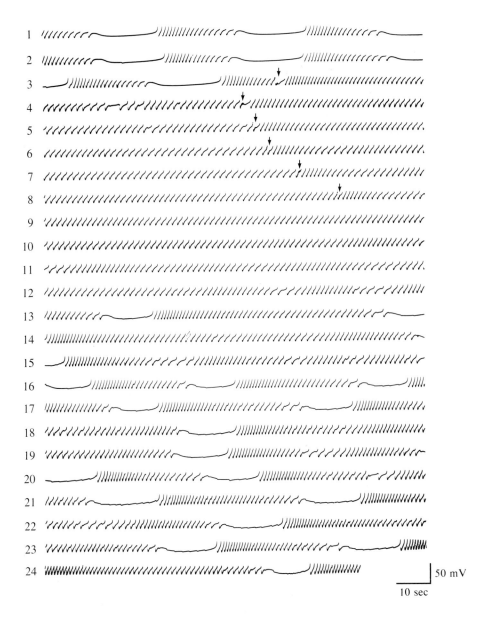

FIGURE 11-20.
Effects of noncontingent stimulation on the regular-bursting rhythm of cell L3. (Continuous record, each line is a 2-min period.) Only the bottom parts of the action potentials are shown. After the control period (lines 1 and 2), six stimuli (6/sec for 1 sec) were applied to the left connective once every 2 min (lines 3–8, arrows). As a result, the burst-onset interval was altered and did not return to control for over 30 min after the last stimulus (lines 9 to 24). The data are graphically illustrated in Figure 11-21. [From Kandel, 1967.]

of spikes alternating with quiet periods (Chapter 7). Although they are not very good cellular analogs of operants because they occur so regularly, the burst activities of cells are a convenient starting point. Stimulation of the cells can be either contingent (all stimuli are presented at the same phase of the burst) or noncontingent (the stimuli are presented at different phases) (Frazier, Waziri, and Kandel, 1965; Pinsker and Kandel, 1967, unpub.; Kandel, 1967; Tauc, 1969; Parnas, Armstrong, and Strumwasser, 1974; Parnas and Strumwasser, 1974).

Noncontingent stimulation produces varied effects on the firing rhythm of autoactive cells, depending upon the cell and the synaptic input. A single noncontingent stimulus train applied to a connective produces in cells L2, L3, L4, and L6 a complex EPSP that causes these regular-bursting cells to fire continuously for a few minutes. Several spaced stimuli produce continuous firing that can last up to 20 minutes (Figures 11-20 and 11-21; Frazier *et al.,* 1965; Kandel, 1967). Repetitive noncontingent activation of the monosynaptic excitatory

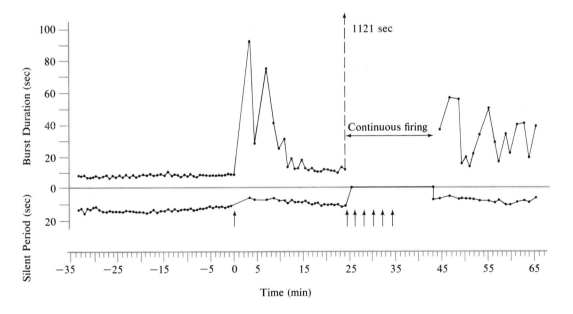

FIGURE 11-21.
Graphic illustration of changes in the burst cycle of L3 due to noncontingent stimulation (data in Figure 11-20 are plotted beginning at +22 min). The data points above the base line (zero) indicate duration of the burst; points below indicate duration of the silent period. After a control period of 35 min (−35 to 0 time) a single stimulus (arrow) was applied to the left connective. The cell's firing pattern was altered for about 20 min. At 25 min, six stimuli were applied to the left connective. This produced continuous firing for 18 min 41 sec (1121 sec), indicated by the line at 24 min, and a prolonged alteration in the burst pattern of the cell. [From Kandel, 1967.]

pathway to cell R15, by stimulation of the right connective (1/sec to 4/sec for 3 min), produces an increase in the number of spikes per burst for as long as 30 minutes to a few hours. This modification can be reversed by hyperpolarizing the cell for several minutes. The synaptic input is critical for this modification, which cannot be produced by simply firing the cell with depolarizing pulses that simulate the EPSP (Figures 11-22 and 11-23). Because this alternation in the firing pattern of R15 is produced by a direct connection, it is very likely that the synaptic action directly alters the burst mechanism (see Figure 7-33, p. 267). Other fibers in the connective produce a prolonged IPSP on this cell. Stimulating these fibers at 3- to 10/sec for 1 sec produces a prolonged inhibition of bursting that lasts up to one hour (Parnas, Armstrong, and Strumwasser, 1974; Tauc, 1969).

When a stimulus is made contingent upon a particular phase of the burst cycle it can produce changes that differ systematically, depending upon where in the burst cycle it is placed (Pinsker and Kandel, 1967; unpub.). Such changes have been examined in the direct IPSP produced in L3 by stimulation of L10 (see Figure 8-9, p. 297). It is useful to divide the burst period into two periods: period A is the burst onset, period B is the late phase of the burst and the post-burst hyperpolarization that lasts from offset of the burst to the onset of the next. If cell L10 is made to fire at the onset of a burst in L3 (contingency A), the IPSPs inhibit the remaining spikes in the burst and shorten the burst-onset interval, thereby increasing the number of bursts per unit time. When cell L10 is made to fire during the end of the quiet period of L3 (contingency B), the IPSPs lead to a lengthening of the interval and a decrease in the number of bursts per unit time (Figure 11-24). Repeated stimulation during contingency

FIGURE 11-22.
Long-lasting effects on bursting rhythm of R15 following repeated (noncontingent) stimulation of the right connective, which activates a direct excitatory connection (see Figure 11-10B). [From Parnas, Armstrong, and Strumwasser, 1974.]

A. Experimental setup for intracellular recording from R15. The monosynaptic EPSP is elicited by stimulating the right connective.

B. *B-1.* Control burst size varies between 17 and 18 spikes per burst. *B-2.* Stimulation of the right connective (10/sec for 4 min) produces EPSPs and continuous firing. *B-3.* Eight minutes after stimulation the burst size has increased from 17–18 to 27–29 spikes per burst and the interburst interval has been prolonged. *B-4.* After 4 min of hyperpolarizing the cell and inhibiting spiking, the cell produces a long train of spikes as a result of rebound excitation. Following the rebound, the burst size decreases to 19–20 spikes and the interburst interval returns to control level. *B-5.* Intracellular depolarization causes 4 min of continuous spiking that simulates firing produced by connective stimulation. *B-6.* Unlike stimulation of the connective, intracellular depolarization does not cause a change in burst size but does prolong the interburst interval.

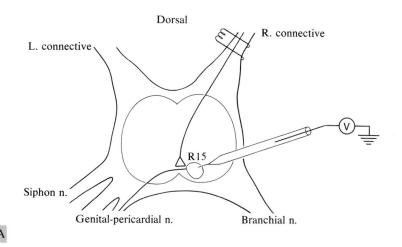

A

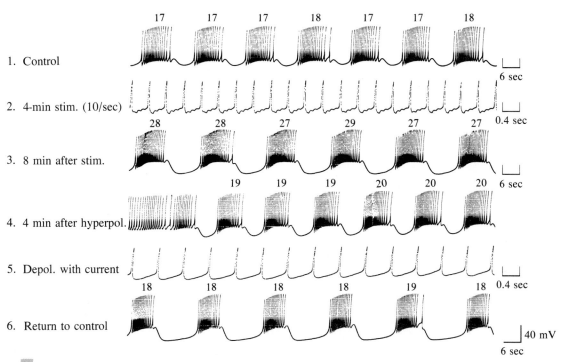

1. Control

2. 4-min stim. (10/sec)

3. 8 min after stim.

4. 4 min after hyperpol.

5. Depol. with current

6. Return to control

6 sec

0.4 sec

6 sec

0.4 sec

40 mV

6 sec

B

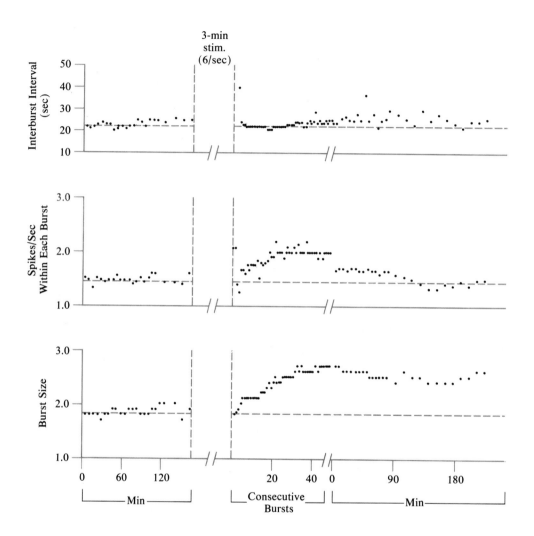

FIGURE 11-23.

Time course of noncontingent modification of the bursting rhythm of R15. Three parameters are plotted: interburst interval, spikes per sec in each burst, and burst size. Control values are plotted over three hours in the left part of each graph. At the break in the curve, the right connective was stimulated (6/sec for 3 min), eliciting an EPSP in R15. For the 40 bursts following stimulation, there is no change in interburst interval but there is an increase in burst size and spikes per sec in each burst. The burst size remains at the post-stimulation level for four hours. [From Parnas, Armstrong, and Strumwasser, 1974.]

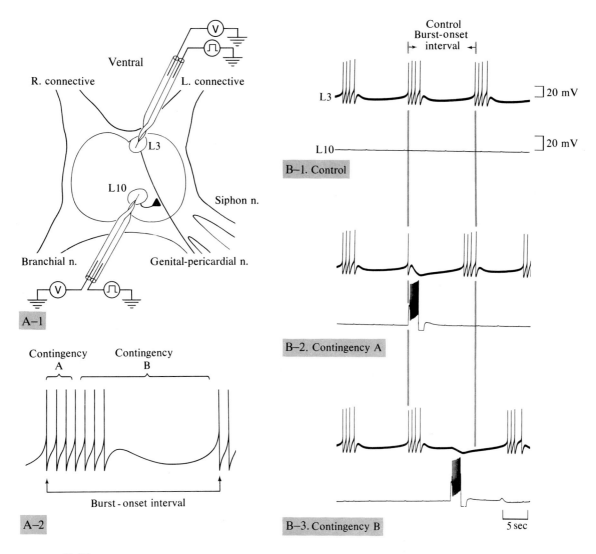

FIGURE 11-24.

Contingent modification of the burst-onset interval in cell L3 in response to a single burst in L10.

A. *A-1.* Experimental arrangement. Intracellular stimulating and recording electrodes are placed in L10 and its inhibitory follower cell L3. The connection between L3 and L10 is monosynaptic and inhibitory. *A-2.* Each burst cycle is divided into two phases (contingencies), A and B. The stimulus can be made contingent on either contingency A or B.

B. Effects of stimulation during contingencies A and B. The same synaptic input produces opposite effects on the burst-onset interval, depending on the phase of the burst cycle in which it is presented. The control burst-onset interval is indicated between the two vertical grey lines. Stimulation during contingency A shortens the interval, whereas stimulation during contingency B lengthens it.

A produces a slight buildup in the number of bursts per unit time. The first of 10 stimuli produces a 10 to 15 percent shortening of the burst-onset interval, whereas the last produces more than a 30 percent shortening. The shortening persists for one or at most a few minutes after the last stimulus. Repeated stimulation during contingency B results in a progressively weaker effect. The first of 10 stimuli produces about a 50 percent lengthening of the interval, whereas the last produces only a 30 percent lengthening. Moreover, following the last stimulus the burst-onset interval shortens for several minutes before returning to control. Although these after-effects of a contingent stimulus are brief and tend to outlast the stimulus for at most a few minutes, they nonetheless provide further evidence that synaptic input can alter the endogenous pacemaker properties of cells (Figure 11-25).

Maximal shortening of the burst-onset interval (during contingency A) occurs when the stimulus is presented immediately following the first spike in the burst, thereby inhibiting the remaining spikes. As the stimulus is presented at progressively later points in the burst cycle, it inhibits fewer spikes and produces less shortening of the burst-onset interval. Beyond a null point the stimulus produces progressively greater lengthening of the interval, with maximal lengthening resulting when the stimulus is placed at the end of the burst cycle. The function relating the burst-onset interval to the point of the stimulus within the burst cycle can thus best be described by two straight lines that intersect at the time when the burst of spikes ends (Figure 11-26).

These results suggest that the stimulus interacts with two overlapping endogenous processes. One process occurs only during the burst of spikes. The

FIGURE 11-25.

Contingent modification of the burst-onset interval in L3 by repeated bursts in L10. [From Pinsker and Kandel, unpub.]

A. Selected records of the effects of repeated stimulation during contingencies A and B. Stimulus number is indicated on the left; burst-onset interval is indicated in parentheses (control = 100 percent, indicated by the vertical grey lines). The short bar under each record indicates the duration of the L10 train of spikes.

B. Graph of an experiment showing the effects of repeated stimulation during contingencies A and B. Ten control burst cycles are followed by stimulation of 10 burst cycles, followed in turn by nonstimulation of 10 burst cycles. When L3 is stimulated during contingency A, the shortening of the burst-onset interval increases; the interval gradually returns to the control value during the last 10 (unstimulated) burst cycles. When L3 is stimulated during contingency B, the lengthening of the burst-onset interval decreases; the interval overshoots (shortens) following stimulation and gradually returns to control levels. Each data point represents an average of seven preparations ±1 SD.

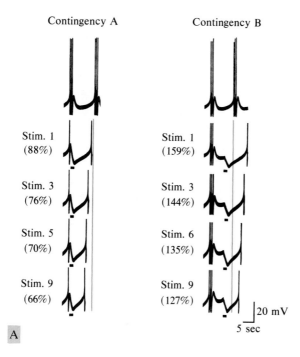

Contingency A   Contingency B

Stim. 1
(88%)

Stim. 3
(76%)

Stim. 5
(70%)

Stim. 9
(66%)

Stim. 1
(159%)

Stim. 3
(144%)

Stim. 6
(135%)

Stim. 9
(127%)

20 mV

5 sec

A

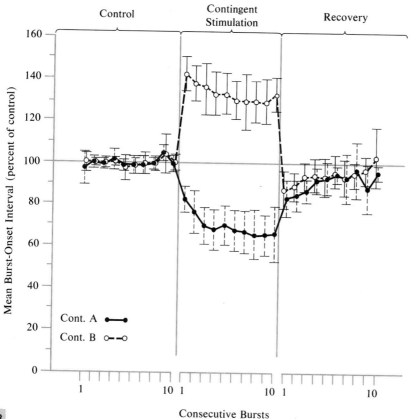

Control   Contingent
Stimulation   Recovery

Mean Burst-Onset Interval (percent of control)

Cont. A ●—●
Cont. B ○--○

Consecutive Bursts

B

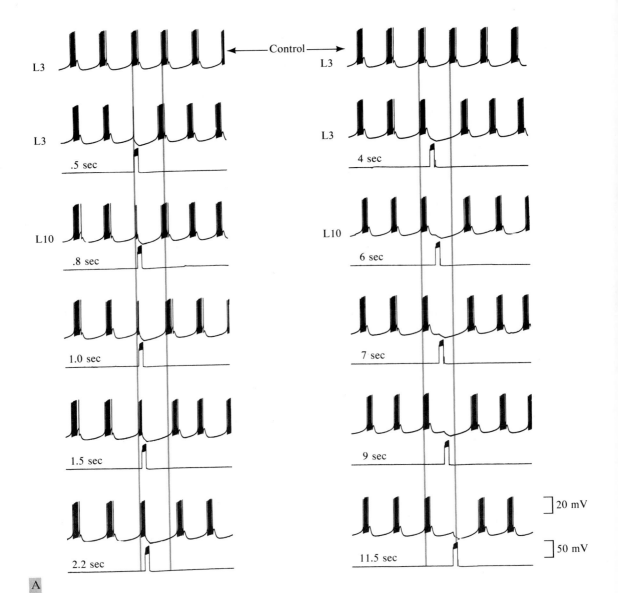

FIGURE 11-26.

Modification of L3 rhythm as a function of stimulation of L10 at different points in the burst cycle of L3. [From Pinsker and Kandel, unpub.]

A. When stimulated during the early part of the burst (0.5 and 0.8 sec after the onset), L10 produces a shortening of the burst-onset interval in L3. At 1 sec after burst onset in L3, L10 produces no effect (null point). Later presentation (1.5 sec to 11.5 sec after burst onset) produces progressive prolongation of the interval. Control interval (indicated by the two vertical grey lines) was 13.4 sec, 8.3 spikes per burst. The time given with each L10 trace is the time after the onset of the burst in L3 that L10 is stimulated.

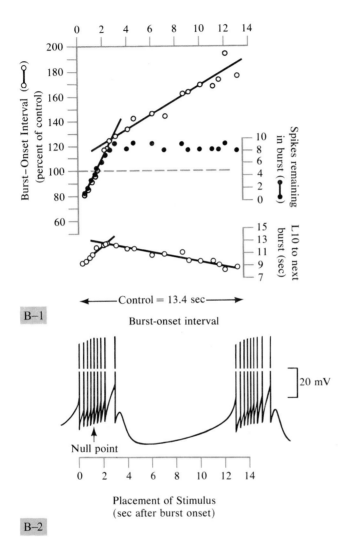

B. Phase response curve. *B-1.* Graph of data illustrated in part *A*, showing the effects of L10 stimulation on both the burst-onset interval with respect to L3 (open circles) and the number of spikes remaining in the L3 burst (solid circles). Lower graph shows that the burst-onset interval between L10 and L3 is a function of the point in the burst cycle of L3 that L10 is stimulated. *B-2.* Control burst-onset interval and null point in the burst cycle for the experiment illustrated in part *A*.

stimulus acts on this process to reduce the number of spikes in the burst and thereby shorten the burst-onset interval. The pacemaker potential has the properties of an afterpotential due to an increased g_K, the amplitude and duration of which are a function of the number of spikes in the burst (see p. 262). Reducing the number of spikes in the burst will therefore reduce the burst-onset interval because fewer spikes will produce less of an afterpotential. Very little g_K will be turned on due to the depolarizing action of both the steady and cyclical g_{Na} (see the model in Figure 7-33, p. 267).

The stimulus also lengthens the burst-onset interval by acting on a second process, perhaps a delay in the recycling of g_{Na}; the lengthening is a linear function of the time after burst onset. The effects of the stimulus on this second process are seen in isolation only when the stimulus is presented after the burst of spikes is over. Thus, an elementary synaptic input can produce opposite effects on the bursting rhythm of spontaneously active cells because stimulation affects primarily one or the other of two different endogenous processes. The monosynaptic input activates both an early fast g_{CL} and a late slow g_K. But contingent synaptic modifications do not seem to depend upon either conductance change. The effects of both contingency A and B modification can be simulated by hyperpolarizing current pulses.

These experiments thus illustrate that autoactive properties of neurons can undergo plastic changes. This mechanism may be important in *Aplysia,* and in nervous systems generally, because many cells are autoactive.

Habituation

Habituation is a gradual decrease in the intensity of a reflex response to a monotonously repeated stimulus (Chapter 1). After surveying a variety of studies on short-term habituation, Thompson and Spencer (1966) listed nine features that characterize this type of behavioral modification.

1. Repeated stimulation leads to a decrease of the response.

2. The more frequent the stimulus presentation, the more rapid and pronounced is the decrease.

3. The weaker the stimulus, the more pronounced is the decrease; strong stimuli may yield insignificant habituation.

4. Recovery may be prolonged by repeated stimulation even after depression has been severe enough to abolish the response (subzero effect).

5. Habituation of the response to a given stimulus can generalize to other stimuli.

6. If the stimulus is withheld, spontaneous recovery of the response occurs.

7. Habituation occurs more rapidly after repeated periods of habituation training and spontaneous recovery.

8. Presentation of a strong stimulus to the same or different pathway produces restoration of the response (dishabituation).

9. With repeated presentations, the effectiveness of dishabituation decreases.

Decrement of response is commonly seen in neuronal aggregates; even very simple neuronal systems show a gradual reduction in the intensity of the system's activation, some features of which are similar to habituation in intact organisms. For example, in the isolated abdominal ganglion of *Aplysia depilans* the EPSP produced in cell R2 by stimulation of a connective undergoes a frequency-dependent homosynaptic depression (Figure 11-27). With repeated stimulation, once every 10 sec, the amplitude of both the complex EPSP produced by whole-nerve stimulation and the unitary, presumably direct, EPSP produced by stimulation of a single axon decreases to about 30 percent of control (Figure 11-28; see also Figure 11-9*B*). Several minutes of rest are required for the EPSP to regain its initial amplitude. Stimulation of another pathway produces heterosynaptic facilitation (analogous to dishabituation).

A similar depression occurs in the giant cell of the left pleural ganglion. Bruner and Tauc (1966) and Bruner and Kehoe (1970) have compared homosynaptic depression in these two cells to behavioral habituation in vertebrates. They found that, although both the complex and the unitary EPSP recover within several minutes, the time for spontaneous recovery becomes progressively longer with repeated presentations of the stimulus (Figure 11-28). The EPSP depression they noted was specific to the stimulated pathway and did not extend from one monosynaptic connection to another (Figure 11-29). However, in their analog, facilitation (analogous to dishabituation) could best be produced by homosynaptic stimulation (posttetanic facilitation).

Bruner and Tauc (1966) and Bruner and Kehoe (1970) also compared these findings from electrically stimulated nerves in isolated ganglia with those made in the left pleural giant cell when the ganglion was still attached to an isolated head, permitting natural stimuli to be used. Repeated stimulation of the tentacles with drops of seawater produced a depression of the polysynaptic EPSPs in the left pleural giant cell similar to the depression produced by electrical stimulation of the nerve (Figure 11-30). This naturally initiated depression was not due to sensory receptor adaptation; rather, it was due to a decrease in excitatory synaptic transmission. Moreover, synaptic depression occurred without a change in the input resistance of the postsynaptic neuron. In addition, the rate of EPSP depression was found to be a function of the rate of

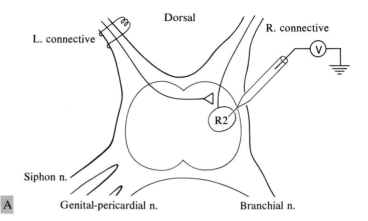

A

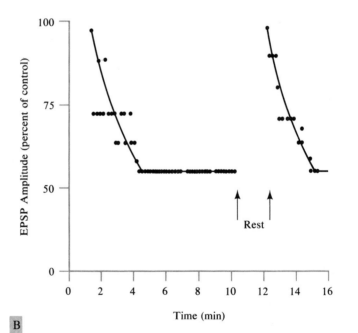

B

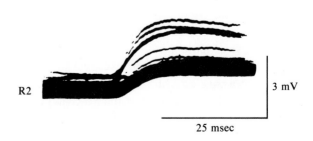

C

FIGURE 11-27. (*Facing Page*).

Neural analog of habituation. [From Kandel and Tauc, 1965a, b.]

A. Experimental setup. Intracellular recording of the complex (polysynaptic) EPSP in R2 produced by stimulation of a connective.

B. Response depression in a complex EPSP. The left connective was stimulated at 1/10 sec. Recovery followed a 2-min rest, at which point stimulation (1/10 sec) was again instituted.

C. Response depression in an elementary (monosynaptic) EPSP produced by stimulating the left connective. Consecutive sweep records have been superimposed to show the initial (large) response followed by a progressive depression to a steady state.

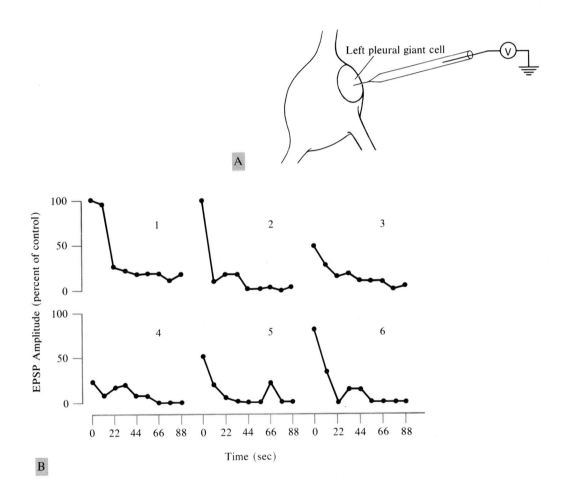

FIGURE 11-28.

Neural analog of habituation in the pleural ganglia. The amplitude of the first EPSP in a series is dependent on the number of stimuli administered in previous series and on the length of rest from previous series. [From Bruner and Kehoe, 1970.]

A. Experimental setup. Intracellular recordings are made from the pleural giant cell LP1 (see Figure 7-9, p. 233).

B. Six series of stimuli are separated by rests of 5 min (graphs 1-4), 20 min (graph 5), and 50 min (graph 6). Recovery is complete after the first series but is only very slight following the next three series of stimuli. Recovery is 50% after a rest of 20 min and 83% after a rest of 50 min.

Siphon

1st stimulus   2nd stim.   3rd stim.   19th stim.   20th stim.

A

Siphon   L. connective   Siphon

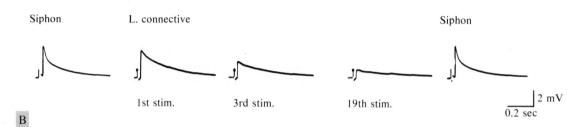

1st stim.   3rd stim.   19th stim.

                                                              ⌐ 2 mV
                                                         └─────
                                                         0.2 sec

B

FIGURE 11-29.

Depression in the EPSP from one nerve does not affect the EPSP from another. [From Bruner and Kehoe, 1970.]

A. Depression of the unitary EPSP in cell R2 in response to a series of 20 stimuli administered to the siphon nerve at 10-sec intervals.

B. After a 5-min rest, the siphon nerve is again stimulated; the response has largely recovered. Nineteen stimuli were then applied at 10-sec intervals to the left connective; the EPSP again shows depression. At the end of this series of stimuli to the left connective a test stimulus was applied to the siphon nerve. The unitary EPSP was unaffected by the previous stimulation of the left connective.

presynaptic stimulation. Faster rates cause greater depression than slower rates (Figure 11-31). However, rates faster than several per second result in posttetanic facilitation. Thus, stimulating the depressed synaptic pathway at 10 per second led to restoration of synaptic transmission.

In addition to depression in excitatory synaptic transmission, a second mechanism for response decrement, the buildup of inhibitory postsynaptic potentials, was proposed by Holmgren and Frenk (1961) based on a hypothesis first developed by Moruzzi (1959). Holmgren and Frenk found that electrical stimulation of the mantle nerve of the land snail at a rate of 1 to 5 per second first produced an excitatory response in the pleural ganglion cells. However, subsequent stimuli produced an oscillating type of hyperpolarizing activity that became more pronounced with repeated stimulation. After stimulation

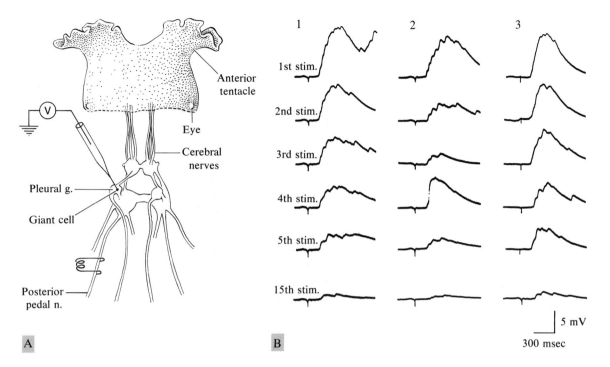

FIGURE 11-30.
Depression of a response in an isolated head preparation. [From Bruner and Tauc, 1966.]

A. Experimental preparation consists of the separated head and perioesophageal ganglia together with most of the connectives and peripheral nerves. The remaining body of the animal is removed. The buccal mass and the buccal ganglia are removed; the connectives between the head and cerebral ganglia are maintained and the peripheral nerves are isolated by removing the posterior part of the head and the rhinophores.

B. Depression of a complex (polysynaptic) EPSP during stimulation of the head by drops of seawater applied every 10 sec. *B-1.* First series of 15 stimuli. *B-2.* Second series of stimuli following 10 min of rest. *B-3.* Continuation of the second series without rest, but following repetitive stimulation (5/sec for 2 sec) of the left posterior pedal nerve. This repetitive stimulation (which itself produced a large complex EPSP, not shown in the figure) restores the initial amplitude of the EPSP (the one observed in the beginning of the first and second series). Similar to natural habituation, the compound action potential due to natural stimuli undergoes spontaneous recovery, and dishabituation.

was discontinued, the hyperpolarizing oscillation and its inhibitory effects subsided in the next 20 to 100 seconds. The nature of the hyperpolarizing oscillations was not specified but seem to be IPSPs recruited by the stimulus. Powerful recruitment of IPSPs also occurs in cells L7 and L10 in the abdominal ganglion of *Aplysia* following repeated stimulation of nerves or connectives (Figure 11-32; Waziri, Kandel, and Frazier, 1969).

Interstimulus interval = 4 min

A

Interstimulus interval = 15 sec

12.5 mV

B

FIGURE 11-31.
EPSP depression in the left pleural giant cell as a function of frequency of stimulation. **A.** The stimulus (drops of seawater on the tentacles) is presented every 4 min and does not produce depression. **B.** The stimulus is presented every 15 sec and produces rapid depression. [From Bruner and Kehoe, 1970.]

Although it does not fit the habituation paradigm, another process that leads to the depression of a neural response by an inhibitory mechanism is presynaptic inhibition, the heterosynaptic depression of the EPSP from one pathway produced by activity in another (Dudel and Kuffler, 1961; Eccles, Schmidt, and Willis, 1962; Frank and Fuortes, 1957). Presynaptic inhibition usually lasts for only a few hundred msec (see Figure 6-20, p. 201), but in *Aplysia* it lasts from several seconds to many minutes (Figure 11-33).

Heuristic Value of Neural Analogs

In addition to providing information about the mechanisms that regulate cellular and synaptic function, neural analogs of learning paradigms are heuristically useful. For example, studies of the cellular analogs illustrate a behaviorally interesting, although neurophysiologically obvious, distinction between nonassociative alpha and nonassociative classical conditioning (Kandel and Tauc, 1965a). In the neural analog, presentation of the priming stimulus increases the amplitude of the EPSP and ultimately produces a new response—the generation of a spike (see Figure 11-13). Thus, as in nonassociative classical conditioning (pseudo-conditioning), the priming stimulus leads to the generation of a new response to the test input.

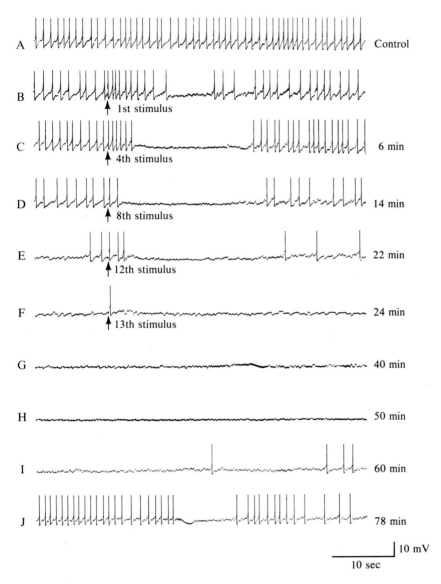

FIGURE 11-32.

Build-up of inhibitory postsynaptic potentials in cell L7 following stimulation of the left connective. **A.** Control. **B-F.** Records of the response to the 1st, 4th, 8th, 12th, and 13th stimuli. The initial stimulus elicits a brief spontaneous acceleration of the cells, followed by a recruitment of IPSPs that becomes more profound as the stimulus is repeated. **G-J.** Gradual decline in the frequency of the recruited IPSP activity over 54 min following the last stimulus. [From Waziri *et al.,* 1969.]

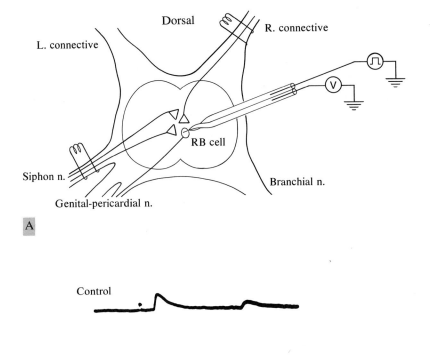

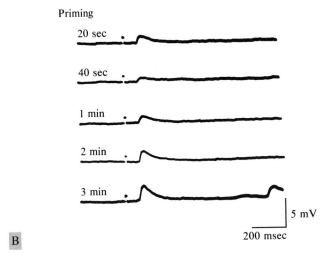

FIGURE 11-33.
Presynaptic inhibition in *Aplysia.* **A.** Experimental setup. **B.** A unitary monosynaptic test EPSP is produced in an RB cell by stimulating the right connective (control). A stimulus train (6/sec for 1 sec) was then applied to the siphon nerve (priming). Following stimulation, the test EPSP is inhibited for 3 min. [From Tauc, 1965.]

From a mechanistic point of view, however, the important change is not the generation of a spike, but the quantitative change in the amplitude of the test EPSP analogous to the quantitative change in a behavioral response during nonassociative alpha conditioning. Spike generation is an obligatory response of the cell; it must occur if the membrane potential reaches firing level. Whether or not a change in EPSP triggers a spike is determined by the amplitude of the test EPSP prior to pairing and by the potential difference between the peak of the EPSP and threshold. If the initial EPSP is small, the same absolute increase will not give rise to an action potential; if the initial EPSP is large the increase will give rise to an action potential. These arguments suggest that nonassociative classical conditioning may in some instances differ only quantitatively from nonassociative alpha conditioning.

BIOCHEMICAL ASPECTS OF NEURONAL PLASTICITY

In contrast to the extensive information available on cellular physiological mechanisms of plasticity, little is known about the biochemical mechanisms. But interest in this area has grown considerably in the last few years. Research has primarily been concerned with the possible role of macromolecular synthesis in memory and learning (Hydén, 1967; Hydén and Lange, 1970; Glassman, 1974; Agranoff, 1967). Several studies indicate that actinomycin D, an agent that blocks DNA-dependent RNA syntheses, or puromycin and acetoxycycloheximide, antibiotics that inhibit protein synthesis by interfering with ribosomal function, produce no impairment of short-term memory but do impair long-term memory (Agranoff, Davis, and Brink, 1966; Agranoff, Davis, Casola and Lim, 1967). However, the behavioral results from experiments in different species are not fully consistent. For example, whereas puromycin interferes with long-term memory in both goldfish and mice, acetoxycycloheximide does not interfere with long-term memory in fully trained or overtrained mice; it only interferes with marginal learning (for review see Barondes, 1970). Moreover, interference with memory induced by puromycin can be blocked by acetoxycycloheximide, although both drugs together still inhibit protein synthesis (Flexner, Flexner, and Stellar, 1963; Flexner and Flexner, 1966; Flexner, Flexner, and Roberts, 1967).

These powerful inhibitors of protein synthesis also have nonspecific effects. For example, when actinomycin D is given to mice in amounts sufficient to inhibit RNA synthesis by more than 75 percent, the mice rapidly become ill and die (Barondes and Cohen, 1967). Furthermore, puromycin produces occult hippocampal seizures in mice, which alone may fully account for the amnesic effects of this drug (Cohen, Ervin, and Barondes, 1966; Cohen and Barondes, 1967). Thus it is difficult to be certain whether the impairment of memory that can be demonstrated with puromycin or actinomycin D results from the inhibition of protein or RNA synthesis or from secondary effects of the inhibitors.

In addition, macromolecular changes cannot simply be related to learning directly. A particular biochemical change must first be causally related to a specific, presumably plastic, change in the activity of specific nerve cells and their interconnections, and these changes must in turn be causally related to the behavioral modification. This demonstration is essential because nonspecific stimuli can also affect the metabolism of neurons (Glassman, 1974). In none of the studies with mammalian organisms have these types of correlations been possible.

To overcome these problems, a number of simple systems are being explored. Three types of problems have been examined: (1) neuronal control of macromolecular synthesis; (2) the role of macromolecular synthesis in the short-term functioning of neurons; and (3) neural determinants of short-term behavioral modification. I shall consider the first two problems here and the third area in the next chapter (see p. 585).

Neuronal Control of Macromolecular Synthesis

Berry (1969) and Peterson and Kernell (1970) have reported that neural activity in *Aplysia* can lead to an increase of RNA synthesis in cell R2. Electrical stimulation of peripheral nerves and connectives for one hour (strong enough to elicit postsynaptic spikes in R2) resulted in an increase in the incorporation of ^{3}H-uridine into RNA in both the nucleus and cytoplasm. Action potentials elicited directly in R2 by depolarizing current pulses produced no effect. Peterson and Kernell (1970) therefore concluded that increased incorporation of the label into RNA resulted only from synaptic activation of the cell and not from the generation of the action potentials. However, whether these changes reflect new synthesis of RNA or only increased uptake of labeled material into the cell (thereby increasing the specific radioactivity of the precursor pool) is not clear (Wilson and Berry, 1972).

Macromolecular Synthesis and the Short-Term Functioning of Neurons

Schwartz, Castellucci, and Kandel (1971) examined the functioning of cells R2, R15, and L7 following inhibition of protein synthesis. They used antibiotics (anisomycin, sparsomycin, and pactamycin) that irreversibly inhibit protein synthesis by up to 97 percent and found that for up to 30 hours inhibition had no effect on the membrane potential or spike generation of these neurons. In autoactive cells (L7 and R15) the capability for spontaneous activity remained. At several synapses that make use of different chemical transmitter substances, synaptic transmission was unimpaired. Even spontaneously occurring EPSPs produced by the action of autoactive interneurons persisted. The available data therefore suggest that the normal short-term function (one day or less) of invertebrate neurons does not depend upon new protein synthesis.

The formation of new protein is also not necessary for certain plastic changes. Schwartz and co-workers (1971) tested the effect of blocking protein synthesis on the posttetanic facilitation of the monosynaptic cholinergic EPSP produced in R15 following a 30-sec tetanization of the right connective (see Figure 11-10). Incubations with anisomycin for periods up to 31 hours did not alter the amplitude and duration of posttetanic facilitation (Figure 11-34). In Chapter 12 I will consider other types of short-term plastic changes that are also independent of new protein synthesis. Thus, although long-term plastic changes may depend on new protein synthesis, the short-term plastic changes so far examined do not. Nor do they depend on the synthesis of proteins that have very fast turnover rates.

Neuronal Control of Small-Molecule Synthesis

The results of experiments with protein-synthesis inhibitors draws attention to the need for new biochemical models of neuronal plasticity not based on alterations in macromolecular synthesis. Because of the central importance of the intracellular messenger cAMP in regulating cellular function,[6] Cedar, Kandel, and Schwartz (1972) examined the effect of neuronal activity on the synthesis of cAMP in the abdominal ganglion of *Aplysia* (for earlier studies in vertebrates see Greengard and Kebabian, 1974). They found that electrical stimulation (1/sec for 15 min) of the ganglion's connectives produced a twofold increase in the synthesis of cAMP that persisted for 45 minutes. The increase seemed to be synaptically mediated; it was blocked by high concentrations of Mg^{++}. The increase appeared to be restricted to the neuropil regions of the ganglion; it could not be detected in the cell bodies of individual neurons nor in the connectives, suggesting that the increase in cAMP brought about by electrical stimulation may be confined to regions of synaptic contact.

[6]cAMP is the abbreviation for 3'5' cyclic adenosine monophosphate. For review see Robison, Butcher, and Sutherland, 1971; see also p. 588.

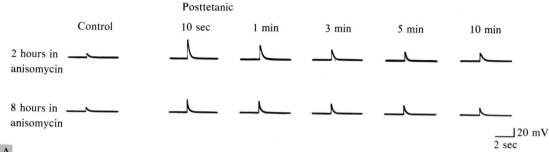

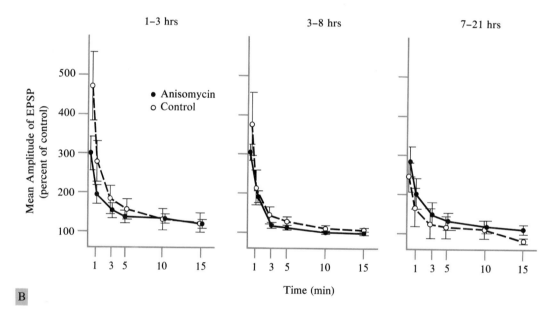

FIGURE 11-34.
Posttetanic facilitation of a cholinergic EPSP in the absence of protein synthesis. [From Schwartz, Castellucci, and Kandel, 1971.]

A. Selected intracellular records from cell R15 (see Figure 11-10 for the experimental setup). The cell was hyperpolarized to prevent firing. The right connective was stimulated every 10 sec to evoke an elementary monosynaptic EPSP. After a brief tetanization (5/sec for 30 sec) the EPSP was again evoked at the control rate. In this experiment posttetanic facilitation was produced every hour. Examples are given of the EPSPs and subsequent potentiation after two and eight hours in the presence of anisomycin. There was a similar increase and duration of the effect in all cases.

B. Amplitude and time course of posttetanic facilitation in R15 in the presence (control) and absence of anisomycin. The mean value of about 10 EPSP amplitudes was taken as control (100%). Ganglia were bathed in artificial seawater for one hour and then incubated either in the presence or absence of anisomycin for additional periods of time as indicated. The mean values ($\pm$ SEM) of the amplitudes of the initial EPSPs at various intervals after tetanization are expressed as percentages of the control value.

In order to determine which transmitter substances might be producing this effect, several candidates—serotonin, dopamine, octopamine, carbamyl-choline, norepinephrine, glutamate, and histamine—were examined. When ganglia were briefly incubated with each of these transmitter substances, only serotonin, dopamine, and octopamine stimulated the formation of cAMP (Cedar and Schwartz, 1972; Levitan and Barondes, 1974). It seems, there-fore, that synaptic activity can control cAMP levels in the postsynaptic cells by means of some, but not all, transmitter substances. We have previously seen that transmitters do not produce synaptic actions of a specific sign. Thus, ACh, serotonin, and dopamine can each produce excitation and inhibition (Chapter 8). However, studies of the effects of synaptic transmission on cAMP synthesis suggest that the biochemical changes resulting from transmitter action may be more specific. But more work needs to be done to allow con-fidence in this argument.

Although the functional role of the biochemical changes resulting from synaptic action is not known, it is inviting to postulate that it may be associ-ated with plastic changes in neural circuits. For example, the biochemical mechanisms underlying short-term plastic change may be composed of a series of reactions that determine the distribution of transmitter substance or its availability. An intracellular messenger such as cAMP could act on a pathway of this kind to stimulate the mobilization of transmitter from one compartment (a long-term store) to another (an immediately releasable store), or it might control the number and availability of release sites. Several recent reports of altered synaptic activity might reasonably be explained by this mechanism (Dudel, 1965; Goldberg and Singer, 1969; Kuba, 1970). Thus, sequential reac-tions that involve the synthesis of degradable small molecules but not new structural receptor or enzymatic components could provide mechanisms for plastic changes.

MORPHOLOGICAL ASPECTS OF NEURONAL PLASTICITY

Whereas biochemical techniques are only beginning to be applied to the study of neuronal plasticity, morphological studies have progressed significantly. These studies have been carried out on two levels. On one level, studies have examined how the collateral axons of a neuron are affected by degeneration (due to lesions) of neighboring neurons. These studies have shown that col-lateral axons can *sprout* into vacated synaptic regions; with reinnervation there is regression of sprouts. Thus, whereas *activity* in one pathway usually produces only relatively brief heterosynaptic changes in another, lasting minutes to

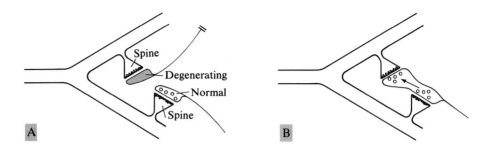

FIGURE 11-35.
A schematic representation of sprouting. Two postsynaptic sites (spines) are illustrated. Each is innervated by a different type of postsynaptic fiber. **A.** Cutting one presynaptic fiber leads to its degeneration. **B.** Degeneration is followed by collateral sprouting and reinnervation of denervated sites by adjacent nondegenerating terminals. [From Raisman and Field, 1973.]

hours, *lesioning* of one pathway causes the sprouting of connections by a parallel pathway, the consequences of which can be long lasting (Figure 11-35; Raisman, 1969; Schneider, 1970; Steward, Cotman, and Lynch, 1973; Grinnell, 1976).

Sprouting has also been studied in several simple neuronal systems. For example, Courtney and Roper (1976) have found that in the cardiac ganglion of the frog most cells (87 percent) receive innervation from branches of the ipsilateral vagus nerve, although some cells (48 percent) also receive connections from the contralateral vagus. Following partial dennervation of these cells by cutting of the ipsilateral vagus, effective synaptic connections begin to sprout from the contralateral vagus within 24 to 48 hours. The number of synaptic terminals per cell seems to stay the same throughout the eight-day period during which sprouting occurs, and the number is the same as in cells of unoperated animals. These results support the idea of sprouting and suggest that the total number of synaptic sites on a cell may be fixed. Moreover, the findings are consistent with the idea developed in the mammalian central nervous system (Hubel and Wiesel, 1965; Grinnell, 1976) that activity in one pathway not only maintains the viability of its synaptic connections but also inhibits the sprouting capability of other pathways that end on that cell.

Nicholls and his colleagues have found even more radical changes in the leech, at synaptic connections within and between ganglia following lesions of nerves and connectives. Isolating a segmental ganglion (by cutting connectives and peripheral nerves) actually causes some connections to change their sign: sensory neurons that normally produce EPSPs will, after the surgical procedure, produce IPSPs in a certain motor neuron. In another motor cell a large chemical EPSP that does not occur in unoperated animals will occur in response to stimulation of sensory neurons. There is indirect evidence that some of the seemingly new synaptic connections that appear after isolation of the ganglion normally occur in a small and less obvious form. The postsurgical synaptic enhancement may represent selective sprouting of some connections as other connections to the motor cells degenerate.[7]

The most remarkable example of neuronal sprouting occurs normally in the vertebrate olfactory system, where the epithelium actually makes new nerve cells throughout the life of the animal. Unlike other nerve cells, the olfactory receptor neurons turn over every 14 to 30 days and are replaced by new ones produced from nonneural precursor cells, the "basal" supporting cells of the olfactory mucosa. The waves of newly differentiated neurons send their axons to the olfactory bulb and establish new synaptic contacts, replacing old ones on the dendrites of the mitral cells (Graziadei and Metcalf, 1971).

On a second, fine structural level, studies have so far only examined the morphological changes within the presynaptic terminal that are associated with short-term high-frequency activity. Preliminary results suggest that prolonged high-frequency activity leads to depletion of synaptic vesicles from the terminals and swelling of the presynaptic terminal membrane (see Figure 6-18B; and Heuser and Reese, 1974). These morphological approaches are very promising. Several key questions concerning synaptic plasticity can best be answered by combining behavioral and morphological studies on the same synapse. What are the morphological concomitants of homosynaptic use? Of disuse? Of heterosynaptic facilitation and depression? Do these involve alterations in the extent of the synaptic area? In the size of the synaptic cleft? In the number or disposition of the synaptic vesicles in the presynaptic terminals? Through a combination of fine structural and electrophysiological analyses, some answers to these questions may be forthcoming. In addition, morphological analyses are likely to clarify the slower, long-term changes that accompany the sprouting of connections.

[7]In addition to sprouting, denervation may involve other factors. In the nerve-muscle synapse of vertebrates and in the vagal synapses on the nerve cells of the cardiac ganglion of the frog, denervation leads to a spread of receptor molecules beyond their normal restricted location just underneath the synaptic terminal. This spread of chemosensitivity contracts with reinnervation (Axelsson and Thesleff, 1959; Miledi, 1960; Kuffler et al., 1971).

SUMMARY AND PERSPECTIVE

The importance of specific neuronal connections for behavior has now been well demonstrated in both invertebrates and vertebrates. To reconcile the notion of a predetermined nervous system with the known modifiability of behavior, the hypothesis of neuronal plasticity has been advanced. A current version of this hypothesis states that even though the anatomical connections between neurons may develop according to a definite plan, the strength or effectiveness of the connections, as well as other properties of nerve cells, are not entirely predetermined and may be altered over time. This hypothesis predicts that neurons, and in particular their synapses, should prove capable of changing their functional properties as a result of repeated or patterned activity.

The plasticity of neurons has been examined both as a function simply of usage and as a function of complex stimulus sequences derived from learning paradigms. These tests of plasticity have revealed that chemical, but not electrical, synapses have considerable plastic potentiality. As a result of previous activity, chemical synapses can undergo short-term alterations of functional effectiveness (lasting from minutes to hours). These changes can be regarded as a form of information storage. Changes in synaptic efficacy are not related to usage in any simple manner. Usage can lead to enhancement at some synapses, depression at others, and to either depression or enhancement at yet others, depending on frequency of use. Moreover, without changing the use of a pathway, effectiveness can be altered by activity in a neighboring pathway.

Changes in synaptic efficacy appear to involve alterations in the regulatory mechanisms within the presynaptic terminals controlling transmitter release (see p. 198 and Figure 6-21). Two sets of mechanisms are thought to be involved. One set, limited to the stimulated, homosynaptic pathway, underlies low-frequency depression, posttetanic depression, and facilitation. A second mechanism is thought to underlie some types of heterosynaptic facilitation and heterosynaptic depression, and involves alterations in one pathway following activity in another.

Other components of neurons, in particular the pattern of the autoactive rhythms of burst-generating cells, are also capable of undergoing plastic change. Here the mediating event is synaptic, but the plasticity of neural action seems to reside in the pacemaker properties of the postsynaptic cell. Table 11-3 summarizes the demonstrated, as well as some possible, sites of plasticity within a single neuron. It appears likely that the different components of the neuron can undergo different and possibly independent plastic changes.

TABLE 11-3
Possible mechanisms for information storage in the central nervous system.

1. Changes in synaptic efficiency directly related to presynaptic transmission:
   A. Homosynaptic changes in synaptic efficacy: EPSPs and IPSPs
   B. Heterosynaptic changes in synaptic efficacy: EPSPs and (?) IPSPs

2. Postsynaptic modulation of the autoactive rhythm of pacemaker cells:
   A. Increments and decrements in pacemaker activity
   B. Turning-on and turning-off of pacemaker activity

3. Changes in the spike-generating capacity of postsynaptic neurons:
   A. Change in the number of remote trigger zones
   B. Change in the extent of the excitable membrane of the cell body or dendrites
   C. Change in the threshold of the trigger zone

4. Changes in the properties of postsynaptic receptors and ionophores:
   A. Spatial extent of the chemosensitive area
   B. Density of receptor sites
   C. Chemical sensitivity of receptor sites
   D. Number and properties of the ionic channels controlled by each receptor site

The study of plasticity is only beginning; many unsolved problems remain. For example, the plastic changes so far described last only minutes and hours. Yet long-term memory lasts days, weeks, and even years. The means by which such long-term changes come about are largely unknown. In addition, although there now exists a catalog of plastic changes, cellular studies do not tell which if any of these mechanisms are relevant to specific instances of behavioral modification in the whole animal.

Apparent similarities between instances of plasticity in an isolated ganglion and behavioral modifications in the whole animal must be viewed with caution. Behavioral modification involves a change in the behavior of the organism, whereas studies of neural analogs and other types of plasticity examine the outputs of neurons, independent of their role in behavior. The relationship between a specific plastic change in neurons and a specific modification of behavior needs in each case to be determined by studies in intact animals. Even at this early juncture it is clear that a given modification of behavior could result from one of several cellular processes. For example, response depression of the sort that might produce habituation could be caused by IPSP recruitment, changes in cellular threshold, increased presynaptic inhibition, or EPSP depression, to list only a few of the most likely mechanisms. Only by analyzing specific instances of habituation can one determine which if any of these processes are operative.

Tanzi and Cajal, early theorists of cellular connectionism, viewed the cellular mechanisms of learning as an extension of the embryological mechanisms involved in synapse formation. A possible relationship between embryological processes and learning is also indicated in studies of imprinting, a type of behavioral modification that occurs only in a critical early phase of development (see p. 20). The sprouting of new connections following the degeneration of older ones also resembles the embryonic development of synapses. However, the work on plasticity reviewed here suggests that at least some changes in neuronal function occur rapidly and may not require an actual, or at least an extensive, physical outgrowth of presynaptic processes. As a result, there seems to be a distinction between ontogenetic and experiential types of plasticity. Whether this distinction is useful and reflects differences in mechanisms will require more detailed comparisons.

SELECTED READING

Eccles, J. C., 1953. *The Neurophysiological Basis of Mind: The Principles of Neurophysiology.* Oxford: Clarendon Press. A remarkable book. The first systematic attempt to think about integrative processes and learning from a cellular perspective.

Hebb, D. O. 1949. *The Organization of Behavior: A Neuropsychological Theory.* New York: Wiley. One of the two most influential books in physiological psychology. It develops a two-process model of memory and emphasizes the importance of early experience.

Konorski, J. 1948. *Conditioned Reflexes and Neuron Organization.* Cambridge University Press. Reprinted 1968. The other highly influential work. An attempt to combine Pavlov's approach to learning with Sherrington's approach to the nervous system. A clear, modern statement on neuronal plasticity and its relation to learning.

CHAPTER 12

Short-Term
Behavioral Modification

One of the most fundamental problems in the neurobiology of behavior concerns the mechanisms of learning and memory. What types of changes does learning produce in the nervous system? How and where is memory stored? Tests of the plasticity hypothesis of learning reveal that chemical, but not electrical, synapses can undergo alterations in functional effectiveness as the result of activity. Other components of the neuron also undergo plastic changes. Are these changes expressed in behavior? Are plastic changes in existing neuronal pathways sufficient to account for the effects of learning? Or does learning produce a more radical rearrangement of the neuronal circuitry, such as an outgrowth of new nerve cells or a sprouting of new connections? To answer these questions, the neuronal circuits of a particular behavior must be examined at the same time the behavior is being modified.

In this and the next chapter I shall consider the cellular mechanisms underlying two simple types of behavioral modification, habituation and dishabituation—the two types that have been studied most successfully at the cellular level. These paradigms were briefly discussed earlier in the book (Chapter 1, p. 5; Chapter 11, p. 494).

GENERAL FEATURES OF HABITUATION AND DISHABITUATION

Habituation is a gradual decrease in the amplitude or probability of a response to repeated presentation of a particular stimulus. Once a response has been habituated, either of two processes can lead to its restoration. *Spontaneous recovery,* a gradual restoration of the response, occurs as a result of withholding the stimulus to which the animal has habituated. *Dishabituation,* a sudden restoration of the response, occurs as a result of a change in the pattern of the stimulus or the introduction of another stimulus.

The habituation paradigm is particularly attractive to pursue at the cellular level for two reasons. First, habituation is ubiquitous in nature. Second, it is the simplest type of learning and therefore easier to study on the cellular level than other, more complex types. Aspects of the cellular mechanisms of habituation have been successfully investigated in several animals: *Aplysia,* crayfish, cockroach, frog, and cat. Habituation also has features in common with other kinds of learning and is sometimes a component of more complex learning. Consequently, neural studies of habituation could provide insights into the mechanisms of other types of learning.

Habituation is involved in many aspects of achieving familiarity with and mastery over the environment and is experienced in one form or another by all animals (including man) in everyday life. When, for example, a sudden noise is perceived, a number of concomitant autonomic changes are triggered, e.g., increases in heart rate and respiratory rate. If the noise is repeated, or if some other stimulus commands attention, response to the noise will abate. Thus, one can become accustomed to initially distracting sounds and learn to work effectively in an otherwise noisy environment. One becomes habituated .to the sound of the clock in the study, and to one's own heart beat, stomach movements, and other bodily sensations. These enter awareness rarely and only under special circumstances. In this sense, habituation is learning to recognize, and ignore as familiar, recurrent external or internal stimuli. Thus, city dwellers may hardly notice the noise of traffic at home but may be awakened by the chirping of crickets in the country. Because of its simplicity as a test for recognizing familiar objects, habituation of autonomic responses to novel visual stimuli is one of the most effective means for studying the development of perception and memory in infants (Cohen, 1973; Jeffrey, 1968).

A related sense in which habituation operates is elimination of inappropriate or exaggerated defensive responses. This is beautifully illustrated in the following fable (apologies to Aesop):

A fox, who had never yet seen a turtle, when he fell in with him for the first time in the forest was so frightened that he was near dying with fear. On his meeting with him for the second time, he was still much alarmed but not to the same extent as at first. On seeing him the third time, he was so increased in boldness that he went up to him and commenced a familiar conversation with him.

The elimination of responses that fail to serve useful functions is as important as the development of new ones. Immature animals often show escape responses to a variety of nonthreatening stimuli. As a result of habituation to common, innocuous stimuli, an animal can put a large number of stimuli that do not affect survival beyond its attention. The animal can focus on stimuli that are novel or that become associated with either satisfying or alarming consequences. Habituation is therefore as important in the development of selective responsiveness to the environment as it is in organizing perception. This is dramatically illustrated in the escape response of birds to a flying hawk. Tinbergen (1951) used hawk models to analyze the responsiveness of birds to the hawk and found that fleeing occurred only when the hawk model flew overhead in a forward, not backward, direction. Schleidt (1961a, b) reexamined this response using turkeys raised in a controlled environment. He found that when first exposed to a circle or square model overhead, the birds fled. With repeated exposure, the turkeys habituated to these stimuli and failed to respond to them. When occasionally exposed to a hawk model they gave fleeing responses. But turkeys repeatedly exposed to hawk models habituated to these and fled in response to circles or squares! The initial response to the novel stimulus was comparable in both groups. Thus, the fact that turkeys make escape responses to hawks but not other birds could result from their having habituated to the shape of birds that they frequently see flying overhead but not to hawks, which they see rarely.

Habituation is not restricted to escape responses. The frequency with which sexual responses are expressed also can be decreased through habituation. Given free access to a receptive female, a male rat will copulate six or seven times over a period of one or two hours, following which he appears sexually exhausted and becomes inactive for 30 minutes or longer. This is sexual habituation, not fatigue. An apparently exhausted male will promptly resume mating if a new female is made available (Fowler and Whalen, 1961).

Frequency of ejaculation as well as time for arousal can also be habituated. A male and female pair of rhesus monkeys first placed together in a cage will couple rapidly and often with little foreplay. After several days the frequency of copulation decreases and every act is preceded by more lengthy preliminaries, each partner examining and stimulating the other. If the male is exposed to a new female partner, he becomes immediately aroused and potent and preliminaries are dropped (Hamilton, 1914).

Aggressive behavior can also be habituated. For example, the three-spine stickleback (*Gasterosteus aculeatus*) defends its territory against intruders. Upon repeated exposure to an intruder, the fish gradually shows less frequent aggressive responses. Habituation of this response is correlated with enhanced responsiveness to sexual stimuli. Thus, habituation to one stimulus may free an animal to respond more effectively to other stimuli that may be necessary for its survival and therefore that of the species (Peeke, Hertz, and Gallagher, 1971).

Habituation is frequently also a component of more complex learning. Animals cannot learn new responses to a stimulus without first eliminating incorrect or inappropriate responses to the stimulus. Habituation is an essential feature of animal training in which undesired responses, such as gun-shyness in dogs or motor-shyness in horses, need to be eliminated (Humphrey, 1933). Habituation serves to economize an animal's energy by restricting behavior to those responses that have rewarding or threatening consequences. According to Wyers, Peeke, and Hertz (1973), habituation serves three adaptive functions.

1. It restricts defensive and escape responses to (infrequent) stimuli most likely to herald a predator's approach.

2. It helps to define for the animal its territory. In a familiar environment an animal habituates to stimuli that in an unfamiliar environment would elicit orienting and exploratory as well as defensive and escape responses. By reducing competing responses, habituation to the environment (the establishment of territory) enhances reproduction.

3. It tends to standardize social behavior of different individuals of the same species and sharpen the response to sign stimuli (see p. 15) by restricting the release of species-specific response patterns to optimal and appropriate situations.

The generality and simplicity of the habituation paradigm have prompted several students of behavior (Schneirla, 1929; Harris, 1943; Thorpe, 1943) to argue that habituation is the most elementary type of learning. This notion has been elaborated by Harlow, who holds that habituation may be the basic modification of behavior from which more complex learning arises.

If one surveys the literature on subvertebrate learning, a single fact stands out in a very unequivocal manner. Learning, all learning which has been adequately described and measured, appears to be the learned inhibition of responses and response tendencies which block the animal or fail to lead it to some terminal response, such as eating or escape from noxious stimulation. Furthermore, it should be remembered that many of these lower animals have reasonably well-developed, somatic-type nervous systems, and there is no reason from an anatomical point of view to suspect that the nature of learning is going to be altered in any subtle, fundamental manner as we progress to higher forms. The law of parsimony requires, at the very least, that we seek as simple a fundamental explanation of vertebrate learning—including human learning—as is consonant with fact. . . . I wish to take the position here that there exists no fundamental difference, other than complexity, between the kinds of learning listed as habituation learning and the kinds listed as conditioned response learning—and, for that matter, the kinds of learning described as reasoning and thinking.

I hold this position in spite of the fact that the Pavlovian-type conditioned response has been taken by many as the paradigm for all learning, with the assumption that a conditioned response involves the formation of some new association. . . . It is however obvious that the animal already possessed the potential capacity to flex to either the (CS) light or sound. . . . From this point of view we may give serious consideration to the fact that conditioning does not produce new stimulus-response connections but that it operates instead to restrict, specify, and channelize stimulus-response potentialities already possessed by the organism. It is entirely possible that a specific visual conditioned stimulus comes to elicit a specific leg flexion response because the presentation of the CS and US in a specific temporal pattern produced inhibition of response to extraneous and distracting external stimuli and inhibition of the other postural responses which the CS was already capable of eliciting.[1]

Harlow has developed a theory of learning that attempts to account for classical conditioning, spatial discrimination, object or cue discrimination, and abstract thought by habituation of inappropriate responses.

Although Harlow greatly overstates the case for habituation, I have quoted him at length only to show that it is possible to establish in theory a hierarchy of learning using habituation as the basic building block. Nevertheless, the habituation paradigm is attractive for cellular analysis not because of this speculation—which quite likely is wrong—but because of its simplicity and ubiquity, which facilitate cellular analysis.

[1]Harlow, 1958, pp. 285–287.

HABITUATION AND DISHABITUATION OF
THE GILL-WITHDRAWAL REFLEX

Like most defensive reflexes, the gill-withdrawal reflex of *Aplysia* undergoes habituation and dishabituation (Figure 12-1).[2] Pinsker and co-workers (1970) found that after a single training session of 10–15 trials, with intervals ranging from 10 seconds to 3 minutes, the response habituates to roughly 30 percent of control. Further stimulation depresses the response even more. Full recovery, as tested with only a single stimulus, requires a period of rest (no stimulation) ranging from 30 minutes to several hours. During the first 10–20 minutes of rest there is a rapid recovery phase that accounts for 75–85 percent of the recovery, followed by a slow and variable return over the next hour or more to the original response level (Figure 12-2). The habituated response is immediately restored, however, if a novel, dishabituatory stimulus, e.g., a strong stimulus to the neck, is presented. On occasion, the dishabituated response is larger than the initial nonhabituated response.

[2]As we have seen (p. 353), molluscs have peripheral neurons that mediate local responses. Some of these responses also habituate when repeatedly elicited (Kandel and Spencer, 1968; Peretz, 1970; Kupfermann, Pinsker, Castellucci, and Kandel, 1971). Whereas the central nervous system is required for habituation of the gill-withdrawal reflex to weak and moderate stimuli to the siphon, the pinnule response to direct stimulation of the gill and its habituation do not require the central nervous system; this response is mediated by independent peripheral neurons (Peretz, 1970; Kupfermann, Carew and Kandel, 1974). The relationship between the centrally mediated gill withdrawal and the peripherally mediated pinnule response is considered in Kandel, *The Behavioral Biology of Aplysia,* forthcoming.

FIGURE 12-1.

Short-term habituation and dishabituation of the gill-withdrawal reflex in *Aplysia*. [From Pinsker, Kupfermann, Castellucci, and Kandel, 1970. Copyright by the American Association for the Advancement of Science.]

A. Experimental arrangement for behavioral studies in the intact, restrained animal. The animal is immobilized in a small aquarium in which cooled, filtered, and aerated seawater is circulated. The edge of the mantle shelf is pinned to a substage and a constant and measurable tactile stimulus (brief jets of seawater from one or more commercially available Water Piks) is applied. The behavioral response (gill contraction) is monitored by means of a photocell placed under the gill; the electrical output of the photocell is recorded on a polygraph.

B. Habituation, spontaneous recovery, and dishabituation of the gill-withdrawal reflex. Records from two training sessions with the same animal. *B-1.* Habituation of the response with 80 repetitions of the stimulus at 3-min intervals. The major decrease occurs within the first 10 stimuli. Following a 122-min rest the response recovered partially. *B-2.* A later training session after the animal had fully recovered; the response was habituated a second time at 1-min intervals. A dishabituatory stimulus, consisting of a strong and prolonged tactile stimulus to the neck region, was presented. Following the dishabituating stimulus, responses were facilitated for several minutes.

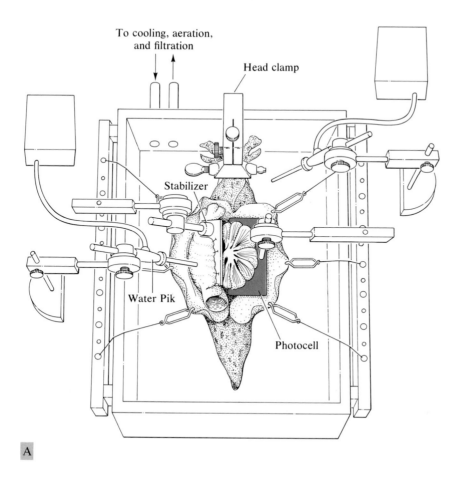

To cooling, aeration, and filtration

Head clamp

Stabilizer

Water Pik

Photocell

A

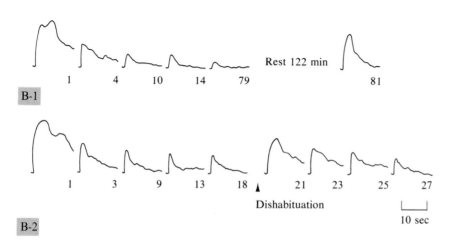

1   4   10   14   79   Rest 122 min   81

B-1

1   3   9   13   18 ▲ 21   23   25   27

Dishabituation

10 sec

B-2

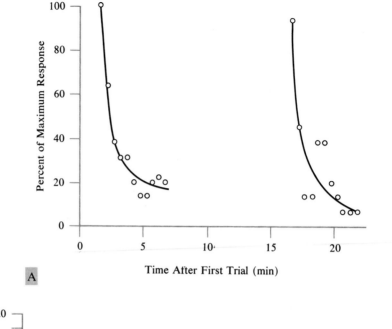

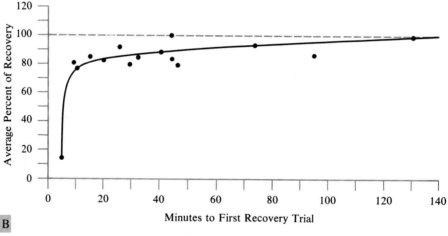

FIGURE 12-2.

Time course of short-term habituation and spontaneous recovery of the gill-withdrawal reflex.

A. Time course of habituation, recovery, and rehabituation. The gill-withdrawal reflex is habituated by presenting a series of tactile stimuli to the siphon once every 30 sec. After a rest of 10 min, the response spontaneously recovers to almost control level. When the response is rehabituated, the depression is slightly more profound than in the first training session. [Modified from Kupfermann, Pinsker, Castellucci, and Kandel, 1971. Copyright by the American Association for the Advancement of Science.]

B. Time course of recovery as tested with a single stimulus following one training session. Recovery was estimated by habituating individual animals and then testing for the percent of recovery of the first response, after different intervals of rest. The curve is based on 44 separate habituation and recovery runs in 27 animals. Each point is the average of three measures (last point based on only two) taken at roughly the same interval. The shortest time in which full recovery occurred was 10 minutes; the longest was 122 minutes. [From Pinsker *et al.,* 1970. Copyright by the American Association for the Advancement of Science.]

Of the nine features that commonly characterize the short-term habituation in vertebrates (see p. 518), eight are found in the gill-withdrawal reflex. These include: (1) response depression; (2) spontaneous recovery; (3) dishabituation; (4) habituation to the dishabituatory stimulus with repeated presentations; (5) less habituation with stronger stimuli; (6) greater habituation with shorter stimulus intervals; (7) more rapid depression with repeated periods of habituation and rest (see p. 610); and (8) prolongation of recovery by continued stimulation after the response has been abolished (subzero effect; see Figure 13-14, p. 614). One feature that is not prominent is generalization of habituation. In *Aplysia*, as with certain responses in man, generalization of habituation is very restricted (see Figure 13-12, p. 628).

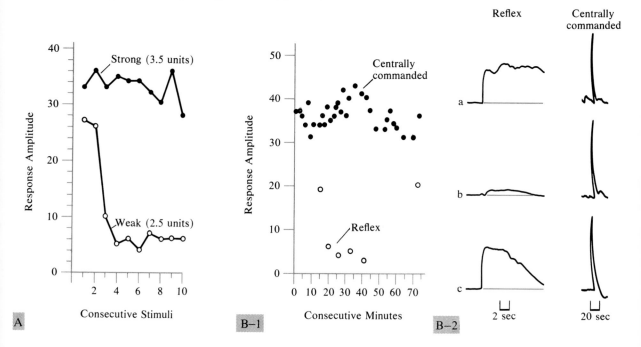

FIGURE 12-3.
Some features of habituation of the gill-withdrawal reflex. [From Pinsker *et al.*, 1970. Copyright by the American Association for the Advancement of Science.]

A. Habituation with weak and strong tactile stimuli. Responses to the strong stimulus (3.5 units on the Water Pik) are larger and show less depression with repetition than responses to the weak stimulus (2.5 units). (The intensity of the stimuli produced by the Water Pik ranges from 1, light touch, to 5, intense pressure.) The interstimulus interval is 1.5 minutes in each case.

B. A comparison of centrally commanded gill contractions and reflex responses of the gill. *B-1.* Centrally commanded gill contractions remain constant in amplitude whereas reflex contractions produced by five tactile stimuli presented to the siphon at 5-min intervals show habituation and then recovery after a 30-min period of rest. *B-2.* Sample records from this experiment compare the amplitudes of reflex and centrally commanded contractions before reflex habituation (B-2a), during maximum habituation (B-2b), and following recovery with rest (B-2c).

NEURAL ANALYSIS OF HABITUATION OF
THE GILL-WITHDRAWAL REFLEX

Localization of Functional Change

Figure 12-4 shows the neural circuit of the gill-withdrawal reflex to weak tactile stimuli applied to the siphon. The possible sites at which a plastic change in neuronal function could account for habituation are indicated. Even in this simple circuit there are nine possible sites for behavioral modifications.

The first sites to consider in specifying the cellular locus of habituation are peripheral. Changes could occur in the sensory receptors of the siphon skin (Figure 12-4B, locus 1), the nerve-muscle synapse (locus 2), or the muscle (locus 3). Do any of these peripheral loci have a role in habituation?

MOTOR FATIGUE AND RECEPTOR ADAPTATION

Contractions of the gill, somewhat similar to the reflex withdrawal response to tactile stimulation, also occur spontaneously as part of the centrally commanded pumping movements (see Figure 9-22B, p. 380). The ability of the gill to respond to these central commands is unaffected by habituation and recovery of the gill-withdrawal reflex (Figure 12-3B). In addition, the habituated gill-withdrawal reflex can be dishabituated. These findings suggest that neuro-

FIGURE 12-4.
Neural circuit of the gill component of the withdrawal reflex to weak stimulation of the siphon (see Figure 9-17, p. 371, for the total reflex).

A. The detailed circuit. The sensory input from the siphon skin to the gill motor neurons is mediated by direct monosynaptic connections from the sensory neurons and via two excitatory interneurons. (Synapses are schematically indicated as being on the cell body; actually they are on the initial segment of the axon and its branches.) The sensory neurons are a population of about 24 cells. The motor neurons and the excitatory and inhibitory interneurons indicated are identified cells. The large and small terminals from gill motor neurons onto the gill represent differences in the output of the different motor neurons; small terminals represent weaker actions than those represented by large terminals.

B. A segment of the total circuit indicating possible loci of plasticity. Locus **1** represents receptor adaptation; locus **2** neuromuscular depression; locus **3** muscle fatigue; locus **4** buildup of post-synaptic inhibition on the motor neuron; locus **5** postsynaptic (or presynaptic) inhibition of the excitatory interneuron; locus **6** presynaptic inhibition of the sensory neuron; locus 7 resistance decrease of the nonsynaptic membrane of the motorneuron; locus **8** synaptic depression of the connection between the excitatory interneuron and the motor neuron; and locus **9** synaptic depression between the sensory neuron and the motor cell. [Modified from Kupfermann and Kandel, 1969; and Kupfermann, Carew, and Kandel, 1974.]

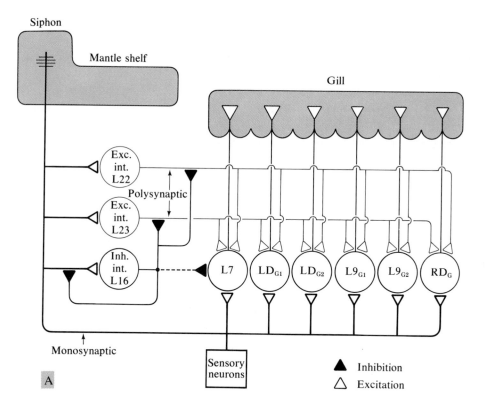

A

Siphon

Mantle shelf

Gill

Exc.
int.
L22

Polysynaptic

Exc.
int.
L23

Inh.
int.
L16

L7   LD$_{G1}$ LD$_{G2}$ L9$_{G1}$ L9$_{G2}$ RD$_{G}$

Monosynaptic

Sensory
neurons

▲ Inhibition
△ Excitation

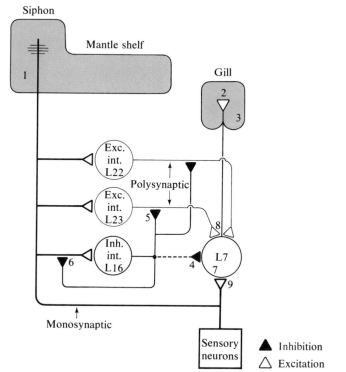

B

Siphon

Mantle shelf

Gill

1

2

3

Exc.
int.
L22

Polysynaptic

Exc.
int.
L23

5

8

Inh.
int.
L16

6

4

L7
7

9

Monosynaptic

Sensory
neurons

▲ Inhibition
△ Excitation

muscular fatigue or fatigue of the gill musculature may not be an important factor in habituation (Pinsker *et al.,* 1970).

To obtain direct evidence that the locus of habituation is not at the nerve–muscle synapse (Figure 12-4*B,* locus 2) or in the muscle itself (locus 3), identified motor neurons innervating the gill musculature need to be stimulated and the effectiveness of these procedures in producing habituation observed. Repeated direct stimulation of the motor neurons to the gill, at intervals that are effective in producing habituation in the intact animal, produces a contraction of the gill that does not attenuate (Figure 12-5*B;* and see Kupfermann, Castellucci, Pinsker, and Kandel, 1970; Carew, Pinsker, Rubinson, and Kandel, 1974). Even when the reflex response to sensory stimuli is depressed by habituation, direct stimulation of gill motor neurons continues to produce responses equivalent to those produced prior to habituation (Figure 12-6*C*). Similarly, following a dishabituatory stimulus, when the reflexive gill contraction increases, the directly evoked gill contraction is unchanged (Kupfermann *et al.,* 1970).

The role of sensory adaptation (Figure 12-4*B,* locus 1) can be investigated by recording from single afferent units in peripheral nerves and intracellularly from individual sensory neurons (Kupfermann *et al.,* 1970; Byrne, Castellucci, and Kandel, 1974a, b, and unpub.). These recordings show that the response of the sensory neurons is unchanged when the skin is stimulated at intervals that produce habituation (Figure 12-5*A*). Additional evidence that changes in receptor responsiveness could not explain habituation stems from experiments (to be considered below) in which stimulation of the receptors is circumvented and a constant electrical stimulus is applied directly to the nerve carrying the afferent fibers. Under these circumstances reflex habituation still occurs.

CENTRAL SYNAPTIC CHANGES

Since neither sensory adaptation (Figure 12-4*B,* locus 1) nor peripheral fatigue (loci 2 and 3) accounts for the changes in reflex response, habituation must result from changes within the central nervous system. This possibility was tested by obtaining intracellular recordings from the major gill motor neurons in the intact animal during the course of habituation (Kupfermann *et al.,* 1970). These recordings revealed a progressive decrease in the discharge of the motor neuron during habituation. The first tactile stimulus of the siphon produced a repetitive discharge of the motor neurons. But when the stimulus was repeated, the gradual decrease in reflex responsiveness was paralleled by a gradual decrease in the number and frequency of evoked action potentials

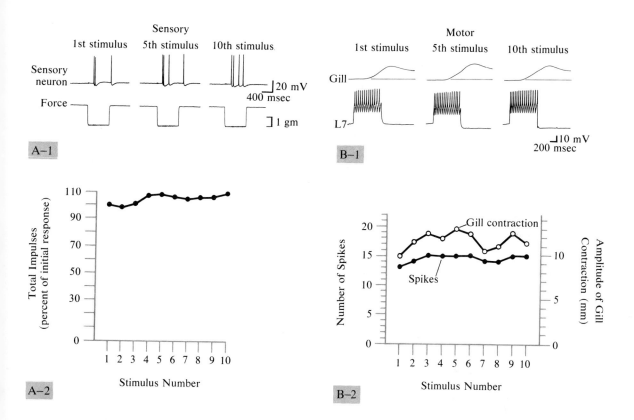

FIGURE 12-5.

Evaluation of the contributions of sensory adaptation and motor fatigue to habituation.

A. Stability of responses of the sensory neuron. *A-1.* Sample records of the responses of a mechanoreceptor neuron to 10 constant-force stimuli delivered to a 0.5 mm point on the siphon skin once every 30 sec in an isolated reflex test system. *A-2.* Average data from 124 training sessions. The total number of spikes elicited by the first stimulus was scored as 100 percent. [From Byrne, Castellucci, and Kandel, unpubl.]

B. Stability of gill contractions produced by motor cell L7. *B-1.* Sample records of the response of cell L7 to 10 constant-current depolarizing pulses injected intracellularly every 30 sec and the concomitantly recorded gill contraction. *B-2.* Graph of the number of spikes per stimulus and the amplitude of the gill contraction as a function of each stimulus. There is little change in the threshold of L7 or the motor response with repeated stimulation. [From Carew, Pinsker, Rubinson, and Kandel, 1974.]

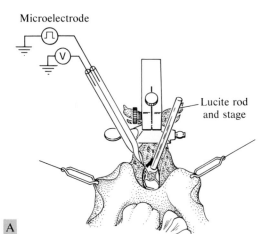

A

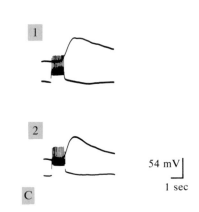

1

2

54 mV |

1 sec

C

1. Habituation

1st stimulus   3rd stim.   8th stim.   15th stim.

2. Rehabituation (9 min rest)

1st stimulus   2nd stim.   5th stim.   8th stim.

3. Dishabituation

13th stimulus   15th stim.   17th stim.   25th stim.

↑

Dishabituation

| 27
| mV

1 sec

B

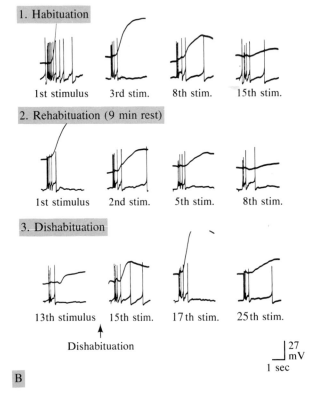

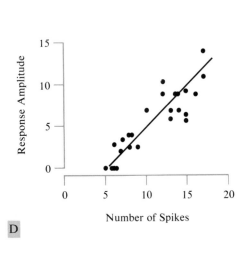

D

in the motor neuron. Recovery of reflex responsiveness, produced either by rest or by a dishabituatory stimulus, was associated with an increase in the number and frequency of action potentials in the motor neurons (Figure 12-6).

Since gill contraction is determined by the activity of three major and three minor motor neurons (see Figure 12-4*A*), the contraction was not correlated perfectly with firing in any one motor cell. Nevertheless, the correlation between motor-neuron firing and gill contraction was reasonably close. Even when repeated tactile stimuli produced a reduction of only one or two spikes in a motor cell (such as L7), there was often a measurable decrease in magnitude of the gill contraction. Changes in the spike out of the motor neuron are therefore causally related to changes in gill contraction (see also Figure 12-6*D*).

The decrease in spike output could be due either to an elevation of the spike-initiation threshold (accommodation) of the motor neuron or to a decrease in synaptic drive. The experiment in which motor responsiveness was examined (Figure 12-5*B*) showed that the motor neurons' threshold to constant-current

FIGURE 12-6.

Central synaptic changes accompanying habituation, spontaneous recovery, and dishabituation. Correlation of the responses of motor neuron L7 and of the amplitude of the reflex responses. [From Kupfermann *et al.*, 1970. Copyright by the American Association for the Advancement of Science.]

A. Experimental arrangement as in Figure 12-1, but in addition a slit has been made in the neck of the animal allowing the abdominal ganglion and its nerves to be externalized, pinned on a lucite stage, and illuminated by means of a lucite-rod light guide. Intracellular recordings are then obtained from identified motor cell L7 during gill withdrawal, allowing one to compare behavioral responses with motor neuron activity.

B. Sample records of gill contractions, measured with a photocell, and simultaneous intracellular recordings from identified motor neuron L7. Tactile stimuli (500 msec in duration) were presented to the mantle shelf every 90 sec. *B-1.* Fifteen stimuli were presented over a period of 21 minutes. The number of action potentials in the first second following the first evoked action potential in each trace is, left to right, 9, 6, 6, and 4. *B-2.* Rehabituation of the reflex following partial recovery due to a 9-min rest. The number of action potentials in the first second following the first evoked action potential in each trace is 7, 6, 5, and 3. *B-3.* Following the last habituation trial shown in the first trace, a strong stimulus was applied to the siphon. The discharge of the motor neuron and the amplitude of the gill contraction progressively increased during the first three stimuli following the dishabituatory stimulus and remained elevated for several minutes. The number of action potentials in the first sec following the first action potential in each trace is 4, 5, 7, and 5.

C. Gill responses produced by intracellular stimulation of the motor neuron L7 obtained before and during the experiment illustrated in part B. *C-1.* Before habituation of the reflex. *C-2.* During maximal habituation. The amplitude of the gill response to intracellular stimulation is not changed by the habituation procedure. This indicates that habituation does not occur at the synapse between the motor neuron and gill muscle.

D. Magnitude of gill contraction produced by motor neuron L7. Same experiment as in parts *B* and *C*. The abscissa indicates the number of spikes produced by depolarizing current pulses of different intensities and 1 sec in duration. The ordinate indicates the amplitude of the gill contraction (in arbitrary units, measuring the output of the photocell). The amplitude of the gill contraction is linearly related to the number of spikes in the motor neuron.

pulses did not change. Therefore the synaptic input must be changing. The synaptic input is normally masked by spike activity. To examine changes in the underlying synaptic potential, one motor cell was hyperpolarized and spike generation prevented. Under these conditions repeated elicitation of the reflex led to a progressive decrease in the amplitude of the complex EPSP produced in the motor neuron by the activity of several sensory neurons and interneurons. Rest and dishabituation restored the amplitude of the EPSP (Castellucci *et al.,* 1970; Castellucci, Byrne, and Kandel, unpubl.). These changes in amplitude can account for the changes in the firing rate of the motor neuron.

To further analyze the mechanisms underlying the change in the complex EPSP, two simplified test systems have been developed. In the first of these, the *isolated reflex test system,* the gill, siphon skin, and abdominal ganglion are removed from the animal and placed in a chamber perfused with aerated, cooled seawater (Figure 12-7*B*). Tactile stimuli are applied to the siphon using a mechanical stimulator that provides constant-force stimuli to discrete points on the skin. Gill movements are recorded with a photocell and intracellular recordings are obtained from gill motor neurons (Byrne, Castellucci, and Kandel, 1974a, b). In a second experimental preparation, the *isolated ganglion test system,* the ganglion only is removed from the animal and examined in isolation from the rest of the nervous system in an experimental chamber perfused with seawater (Figure 12-7*C*). The intracellular responses of

FIGURE 12-7.
Comparison of recordings from an intact preparation with a slit in the neck and those from two simpler test systems. [Modified from Carew *et al.,* 1971, and Byrne *et al.,* unpubl.].

A. Intact preparation. *A-1.* Experimental arrangement as in Figure 12-6. Intracellular recordings are obtained from major gill motor cell L7. *A-2.* Simultaneous recordings of gill contraction (obtained with a photocell) and a complex EPSP in L7 in response to an 800-msec jet of seawater applied to the siphon. Cell L7 was hyperpolarized about 50 mV to prevent spike generation. The remaining gill contraction is due to the activity of the other gill motor neurons in the reflex (see Figure 9-12*A,* p. 364).

B. Isolated reflex test system. *B-1.* Diagram of the experimental arrangement. The gill, siphon, and abdominal ganglion have been removed from the animal. The siphon skin is pinned to a lucite stage and stimulated with a mechanically driven constant-force probe having a tip diameter of 0.5 mm. The gill response is recorded with a photocell; intracellular recordings are obtained from L7. *B-2.* Simultaneous records of gill contraction, the complex EPSP in L7, and the constant-force probe.

C. Isolated ganglion test system. *C-1.* Diagram of the isolated ganglion. The siphon nerve, which innervates the siphon, is electrically stimulated. Stimulation of the right connective (carrying fibers from the head) provides input for heterosynaptic facilitation and dishabituation. Intracellular recordings from L7. *C-2.* An EPSP is produced in L7 by a 1.5-msec electrical shock to the siphon nerve. The cell was hyperpolarized about 50 mV to prevent firing. The general configurations and time courses of the PSPs produced with natural stimulation of the gill (*A-2, B-2*) and with electrical stimulation of a nerve (*C-2*) are roughly comparable.

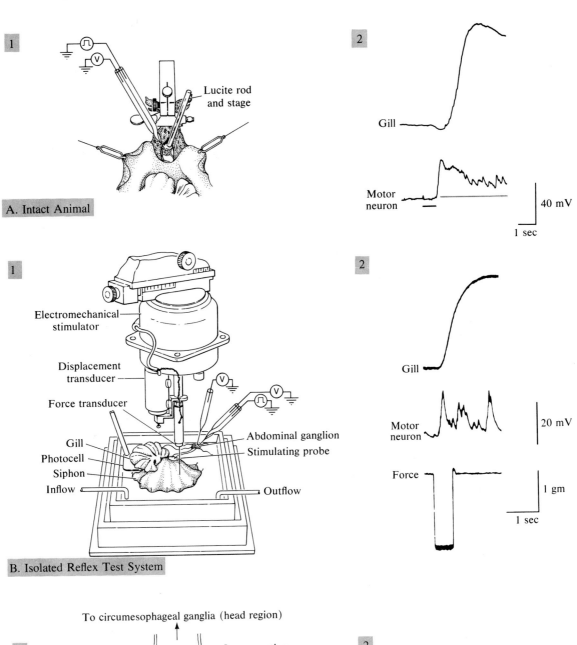

1

Lucite rod
and stage

A. Intact Animal

2

Gill

Motor
neuron

40 mV

1 sec

1

Electromechanical
stimulator

Displacement
transducer

Force transducer

Gill
Photocell
Siphon
Inflow

Abdominal ganglion
Stimulating probe

Outflow

B. Isolated Reflex Test System

2

Gill

Motor
neuron

20 mV

Force

1 gm

1 sec

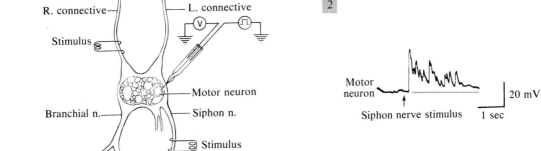

To circumesophageal ganglia (head region)

1

R. connective

L. connective

Stimulus

Motor neuron

Branchial n.

Siphon n.

Stimulus

C. Isolated Ganglion Test System

2

Motor
neuron

20 mV

Siphon nerve stimulus

1 sec

motor neuron L7 are used to monitor the output of the reflex. The siphon nerve, which carries the afferent input from the siphon skin, is electrically stimulated to elicit the reflex response (Castellucci *et al.*, 1970).

In both test systems the complex EPSPs elicited in L7 by either natural or electrical stimuli resemble those produced in L7 by natural stimuli to the siphon in the intact animal (Figure 12-7). By means of these experimental preparations it becomes possible to analyze the behavior on a number of different levels, ranging from the intact animal to simplified test systems that resemble the neural analogs of learning considered in Chapter 11 (p. 497). However, the test systems differ from neural analogs in an important way. Rather than being arbitrarily selected neuronal systems of unknown behavioral relevance, the test systems represent an actual microcircuit that contains critical components of the full neural circuit producing the behavior.

In these test systems, as in the intact animal, repeated mechanical stimulation of the siphon skin or electrical stimulation of the siphon nerve at intervals that produce habituation causes a decrease in the complex EPSP in the gill motor neuron L7. With rest, the depressed EPSP recovers (Figures 12-6, 12-8, 12-14).

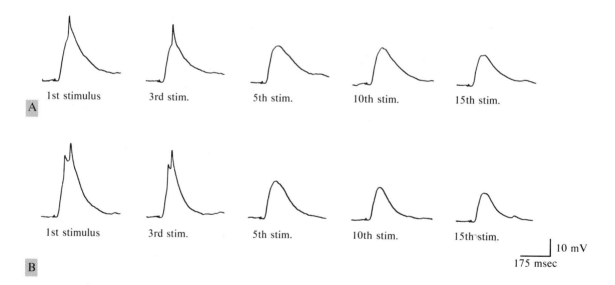

FIGURE 12-8.

Response depression in the isolated ganglion test system. Sample records from two consecutive training sessions illustrate the changes in the complex EPSP produced in motor neuron L7 by 15 electrical stimuli applied to the siphon nerve at 10-sec intervals. The membrane potential of L7 was hyperpolarized beyond the reversal level of the IPSP in order to prevent firing and thus see the PSPs more clearly. This procedure also makes it unlikely that depression is due to an IPSP that becomes progressively more prominent. **A.** With repeated stimulation of the siphon nerve, the EPSP decreases in amplitude. **B.** Following a 15-min rest, the EPSP recovers, but the next 15 stimuli produce a slightly more profound depression. [From Castellucci, Byrne, and Kandel, unpubl.]

CHANGE IN INPUT RESISTANCE VERSUS SYNAPTIC DRIVE

The changes in the amplitude of the complex EPSP could be produced either by a prolonged change in the input resistance of the motor neuron (Figure 12-4B, locus 7) or by a change in synaptic input to the motor neuron (loci 8, 9). Castellucci and co-workers (1970) examined the input resistance of the motor neuron by injecting a brief constant-current pulse into the cell and found no changes in the resulting electrotonic potential during the depression of the complex EPSP (Figure 12-9A).

Because these measurements of input resistance are made in the cell body of L7 and the synapses occur on the axon and its ramifications, it is difficult to know to what degree they provide an index of resistance changes in the synaptic region. Carew, Castellucci, and Kandel (1971) therefore examined the effect of spontaneous synaptic activity on the input resistance in the cell body and found that the characteristic burst of elementary IPSPs produced in L7 by Interneuron II led to a large reduction in the amplitude of the electrotonic potential (Figure 12-9B). Measurements of electrotonic potential in the cell body can thus detect resistance changes due to synaptic activity in at least some parts of the synaptic region. Moreover, as we shall see in Chapter 13 (p. 630), depression of the EPSP caused by stimulation of one part of the receptive field does not alter the amplitude of the EPSP produced by stimulating another part. This finding also argues against a resistance change in the synaptic region of the motor neuron. Conceivably, the synaptic depression may have produced a change in resistance that could not be detected because it occurred in a region remote from the cell body. The weight of the evidence, however, suggests that the synaptic depression accompanying habituation is due to change in synaptic input.

POSTSYNAPTIC INHIBITORY BUILDUP VERSUS EXCITATORY SYNAPTIC DEPRESSION

Changes in synaptic input to the motor neuron could be caused by one of two mechanisms: (1) A progressive increase in the central postsynaptic inhibitory drive that alters the number, the firing frequency, or the effect of the active afferent neurons contributing to the complex EPSP; this could be achieved by an inhibitory action on the interneurons or on the motor neurons (Figure 12-4B, loci 4 and 5). (2) A progressive decrease in the excitatory synaptic input to the motor neurons due to the reduced synaptic efficacy of individual sensory neurons or inter-neurons (Figure 12-4B, loci 6 or 9).

As a first step toward distinguishing between these alternatives, the membrane potential of the motor neuron was hyperpolarized well beyond the reversal potential of the spontaneous IPSPs (Castellucci et al., 1970). If the decrease in the complex EPSP is due to buildup in postsynaptic inhibition,

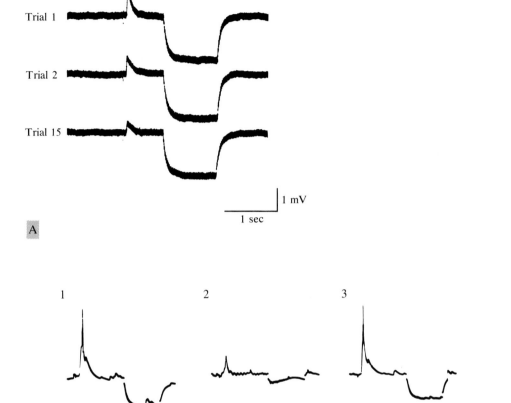

Trial 1

Trial 2

Trial 15

1 mV

1 sec

A

1   2   3

10 mV

1 sec

B

FIGURE 12-9.

Measurements of input resistance during synaptic depression.

A. No change in input resistance in the cell body of motor neuron L7 during EPSP depression. An EPSP was elicited by electrical stimulation of the siphon skin every 10 sec. Following every EPSP, the input resistance of L7 was measured by the amplitude of an electrotonic potential produced by a constant-current pulse. [From Pinsker, Kandel, Castellucci, and Kupfermann, 1970.]

B. Changes in the input resistance of the synaptic region can be detected by measuring the input resistance of the cell body. When L7 receives a massive burst of PSPs, such as that from Interneuron II, its input resistance decreases. (*B-1* to *B-3* are consecutive traces, 10 sec apart. In each trace an EPSP was elicited in L7 by stimulating the siphon nerve. Hyperpolarizing pulses were injected into the cell body of L7 to monitor the input resistance.) *B-1.* Control. *B-2.* A spontaneously occurring burst of PSPs attributed to Interneuron II produced a decrease (short circuiting) of the EPSP and a large decrease in the amplitude of the electrotonic potential. *B-3.* Control. [From Carew, Castellucci, and Kandel, 1971.]

then the complex EPSP must include, in addition to the evident EPSPs, some occult IPSPs that increase progressively. As they increase, the conductance change produced by the IPSPs would reduce (short-circuit) the EPSPs. With the cell hyperpolarized well beyond the reversal level of the IPSPs, the IPSP will produce depolarizing synaptic actions. An increase in inhibition should then result in an increase in total EPSP amplitude, or at least a much less dramatic decrease. Yet with repetitive stimulation at this hyperpolarized level, the PSP still decreased in amplitude and the kinetics of the depression were similar to those at the resting potential (Figure 12-8). These experiments suggest that increased postsynaptic inhibition does not produce the synaptic depression.

To determine whether habituation is due to a decrease in excitatory synaptic efficacy, individual sensory neurons were examined in isolation to see whether the direct monosynaptic EPSPs they produced in the motor neuron undergo a depression that parallels the habituation. Castellucci and co-workers (1970) therefore further simplified the afferent pathway of the reflex and examined the EPSP produced in the motor cell by electrical stimulation of a single mechanoreceptor neuron in the siphon skin. Repeated stimulation of this monosynaptic pathway at intervals that produce habituation results in a synaptic depression (Figure 12-10).

PRESYNAPTIC INHIBITION VERSUS HOMOSYNAPTIC DEPRESSION

The electrical stimulus to the skin that activates one sensory neuron could however also excite nearby sensory neurons that in turn might activate interneurons synapsing on the terminals of the test sensory neuron. Consequently, the depression in the EPSP could be due to presynaptic inhibition from parallel afferent fibers (Figure 12-4B, locus 6) that depress the sensory neuron's ability to release transmitter (see Figure 6-20, p. 201). Alternatively, the depression could result from a homosynaptic change in the effectiveness of the excitatory connections themselves. To distinguish between these possibilities, Castellucci and co-workers (1970; and Castellucci and Kandel, 1974) impaled individual mechanoreceptor neurons and stimulated them in isolation. Each spike in the sensory neuron produces a direct EPSP in the gill motor neuron similar to that caused by localized stimulation of the skin (Figure 9-15). This EPSP also undergoes depression with repeated stimulation and recovers with rest. The depression occurs even when the ganglion is bathed in a solution with a high concentration of divalent cations, which elevate the firing threshold of neurons and reduce the likelihood of a presynaptic inhibitory neuron being recruited. These observations tend to exclude presynaptic inhibition and suggest that depression of the EPSP results from a change in excitatory synaptic efficacy (Figure 12-11).

558

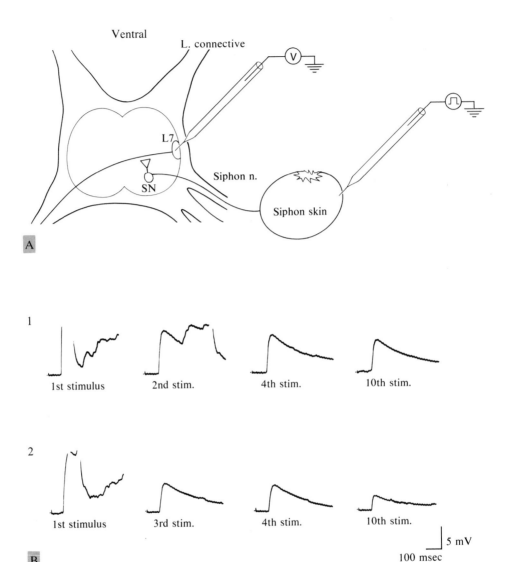

A

1

1st stimulus   2nd stim.   4th stim.   10th stim.

2

1st stimulus   3rd stim.   4th stim.   10th stim.

5 mV

100 msec

B

FIGURE 12-10.
Response depression of a unitary (presumably monosynaptic) EPSP in motor neuron L7, produced by electrical stimulation of a 5–10 μm point on the siphon, which activates a single mechanoreceptor. **A.** Experimental arrangement. **B.** Selected responses to stimulation at an interstimulus interval of 10 sec. *B-1.* Progressive depression of the EPSP during 10 stimuli. *B-2.* Recovery following a 12-min rest. Repetition of training leads to more profound depression. The late responses to stimuli 1 and 2 in the first session (*B-1*) and to stimulus 1 in the second session (*B-2*) are presumably due to interneuronal activity. [From Castellucci *et al.,* 1970.]

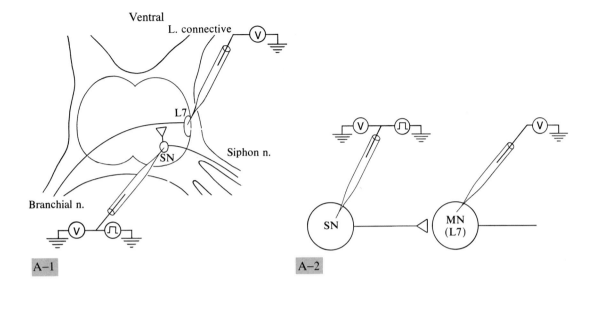

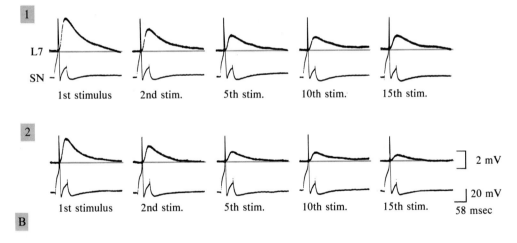

FIGURE 12-11.
Response depression of the monosynaptic EPSP produced in L7 by stimulating an individual sensory neuron. [From Castellucci and Kandel, 1974.]

A. Experimental arrangement. *A-1.* Isolated ganglion, illustrating the position of the electrode for recording from L7 and the sensory neuron. *A-2.* Schematic drawing of direct connections between the sensory neuron and motor neuron L7.

B. Changes in the monosynaptic connections between the sensory and motor neurons as a result of repeated stimulation every 10 sec. *B-1.* Selected records from responses to 15 consecutive stimuli. *B-2.* A second training session of 15 stimuli following a 15-min rest produces a more profound depression. The ganglion was bathed in solutions with high Mg^{++} and Ca^{++} concentrations to block interneuronal activity.

Synaptic depression is not restricted to the synapses that the sensory neurons make on gill motor neuron L7. A similar depression occurs in the EPSPs produced by the various branches of the sensory neurons: on the two types of excitatory interneurons, on the inhibitory neuron, and on other motor neurons of the gill. A similar depression also occurs in the EPSPs produced by the sensory neurons in both the central and peripheral siphon motor neurons. As in the reflex response, most of the depression in the monosynaptic EPSPs occurs within the first five stimuli (Figure 12-12). Recovery is obtained when the stimulus is withheld for 15 or more minutes. These results are summarized in the schematic circuit in Figure 12-13, which shows that the major loci for habituation are the synapses made by the branches of the sensory neurons on the three interneurons and the various gill and siphon motor neurons (Castellucci, Byrne, and Kandel, unpubl.; Bailey *et al.*, 1975).

The contribution of the connections between the interneurons and the motor neurons to habituation has not yet been adequately evaluated, but it is thought to be small. A large component of the complex EPSP recorded in the motor neurons is caused by direct monosynaptic actions from sensory neurons (see Figure 9-18, p. 373). Since the complex EPSP is mostly monosynaptic, one should be able to synthesize the behavior in a microreflex (p. 374) and show that the firing of an *individual* sensory neuron can bring about habituation. Using the isolated reflex test system (Figure 12-7), Byrne and co-workers (unpubl.) applied mechanical stimuli to the skin while recording simultaneously from one of the sensory neurons activated by the stimulus, from a major gill motor neuron, and from the gill. The first stimulus produced a discharge in the sensory and motor neurons and a gill contraction. Repeated stimulation did not alter the responsiveness of the sensory neuron, but caused the complex EPSP in the motor neuron and the gill contraction to decrease progressively (Figure 12-14).

FIGURE 12-12.
Synaptic depression in complex and monosynaptic EPSPs at two interstimulus intervals. **A.** Complex EPSP elicited in the reflex test system at 30-sec intervals using natural stimuli. (Data based on 33 training sessions.) **B.** Complex EPSP elicited at 30-sec intervals in the isolated ganglion test system using electrical stimuli. (Data based on 10 sessions). **C.** Complex EPSP elicited at 10-sec intervals in the isolated ganglion test system using electrical stimuli. (Data based on 4 sessions.) **D.** Monosynaptic EPSP elicited at 10-sec intervals. (Data based on 11 sessions.) The similarity of the graphs suggests that the kinetics of depression of the complex EPSP is similar whether produced by a natural stimulus to the siphon (graph A) or an electrical stimulus to the siphon nerve (graph B). The kinetics of depression of the complex EPSP (graphs A to C) are in turn comparable to the monosynaptic EPSP (graph D). Moreover, the depression is similar whether stimuli are presented every 30 sec (A and B) or every 10 sec (C and D). [Based on an unpublished experiment by Byrne, Carew, Castellucci, and Kandel.]

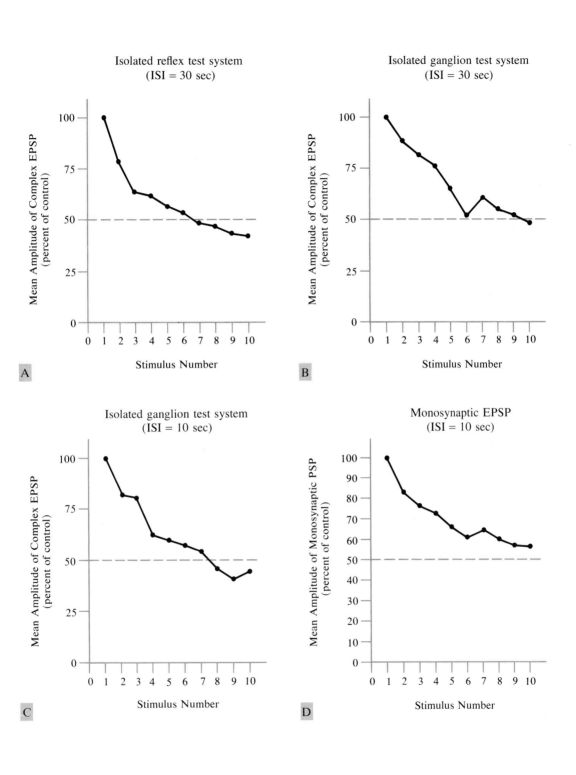

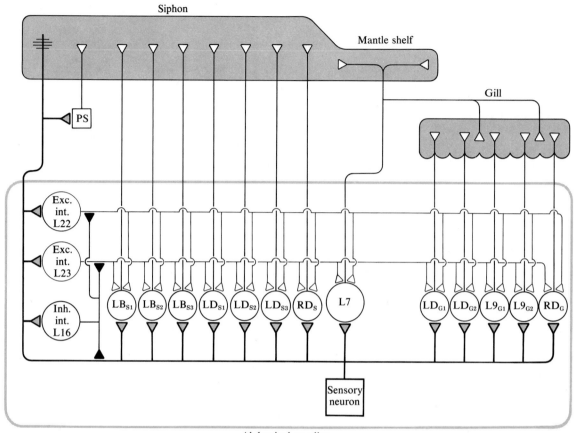

Abdominal ganglion

FIGURE 12-13

Sites of neuron plasticity that mediate habituation. Schematic illustration of the identified neurons involved in the defensive withdrawal reflex of the mantle organs elicited by weak tactile stimulation of the siphon. All motor neurons and interneurons receive excitatory synaptic input from the sensory neurons. These excitatory connections have plastic capabilities (hatched) and undergo depression with repeated stimulation. The peripheral siphon motor neuron (PS) cluster includes about 20 cells; the sensory neuron (cluster includes about 24 cells.

FIGURE 12-14 (*Facing Page*).
Habituation in the isolated reflex test system. [From Byrne, Castellucci, and Kandel, unpubl.]

A. Experimental arrangement as in Figure 12-7*B*. Intracellular recordings are obtained from a motor neuron and a sensory neuron. The sensory neuron is stimulated with a direct intracellular current pulse. A constant-force stimulus is applied to the siphon skin and photocell records are taken from the gill.

B. Response depression to stimulation of the skin and a sensory neuron. *B-1*. Selected records from a training session in which the tactile stimuli were repeatedly presented to the skin every 30 sec. The first stimulus produces a discharge in the sensory neuron, motor neuron, and a gill contraction. Subsequent stimuli produce no depression in the sensory discharge but the motor neuron response and gill contraction decrease. *B-2*. After a 15-min rest the single sensory neuron was repetitively fired (8 spikes per trial) to simulate the response of the population of three to eight sensory neurons usually activated by the natural stimulus used in *B-1*. The first sensory discharge produces a motor discharge and gill contraction. But subsequent sensory discharges produce progressively less response in the motor neuron and progressively smaller gill contractions.

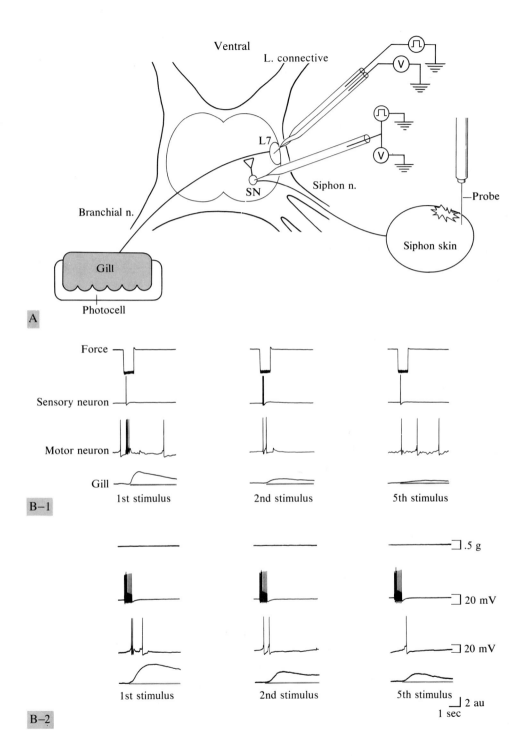

Ventral

L. connective

L7

SN

Siphon n.

Branchial n.

—Probe

Siphon skin

Gill

Photocell

A

Force

Sensory neuron

Motor neuron

Gill

1st stimulus

2nd stimulus

5th stimulus

B–1

.5 g

20 mV

20 mV

1st stimulus

2nd stimulus

5th stimulus

2 au

1 sec

B–2

A weak stimulus to a 0.5 mm point on the siphon activates about 8 sensory cells, each of which fires one to two spikes (see p. 375). As we have seen, one can simulate activation of the eight sensory neurons (transforming the spatial pattern of activation into a temporal pattern) by firing a single sensory neuron repetitively so as to produce about eight spikes. This transformation works well. A burst of spikes in the sensory neuron produces a motor discharge and gill contraction comparable to that evoked by the natural stimulus. When the burst of spikes in the sensory neuron is repeatedly elicited, at rates that cause habituation, the EPSP produced by the sensory neuron in the motor neuron decreases progressively, as does the motor response and gill contraction (Figure 12-14). By contrast, when the motor neuron alone is stimulated repeatedly (with an intracellular current pulse) no depression in the gill contraction occurs. Thus, one can show in several ways that the major locus for reflex habituation is in the synapses between the sensory neurons and the motor neurons and interneurons.

Mechanism of Habituation

SYNAPTIC DEPRESSION AND THE LEVEL OF TRANSMITTER RELEASE

The depression of synaptic transmission between the sensory neurons and the inter- and motor neurons superficially resembles homosynaptic (low-frequency) depression that occurs at peripheral junctions and central neurons (Chapter 11). But the changes differ in detail from most forms of homosynaptic depression. First, the depression at the sensory-neuron synapses is unusually profound and long lasting. Second, at most synapses the degree of depression is related to the amount of transmitter release. Depression is increased when transmitter release is increased and is decreased when transmitter release is decreased. At low levels of release most synapses actually show facilitation rather than depression (for an exception see Bruner and Kennedy, 1970). This has been interpreted to mean that each action potential in the terminal activates two processes: (1) a depletion of some transmitter from a readily releasable pool, perhaps from active release sites; and (2) a mobilization of transmitter into the active release sites from the larger storage pool.

With low rates of stimulation depletion dominates over mobilization at high levels of release; at low levels of release the reverse holds. However, at the sensory-neuron synapses depression occurs at all levels of release—at normal levels, at increased levels (achieved by broadening the presynaptic action potential by injecting TEA into the presynaptic neuron), and at reduced levels (achieved by increasing the Mg^{++} to Ca^{++} ratio of the external medium). This finding suggests that synapses of the sensory neurons are not as capable of mobilizing transmitter in response to repeated stimulation as are most other synapses.

Finally most synapses that show homosynaptic depression at low rates of stimulation show facilitation if stimulated to fire at high rates, because high rates augment mobilization. The synapses of the sensory cells, however, do not facilitate at any frequency of stimulation. This suggests that the mobilization process in the terminals of the sensory neurons is not also sensitive to homosynaptic activation at any frequency of stimulation. Rather, each impulse in the sensory neuron seems to discharge a fixed fraction of the releasable sites. The fraction discharged is independent of the size of the releasable pool, the level of release (an indirect measure of the size of the releasable pool), or the rate of activation.

These experiments suggest that the mechanism for synaptic depression is presynaptic; they cannot, however, rule out a possible change in the sensitivity of the postsynaptic receptors.

POSTSYNAPTIC VERSUS PRESYNAPTIC MECHANISMS

To distinguish between a presynaptic mechanism (a decrease in transmitter release) and a postsynaptic mechanism (a decrease in the sensitivity of the postsynaptic receptors), Castellucci and Kandel (1974) carried out a quantal analysis of synaptic transmission between sensory and motor neurons. A quantal analysis provides a means for estimating the number and size of quanta released by an action potential in the terminals (p. 184). The average amplitude of an elementary excitatory synaptic potential $\bar{E}$ produced by an action potential is given by

$$\bar{E} = m \times \bar{q}$$

where m, the average quantal content (or quantal output), is the average number of quanta released in a series of stimuli, and $\bar{q}$ is the average amplitude of the unit synaptic potential, the potential produced by a single quantum.

Both m and $\bar{q}$ can be determined experimentally and provide the most direct way to distinguish between pre- and postsynaptic processes in synaptic transmission. The value of m is an indicator of the number of transmitter quanta released by the presynaptic terminal. Changes in m therefore indicate alterations in the presynaptic processes leading to transmitter release. The value of

$\bar{q}$ reflects the response of the receptor to a transmitter quantum. A constant $\bar{q}$ value indicates constant sensitivity of the postsynaptic receptor. Given normal filling of vesicles, and therefore a constant amount of transmitter per quantum, a decrease in $\bar{q}$ indicates a decrease in the responsiveness of the receptor.

The first suggestion that transmission at the synapses between the sensory and motor neurons is quantized came from a comparison between normal and low levels of transmitter release (Figure 12-15). At normal levels of release repetitive stimulation of the sensory neuron produces only small fluctuations in the amplitude of the EPSPs evoked in the motor neuron. But when transmitter output is reduced, repetitive stimulation of the sensory neuron produces marked fluctuations in the amplitude of the EPSP in an apparently quantal fashion, and one observes occasional failures (Figure 12-15). These findings alone suggest that transmission is probably quantized and that the synaptic potential is made up of an integral number of quanta. The existence of quantal transmission at this synapse was further supported by amplitude histograms of blocks of consecutive (30 to 100) responses in stable regions where the EPSP changed by less than 15 percent (Figure 12-16). These histograms reveal a peak of failures (black bar) followed by a multimodal distribution with the mean voltage value of each subsequent peak (on the abscissa) being an integral multiple of the first response peak. For example, in the histogram of Figure 12-16A the first response peak is at 34 μV, the second at 68, the third at 102, and the fourth at about 136. These response peaks suggest that transmitter is released in packets that are integral multiples of a unit packet.

The first response peak was assumed to be the amplitude of $\bar{q}$, the unit EPSP. The mean q value estimated in this way (called $\bar{q}_1$) was 33.8 μV (with range of 14 μV to 70 μV). These values are similar to those found by Miledi (1967) at another molluscan synapse, the giant synapse of the squid.

To improve the signal-to-noise ratio and insure that failures were correctly identified, all failures were averaged by computer. The averaged failures were indistinguishable from the averaged background fluctuations. Averaging did not reveal a smaller, occult unit EPSP. The unit EPSPs were also averaged and found to have a similar configuration to the average of the first few large evoked EPSPs. The results indicate that one can also successfully recognize the unit quantal signal and carry out a quantal analysis at this synapse (Figure 12-17).

Knowing $\bar{E}$, the average amplitude of the evoked EPSPs, and $\bar{q}_1$, the value of m_1 can be estimated from the amplitude histogram (since $m_1 = \bar{E}/\bar{q}_1$). In the experiment in Figure 12-16A, m_1 was 1.7.

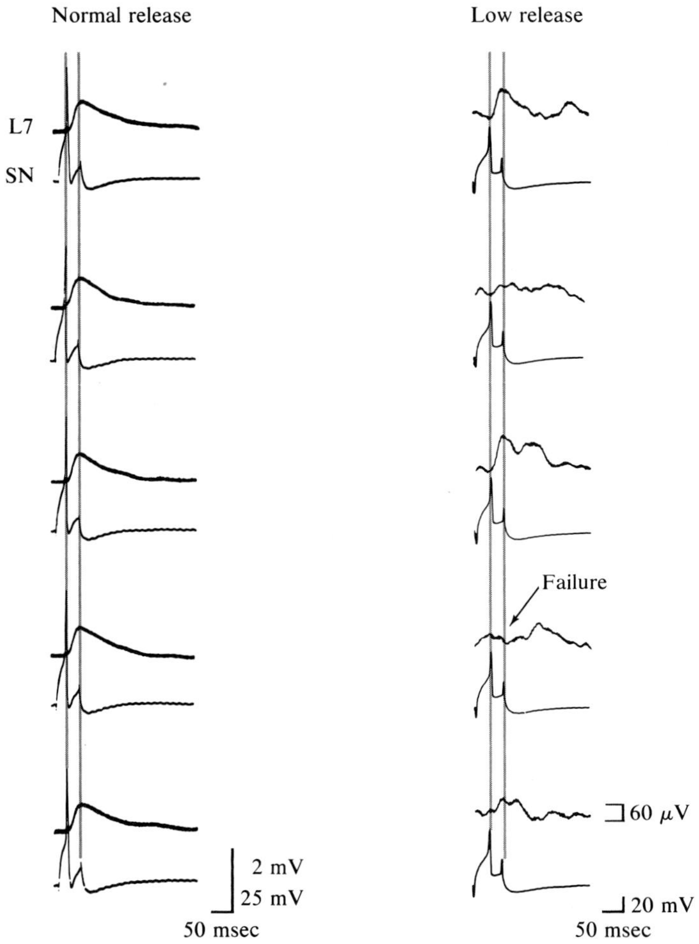

FIGURE 12-15.
Fluctuations in synaptic potential at normal and low levels of transmitter release. Sample records of successive EPSPs evoked every 10 sec in cell L7 by a series of intracellular stimuli to the sensory neuron. At normal levels of release there are no marked fluctuations in consecutive responses. At low levels of release there are marked fluctuations in amplitude in consecutive responses and an occasional failure (arrow). The magnitude of the response in time intervals (the window, indicated by the vertical lines between the peak of the presynaptic spike and the peak of the first EPSP) was used to obtain amplitude histograms of the EPSPs (see Figure 12-16A). [From Castellucci and Kandel, 1974.]

One can obtain a second and independent estimate of m at low levels of release by assuming a Poisson distribution (Equation 6-24, p. 193) and examining the ratio of failures (N_0) to the total number of trials (N) where

$$m_0 = \ln \frac{N}{N_0}$$

As we have seen in Chapter 6, transmitter release in response to an action potential in the presynaptic terminals follows a binomial distribution, where $m = n \times p$ (see Equation 6-16, p. 189); p is the average probability that a quantum is released and n is the pool from which the quanta are released. If p is small compared to n, under conditions of low release, the two-parameter binomial distribution can be approximated by the one-parameter Poisson distribution. Here P_x, the probability of any number (x) of quanta being released,

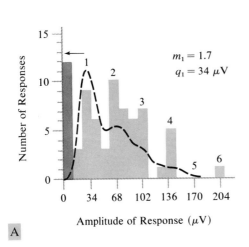

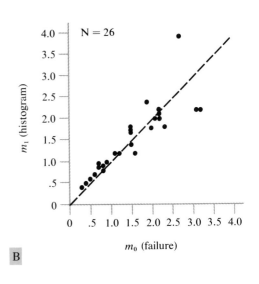

A

B

FIGURE 12-16.
Quantal analysis. [From Castellucci and Kandel, 1974; and Kandel, Brunelli, Byrne, and Castellucci, 1976.]
A. Amplitude histogram of 70 consecutive responses in cell L7 (interstimulus interval 10 sec). The peak of failures (dark bar, 0 mV) is followed by a multimodal distribution of responses. The first peak of responses was assumed to be the amplitude of a unit potential (q_1). Numbers at the top of the histogram indicate successive multiples of the estimated unit potential; peaks 1 to 4 were roughly integral multiples of the unit peak. From the value of m_1 obtained from these amplitude histograms, a predicted curve (dotted line) was generated assuming a Poisson distribution and a coefficient of variation for q_1 of 30 percent. The arrow on the ordinate refers to the predicted number of failures.

B. Correlation between estimates of m from the histogram method (m_1) and from the method of failures (m_0) derived from the Poisson equation.

is simply given by

$$P_x = \frac{m^x e^{-m}}{x!}$$

If one knows m, one knows the entire distribution.

To justify the use of the Poisson distribution, Castellucci and Kandel (1974; and Kandel *et al.*, 1976) first estimated p from the number of failures and from m_1 (see Chapter 6, footnote 7, p. 191) and found it to be reasonably low (0.16). They next used the Poisson distribution to predict the number of responses found in each quantal class. In 31 of 38 plateau regions the predicted values were not statistically different from the observed values. Moreover, the values predicted from a binomial distribution did not provide a better fit to the data than that obtained with the Poisson distribution.

Castellucci and Kandel assumed a Poisson distribution and calculated a theoretical curve based upon knowledge of the number of quantal events in each class and the coefficient of variation of the quantal unit. The distribution of the quantal unit is best obtained from the distribution of the spontaneous

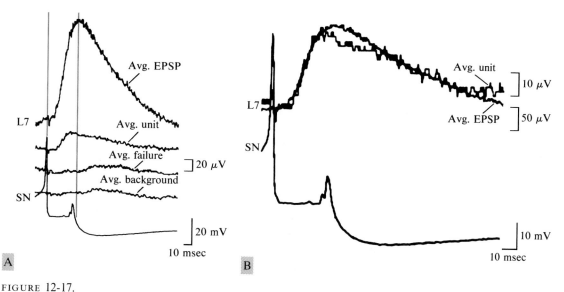

FIGURE 12-17.

Computer-averaged failures and unit EPSPs. [From Castellucci and Kandel, 1974.]

A. The first five EPSPs of the series were averaged to establish the interval window (cf. Figure 12-15). Failures (N = 21) were averaged separately; the background noise (N = 21) was averaged 2 sec before the occurrence of a presynaptic spike. The failure average was indistinguishable from the background-noise average; there was no time-locked depolarization. This indicates that no small responses were missed.

B. Time courses of the averaged EPSPs and unit peaks. The unit peak is superimposed on the average EPSP to illustrate the similarity of their time course.

miniature synaptic potentials (see Figure 6-17B).[3] Theoretical curves generated in this way fit the observed data reasonably well and suggest that a Poisson distribution can serve as an approximation of the PSP distribution.

Using the Poisson equation, a second estimate of m (m_0) can be obtained. The estimates obtained from the Poisson equation and those obtained from the histograms proved highly correlated (Figure 12-16B).

Using two independent estimates, quantal output and average quantal size could now be estimated with repeated stimulation. A habituation training session at normal levels of release consists of 10–15 stimuli. This small number precludes a meaningful analysis because Poisson statistics require many responses. However, at low levels of transmitter release, where each EPSP is only 1 to 5 percent of the amplitude of the EPSP at normal levels of release, 200 stimuli release roughly the same amount of transmitter as do 15 stimuli at normal release levels. Poisson statistics also require that the mean response be relatively stable. By discarding the first three responses generated in a habituation training session at normal levels of release, one can distinguish three relatively stable regions during the next 12 evoked EPSPs. Similarly, during 200 consecutive EPSPs at low release, one can (after discarding the first few responses) also observe three successive, relatively stable (plateau) regions, containing about 50 responses each, in which EPSP depression is less than 15 percent. Each plateau region can be compared to the consecutive regions (of four EPSPs each) at high release (Figure 12-18). Thus, the possible mechanism of the synaptic depression accompanying a single habituation training session can be analyzed by comparing the first plateau region of about 50 EPSPs (roughly equivalent to EPSPs 4, 5, 6, and 7 of the 15 responses elicited at high levels of transmitter release) to the two subsequent plateau regions (equivalent at high release to the next two regions of four responses each). The m and $\bar{q}$ values can then be calculated and compared in the consecutive plateau regions and used as an index of the synaptic depression accompanying repeated stimulation. Estimates of m and $\bar{q}$ were obtained in three independent ways: at low levels of release by means of amplitude histograms and failure analysis, and at normal levels of release by estimating the coefficient of variation of the amplitude of the EPSP (see p. 194).

The amplitude histograms of each plateau region showed successive peaks that again corresponded roughly to integral multiples of the unit peak (Figure 12-19). With repetition (compare regions 1, 2, and 3 in Figure 12-19), the

---

[3]Since the miniature synaptic potentials associated with a particular presynaptic neuron cannot be measured in central nervous systems (where each cell receives connections from many presynaptic neurons), Castellucci and Kandel assumed a commonly used coefficient of variation of 30 percent (Martin, 1966).

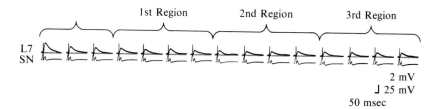

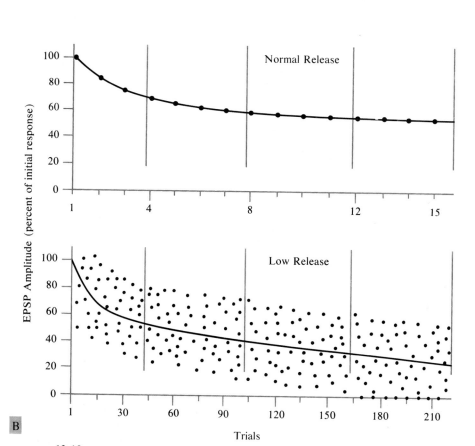

FIGURE 12-18.

Schematic diagram of three plateau regions used to study synaptic depression at low release.

A. During the course of 15 consecutive stimuli the elementary EPSP produced by the sensory neurons decreases in amplitude, parallel with the time course of habituation of the gill-withdrawal reflex.

B. Upper graph plots the amplitude of the EPSP illustrated in part *A*. By disregarding the first three responses, one can distinguish three successive regions of relative stability during the decrement. These plateau regions are compared to one another in the analysis of the mechanism of synaptic depression. Lower graph plots the amplitude of the EPSP at reduced levels of transmitter release; the amplitude of the initial EPSP is 1 to 5 percent that of the initial EPSP at normal release. Thus the decline in amplitude over 200 consecutively evoked EPSPs at low levels of transmitter release is roughly comparable to the decline in amplitude over 15 evoked EPSPs that occur during a behavioral habituation training session. Each consecutive plateau region at low release (during which the EPSP decrement was less than 15 percent) was roughly comparable to consecutive groups of 5 EPSPs at normal release.

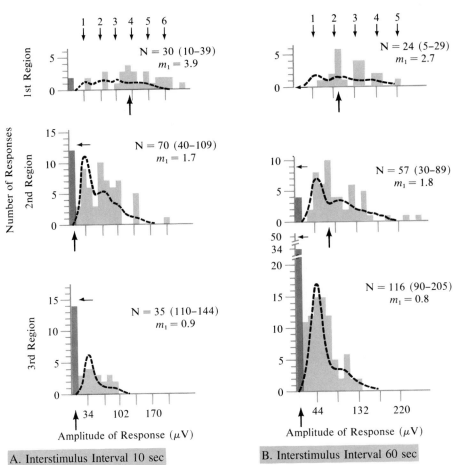

FIGURE 12-19.

Amplitude histograms of consecutive plateau regions during synaptic depression produced at interstimulus intervals of 10 and 60 sec. [From Castellucci and Kandel, 1974.]

A. Histograms of successive plateau regions during the synaptic depression accompanying 144 intracellular stimuli to a sensory neuron at 10-sec intervals. Arrows on the abscissa refer to the mode, the highest peak in the histogram. With repeated stimulation the percent of failures increases and the mode shifts to the left. N = number of stimuli; the two values in parentheses refer to the first and last EPSPs of the plateau region.

B. Histograms of successive plateau regions of EPSPs produced by 205 stimuli at 60-sec intervals.

value of the unit peak, the measure of quantal size, did not change but the incidence of failures increased about sevenfold from the first to the third region. For example in Figure 12-19A the 30 stimuli in the first region produced 2 failures (or 7%), the next 70 stimuli produced 12 failures (or 17%), and the 35 stimuli in the last region produced 14 failures (or 40%). In addition, the mode of the evoked EPSPs (the value of the major peak) shifted from the third or fourth response peak (in region 1) to either the quantal unit peak or the failure peak (in region 3). This analysis thus indicates that the EPSP depression that occurs during continued stimulation is due to a decrease in quantal output; the quantal size remains constant.

Similar results were obtained with failure analysis (Figure 12-20). With continued stimulation, the decrease in the size of the average EPSP is paralleled by a decrease in quantal output while quantal size does not change significantly. Thus, at interstimulus intervals of 10 seconds the mean normalized value of q in the three plateau regions was 100, 104, and 98 percent, respectively, while the mean normalized value for m (and that of $\bar{E}$) decreased progressively over the three regions from 100 to 63 to 46 percent. Similar trends occurred at interstimulus intervals of 30 and 60 seconds.

A third estimate of m and $\bar{q}$ can be obtained from the coefficient of variation of the EPSPs. In a Poisson distribution the variance (σ^2) equals the mean, which in turn equals m. Therefore the standard deviation (σ_m), which is the square root of the variance of m_1, will equal $\sqrt{m}$. The coefficient of variation (CV), which is the standard deviation divided by the mean, will be

$$CV = \frac{\sqrt{m}}{m}$$

Squaring both sides and solving for m gives

$$m_{\mathrm{CV}} = \frac{1}{(CV)^2}$$

Since

$$CV = \frac{\sigma_m}{m} = \frac{\sigma_{\mathrm{E}}}{\bar{E}}$$

where σ_{E} is the standard deviation of the amplitudes of all EPSPs and $\bar{E}$ is their mean amplitude, the expression for m_{CV} may therefore be transformed to

$$m_{\mathrm{CV}} = \frac{\bar{E}^2}{\sigma_{\mathrm{E}}^2}$$

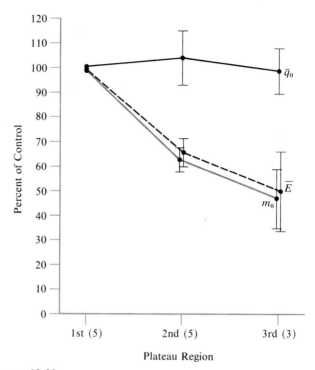

FIGURE 12-20.

Comparison of an average EPSP amplitude ($\overline{E}$) during repeated stimulation (interstimulus interval 10 sec) and quantal size (q_0) and quantal output (m_0) estimated by the method of failures. Each point in each of the three consecutive plateau regions represents the mean of either three or five experiments (indicated in parentheses). The quantal output per stimulus decreases in parallel with the decrease of EPSP amplitude; quantal size does not change significantly.

where σ_E^2 is the variance of the amplitudes of all EPSPs. Estimates based on coefficient of variation are useful because they can be obtained at normal levels of release, but this method is the least reliable of the three techniques and tends to overestimate m. Estimates of m based on coefficient of variation were consistently higher (and those of q consistently lower) than those derived using the other two methods. Nonetheless, the trend revealed by this method was similar to that produced by the others: with repeated stimulation, quantal size remains constant but quantal output decreases.

The quantal analysis, at both low and normal levels of transmitter release, indicates that the critical change accompanying habituation of the gill-withdrawal reflex is a decrease in the number of transmitter quanta released by the sensory neuron. The sensitivity of the postsynaptic receptor on the motor neuron remains unaffected.

LOCUS AND MECHANISM OF DISHABITUATION OF THE GILL-WITHDRAWAL REFLEX

In an intact animal habituation of the gill-withdrawal reflex to repeated stimulation of the siphon can be dishabituated by applying a strong tactile stimulus to the head. Similarly, in the isolated ganglion test system, a train of strong stimuli to the connective (that carries the fibers from the head) restores a depressed EPSP (Figure 12-21). Dishabituation leads to the discharge of the gill motor neuron and could increase the EPSP by feedback to the sensory neurons from action potentials in the motor neuron. However, this is not the case. Directly firing the motor neuron with intracellular pulses does not increase the depressed complex EPSP (Figure 12-22).

The increase occurs in an elementary monosynaptic EPSP produced by natural stimulation of a single receptor fiber or by electrical stimulation of a single sensory neuron (Figures 12-23 and 12-24). Such an increase could result from posttetanic facilitation due to a repetitive discharge of the sensory neuron produced by connective stimulation. However, the connective stimulus that increases the EPSP to control level (Figure 12-24) and even beyond (Figure 12-25) does not fire the sensory neuron. Moreover, as we have seen, the sensory neuron does not give rise to posttetanic facilitation even when it is fired repetitively by intracellular stimulation. Thus, posttetanic facilitation does not contribute to restoring synaptic effectiveness. Rather, the connective stimulus enhances the depressed EPSP by means of *heterosynaptic facilitation* (see p. 500).

Quantal analysis of the heterosynaptic facilitation based on amplitude histograms, failure analysis, and coefficient of variation indicates that the connective stimulus leads to a sudden increase in quantal output without a change in quantal size (Figures 12-26, 12-27). The heterosynaptic facilitation mediating dishabituation therefore results from a presynaptic facilitation. There is an increase in quantal output by the sensory neuron; receptor responsiveness is unaffected (Castellucci and Kandel, 1975; Kandel et al., 1976).

1. Stimulus to siphon nerve

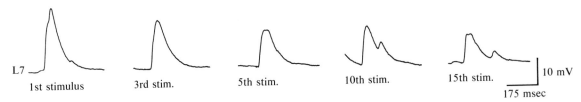

L7

1st stimulus   3rd stim.   5th stim.   10th stim.   15th stim.

10 mV

175 msec

2. Stimulus to left connective

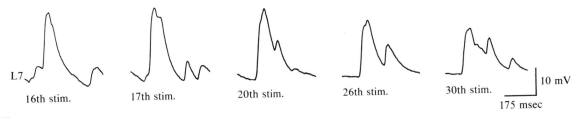

L7

15th stim.   16th stim.

25 mV

1 sec

3. Stimulus to siphon nerve

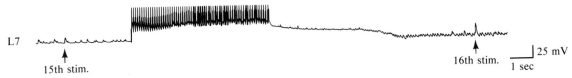

L7

16th stim.   17th stim.   20th stim.   26th stim.   30th stim.

10 mV

175 msec

A

1. Stimulus to siphon nerve

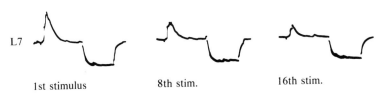

L7

1st stimulus   8th stim.   16th stim.

2. Stimulus to left connective

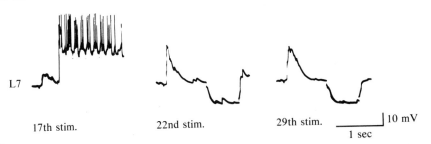

L7

17th stim.   22nd stim.   29th stim.

10 mV

1 sec

B

FIGURE 12-21 (*Facing Page*).
Synaptic changes underlying dishabituation.

A. Heterosynaptic facilitation of the complex EPSP in cell L7 in the isolated ganglion test system. *A-1.* Sample records illustrating the synaptic depression produced during a training session consisting of 15 consecutive stimuli to the siphon nerve. *A-2.* A strong stimulus applied to the left connective (6/sec for 6 sec) produces discharge of the cell and facilitation of the PSP. Arrows at the beginning and end of the trace point to complex EPSPs before and after the stimulus to the connective; these EPSPs are number 15 in part *A-1* and number 16 in part *A-3*. Sample records from the next 15 consecutive stimuli to the siphon nerve (stimuli 16-30) following the stimulus to the connective illustrated in part *A-2*. [From Castellucci and Kandel, unpubl.]

B. Absence of input resistance change in the soma of motor neuron L7 during heterosynaptic facilitation in the isolated ganglion test system. The siphon nerve was stimulated at an interstimulus interval of 10 sec and the EPSP recorded. A constant-current hyperpolarizing pulse was applied through another intracellular electrode and the resulting electrotonic potential was used to measure the input resistance. *B-1.* With repeated stimulation the EPSP decreased in amplitude. *B-2.* Stimulation of the left connective (6/sec for 6 sec) produced facilitation of the EPSP. There is no significant change in input resistance in the soma either during depression or during facilitation, as indicated by the relative constancy of the electrotonic potential in L7 during facilitation. [From Carew, Castellucci, and Kandel, 1971.]

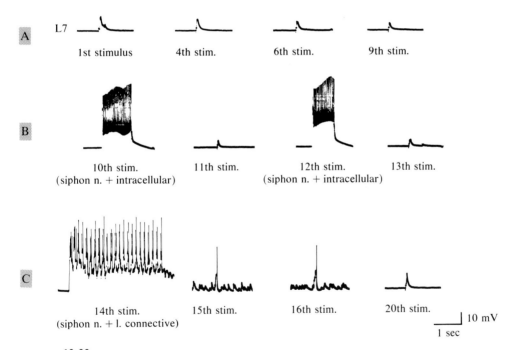

FIGURE 12-22.
Failure of a directly initiated train of action potentials in gill motor neuron L7 to produce facilitation of a depressed EPSP (sample records from continuous experiments). [From Carew, Castellucci, and Kandel, 1971.]

A. Records illustrating depression of the complex EPSP produced by stimulation of the siphon nerve every 10 sec.

B. During the tenth and twelfth stimuli to the siphon nerve, the motor cell was also stimulated intracellularly. The high-frequency burst of action potentials did not facilitate the depressed EPSP in trials 11 or 13.

C. On trial 14 a train of stimuli applied to the left connective (7/sec for 3.3 sec) facilitated the EPSP in trial 15 and in the next few trials.

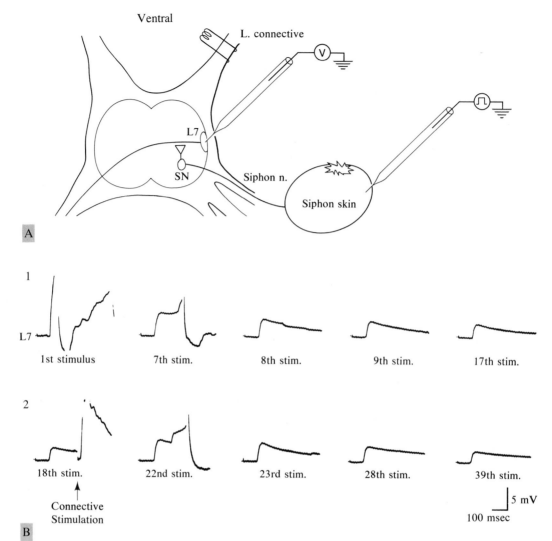

Ventral

L. connective

L7

SN

Siphon n.

Siphon skin

A

1

L7

1st stimulus   7th stim.   8th stim.   9th stim.   17th stim.

2

18th stim.   22nd stim.   23rd stim.   28th stim.   39th stim.

Connective
Stimulation

5 mV

100 msec

B

FIGURE 12-23.
Depression and heterosynaptic facilitation of elementary (presumably monosynaptic) EPSPs produced in cell L7 by stimulating the siphon skin. **A.** Experimental arrangement. **B.** Synaptic depression and facilitation (selected records from consecutive responses). *B-1.* Depression of responses to a series of 17 electrical stimuli to the siphon skin, one every 10 sec. *B-2.* Facilitation of the depressed EPSP following a train of stimuli applied to the left connective (9/sec for 2 sec) after the 18th stimulus to the siphon skin. The tops of the spikes have been cut off in several traces. [From Castellucci *et al.,* 1970.]

FIGURE 12-24 (*Facing Page*).
Depression and facilitation of monosynaptic EPSP produced in the motor neuron by stimulating a single sensory neuron. **A.** Experimental arrangement. **B.** Synaptic depression and facilitation. *B-1.* Selected records of depression during a series of 15 consecutive stimuli to the sensory neuron, one every 10 sec. *B-2.* The left connective was stimulated between the 15th and 16th stimulus to the sensory neuron. Connective stimulation did not fire the sensory neuron. (Vertical lines in the sensory neuron trace between spike number 15 and 16 are shock artifacts.) *B-3.* Selected records of the synaptic facilitation of the response to the 17th stimulus and the next 20 responses following connective stimulation. Facilitation lasted several minutes. Interneuronal activity was reduced by using a solution with high Mg^{++} and Ca^{++} concentrations. [From Kandel, Brunelli, Byrne, and Castellucci, 1976.]

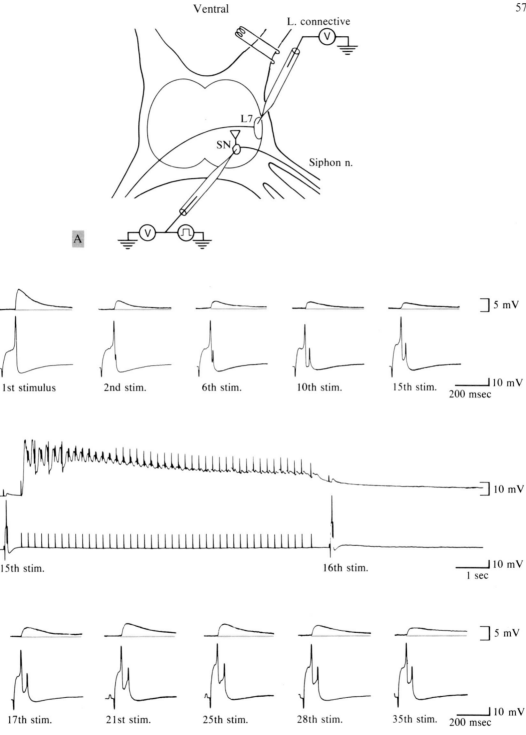

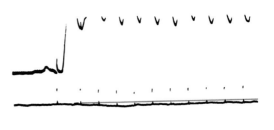

L7

SN

1st stimulus 2nd stim. 3rd stim.

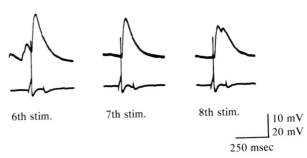

FIGURE 12-25.
Synaptic hyperpolarization of the sensory neuron during onset of heterosynaptic facilitation of the monosynaptic EPSP.

A. Depression of the monosynaptic EPSP in gill motor neuron L7 produced by intracellular stimulation of a sensory cell every 10 sec. Heterosynaptic stimulation of the left connective (7/sec for 5.5 sec) causes sustained hyperpolarization of the sensory cell (note the stimulus artifacts in the sensory neuron). The resulting EPSP produced in L7 by the sensory neuron is facilitated. [From Carew, Castellucci, and Kandel, 1971.]

B. The hyperpolarizing response in the sensory neuron is an increased-conductance IPSP. *B-1.* Slow hyperpolarization is obtained when a peripheral nerve or connective is stimulated at 10/sec for 0.5 sec. *B-2.* When the membrane potential is hyperpolarized, the hyperpolarizing PSP is reversed to a depolarizing PSP, one consistent with an increased-conductance EPSP. [From Byrne *et al.*, 1974b.]

6th stim. 7th stim. 8th stim.

10 mV
20 mV

250 msec

A

1

2

10 mV

2 sec

B

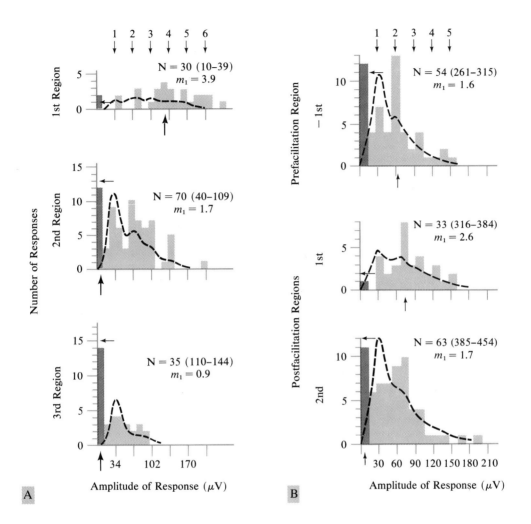

FIGURE 12-26.
Amplitude histograms comparing increased quantal output during heterosynaptic facilitation and decreased quantal output during synaptic depressions. [From Kandel, Brunelli, Byrne, and Castellucci, 1976.]

A. Synaptic depression. With repeated stimulation the number of failures increases from 2 per 30 responses in the first region to 14 per 35 responses in the third region. The position of the unit peak does not change (see the legend for Figure 12-19A for further details).

B. Synaptic facilitation. In the region preceding the facilitating stimulus there are many failures. In the two regions following the facilitating stimulus there are proportionately fewer failures but the unit peak (or its multiples) is not altered, indicating that the estimate of m increases but $\bar{q}$ does not change. With continued stimulation (2nd region) the number of failures again increases but the unit peak remains the same. Interstimulus interval 10 sec.

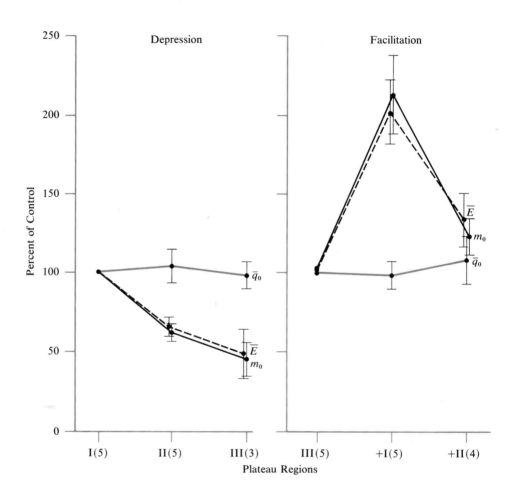

FIGURE 12-27.

Comparison of the average EPSP during depression and facilitation with quantal size and quantal output estimated by the failure method. Five experiments on synaptic depression were normalized to the first region, where failures were encountered. Another five experiments on synaptic facilitation were normalized to the first region prior to facilitation. During both depression and facilitation the estimated value of $\bar{q}$ did not change significantly between successive regions, whereas $\bar{E}$, the average EPSP amplitude, and m, the quantal output decreased by 50 percent during the depression and increased by 100 percent in relation to the control region during facilitation. [From Kandel, Brunelli, Byrne, and Castellucci, 1976.]

A CELLULAR ELECTROPHYSIOLOGICAL MODEL OF SHORT-TERM HABITUATION AND DISHABITUATION

The simplified circuit diagram in Figure 12-28 illustrates the model proposed by Castellucci and co-workers to account for the locus and mechanism of habituation and dishabituation (Castellucci *et al.,* 1970; Castellucci and Kandel, 1974; 1975). According to this model, habituation and dishabituation involve a common locus and mechanism: the presynaptic connections made by the sensory neurons onto the motor neurons and interneurons. Habituation results from a homosynaptic depression (self-generating), dishabituation from presynaptic facilitation. The interneurons mediating dishabituation are postulated to synapse on or near the presynaptic terminals of the sensory neurons and

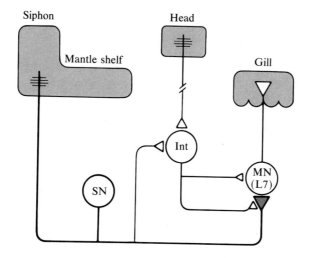

FIGURE 12-28.
Postulated plastic changes in the circuit underlying habituation and dishabituation of the gill reflex. For schematic purposes only one sensory neuron (SN) and one motor neuron (L7) are illustrated. Similar processes are thought to be operative at all the synaptic contacts made by the population of sensory neurons on the other motor neurons and interneurons. The terminal of a sensory neuron (SN) is the common locus of both the synaptic depression underlying habituation and the presynaptic facilitation underlying dishabituation. The pathway from the head mediating presynaptic facilitation ends on an excitatory interneuron (L22 or L23) that synapses on the sensory neuron terminal. The plastic synapse from the sensory to motor neuron is indicated in grey. [From Kandel, Brunelli, Byrne, and Castellucci, 1976.]

facilitate their transmitter release. There is as yet no direct evidence for this postulate. However, electron microscopic studies of sensory neurons labeled with the electron-dense marker horseradish peroxidase suggest that some apparent presynaptic terminals end on the presynaptic terminals of the sensory neurons (Bailey *et al.*, 1976; and see the frontispiece here).

Habituation involves a decrease in the quantal output from the presynaptic terminals of the sensory neurons. It therefore resembles the brief low-frequency depression seen at many synapses (Figure 11-9, p. 490; Figure 11-28, p. 521). As we have seen, however, the synapses in the habituating pathway differ from the usual synapses in that they undergo remarkably large and prolonged depression and do not facilitate at any frequency of firing or at any level of transmitter release. These differences suggest that additional features are operative in synapses of this habituating pathway. What these factors are is not known. One possibility is that mobilization of transmitter from the storage pool into release sites is much slower at the sensory-to-motor neuron synapse than at other synapses and is not enhanced by impulse activity. As a result action potentials in presynaptic terminals rapidly lead to depletion of the releasable pool of transmitter.

Mobilization of transmitter by impulse activity is thought to be due to an accumulation of Ca^{++} in the terminals following its entry with impulse activity (Katz and Miledi, 1968; Rahamimoff, 1974). In turn, the level of free Ca^{++} is thought to be largely regulated by mitochondrial uptake (Lehninger, 1970; Rahamimoff and Alnaes, 1973). Perhaps Ca^{++}-uptake mechanisms are particularly effective in the terminals of the sensory neurons because of a dense packing of mitochondria, so that Ca^{++} can not accumulate and therefore will not lead to the transmitter mobilization necessary for sustained activity. A progressive increase in Ca^{++}-uptake by mitochondria could slow the mobilizing of transmitter into release sites from a storage pool and also decrease the probability that an impulse will reload a release site. Alternatively, the Ca^{++} level may be controlled by the external membrane. With repeated impulse activity, less Ca^{++} may move into the terminals.[4]

[4]One might be able to test some of these possibilities with quantitative electron microscopic studies of the terminals of the sensory neurons following injection of electron-opaque or radio-labeled markers. Morphological studies could indicate whether the terminals contain an unusually high density of mitochondria. These studies could also reveal whether habituation leads to a reduction in the number of vesicles bound to the surface membrane of the terminal and perhaps even a reduction in total number of vesicles present in the terminal. Depletion of synaptic vesicles has been found in one vertebrate central synapse following unusually prolonged stimulation, 10–20/sec for 10 minutes, a rate that produced profound synaptic depression at this synapse (Model, Highstein and Bennett, 1975). In synapses designed to mediate habituation, 15 to 20 action potentials might be sufficient to produce comparable depletion. Aspects of the Ca^{++} accu-

Presynaptic facilitation and the accompanying synaptic hyperpolarization of the sensory neurons is mediated, at least in part, by the interneurons of the reflex pathway (Figure 12-4). Each of these interneurons receives input from the head ganglia, in addition to excitatory synaptic convergence from the other sensory neurons. The inhibitory interneuron L16 produces hyperpolarization that accompanies the facilitation of the sensory neurons (Figure 12-25*B*). Preventing L16 from firing blocks the synaptic hyperpolarization of the sensory neurons. However, direct firing of L16 does not produce facilitation. The facilitation is independent of the hyperpolarization; it is produced by the excitatory interneurons L22 and L23. Intracellular stimulation of either of these interneurons facilitates the synaptic actions of the sensory neurons on the motor neurons (Hawkins, Castellucci, and Kandel, unpubl.).

Dishabituation is typically associated with arousal of the organism (see p. 635). Although it is not as yet known whether the actions of L22 and L23 quantitatively account for all of the facilitation, the finding that the two cells contribute importantly to it suggests that at least part of the arousal system for this reflex is located in the abdominal ganglion and is an integral part of the neural circuit of this behavior. A corollary to this finding, which we will consider again later (Chapter 15, p. 659) is that arousal is not a unitary process localized in one neuronal group, but is multifaceted and anatomically distributed. Different behavioral systems probably have anatomically distinct neuronal systems for mediating arousal of the organism.

A MOLECULAR MODEL OF DISHABITUATION

In an attempt to analyze the molecular mechanisms underlying habituation and dishabituation of the gill-withdrawal reflex, Schwartz, Castellucci, and Kandel (1971) examined the effect of inhibiting protein synthesis in the isolated abdominal ganglion test system. The neuronal correlates of habituation and dishabituation (in response to electrical stimulation of the siphon nerve) were monitored in a gill motor neuron (L7) in the presence (and absence) of anisomycin, an antibiotic that interferes with protein synthesis in *Aplysia* neurons. Inhibition of protein synthesis did not block either the decrease in synaptic efficacy associated with habituation or the increase associated with dishabituation (Figure 12-29*A*).

---

mulation hypothesis could be tested by examining how impulse activity affects the level of free Ca^{++} as determined with intracellularly injected aequorin (see Note 6, p. 591). Alternatively, one could determine how inhibitors of mitochondrial uptake of Ca^{++} affect synaptic depression and facilitation (see Rahamimoff and Alnaes, 1973).

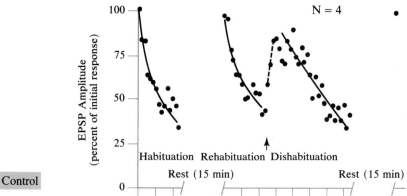

A-1. Control

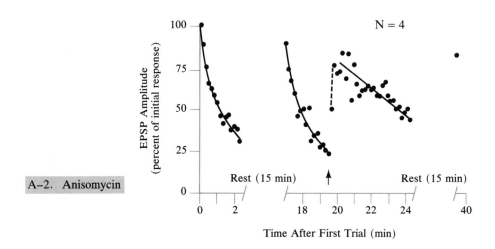

A-2. Anisomycin

FIGURE 12-29.
Biochemical studies of the synaptic depression and facilitation underlying habituation and dishabituation.

A. Changes in the complex EPSP produced in cell L7 by stimulation of the siphon nerve, both in the absence and in the presence of the protein-synthesis inhibitor anisomycin (18 μM). Fifteen consecutive stimuli produced a decrease in the amplitude of the EPSP of about 30 percent. A rest of 15 min resulted in almost complete recovery under both conditions. Fifteen stimuli were then applied, and this procedure again produced a depression in the EPSP amplitude. Stimulation of the left connective (arrows) produced facilitation lasting several minutes. After a second 15-min rest, one stimulus was given to test recovery. In both experiments the isolated ganglia were incubated for 30 to 50 min. *A-1.* Control. *A-2.* Stimulation in the presence of anisomycin. No differences between the two sets of data were observed. This indicates that protein synthesis is not necessary for short-term habituation or facilitation to occur. [From Schwartz, Castellucci, and Kandel, 1971.]

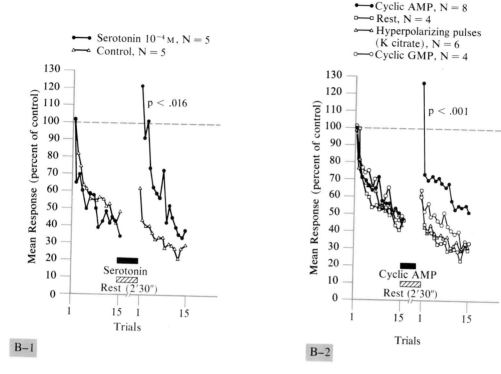

B. Effects of serotonin and intracellular injection of cyclic AMP on the monosynaptic EPSP produced by a sensory neuron in motor neuron L7. [From Brunelli, Castellucci, and Kandel, 1976.]

B-1. Effects of serotonin. A sensory neuron was stimulated once every 10 sec and produced a monosynaptic EPSP in a gill or siphon motor neuron. In each experiment the EPSP amplitudes were normalized to initial control EPSP. In the control group 15 stimuli produced synaptic depression. After a rest of 2½ min (and a slight recovery of the EPSP) a second series of 15 stimuli was applied to the sensory neurons, producing further depression. In the experimental group the initial series of 15 stimuli produced similar synaptic depression. During the 2½ min rest period the ganglion was perfused with a solution containing serotonin (2×10^{-4} M). During the initial EPSP depression there is complete overlap of the control and experimental curves. But after treatment with serotonin the experimental curve is significantly higher than the control curve. To determine the statistical differences between the experimental and control curves, the amplitude of the responses to the series of 15 stimuli were summed to obtain a single score and a two-tailed Mann-Whitney U test was performed ($p < 0.016$, •—•).

B-2. Effects of cyclic AMP. The protocol was similar to that used in part *A*. A sensory neuron was stimulated 15 times, once every 10 sec, and EPSP amplitudes were normalized to initial control. During the 2-min rest, cyclic AMP was iontophoresed into the body cell of a sensory neuron in the experimental group (•—•, N = 8). One-sec hyperpolarizing current pulses (3 to 4.5×10^{-7} A) were applied every 2 sec for 2 min 30 sec from one barrel of a double electrode filled with 1.5 M cyclic AMP; the second (recording) barrel was filled with 3 M potassium citrate. Thirty seconds after the end of the injection a second series of 15 stimuli was given. In the first control group the 2½ min rest was simply followed by 15 additional stimuli without any current injection. In a second group a hyperpolarizing current pulse, comparable to that used to inject cAMP, was injected into the sensory neuron but the pulses were applied through an electrode filled with 3 M potassium citrate. In a third group 1.5 M cyclic GMP was injected intracellularly. There is a complete overlap of the four curves during the initial EPSP depression produced by the first series of 15 stimuli. But after injection of cAMP the experimental group was significantly higher than either the first control group (□—□, $p < 0.001$, two tailed Mann-Whitney U test) or the two other control groups (▵—▵, $p < 0.001$ and ○—○, $p < 0.014$).

Short-term habituation and dishabituation thus appear not to depend on the synthesis of new proteins or on proteins that have remarkably fast turn-over rates. This conclusion is consistent with experiments in vertebrates that indicate that new protein synthesis is not required for short-term learning (Agranoff, 1967; Barondes, 1970). As indicated in Chapter 11 (p. 529), the biochemical mechanisms underlying these short-term plastic changes might be composed of a sequence of reactions that involve the synthesis of small molecules but not the production of new structural, receptor, or enzymatic components. For example, the presynaptic facilitation underlying dishabituation might involve a small regulatory molecule, perhaps a second messenger, that regulates the mobilization of transmitter substance or its availability by simply altering the accumulation of free Ca^{++} in the presynaptic terminals.

As a first step in analyzing possible molecular mechanisms underlying dishabituation, Cedar, Kandel, and Schwartz (1972) examined the effects of prolonged stimulation of the connectives from the head ganglia and found that it leads to an increase in cAMP (3'5' cyclic adenosine monophosphate) in the entire ganglion. This increase is produced by synaptic action and blocked by blocking synaptic transmission.

Cyclic AMP is one of several intracellular messengers (cyclic GMP is another) involved in mediating the actions produced by hormones (and biogenic amines) on muscle and other target cells in the body. With the exception of a few fat-soluble molecules (such as the steroids) most hormones are water soluble and do not dissolve in the lipid phase of the membrane and therefore do not enter the cell whose function they regulate. Most hormones act on receptors on the external surface of the membrane. This interaction initiates the synthesis (or accumulation) of certain small molecules in the cell, called *second messengers,* that internalize the message of the hormone (the first messenger). Second messengers trigger a series of biochemical reactions that induce the appropriate change in cellular function.

Second messengers are thought to serve at least two functions: amplification and regulation. By means of a second messenger a single hormone molecule can trigger a cascading series of biochemical steps. For example, by acting through cAMP, one molecule of the hormone epinephrine triggers the release of 10^8 glucose molecules from a liver cell (Goldberg, 1975). In addition, the intricate system of intracellular reactions necessary to mediate the actions of a second messenger can also be used by other intracellular messengers, for example, cGMP, prostaglandins, or Ca^{++}. These other messengers can regulate the activity of the initial second messenger, enhancing or counteracting its action.

In the abdominal ganglion of *Aplysia* the increase in cAMP outlasts the stimulation. A few minutes of synaptic stimulation leads to an increase in cAMP that is maintained for almost an hour. This synaptic action is simulated by three biogenic amines, serotonin, dopamine, and octopamine (Cedar and Schwartz, 1972; Levitan and Barondes, 1974). For example, a transient (5 minute) exposure to serotonin produces a sustained increase in cAMP that persists for 45 minutes. This increase parallels the time course of presynaptic facilitation produced by a strong stimulus. These findings suggest the specific hypothesis that dishabituation may be mediated by one of the three amines that enhance transmitter release by increasing cAMP in the presynaptic terminals.[5]

To test this hypothesis Brunelli, Castellucci, and Kandel (unpub.) examined the effects of the three amines and cAMP on the monosynaptic EPSP produced in gill motor neuron L7 by stimulation of individual sensory neurons. They found that serotonin (but not dopamine or octopamine) simulates presynaptic facilitation (Figure 12-29B). Introducing serotonin into the bathing solution increases the amplitude of the EPSP and retards its depression. (A similar enhancement of synaptic transmission in the pleural ganglia of *Aplysia* has been described by Shimahara and Tauc, 1972, 1975b.) Moreover, presynaptic facilitation, produced by nerve stimulation, is reversibly blocked by cinanserin, a blocking agent for serotonin-mediated synaptic transmission in *Aplysia*. Cinanserin does not interfere with the EPSP, nor does it block the synaptic depression due to repeated stimulation; it selectively affects facilitation.

To test whether the action of serotonin is mediated by cAMP, dibutyryl cAMP, a lipid soluble analog of cAMP that can cross the cell membrane, was applied to the bathing solution. It was found to enhance excitatory synaptic transmission between the sensory and motor neurons but not at two other excitatory synapses examined. Moreover, when cAMP was injected intracellularly into the sensory neuron it produced enhancement of synaptic transmission that simulated presynaptic facilitation. By contrast, intracellular injection of other nucleotides (5' AMP and cyclic GMP) was ineffective (Figure 12-29B).

Based on these experiments Kandel and co-workers (1976) have developed a molecular hypothesis about presynaptic facilitation. Figure 12-30 summarizes

---

[5]Some indirect support for this idea comes from experiments on homosynaptic (low-frequency) depression at the nerve–muscle synapse of vertebrates, where synaptic depression is counteracted by norepinephrine. The norepinephrine action appears to be mediated by cAMP, which acts to increase the quantal output without altering receptor sensitivity (Goldberg and Singer, 1969; Miyamoto and Breckenridge, 1974).

these speculations and relates them to habituation. Repeated stimulation of the sensory neurons leads to a progressive decrease in the amount of free Ca^{++} associated with each action potential in the terminal (due either to a decrease in Ca^{++} influx or to uptake of free Ca^{++} by mitochondria). A low level of free Ca^{++} could account for the slow mobilization and reloading of the release site, the features that seem to characterize the terminals of this habituating pathway. Because of slow mobilization, the nerve terminals of the sensory neurons normally only load a small number of vesicles into release sites following impulse activity. This small reservoir is rapidly reduced and may even be depleted because, unlike most other synapses, repeated impulse activity in the terminals does not enhance mobilization (Figure 12-27). As a result, at least several minutes of rest are required for recovery of normal transmission following a habituation training session of 10 to 15 stimuli. However the rate of mobilization can be speeded up by presynaptic facilitation. Facilitation is postulated to be produced by one or more serotonergic cells, perhaps L22 and

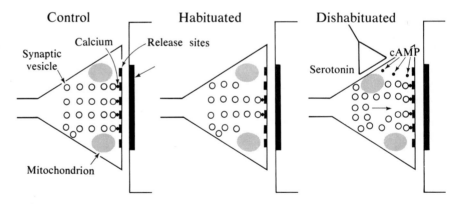

FIGURE 12-30.
Suggested mechanisms for depression and facilitation based on vesicle mobilization. Each consecutive action potential in the terminal of a sensory neuron may lead to a progressively smaller increase in free Ca^{++} either because of a high density of mitochondria that take up free Ca^{++} as it comes in with the action potential or because of a progressive decrease in the Ca^{++}-permeability of the terminal membrane. As a result, repeated stimulation does not lead to effective mobilization of vesicles to the release sites at the terminal and a partial depletion of vesicles is achieved after 5 to 15 impulses. This could account for habituation. Dishabituation is mediated by serotonergic interneurons, and produces two complementary actions: (1) a synaptic action, which increases the Ca^{++} concentration and briefly enhances mobilization, and (2) a prolonged increase in cAMP, which increases Ca^{++} levels by inhibiting Ca^{++}-uptake by the mitochondria or increasing Ca^{++} influx, thereby enhancing mobilization of transmitter vesicles.

L23. These cells are assumed to briefly alter the permeability of the presynaptic terminals of the sensory neuron, thereby transiently increasing Ca^{++} entry and speeding up mobilization. The action of serotonin is followed by a second action: increasing the level of cAMP in the sensory terminal, which in turn produces a more sustained increase in transmitter release by a long-term buildup in the Ca^{++} level either through inhibition of its uptake by mitochondria or by increasing Ca^{++} influx.[6]

There is as yet no direct evidence that L22 and L23 are serotonergic. Similarly, the suggestion that cAMP participates in presynaptic facilitation is based on pharmacological evidence and is not yet compelling. I have indulged in this exercise only to indicate that it is now possible to develop a testable molecular hypothesis of a plastic neural change related to a behavioral modification.

CELLULAR ANALYSES OF HABITUATION IN OTHER INVERTEBRATES

Escape Swimming of the Crayfish

Crayfish evade capture by darting backward (escape swimming) when perturbed. They do this by means of a stereotypic fixed act, a tail-flip response consisting of a sudden flexion of the abdomen. In this movement the tail fin is used as a paddle that can exert a backward and sometimes upward thrust against the water. Krasne and Woodsmall (1969) have found that 5 to 10 elicitations of the escape reflex, with mechanical stimuli at 5-minute intervals, cause an all-or-none reflex failure in most animals. After two hours of rest only a small number of animals recover; full recovery of the population requires more than six hours (Figure 12-31).

The tail-flip response can be elicited in a number of ways. When elicited by a phasic mechanical stimulus to the tail, it involves identified symmetrical

[6]Stinnakre and Tauc (1973) injected the Ca^{++}-dependent fluorescent protein aequorin into the cell body of individual *Aplysia* neurons and found measurable light emission (a measure of Ca^{++} influx) associated with each action potential (see Figure 7-22, p. 250). The control of Ca^{++} influx by the membrane potential in *Aplysia* cells corresponds remarkably well with that found by Katz and Miledi in the presynaptic terminal membrane at the giant synapse of the squid (see p. 197). This led Stinnakre and Tauc to suggest that changes in Ca^{++} influx in the cell body might mirror the changes in Ca^{++} influx that occur in the presynaptic terminal membrane. By injecting aequorin into the cell body of the sensory neurons, one might obtain an idea of how Ca^{++} influx might be controlled by serotonin, hyperpolarization, and cAMP.

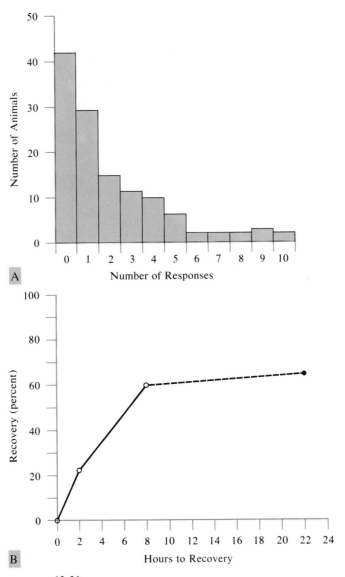

FIGURE 12-31.
Habituation of the tail-flip response of the crayfish. [From Krasne and Woodsmall, 1969.]

A. Bar graph of the number of tail-flip responses to 10 consecutive stimuli based on data obtained from 122 previously rested crayfish. Each animal was given 10 trials with 5-sec intervals. The data show individual differences in excitability.

B. The time course of recovery from reflex failure. Computation of recovery was based on 16 measurements at 0 hr, 13 at 2 hr, 14 at 8 hr, and 11 at 22 hr. The 22-hr point is not entirely comparable with the other points because it was obtained in another experiment.

interneurons, the left and right lateral giant interneurons (Wine and Krasne, 1972).[7] The two lateral giant fibers are tightly interconnected by means of an electrical synapse and function as a single axon. They in turn excite a giant motor neuron as well as a population of smaller fast-flexor motor neurons that innervate the phasic flexor muscles of the abdomen (Kennedy and Takeda, 1965; Mittenthal and Wine, 1973). Each lateral giant is a command neuron. A single action potential in one cell produces the tail-flip escape response.

The lateral giant fibers are activated by tactile stimulation of the pit-hair receptors on the surface of the abdominal carapace. The sensory neurons have direct connections to the giant fibers and indirect connections to them via at least three (and probably more) interneurons. All the connections to the giant fibers, both direct and indirect, are thought to be mediated by electrical synapses. Thus, here as in the two other fixed acts (inking and egg-laying) that we considered previously (Chapter 9, p. 415) electrically interconnected cells are prominently involved in generating an all-or-none response. By contrast, the synapses made by the tactile sensory neurons on the interneuron are chemically mediated (Figure 12-33; see Zucker, Kennedy, and Selverston, 1971; Zucker, 1972a, c).

Various points of this neural circuit have been examined to determine the locus for habituation (Krasne and Woodsmall, 1969; Krasne, 1969; and Zucker, 1972a, b, c). Krasne and Woodsmall found that prolonged stimulation of the lateral giant, at frequencies up to 1/sec, results in continued appearance of tail flips, suggesting that the locus of habituation must be afferent to the giant fiber. Using an isolated abdomen preparation, Krasne found that habituation involved a failure of transmission of the synaptic input to the lateral giants. Stimulation of the afferent root produced a decrease in the complex EPSP in the lateral giant, the time course of which paralleled that of behavioral habituation (Figure 12-32*A*). Zucker (1972b) found that the tactile receptors do not adapt to repeated stimuli, and neither the lateral giant fibers nor the tactile interneurons show long-lasting refractoriness. The threshold and input resistance of the giant fibers also do not change following repeated stimulation. The decrease in the complex EPSP in the lateral giant is attributable to a decline in the chemically mediated EPSP produced in the interneurons by the sensory neurons. The elementary EPSPs produced by the sensory neurons in the interneurons undergo homosynaptic depression when stimulated at fre-

---

[7]The lateral giant neurons are segmented, being composed of units that are tightly coupled by electrical synapses. Each axon segment has its own large cell body, contralateral to the axon (Remler, Selverston, and Kennedy, 1968). The lateral giants are therefore actually several fused neurons, but their electrical synaptic connections make them behave functionally as one unit (Kao, 1960; Watanabe and Grundfest, 1961).

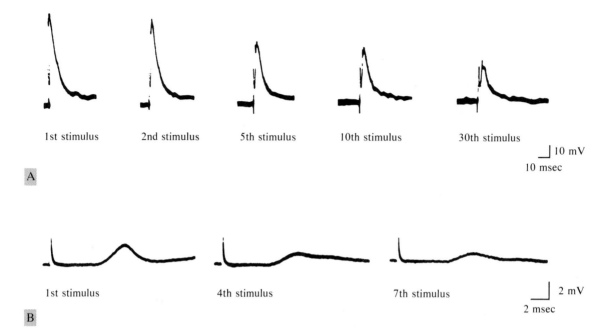

1st stimulus   2nd stimulus   5th stimulus   10th stimulus   30th stimulus

┘ 10 mV

10 msec

A

1st stimulus   4th stimulus   7th stimulus

┘ 2 mV

2 msec

B

FIGURE 12-32.
EPSP depression accompanying habituation of the tail-flip response of the crayfish. **A.** A complex EPSP in the lateral giant neuron (recorded in the fifth ganglion) in response to stimulation (interstimulus interval 10 sec) of the second root of the same segment. **B.** Properties of monosynaptic EPSPs produced in identified tactile interneuron in response to stimulation (interstimulus interval 2 sec) of individual tactile neurons. [From Zucker, 1972b.]

quencies ranging from 1/sec to 1/min (Figure 12-32*B*). Thus, crayfish escape, an all-or-none fixed act, and gill-withdrawal, a graded reflex act, share a common locus and mechanism for habituation. In each case the locus for habituation is the site of the central termination of the primary sensory neurons, and the mechanism is homosynaptic depression (Figure 12-33, compare with Figure 12-13). On the basis of a preliminary analysis of the coefficient of variation of the complex EPSP, Zucker (1972b) suggested that the depression is presynaptic, involving a decrease in the number of quanta released by each impulse. It is still necessary, however, to carry out an analysis on the elementary EPSP produced by stimulation of single sensory neurons, at both normal and low levels of transmitter release (using failure analysis) before one can be certain of the mechanism.

A fascinating aspect of habituation of the tail-flip response has been studied by Krasne and Bryant (1973). They addressed themselves to the question of

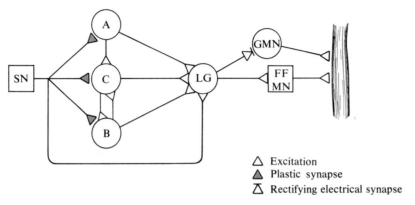

△ Excitation
▲ Plastic synapse
◬ Rectifying electrical synapse

FIGURE 12-33.
Neural circuit for generating the tail-flip response of the crayfish in response to phasic mechanical stimulation of the abdomen. Circles represent single neurons; squares represent populations of similar neurons. Shaded terminals are chemical synapses that undergo depression with repeated stimulation. All contacts onto the lateral giant (LG) are electrical junctions. The electrical synapse between the lateral giant and the giant motor neuron (GMN) is unidirectional (rectifying). SN, tactile sensory neuron; A, B, C, identified tactile interneurons; LG, lateral giant; GMN, giant motor neuron. FFMN, fast flexor motor neurons. (Compare with Figure 12-13, p. 562). [From Zucker, 1972a.]

how the animal adjusts for possible maladaptive usage of habituation. Can an animal protect itself from self-induced tendencies to habituate? Animals will swim with rapid abdominal flexion not only in response to abdominal contact but also in response to a variety of other stimuli. These tail flips in turn can fire the mechanosensitive sensory neurons from the abdomen. However, the synapses of these sensory neurons readily undergo depression, which might leave the animal unable to make an escape response to other noxious stimuli. Krasne and Bryant found that during swimming produced by other stimuli the synapses between sensory neurons and interneurons are protected from habituation by presynaptic inhibition (see also, Kennedy, Calabrese, and Wine, 1974). These findings suggest that the use of these highly plastic synapses is regulated by a control circuit in the central nervous system.

Escape Response of the Cockroach

A puff of air applied to the anal antennae (cerci) of a cockroach triggers an escape reflex (Pumphrey and Rawdon-Smith, 1937) that rapidly habituates with repeated stimulation (Roeder, 1948). The reflex is thought to be mediated by cercal afferent fibers that synapse on the interneurons in the last (sixth) abdominal segment. The postsynaptic interneurons in the abdominal ganglion terminate on motor neurons in the thoracic ganglia that innervate the legs of the corresponding segment (Roeder, 1948; Dagan and Parnas, 1970). One group of interneurons in the pathway from the cercal afferent fibers to the thoracic ganglia is a giant interneuron system (Roeder, 1948; Zilber-Gachelin and Chartier, 1973a). These neurons receive direct, apparently monosynaptic connections from the cercal mechanoreceptors. Stimulation of the anal cerci even at the low rate of one every five minutes leads to decreased firing of the giant fibers without a concomitant change in the firing of the sensory cells (Niklaus, 1965; Zilber-Gachelin and Chartier, 1973a). Part of this decrease appears to be due to homosynaptic depression of the chemical synapses between sensory neurons and giant neurons (Callec, Guillet, Pichon, and Boistel, 1971). Repeated stimulation (at least 1/sec) of a single receptor results in depression of the elementary EPSP that it produces in the giant neurons (Figure 12-34). In addition, the cercal input to the other interneurons also decreases (Zilber-Gachelin and Chartier, 1973a). Thus, as in the crayfish and *Aplysia*, the central synapse of the sensory cells may prove a critical site for habituation. However, there appears to be a second site of depression in the thoracic ganglia, where the interneurons in the abdominal ganglion engage the motor cells (Zilber-Gachelin and Chartier, 1973b). The thoracic ganglia also seem to be the site for dishabituation (Figure 12-35).

CELLULAR ANALYSIS OF HABITUATION IN VERTEBRATES

It would obviously be of great interest to know whether vertebrates and invertebrates share aspects of a common mechanism for habituation and dishabituation. Fortunately, as a result of the pioneering work of Spencer, Thompson, and Neilson a great deal is known about one form of mammalian habituation: flexion withdrawal of the cat. I will consider here some similarities of habituation in vertebrates and invertebrates; similarities of dishabituation will be taken up in Chapter 13.

The reflex responses of the spinal cord of vertebrates are important for posture and locomotion. In the course of analyzing the neural mechanism of spinal reflexes, Sherrington found that certain reflex responses, such as the flexion withdrawal of a limb in response to stimulation of the skin, decreased with repeated stimulation and recovered only after many seconds of rest. Sherrington was greatly influenced by Cajal's work. In fact, it was Sherrington

Stimulus 3/sec

1 sec

Stimulus 25/sec

100 msec

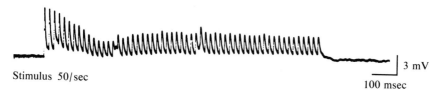

Stimulus 50/sec

3 mV

100 msec

FIGURE 12-34.

Synaptic depression of the monosynaptic connections of sensory neurons in the cockroach recorded from the interneuron of the abdominal ganglion. Repetitive electrical stimulation of the cercal nerve at different frequencies. In the second trace (25/sec) the first EPSP gives rise to an action potential, the rapid phase of which is shown. [From Callec, Guillet, Pichon, and Boistel, 1971.]

who coined the term "synapse" for the zone of apposition between neurons described by Cajal. It was therefore perhaps only natural for him to attribute the decrease in the responsiveness of the withdrawal reflex to a decrease in function at the synapses through which the motor neuron responsible for flexion was repeatedly activated. Thus, as early as 1906 Sherrington had suggested that a plastic change at central synapses could underlie response decrement. Sherrington found that the depression was not due to fatigue of either the muscles or the sensory receptors but, with the neurophysiological tech-

598

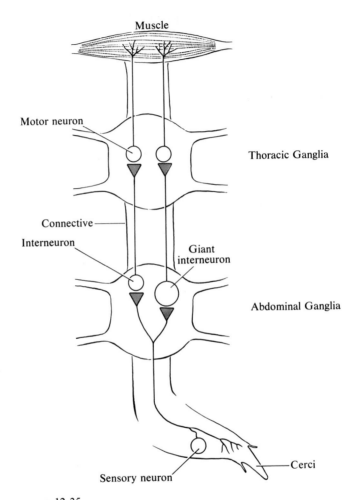

FIGURE 12-35.
Diagram of cercal escape response illustrating two sites of plasticity, one
in the abdominal ganglia, where sensory neurons synapse on interneurons,
and a second in the thoracic ganglia, where the interneurons synapse on the
motor cells. Grey terminals indicate plastic synapses.

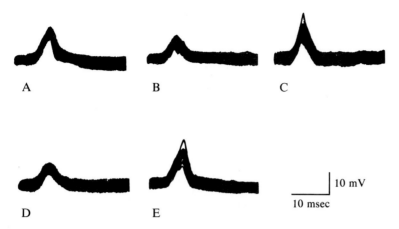

FIGURE 12-36.

Depression and restoration of a polysynaptic, complex EPSP in a motor neuron in the spinal cord of a spinalized cat. Intracellular recordings from a deep peroneal motoneuron. **A.** Superimposed records during the control period, interstimulus interval 3 sec. **B.** Superimposed records of synaptic depression, interstimulus interval 1 sec. **C.** Superimposed records of response restoration after tetanization of the tibial nerve (100/sec for 4 sec). **D.** Subsequent depression produced by continued stimulation, interstimulus interval 1 sec. **E.** Period of recovery, interstimulus interval 3 sec. [From Spencer, Thompson, and Neilson, 1966b.]

niques then available, he could not fully test his intriguing synaptic hypothesis.

The problem was reinvestigated in 1936 by Prosser and Hunter, who found that the habituated flexion withdrawal can be restored to full size (dishabituated) by the application of a strong new stimulus to another part of the skin. In 1966 Spencer, Thompson, and Neilson recorded intracellularly from the motor neurons that mediate the flexion reflex in the cat and found that depression of neuronal responses did not involve a change in the input resistance of the motor neurons but did involve a change in the excitatory synaptic impingement on them (Figure 12-36). Because the central synaptic pathways of the flexion withdrawal reflex in the cat are complex, involving many connections between sensory and motor cells through interneurons that have not yet been worked out, the critical change underlying habituation has not yet been localized. But on the basis of the data available to them, Spencer and his colleagues postulated that habituation of this reflex involved a homosynaptic depression of the synaptic input to the interneurons.

Subsequent work by Groves and Thompson (1970) provided further support for this idea. They examined several populations of interneurons and found that only certain classes showed a depression of response. Although the indirect evidence for a plastic change at excitatory synapses onto interneurons is good, the complexity of the spinal pathways has prevented investigators from distinguishing between plastic changes at specific synapses or dynamic changes in the circulation of impulses in loops of neurons. Also, it has not yet been possible to specify directly whether inhibitory or excitatory synaptic transmission is altered (see for example Wickelgren, 1967; Wall, 1970).

In an interesting extension of this work, Farel, Glanzman, and Thompson (1973) examined a monosynaptic analog of habituation in the frog spinal cord. They stimulated a descending pathway, the lateral columns, that produces a direct excitatory connection onto motor neurons and found that at low frequencies of stimulation this input shows a depression that shares the features of habituation of the flexion reflex in the frog. Since the lateral columns carry the axons of interneurons that may participate in a defensive reflex (jumping), the findings raise the possibility that habituation of a vertebrate defensive reflex may also involve a plastic change in excitatory synaptic connections. This finding is particularly interesting in view of the fact that human spinal reflexes also habituate (see Chapter 13, p. 633).

HIGHER-ORDER FEATURES OF HABITUATION

There is now suggestive evidence that homosynaptic depression is a common mechanism for habituation. However, certain types of habituation in higher vertebrates and man have two features that seem, on the surface, inconsistent with this mechanism. First, during habituation of the orienting response of the cat (or man) to a repeated tone, a transient decrease in the intensity of the tone will lead to dishabituation (Sokolov, 1963a, b). Second, certain reflex responses (such as changes in heart rate in response to localized cooling) become more resistant to habituation following lesions of the frontal cortex (Glaser, 1966). In each of these cases it has been thought that habituation must be due to some active process—the activation of a neural comparator circuit in the case of dishabituation of the orienting response (Sokolov, 1963a, b), and a buildup of inhibition in the case of increased resistance to habituation following lesions (Glaser, 1966). Actually, both features can also be explained by a homosynaptic depression. Dishabituation due to a decrease in intensity or change in pattern could reflect the activation of a different pathway. In the highly organized nervous systems of vertebrates a weaker tone may not simply activate fewer fibers but probably activates different fibers. The apparent dishabituation may therefore reflect the activation of a new population of sensory neurons or interneurons that were not previously activated. Vertebrates, having more sophisticated perceptual capabilities than invertebrates, might utilize the same elementary mechanism for habituation as invertebrates and simply superimpose upon this mechanism additional discriminative capabilities (for discussion see Krasne, 1976).

Similarly, a brain lesion might not act on the spinal mechanisms of habituation but could remove an inhibitory system that acts on neurons capable of tonically dishabituating a reflex. With the removal of this inhibitory system the reflex becomes disinhibited and habituation of the response becomes difficult to achieve.

These arguments are clearly indirect and they should not be taken to indicate that all forms of habituation are likely to involve homosynaptic depression. That is improbable. However, the arguments do make clear that one cannot *exclude* the possibility of a homosynaptic mechanism for depression for *any* form of habituation, no matter how complex, on the basis of behavioral or lesions studies alone.

SUMMARY AND PERSPECTIVE

Habituation and dishabituation are often considered the most elementary forms of learning. Habituation is a decrease in a behavioral response that occurs when a stimulus is repeatedly presented. The habituated response recovers spontaneously if the stimulus is withheld for some time. But immediate restoration (dishabituation) of a previously habituated response can be produced by a change in the pattern of the stimulus or by the presentation of another stimulus. Habituation is widespread throughout the animal kingdom and is probably the most common form of behavior modification, even in humans. Because it is also simple, habituation has been more fully explored on the cellular level than any other behavioral modification.

The gill-withdrawal reflex of *Aplysia* habituates when elicited 10 to 15 times at intervals of 10 minutes or less. Following a single habituation session of 10 to 15 stimuli, recovery of reflex responsiveness to a single test stimulus requires a period of rest ranging from 30 minutes to several hours. But a strong stimulus to the head of the animal produces immediate dishabituation. Habituation of this reflex shows all of the nine characteristics of habituation in vertebrates.

Habituation of the gill-withdrawal reflex is associated with a decrease in the amplitude of the complex EPSP produced in the gill motor neurons by the activation of sensory neurons and interneurons. This decrease in the EPSP leads to a reduction in the number of spikes in the motor neuron, which in turn reduces the amplitude of the gill contractions. The synaptic depression is due to a self-generated decrease in the effectiveness of the elementary excitatory connections made by the sensory neurons on their central target cells, the motor neurons, and the interneurons. The depression results from a decrease in the quantal output of transmitter substance produced by each action potential in the sensory neurons. Dishabituation involves a sudden increase in effectiveness (presynaptic facilitation) of these direct connections from the sensory neurons by increasing transmitter quantal output. With both habituation and dishabituation the sensitivity of the postsynaptic receptor is unaffected.

Habituation and dishabituation of the gill-withdrawal reflex involve only plastic changes in the excitatory transmission of previously existing connections. The properties of the neurons are not altered, nor is there any fundamental change in the circuit. Since no new connections are added and no existing connections disappear, new growth of connections need not be postulated to explain the behavioral modification. Nor does short-term habituation involve dynamic reverberatory activity in an interconnected chain of neurons. Rather, one class of synapses, that between the sensory neuron and its central target cells, undergoes a profound low-frequency homosynaptic depression during habituation and a presynaptic facilitation during dishabituation. A similar locus and mechanism underlies habituation of escape swimming in the crayfish, an all-or-none fixed act, and the escape behavior of the cockroach. Homosynaptic depression, but at another locus, may also underlie habituation of the flexion reflex in mammals.

These analyses of habituation and dishabituation are consistent with the notion that genetic and developmental processes determine the properties of individual cells and the specific anatomical interconnections between cells. Nevertheless, these processes leave unspecified the degree of effectiveness of certain of these connections and thus allow experience to modify behavior. Moreover, in the gill-withdrawal reflex of *Aplysia,* for which the data is most complete, the neural system for modifying behavior is an integral part of, and not superimposed upon, the neural circuit of the behavior. These findings help to clarify the paradox that plagued Karl Lashley when he discovered that removal of large parts of the cortex did not interfere with learning (see p. 612). The failure of a lesion to produce any manifest effect on learning is not due to equipotentiality or mass action, but to the extensively distributed, overlapping organization of the sensory, motor, and interneuronal systems in the brain (for examples in *Aplysia,* see Fig. 12-14*A*). Individual elements can often be removed without complete loss of a behavior because the remaining cells continue to mediate the behavior. Since it is built directly into the neuronal circuitry of the behavioral response, the capability for modification is also extensively distributed, and hence is not erased by lesions involving only some of the cells.

The finding that a profound form of homosynaptic depression underlies the habituation of a defensive reflex in *Aplysia,* crayfish, and perhaps also in vertebrates, makes it likely that this mechanism is involved in simple forms of habituation in many animals. Nevertheless, other forms of short-term habituation may exist (see for example, Sokolov, 1963a, b). Also, habituation varies not only in its complexity but also in its duration. It would therefore be incorrect, on the basis of the limited evidence available, to assume that all forms of habituation necessarily involve the same physiological mechanisms. Other types of habituation should be examined on the cellular level to obtain a broader view of the possible mechanisms involved.

SELECTED READING

Horn, G., and R. A. Hinde, eds. 1970. *Short-Term Changes in Neural Activity and Behaviour.* Cambridge, England: The University Press. A collection of papers on habituation and other short-term processes.

Jeffrey, W. E. 1968. The orienting reflex and attention in cognitive development. *Psychological Reviews,* 75:323–334. Reviews the importance of habituation for the development of perception and memory in newborn infants.

Krasne, F. B. 1973. Learning in crustacea. In *Invertebrate Learning,* Vol. 2, edited by W. C. Corning, J. A. Dyal, and A. O. D. Willows. New York: Plenum Press, pp. 49–130. An excellent discussion of habituation in the crayfish.

Krasne, F. B. 1976. Recent developments in the neurobiology of memory. In Society for Neuroscience 5th Annual Meeting. Brain Information Service Conference Report #43, pp. 27–30. A good discussion of the relation of homosynaptic depression to higher-order features of habituation.

Spencer, W. A., R. F. Thompson, and D. R. Neilson, Jr. 1966. Response decrement of the flexion reflex in the acute spinal cat and transient restoration by strong stimuli, *Journal of Neurophysiology,* 29: 221–239. Decrement of ventral root electrotonus and intracellularly recorded PSPs produced by iterated cutaneous afferent volleys, *ibid,* 29: 253–274. The classic studies on the cellular mechanisms of habituation in the isolated cat spinal cord.

Thompson, R. F., and W. A. Spencer. 1966. Habituation: a model phenomenon for the study of neuronal substrates of behavior. *Psychological Reviews,* 73:16–43. A highly influential review summarizing the critical features of short-term habituation.

Thorpe, W. H. 1956. *Learning and Instinct in Animals.* Cambridge, Mass.: Harvard University Press. Perhaps the single most important source of the current interest in habituation.

CHAPTER 13

Relationships Between Short-Term and Long-Term Behavioral Modifications

A key question in the study of learning is: How are various types of behavioral modifications interrelated at a mechanistic level? What is the relationship of short-term and long-term habituation? Is long-term habituation an extension of the process underlying short-term habituation, or does it involve different cellular mechanisms operating at a different anatomical locus? Following the discovery that most forms of learning give rise to both short- and long-term memory there were repeated attempts, all rather unsuccessful, to determine whether these two apparently different forms of memory represent separate processes or merely two different phases of a single process.

Similar questions have dominated the debate and the research on the relationship between habituation and dishabituation, between dishabituation and sensitization, between sensitization and classical conditioning, and between classical and instrumental conditioning. In each case the debate is over one-process versus two-process theories.[1] Is dishabituation a wiping out of habituation or is it an independent facilitatory process, a form of sensitization? Is sensitization a component of classical conditioning? Are classical and instrumental conditioning two aspects of a common learning mechanism? To resolve these questions, the relationships between types of behavioral modification

[1]See Chapter 1, p. 9. See also Pavlov, 1927; Guthrie, 1935; Skinner, 1938; Hull, 1943; Konorski, 1948; Rescorla and Solomon, 1967; and Groves and Thompson, 1970.

must be analyzed on a more fundamental level than that of behavior. This requires that one determine the locus in the nervous system and the mechanism for each type of modification.

In this chapter I shall consider three relationships: (1) short-term and long-term memory (habituation); (2) habituation, dishabituation, and sensitization; and (3) short-term and long-term sensitization. In the course of discussing these relationships I shall also consider, in a more speculative way, new data on the relationship of habituation and sensitization to more complex forms of associative learning.

SHORT-TERM AND LONG-TERM MEMORY

In the broadest sense *memory* is defined as the ability of an animal to retain or store a behavioral modification (Hilgard and Bower 1975). Memory is typically measured by first exposing a subject to a training session and then determining at various times how much the subject has retained (or forgotten). Thus habituation, or any learning process, involves two distinct phases: *acquisition,* the mastery of learning, and *retention,* the time course over which memory is retained. Although many scientists believe that all memories are produced by a single mechanism, the evidence is not compelling. Different types of learning may well give rise to different types of storage processes. In addition, it is not known how various short-term forms of a memory, lasting minutes and hours, and long-term forms, lasting days and weeks, relate to each other. These questions address one of the major issues in the investigation of the neural structure of behavior. The study of memory processes therefore provides one of the best instances in which cellular analyses are likely to clarify questions about intervening variables (acquisition, memory), questions that cannot be resolved with behavioral analyses alone (see p. 23). I will therefore consider behavioral studies of memory in some detail.

That there may be different processes forming short- and long-term memories was first suggested by William James in 1890: James's suggestion was subsequently supported by the experiments of Müller and Pilzecker in 1900. They found that a subject would forget recently learned material if he actively engaged in mental activity immediately after learning; they called this forgetting *retroactive interference.* Many subsequent studies in animals and humans confirmed Müller and Pilzecker's observation that newly formed memories are susceptible to disruption for a short period of time after formation. Later, memories become more stable (consolidated) and less capable of being disrupted (Minami and Dallenbach, 1946; Postman and Alper, 1946; McGeoch, 1952).

Early thinking about memory storage was also influenced by the clinical observation that head trauma or convulsions could produce amnesia for events that recently preceded the trauma without affecting memory of older events. This phenomenon, called *retrograde amnesia,* is found in epileptics, who frequently have amnesia for events immediately preceding a seizure. McDougall (1901) suggested that the retroactive interference found by Müller and Pilzecker might explain the retrograde amnesia.

In 1949 Duncan produced retrograde amnesia experimentally in animals. He found that memory was disrupted if a learning trial was closely followed by a convulsive seizure. In the same year Hebb (1949) developed a dual-trace hypothesis (see p. 478). Hebb suggested that short-term memory consisted of a dynamic, easily disrupted reverberating electrical activity in closed neural circuits in the brain. Long-term memory involved more permanent structural change in the brain.

During the next 20 years many experiments examined the conditions that impaired or facilitated short- and long-term memory (for review see McGaugh and Herz, 1972). A variety of experimental methods, including spreading depression, hypothermia, and anoxia—all usually injurious to brain function—were found to interfere with the short-term process. On the other hand, several antibiotics that inhibit protein synthesis (puromycin, cycloheximide) were found to interfere with long-term memories (Flexner, Flexner, and Stellar, 1963; for review see Agranoff, 1967, and Barondes, 1970). From these studies it was concluded that short-term memories involve a labile process that, if allowed to run its course, is converted to a more stable neural process by a mechanism called *consolidation* (see Russell, 1959; McGaugh, 1966).

Based in part on these findings, several students of human learning have developed multiprocess models of memory. For example, a common model of verbal memory combines the concept of short- and long-term memory with ideas derived from computer processing, storage, and retrieval (Figure 13-1; Bower, 1967; Atkinson and Shiffrin, 1968). According to this model, new information (a telephone number or a nonsense syllable) is held in a sensory store for a few seconds and then either attended to or ignored. If ignored, the trace decays and disappears; if attended to, it is conveyed to a short-term store, which has a limited capacity. Unless operated upon by various mechanisms, such as rehearsal, the trace decays from the short-term store, usually within 30 seconds. Also, if too much information is transmitted to the short-term store and its capacity is exceeded, some of the information in the short-term store is displaced and lost. A memory trace that is acted upon in the short-term store by rehearsal gradually gains access to the long-term store. Access of information to the long-term store is usually not accomplished in one step, but

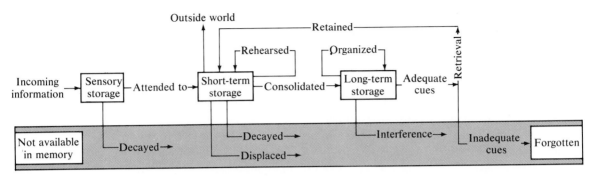

FIGURE 13-1.
Multistage model of human verbal learning. [Modified from Bower, 1967.]

in segments or chunks. The longer the information remains in the short-term store, i.e., the more the opportunity for rehearsal, the greater is the number of segments of information that can gain access to the long-term store. The long-term store has unlimited capacity and can hold information permanently. For recall and action, information must be transferred to the short-term store, which is the interface with the outside world. Forgetting is assumed to occur through inaccessibility or retrieval failure, not through decay.

This multistage verbal memory model is similar to and derives from the two-process model based on animal studies. Both models postulate two stores in series, with rules that determine how a trace moves from one to another. In each case a mechanism (consolidation or rehearsal) is thought to be necessary during the early and labile phase for transfer from short- to long-term store. But the model of human memory, based on verbal learning, specifies some features more completely than the model of animal memory based on non-verbal learning. For example, the short-term store is conceived of as being in contact with both the outside world and with long-term memory. All memory must be returned to the short-term store before it can be utilized. Similarly, the rehearsal process, in addition to having the features of consolidation, prepares information for memory. Finally, structural and organizational features of long-term storage transform the memory trace. Thus, in human verbal learning the trace, once having been stored in long-term memory, is thought to assume a different character than it had when it first appeared in the short-term store.

Despite their widespread use, multiprocess models have been criticized on several grounds. For instance, it has proven difficult to define a time course for short-term memory. Based on studies of retrograde amnesia in animals, the time course of short-term memory varies from seconds to hours or even days depending upon the disruptive procedure and the learning task (Chorover and Schiller, 1965; Quartermain, Paolino, and Miller, 1965; Kopp, Bohdanecky, and Jarvik, 1966). In some experiments the recently learned material that at first appeared lost, as a result of retrograde amnesia, returned days or weeks later (Weiskrantz, 1966). Also, most procedures that interfere with memory are traumatic, some devastatingly so, and the data obtained by different memory-blocking agents are not consistent. This has led some animal and human psychologists to suggest that what appears as short-term memory may only be an early, sensitive phase of long-term memory (Weiskrantz, 1970; Craik and Lockhart, 1972; Wickelgren, 1973). According to this view, memory consists of a single trace that changes in character with time and with processing.

The conflict between the one- and two-process models is central to the renewed interest that psychologists have shown in the neural analysis of learning, and poses a specific challenge to the cellular analysis of memory processes. Do short- and long-term memories (for a given learning task) represent extensions of the same process or two different processes? If they represent two different processes, where is each located and how do the two interact?

The Transition from Short-Term to Long-Term Habituation

One feature of habituation, long overlooked, is that repeated training often leads to long-term habituation lasting days and even weeks (Thorpe, 1956; Hinde, 1954a, b; Jeffrey, 1968). The habituation paradigm can therefore serve as a useful tool for studying both short- and long-term memory. For example, in *Aplysia* a single habituation training session of 10 tactile stimuli to the siphon produces short-term habituation of the siphon-withdrawal reflex that lasts only several hours (Figure 12-1*B*, p. 543), but four repeated sessions (10 trials a day) produce long-term habituation. In each of the four sessions there is a progressive buildup of habituation so that by the fourth day the median reflex response (sum of trials 1 to 10) is only 20 percent of what it had been on the first day. Three weeks (day 26) after training, experimental animals still show significantly greater retention of habituation than a control group, which is given no training (Figure 13-2). The training sessions need not be separated by 24 hours. As we shall see, equally good retention follows when the sessions are separated by as little as 1.5 hours.

Although the siphon-withdrawal component of the defensive reflex has the advantage that it can be studied in unrestrained animals (because the siphon

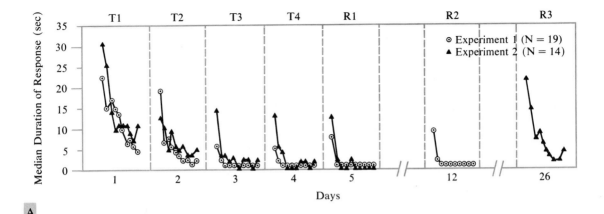

A

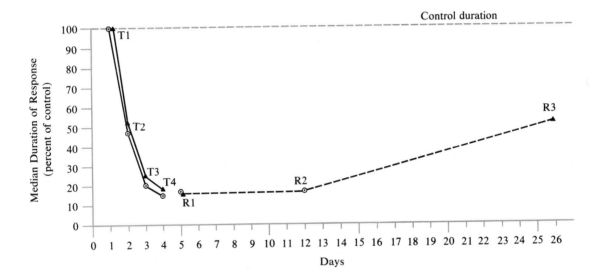

FIGURE 13-2.
Long-term habituation of siphon withdrawal in *Aplysia*. [From Carew, Pinsker, and Kandel, 1972. Copyright by the American Association for the Advancement of Science.]

A. Habituation during four daily sessions of training (T1 to T4), and retention one day (R1), one week (R2), and three weeks (R3) after training. Each session consisted of 10 trials (10 stimuli). Data from two experiments are compared. In experiment 1 retention was tested one day (R1) and one week (R2) after training. In experiment 2 retention was tested one day (R1) and three weeks (R3) after training. Each data point is the median response of the population in a single trial.

B. Time course of habituation, based on the experiments illustrated in part *A*. The score for each daily session is the median of the sum of 10 trials. Control duration (100 percent) is the response time during the first day of training.

extends out of the mantle cavity and can be directly observed), most of the neural analysis of short-term habituation has been carried out on the gill-withdrawal component, which can only be studied in restrained animals (Chapter 12). Carew and co-workers (1972) therefore examined whether long-term habituation of the siphon-withdrawal component of the defensive reflex was accompanied by long-term habituation of the gill-withdrawal component.[2] Animals from an experimental group (which received four days of habituation training) and from a control group (which received no habituation training) were restrained to expose the gill and the gill-withdrawal reflex was measured using a blind procedure. One day and one week after training the experimental animals exhibited significantly greater habituation of gill withdrawal than did the controls (Figure 13-3). Repeated training with siphon stimulation therefore gives rise to long-term habituation of both gill withdrawal and siphon withdrawal.

The finding that long-term habituation occurs in the gill-withdrawal component thus provides an opportunity for studying the synaptic changes that accompany the transition from short- to long-term habituation.

Neural Studies of the Transition from Short-Term to Long-Term Habituation

As a first step in the study of the neural mechanisms of long-term habituation, Carew and Kandel (1973) used a shortened training schedule (Figure 13-4 and 13-9C). They separated four training sessions (of 10 stimuli each) by only 1.5 hours and still obtained long-term habituation. This brief schedule in the behavioral experiments allowed them to study the cellular changes that accompany the entire acquisition of long-term habituation in the *isolated ganglion test system* (see Figure 12-7C, p. 553).

In the isolated ganglion test system the siphon or branchial nerve was stimulated electrically. These two nerves carry the axons of the LE and RE clusters of sensory neurons that innervate the siphon and purple gland, the two regions that make up the receptive field of the defensive withdrawal reflex. Intracellular recordings were obtained from L7, one of the major gill motor

---

[2]Although the siphon component of the withdrawal reflex is largely (55 percent) centrally mediated, it is in part mediated by peripherally located motor neurons (Figure 12-13, p. 562). The sensory input to the peripheral and central siphon motor neurons is mediated by different branches of the sensory neurons in the LE cluster in the abdominal ganglion. When examined simultaneously, both the central and peripheral branches of individual sensory neurons show similar kinetics of short-term depression with repeated stimulation (Bailey et al., 1975). The existence of common sensory neurons for the peripheral and central siphon motor neurons on the one hand, and the siphon and gill motor neurons on the other, probably explains why repeated stimulation of the siphon leads to parallel habituation of both siphon and gill.

Experimental

Anterior mantle (test)   Siphon habituation   Purple gland (test)

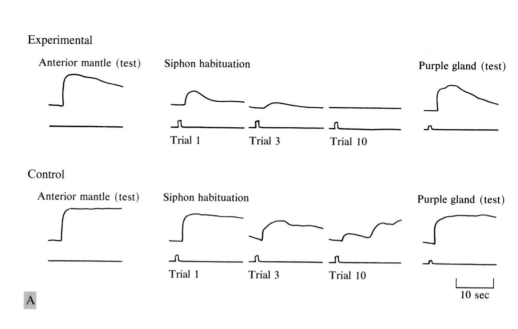

Trial 1   Trial 3   Trial 10

Control

Anterior mantle (test)   Siphon habituation   Purple gland (test)

Trial 1   Trial 3   Trial 10

10 sec

A

1. One-Day Retention

2. One-Week Retention

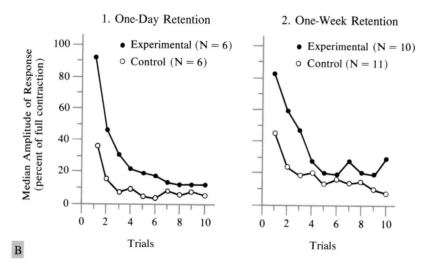

Median Amplitude of Response
(percent of full contraction)

• Experimental (N = 6)
○ Control (N = 6)

• Experimental (N = 10)
○ Control (N = 11)

Trials

Trials

B

neurons. The complex EPSP elicited in L7 by electrical stimulation of either the branchial or siphon nerve resembled the one produced by natural stimulation. The siphon and branchial nerves were alternated as experimental and control nerves in different experiments.

In the neural experiments, as in the behavioral experiments, a training schedule of four sessions of 10 stimuli each was followed. The sessions were separated by 1.5 hours. In the first session both the experimental and control nerves were stimulated. The resulting EPSP depressions were comparable, indicating that the depression produced by stimulation of one nerve does not generalize to the other nerve (a point we will consider again later). Each ganglion can therefore serve as its own control. This is essential for distinguishing a real decrease in synaptic effectiveness (due to repeated stimulation) from deterioration of the preparation (due to prolonged isolation). The experimental nerve was then stimulated during the three remaining sessions. The EPSP depression built up across sessions, so that by the fourth session the first response was significantly more depressed than a single test stimulus presented to the control nerve at the end of the session. Moreover, the median amplitude of all 10 responses in the fourth session was significantly less than the median of the 10 responses in the first session (Figure 13-4B). This significant build-up in synaptic depression parallels the buildup of behavioral habituation produced by an identical sequence of tactile stimulation of the intact animal (Figure 13-4).

FIGURE 13-3.
Long-term habituation of gill withdrawal. [From Carew, Pinsker, and Kandel, 1972. Copyright by the American Association for the Advancement of Science.]

A. Sample records from one experiment. One week prior to the experiment the experimental animals received 10 stimuli to the siphon each day for four days and exhibited significantly greater retention of habituation of siphon withdrawal than did the control animals. One week following this test retention of habituation of the gill-withdrawal component of the defensive reflex was measured. A single test gill contraction was first produced by a strong tactile stimulus to the anterior neck region. After a brief period of rest, habituation of gill withdrawal was produced by stimulating the siphon with jets of water (10 trials; intertrial interval = 30 sec). This was followed by another single test stimulus to the purple gland. Stimulation of the anterior mantle and purple gland produced comparable responses in both experimental and control animals. However, stimulation of the siphon produced significantly greater habituation of the gill-withdrawal reflex in experimental animals than it did in the controls.

B. Comparison of the gill-withdrawal reflex in experimental and control animals one day and one week after long-term habituation training produced by siphon stimulation. Median amplitude of the gill responses is expressed as a percentage of a full contraction produced by a strong stimulus to the neck. *B-1.* Twenty-four hours after siphon habituation experimental animals exhibited significantly greater habituation ($p < 0.001$) of the gill-withdrawal reflex than did the controls (when the sums of trials 1–10 were compared). *B-2.* One week after siphon habituation experimental animals still exhibited significantly greater habituation ($p < 0.001$) than the controls. These data indicate that stimulation of the siphon produces long-term habituation of both gill and siphon withdrawal.

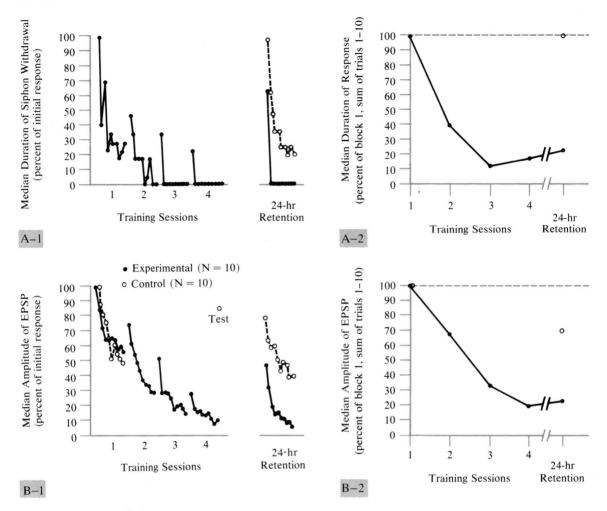

FIGURE 13-4.

Comparison of reflex habituation and EPSP depression. [From Carew and Kandel, 1973. Copyright by the American Association for the Advancement of Science.]

A. Acquisition and retention of long-term habituation of siphon withdrawal following four training sessions of 10 trials each (intertrial interval = 30 sec; intersession interval = 1.5 hr). *A-1.* Buildup during four training sessions and retention one day later. (Data are from Figure 13–9C; see that figure for the complete run, including one-week retention.) *A-2.* Time course of acquisition and retention. For each training session and the 24-hour retention test the responses have been summed and expressed as a single score, the median of 10 trials.

B. Acquisition and retention of EPSP depression. *B-1.* The EPSP amplitudes from both experimental and control nerves are expressed as a percentage of the initial amplitude. The siphon nerve served as the experimental nerve in six experiments and the branchial nerve was experimental in four experiments. In the first session 10 stimuli were applied to the experimental nerve and then to the control nerve; the EPSP depression produced in cell L7 by both nerves was comparable, indicating a lack of EPSP generalization. Subsequent sessions of stimulation of the experimental nerve produced progressive buildup of EPSP depression. Following the fourth session with the experimental nerve a singe test stimulus to the control nerve produced an EPSP that had recovered to 85 percent of control, indicating that deterioration cannot account for experimental EPSP depression. The cell was reimpaled 24 hours after training and repolarized to the membrane potential maintained for training. In the retention test, stimulation produced significantly greater ($p < 0.001$) EPSP depression in the experimental nerve than in the control nerve. *B-2.* Time course of acquisition and retention. For each training session and the 24-hour retention test the responses have been summed and expressed as a single score, the median of 10 trials.

At the end of the four training sessions the microelectrode was removed from the motor neuron and the ganglion was maintained in an enriched culture medium (Strumwasser and Bahr, 1966). Twenty-four hours after the last training session the motor neuron was reimpaled to determine retention of EPSP depression. The membrane potential was restored to approximately the same voltage as during training. A series of 10 stimuli was applied to both the experimental and control nerves in turn (the parameters of stimulation were identical to those used during training). The EPSP produced by the first stimulus to the experimental nerve was significantly smaller than the EPSP produced by the first stimulus to the control nerve (Figure 13-4B1). Moreover, the *sum* of all 10 responses to stimulation of the experimental nerve was significantly smaller than the sum of all 10 responses to stimulation of the control nerve (Figure 13-4B2). Examples of recordings from a single animal in the above experiment are shown in Figure 13-5.

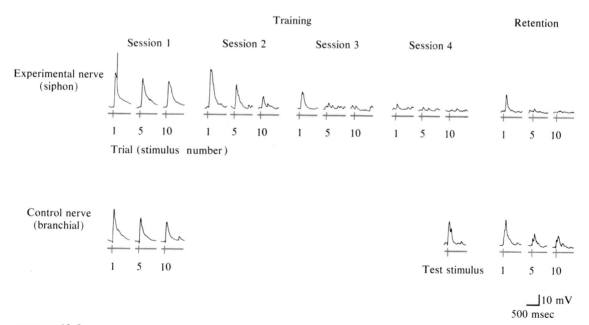

FIGURE 13-5.
Acquisition and retention after 24 hours of the complex EPSP depression. Sample records from responses in cell L7 following stimulation of the siphon (experimental) and branchial (control) nerves. Each training session consisted of 10 trials. The control nerve was stimulated for one session at the beginning of the experiment. Single test stimuli were also applied to the control nerve at the end of session 4. Twenty-four hours later retention of EPSP depression in the experimental and control nerves was tested. The experimental EPSPs showed significantly greater ($p < 0.001$) and more rapid depression than the control EPSPs. [From Carew and Kandel, 1973. Copyright by the American Association for the Advancement of Science.]

The acquisition of long-term habituation of the gill-withdrawal reflex therefore involves a gradual but profound decline in the synaptic potential produced in the gill motor neuron L7 by stimulation of the afferent nerve. These changes are retained for at least 24 hours. That such limited training (40 stimuli) should lead to prolonged plastic changes in neuronal function is interesting in view of the general difficulty (in both vertebrates and invertebrates) of producing 24-hour long changes with intensive, continuous, stimulation, even after thousands of stimuli (Fig. 11-4B, p. 483). The ease with which prolonged neuronal changes were produced in these experiments can probably be attributed to the use of a stimulus pattern of known behavioral effectiveness and the selection of a synaptic pathway known to be involved in a modifiable reflex.

Since the test system lacks receptor and effector components, these synaptic changes obviously occurred centrally and did not involve sensory adaptation or motor fatigue. The alterations also did not result from a change in the input resistance of the cell body, although one cannot exclude a resistance change at a site remote from the cell body. Since the complex EPSP is largely made up of the monosynaptic EPSP produced by the sensory neurons (see p. 372), some of the synaptic changes underlying long-term habituation must occur at the same site and at least superficially resemble those found with short-term habituation.

Transmitter Release and the Acquisition of Long-Term Habituation

The finding that short- and long-term habituation have a common locus permits one to examine further the mechanisms for acquisition of the long-term process. Multistage models of memory postulate two memory stores in series—a short-term store and a long-term store. Are the stores for short- and long-term habituation arranged in series or in parallel? Is the transmitter release that accompanies short-term habituation a necessary prerequisite for the long-term synaptic depression that accompanies long-term habituation? Or is the long-term depression due to an independent and parallel process that depends only on impulse activity and the concommitant ion movements (Na^+ influx and K^+ efflux) in the presynaptic terminals? These questions have been examined in the isolated ganglion test system by Castellucci and Kandel (unpubl.).

The shortened training schedule was used (four sessions of 10 stimuli each, with a 1.5 hour interval between sessions). During the first session both the experimental and control nerves were stimulated. The ganglion was then bathed in a high-Mg^{++} solution to block transmitter release (see p. 188). During the second and third training sessions only the experimental nerve was stimulated; the control nerve was not. The high-Mg^{++} solution was then removed and normal seawater was reintroduced, so that transmitter release was restored. Experimental and control nerves were then both examined during the fourth session. If the buildup of synaptic depression did not require transmitter release, then the responses of the experimental nerve in the fourth training session should resemble those normally found with the acquisition of long-term habituation (Figure 13-6A). This proved not to be the case. The synaptic responses to stimulation of the experimental nerve did not differ from control and failed to show the profound buildup of depression characteristic of the acquisition of long-term habituation (Figure 13-6B). These findings (although indirect) suggest that Na^+ influx or K^+ efflux are insufficient to produce synaptic depression. One or more steps in the activation of transmitter release seem to be essential for long-term habituation. The mechanisms for short- and long-term habituation therefore appear to be in series.

Comparison of Short-Term and Long-Term Habituation

A large component of the complex EPSP produced in the motor neurons by stimulation of the siphon nerve is produced by the monosynaptic connections of the sensory neurons (see p. 372). Short- and long-term habituation therefore seem to share a common neural locus: the synapses that the sensory neurons make on the motor neurons and interneurons. It is also likely that short- and long-term habituation share aspects of a common cellular mechanism, homosynaptic depression. By these criteria alone, long-term habituation would seem to be an extension of the short-term process.

However, by using better criteria, finer distinctions may emerge. For example, the short-term synaptic depression that underlies short-term habituation is presynaptic, involving a decrease in the number of quanta released per impulse (see p. 574). The factors producing the long-term synaptic depression underlying long-term habituation are not yet known, and may prove to be postsynaptic. If long-term depression proves to be presynaptic, however, this finding would support, on a more fundamental level, the notion that short- and long-term memory are stages of a single memory process. Nevertheless, the time courses of the two types of memory are very different, and this fact alone implies two different processes in the presynaptic terminals (Figure 13-7). Short-term memory may be the result of depletion of the *releasable* pool of transmitter (due to inadequate mobilization); long-term memory may be the result of depletion of the *storage* pool of transmitter (due to either

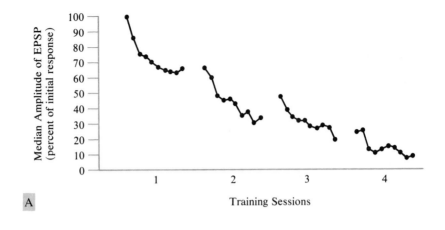

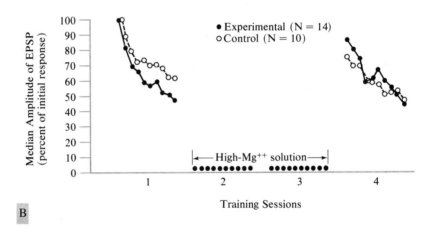

FIGURE 13-6.

The role of transmitter release in the buildup of long-term synaptic depression. [Castellucci and Kandel, unpublished observations.]

A. In normal seawater four training sessions, each consisting of 10 stimuli applied to the siphon nerve, produce a progressive buildup of the depression of the complex EPSP in motor neuron L7 (interstimulus interval = 30 sec; intersession interval = 90 min; N = 14).

B. Experimental and control nerves are both stimulated 10 times during the first training session. The bathing solution is then changed to a high-Mg^{++} solution to block transmitter release and two more sets of stimuli are applied to the experimental nerve only. Normal seawater is then reintroduced and both the experimental and control nerves are stimulated during the fourth training session. The expected buildup of EPSP depression does not occur. These findings suggest that transmitter release accompanying the short-term depression is a necessary condition for the acquisition of long-term depression.

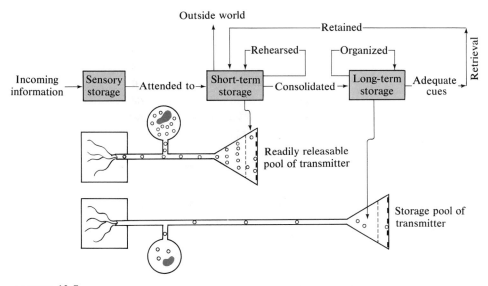

FIGURE 13-7.
A cellular model of a two-stage memory trace assuming a single cellular locus and a common plastic mechanism, synaptic depression (based on the model in Figure 13-1). Short-term storage involves mobilization of transmitter substance from a storage pool of vesicles to a readily releasable pool of vesicles. One training session leads to partial depletion of the readily releasable pool (short-term habituation). Repeated training sessions produce depletion of the storage pool, which leads to reduced synthesis and transport of transmitter substance.

inhibition of synthesis or transport of transmitter from the cell body). Although obviously speculative, this hypothesis is simple and not overly far-fetched. If the hypothesis is accurate, the debate over one-process versus two-process memory is an empty one. In this model short- and long-term memory can be regarded *either* as phases of a single process (homosynaptic depression at the terminal on the sensory neurons) or as two discrete processes (depletion of the releasable pool of transmitter substance due to inadequate mobilization followed by depletion of the storage pool by inhibition of synthesis). Moreover, both short- and long-term synaptic changes could be caused by a common mechanism, for example, a progressive decrease in the free Ca^{++} concentrations in the terminals. This decrease leads in the short term to inadequate mobilization and a reduction in the releasable pool of transmitter, and in the long-term to depletion of the storage pool. Depletion of the storage pool could give rise to a biochemical signal from the presynaptic terminals to the cell body, leading to inhibition of transmitter synthesis or transport.

It should be emphasized, however, that different learning processes are likely to involve different types of memory—no single rule is likely to apply for all types.

Relationship of Synaptic and Behavioral Changes

With repeated training the buildup of behavioral habituation reaches asymptote at zero (no response) by the third session, with the exception of the first response of the session (Figure 13-2*A* and 13-4*A*). One might assume therefore that because there is no further change in the behavior no further changes are occurring in the nervous system. Actually, in experiments measuring long-term synaptic depression the complex EPSP continues to decrease throughout the third and fourth sessions (compare Figures 13-4*A* and *B*). Although undetectable behaviorally, this continued synaptic depression is likely to affect the retention of habituation or the ease with which habituation might reoccur. Such a mechanism probably underlies a commonly described feature of habituation, the subzero effect, whereby continued stimulation of a completely habituated response prolongs the retention of habituation (Figure 13-2). The point at which the synaptic input first becomes capable of triggering the motor cell constitutes the behavioral threshold. As long as the EPSPs are above this threshold, depressions in EPSP amplitude produce a concomitant depression in behavioral response. However, synaptic depression below this threshold produces no further observable change, even though changes in synaptic transmission may continue to occur.[3]

Relationship of Long-Term Habituation to Complex Learning

Although long-term habituation has been examined in surprisingly few behavioral responses, the gill-withdrawal reflex is not unique. The mobbing response of chaffinches undergoes long-term habituation (Hinde, 1954b), as does the flexion withdrawal response of frogs (Farel, 1971) and the cold-pressure (Glaser, 1966) and galvanic skin responses of man (Kimmel and Goldstein, 1967). In all four of these response systems the time course for acquisition of long-term habituation is similar (Figure 13-8*A*). The rapid recovery of responsiveness after habituation has often been the basis for distinguishing habituation and learning (for example see Hilgard and Marquis, 1940; Rosvold, 1959). However, the finding that habituation can be prolonged by repeated training indicates that habituation shares at least one feature with complex learning, namely long-term retention. Does habituation share other features with complex forms of learning? The learning curve of long-term habituation does in fact resemble (at least superficially) the learning of simple motor tasks. For example, human subjects instructed to hold a telegraph key

[3]See p. 524 for discussion of a similar threshold relationship between alpha and classical conditioning.

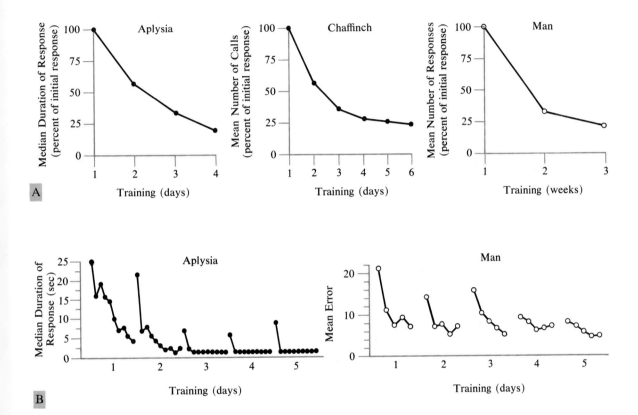

FIGURE 13-8.

Comparison of long-term habituation in *Aplysia* with vertebrate habituation and human learning.

A. Long-term habituation. *A-1.* Long-term habituation of siphon withdrawal in *Aplysia* (data from Figure 13-2*B*). *A-2.* Long-term habituation of the calling response of the chaffinch while mobbing a live owl presented three minutes each day. [From Hinde, 1954b.] *A-3.* Long-term habituation of the galvanic skin response of man to a tone. Each data point is the mean result of a weekly session. In each session the subject received at most 50 repeated tone presentations. The training was stopped if at any point the subject showed no response to two consecutive stimuli. [Replotted from Kimmel and Goldstein, 1967.]

B. Comparison of the acquisition of long-term habituation in *Aplysia* and learning in man. *B-1.* Buildup of habituation in *Aplysia* during four days of training (data replotted from experiment 1 in Figure 13-2*A*). *B-2.* Buildup of learning (reduction of errors) in human subjects during four days of training. The subjects were asked to depress a telegraph key for 0.7 sec. [From MacPherson, Dees, and Grindley, 1949.]

for 0.7 seconds improve their timing during the first practice session and also progressively over successive days of practice (MacPherson, Dees, and Grindley, 1949). As with long-term habituation, there is a slight forgetting between the last performance of one day and the first performance of the next, despite the progressive improvement over days (Figure 13-8*B*).

In the learning of motor tasks (Kling, 1971) and in verbal learning (Bjork, 1970) short-term retention is as good or even better with massed training than with distributed training. By contrast, long-term retention is much better with distributed training than with massed training.

To compare the effects of massed and distributed training on short- and long-term retention of habituation, Carew and co-workers (1972) randomly divided animals into three groups: distributed training, massed training, and a control group that received no training. Animals receiving distributed training were given 10 stimuli per day over four days. Animals receiving massed training were given no stimulation on days 1 to 3; on day 4 they were given 40 consecutive stimuli. Despite the fact that both groups had received the same number of training trials, the animals receiving distributed training exhibited significantly greater retention of habituation one day and one week after training than did the animals receiving massed training. As is the case for massed learning in man (Bjork, 1970), animals in the massed training group showed significant retention of habituation one day after training (compared to their responses on trials 1 to 10 of the previous day). But on the one-week retention test the massed-training group showed no significant retention while the distributed-training group did exhibit significant retention. Thus, although both massed and distributed training groups showed identical short-term habituation at the end of the fortieth trial, massed training is not retained as long as distributed training (Figure 13-9).

What features of distributed training lead to the conversion from short- to long-term retention? Must the training sessions be separated by a complete day, or can shorter intervals also be effective? To examine this question, the interval between training sessions was reduced to 1.5 hours (Carew and Kandel, 1973). As we have seen, this shorter interval also leads to a buildup of habituation across sessions and significantly greater habituation one day and one week after training (Figures 13-9 and 13-10).

The failure to obtain good long-term retention with massed training indicates that the conversion from short- to long-term habituation is critically dependent upon the intervals between training sessions. But the experiments do not provide direct insights into the mechanism of the conversion. An important feature of distributed learning might be the partial forgetting (in the case of habituation, the recovery of reflex responsiveness) that occurs with interruptions in training. Rehabituation of a partially recovered response may be the

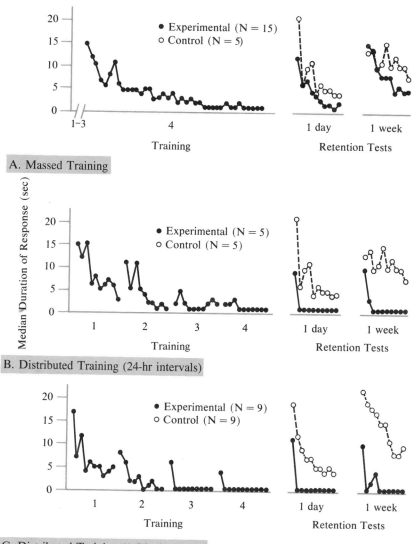

A. Massed Training

B. Distributed Training (24-hr intervals)

C. Distributed Training (1.5 hr intervals)

FIGURE 13-9.
Comparison of massed and distributed training of habituation of siphon withdrawal in *Aplysia*. [Parts *A* and *B* are from Carew *et al.,* 1972; part *C* is from Carew and Kandel, 1973. Copyright by the American Association for the Advancement of Science.]

A. Massed training. Experimental animals received no stimulation for three days. On the fourth day 40 consecutive stimuli (trials) were applied to the siphon (intertrial interval = 30 sec). Control animals received no training. Retention was tested one day and one week after training, and was not significantly different from control.

B. Distributed training (24 hours between training sessions). Experimental animals were given four sessions of training (10 trials per session, each session was separated by one day); control animals received no training. Retention of habituation was tested one day and one week after training. (Studies illustrated in parts *A* and *B* were carried out concurrently; control animals were the same in both experiments.)

C. Distributed training (1.5 hours between training periods). Experimental animals were given training; controls received no training. Retention was tested one day and one week after training.

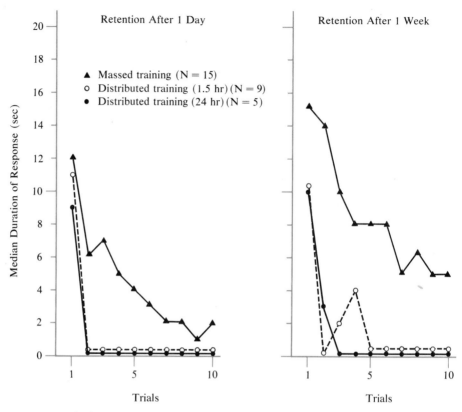

FIGURE 13-10.
Detailed comparison of retention of habituation one day and one week following massed and distributed training. Graphs are based on data from experiments described in Figure 13-9. Distributed training produces greater retention of habituation than does massed training. [Based on Carew and Kandel, 1973.]

crucial feature for the development of long-term habituation. As Peterson (1966) has pointed out in discussing the effectiveness of distributed training for verbal learning, "It remains a paradox that in order to remember something better, you should allow some forgetting to occur." The solution to this apparent paradox will undoubtedly bring about important insights into long-term memory. Following the model illustrated in Figure 13-7, under conditions of distributed training the slow mobilization process characteristic of synapses in a habituation pathway would continuously deplete the releasable pool and thus eventually deplete the storage pool. Depletion of the storage pool might give rise to a signal transmitted back to the cell body, inhibiting the synthesis or transport of additional transmitter substances.

HABITUATION, DISHABITUATION, AND SENSITIZATION

Pavlov (1927) found that after extinction of a classically conditioned response the presentation of a strong or novel stimulus could enhance an animal's responsiveness, so that a conditioned stimulus that had become ineffective could again trigger the conditioned responses. This phenomenon was later studied in detail by Grether (1938) and came to be known as *sensitization* (see p. 12). Pavlov, and later Humphrey (1933) and others, also found that a habituated response can be rapidly reestablished by a stimulus stronger than the one to which the animal had become habituated. This process was called *dishabituation.* Sensitization and dishabituation were initially regarded as different processes: sensitization was thought of as an excitatory process, while dishabituation was thought to be a removal of inhibition.[4] Pavlov and subsequently Humphrey (1930a, b) thought that habituation was the building up of an inhibitory process, so they attributed dishabituation—the restoration of reflex responsiveness following habituation—to a wiping out of habituation by *removal of inhibition.*

The first clear suggestion that dishabituation is not related to habituation came from Sharpless and Jasper (1956). They proposed that dishabituation is not merely a wiping out of habituation but a separate process, a type of sensitization in which a strong stimulus enhances the responsiveness of many reflexes. A similar conclusion was reached by Hagbarth and Kugelberg (1958) from studies of human abdominal reflexes. Direct evidence in support of this idea was provided by Spencer, Thompson, and Neilson (1966a) from studies of dishabituation of the flexion reflex in the cat. They found that a strong stimulus enhanced nonhabituated responses as well as habituated ones, and further, that habituated responses could be facilitated beyond their prehabituation levels. If dishabituation were simply the removal of habituation, a strong stimulus would at most only restore a response to its prehabituation level.

The relationship between habituation, dishabituation, and sensitization can be clearly demonstrated in the gill-withdrawal reflex of *Aplysia.* This reflex can be elicited by one of two anatomically distinct and functionally independent afferent pathways: (1) the siphon skin, innervated by the LE cluster of mechanoreceptor neurons, the axons of which run in the siphon nerve, and (2) the anterior third of the mantle shelf (which includes the purple gland), innervated by the RE cluster of sensory neurons, the axons of which run in the branchial nerve (Figure 13-11). In the intact animal habituation of the gill-withdrawal reflex by stimulation of the siphon does not affect the animal's responsiveness to stimulation of the purple gland and vice versa—there is no generalization of

---

[4]Recent examples of this theoretical position are found in Konorski, 1948, 1967.

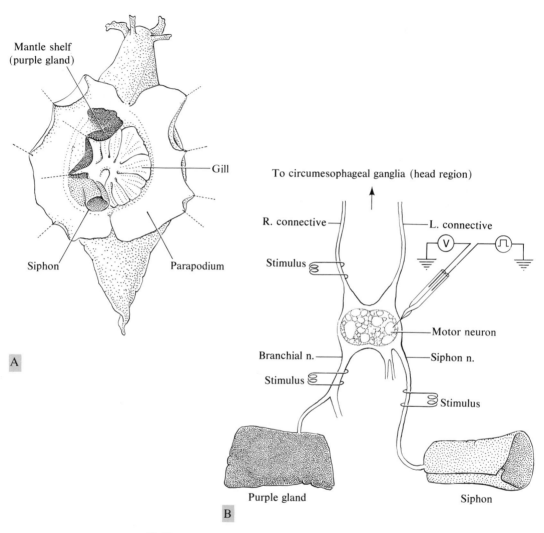

FIGURE 13-11.

Two independent pathways for gill withdrawal. [From Carew, Castellucci, and Kandel, 1971.]

A. The intact animal, illustrating the two reflex pathways, one from the anterior part of the mantle shelf and its edge (dark stipple), the other from the siphon and the posterior part of the mantle shelf (light stipple).

B. Diagram of the isolated ganglion test system of the gill-withdrawal reflex, illustrating innervation of the siphon skin by the siphon nerve and innervation of the purple gland by the branchial nerve. The siphon and branchial nerves are stimulated electrically at rates that produce habituation in the animal. Stimulation of either connective produces heterosynaptic facilitation.

habituation from one pathway to the other (Figure 13-12*A*). Moreover, in the isolated ganglion test system repeated stimulation of the siphon nerve produces a depression in the EPSP recorded in the gill motor neuron L7 but does not alter the amplitude of the EPSP produced by the branchial nerve (Figure 13-12*B*). Thus, synaptic depression also does not generalize from the branchial to the siphon nerve.

Carew and co-workers (1971) used the two reflex pathways to study the effects of a strong stimulus on both habituated and nonhabituated responses. If dishabituation were simply the removal of habituation, the strong stimulus would facilitate only the habituated pathway and restore it to its prehabituation values. If dishabituation were a form of sensitization, then a strong stimulus would facilitate other (nonhabituated) responses as well as the habituated one and the habituated response might even be facilitated beyond its prehabituation value.

In restrained animals a single test stimulus was presented first to the purple gland and then to the siphon, producing in each case a gill-withdrawal reflex. One of the two reflex pathways was then habituated by repeated stimulation and the other was used as a control. Following habituation of one pathway, a strong tactile stimulus was presented to the head. This stimulus produced sensitization of the nonhabituated (control) pathway, as well as dishabituation of the habituated pathway. The amplitude of the response to the dishabituated pathway was sometimes greater than the prehabituation value (Figure 13-13).

The neural correlates of dishabituation and sensitization were examined in the isolated ganglion test system with similar results (Figures 13-14, 13-15). A strong stimulus to the connective significantly facilitated both the depressed EPSPs produced by a previously stimulated pathway and the nondepressed EPSPs produced by a nonstimulated pathway. In most cases (67 percent) the depressed EPSPs were facilitated beyond the prestimulation level; the non-depressed EPSPs showed similar increases. But the depressed EPSPs showed proportionately more facilitation when the last observed depressed response was used for comparison. These data suggest that the facilitating stimulus facilitates both depressed and nondepressed responses toward a common maximum or ceiling (perhaps determined by the reversal potential of the EPSP; see Figure 6-10, p. 167) and that depressed EPSPs are proportionately more facilitated because they are smaller and farther away from this ceiling.

Thus, in the intact animal a strong stimulus capable of dishabituating one pathway can also sensitize a nonhabituated pathway. Similarly, in the isolated ganglion a strong stimulus that facilitates a depressed EPSP usually also facilitates a nondepressed EPSP. Taken together, the data prove that dishabituation

1. Purple Gland Habituation

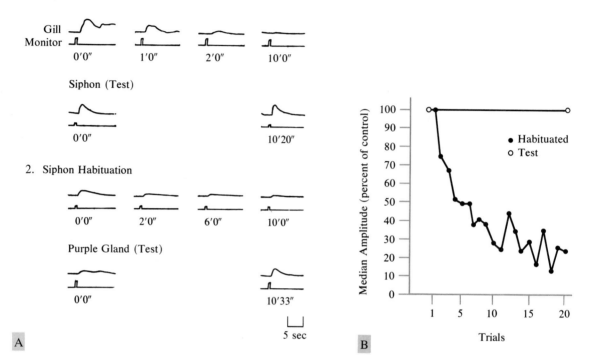

FIGURE 13-12.

Lack of generalization of habituation from the purple gland to the siphon. [From Carew *et al.*, 1971.]

A. Selected responses from a behavioral experiment using a preparation similar to that illustrated in Figure 13–11*A*. Stimuli were applied to both the purple gland and the siphon to produce a gill-withdrawal reflex (recorded with a photocell). *A-1.* A single test stimulus was first presented to the siphon (0′ 0″). Stimuli were then applied to the purple gland every 60 sec; this habituated the reflex response to stimulation of the purple gland. A second test stimulus to the siphon (10′ 20″) elicited an undiminished response, similar in amplitude to the response to the first test stimulus (0′ 0″). Habituation of the gill-withdrawal response to stimulation of the purple gland therefore does not generalize to stimulation of the siphon. *A-2.* After a 40-min rest the experimental procedure was reversed. A test stimulus to the purple gland (0′ 0″) elicited gill withdrawal that was somewhat less intense than normal, indicating that recovery from the earlier habituation was incomplete. The gill-withdrawal reflex was then habituated by repetitive stimulation (once every 60 sec) to the siphon. A second test stimulus to the purple gland (10′ 33″) elicited more normal gill withdrawal, indicating again that habituation of gill withdrawal by means of one pathway (the siphon) does not generalize to another (the purple gland).

B. Summary of eight runs in four experiments designed to examine generalization of habituation between the habituated and test pathways. A sign test revealed no significant difference between the second and first test stimuli to the test pathway, even though significant depression ($p < 0.001$) was obtained following stimulation of the habituated pathway. (Intertrial interval 2 min).

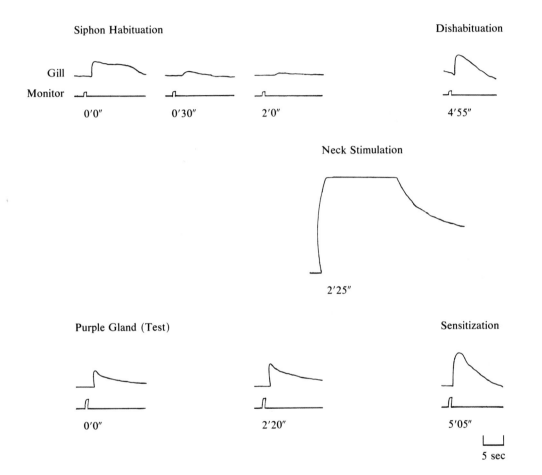

FIGURE 13-13.
Facilitation of habituated and nonhabituated responses following presentation of a single strong stimulus to the neck. After a single test stimulus to the purple gland (0′ 0″), repeated stimuli were delivered to the siphon every 30 sec, producing habituation of gill withdrawal. A second test to the purple gland (2′ 20″) indicated that no generalization of habituation had occurred. A strong stimulus was then applied to the neck region, producing prolonged gill contraction. Subsequent stimulation of the purple gland (2′ 40″, after neck stimulation) produced a facilitated nonhabituated response (sensitization); stimulation of the siphon (2′ 30″ after neck stimulation) produced a facilitated habituated response (dishabituation). [From Carew, Castellucci, and Kandel, 1971.]

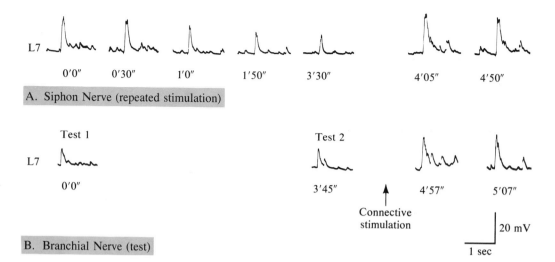

A. Siphon Nerve (repeated stimulation)

B. Branchial Nerve (test)

FIGURE 13-14.

Lack of generalization of EPSP depression and facilitation. Depressed and nondepressed complex EPSPs following a single train of stimuli to a connective. A single test stimulus delivered to the branchial nerve produced a complex EPSP in L7. Stimulation (intertrial interval = 10 sec) of the siphon nerve produced EPSP depression. A second test to the branchial nerve (3′ 45″) revealed that EPSP depression had not generalized from the siphon nerve to the branchial nerve. A train of stimuli (6/sec for 6 sec) was then delivered to the left connective. Subsequent stimulation first of the siphon nerve and then the branchial nerve revealed facilitation of both depressed (part A) and nondepressed (part B) EPSPs, respectively. See Figure 13–15 for pooled data from several experiments. [From Carew, Castellucci, and Kandel, 1971.]

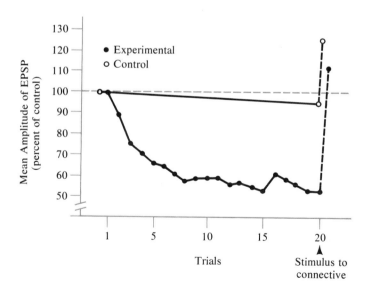

FIGURE 13-15.

Pooled data from nine experiments described in Figure 13-14. A single test EPSP was produced by stimulation of the control nerve (N = 9). The experimental nerve (N = 9) was then repeatedly stimulated (ISI-10 sec), producing significant ($p < 0.005$) depression of the EPSP in the monitor neuron. A second stimulus was then applied to the control nerve, producing a test EPSP that was not significantly different from the first test EPSP. This demonstrated that depression had not generalized to the test nerve. After a train of stimuli to the connective (6/sec for 6 sec), both depressed and nondepressed EPSPs were significantly facilitated ($p < 0.005$ and $p < 0.025$, respectively).

of the gill-withdrawal reflex is not simply the removal of habituation but is an independent process—a special case of sensitization. The findings also support the idea that habituation and dishabituation involve separate mechanisms regulating transmitter release at common synapses (see Figure 12-28, p. 583). Habituation leads to a homosynaptic depression that is limited to the stimulated pathway; sensitization leads to a widespread presynaptic facilitation involving both stimulated and unstimulated pathways.

Differences in the extensiveness of the pathways mediating habituation and sensitization explain why sensitization generalizes but habituation does not. The major plastic changes in the pathway of the gill-withdrawal response occur at the synapses between the primary sensory neurons and the motor neurons. Because the receptive field of most of these sensory neurons is restricted to either the siphon or the mantle shelf, a stimulus to the siphon would not be expected to—and in fact does not—generalize to the mantle shelf. However, a strong stimulus from a single site, such as the head, produces much more extensive presynaptic facilitation and affects a number of closely related reflexes. A model of the spatial relationships of the pathways mediating habituation and sensitization and their synaptic properties is shown in Figure 13-16. Although dishabituation is a form of sensitization, the difference between the two is not trivial. A strong stimulus tends to bring both dishabituation and sensitization toward a common maximum. As a result, the dishabituating effects of a strong stimulus are always likely to be more powerful than its sensitizing effects.

Habituation and dishabituation are separate processes in vertebrates as well, suggesting that this may be quite general (Hagbarth and Finer, 1963; Spencer, Thompson, and Neilson, 1966a). For example, in studies of human abdominal reflexes, Hagbarth and Kugelberg (1958) found that habituation is restricted to the stimulated site; little generalization occurs. Habituation of the response to repeated stimulation of one abdominal site has no effect on the response elicited by stimulation of a neighboring region 3 to 4 cm away. Stimuli separated by this distance presumably activate separate pathways because the separation represents the limit of the subject's capacity for two-point discrimination (Figure 13-17). A strong stimulus will facilitate a habituated abdominal reflex beyond the prehabituation level (Figure 13-18) and will also facilitate nonhabituated responses. Although detailed psychophysical studies were not done, Hagbarth and Kugelberg report that subjects sometimes had the illusion of declining stimulus intensity during habituation and of increasing stimulus intensity during sensitization.

632

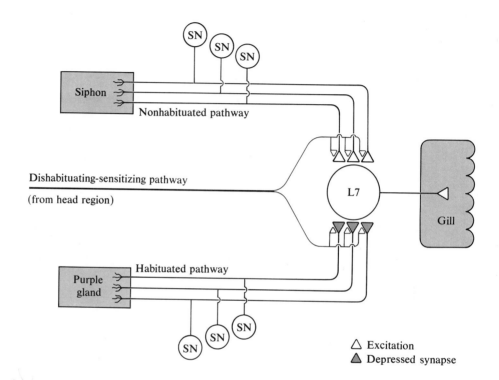

FIGURE 13-16.
Schematic model of habituation and sensitization of the gill-withdrawal reflex. Following stimulation of either the purple gland or the siphon, only the stimulated pathway is habituated; there is homosynaptic (self-generated) depression of the synapses between the sensory neurons (SN) and motor neuron L7. By contrast, a strong stimulus from the head sensitizes both the habituated and nonhabituated pathways due to presynaptic facilitation of all the synapses made by primary sensory neurons on the motor neuron L7. The difference in the spatial distribution of the pathways mediating habituation and sensitization explains the lack of generalization of habituation as well as the evident generalization of sensitization. The model accounts only for the monosynaptic components of the gill-withdrawal reflex. The effects of sensitization on the polysynaptic components of the reflex have been analyzed. Grey terminals indicate depressed synapses. [From Carew, Castellucci, and Kandel, 1971.]

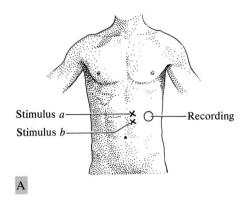

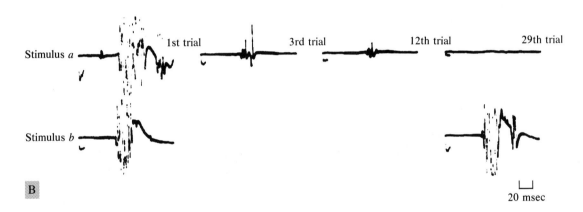

FIGURE 13-17.

Lack of generalization of habituation of human abdominal reflexes. [From Hagbarth and Kugelberg, 1958.]

A. Experimental arrangement. Electrical stimuli are applied to two sites of the abdomen (*a* and *b*), 4 to 5 cm apart. Recordings were obtained by inserting electrodes in the left external oblique muscles at the site indicated. Stimuli were applied every 5 to 10 sec.

B. A single stimulus was first applied to site *b*. Site *a* was then stimulated repeatedly for 29 trials (trials 1, 3, 12, and 29 are illustrated), producing habituation of the abdominal reflex. A single stimulus was then applied to site *b*. Full reflex responsiveness was obtained, indicating specificity of habituation to the stimulated pathway. Repeated stimulation at site *b* also produced habituation that did not generalize to *a*.

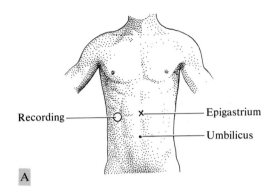

A

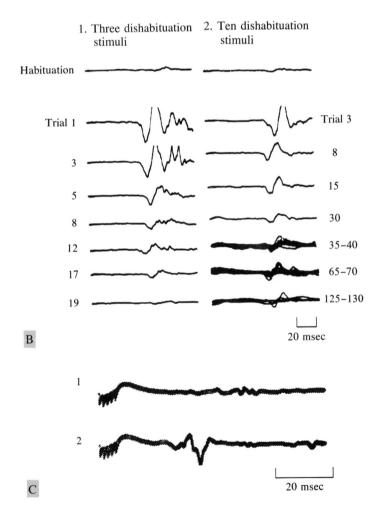

1. Three dishabituation stimuli

2. Ten dishabituation stimuli

Habituation

Trial 1

3

5

8

12

17

19

Trial 3

8

15

30

35–40

65–70

125–130

20 msec

B

1

2

20 msec

C

FIGURE 13-18.

Sensitization of human abdominal reflex. [From Hagbarth and Kugelberg, 1958.]

A. Experimental arrangement. Recordings obtained from right external oblique muscles. Both the test stimulus (a weak mechanical stimulus) and the sensitizing stimulus (a strong blow) were applied to the epigastrium.

B. Sensitization. *B-1.* Repeated weak mechanical stimuli to the epigastrium (ISI = 5–10 sec) produced habituation of reflex (top trace). Three sensitizing stimuli (blows to epigastrium) were then applied. These produced sensitization of the previously depressed response that lasted for the next 17 trials and the response returned to control by the 19th trial. *B-2.* Following the 19th stimulus in part *B-1*, 10 sensitizing stimuli were applied and produced more prolonged sensitization than in *B-1* (lasting for the next 125 trials).

C. Sensitization by verbal suggestion. *C-1.* A weak electrical stimulus applied to the epigastrium produced no response. *C-2.* The stimulus became effective after the subject was told that the next stimulus would be painful.

SENSITIZATION AND AROUSAL

Groves and Thompson (1970) have proposed that just as dishabituation is a special case of sensitization, so might sensitization be a component of an even more general state, behavioral arousal. Behavioral arousal can be operationally analyzed into at least two distinct components: (1) orientation toward the stimulus, and (2) a generalized alteration of behavioral responsiveness, consisting of heightened responsiveness in certain behavioral systems and concomitant depression in others. Sensitization is related to this second component; it is thought to mediate the heightened responsiveness that accompanies behavioral arousal. Since cyclic AMP seems to have a role in sensitization in *Aplysia* (p. 585) it may, by extension, mediate this component of arousal in vertebrates.

To the extent that arousal may include heightened responsiveness, the study of sensitization provides an opportunity to substitute a more restricted and testable concept (sensitization) for the somewhat vague, albeit useful, behavioral concept (arousal). Arousal can be produced by a number of different factors: attention, fear, or a high level of sensory distraction (background stimulation), as well as by motivational factors, such as hunger and thirst. Some of these have been shown to impair reflex habituation and enhance responsiveness in vertebrates. For example, in man attention and fear can sensitize (dishabituate) habituated responses (Figure 13-18C). It would therefore be of interest to see whether these and other variables known to alter behavioral responsiveness in vertebrates, such as motivation, can affect habituation in *Aplysia*. I will later consider how arousal due to motivational factors relates to arousal due to noxious stimuli (see Chapter 15, p. 659).

SHORT-TERM AND LONG-TERM SENSITIZATION

Because habituation can be prolonged by repeated training sessions, Pinsker and co-workers (1973) studied sensitization to see whether it too might be prolonged. Experimental animals received sensitization training consisting of an electric shock to the head four times per day for four days; control animals received no shocks. One day and one week after training, the sensitized animals showed significantly longer siphon withdrawal than did the controls. The experimental group also showed significantly longer siphon withdrawal one day and one week after training compared to their own initial values, whereas the control group's scores were not changed from their initial values. Thus, one day after training the sum of all 10 trials of the experimental animals was increased to 720 percent of control; one week later it was still 350 percent above control. In the retention test three weeks after training the experimental group was no longer significantly different from the controls, indicating that the sensitized animals had largely recovered. However, even at three weeks the experimental animals showed some effects of sensitization compared with their own initial scores (Figure 13-19).

The amplitude and duration of reflex responsiveness can be enhanced by spacing the training sessions over a period of 10 days (Figure 15-1, p. 655; Pinsker *et al.*, 1973). Thus, as with habituation, sensitization can be prolonged for days and even weeks, and distributed training appears to be particularly effective in prolonging sensitization. Moreover, long-term enhancement of reflex responsiveness occurs whether or not there has been prior habituation. These results indicate that, like short-term dishabituation, long-term dishabituation is not merely the removal of habituation but is an independent facilitory process (sensitization). There are no studies as yet, however, of the cellular concomitants of long-term sensitization. It will be interesting to see whether an increase in cyclic AMP in the presynaptic terminals of the sensory neurons, thought to be important for short-term sensitization, is also involved in long-term sensitization.

SENSITIZATION AND ASSOCIATIVE CLASSICAL CONDITIONING

Because sensitization produces enhancement in one pathway as a result of activity in another pathway, it provides a natural bridge between the habituation paradigm and the more complex types of associative learning, such as classical conditioning. Sensitization is different from conditioning, however, in that it does not depend on pairing of activity in two pathways. Classical conditioning is dependent on the temporal pairing of two stimuli in a precise sequence; sensitization is not dependent on paired presentation. The similarity

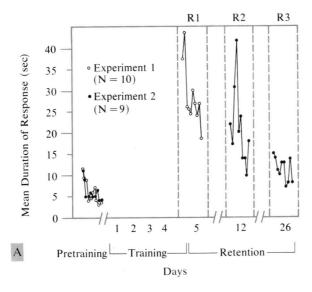

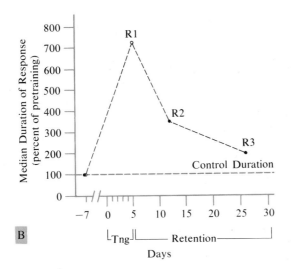

FIGURE 13-19.
Long-term sensitization of siphon withdrawal. [From Pinsker, Hening, Carew, and Kandel, 1973. Copyright by the American Association for the Advancement of Science.]

A. The median duration of siphon withdrawal is shown for each trial of a 10-trial block (minimum intertrial interval was 30 sec). The results of two independent experiments are shown. Experimental animals were given four electrical shocks per day for four days; the controls received no shocks. In experiment 1 retention of dishabituation was tested one day (R1) after the last shock. In experiment 2 retention was tested one week (R2) and three weeks (R3) after the last shock. Significant sensitization was evident in the one-day and one-week retention tests, while there was almost complete recovery in the three-week retention test.

B. Time course of sensitization (data from part *A*). Responses for each daily session have been summed and expressed as a single (median) score. Twenty-four hours and one week after the end of training the experimental animals showed significantly longer siphon withdrawal compared to their own pretraining score and the score of the controls, both of which are expressed in the graph as 100 percent (Mann-Whitney U tests used for intergroup comparisons). In the three-week retention test there was no longer a significant difference between the experimental and control groups, but the responses of the experimental animals were still significantly prolonged compared to their own pretraining levels (Wilcoxon matched pairs signed-ranks test; $p < 0.005$); the control animals were unchanged.

of the two processes nonetheless suggests that they may be related (Kimmel, 1964; Kandel and Tauc, 1965b; Kandel, 1967; Wells, 1968).

An attractive possibility is that the mechanisms of sensitization form part of the mechanism of associative classical conditioning. Prior to pairing with an unconditioned stimulus, a conditioned stimulus may be capable of eliciting, if only in subthreshold form, all the responses to which it can subsequently become conditioned. Whether or not this neural potentiality becomes realized in behavior (or is even detected) depends on a number of variables, including the nature and strength of the CS, the context in which it is examined, and the sensitivity of the measuring instrument. The effect of the US may be twofold. The first, a nonassociative effect, is to enhance a number of preexisting (alpha) neural responses to the CS. Depending on the nature of the US and the CS, some neural responses become enhanced more than others. This enhancement may bring to threshold an effector response that is qualitatively different from the response to the CS. The second effect is associative; it depends upon the reinforcing stimulus and, even more, on the pattern of reinforcement, on how the UR is paired with the CS.

According to this simplified model, all conditioning is at the outset a form of sensitization (nonassociative alpha and nonassociative classical conditioning). More complex forms of learning, such as associative alpha and associative classical conditioning, are subsequent modifications that depend more on the pattern of reinforcement than on the initial, nonspecific facilitation of the CS by the US.[5]

Figure 13-20 compares this differential facilitation model of classical conditioning with the conventional view based on stimulus substitution. In order to test the differential facilitation model it will be useful to develop and analyze instances of associative alpha and associative classical conditioning and to determine how these processes relate to nonassociative conditioning. An attractive possibility is that associative specificity might be mediated by a form of presynaptic facilitation that is critically dependent upon previous activity in the test pathway (Kandel and Tauc, 1965b). In certain synapses presynaptic facilitation might be more effective if the invasion of the presynaptic terminals of the test pathway is followed within 200 to 500 msec by a presynaptic input from the US pathway. Other models for specificity have also been proposed (see for example Kupfermann and Pinsker, 1969).

---

[5]For earlier and somewhat parallel views see Grant, 1943; Sokolov, 1963a, b; and particularly Sperry, 1965a.

Stimulus Substitution Model

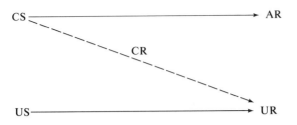

Differential Facilitation Model

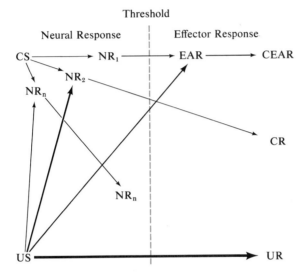

FIGURE 13-20.

Models of classical conditioning based on stimulus substitution and differential facilitation.

According to the stimulus-substitution hypothesis (based on Skinner, 1938) the conditioned stimulus (CS) produces a behavior, the alpha response (AR). The unconditioned stimulus (US) produces an unconditioned response (UR). Pairing of the CS with the US leads to a new behavioral response, the conditioned response (CR), which resembles the unconditioned response. Presumably this involves the development of a new neural pathway (dotted line) through which the CS (which previously did not have access to the UR) now produces responses similar to the UR.

According to the differential-facilitation model the CS gives rise to a family of neural responses, each of which is designated NR_1, NR_2, etc. The efficacy of each is indicated by the length of the arrow originating from the CS. Some of the neural responses reach effector threshold (NR_1) and produce an effector alpha response (EAR). Others remain below threshold; of these, some might be capable of producing responses if appropriately facilitated (NR_2), others might never reach behavioral threshold or might not involve effector systems (NR_n). The type of effector response produced by the neural response varies; some will resemble the UR more than others. The US also produces a family of neural responses; these interact with the neural responses to the CS, facilitating some and inhibiting others. In this scheme only the simplest case of differential facilitation is considered. The varying ability of the US to facilitate different neural responses is indicated by the thickness of the arrow. When NR_1 is facilitated, it gives rise to a specific conditioned effector alpha response (CEAR). Similarly, when NR_2 is facilitated, it produces an effector response. Since this response to the CS was not previously evident, it constitutes a new response, a CR. The degree of resemblance between the CR and UR depends upon the nature of the CS and US. The CR will resemble the UR if the neural response facilitated by the US shares a large number of common motor pathways with the UR.

SUMMARY AND PERSPECTIVE

The relationships of three types of behavioral modification can be examined in the defensive-withdrawal reflex of *Aplysia*: short- and long-term habituation; habituation, dishabituation, and sensitization; and short- and long-term sensitization.

Short-term habituation and sensitization are readily prolonged, up to three weeks, by repeated training. In both cases distributed training is more effective than massed training. The ease with which habituation can be prolonged is particularly interesting in view of the long-standing controversy among experimental psychologists as to whether habituation should be considered a form of learning. Because of the simplicity of habituation, psychologists have not studied it as extensively as associative learning and have therefore not appreciated that there appear to be two forms of habituation: a restricted and an extended form.

In its restricted form habituation consists of (1) short-term depression of the behavioral response, lasting minutes or hours; (2) no capability for dishabituation, or only homosynaptic dishabituation; and (3) no differences between massed and distributed training. This is the form that psychologists have commonly encountered, and have considered a form of adaptation. In its extended form habituation can be prolonged for days and weeks by repeated training; the duration is dependent upon the pattern of stimulation. In addition, habituation can be dishabituated heterosynaptically, for hours or days, by stimulating a site or pathway not involved in the habituation.

Three features are usually thought to distinguish the more interesting from less interesting forms of learning: (1) the duration of retention; (2) the specificity of the pattern of stimulation; and (3) the manipulability of the response (so that training can lead to an increase as well as a decrease in response strength). Thus what distinguishes the extended form of habituation from the restricted are features that generally distinguish the more interesting from less interesting forms of learning. Consequently, an effective argument can be made that the extended form of habituation blends, almost imperceptibly, with other types of learning. As further analysis of the neural mechanisms of associative forms of learning is carried out, a more direct appraisal of their relationship to habituation will become possible.

The extended form of habituation provides an unusually powerful paradigm for studying memory mechanisms: how an animal retains behavioral modification. These studies can answer the question of how long-term habituation differs from the short-term form. Are there two types of memory processes or merely two stages in a single process?

The possibility that there are two types of memory processes has been inferred from studies of learning in humans and lower animals. Experiments in animals using protein- and RNA-synthesis inhibitors suggest that a labile, dynamic, short-term process (circulation of impulses) gives rise to a long-term structural change, which requires the synthesis of new protein and perhaps even of new RNA. The interpretation of these results is difficult, however, because the inhibitors produce other physiological side effects and most of the studies rely solely on behavioral criteria for the assessment of memory impairment. In *Aplysia,* in which the neural correlates of habituation can be studied directly, the short-term process is not dynamic—it does not depend upon the circulation of impulses. Instead, it is plastic, i.e., it depends upon changes in synaptic efficacy. The long-term process does not seem to require a gross structural change, only a progressive decline in the excitatory synaptic potential of the motor neurons. Since there is good evidence that a significant component of this synaptic potential is produced by the direct connections from the sensory neurons to the motor neurons (see Chapter 9, p. 372), it seems likely that both short- and long-term habituation have a common locus, namely, the synapses of the sensory neurons on the motor cells. It also appears likely that both processes share features of a common cellular mechanism, homosynaptic depression.

However, the subcellular mechanisms for short- and long-term habituation may not necessarily be the same. Short-term habituation involves a change in the number of quanta released from the presynaptic terminal, due perhaps to the slow mobilization of transmitter substance into release sites, which results in a depletion of the releasable pool of transmitter substance (the vesicles that are loaded at release sites). The subcellular changes underlying long-term habituation are not yet known, but repeated and distributed activation of a synapse in the face of inadequate mobilization might, over the long term, lead to secondary changes, perhaps to an inhibition of transmitter synthesis, which would reduce the storage pool (the other, unloaded, vesicles in the terminal). The operation of a single cellular process (inadequate mobilization due to a progressive reduction in the free Ca^{++} of the terminals) with secondary long-term consequences might explain why some aspects of the data on memory fit better with a single-process model, whereas other aspects fit better with a two-process model.

Studies of the relation of habituation, dishabituation, and sensitization indicate that habituation of one pathway does not alter the reflex responsiveness of another pathway. However, a dishabituating stimulus facilitates the responses of both habituated and nonhabituated pathways. Experiments indicate that dishabituation is not a removal of habituation but a form of sensitization by which a strong stimulus enhances a family of related reflex responses.

There are as yet no studies of the cellular changes underlying long-term sensitization. Such studies are now particularly important in order to test the generality of the ideas derived from studies of habituation. Studies of sensitization may also provide a crucial bridge for analyzing the mechanisms underlying learning by association.

SELECTED READING

Agranoff, B. W., 1967. Agents that block memory. In *The Neurosciences: A Study Program,* edited by G. C. Quarton, T. Melnechuk, and F. O. Schmitt. New York: Rockefeller University Press. Biochemical studies of learning in the goldfish.

Barondes, S. H. 1970. Multiple steps in the biology of memory. In *The Neurosciences, Second Study Program,* edited by F. O. Schmitt, G. C. Quarton, T. Melnechuk, and G. Adelman. New York: Rockefeller University Press. Biochemical studies of learning in the rat.

Groves, P. M., and R. F. Thompson. 1970. Habituation: A dual-process theory. *Psychological Reviews,* 77:419–450. A dual-process hypothesis for habituation and dishabituation.

Spencer, W. A., R. F. Thompson, and D. R. Neilson, Jr. 1966. Response decrement of the flexion reflex in the acute spinal cat and transient restoration by strong stimuli, *Journal of Neurophysiology,* 29:221–239. First clear demonstration that dishabituation is a special case of sensitization.

CHAPTER 14

Cellular Studies of Behavior
In Perspective

In the preceding chapters I outlined an approach to the cellular mechanisms of behavior and learning based on investigations of opisthobranch molluscs. Several elementary forms of behavior have now been at least partially analyzed in opisthobranch molluscs and the analysis of two complex behavioral responses, feeding and locomotion, has been initiated. Moreover, within the last few years progress has perceptibly quickened. It is probable that in the not too distant future cellular analysis will be applied to other, more complex problems in the study of behavior, perhaps even to abnormal behavior. I shall discuss some of the implications of these new directions in the next chapter. Before doing so, however, I should like to relate the main principles that have emerged from the study of *Aplysia* and other opisthobranch molluscs to findings from cellular studies of behavior in other invertebrates, and consider some of the limitations of using simple systems to study behavior.

COMPLEMENTARITY OF CELLULAR STUDIES OF BEHAVIOR IN *APLYSIA* AND OTHER INVERTEBRATES

Although the cellular study of behavior in invertebrates is only beginning, a surprisingly good fit exists between the findings from studies of widely different animals. A number of the principles described here for *Aplysia* were first found in arthropods, and several of the cellular mechanisms described in *Aplysia* have now been found in insects and crayfish. This generality presumably reflects a basic similarity in the organizational principles of behavior in different invertebrate phyla and encourages the thought that these principles may also be found among vertebrates. It also argues for the complementarity of the two very different methods used to study the cellular mechanisms of behavior: Adrian's single-fiber dissection technique used so extensively in arthropods, and the intracellular recording techniques used in *Aplysia* and other opisthobranch molluscs. To illustrate these points I shall consider here only a few examples of behavior in various nonmolluscan invertebrates.

Neural Control of Locomotion

A major question in the study of motor systems has been the degree to which sequences of behavior are peripherally or centrally controlled. Do recurrent motor patterns, such as walking, swimming, or flying, require proprioceptive feedback from peripheral structures for the timing of the motor pattern? Or can the central nervous system produce these patterns without information from the periphery? Eighty years ago Mott and Sherrington (1895) suggested that the sequential movements of a motor pattern are due to a series of chained reflexes. According to this view, each phase of a movement is triggered by proprioceptive input from peripheral receptors activated by the previous movement. Thus, although locomotion may be initiated by input from either peripheral receptors or a central command, once initiated the motor pattern would require proprioceptive feedback from the limbs for timing of the several phases of locomotion. However, recent studies have shown this view to be incorrect for locomotion in several arthropods. The most detailed and important of these studies are those by Donald Wilson on the flight of locusts.

Locust flight is controlled by three thoracic ganglia (Wilson, 1961, 1965). Flight continues even when these ganglia are isolated from the rest of the nervous system. The thoracic ganglia receive two types of signals from the periphery: one consists of steady signals (from head hairs) and the other of phasic signals (from wing receptors). But the output pattern is independent of these input signals. Most of the detailed information about wing position, frequency, and amplitude is discarded by the neurons. Information about timing, even from the phasic receptors, is also not used. The thoracic ganglia simply extract an average of the excitatory input from the stretch receptors in each wing and from other sensory receptors involved in flight. This average level determines the wingbeat frequency. Moreover, the average level of excitatory input does not itself determine the details of the wingbeat pattern; it only activates the central program for flight.

Wilson has also shown how sensory stimuli interact with the central program for flight. He found that the stereotypic program for flight is produced by the neurons of the thoracic ganglia and their interconnection. Sensory input supplements this central motor program with information about body size and environmental conditions that cannot be genetically preprogrammed. This information is needed to compensate for minor genetic or developmental errors, such as a fundamental mistake in the central program, an inherent asymmetry in the output pattern, or a transient change in the environmental flight conditions (a change in wind direction). For example, phasic sensory feedback can compensate for variations in load during locomotion.

Other types of locomotor sequences in invertebrates, such as walking by the milkweed bug and the cockroach and swimming by the leech, are also centrally controlled (Hoy and Wilson, 1969; Pearson, 1972; Kristan and Stent, 1976). In each of these cases proprioceptive reflexes reinforce and correct the central program. Some nonlocomotor sequences in arthropods are also centrally controlled. The rhythmic movements of the thoracic and abdominal appendages (the swimmerettes) of the crayfish are under central control of the first five abdominal ganglia (Hughes and Wiersma, 1960b; and Ikeda and Wiersma, 1964). The rhythmic, coordinated motor discharge that mediates these movements persists even after complete elimination of sensory inflow to these five ganglia (deafferentation) and their isolation from the rest of the central nervous system (Ikeda and Wiersma, 1964).

There is also good evidence for central control of the motor sequence of the song of the leafhopper (Hagiwara and Watanabe, 1956) and the cricket (Huber, 1962), the heartbeat of crustacea (Hagiwara, 1961; Maynard, 1955, 1961), the copulatory movements of the praying mantis (Roeder, Tozian, and Weiant, 1960), the respiration of insects (Miller, 1965), the eye movement of crabs (Burrows and Horridge, 1968; Horridge, 1966), and the grinding stomach movements of the lobster (Maynard, 1972; Mulloney and Selverston, 1974b). Deafferentation experiments indicate that proprioceptive feedback also may not be essential for certain types of voluntary movements in vertebrates (Taub and Berman, 1963, 1968; Grillner, 1975). For example, locomotion in vertebrates is thought to be produced by the coordinated activity of independent central pattern generators—one for each point of each limb—located in the spinal cord (Grillner, 1975).

Most of the pioneer studies of locomotor sequences in arthropods were based on extracellular recording and stimulation of single axons, and did not define the interconnection and cellular properties of the neurons involved in generating the central program. Nevertheless, the data from these experiments are consistent with a model proposed by Wilson (1966) that postulates that the neurons generating the central program are autoactive. However, with the exception of the cardiac ganglion and the stomatogastric ganglion of the lobster (Hagiwara, 1961; Maynard, 1972; Mulloney and Selverston, 1974b) it has not been possible to make the kind of *intra*cellular recordings necessary to verify the endogenous activity of the cells involved in the motor sequences. As a result, certain questions remain. Are single neurons capable of spontaneous activity, or is a network of neurons required? How general a mechanism is autoactivity? Where does the central program for a motor sequence reside? What types of cellular mechanisms are involved?

Intracellular studies of motor systems in opisthobranch molluscs have provided some answers to these questions. Many molluscan nerve cells have been shown to be autoactive. Preliminary analyses of a few central programs in *Aplysia* indicate that they can reside either in the motor cells themselves or in the command cells (single neurons that trigger complete motor sequences), depending on whether the effector system is used for only one or for several types of behavior. When located in the motor cells the program can involve the capability of repetitive firing among electrically interconnected cells. When located in the command cells the program can involve a burst-generating capability in the command elements.

Command Cells

Command cells were discovered by Wiersma in his studies of the giant fiber system in the crayfish. The giant fiber system consists of two pairs of giant interneurons, the medial giants and the lateral giants, and many pairs of giant motor neurons. In 1938 Wiersma found that a single electrical stimulus applied to the axons of any one of the four giant interneurons could produce a complete escape response consisting of a strong tail flip accompanied by eyestalk, antennae, leg, and swimmeret movement (Wiersma, 1938, 1952; Roberts, 1968; for earlier work see Johnson, 1924, 1926). Each of the four giant interneurons apparently makes synaptic connections with every ganglion of the nerve cord, thereby producing a similar although not identical behavior.

In 1952 Wiersma found that individual neurons in the esophageal commissure triggered a defensive response consisting of the animal raising its head and claws while supporting its body on the thoracic legs. A decade later Wiersma, in collaboration with Hughes and Ikeda, described five command neurons, each of which can cause rhythmic beating of the swimmerettes

(Hughes and Wiersma, 1960). Atwood and Wiersma (1967) described eight additional command neurons in the esophageal commissures, each of which can control claw, leg, and abdominal movements; and Kennedy and his colleagues provided detailed evidence that single neurons command abdominal postural adjustments (Evoy and Kennedy, 1967; Kennedy, Evoy and Hanawalt, 1966; Kennedy, Evoy, Dane and Hanawalt, 1967).

As Atwood and Wiersma (1967) have pointed out, the term "command cell" indicates only that these cells call into play a readily recognizable behavioral act. While the term is descriptively convenient, it implies nothing about the mechanisms by which these cells engage their respective motor neurons.

In arthropods it has often been difficult to delineate patterns of connection between command elements and the neuronal populations they innervate. However, command cells have also been found in opisthobranchs, and because of the size of the cells, intracellular analysis of the connections made by some command cells has become possible. Several of the command cells studied are multiaction cells whose range of innervation is extensive and whose control over behavior is exercised by direct connections to motor cells. Some cells, such as the metacerebral cells of *Aplysia,* have peripheral as well as central actions (Weiss, Cohen, and Kupfermann, 1975).

Organization of Sensory and Motor Systems

The functional organization of sensory and motor systems in arthropods has been thoroughly explored, and several principles have emerged. Neurons that signal position have been found in both the motor and the sensory systems. Those in the motor system signal position of movement; those in the sensory system signal the position of the sensory stimulus on the body surface.

The most detailed studies of sensory systems so far have been those of sensory interneurons in the somatic sensory and visual systems of the crayfish by Wiersma and his colleagues. Their findings are probably representative of most arthropods. For example, a given mechanical stimulus to a point on the abdomen activates a large number of interneurons, each with different size receptive fields. Some interneurons respond only to stimulation of that particular abdominal segment; others respond to stimulation of that segment and one or more adjacent segments; still others may respond to tactile stimulation of any segment of the entire abdomen, and so forth. Thus, the receptor surface is represented by a series of unique cells, some with very localized receptive fields and others with broad and overlapping receptive fields.

A similar organization exists in the crayfish visual system, where the optic nerve contains a number of identifiable interneurons, each with a unique receptive field. Some interneurons respond to stimulation in a relatively restricted field, such as a small segment of a retinal quadrant, while others respond to stimulation in wider and overlapping fields, such as a hemiretina or even an entire retina (Wiersma and Roach, 1976). A hierarchical organization of the sensory cells based on size of the sensory receptive field or motor innervation has also been found in *Aplysia* for the primary mechanoreceptors innervating the siphon skin and for the motor cells for respiration, circulation, and inking.

Thus, like command interneurons that code for specific features of motor output, sensory interneurons code for specific features of the sensory input. Appropriate connections between sensory neurons responding to specific features of the environment and command cells that trigger complex behavioral sequences could in principle explain why only a complex sensory stimulus (a sign stimulus) can trigger a given fixed-action pattern.

Locus and Mechanism for Habituation

The critical change underlying habituation of the crayfish tail-flip response, a fixed act, and the *Aplysia* gill-withdrawal reflex, a graded reflex, involves a homosynaptic depression at the central synapse of the primary sensory neurons. Although the evidence from the crayfish is less complete than that from *Aplysia,* nevertheless it also points to a presynaptic decrease in the number of transmitter quanta released per impulse.

RESTRICTIONS AND LIMITATIONS OF A SIMPLE-SYSTEMS APPROACH

There are three areas in which limitations to the simple-systems approach to behavior have become evident in the course of work on higher invertebrates: methodology, fit between an established neuronal circuit and a behavior, and the applicability of findings across species.

Methodological Limitations

The two most glaring methodological limitations are the lack of morphological and biochemical insights into the cellular mechanisms of behavior. For example, in *Aplysia* and other invertebrates it is now possible to show how specific anatomical loci within a neural circuit become functionally modified during modification of a particular behavior. With some degree of certainty one can also specify the synaptic mechanism. But it would be valuable to have a more profound understanding of the mechanism that produces the alterations of transmitter release. Is it a change in mobilization? In the probability of release? In the number of release sites? Much work is required to advance from the present limited cellular explanation to a more fundamental molecular analysis. Some of these questions might be better examined with morphological techniques that allow one to visualize the synaptic terminals of identified cells.

However, there are at present relatively few good morphological techniques for studying synapses. In contrast to our rapidly increasing understanding of the organization of the cell-body layer of invertebrate ganglia, analysis of the neuropil is proceeding slowly. This is due to the complicated structure of even the simplest neuropil in the nervous system of invertebrates, which seems to offer to morphologists no particular advantage over the more orderly organization of the neuropil of the vertebrate brain. However, several techniques have recently been developed for marking the synapses of individual identified cells so that they can be examined under the electron microscope (see Kater and Nicholson, 1973; Muller and McMahan, 1975; Thompson *et al.*, 1976; and see frontispiece). These techniques are likely to bring important advances in correlating synaptic structure to function. However, even good marking techniques will not provide a molecular explanation. Ultimately the synapses of these systems will have to be examined biochemically as well as morphologically, perhaps through the use of histochemical and immunochemical techniques that permit one to visualize specific species of molecules in the terminals.

Other technical problems need solution. For example, molluscs have an extensive peripheral nervous system that can mediate local responses in the absence of the central nervous system. The analysis of any behavior must take into consideration possible contributions from the peripheral nervous system (Kandel and Spencer, 1968; Peretz, 1970; Kupfermann *et al.*, 1971, 1974; and Lukowiak and Jacklet, 1972). Techniques for studying the peripheral nervous system and distinguishing central from peripheral effects are beginning to be developed. Particularly encouraging has been the finding that certain peripheral responses, such as the peripheral component of siphon withdrawal in *Aplysia*, are mediated by a group of peripheral motor cells that have properties similar to the central motor neurons, except for more restricted fields of innervation. Moreover, these peripheral motor neurons are innervated by the same centrally located sensory neurons that innervate the central motor neurons. However, much progress is necessary before a detailed understanding of the peripheral component of most other systems is worked out.

Closeness of Fit

How well does a given neuronal circuit account for a behavior and how well do observed plastic changes in the circuit account for modification of that behavior? In *Aplysia* there is now fairly compelling evidence that the motor cells so far discovered account in large measure for the motor components of the defensive-withdrawal reflex, respiratory pumping, inking, and heart rate regulation. However, except for gill withdrawal, understanding of the sensory components is less complete. Once all the sensory neurons and interneurons for each of these behavioral responses have been identified, it will then have to be shown that the established neuronal circuit accounts entirely for the behavior and its modification. Models are needed that will permit one to test quantitatively the fit between an established circuit and a behavior.

It is very likely that simple reflexes in invertebrates can be fully accounted for by neural elements. Nevertheless, the important test of fit will come with cellular analyses of more complex behavior. The question of how well cellular analysis can account for complex behavior has often centered on the concept of *emergent properties*. According to one view, usually restricted to higher mental processes, the complex behavior of an animal cannot be accounted for by the sum of the properties of individual neurons. The improbability of accounting for a behavior in terms of its component cells is attributable to noncellular features, such as extracellular electrical or chemical fields, or even to nonbiological features, such as "mind."[1] The alternative view, to which I subscribe, is that emergent properties arise from the ability of interacting systems of cells to manifest properties that could not or might not be readily inferred from studies of individual cells in isolation. By developing quantitative cellular models of complex behavior it should be possible to deny the plausibility of noncellular contributions and also to make a number of biological discoveries, including some that may be quite counter-intuitive.

Generality of Findings

Although animals in all phyla share certain neuroanatomical features, the demonstration of a principle of cellular organization in one phylum does not guarantee that it will be found in others. To what degree are the principles encountered in *Aplysia* likely to be applicable to vertebrates, and specifically

---

[1]For different versions of these views, in relation to conscious behavior, see Lashley, 1929; Eccles, 1953; Pribram, 1971.

to man? Does the fact that habituation is similar in *Aplysia* and vertebrates mean that the mechanism for habituation is universal? Is the increased complexity in vertebrate behavior merely due to an increased complexity of perceptual and motor apparatuses (the basic mechanism for habituation being the same), or does increased complexity bring with it new kinds of plastic change? At present these questions cannot be answered because little is known about the neural mechanisms of behavior in vertebrates in general, and next to nothing is known about man. We can only fall back on formal, indirect comparisons between man and invertebrates.

For example, of the nine features of short-term habituation in vertebrates (p. 518), eight also apply to *Aplysia*, and the one exception, generalization of habituation, does not always apply in man. In addition, three more complex features recently discovered in vertebrates are particularly interesting. First, dishabituation is not related to habituation; it is a separate process, a form of sensitization, (Sharpless and Jasper, 1956; Spencer, Thompson, and Neilson, 1966a; Hagbarth and Kugelberg, 1958). Second, habituation can be made to endure for days and weeks (Hinde, 1954b). Third, habituation is sensitive to the pattern of stimulation (Hinde, 1954b). All three features have been found in the gill-withdrawal reflex of *Aplysia*.

Preliminary cellular analyses of habituation in the vertebrate nervous system are quite consistent with the picture that has emerged from *Aplysia* and crayfish. The studies of Spencer, Thompson, and Neilson (1966b) have provided indirect evidence that habituation is a type of homosynaptic mechanism of excitatory synaptic depression. The recent studies of Farel, Glanzman, and Thompson (1973) suggest that in the frog synaptic depression may involve a monosynaptic excitatory system. Once it becomes possible to work out monosynaptic pathways in the cat and thus to examine their plastic capabilities, it may then become possible to determine whether the mechanism of the flexion reflex of mammals resembles that found in invertebrate defensive responses.

To document the generality of a mechanism found in invertebrates, a range of experiments is needed in progressively more complex lower vertebrates and mammals. Primitive vertebrates, such as lamprey, are likely to be particularly useful. The neuronal organization of simple behavior in these lower vertebrates is not beyond the power of present techniques to investigate. Individual cells or cell groups in the lamprey can be identified and their connections studied (Rovainen, 1967; Martin and Wickelgren, 1971). Extending cellular analysis from invertebrates to simple vertebrates would strengthen the argument that these two major phylogenetic divisions share similar nervous systems and that the properties of their cellular elements and the principles relating patterns of connection and types of behavior are common to both. Much is likely to be learned even in instances in which cellular findings do not prove general. Cellular analysis of a behavior forces one to defend the logic of a behavioral definition and to break down the behavior into its component parts. The very attempt to analyze a behavior on the cellular level therefore has considerable heuristic value for other behavioral studies. In addition, analyses that do not provide general solutions may nevertheless illustrate and solve technical problems necessary for the analyses of vertebrate behavior.

In working with invertebrates, however, one important consideration should be kept in mind. Cellular studies of gill withdrawal and escape swimming are interesting only insofar as they can produce general models of defensive reflexes and locomotion, and thus serve as useful starting points for the systematic examination of the neurobiology of vertebrate behavior. Deprived of this objective, the research can lose interest in a way that neural research in vertebrates does not. Stripped of its behavioral relevance, looked at simply as a neural machine, there is something grand about the wiring diagram of the cerebellum or the hippocampus that is lacking in the wiring diagram of a molluscan or arthropod ganglion. It does not take a great leap of faith to see that the functioning of the cat cerebellum is like our own, but studies of invertebrates must aim at generality to be interesting for human behavior.

SELECTED READING

Kennedy, D., and W. J. Davis. 1976. The organization of invertebrate motor systems. In "Cellular Biology of Neurons" (Vol. 1, Sect. 1, *Handbook of Physiology*), edited by E. R. Kandel. Baltimore: Williams and Wilkins. Discusses command cells and their role in behavior.

Wiersma, C. A. G., and J. L. M. Roach. 1976. Principles in the organization of invertebrate sensory systems. In "Cellular Biology of Nuerons" (Vol. I, Sect. 1, *Handbook of Physiology*). A review of invertebrate sensory systems with emphasis on somatic and visual sensation.

Wilson, D. M. 1968. The flight-control system of the locust. *Scientific American,* 218(5):83–90. A good introduction to the classic work on the central control of insect flight.

CHAPTER 15

Implications for the Study of Abnormal Behavior

Claude Bernard (1865), one of the founders of modern experimental medicine, was the first to appreciate that disease states are often extreme manifestations of normal processes and follow lawful patterns that can be successfully analyzed with biological techniques. In part as an outgrowth of Bernard's influence, the efforts of biology and medicine have often become fused in a common purpose—the biological analysis of disease. In many areas of medicine this analysis can now be accomplished on the cellular and even molecular level. As a result, normal cellular functioning and its alteration in disease have come to be viewed within a common, biological framework. This framework, which encompasses both normal and abnormal behavior, has not yet been firmly established within psychiatry. Although it has long been recognized that some disturbances of behavior originate in normal behavior, few attempts have been made to develop simple animal models of abnormal behavior that would allow a cellular analysis of how normal behavior can be distorted by experiential or genetic factors.

It is unlikely that invertebrates will yield models of abnormal behavior that are comparable to the disease states of man. Nevertheless, it may be possible to develop models in invertebrates that include essential components of human abnormalities. Moreover, these models may illustrate some of the mechanisms by which environmental factors exaggerate of disrupt normal adaptive responses.

In this final chapter I shall outline studies of invertebrates that might throw light on how alterations in learning and motivation produce abnormal behavior. Because many abnormalities in the adult are thought to originate in infancy, I shall also outline experiments designed to examine how experiences at an early stage in development may have devastating consequences for later stages.

ABNORMAL BEHAVIOR PRODUCED BY STRESS

The psychological disorders of man are commonly divided, according to severity, into two large classes: (1) mild disturbances, or *neuroses,* and (2) severe disturbances, or *psychoses.* Each of these classes (and their numerous subclasses) involves one or more of seven common psychological states: anxiety, depression, mania, paranoia, delusion, hallucination, and thought disorders. The first four of these are often experienced in mild form by normal people and more frequently and more severely by neurotically disturbed people. In addition, people who are psychotically disturbed experience one or more of the other three disorders: delusions, hallucinations, and thought disorders (see for example Coleman, 1972; Solomon and Patch, 1971).

Those disturbed states in which there is a cognitive component—paranoia, delusion, hallucination, and thought disorders—can obviously not be studied in simple animal models, although it is sometimes possible to recognize aspects of these disorders in primates (Harlow and Harlow, 1962; Kaufmann and Rosenblum, 1967). Anxiety (excessive fear) and depression, however, have been well demonstrated in primates (see for example Kaufmann and Rosenblum, 1967). The mechanisms of these abnormal states are also those most likely to be found in simpler animals. All animals show fear responses (withdrawal and escape); anxiety is therefore the affective state that can perhaps best be studied in invertebrates.

Anxiety is a normal response to threat, whether to one's person, attitudes, or self-esteem. Normal anxiety is adaptive; it signals potential danger and can contribute to the mastery of a difficult situation and thus to personal growth. Excessive anxiety, on the other hand, can be dysfunctional. Anxiety has subjective as well as objective manifestations. The subjective manifestations range from a heightened sense of awareness to a deep fear of impending disaster. The objective manifestations consist of heightened responsiveness (hyperarousal), restlessness, and autonomic changes (increased heart rate and blood pressure). Anxiety is considered pathological when it becomes unduly severe and persistent, when it no longer serves only to signal danger but becomes disorganizing.

Excessive fear responses are also seen in invertebrates. For example, in *Aplysia* an extreme form of long-term sensitization follows many days of repeated noxious stimuli (Figure 15-1). The extreme form of sensitization provides a simple example of how a normal fear response can evolve into one that is exaggerated and therefore potentially abnormal. As I have noted earlier, if an *Aplysia* receives four painful stimuli to the siphon a day for four days it will show a heightened responsiveness (long-term sensitization) of siphon withdrawal that persists for about a week (see Figure 13-19, p. 637). Pinsker

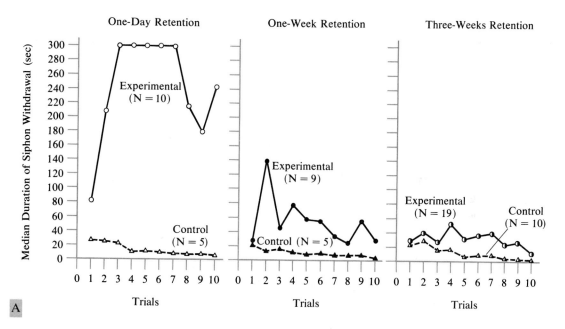

FIGURE 15-1.
The extreme form of sensitization of siphon withdrawal. The experimental group was given two shocks to the neck each day for 10 days. Half of the experimental animals were then tested for retention of sensitization one day after training and the other half one week after training. Three weeks after training all animals were tested for retention. The median duration of siphon withdrawal is shown for each trial of a 10-trial training session (intertrial interval = 30 sec). In all comparisons the sensitized animals showed significantly greater siphon withdrawal than the controls. Note the difference in the duration of siphon withdrawal in these animals compared to those illustrated in Figure 13-19. The flat peak of the experimental group in the one-day retention test results from the fact that more than half of the animals exceeded a predetermined 300-sec limit on those trials [From Pinsker *et al.*, 1973. Copyright by the American Association for the Advancement of Science.]

and co-workers (1973) have found that if training is continued for 10 days, the heightened responsiveness becomes much more intense (about 10 times that seen after four days of training). Not only is the reflex response dramatically enhanced, but the time course of sensitization in these animals (tested by 10 stimuli to the siphon) is radically different from that of normal animals. Instead of showing the monotonic decline in responsiveness typical of normal habituation, animals that have been extremely sensitized exhibit a progressive buildup of responsiveness (compare Figures 15-1 and 13-19). This extreme form of long-term sensitization is analogous to one component of a chronic hyper-arousal state, or in loosely anthropomorphic terms, to a component of chronic anxiety. The parallel between sensitization and arousal has also been pointed out by Hagbarth and Kugelberg (1958) in their studies of the abdominal reflex in man:

> Sensitization also increases the spatial irradiation of the reflex in the musculature of the abdominal wall. . . . It often irradiates to the flexor muscles of the extremities and causes general crouching of the body. . . . With increasing strength of sensitization the abdominal skin reflex gradually emerges into a more widespread reaction which has a striking resemblance not only to startle responses but also to the pathological abdominal mass reflexes sometimes encountered in spinal man.

The extreme form of sensitization in *Aplysia* has other resemblances to chronic anxiety. The sensitization persists for many days after the stressful stimulus has stopped. Also, the animals are restless, locomote more, and try again and again to escape from their cages, which normally they do not do. Finally, the animals show autonomic changes, including a prolonged increase in heart rate (Dieringer and Koester, unpub.). In man some autonomic changes due to stress, such as repeated elevations of blood pressure, are thought to lead to the development of hypertension in genetically susceptible individuals.

Extreme sensitization is a useful model of behavioral abnormality because one can readily trace its origins in normal adaptive behavior. Animals habituate readily in a safe environment, where they learn quickly to ignore a novel but nonharmful stimulus that is repeatedly presented. An animal that has received a *few* harmful stimuli becomes sensitized; it learns that, for the moment, its environment is no longer completely safe and it will not habituate as readily as an unperturbed animal. But an animal that has received *repeated* harmful stimuli learns that its environment is dangerous. In the presence of danger, any tendency for habituation of withdrawal reflexes is maladaptive, whereas heightened responsiveness is adaptive. However, the persistence of heightened responsiveness for days after the noxious stimulation has stopped is no longer adaptive. Thus, the ability to respond effectively to extreme danger carries with it the undesirable consequence that the heightened responsiveness cannot be readily shut off; it persists after the danger is gone. Insofar as an animal's responsiveness relates more to stimuli presented earlier in its history than to current stimuli, the animal can be thought of as responding abnormally. The difference between an appropriate response (to an actual threat) and an inappropriate response (to a threat that has passed) is one of the features that distinguishes appropriate anxiety from inappropriate or chronic anxiety.

Because of the possible value of the extreme form of sensitization as a model of chronic anxiety, it would be of interest to explore the physiological and biochemical mechanisms that account for both the long-term and extreme forms. How do long-term and extreme sensitization relate to each other and to short-term sensitization? Do the long-term and extreme forms represent progressively longer and more profound forms of heterosynaptic facilitation? Is the extreme form mediated by serotonin through cyclic AMP, as may be the case for short-term sensitization? Many forms of anxiety in man respond well to pharmacological treatment with antianxiety agents, such as diazepam (Valium) and chlordiazepoxide (Librium). It would be of interest to examine the effects of these drugs on the extreme form of long-term sensitization in *Aplysia*.

I would again emphasize that the analogy I have drawn here between long-term sensitization and one component of chronic anxiety, neurotic arousal, should not be taken too literally. The responses of a chronically anxious individual do not usually habituate, whereas after some time the responses of a chronically sensitized *Aplysia* do. Nevertheless, some components of chronic anxiety can be viewed as pathological forms of sensitization. Insofar as this is so, understanding of the normal forms of sensitization throws some light on the pathological forms of both sensitization and anxiety. Moreover, as we gain better understanding of the cellular mechanisms that underlie learning, we are likely to be in a better position to develop animal models that permit insight into the cellular mechanisms underlying behavioral abnormalities related to learning.

MOTIVATIONAL STATE AND STRESS

I have so far considered only simple learned behavior (sensitization) and the effects of extreme stress on that behavior (extreme sensitization). But behavior, whether appropriate or disturbed, reflects not only what the animal has learned but what the animal needs or wants. While in earlier chapters I have considered learning in some detail, I have said little about the factors that determine the needs and wants of an animal. These factors are necessarily more vaguely defined than those involved in learning and cannot be studied directly; they are thought to reflect an intervening variable variously called *drive* or *motivational state* (see p. 22). The concept of motivational state was developed to explain variations in a given behavioral performance.

For example, a strong stimulus that usually produces a large response may fail to do so; a constant stimulus that normally produces an unvarying response may produce variations of that response; or a change in the stimulus may not be reflected by a change in the response. These common deviations from expected performance have been explained by postulating that the animal's motivational state varies. The motivational state of an organism is thought to be controlled by basic homeostatic regulatory processes essential for survival: feeding, respiration, sex, temperature regulation, and self-protection. Alterations of the motivational state are therefore produced by changes in the internal condition of the animal relative to a given set point of the regulatory process. Thus, motivation varies as a function of food deprivation (hunger), water deprivation (thirst), and sexual deprivation. Insofar as behavior is controlled by *internal* rather than *external* variables it is said to be controlled by a motivational state.

Three features are common to an increase in motivational states: (1) increased arousal and motor activity; (2) a lowering of the behavioral threshold, which is both *selective* (favoring one response system) and *general* (affecting nonrelated response systems as well); (3) an organization of individual behavioral components into specific (goal-oriented) sequences, e.g., appetitive behavior. Attainment of a goal decreases the intensity of the motivational state and increases the behavioral threshold, which leads to a dropping out of one or more components of the behavioral sequence and finally to cessation of response.

In the past the cellular mechanisms of motivational states were not accessible to study. Recently, however, two factors have been found that have facilitated their study. First, features of motivational state have been recognized in invertebrates (Dethier, 1966), including *Aplysia* (Kupfermann, 1974a). Second, alterations in motivational state have been found to initiate *arousal functions* as well as *cue functions* (for review see Brown, 1961). For example, an increase in hunger due to food deprivation selectively activates a cue function, feeding behavior. But, in addition, food deprivation has a broader arousal action and alters reflex responses not concerned with feeding. Satiety, on the other hand, not only inhibits feeding behavior but also reduces the arousal level of the animal; the animal will become behaviorally tranquilized (less reactive to environmental stimuli) and may go to sleep. The general arousal function controls the animal's activity level and responsiveness to stimuli without specifying the kind of activity or what stimulus the animal will respond to. Specific cue functions tend to channel the arousal toward certain response systems, e.g., escape in response to pain, feeding in response to food deprivation.

Cue functions are expressed only in behavioral responses associated with complex neural circuits, such as feeding, drinking, and sex. By contrast, arousal functions of motivational states are manifested in elementary as well as complex behavior, and are often readily analyzed in each. For example, *Aplysia* exposed to food show enhanced cue functions: they orient to the food and make spontaneous biting movements (Kupfermann, 1974a). In addition, the animals show altered arousal functions: they stop locomoting, their biting reflex is enhanced, and the siphon-withdrawal reflex is dramatically depressed (Advokat, Carew, and Kandel, unpubl.). Studies of motivational state have provided important insights into the nature of behavioral arousal. For example, comparison of arousal produced in simple behavior by motivational state (food) with that produced by noxious stimuli indicates that arousal is not a unidimensional process. Whereas arousal by food stimuli depresses siphon withdrawal, arousal by noxious stimuli sensitizes siphon withdrawal (Figure 13-9, p. 623). A dichotomy is also seen in complex behavior. Noxious stimuli depress the biting response; food sensitizes it (Kupfermann and Pinsker, 1968; Susswein, Kupfermann, and Weiss, unpubl.).

Recent studies suggest that both elementary and complex behavior seem to utilize a similar mechanism for sensitization: serotonergic interneurons act on target cells to increase the level of cyclic AMP (see Figure 10-31*B* and 12-28). Even though the *mechanisms* are similar, the loci for sensitization are nonetheless separate. Excitatory interneurons mediate sensitization of siphon and gill withdrawal; the metacerebral cells mediate sensitization of biting (Hawkins, Castellucci, and Kandel, unpubl.; Weiss, Schonberg, Cohen, Mandelbaum, and Kupfermann, unpubl.). In each case the cells mediating, sensitization are an integral part of the neural circuit for the behavior. As a result of this anatomical independence of the sensitizing system, arousal produced by noxious and appetitive stimuli can elicit opposite effects on defensive and appetitive response systems.

Whereas the sensitizing component of arousal is now beginning to be understood, there are no data on how arousal produces depression of response. Studies of the effects of motivational state on siphon withdrawal could indicate whether the response depression results from inhibition of a tonically active sensitizing system or is due to an independent mechanism.

In man, abnormalities in motivation are reflected in many disease states.

They are perhaps seen most clearly in a group of neurotic disorders, the *consummatory diseases,* in which the regulatory processes that control eating and drinking are disturbed (Schachter, 1971). These states include obesity due to overeating and *anorexia nervosa* (pathological undereating). Since the neuronal mechanisms of feeding behavior in *Aplysia* and other invertebrates are beginning to be understood, it might be possible to examine aspects of these disorders in invertebrates, by disturbing normal feeding patterns and determining whether this produces alterations in the motivation. Is there an alteration in the motivational effects of feeding if animals are fed out of phase with their circadian activity pattern? What happens if animals are given large meals infrequently rather than the more normal pattern of small frequent feedings? By examining various arousal and cue functions of motivational states in animals exposed to altered feeding patterns, it might be possible to detect disturbances of the motivational state.

BEHAVIORAL MODIFICATION AND DEVELOPMENT

Studies of humans as well as other animals indicate that normal behavioral development follows a definite course that proceeds through well-defined stages. Infants crawl before they walk, they have diffuse emotional responses before they develop specific ones, and even the acquisition of language and reasoning follows a well-defined sequence. Mastery of tasks requires a specific interaction between the developing organism and its environment. If this interaction is disturbed, normal development may be disrupted. For example, newborn kittens deprived of visual patterns throughout the first 16 weeks of life become permanently blind, and stimuli that normally excite the cortical cells become much less effective (Wiesel and Hubel, 1965). Similar deprivation in an adult cat has no effect. Infant monkeys reared in social isolation for the first year of life fail to achieve normal social and sexual adjustment (Harlow and Harlow, 1962). Similar social isolation has few consequences for adult monkeys. These instructive studies stand as landmarks in the attempt to study in animals the mechanisms whereby alterations in the normal environment of the newborn leads to altered behavior in later life.

The ability to rear *Aplysia* and other gastropods in the laboratory might make it possible to carry developmental studies to a more mechanistic level. Before studying abnormal behavior, however, it will be important to determine the normal pattern of development of behavior, about which surprisingly little is known. Using the behavioral responses of *Aplysia* as a model, several questions may be explored. How does the developmental sequence for reflex patterns compare to the developmental sequence of fixed-action patterns? What is the correlation of the development of a behavioral response with that of its neural architecture? How do environmental and social factors (the presence or absence of other animals) affect the development of behavior? What environmental conditions maximize, retard, or disturb normal development?

Aplysia develops through four stages: premetamorphic, metamorphic, juvenile, and reproductive adult (see p. 74). Having once established the normal timetable for the emergence of a behavior within this developmental sequence, it would be interesting to see how learning and memory vary as functions of development. Habituation and sensitization are particularly good learning paradigms for this purpose and have been used extensively for tracing the development of memory in human infants (Cohen, 1973; Jeffrey, 1968). For example, newborn infants do not habituate to novel stimuli; the ability to habituate develops only in the second to third month of life. If a similar relationship between developmental stages and habituation is found to obtain for the gill-withdrawal reflex in *Aplysia*, one could determine the physiological changes that lead to the capability for habituation. Is the nervous system at birth incapable of habituation? Or is this capability present but overridden by sensitization because the entire environment is novel and thus exerts a powerful arousing effect? Finally, it would be interesting to determine how the animal's early training at one stage affects performance at a later stage. What are the consequences for adult habituation of repeated habituation training in the premetamorphic or metamorphic phase? Does extreme sensitization training early in life produce irreversible effects? How do alterations in the feeding pattern of the early juvenile, e.g., irregular meals, affect adult feeding behavior?

I have considered only abnormal behavior that can be analyzed by cellular biological approaches. Much can also be learned in the future by combining behavioral and genetic approaches (see p. 59). For example, single gene mutations in *Drosophila* have been used to develop a variety of behavioral abnormalities including visual disturbances, abnormalities in circadian rhythm and sexual behavior, muscular dystrophies, and sudden cessation of development (Benzer, 1973). Analysis of *Drosophila* has provided a powerful methodology for examining the genetic components of normal behavioral systems. This approach promises to revolutionize behavioral genetics and in so doing to shed much additional light on the neural mechanisms of both normal and abnormal behavior.

In this final chapter I have argued that, as in other areas of behavior, the cellular mechanisms of abnormal behavior are likely to be general. The key to their analysis will often lie in the selection of an appropriate experimental animal. Most abnormalities will be better analyzed in man than in invertebrates, but certain aspects of abnormal behavior can perhaps best be understood in simpler animals.

In attempting to extend the generalities of neuronal function to abnormal behavior I have taken chances and speculated freely. But only insofar as such speculations can be tested empirically will a true understanding of animal behavior in general, and in the long run our own behavior, be advanced.

SELECTED READING

Benzer, S. 1973. Genetic dissection of behavior. *Scientific American,* 229(6): 24–37. A good discussion of modern behavioral genetics in *Drosophila.*

Dethier, V. G. 1966. Insects and the concept of motivation. *Nebraska Symposium on Motivation.* Reviews the usefulness of invertebrates for motivational studies.

Hagbarth, K. E., and E. Kugelberg. 1958. Plasticity of the human abdominal skin reflex. *Brain,* 81:305–318. A study of sensitization in man.

Kety, S. S. 1974. From rationalization to reason. *American Journal of Psychiatry,* 131:957–963. A critique of the unnecessarily conflicted relationship between psychiatry and medical biology.

Sidman, M. 1960. Normal sources of pathological behavior. *Science,* 132:61–68. A behaviorist looks at abnormal behavior.

Appendix

TABLE I. Identified neurons and cell clusters in the abdominal ganglion of *Aplysia*.

TABLE II. Known connections made by interneurons located in the abdominal ganglion of *Aplysia*.

TABLE I.

Identified neurons and cell clusters in the abdominal ganglion of *Aplysia*. (See Figures 7-5 and 7-6, pp. 226–227, for anatomical maps; see p. 228 for a description of the nomenclature of neurons in *Aplysia*.)

Cell	Firing pattern*	Function	Transmitter
L1	Silent		
L2	Bursting		
L3	Bursting		
L4	Bursting		
L5	Silent		
L6	Bursting		
L7	Irregular	Gill motor neuron	
L8	Irregular		
L9 (see $L9_{G1}$ below)			
L10	Irregular with burst tendency	Command cell; increases cardiac output	ACh
L11	Irregular with burst tendency		
L12	Irregular		
L13	Irregular		
L14A	Silent	Neuroglandular cell; causes inking	
L14B	Silent	Neuroglandular cell; causes inking	
L14C	Silent	Neuroglandular cell; causes inking	
L15		Excitatory interneuron	
L16	Silent	Inhibitory interneuron	
L17			
L18	Silent	Sensory cell responding to noxious stimuli	
L19	Irregular	Interneuron	
L20A			
L20B			
L21	Silent	Interneuron	
L22	Silent	Interneuron	
L23	Silent	Interneuron	
L24	Regular firing	Interneuron	
LA cluster	Silent	Neurosecretory; egg-laying	Polypeptide hormone

*See Figure 7-14, p. 240, for maps of the distribution of the abdominal neurons according to firing pattern.

Cell	Firing pattern*	Function	Transmitter
LB_{S1}	Irregular	Siphon motor neuron	
LB_{S2}	Irregular	Siphon motor neuron	
LB_{S3}	Irregular	Siphon motor neuron	
LB_{VC1}	Irregular	Vasoconstrictor motor neuron	ACh
LB_{VC2}	Irregular	Vasoconstrictor motor neuron	ACh
LB_{VC3}	Irregular	Vasoconstrictor motor neuron	ACh
$L9_{G1}$	Irregular	Gill motor neuron	
$L9_{G2}$	Irregular	Gill motor neuron	
LC_P	Irregular	Pericardial motor neuron	
LC_K	Irregular	Kidney motor neuron	
LD_{G1}	Irregular	Gill motor neuron	ACh
LD_{G2}	Irregular	Gill motor neuron	ACh
LD_{S1}	Irregular	Siphon motor neuron	
LD_{S2}	Irregular	Siphon motor neuron	
LD_{S3}	Irregular	Siphon motor neuron	
LD_{HI1}	Irregular	Heart inhibitory motor neuron	ACh
LD_{HI2}	Irregular	Heart inhibitory motor neuron	ACh
LD_{HE}	Silent	Heart excitatory motor neuron	
LE cluster	Silent	Siphon sensory cells	
R1	Silent		
R2	Silent		ACh
R3–R13	Regular	Neurosecretory	
R14	Regular or silent	Neurosecretory	
R15	Bursting	Neurosecretory; water balance	Polypeptide hormone
R16	Silent		
RA cluster	Silent	Neurosecretory; egg-laying	Polypeptide hormone
RB_{HE}	Irregular	Heart excitatory motor neuron	5-HT
RC cluster	Irregular		
RD_G	Irregular	Gill motor neuron	
RD_S	Irregular	Siphon motor neuron	
RE cluster	Silent	Siphon sensory cells	

TABLE II.
Known connections made by interneurons located in the abdominal ganglion of *Aplysia*.

Interneuron number	Identified cell number	Identified follower cells	Synaptic action
I	L10	L1, L2, L3, L4, L5 L6, L8, L11, L12 L13 LB_{S1}, $LB_{VC1,2,3}$, LC_P, $LD_{HI1,2}$ L9, R15, R16, RB clusters L7 L20A, L20B, L21	Inhibitory* Excitatory Excitatory-inhibitory* Electrical
II	Unidentified	L2, L3, L4, L5, L6, L7, L8 L9, L10, L20A, L20B, L21, L24, $LB_{VC1,2,3}$ $LB_{S1,2,3}$ $L9_{G1,2}$, RB cluster L11, L14A, L14B, L14C, $LD_{G1,2}$ $LD_{S1,2,3}$, $LD_{HI1,2}$, LD_{HE} R2, R16 RD_G, RD_S R15	Inhibitory Excitatory Excitatory-inhibitory
III	Unidentified	L10	Inhibitory
IV	Unidentified	R16	Excitatory
V	Unidentified	L10 L2, L3, L4, L5, L6	Inhibitory Excitatory
VI	L15	L5, L3	Excitatory
VII	Unidentified	L2, L3, L4, L6	Inhibitory
VIII	Unidentified	L12, L13	Excitatory
IX	Unidentified	R1, R2	Excitatory
X	Unidentified	L1, R1, R2	Excitatory
XI	L24	LD_{HE}, $LD_{HI1,2}$, $LD_{G1,2}$ $LD_{S1,2,3}$, RD_G, RD_S, L7 RB cluster†	Inhibitory Excitatory
XII	Unidentified	RB cluster	Excitatory
XIII	Unidentified	R15, RB cluster	Excitatory
XIV	L16	LE sensory neurons	Inhibitory
XV	L19	L7, $LB_{S1,2}$	Inhibitory
XVI	L21	L3, L4‡ L10	Inhibitory Electrical
XVII	L22	L7	Excitatory
XVIII	L23	L7	Excitatory

*Unidentified inhibitory and excitatory-inhibitory followers have also been observed in the right caudal quarter-ganglion.

†With the exception of cell RB_{HE}.

‡Cell L21 also produced IPSPs in several unidentified follower cells in the left rostral quarter-ganglion.

Literature Cited

Adey, W. R.
  1972 Organization of brain tissue: Is the brain a noisy processor? *Int. J. Neurosci.,* **3:**271–284.
Adrian, E. D.
  1930 The activity of the nervous system in the caterpillar. *J. Physiol. (London),* **70:**xxxiv–xxxv.
  1932a The activity of the optic ganglion of *Dytiscus marginalis. J. Physiol. (London),* **75:**26P–27P.
  1932b *The Mechanism of Nervous Action; Electrical Studies of the Neurone.* Philadelphia: University of Pennsylvania Press.
  1937 Synchronized reactions in the optic ganglion of *Dytiscus. J. Physiol. (London),* **91:**66–89.
Adrian, E. D., and D. W. Bronk.
  1928 The discharge of impulses in motor nerve fibres. Part I: Impulses in single fibres of the phrenic nerve. *J. Physiol. (London),* **66:**81–101.
Adrian, E. D., and B. H. C. Matthews.
  1934 The Berger rhythm: Potential changes from the occipital lobes in man. *Brain,* **57:**355–385.
Adrian, E. D., and Y. Zotterman.
  1926 The impulses produced by sensory nerve endings. Part II: The response of a single end-organ, *J. Physiol. (London),* **61:**151–171.
Advokat, C., T. J. Carew, and E. R. Kandel.
  1976 Modulation of a simple reflex in *Aplysia californica* by arousal with food stimuli. *Neuroscience Abstracts* 2.
Agranoff, B. W.
  1967 Agents that block memory. In *The Neurosciences: A Study Program,* edited by G. C. Quarton, T. Melnechuk, and F. O. Schmitt. New York: Rockefeller University Press, pp. 756–764.
Agranoff, B. W., R. E. Davis, and J. J. Brink.
  1966 Chemical studies on memory fixation in goldfish. *Brain Res.,* **1:**303–309.
Agranoff, B. W., R. E. Davis, L. Casola, and R. Lim.
  1967 Actinomycin D blocks formation of memory of shock-avoidance in goldfish. *Science (Wash., D.C.),* **158:**1600–1601.
Alexander, R. D.
  1962 Evolutionary change in cricket acoustical communication. *Evolution,* **16:**443–467.

Alving, B. O.

1968 Spontaneous activity in isolated somata of *Aplysia* pacemaker neurons. *J. Gen. Physiol.*, **51**:29–45.

Apáthy, S.

1897 Das leitende Element des Nervensystems und seine topographischen Beziehungen zu den Zellen. *Mitt. Zool. Stn. Neapel*, **12**:495–748.

Arch, S.

1972a Polypeptide secretion from the isolated parietovisceral ganglion of *Aplysia californica*. *J. Gen Physiol.*, **59**:47–59.

1972b Biosynthesis of the egg-laying hormone (ELH) in the bag cell neurons of *Aplysia californica*. *J. Gen. Physiol.*, **60**:102–119.

Armstrong, C. M.

1975 Ionic pores, gates and gating current. *Quart. Rev. Biophys.*, **7**:179–210.

Arvanitaki, A.

1938 *Propriétés rythmiques de la matière vivante. Variations graduées de la polarisation et rythmicités*, 2 v. Paris: Hermann.

1939 Recherches sur la réponse oscillatoire locale de l'axone géant isolé de 'Sepia.' *Arch. Int. Physiol.*, **49**:209–256.

Arvanitaki, A., and H. Cardot.

1941a Contribution à la morphologie du système nerveux des Gastéropodes. Isolement, à l'état vivant, de corps neuroniques. *C. R. Séances Soc. Biol. Fil.*, **135**:965–968.

1941b Les caractéristiques de l'activité rythmique ganglionnaire "spontanée" chez l'Aplysie. *C. R. Séances Soc. Biol. Fil.*, **135**:1207–1211.

1941c Réponses rythmiques ganglionnaires, graduées en fonction de la polarisation appliquée. Lois des latences et des fréquences. *C. R. Séances Soc. Biol. Fil.*, **135**:1211–1216.

1941d Réponses autonomes ganglionnaires à une polarisation appliquée. *C. R. Séances Soc. Biol. Fil.*, **135**:1216–1221.

Arvanitaki, A., and N. Chalazonitis.

1958 Configurations modales de l'activité, propres à différents neurones d'un même centre. *J. Physiol. (Paris)* **50**:122–125.

1964 Processus d'excitation d'un neurone autoactif sur ondes lentes. *C. R. Séances Soc. Biol. Fil.*, **158**:1119–1122.

1967 Electrical properties and temporal organization in oscillating neurons (*Aplysia*). In *Symposium on Neurobiology of Invertebrates*. Hungarian Academy of Sciences, pp. 169–199.

Arvanitaki, A., and S. H. Tchou.

1942 Les lois de la croissance relative individuelle des cellules nerveuses chez l'Aplysie. *Bull. Histol. Appl. Physiol. Pathol.*, **19**:224–256.

Ascher, P.

1968 Electrophoretic injections of dopamine on *Aplysia* neurones. *J. Physiol. (London)*, **198**:48P–49P.

1972 Inhibitory and excitatory effects of dopamine on *Aplysia* neurones. *J. Physiol. (London)*, **225**:173–209.

Ascher, P. and J. S. Kehoe.

1975 Amine and amino receptors in gastropods. In *Handbook of Psychopharmacology*, Vol. 14, edited by L. Inverson, S. Inverson and S. Snyder. New York: Plenum, pp. 265–310.

Ashby, W. R.

1952 *Design for a Brain*. New York: Wiley.

Atkinson, R. C., and R. M. Shiffrin.

1968 Human memory: A proposed system and its control processes. In *The Psychology of Learning and Motivation*, Vol. 2, edited by K. W. Spence and J. T. Spence. New York: Academic Press, pp. 89–195.

Attardi, D. G., and R. W. Sperry.
  1963 Preferential selection of central pathways by regenerating optic fibers. *Exp. Neurol.,* **7:**46–64.
Atwood, H. L., and C. A. G. Wiersma.
  1967 Command interneurons in the crayfish central nervous system. *J. Exp. Biol.,* **46:**249–261.
Atz, J. W.
  1970 The application of the idea of homology to behavior. In *Development and Evolution of Behavior: Essays in Memory of T. C. Schneirla,* edited by L. R. Aronson, E. Tobach, D. S. Lehrman, and J. S. Rosenblatt. San Francisco: Freeman, pp. 53–74.
Axelsson, J., and S. Thesleff.
  1959 A study of supersensitivity in denervated mammalian skeletal muscle. *J. Physiol. (London),* **147:**178–193.
Bailey, C. H., V. F. Castellucci, J. D. Koester, and E. R. Kandel.
  1975 Central mechanoreceptor neurons in *Aplysia* connect to peripheral siphon motor neurons: A simple system for the morphological study of the synaptic mechanism underlying habituation. *Neuroscience Abstracts,* **1:**588.
Bailey, C. H., Thompson, E. B., Castellucci, V., and Kandel, E. R.
  1976 Fine structure of synapses of identified sensory cells which mediate the gill-withdrawal reflex in *Aplysia californica. Neuroscience Abstracts* 2.
Baker, P. F., A. L. Hodgkin, and T. I. Shaw.
  1962 Replacement of the axoplasm of the giant nerve fibres with artificial solutions. *J. Physiol. (London),* **164:**330–354.
Barlow, G. W.
  1968 Ethological units of behavior. In *The Central Nervous System and Fish Behavior,* edited by D. Ingle. Chicago: University of Chicago Press, pp. 217–232.
Barnett, S. A.
  1963 *The Rat: A Study in Behaviour.* Chicago: Aldine.
Barondes, S. H.
  1970 Multiple steps in the biology of memory. In *The Neurosciences: Second Study Program,* edited by F. O. Schmitt, G. C. Quarton, T. Melnechuk, and G. Adelman. New York: Rockefeller University Press, pp. 272–278.
Barrett, J. N., and W. E. Crill.
  1972 Voltage clamp analysis of conductances underlying cat motoneuron action potentials. *Fed. Proc.,* **31:**305, abstract 509.
Beach, F. A.
  1950 The snark was a boojum. *Am. Psychol.,* **5:**115–124.
Bennett, M. V. L.
  1968 Similarities between chemically and electrically mediated transmission. In *Physiological and Biochemical Aspects of Nervous Integration,* edited by F. D. Carlson. Englewood, N.J.: Prentice-Hall, pp. 73–128.
  1971 Analysis of parallel excitatory and inhibitory synaptic channels. *J. Neurophysiol.,* **34:**69–75.
  1972 Comparison of electrically and chemically mediated synaptic transmission. In *Structure and Function of Synapses,* edited by G. D. Pappas and D. P. Purpura. New York: Raven Press, pp. 221–256.
  1973 Function of electrotonic junctions in embryonic and adult tissues. *Fed. Proc.,* **32:**65–75.
  1976 Electrical transmission: A functional analysis and comparison to chemical transmission. In "Cellular Biology of Neurons" (Vol. 1, Sect. 1, *Handbook of Physiology*), edited by E. R. Kandel. Baltimore: Williams and Wilkins.

Bentley, D. R.
    1971 Genetic control of an insect neuronal network. *Science (Wash., D.C.)*, **174:**1139–1141.
Bentley, D., and R. R. Hoy.
    1974 The neurobiology of cricket song. *Sci. Am.*, **231**(2):34–44.
Benzer, S.
    1967 Behavioral mutants of *Drosophila* isolated by counter-current distribution. *Proc. Nat'l. Acad. Sci. U.S.A.*, **58:**1112–1119.
    1973 Genetic dissection of behavior. *Sci. Am.*, **229**(6):24–37.
Bernard, C.
    1865 Introduction à l'étude de la médecine expérimentale. Paris: J. B. Baillière et fils. [English translation by H. C. Green, *An Introduction to the Study of Experimental Medicine*, 1957. New York: Dover.]
Bernstein, J.
    1902 Untersuchungen zur Thermodynamik der bioelektrischen Ströme. Erster Theil. *Pflügers Archiv gesamte Physiol. Menschen Tiere*, **92:**521–562.
Berry, M. S., and G. A. Cottrell.
    1975 Excitatory, inhibitory and biphasic synaptic potentials mediated by an identified dopamine-containing neurone. *J. Physiol.*, **244:**589–612.
Berry, R. W.
    1969 Ribonucleic acid metabolism of a single neuron: Correlation with electrical activity. *Science (Wash., D.C.)*, **116:**1021–1023.
Beswick, F. G., and R. T. W. L. Conroy.
    1965 Optimal tetanic conditioning of heteronymous monosynaptic reflexes. *J. Physiol. (London)*, **180:**134–146.
Beurle, R. L.
    1956 Properties of a mass of cells capable of regenerating pulses. *Phil. Trans. R. Soc. London (B)*, **240:**55–94.
Bezanilla, F., and C. M. Armstrong.
    1974 Gating currents of the sodium channels: Three ways to block them. *Science (Wash., D.C.)*, **183:**753–754.
Binstock, L., and L. Goldman.
    1969 Current- and voltage-clamped studies on *Myxicola* giant axons: Effect of tetrodotoxin. *J. Gen. Physiol.*, **54:**730–740.
Bishop, P. D., and H. D. Kimmel.
    1969 Retention of habituation and conditioning. *J. Exp. Psychol.*, **81:**317–321.
Bittner, G. D., and D. Kennedy.
    1970 Quantitative aspects of transmitter release. *J. Cell Biol.*, **47:**585–592.
Bjork, R. A.
    1970 Repetition and rehearsal mechanisms in models for short-term memory. In *Models of Human Memory*, edited by D. Norman. New York: Academic Press, pp. 307–330.
Blankenship, J. E., and R. E. Coggeshall.
    1973 The abdominal ganglion of *Aplysia willcoxi*. *Society for Neuroscience, Program and Abstracts*, **3:**287.
Blankenship, J. E., H. Wachtel, and E. R. Kandel.
    1971 Ionic mechanisms of excitatory, inhibitory, and dual synaptic actions mediated by an identified interneuron in abdominal ganglion of *Aplysia*. *J. Neurophysiol.*, **34:**76–92.
Blodgett, H. C.
    1929 The effect of the introduction of reward upon the maze performance of rats. *Univ. Calif. Publ. Psychol.*, **4:**113–134.
Bloedel, J., P. W. Gage, R. Llinas, and D. M. J. Quastel.
    1966 Transmitter release at the squid giant synapse in the presence of tetrodotoxin. *Nature (London)*, **212:**49–50.

Bodian, D.

1967 Neurons, circuits, and neuroglia. In *The Neurosciences: A Study Program,* edited by G. C. Quarton, T. Melnechuk, and F. O. Schmitt. New York: Rockefeller University Press, pp. 6–24.

Boring, E. G.

1950 *A History of Experimental Psychology,* 2nd ed. New York: Appleton-Century-Crofts,

Bottazzi, F.

1899 Ricerche fisiologiche sul Sistema nervoso viscerale delle Aplisie e di alcuni Cefalopodi. *Riv. Sci. Biol.,* **1:**837–924.

Bower, G.

1967 A multicomponent theory of the memory trace. In *The Psychology of Learning and Motivation,* Vol. 1, edited by K. W. Spence and J. T. Spence. New York: Academic Press, pp. 229–325.

Boycott, B. B., and J. Z. Young.

1950 The comparative study of learning. *Symp. Soc. Exp. Biol.,* **4:**432–453.

1955 A memory system in *Octopus vulgaris* Lamarck. *Proc. R. Soc. London* (B), **143:**449–480.

Boyd, I. A., and A. R. Martin.

1956 The end-plate potential in mammalian muscle. *J. Physiol. (London),* **132:**74–91.

Brenner, S.

1973 The genetics of behaviour. *Br. Med. Bull.,* **29:**269–271.

Bridgman, P. W.

1927 *The Logic of Modern Physics.* New York: Macmillan.

Brinley, F. J., Jr.

1974 Excitation and conduction in nerve fibers. In *Medical Physiology,* 13th ed., Vol. 1, edited by V. B. Mountcastle, St. Louis: C. V. Mosby, pp. 34–76.

Brock, L. G., J. S. Coombs, and J. C. Eccles.

1952 The recording of potentials from motoneurones with an intracellular electrode. *J. Physiol. (London),* **117:**431–460.

Brodwick, M. S., and D. Junge.

1972 Post-stimulus hyperpolarization and slow potassium conductance increase in *Aplysia* giant neurone. *J. Physiol. (London),* **223:**549–570.

Brogden, W. J.

1939 Sensory pre-conditioning. *J. Exp. Psychol.,* **25:**323–332.

Bronk, D. W., J. Z. Young, and R. W. Gerard.

1936 *Cold Spring Harbor Symp. Quant. Biol.* **4:**293.

Brown, J. S.

1961 *The Motivation of Behavior.* New York: McGraw-Hill.

Brown, P. L., and H. M. Jenkins.

1968 Auto-shaping of the pigeon's key-peck. *J. Exp. Anal. Behav.,* **11:**1–8.

Brunelli, M., V. Castellucci, and E. R. Kandel.

1975 A possible role for serotonin and cyclic AMP in heterosynaptic facilitation. *Neuroscience Abstracts,* **1:**568.

Bruner, J., and J. Kehoe.

1970 Long-term decrements in the efficacy of synaptic transmission in molluscs and crustaceans. In *Short-Term Changes in Neural Activity and Behaviour,* edited by G. Horn and R. A. Hinde. Cambridge, England: Cambridge University Press, pp. 323–359.

Bruner, J., and D. Kennedy.

1970 Habituation: occurrence at a neuromuscular junction. *Science (Wash., D.C.).* **169:**92–94.

Bruner, J., and L. Tauc.

1966 Long-lasting phenomena in the molluscan nervous system. *Symp. Soc. Exp. Biol.,* **20:**457–475.

Bryant, H., and D. Weinreich.
  1975 Transsynaptic connections in *Aplysia* examined with tetraethylammonium. *J. Physiol. (London),* **244:**181–195.

Bullock, T. H.
  1947 Problems in invertebrate electrophysiology. *Physiol. Rev.,* **27:**643–664.

Bullock, T. H., and G. A. Horridge.
  1965 *Structure and Function in the Nervous Systems of Invertebrates,* 2 vols. San Francisco: W. H. Freeman.

Bürger, O.
  1890 Untersuchungen über die Anatomie und Histologie der Nemertinen nebst Beiträgen zur Systematik. *Z. Wiss. Zool.,* **50:**1–277.

Burrows, M., and G. A. Horridge.
  1968 The action of the eyecup muscles of the crab, *Carcinus,* during optokinetic movements. *J. Exp. Biol.,* **49:**223–250.

Burrows, M., and G. Hoyle.
  1973 Neural mechanisms underlying behavior in the locust *Schistocerca Gregaria.* III: Topography of limb motor neurons in the metathoracic ganglion. *J. Neurobiol.,* **4:**167–186.

Byrne, J.
  1975a Input-output characteristics of the defensive gill-withdrawal reflex in *Aplysia californica. Neuroscience Abstracts,* **1:**591.
  1975b A feedback controlled stimulator that delivers controlled displacement of forces to cutaneous mechanoreceptors. *IEEE Trans. Bio-Med. Eng.* BME **22:**66–69.

Byrne, J., V. Castellucci, and E. R. Kandel.
  1974a Quantitative aspects of the sensory component of the gill-withdrawal reflex in *Aplysia. Progr. and Abst. Soc. Neuro. Sci.,* Fourth Annual Meeting, 160.
  1974b Receptive fields and response properties of mechanoreceptor neurons innervating siphon skin and mantle shelf in *Aplysia. J. Neurophysiol.,* **37:**1041–1064.
  UNPUBL. Stimulus-response relations and stability of *Aplysia* mechanoreceptor neurons.
  UNPUBL. Quantitative aspects of the contribution of mechanoreceptor sensory neurons to the defensive gill-withdrawal reflex in *Aplysia.*

Byrne, J., N. Dieringer, J. Koester, and E. Shapiro.
  1976 Ionic mechanisms contributing to inking behavior in *Aplysia, Neuroscience Abstracts* 2.

Cajal, S. R.
  1894 La fine structure des centres nerveux. *Proc. R. Soc. London* (B), **55:**444–468.
  1911 *Histologie du Système Nerveux de L'homme et des Vertébrés.* Vol. 2. Paris: Maloine. [Republished 1955, *Histologie du Système Nerveux.* Translated by L. Azoulay. Madrid: Instituto Ramón y Cajal.]

Callec, J. J., J. C. Guillet, Y. Pichon, and J. Boistel.
  1971 Further studies on synaptic transmission in insects. II: Relations between sensory information and its synaptic integration at the level of a single giant axon in the cockroach. *J. Exp. Biol.,* **55:**123–149.

Carew, T. J., V. Castellucci, J. Byrne, and E. R. Kandel.
  1976 Quantitative analysis of the contribution of central and peripheral nervous systems to the gill-withdrawal reflex in *Aplysia californica. Neuroscience Abstract* 2.

Carew, T. J., V. F. Castellucci, and E. R. Kandel.
  1971 An analysis of dishabituation and sensitization of the gill-withdrawal reflex in *Aplysia. Int. J. Neurosci.,* **2:**79–98.

Carew, T. J., and E. R. Kandel.
  1973 Acquisition and retention of long-term habituation in *Aplysia:* Correlation of behavioral and cellular processes. *Science (Wash., D.C.),* **182:**1158–1160.

1976 Two functional effects of decreased-conductance EPSP's: synaptic augmentation and increased electrotonic coupling. *Science (Wash., D.C.),* **192:**150–153.

UNPUBL. Inking in *Aplysia californica:* I The neural circuit of an all-or-none behavioral response.

UNPUBL. Inking in *Aplysia californica:* II The central program for inking.

Carew, T. J., H. M. Pinsker, and E. R. Kandel.

1972 Long-term habituation of a defensive withdrawal reflex in Aplysia. *Science (Wash., D.C.),* **175:**451–454.

Carew, T. J., H. Pinsker, K. Rubinson, and E. R. Kandel.

1974 Physiological and biochemical properties of neuromuscular transmission between identified motoneurons and gill muscle in *Aplysia. J. Neurophysiol.,* **37:**1020–1040.

Carpenter, D. O.

1970 Membrane potential produced directly by the Na+ pump in *Aplysia* neurons. *Comp. Biochem. Physiol.,* **35:**371–385.

1973a Electrogenic sodium pump and high specific resistance in nerve cell bodies of the squid. *Science (Wash., D.C.),* **179:**1336–1338.

1973b Ionic mechanisms and models of endogenous discharge of *Aplysia* neurones. *Symposium Neurobiol. Invert.,* Tihany 1971, pp. 35–58.

Carpenter, D. O., and B. O. Alving.

1968 A contribution of an electrogenic Na+ pump to membrane potential in *Aplysia* neurons. *J. Gen. Physiol.,* **52:**1–21.

Carpenter, D., and R. Gunn,

1970 The dependence of pacemaker discharge of *Aplysia* neurons upon Na+ and Ca++. *J. Cell. Physiol.,* **75:**121–128.

Castellucci, V., J. Byrne, and E. R. Kandel.

UNPUBL. Mechanisms of synaptic plasticity associated with habituation of gill and siphon withdrawal reflex in *Aplysia.*

Castellucci, V., and E. R. Kandel.

1974 A quantal analysis of the synaptic depression underlying habituation of the gill-withdrawal reflex in *Aplysia. Proc. Nat'l. Acad. Sci. U.S.A.,* **71:**5004–5008.

1975 Quantal analysis of heterosynaptic facilitation underlying dishabituation in *Aplysia. Fed. Proc.,* **34:**418, abstract 1117.

IN PRESS An invertebrate system for the cellular study of habituation and sensitization. In *Habituation: Perspectives from Child Development, Animal Behavior, and Neurophysiology,* edited by T. Tighe and R. N. Leaton. Hillsdale, N.J.: Lawrence Erlbaum Associates.

Castellucci, V., E. R. Kandel, and J. H. Schwartz.

1972 Macromolecular synthesis and the functioning of neurons and synapses. In *Structure and Function of Synapses,* edited by G. D. Pappas and D. P. Purpura. New York: Raven Press, pp. 193–219.

Castellucci, V., H. Pinsker, H. I. Kupfermann, and E. R. Kandel.

1970 Neuronal mechanisms of habituation and dishabituation of the gill-withdrawal reflex in *Aplysia. Science (Wash., D.C.),* **167:**1745–1748.

Cedar, H., E. R. Kandel, and J. H. Schwartz.

1972 Cyclic adenosine monophosphate in the nervous system of *Aplysia californica.* I: Increased synthesis in response to synaptic stimulation. *J. Gen. Physiol.,* **60:**558–569.

Cedar, H., and J. H. Schwartz.

1972 Cyclic adenosine monophosphate in the nervous system of *Aplysia californica.* II: Effect of serotonin and dopamine. *J. Gen. Physiol.,* **60:**570–587.

Chen, C. F., R. von Baumgarten, and R. Takeda.

1971 Pacemaker properties of completely isolated neurones in *Aplysia californica. Nature New Biol.,* **233:**27–29.

Chiarandini, D. J., E. Stefani, and H. M. Gerschenfeld.
　1967 Ionic mechanisms of cholinergic excitation in molluscan neurons. *Science (Wash., D.C.)*, **156:**1597–1599.

Chorover, S. L., and P. H. Schiller.
　1965 Short-term retrograde amnesia in rats. *J. Comp. Physiol. Psychol.*, **59:**73–78.

Christoffersen, G. R. J.
　1972 Steady state contribution of the Na–K pump to the membrane potential in identified neurons of terrestrial snail, *Helix pomatia. Acta. Physiol. Scand.*, **86:**498–514.

Coggeshall, R. E.
　1967 A light- and electron-microscope study of the abdominal ganglion of *Aplysia californica. J. Neurophysiol.* **30:**1263–1287.

Coggeshall, R. E., E. R. Kandel, I. Kupfermann, and R. Waziri.
　1966 A morphological and functional study on a cluster of identifiable neurosecretory cells in the abdominal ganglion of *Aplysia californica. J. Cell Biol.*, **31:**363–368.

Cohen, H. D., and S. H. Barondes.
　1967 Puromycin effect on memory may be due to occult seizures. *Science (Wash., D.C.)*, **157:**333–334.

Cohen, H. D., F. Ervin, and S. H. Barondes.
　1966 Puromycin and cycloheximide: Different effects on hippocampal electrical activity. *Science (Wash., D.C.)*, **154:**1557–1558.

Cohen, J., K. Weiss, and I. Kupfermann.
　1974 Physiology of the neuromuscular system of buccal muscle of *Aplysia. The Physiologist*, **17:**198 (abstract).

Cohen, L. B.
　1973 Two processes in visual attention. *Merrill-Palmer Q.*, **19:**157–180.

Cohen, M. J.
　1965 The dual role of sensory systems: Detection and setting central excitability. *Cold Spring Harbor Symp. Quant. Biol.*, **30:**587–599.

Cohen, M. J., and J. W. Jacklet.
　1967 The functional organization of motor neurons in an insect ganglion. *Philos. Trans. R. Soc. London* (B), **252:**561–572.

Cole, K. S., and H. J. Curtis.
　1939 Electric impedance of the squid giant axon during activity. *J. Gen. Physiol.*, **22:**649–670.

Cole, K. S., and A. L. Hodgkin.
　1939 Membrane and protoplasm resistance in the squid giant axon. *J. Gen. Physiol.*, **22:**671–687.

Coleman, J. C.
　1972 *Abnormal Psychology and Modern Life*, 4th ed. (W. E. Broen, Jr., contributing author). Glenview, Ill.: Scott, Foresman.

Connor, J. A., and C. F. Stevens.
　1971a Voltage clamp studies of a transient outward membrane current in gastropod neural somata. *J. Physiol. (London)*, **213:**21–30.
　1971b Prediction of repetitive firing behaviour from voltage clamp data on an isolated neurone soma. *J. Physiol (London)*, **213:**31–53.

Cooke, I. M., G. Leblanc, and L. Tauc.
　1974 Sodium pump stoichiometry in *Aplysia* neurones from simultaneous current and tracer measurements. *Nature (London)*, **251:**254–256.

Coombs, J. S., J. C. Eccles, and P. Fatt.
　1955 The specific ionic conductances and the ionic movements across the motoneural membrane that produce the inhibitory post-synaptic potential. *J. Physiol. (London)*, **130:**326–373.

Courtney, K., and S. Roper.
    1976 Sprouting of synapses after partial denervation of frog cardiac ganglion. *Nature*, **259**:317–319.

Craig, W.
    1918 Appetites and aversions as constituents of instincts. *Biol. Bull. Mar. Biol. Lab (Woods Hole)*, **34**:91–107.

Craik, F. I. M., and R. S. Lockhart.
    1972 Levels of processing: A framework for memory research. *J. Verb. Learn. Verb. Behav.*, **11**:671–684.

Culbertson, J. T.
    1963 *The Minds of Robots. Sense Data, Memory Images, and Behavior in Conscious Automatons.* Urbana: University of Illinois Press.

Curtis, D. R. and R. W. Ryall.
    1966 The synaptic excitation of Renshaw cells. *Exp. Brain Res.*, **2**:81–96.

Curtis, H. J., and K. S. Cole.
    1942 Membrane resting and action potentials from the squid giant axon. *J. Cell. Comp. Physiol.*, **19**:135–144.

Dagan, D., and I. Parnas.
    1970 Giant fibre and small fibre pathways involved in the evasive response of the cockroach *Periplaneta americana. J. Exp. Biol.*, **52**:313–324.

Dale, H. H.
    1935 Pharmacology and nerve-endings. *Proc. R. Soc. Med.*, **28**:319–332.

Darnell, J. E., Jr.
    1968 Ribonucleic acids from animal cells. *Bacteriol. Rev.*, **32**:262–290.

Darwin, C.
    1860 *On the Origin of Species by Means of Natural Selection,* New York: Appleton.
    1871 *The Descent of Man, and Selection in Relation to Sex.* 2 Vols. New York: Appleton.
    1873 *The Expression of the Emotions in Man and Animals.* New York: Appleton.
    1884 Instinct: A posthumous essay. In *Mental Evolution in Animals,* edited by G. J. Romanes. New York: Appleton, pp. 159–317.

Davis, W. J., G. J. Mpitsos, and J. M. Pinneo.
    1974a The behavioral hierarchy of the mollusk *Pleurobranchaea.* I: The dominant position of the feeding behavior. *J. Comp. Physiol.*, **90**:207–224.
    1974b The behavioral hierarchy of the mollusk *Pleurobranchaea.* II: Hormonal suppression of feeding associated with egg-laying. *J. Comp. Physiol.*, **90**:225–243.

Davis, W. J., M. V. S. Siegler, and G. J. Mpitsos.
    1973 Distributed neuronal oscillators and efference copy in the feeding system of *Pleurobranchaea. J. Neurophysiol.*, **36**:258–274.

Del Castillo, J., and B. Katz.
    1954a The effect of magnesium on the activity of motor nerve endings. *J. Physiol. (London)*, **124**:553–559.
    1954b Quantal components of the end-plate potential. *J. Physiol. (London)*, **124**:560–573.
    1954c Statistical factors involved in neuromuscular facilitation and depression. *J. Physiol. (London)*, **124**:574–585.
    1954d Changes in end-plate activity produced by pre-synaptic polarization. *J. Physiol. (London)*, **124**:586–604.
    1955 On the localization of acetylcholine receptors. *J. Physiol. (London)*, **128**:157–181.
    1957 La base 'quantale' de la transmission neuro-musculaire. In "Microphysiolgie comparée des éléments excitables," *Colloq. Int. Natl. Rech. Sci.*, **67**:245–258.

De Robertis, E.
    1971 Molecular biology of synaptic receptors. *Science (Wash., D.C.)*, **171**:963–971.

Deschênes, M., and M. V. L. Bennett.

1974 A qualification to the use of TEA as a tracer for monosynaptic pathways. *Brain Res.,* **77:**169–172.

Dethier, V. G.

1966 Insects and the concept of motivation. *Nebraska Symposium on Motivation,* pp. 105–136.

De Weer, P., and D. Geduldig.

1973 Electrogenic sodium pump in squid giant axon. *Science (Wash., D.C.),* **179:**1326–1328.

Diamond, J.

1971 The Mauthner cell. In *Fish Physiology,* vol. 5, edited by W. S. Hoar and D. J. Randall. New York: Academic Press, pp. 265–346.

Dieringer, N., and J. Koester.

UNPUBL. The behavioral control of heart rate in *Aplysia.*

Dodge, R.

1923a Habituation to rotation. *J. Exp. Psychol.,* **6:**1–35.

1923b Thresholds of rotation. *J. Exp. Psychol.,* **6:**107–137.

Dorsett, D. A.

1967 Giant neurons and axon pathways in the brain of *Tritonia. J. Exp. Biol.,* **46:**137–151.

Dudel, J.

1965 Facilitatory effects of 5-hydroxy-tryptamine on the crayfish, neuromuscular [*Astacus fluviatilis, Orconectes virilis*] junction. *Arch. Exp. Pathol. Pharmakol.,* **249:**515–528.

Dudel, J., and S. W. Kuffler.

1961 Presynaptic inhibition at the crayfish neuromuscular junction. *J. Physiol. (London),* **155:**543–562.

Duncan, C. P.

1949 The retroactive effect of electroshock on learning. *J. Comp. Physiol. Psychol.,* **42:**32–44.

Eales, N. B.

1921 *Aplysia.* Liverpool Marine Biology Committee, Memoir #24, *Proc. and Trans., Liverpool Biological Society,* **35:**183–266.

1950 Torsion in Gastropoda. *Proc. Malacol. Soc. London,* **28:**53–61.

1960 Revision of the world species of *Aplysia* (Gastropoda, Opisthobranchia). *Bull. Br. Mus. (Nat. Hist.) Zool.,* **5:**267–404.

Eccles, J. C.

1953 *The Neurophysiological Basis of Mind. The Principles of Neurophysiology.* Oxford: Clarendon Press.

1957 *The Physiology of Nerve Cells.* Baltimore: Johns Hopkins Press.

1964 *The Physiology of Synapses.* Berlin: Springer.

1969 *The Inhibitory Pathways of the Central Nervous System.* Springfield, Ill.: Charles C. Thomas.

1973 *The Understanding of the Brain.* New York: McGraw-Hill.

Eccles, J. C., P. Fatt, and K. Koketsu.

1954 Cholinergic and inhibitory synapses in a pathway from motor-axon collaterals to motoneurones. *J. Physiol. (London),* **126:**524–562.

Eccles, J. C., P. Fatt, and S. Landgren.

1956 Central pathway for direct inhibitory action of impulses in largest afferent nerve fibers to muscle. *J. Neurophysiol.,* **19:**75–98.

Eccles, J. C., P. Fatt, S. Landgren, and G. J. Winsbury.

1954 Spinal cord potentials generated by volleys in the large muscle afferents. *J. Physiol. (London),* **125:**590–606.

Eccles, J. C., B. Katz, and S. W. Kuffler.

1941 Nature of the "end plate potential" in curarized muscle. *J. Neurophysiol.,* **4:**362–387.

Eccles, J. C., R. F. Schmidt, and W. D. Willis.
    1962 Presynaptic inhibition of the spinal monosynaptic reflex pathway. *J. Physiol. (London)*, **161**:282–297.
Eckert, R., and H. D. Lux.
    1976 A voltage sensitive persistent calcium conductance in neuronal somata of *Helix*. *J. Physiol.*, **254**:129–151.
Ehrenberg, C. G.
    1836 *Beobachtung einer auffallenden bisher unbekannten Struktur des Seelenorgans bei Menschen und Tieren*. Berlin.
Eibl-Eibesfeldt, I.
    1970 *Ethology: The Biology of Behavior*, translated by E. Klinghammer. New York: Holt, Rinehart and Winston.
Eisenstadt, M., J. E. Goldman, E. R. Kandel, H. Koike, J. Koester, and J. H. Schwartz.
    1973 Intrasomatic injection of radioactive precursors for studying transmitter synthesis in identified neurons of *Aplysia californica. Proc. Natl. Acad. Sci. U.S.A.*, **70**:3371–3375.
Elliot, T. R.
    1905 The action of adrenalin. *J. Physiol., (London)*, **32**:401–467.
Epstein, R., and L. Tauc.
    1970 Heterosynaptic facilitation and post-tetanic potentiation in *Aplysia* nervous system. *J. Physiol. (London)*, **209**:1–23.
Evarts, E. V., and J. R. Hughes.
    1957 Relation of posttetanic potentiation to subnormality of lateral geniculate potentials. *Am. J. Physiol.*, **188**:238–244.
Evoy, W. H., and D. Kennedy.
    1967 The central nervous organization underlying control of antagonistic muscles in the crayfish. I: Types of command fibers. *J. Exp. Zool.*, **165**:223–238.
Faber, D. S., and M. R. Klee.
    1972 Membrane characteristics of bursting pacemaker neurones in *Aplysia. Nature New Biol.*, **240**:29–31.
    1974 Strychnine interactions with acetylcholine, dopamine, and serotonin receptors in Aplysia neurons. *Brain Res.*, **65**:109–126.
Fabre, J. H.
    1897–1907 *Souvenirs Entomologiques. Études sur l'instinct et les moeurs des Insectes*. Paris.
    1921 *Fabre's Book of Insects*, retold from A. Teixeira de Mattos' translation of Fabre's *Souvenirs Entomologiques*, by Mrs. R. Stawell. New York: Dodd, Mead.
Farel, P. B.
    1971 Long-lasting habituation in spinal frogs. *Brain Res.*, **33**:405–417.
Farel, P. B., D. L. Glanzman, and R. F. Thompson.
    1973 Habituation of a monosynaptic response in vertebrate central nervous system: lateral column–motoneuron pathway in isolated frog spinal cord. *J. Neurophysiol.*, **36**:1117–1130.
Fatt, P. and B. Katz
    1951 An analysis of the end-plate potential recorded with an intra-cellular electrode. *J. Physiol. (London)*, **115**:320–370.
    1953 The effect of inhibitory nerve impulses on a crustacean nerve fibre. *J. Physiol. (London)*, **121**:374–389.
Fechner, G. T.
    1860 *Elemente der Psychophysik*, 2 vols. Leipzig: Breitkopf und Härtel.
Fentress, J. C., and R. W. Doty.
    1971 Effect of tetanization and enucleation upon excitability of visual pathways in squirrel monkeys and cats. *Exp. Neurol.*, **30**(3):535–554.

Fischbach, G. D.
1972 Synapse formation between dissociated nerve and muscle cells in low density cell cultures, *Dev. Biol.,* **28:**407–429.

Flexner, J. B., L. B. Flexner, and E. Stellar.
1963 Memory in mice as affected by intracerebral puromycin. *Science (Wash., D.C.),* **141:**57–59.

Flexner, L. B., and J. B. Flexner.
1966 Effect of acetoxycycloheximide and of an acetoxycycloheximide–puromycin mixture on cerebral protein synthesis and memory in mice. *Proc. Natl. Acad. Sci. U.S.A.,* **55:**369–374.

Flexner, L. B., J. B. Flexner, and R. B. Roberts.
1967 Memory in mice analyzed with antibiotics. *Science (Wash., D.C.),* **155:**1377–1382.

Forbes, A.
1922 The interpretation of spinal reflexes in terms of present knowledge of nerve conduction. *Physiol. Rev.,* **2:**361–414.

Fowler, H., and R. E. Whalen.
1961 Variation in incentive stimulus and sexual behavior in the male rat. *J. Comp. Physiol. Psychol.,* **54:**68–71.

Frank, K., and M. G. F. Fuortes.
1957 Presynaptic and postsynaptic inhibition of monosynaptic reflexes. *Fed. Proc.,* **16:**39–40.

Frazier, W. T., E. R. Kandel, I. Kupfermann, R. Waziri, and R. E. Coggeshall.
1967 Morphological and functional properties of identified neurons in the abdominal ganglion of *Aplysia californica. J. Neurophysiol.,* **30:**1288–1351.

Frazier, W. T., R. Waziri, and E. R. Kandel.
1965 Alterations in the frequency of spontaneous activity in *Aplysia* neurons with contingent and non-contingent nerve stimulation. *Fed. Proc.,* **24:**522, abstract 2171.

Freud, S.
1895 Project for a Scientific Psychology. In *The Origins of Psycho-analysis,* edited by M. Bonaparte, A. Freud, E. Kris, translated by E. Mosbacher and J. Strachey. London: Imago, 1954.

Friedländer, B.
1888 Beiträge zur Kenntnis des Centralnervensystems von *Lumbricus. Z. Wiss Zool.,* **47:**47–84.

Frings, H., and C. Frings.
1965 Chemosensory bases of food-finding and feeding in *Aplysia juliana* (Mollusca, Opisthobranchia). *Biol. Bull. (Woods Hole),* **128:**211–217.

Frisch, K. von.
1950 *Bees, Their Vision, Chemical Senses, and Language.* Ithaca, New York: Cornell University Press.

Furshpan, E. J., and D. D. Potter.
1957 Mechanism of nerve-impulse transmission at a crayfish synapse. *Nature (London),* **180:**342–343.
1959 Transmission at the giant motor synapses of the crayfish. *J. Physiol. (London),* **145:**289–325.
1968 Low-resistance junctions between cells in embryos and tissue culture. *Curr. Top. Dev. Biol.,* **3:**95–127.

Furukawa, T., and E. J. Furshpan.
1963 Two inhibitory mechanisms in the Mauthner neurons of goldfish. *J. Neurophysiol.,* **26:**140–176.

Gainer, H.
1972 Effects of experimentally induced dispause on the electrophysiology and protein synthesis patterns of identified molluscan neurons. *Brain Res.,* **39:**387–402.

Gainer, H., and Z. Wollberg.
  1974 Specific protein metabolism in identifiable neurons of *Aplysia californica. J. Neurobiol.,* **5:**243–261.
Garcia, J., W. G. Hankins, and K. W. Rusiniak.
  1974 Behavioral regulation of the milieu interne in man and rat. *Science (Wash., D.C.),* **185:**824–831.
Garcia, J., B. K. McGowan and K. F. Green
  1972 Biological constraints on conditioning. In *Classical Conditioning. II: Current Research and Theory,* edited by A. H. Black and W. F. Prokasy. New York: Appleton-Century-Crofts, pp. 3–27.
Gardner, D.
  1971 Bilateral symmetry and interneuronal organization in the buccal ganglia of *Aplysia. Science (Wash., D.C.),* **173:**550–553.
Gardner, D., and E. R. Kandel.
  1972 Diphasic postsynaptic potential: A chemical synapse capable of mediating conjoint excitation and inhibition. *Science (Wash., D.C.),* **176:**675–678.
Geduldig, D., and R. Gruener.
  1970 Voltage clamp of the *Aplysia* giant neurone: Early sodium and calcium currents. *J. Physiol. (London),* **211:**217–244.
Geduldig, D., and D. Junge.
  1968 Sodium and calcium components of action potentials in the *Aplysia* giant neurone. *J. Physiol. (London),* **199:**347–365.
Geffen, L. B., and B. Jarrott.
  1976 Physiology and biochemistry of cholinergic transmission. In "Cellular Biology of Neurons" (Vol. 1, Sec. 1, *Handbook of Physiology, The Nervous System*), edited by E. R. Kandel. Baltimore: Williams and Wilkins.
Gerschenfeld, H. M.
  1966 Chemical transmitters in invertebrate nervous systems. *Symp. Soc. Exp. Biol.,* **20:**299–323.
  1973 Chemical transmission in invertebrate central nervous systems and neuromuscular junctions. *Physiol. Rev.,* **53:**1–119.
Gerschenfeld, H. M., P. Ascher, and L. Tauc.
  1967 Two different excitatory transmitters acting on a single molluscan neurone. *Nature (London),* **213:**358–359.
Gerschenfeld, H. M., and D. Paupardin-Tritsch.
  1974a Ionic mechanisms and receptor properties underlying the responses of molluscan neurones to 5-hydroxytryptamine. *J. Physiol. (London),* **243:**427–456.
  1974b On the transmitter function of 5-hydroxytryptamine at excitatory and inhibitory monosynaptic junctions. *J. Physiol. (London),* **243:**457–481.
Getting, P. A.
  1975 *Tritonia* swimming: Triggering of a fixed-action pattern. *Brain Res.,* **96:**128-133.
Getting, P., and A. O. D. Willows.
  1974 Modification of neuron properties by electrotonic synapses. II. Burst formation by electrotonic synapses. *J. Neurophysiol.,* **37:**858–868.
Giller, E., Jr., and J. H. Schwartz.
  1968 Choline acetyltransferase: Regional distribution in the abdominal ganglion of *Aplysia. Science (Wash., D.C.),* **161:**908–911.
  1971 Choline acetyltransferase in identified neurons of abdominal ganglion of *Aplysia californica. J. Neurophysiol.,* **34:**93–107.
Ginsborg, B. L.
  1967 Ion movements in junctional transmission. *Pharmacol. Rev.,* **19:**289–316.
Glaser, E. M.
  1966 *The Physiological Basis of Habituation.* London: Oxford University Press.

Glassman, E.
  1974 Macromolecules and behavior: a commentary. In *The Neurosciences: Third Study Program,* edited by F. O. Schmitt and F. G. Worden. Cambridge, Mass.: M.I.T. Press. pp 667–678.
Gola, M.
  1974 Neurones à ondes-salves des mollusques. Variations cycliques lentes des conductances ioniques. *Pflügers Archiv gesamte Physiol. Menschen Tiere,* **352:**17–36.
Goldberg, A. L., and J. J. Singer.
  1969 Evidence for a role of cyclic AMP in neuromuscular transmission. *Proc. Natl. Acad. Sci. U.S.A.,* **64:**134–141.
Goldberg, N. D.
  1975 Cyclic nucleotides and cell function. In *Cell Membranes: Biochemistry, Cell Biology and Pathology,* edited by G. Weissmann. New York: HP, pp. 185–201.
Goldman, D. E.
  1943 Potential, impedance, and rectification in membranes. *J. Gen. Physiol.,* **27:**37–60.
Goldschmidt, R.
  1907 Einiges vom feineren Bau des Nervensystems. *Verh. Dtsch. Zool. Ges.,* pp. 130–132.
  1908 Das Nervensystem von *Ascaris* lumbricoides und megalocephala: Ein Versuch, in den Aufbau eines einfachen Nervensystems einzudringen. Erster Teil. *Z. Wiss. Zool.,* **90:**73–136.
Gorman, A. L. F., and M. F. Marmor.
  1970 Contributions of the sodium pump and ionic gradients to the membrane potential of a molluscan neurone. *J. Physiol. (London),* **210:**897–917.
Gorman, A. L. F., and M. Mirolli.
  1972 The passive electrical properties of the membrane of a molluscan neurone. *J. Physiol. (London),* **227:**35–49.
Goronowitsch, N.
  1888 Das Gehirn und die Cranialnerven von *Acipenser ruthenus.* Ein Beitrag zur Morphologie des Wirbelthierkopfes. *Morphol. Jahrb.,* **13:**427–574.
Graham, J., and R. W. Gerard.
  1946 Membrane potentials and excitation of impaled single muscle fibers. *J. Cell. Comp. Physiol.,* **28:**99–117.
Grant, D. A.
  1943 Sensitization and association in eyelid conditioning. *J. Exp. Psychol.,* **32:**201–212.
Graubard, K.
  1973 *Morphological and Electrotonic Properties of Identified Neurons of the Mollusc* Aplysia californica. Dissertation, University of Washington.
Graziadei, P. P. C., and J. F. Metcalf.
  1971 Autoradiographic and ultrastructural observations of the frog's olfactory mucosa. *Z. Zellforsch. Mikrask. Anat.* **116:**305.
Greengard, P., and J. W. Kebabian.
  1974 Role of cyclic AMP in synaptic transmission in the mammalian peripheral nervous system. *Fed. Proc.,* **33:**1059–1067.
Grether, W. F.
  1938 Pseudo-conditioning without paired stimulation encountered in attempted backward conditioning. *J. Comp. Psychol.,* **25:**91–96.
Grillner, S.
  1975 Locomotion in vertebrates: central mechanisms and reflex interaction. *Physiol. Rev.* **55**(2):247–304.
Grinnell, A.
  1976 Specificity of neurons and their interconnections. In "Cellular Biology of Neurons" (Vol. 1, Sect. 1, *Handbook of Physiology, The Nervous System*), edited by E. R. Kandel. Baltimore: Williams and Wilkins.

Groves, P. M., and R. F. Thompson.
1970 Habituation: A dual-process theory. *Psychol. Rev.,* **77**:419–450.

Grundfest, H.
1957 Electrical inexcitability of synapses and some consequences in the central nervous system. *Physiol. Rev.,* **37**:337–361.

Guthrie, E. R.
1935 *The Psychology of Learning.* New York: Harper and Brothers.

Hagbarth, K. E., and R. L. Finer.
1963 The plasticity of human withdrawal reflexes to noxious skin stimuli in lower limbs. *Progress in Brain Research,* **1**:65–81.

Hagbarth, K. E., and E. Kugelberg.
1958 Plasticity of the human abdominal skin reflex. *Brain,* **81**:305–318.

Hagiwara, S.
1961 Nervous activities of the heart in *Crustacea. Ergeb. Biol.,* **24**:287–311.

Hagiwara, S., K. Kusano, and N. Saito.
1961 Membrane changes of *Onchidium* nerve cell in potassium-rich media. *J. Physiol. (London),* **155**:470–489.

Hagiwara, S., and K. Takahashi.
1967 Surface density of calcium ions and calcium spikes in the barnacle muscle fiber membrane. *J. Gen. Physiol.,* **50**:583–601.

Hagiwara, S., and A. Watanabe.
1956 Discharges in motoneurons of cicada. *J. Cell. Comp. Physiol.,* **47**:415–428.

Hale, E. B., and J. O. Almquist.
1960 Relation of sexual behavior to germ cell output in farm animals. *J. Dairy Sci.,* **43**(Suppl.):145–469.

Hamilton, G. V.
1914 A study of sexual tendencies in monkeys and baboons. *J. Anim. Behav.,* **4**:295–318.

Hardy, A.
1959 *The Open Sea: Its Natural History.* Part II: *Fish and Fisheries.* London: Collins.

Harlow, H. F.
1958 The evolution of learning. In *Behavior and Evolution,* edited by A. Roe and G. G. Simpson. New Haven: Yale University Press, pp. 269–290.

Harlow, H. F., and M. K. Harlow.
1962 Social deprivation in monkeys. *Sci. Am.,* **207**(5):136–146.

Harris, J. D.
1943 Habituatory response decrement in the intact organism. *Psychol. Bull.,* **40**:385–422.

Hartline, H. K.
1934 Intensity and duration in the excitation of single photoreceptor units. *J. Cell. Comp. Physiol.,* **5**:229–247.

Hartline, H. K., and C. H. Graham.
1932 Nerve impulses from single receptors in the eye. *J. Cell Comp. Physiol.,* **1**:277–295.

Hartline, H. K., and F. Ratliff.
1957 Inhibitory interaction of receptor units in the eye of *Limulus. J. Gen. Physiol.,* **40**:357–376.

Hebb, D. O.
1949 *The Organization of Behavior: A Neuropsychological Theory.* New York: Wiley.
1958 *Textbook of Psychology,* Philadelphia: Saunders.

Heinroth, O.
1911 Beiträge zur Biologie, namentlich Ethologie und Psychologie der Anatiden. *Verh. V. Int. Ornithol. Konger. (Berlin),* **1910**:589–702.

Herrnstein, R. J., and E. G. Boring, eds.
1965 *A Source Book in the History of Psychology,* Cambridge, Mass.: Harvard University Press.

Heuser, J. E., and T. S. Reese.

1974 Morphology of synaptic vesicle discharge and reformation at the frog neuromuscular junction. In *Synaptic Transmission and Neuronal Interaction* (Society of General Physiologists Series, vol. 28), edited by M. V. L. Bennett. New York: Raven Press, pp. 59–77.

Hilgard, E. R.

1956 *Theories of Learning,* 2nd ed. New York: Appleton-Century-Crofts.

Hilgard, E. R., and G. H. Bower.

1975 *Theories of Learning,* 4th ed. Englewood Cliffs, N.J.: Prentice-Hall.

Hilgard, E. R., and D. G. Marquis.

1940 *Conditioning and Learning.* New York: Appleton-Century.

Hille, B.

1967 The selective inhibition of delayed potassium currents in nerve by tetraethylammonium ion. *J. Gen. Physiol.,* **50:**1287–1302.

1976 Ionic basis of resting potentials and action potentials. In "Cellular Biology of Neurons" (Vol. 1, Sect. 1, *Handbook of Physiology, The Nervous System*), edited by E. R. Kandel. Baltimore: Williams and Wilkins.

Hinde, R. A.

1954a Factors governing the changes in strength of a partially inborn response, as shown by the mobbing behaviour of the chaffinch (*Fringilla coelebs*). I. The nature of the response, and an examination of its course. *Proc. R. Soc. London* (B), **142:**306–331.

1954b Factors governing the changes in strength of a partially inborn response, as shown by the mobbing behaviour of the chaffinch (*Fringilla coelebs*). II. The waning of the response. *Proc. Roy. Soc. London* (B), **142:**331–358.

1970 *Animal Behaviour. A Synthesis of Ethology and Comparative Psychology,* 2nd ed. New York: McGraw-Hill.

Hodgkin, A. L.

1964 *The Conduction of the Nervous Impulse.* Liverpool: Liverpool University Press.

Hodgkin, A. L., and A. F. Huxley.

1939 Action potentials recorded from inside a nerve fibre. *Nature (London),* **144:**710–711.

1945 Resting and action potentials in single nerve fibres. *J. Physiol. (London),* **104:**176–195.

1952a Currents carried by sodium and potassium ions through the membrane of the giant axon of *Loligo. J. Physiol. (London),* **116:**449–472.

1952b The components of membrane conductance in the giant axon of *Logilo. J. Physiol. (London),* **116:**473–496.

1952c The dual effect of membrane potential on sodium conductance in the giant axon of *Logilo. J. Physiol. (London),* **116:**497–506.

1952d A quantitative description of membrane current and its application to conduction and excitation in nerve. *J. Physiol. (London),* **117:**500–544.

Hodgkin, A. L., A. F. Huxley, and B. Katz.

1949 Ionic currents underlying activity in the giant axon of the squid. *Arch. Sci. Physiol.,* **3:**129–150.

1952 Measurement of current-voltage relations in the membrane of the giant axon of *Loligo. J. Physiol. (London),* **116:**424–448.

Hodgkin, A. L., and B. Katz.

1949 The effect of sodium ions on the electrical activity of the giant axon of the squid. *J. Physiol. (London),* **108:**37–77.

Hodgkin, A. L., and R. D. Keynes.

1955 Active transport of cations in giant axons from *Sepia* and *Loligo. J. Physiol. (London),* **128:**28–60.

Hodos, W., and C. B. G. Campbell.

1969 *Scala Naturae:* Why there is no theory in comparative psychology. *Psychol. Rev.,* **76:**337–350.

Holmgren, B., and S. Frenk.
1961 Inhibitory phenomena and 'habituation' at the neuronal level. *Nature (London)*, **192**:1294–1295.

Horn, G., and R. A. Hinde, eds.
1970 *Short-term Changes in Neural Activity and Behaviour.* Cambridge, Eng.: University Press.

Horridge, G. A.
1966 Optokinetic memory in the crab, *Carcinus. J. Exp. Biol.*, **44**:233–245.

Hotta, Y., and S. Benzer.
1972 Mapping of behaviour in *Drosophila* mosaics. *Nature (London)*, **240**:527–535.

Hoy, R. R., and D. M. Wilson.
1969 Rhythmic motor output in the leg motor neurons of the milkweed bug *Oncopeltus. Fed. Proc.* **28**:1837.

Hoyle, G.
1964 Exploration of neuronal mechanisms underlying behavior in insects. In *Neural Theory and Modeling*, edited by R. F. Reiss. Stanford, Calif.: Stanford University Press, pp. 346–376.
1973 Neural machinery underlying behavior in insects. In *The Neurosciences: Third Study Program*, edited by F. O. Schmitt and F. G. Worden. Cambridge, Mass.: M.I.T. Press,. pp. 397–410.

Hubel, D. H., and T. N. Wiesel.
1962 Receptive fields, binocular interaction and functional architecture in the cat's visual cortex. *J. Physiol. (London)*, **160**:106–154.
1963 Receptive fields of cells in striate cortex of very young, visually inexperienced kittens. *J. Neurophysiol.*, **26**:994–1002.
1965 Binocular interaction in striate cortex of kittens reared with artificial squint. *J. Neurophysiol.*, **28**:1041–1059.
1970 The period of susceptibility to the physiological effects of unilateral eye closure in kittens. *J. Physiol. (London)*, **206**:419–436.

Huber, F.
1962 Central nervous control of sound productions in crickets and some speculations on its evolution. *Evolution*, **16**:429–442.

Hughes, G. M., and L. Tauc.
1961 The path of the giant cell axons in *Aplysia depilans. Nature (London)*, **191**:404–405.
1963 An electrophysiological study of the anatomical relations of two giant nerve cells in *Aplysia depilans. J. Exp. Biol.* **40**:469–486.

Hughes, G. M., and C. A. G. Wiersma.
1960 The co-ordination of swimmeret movements in the crayfish. *Procambarus clarkii* (Girard). *J. Exp. Biol.*, **37**:657–670.

Hughes, J. R., E. V. Evarts, and W. H. Marshall.
1956 Post-tetanic potentiation in the visual system of cats. *Am. J. Physiol.*, **186**:483–487.

Hull, C. L.
1943 *Principles of Behavior. An Introduction to Behavior Theory.* New York: Appleton-Century-Crofts.

Hultborn, H., E. Jankowska, and S. Lindström.
1968 Inhibition in Ia inhibitory pathway by impulses in recurrent motor axon collaterals. *Life Sci.*, **7**:337–339.

Humphrey, G.
1930a Le Chatelier's rule, and the problem of habituation and dehabituation in *Helix albolabris. Psychol. Forsch.*, **13**:113–127.
1930b Extinction and negative adaptation. *Psychol. Rev.*, **37**:361–363.
1933 *The Nature of Learning in Its Relation to the Living System.* New York: Harcourt, Brace.

Hurlbut, W. P., and B. Ceccarelli.
1974 Transmitter release and recycling of synaptic vesicle membrane at the neuromuscular junction. *Advances in Cytopharmacology,* **2:**141–154.

Huxley, T. H.
1880 *The Crayfish. An Introduction to the Study of Zoology.* New York: Appleton.

Hydén, H.
1967 Biochemical changes accompanying learning. In *The Neurosciences: A Study Program,* edited by G. C. Quarton, T. Melnechuk, and F. O. Schmitt. New York: Rockefeller University Press, pp. 765–771.

Hydén, H., and P. W. Lange.
1970 Do specific biochemical correlates to learning processes exist in brain cells? In "Biochemistry of Simple Neuronal Models," (*Advances in Biochemical Psychopharmacology* 2), edited by E. Costa and E. Giacobini. New York: Raven Press, pp. 317–338.

Ihering, H. von.
1877 *Vergleichende Anatomie des Nervensystemes und Phylogenie der Mollusken.* Leipzig: Engelmann.

Ikeda, J., and C. A. G. Wiersma.
1964 Autogenic rhythmicity in the abdominal ganglia of the crayfish: The control of swimmeret movements. *Comp. Biochem. Physiol.,* **12:**107–115.

Immelmann, K.
1972 Sexual and other long-term aspects of imprinting in birds and other species. *Adv. Study Behav.,* **4:**147–174.

Jacklet, J. W., B. Peretz, and K. Lukowiak.
1975 Habituation of the gill reflex in *Aplysia:* sites in the peripheral neural system. *Fed. Proc.,* **34:**359 No. 787.

Jahan-Parwar, B.
1970 Conditioned response in *Aplysia californica. Am. Zool.,* **10:**287 (abstract).

Jahan-Pawar, B. and R. von Baumgarten
1967 Untersuchungen zur Spezifitaetsfrage der heterosynaptischen Facilitation bei *Aplysia californica. Pflügers Archiv gesamte Physiol. Menschen Tiere.,* **295:**347–360.

James, W.
1890 *The Principles of Psychology,* 2 vols. New York: Holt.

Jansen, J. K. S., K. J. Muller, and J. G. Nicholls.
1974 Persistent modification of synaptic interactions between sensory and motor nerve cells following discrete lesions in the central nervous system of the leech. *J. Physiol. (London),* **242:**289–305.

Jansen, J. K. S., and J. G. Nicholls.
1973 Conductance changes, an electrogenic pump and hyperpolarization of leech neurones following impulses. *J. Physiol. (London),* **229:**635–655.

Jeffrey, W. E.
1968 The orienting reflex and attention in cognitive development. *Psychol. Rev.,* **75:**323–334.

Jennings, H. S.
1906 *Behavior of the Lower Organisms.* New York: Columbia University Press.

John, E. R.
1967 *Mechanisms of Memory.* New York: Academic Press.
1972 Switchboard versus statistical theories of learning and memory. *Science (Wash., D.C.),* **177:**850–864.

Johnson, G. E.
1924 Giant nerve fibers in crustaceans, with special reference to *Cambarus* and *Palaemonetes. J. Comp. Neurol.,* **36:**323–373.
1926 Studies on the functions of the giant nerve fibers of crustaceans with special reference to *Cambarus* and *Palaemonetes. J. Comp. Neurol.* **42:**19–33.

Jordan, H.
1917 Das Wahrnehmen der Nahrung bei *Aplysia limacina* und *Aplysia depilans. Biol. Zentralbl.,* **37:**2–9.

Junge, D., and J. Miller.
1974 Different spike mechanisms in axon and soma of molluscan neurons. *Nature (London),* **252:**155–156.

Junge, D., and C. L. Stephens.
1973 Cyclical variation of potassium conductance in a burst-generating neurone in *Aplysia. J. Physiol. (London),* **235:**155–181.

Kado, R. T.
1973 *Aplysia* giant cell: soma-axon voltage clamp current differences. *Science (Wash., D.C.),* **182:**843–845.

Kandel, E. R.
1967 Cellular studies of learning. In *The Neurosciences: A Study Program,* edited by G. C. Quarton, T. Melnechuk, and F. O. Schmitt. New York: Rockefeller University Press, pp. 666–689.

1968 Dale's principle and the functional specificity of neurons. In *Psychopharmacology: A Review of Progress 1957–1967* (Public Health Service Publ. No. 1836), edited by D. F. Efron. Wash., D.C.: Government Printing Office, pp. 385–398.

1970 Nerve cells and behavior. *Sci. Am.,* **223**(1):57–67, 70.

FORTHCOMING *The Behavioral Biology of Aplysia.* San Francisco: Freeman.

Kandel, E. R., T. J. Carew, and J. Koester.
1976 Some rules relating the biophysical properties of neurons and their patterns of interconnections to behavior. In *Electrobiology of Nerve, Synapse and Muscle,* edited by J. B. Reuben, D. P. Purpura, M. V. L. Bennett, and E. R. Kandel. New York: Raven Press.

Kandel, E. R., M. Brunelli, J. Byrne, and V. Castellucci.
1976 A common presynaptic locus for the synaptic changes underlying short-term habituation and sensitization of the gill-withdrawal reflex in *Aplysia. Cold Spring Harbor Laboratory Symposium on Quantitative Biology LX: The Synapse,* pp. 465–482.

Kandel, E. R., W. T. Frazier, and H. Wachtel.
1969 Organization of inhibition in abdominal ganglion of *Aplysia.* I: Role of inhibition and disinhibition in transforming neural activity. *J. Neurophysiol.,* **32:**496–508.

Kandel, E. R., W. T. Frazier, R. Waziri, and R. E. Coggeshall.
1967 Direct and common connections among identified neurons in *Aplysia. J. Neurophysiol.,* **30:**1352–1376.

Kandel, E. R., and D. Gardner.
1972 The synaptic actions mediated by the different branches of a single neuron. *Neurotransmitters Res. Publ. A. R. N. M. D.,* **50:**91–144.

Kandel, E. R., and I. Kupfermann.
1970 The functional organization of invertebrate ganglia. *Annu. Rev. Physiol.,* **32:**193–258.

Kandel, E. R., and W. A. Spencer.
1968 Cellular neurophysiological approaches in the study of learning. *Physiol. Rev.,* **48:**65–134.

Kandel, E. R., and L. Tauc.
1963 Augmentation prolongée de l'efficacité d'une voie afférente d'un ganglion isolé après l'activation couplée d'une voie plus efficace. *J. Physiol. (Paris),* **55:**271–272.

1964 Mechanism of prolonged heterosynaptic facilitation. *Nature (London),* **202:**145–147.

1965a Heterosynaptic facilitation in neurones of the abdominal ganglion of *Aplysia depilans. J. Physiol. (London),* **181:**1–27.

1965b Mechanism of heterosynaptic facilitation in the giant cell of the abdominal ganglion of *Aplysia depilans. J. Physiol. (London),* **181:**28–47.

1966a Input organization of two symmetrical giant cells in the snail brain. *J. Physiol. (London)*, **183**:269–286.

1966b Anomalous rectification in the metacerebral giant cells and its consequences for synaptic transmission. *J. Physiol. (London)*, **183**:287–304.

Kandel, E. R., and H. Wachtel.

1968 The functional organization of neural aggregates in *Aplysia*. In *Physiological and Biochemical Aspects of Nervous Integration,* edited by F. D. Carlson. Englewood Cliffs, N.J.: Prentice-Hall, pp. 17–65.

Kao, C. Y.

1960 Postsynaptic electrogenesis in septate giant axons. II: Comparison of medial and lateral giant axons of crayfish. *J. Neurophysiol.,* **23**:618–635.

Karlin, A.

1974 The acetylcholine receptor: Progress report. *Life Sciences,* **14**:1385–1415.

Katayama, Y.

1970 Identification of neuronal connections in the central nervous system of *Onchidium verruculatum. Jap. J. Physiol.,* **20**:711–724.

1973 Electrophysiological identification of neurones and neural networks in the perioesophageal ganglion complex of the marine pulmonate mollusc, *Onchidium verruculatum. J. Exp. Biol.,* **59**:739–751.

Kater, S. B.

1974 Feeding in *Helisoma trivolvis:* The morphological and physiological bases of a fixed action pattern. *Am Zool.,* **14**:1017–1036.

Kater, S. B., and C. H. Fraser Rowell.

1973 Integration of sensory and centrally programmed components in generation of cyclical feeding activity of *Helisoma trivolvis. J. Neurophysiol.,* **36**:142–155.

Kater, S. B., and C. Nicholson, eds.

1973 *Intracellular Staining in Neurobiology.* New York: Springer-Verlag.

Katz, B.

1966 *Nerve, Muscle and Synapse.* New York: McGraw-Hill.

1969 *The Release of Neural Transmitter Substances.* Springfield, Ill: Charles C. Thomas.

Katz, B., and R. Miledi.

1963 A study of spontaneous miniature potentials in spinal motoneurones. *J. Physiol. (London),* **168**:389–422.

1965a The effect of calcium on acetylcholine release from motor nerve terminals. *Proc. R. Soc. London* (B), **161**:496–503.

1965b The effects of temperature on the synaptic delay at the neuromuscular junction. *J. Physiol.,* **181**:656–670.

1967a The timing of calcium action during neuromuscular transmission. *J. Physiol. (London),* **189**:535–544.

1967b The release of acetylcholine from nerve endings by graded electric pulses. *Proc. R. Soc. London* (B), **167**:23–38.

1967c A study of synaptic transmission in the absence of nerve impulses. *J. Physiol. (London),* **192**:407–436.

1968 The role of calcium in neuromuscular faciliation. *J. Physiol. (London),* **195**:481–491.

1969a Tetrodotoxin-resistant electric activity in presynaptic terminals. *J. Physiol. (London),* **203**:459–487.

1969b The effects of divalent cations on transmission in the squid giant synapse. *Pubbl. Stn. Zool. Napoli,* **37**:303–310.

1970 Further study of the role of calcium in synaptic transmission. *J. Physiol. (London),* **207**:789–801.

1971 Further observations on acetylcholine noise. *Nature New Biol.,* **232**:124–126.

1972 The statistical nature of the acetylcholine potential and its molecular components. *J. Physiol. (London)*, **224:**665–699.

Kaufman, I. C., and L. A. Rosenblum.
1967 Depression in infant monkeys separated from their mothers. *Science (Wash., D.C.),* **155:**1030–1031.

Kehoe, J. S.
1967 Pharmacological characteristics and ionic bases of a two component postsynaptic inhibition. *Nature (London),* **215:**1503–1505.
1969a Single presynaptic neurone mediates a two component postsynaptic inhibition. *Nature (London),* **221:**866–868.
1969b Suppression sélective par l'ion tétraéthylammonium d'une inhibition cholinergique résistant au curare. *C. R. Acad. Sci. (Paris)* (D), **268:**111–114.
1972a Ionic mechanisms of a two-component cholinergic inhibition in *Aplysia* neurones. *J. Physiol. (London),* **225:**85–114.
1972b Three acetylcholine receptors in *Aplysia* neurones. *J. Physiol. (London),* **225:**115–146.
1972c The physiological role of three acetylcholine receptors in synaptic transmission in *Aplysia. J. Physiol. (London),* **225:**147–172.
1975 Analysis of a 'resting' potassium permeability that can be synaptically reduced. *J. Physiol. (London),* **244:**23P–24P.

Kehoe, J. S., and P. Ascher.
1970 Re-evaluation of the synaptic activation of an electrogenic sodium pump. *Nature (London),* **225:**820–823.

Kehoe, J. S., R. Sealock, and C. Bon.
1976 Effects of α-toxins from *Bungarus multicinctus* and *Bungarus coeruleus* on cholinergic responses in *Aplysia* neurone. *Brain Research.*

Kendler, H. H.
1974 *Basic Psychology,* 3rd ed. Menlo Park, Ca.: W. A. Benjamin.

Kennedy, D.
1971 Fifteenth Bowditch Lecture: Crayfish Interneurons. *Physiologist,* **14:**5–30.

Kennedy, D., R. L. Calabrese, and J. J. Wine.
1974 Presynaptic inhibition: Primary afferent depolarization in crayfish neurons. *Science (Wash., D.C.),* **185:**451–454.

Kennedy, D., and W. J. Davis.
1976 The organization of invertebrate motor systems. In "Cellular Biology of Neurons" (Vol. 1, Sec. 1, *Handbook of Physiology, The Nervous System*). Edited by E. R. Kandel. Baltimore: Williams and Wilkins.

Kennedy, D., W. H. Evoy, B. Dane, and J. T. Hanawalt.
1967 The central nervous organization underlying control of antagonistic muscles in the crayfish. II: Coding of position by command fibers. *J. Exp. Zool.,* **165:**239–248.

Kennedy, D., W. H. Evoy, and J. T. Hanawalt.
1966 Release of coordinated behavior in crayfish by single central neurons. *Science (Wash., D.C.),* **154:**917–919.

Kennedy, D., A. I. Selverston, and M. P. Remler.
1969 Analysis of restricted neural networks. *Science (Wash., D.C.),* **164:**1488–1496.

Kennedy, D., and K. Takeda.
1965 Reflex control of abdominal flexor muscles in the crayfish. II: The tonic system. *J. Exp. Biol.,* **43:**229–246.

Kerkut, G. A., M. C. French, and R. J. Walker.
1970 The location of axonal pathways of identifiable neurones of *Helix aspersa* using the dye Procion yellow M-4R. *Comp. Biochem. Physiol.,* **32:**681–690.

Kerkut, G. A., and R. W. Meech.

1966 The internal chloride concentration of H and D cells in the snail brain. *Comp. Biochem. Physiol.,* **19:**819–832.

Kerkut, G. A., and R. C. Thomas.

1964 The effect of anion injection and changes in the external potassium and chloride concentration on the reversal potentials of the IPSP and acetylcholine. *Comp. Biochem. Physiol.,* **11:**199–213.

Kety, S. S.

1974 From rationalization to reason. *Am. J. Psychiatry,* **131:**957–963.

Keynes, R. D., and E. Rojas.

1973 Characteristics of the sodium gating current in the squid giant axon. *J. Physiol. (London),* **233:**28P–30P.

1974 Kinetics and steady-state properties of the charged system controlling sodium conductance in the squid giant axon. *J. Physiol. (London),* **239:**393–434.

Kimble, G. A.

1961 *Hilgard and Marquis' Conditioning and Learning,* rev. ed. New York: Appleton-Century-Crofts.

Kimmel, H. D.

1964 Theoretical note: A further analysis of GSR conditioning: A reply to Stewart, Stern, Winokur and Fredman. *Psychol. Rev.,* **71:**160–166.

1973 Habituation, Habituability, and Conditioning. In *Habituation,* Vol. 1 Behavioral Studies, edited by H. V. S. Peeke and M. J. Herz. New York: Academic Press, pp. 219–239.

Kimmel, H. D., and A. J. Goldstein.

1967 Retention of habituation of the GSR to visual and auditory stimulation. *J. Exp. Psychol.,* **73:**401–404.

Klett, R. P., B. W. Fulpius, D. Cooper, and E. Reich.

1974 The nicotinic acetylcholine receptor: Characterization and properties of a macromolecule isolated from *Electrophorus electricus.* In *Synaptic Transmission and Neuronal Interaction* (Society of General Physiologists Series, vol. 28), edited by M. V. L. Bennett. New York: Raven Press, pp. 179–190.

Kling, J. W.

1971 Learning: Introductory survey. In *Woodworth and Schlosberg's Experimental Psychology,* 3rd ed., edited by J. W. Kling and L. A. Riggs. New York: Holt, Rinehart and Winston, pp. 551–613.

Koester, J., N. Dieringer, and E. R. Kandel.

UNPUBL. Further studies of the conversion from excitation to inhibition in a dual synapse.

Koester, J., and E. R. Kandel.

UNPUBL. Further identification of neurons in the abdominal ganglion of *Aplysia* using behavioral criteria.

Koester, J., E. Mayeri, G. Liebeswar, and E. R. Kandel.

1974 Neural control of circulation in *Aplysia.* II: Interneurons. *J. Neurophysiol.,* **37:**476–496.

Koike, H., E. R. Kandel, and J. H. Schwartz.

1974 Synaptic release of radioactivity after intrasomatic injection of choline-^{3}H into an identified cholinergic interneuron in abdominal ganglion of *Aplysia californica. J. Neurophysiol.,* **37:**815–827.

Koketsu, K.

1969 Cholinergic synaptic potentials and the underlying ionic mechanisms. *Fed. Proc.,* **28:**101–112.

Konorski, J.

1948 *Conditioned Reflexes and Neuron Organization.* Cambridge, Eng.: Cambridge University Press. (Reprinted 1968.)

1967 *Integrative Activity of the Brain: An Interdisciplinary Approach.* Chicago: University of Chicago Press.

Kopp, R., Z. Bohdanecky, and M. E. Jarvik.
1966 Long temporal gradient of retrograde amnesia for a well-discriminated stimulus. *Science (Wash., D.C.),* **153:**1547–1549.

Krasne, F. B.
1969 Excitation and habituation of the crayfish escape reflex: The depolarizing response in lateral giant fibres of the isolated abdomen. *J. Exp. Biol.,* **50:**29–46.
1973 Learning in Crustacea. In *Invertebrate Learning,* Vol. 2, edited by W. C. Corning, J. A. Dyal, and A. O. D. Willows. New York: Plenum, pp. 49–130.
1976 Recent developments in the neurobiology of memory. In *Society for Neuroscience 5th Annual meeting. B. I. S., Conference Report #43,* pp. 27–30.

Krasne, F. B. and J. S. Bryant.
1973 Habituation: Regulation through presynaptic inhibition. *Science (Wash., D.C.),* **182:**590–592.

Krasne, F. B., and K. S. Woodsmall.
1969 Waning of the crayfish escape response as a result of repeated stimulation. *Anim. Behav.,* **17:**416–424.

Krech, D.
1950 Dynamic systems, psychological fields, and hypothetical constructs. *Psychol. Rev.,* **57:**283–290.

Kriegstein, A. R., V. Castellucci, and E. R. Kandel.
1974 Metamorphosis of *Aplysia californica* in laboratory culture. *Proc. Natl. Acad. Sci. U.S.A.,* **71:**3654–3658.

Kristan, W. B., Jr., and G. S. Stent.
1976 Peripheral feedback in the leech swimming rhythm. In *Cold Spring Harbor Symposium on Quantitative Biology,* Vol. XL.

Kristan, W. B., Jr., G. S. Stent, and C. A. Ort.
1974a Neuronal control of swimming in the medicinal leech. I. Dynamics of the swimming rhythm. *J. Comp. Physiol.* **94:**97–119.
1974b Neuronal control of swimming in the medicinal leech. III. Impulse patterns of the motor neurons. *J. Comp. Physiol.* **94:**155–176.

Kuba, K.
1970 Effects of catecholamines on the neuromuscular junction in the rat diaphragm. *J. Physiol. (London),* **211:**551–570.

Kuffler, S. W.
1948 Physiology of neuro-muscular junctions: electrical aspects. *Fed. Proc.,* **7:**437–446.

Kuffler, S. W., M. J. Dennis, and A. J. Harris.
1971 The development of chemosensitivity in extrasynaptic areas of the neuronal surface after denervation of para-sympathetic ganglion cells in the heart of the frog. *Proc. R. Soc.* (B) **177:**555–563.

Kuffler, S. W., and J. G. Nicholls.
1966 The physiology of neuroglial cells. *Ergeb, Physiol. Biol. Chem. Exp. Pharmakol.,* **57:**1–90.

Kuffler, S. W., and D. D. Potter.
1964 Glia in the leech central nervous system: physiological properties and neuron-glia relationship. *J. Neurophysiol.,* **27:**290–320.

Kükenthal, W.
1887 Uber das Nervensystem der Opheliaceen. *Jena Z. Naturwiss.,* **20:**511–580.

Kuno, M.
1964a Quantal components of excitatory synaptic potentials in spinal motoneurones. *J. Physiol. (London),* **175:**81–99.

1964b Mechanism of facilitation and depression of the excitatory synaptic potential in spinal motoneurones. *J. Physiol. (London),* **175:**100–112.

Kuno, M., and J. T. Miyahara.

1969 Analysis of synaptic efficacy in spinal motoneurones from "quantum" aspects. *J. Physiol. (London),* **201:**479–493.

Kunze, D. L., and A. M. Brown.

1971 Internal potassium and chloride activities and the effects of acetylcholine on identifiable *Aplysia* neurones. *Nature New Biol.,* **229:**229–231.

Kupfermann, I.

1967 Stimulation of egg-laying: Possible neuroendocrine function of bag cells of abdominal ganglion of *Aplysia californica. Nature (London),* **216:**814–815.

1968 A circadian locomotor rhythm in *Aplysia californica. Physiol. Behav.,* **3:**179–181.

1970 Stimulation of egg-laying by extracts of neuroendocrine cells (bag cells) of abdominal ganglion of *Aplysia. J. Neurophysiol.,* **33:**877–881.

1974a Feeding behavior in Aplysia: A simple system for the study of motivation. *Behav. Biol.,* **10:**1–26.

1974b Dissociation of the appetitive and consummatory phases of feeding behavior in *Aplysia:* a lesion study. *Behav. Biol.,* **10:**89–97.

Kupfermann, I., and T. J. Carew.

1974 Behavior patterns of *Aplysia californica* in its natural environment. *Behav. Biol.,* **12:**317–337.

Kupfermann, I., T. J. Carew, and E. R. Kandel.

1974 Local, reflex, and central commands controlling gill and siphon movements in *Aplysia. J. Neurophysiol.,* **37:**996–1019.

Kupfermann, I., V. Castellucci, H. Pinsker, and E. R. Kandel.

1970 Neuronal correlates of habituation and dishabituation of the gill-withdrawal reflex in *Aplysia. Science (Wash., D.C.),* **167:**1743–1745.

Kupfermann, I., and J. Cohen.

1971 The control of feeding by identified neurons in the buccal ganglion of *Aplysia. Am. Zool.,* **11:**667, abstract 243.

Kupfermann, I., and E. R. Kandel.

1968 Reflex function of some identified cells in *Aplysia. Fed. Proc.,* **27:**277, abstract 348.

1969 Neuronal controls of a behavioral response mediated by the abdominal ganglion of *Aplysia. Science (Wash., D.C.),* **164:**847–850.

1970 Electrophysiological properties and functional interconnections of two symmetrical neurosecretory clusters (bag cells) in abdominal ganglion of *Aplysia. J. Neurophysiol.* **33:**865–876.

Kupfermann, I., and H. Pinsker.

1968 A behavioral modification of the feeding reflex in *Aplysia californica. Commun. Behav. Biol.* (A), **2:**13–17.

1969 Plasticity in *Aplysia* neurons and some simple neuronal models of learning. In *Reinforcement and Behavior,* edited by J. Tapp. New York: Academic Press, pp. 356–386.

Kupfermann, I., H. Pinsker, V. Castellucci, and E. R. Kandel.

1971 Central and peripheral control of gill movements in *Aplysia. Science (Wash., D.C.),* **174:**1252–1256.

Kupfermann, I., and K. R. Weiss.

1974 Functional studies on the metacerebral cells in *Aplysia. Abstr. and Proc. Soc. of Neurosci.,* **3:**375.

1976 Water regulation by a presumptive hormone contained in identified neurosecretory cell R15 of *Aplysia. J. Gen. Physiol.,* **67:**113–123.

Kusano, K., D. R. Livengood, and R. Werman.

1967 Correlation of transmitter release with membrane properties of the presynaptic fiber of the squid giant synapse. *J. Gen. Physiol.,* **50:**2579–2601.

Langley, J. N.
  1906 Croonian Lecture 'On nerve endings and on special excitable substances in cells.'
      Proc. R. Soc. London (B) **78**:170–94.
Larrabee, M. G., and D. W. Bronk.
  1947 Prolonged facilitation of synaptic excitation in sympathetic ganglia. *J. Neurophysiol.*,
      10:139–154.
Lashley, K. S.
  1929 *Brain Mechanisms and Intelligence: A Quantitative Study of Injuries to the Brain.*
      Chicago: Chicago University Press.
  1938 Experimental analysis of instinctive behavior. *Psychol. Rev.,* **45**:445–471.
Lehninger, A. L.
  1970 Mitochondria and calcium ion transport. *Biochem. J.,* **119**:129–138.
Lehrman, D. S.
  1953 A critique of Konrad Lorenz's theory of instinctive behavior. *Q. Rev. Biol.,* **28**:337–363.
Lettvin, J. Y., H. R. Maturana, W. S. McCulloch, and W. H. Pitts.
  1959 What the frog's eye tells the frog's brain. *Proc. I. R. E.,* **47**:1940–51.
Levitan, H., and L. Tauc.
  1972 Acetylcholine receptors: Topographic distribution and pharmacological properties
      of two receptor types on a single molluscan neurone. *J. Physiol. (London),* **222**:537–558.
Levitan, H., L. Tauc, and J. P. Segundo.
  1970 Electrical transmission among neurons in the buccal ganglion of a mollusc, *Navanax
      inermis. J. Gen. Physiol.,* **55**:484–496.
Levitan, I. B., and S. H. Barondes.
  1974 Octopamine and serotonin-stimulated phosphorylation of specific protein in the abdom-
      inal ganglion of *Aplysia californica. Proc. Nat. Acad. Sci. U.S.A.* **71**:1145–1148.
Leydig, F.
  1864 *Vom Bau des thierischen Körpers. Handbuch der vergleichenden Anatomie,* Vol. 1.
      Tübingen.
  1886 Die riesigen Nervenröhren im Bauchmark der Ringelwürmer. *Zool. Anz.,* **9**:591–597.
Libet, B.
  1970 Generation of slow inhibitory and excitatory postsynaptic potentials. *Fed. Proc.,*
      29:1945–1956.
Lickey, M. E.
  1968 Learned behavior in *Aplysia vaccaria. J. Comp. Physiol. Psychol.,* **66**:712–718.
Liebeswar, G., J. E. Goldman, J. Koester, and E. Mayeri.
  1975 Neural control of circulation in *Aplysia.* III. Neurotransmitters. *J. Neurophysiol.,*
      38:767–779.
Liebeswar, G., J. Koester, and J. E. Goldman.
  1973 Identified motor neurons controlling the circulation in *Aplysia:* Synthesis of transmitters
      and synaptic pharmacology. *Program Abstr. Soc. Neurosci., 3rd Annu. Meeting.,* p. 288.
Liley, A. W.
  1956a An investigation of spontaneous activity at the neuromuscular junction of the rat.
      J. Physiol. (London), **132**:650–666.
  1956b The quantal components of the mammalian end-plate potential. *J. Physiol. (London),*
      133:571–587.
Liley, A. W., and K. A. K. North.
  1953 An electrical investigation of effects of repetitive stimulation on mammalian neuro-
      muscular junction. *J. Neurophysiol.* **16**:509–527.
Llinás, R., and C. Nicholson.
  1975 Calcium role in depolarization-secretion coupling: An aequorin study in squid giant
      synapse. *Proc. Nat. Acad. Sci. U.S.A.* **72**:187–190.

Lloyd, D. P. C.

1941 A direct central inhibitory action of dromically conducted impulses. *J. Neurophysiol.*, **4**:184–190.

1949 Post-tetanic potentiation of response in monosynaptic reflex pathways of the spinal cord. *J. Gen. Physiol.*, **33**:147–170.

Lloyd, D. P. C., and V. J. Wilson.

1957 Reflex depression in rhythmically active monosynaptic reflex pathways. *J. Gen. Physiol.*, **40**:409–426.

Loeb, J.

1918 *Forced Movements, Tropisms, and Animal Conduct.* Philadelphia: Lippincott.

Loewi, O.

1921 Ueber humorale Uebertragbarkeit der Herznervenwirkung. I. Mitteilung. *Pflügers Archiv gesamte Physiol. Menschen Tiere* **189**:239–242. (English translation "On the humoral propagation of cardiac nerve action," in *Cellular Neurophysics*, edited by I. Cooke and M. Lipkin, Jr. New York: Holt, Rinehart and Winston, 1972.)

Loewi, O., and E. Navratil.

1926 Ueber humorale Uebertragbarkeit der Herznervenwirkung. X. Mitteilung: Ueber das Schicksal des Vagusstoffs. *Pflügers Archiv gesamte Physiol. Menschen Tiere* **214**:678–688. (English translation "On the humoral propagation of cardiac nerve action," in *Cellular Neurophysics*, edited by I. Cooke and M. Lipkin, Jr. New York: Holt, Rinehart and Winston, 1972.)

Loh, Y. P., Y. Sarne, and H. Gainer.

1975 Heterogeniety of proteins synthesized, stored, and released by the bag cells of *Aplysia californica. J. Comp. Physiol.*, **100**:283–285.

Lopresti, V., E. R. Macagno, and C. Levinthal.

1973 Structure and development of neuronal connections in isogenic organisms: Cellular interactions in the development of the optic lamina of *Daphina. Proc. Natl. Acad. Sci. U.S.A.*, **70**:433–437.

Lorente de Nó, R.

1934 Studies on the structure of the cerebral cortex. II: Continuation of the study of the ammonic system. *J. Psychol. Neurol.*, **46**:113–177.

1938 Analysis of the activity of the chains of internuncial neurons. *J. Neurophysiol.*, **1**:207–244.

Lorenz, K.

1935 Der Kumpan in der Umwelt des Vogels. *J. Ornithol.*, **83**:137–213; 289–413. (English translation by R. Martin, "Companions as factors in the bird's environment." In *Studies in Animal and Human Behaviour*, Vol. 1, by K. Lorenz. London: Methuen, 1970.)

1937 Ueber die Bildung des Instinktbegriffes. *Naturwissenschaften,* **25**:289–300, 307–318, 324–331. (English translation by R. Martin, "The establishment of the instinct concept." In *Studies in Animal and Human Behaviour*, Vol. 1, by K. Lorenz. London: Methuen, 1970.)

1950 The comparative method in studying innate behaviour patterns. *Symp. Soc. Exp. Biol.*, **4**:221–268.

1965 *Evolution and Modification of Behavior.* Chicago: University of Chicago Press.

Lorenz, K., and N. Tinbergen.

1938 Taxis und Instinkthandlung in der Eirollbewegung der Graugans. *Z. Tierpsychol.*, **2**(1):1–29. (English translation by R. Martin, "Taxis and instinctive behavior pattern in egg-rolling in the Greylag goose." In *Studies in Animal and Human Behaviour*, Vol. 1, edited by K. Lorenz. London: Methuen, 1970.)

Lubbock, J.

1882 *Ants, Bees, and Wasps. A Record of Observations on the Habits of the Social Hymenoptera.* New York: D. Appleton.

1888 *On the Senses, Instincts, and Intelligence of Animals with Special Reference to Insects.* New York: D. Appleton.

Lugaro, E.

1899 I recenti progressi dell' anatomia del sistema nervoso in rapporto alla psicologia ed alla psichiatria. *Riv. Patol. nerv. ment., t.* IV, fasc. 11–12. (Cited in Cajal, 1911).

Lukowiak, K., and J. W. Jacklet.

1972 Habituation and dishabituation: Interactions between peripheral and central nervous systems in *Aplysia. Science (Wash., D.C.),* **178:**1306–1308.

MacGinitie, G. E.

1934 The egg-laying activities of the sea hare, *Tethys californicus* (Cooper). *Biol. Bull. (Woods Hole),* **67:**300–303.

MacPherson, S. J., V. Dees, and G. C. Grindley.

1949 The effect of knowledge of results on learning and performance. III: The influence of the time interval between trials. *Q. J. Exp. Psychol.,* **1:**167–174.

McCaman, R. E., and S. A. Dewhurst.

1970 Choline acetyltransferase in individual neurons of *Aplysia californica. J. Neurochem.,* **17:**1421–1426.

McCleary, R. A.

1961 Response specificity in the behavioral effects of limbic system lesions of the cat. *J. Comp. Physiol. Psychol.,* **54:**605–613.

McDougall, W.

1901 Experimentelle Beiträge zur Lehre vom Gedächtniss. *Mind,* **10:**388–394.

McGaugh, J. L.

1966 Time-dependent processes in memory storage. *Science (Wash., D.C.),* **153:**1351–1358.

McGaugh, J. L., and M. J. Herz.

1972 *Memory Consolidation.* San Francisco: Albion.

McGeoch, J. A.

1952 *The Psychology of Human Learning,* 2nd ed., revised by A. L. Irion. New York: Longmans, Green.

Macagno, E. R., V. Lopresti, and C. Levinthal.

1973 Structure and development of neuronal connections in isogenic organisms: Variations and similarities in the optic system of *Daphnia magna. Proc. Natl. Acad. Sci. U.S.A.,* **70:**47–61.

Maier, N. R. F.

1932 A study of orientation in the rat. *J. Comp. Psychol.,* **14:**387–399.

Manning, A.

1972 *Introduction to Animal Behavior,* 2nd ed. Reading, Mass.: Addison-Wesley.

Marler, P., and W. J. Hamilton, III.

1966 *Mechanisms of Animal Behavior.* New York: Wiley.

Marmor, M. F., and A. L. F. Gorman.

1970 Membrane potential as the sum of ionic and metabolic components. *Science (Wash., D.C.),* **167:**65–67.

Martin, A. R.

1966 Quantal nature of synaptic transmission. *Physiol. Rev.,* **46:**51–66.

1976 Junctional transmission. II: Presynaptic mechanisms. In "Cellular Biology of Neurons" (Vol. 1, Sect. 1, *Handbook of Physiology, The Nervous System*), edited by E. R. Kandel. Baltimore: Williams and Wilkins.

Martin, A. R. and G. Pilar.

1963 Dual mode of synaptic transmission in the avian ciliary ganglion. *J. Physiol. (London),* **168:**443–463.

1964 Pre-synaptic and post-synaptic events during post-tetanic potentiation and facilitation in the avian ciliary ganglion. *J. Physiol (London),* **175:**17–30.

Martin, A. R., and W. O. Wickelgren.

1971 Sensory cells in the spinal cord of the sea lamprey. *J. Physiol. (London)*, **212**:65–83.

Martini, E.

1908 Die Konstanz histologischer Elemente bei Nematoden nach Abschluss der Entwickelungsperiode. *Verh. Anat. Ges.*, 1908:132–134.

Mathieu, P. A., and F. A. Roberge.

1971 Characteristics of pacemaker oscillations in *Aplysia* neurons. *Can. J. Physiol. Pharmacol.*, **49**:787–795.

1973 Interval analysis of bursting discharges in *Aplysia* neurons. *BioSystems*, **5**:160–169.

Mauthner, L.

1859 Untersuchungen über den Bau des Rückenmarkes der Fische. *Sitzungsber. K. Akad. Wiss. Math.- Naturwiss, Cl. (Vienna)*, **34**:31–36.

Mayeri, E.

1973 Functional organization of the cardiac ganglion of the lobster, *Homarus americanus*. *J. Gen. Physiol.*, **62**:448–472.

Mayeri, E., J. Koester, I. Kupfermann, G. Liebeswar, and E. R. Kandel.

1974 Neural control of circulation in *Aplysia*. I. Motoneurons. *J. Neurophysiol.*, **37**:458–475.

Maynard, D. M.

1955 Activity in a crustacean ganglion. II. Pattern and interaction in burst formation. *Biol. Bull. (Woods Hole)*, **109**:420–436.

1961 Cardiac inhibition in decapod crustacea. In *Nervous Inhibition: International Symposium on Nervous Inhibition, 2nd,* edited by E. Florey, New York: Pergamon, pp. 144–178.

1972 Simpler networks. *Ann. N.Y. Acad. Sci.*, **193**:59–72.

Maynard, D. M., and H. A. Atwood.

1969 Divergent post-synaptic effects produced by single motor neurons of the lobster stomatogastric ganglion. *Am. Zool.*, **9**:1107, abst. 248.

Mayr, E.

1942 *Systematics and the Origin of Species from the Viewpoint of a Zoologist.* New York: Columbia University Press. (Corrected edition, with new preface, published 1964, New York: Dover.)

1963 *Animal Species and Evolution.* Cambridge, Mass.: Harvard University Press.

Mazzarelli, G.

1893 Monografia delle Aplysiidae del Golfo di Napoli. *Mem. Soc. Ital. Sci. (Rome)*, **9**:1–222.

Meech, R. W. and N. B. Standen.

1975 Potassium activation in *Helix aspersa* under voltage clamp: A component mediated by calcium influx. *J. Physiol. (London)*, **249**:211–239.

Mendel, G.

1866 Versuche über Pflanzen-Hybriden. *Verh. Naturforsch. Ver. Brünn*, **4**:3–47. (English translation in *The Origin of Genetics: A Mendel Source Book,* edited by C. Stern and E. R. Sherwood. San Francisco: Freeman, 1966, pp. 1–48).

Mendelson, M.

1971 Oscillator neurons in crustacean ganglia. *Science (Wash., D.C.)*, **171**:1170–1173.

Michaelson, D. M., and M. A. Raftery.

1974 Purified acetylcholine receptor: Its reconstitution to a chemically excitable membrane. *Proc. Natl. Acad. Sci. U.S.A.*, **71**:4768–4772.

Miledi, R.

1960 The acetylcholine sensitivity of frog muscle fibers after complete or partial denervation. *J. Physiol. (London)*, **151**:1–23.

1967 Spontaneous synaptic potentials and quantal, release of transmitter in the stellate ganglion of the squid. *J. Physiol. (London)*, **192**:379–406.

1973 Transmitter release induced by injection of calcium ions into nerve terminals. *Proc. R. Soc. London* (B), **183**:421–425.

Miller, N. E.

1967 Certain facts of learning relevant to the search for its physical basis. In *The Neurosciences: A Study Program,* edited by G. C. Quarton, T. Melnechuk, and F. O. Schmitt. New York: Rockefeller University Press, pp. 643–652.

Miller, P. L.

1965 The central nervous control of respiratory movements. In *The Physiology of the Insect Central Nervous System,* edited by J. E. Treherne and J. W. L. Beament. New York: Academic Press, pp. 141–155.

Miller, S., and J. Konorski.

1928 Sur une forme particulière des réflexes conditionnels. *C. R. Hebd. Séances Mém. Soc. Biol.,* **99:**1155–1157.

Minami, H., and K. M. Dallenbach.

1946 The effect of activity upon learning and retention in the cockroach. *Amer. J. Psychol.,* **59:**1–58.

Mittenthal, J. E., and J. J. Wine.

1973 Connectivity patterns of crayfish giant interneurons: Visualization of synaptic regions with cobalt dye. *Science (Wash., D.C.),* **179:**182–184.

Miyamoto, M., and B. McL. Breckenridge.

1974 A cyclic adenosine monophosphate link in the catecholamine enhancement of transmitter release at the neuromuscular junction. *J. Gen. Physiol.* **63:**609–624.

Model. P. G., S. M. Highstein, and M. V. L. Bennett.

1975 Depletion of vesicles and fatigue of transmission at a vertebrate central synapse. *Brain Res.,* **98:**209–228.

Morgan, C. L.

1894 *An Introduction to Comparative Psychology.* New York: Charles Scribner's Sons.

Morgan, T. H.

1927 *Experimental Embryology.* New York: Columbia University Press.

Morris, D.

1957 "Typical intensity" and its relation to the problem of ritualisation. *Behaviour,* **11:**1–12.

Moruzzi, G.

1959 The ascending reticular system, the regulation of the sensory inflow and the problem of visual habituation. *Il Nuovo Cimento,* **13**(Suppl. 2):532–538.

Mott, F. W., and C. S. Sherrington.

1895 Experiments upon the influence of sensory nerves upon movement and nutrition of the limbs. Preliminary communication. *Proc. R. Soc. London* (B), **57:**481–488.

Mountcastle, V. B.

1957 Modality and topographic properties of single neurons of cat's somatic sensory cortex. *J. Neurophysiol.,* **20:**408–434.

Mountcastle, V. B., and R. J. Baldessarini.

1974 Synaptic transmission. In *Medical Physiology,* 13th ed., Vol. 1, edited by V. B. Mountcastle. St. Louis: Mosby, pp. 182–223.

Müller, G. E., and A. Pilzecker.

1900 Experimentelle Beiträge zur Lehre vom Gedächtnis. *Z. Psychol. Physiol. Sinnesorg. Ergänzungsband.,* **1:**1–300.

Muller, H. J.

1927 Artificial transmutation of the gene. *Science (Wash., D.C.),* **66:**84–87.

Muller, K. J. and U. J. McMahan.

1975 The arrangement and structure of synapses formed by specific sensory and motor neurons in segmental ganglia of the leech. *Proc. Am. Assoc. of Anatomists* 88th session, abstract 231.

Mulloney, B., and A. I. Selverston.

1974a Organization of the stomatogastric ganglion of the spiny lobster. I: Neurons driving the lateral teeth. *J. Comp. Physiol.*, **91**:1–32.

1974b Organization of the stomatogastric ganglion of the spiny lobster. III: Coordination of the two subsets of the gastric system. *J. Comp. Physiol.*, **91**:53–78.

Murray, M. J., Jr.

1971 *The Biology of a Carnivorous Mollusc: Anatomical, Behavioral, Electrophysiological Observations in* Navanax inermis. Dissertation, University of California, Berkeley.

Myers, R. E., and R. W. Sperry.

1958 Interhemispheric communication through the corpus callosum: Mnemonic carry-over between the hemispheres *A.M.A. Arch. Neurol. Psychiatry,* **80**:298–303.

Nabias, B. de.

1894 Recherches histologiques et organologiques sur les centres nerveux des Gastéropodes. *Actes Soc. Linn. Bordeaux,* **47**:11–202.

Nakajima, S., and K. Takahashi.

1966 Post-tetanic hyperpolarization and electrogenic Na pump in stretch receptor neurone crayfish. *J. Physiol. (London),* **187**:105–127.

Nakamura, Y., S. Nakajima, and H. Grundfest.

1965 The action of tetrodotoxin on electrogenic components of squid giant axons. *J. Gen. Physiol.,* **48**:985–996.

Neher, E.

1971 Two fast transient current components during voltage clamp on snail neurons. *J. Gen. Physiol.,* **58**:36–53.

Newby, N. A.

1973 Habituation to light and spontaneous activity in isolated siphon of *Aplysia:* pharmacological observations. *Comp. Gen. Pharmacol.* **4**:91–100.

Nicholls, J. G., and D. A. Baylor.

1968 Specific modalities and receptive fields of sensory neurons in CNS of the leech. IJ. Neurophysiol., **31**:740–756.

Nicholls, J. G., and D. Purves.

1970 Monosynaptic chemical and electrical connexions between sensory and motor cells in the central nervous system of the leech. *J. Physiol. (London),* **209**:647–667.

Nicholls, J. G., and D. Van Essen.

1974 The nervous system of the leech. *Sci. Am.,* **230**(1):38–48.

Niklaus, R.

1965 Die Erregung einzelner Fadenhaare von *Periplaneta americana* in Abhängigkeit von der Grösse und Richtung der Auslenkung. *Z. vgl. Physiol.,* **50**:331–362.

Ort, C. A., W. B. Kristan, Jr. and G. S. Stent.

1974 Neuronal control of swimming in the medicinal leech. II. Identification and connections of motor neurons. *J. Comp. Physiol.* **94**(A):121–154.

Otsuka, M., E. A. Kravitz, and D. D. Potter.

1967 Physiological and chemical architecture of a lobster ganglion with particular reference to gamma-aminobutyrate and glutamate. *J. Neurophysiol.,* **30**:725–752.

Paine, R. T.

1963 Food recognition and predation on opisthobranchs by *Navanax inermis* Gastropoda: opisthobranchia). *Veliger,* **6**:1–9.

Parmentier, J., and J. Case.

1972 Structure-activity relationships of amino acid receptor sites on an identifiable cell body in the brain of the land snail *Helix Aspersa. Comp. Biochem. Physiol.,* **43**(A):511–518.

Parnas, I., D. Armstrong, and F. Strumwasser.

1974 Prolonged excitatory and inhibitory synaptic modulation of a bursting pacemaker neuron. *J. Neurophysiol.,* **37**:594–608.

Parnas, I., and F. Strumwasser.
 1974 Mechanisms of long lasting inhibition of a bursting pacemaker neuron. *J. Neurophysiol.*, **37**:609–620.

Pavlov, I. P.
 1906 The scientific investigation of the psychical faculties of processes in the higher animals. *Science, (Wash., D.C.)* **24**:613–619.
 1927 *Conditioned Reflexes: An Investigation of the Physiological Activity of the Cerebral Cortex*, translated and edited by G. V. Anrep. London: Oxford University Press.
 1928 *Lectures on Conditioned Reflexes*, translated by W. H. Gantt. New York: International Publishers.

Pearson, K. G.
 1972 Central programming and reflex control of walking in the cockroach. *J. Exp. Biol.* **56**:173–193.

Pearson, K. G., and J. F. Iles.
 1973 Nervous mechanisms underlying intersegmental coordination of leg movements during walking in the cockroach. *J. Exp. Biol.* **58**:725–744.

Peckham, G. W., and E. G. Peckham.
 1887 Some observations on the mental powers of spiders. *J. Morphol.*, **1**:383–419.

Peeke, H. V. S., M. J. Herz, and J. E. Gallagher.
 1971 Changes in aggressive interaction in adjacently territorial Convict Cichlids (*Cichlasoma nigrofasciatum*): A study of habituation. *Behaviour*, **40**:43–54.

Peretz, B.
 1969 Central neuron initiation of periodic gill movements. *Science (Wash., D.C.)*, **166**:1167–1172.
 1970 Habituation and dishabituation in the absence of a central nervous system. *Science (Wash., D.C.)*, **169**:379–381.

Peretz B., J. W. Jacklet, and K. Lukowiak.
 1976 Habituation of reflexes in *Aplysia*: contribution of the peripheral and central nervous systems. *Science*, **191**:396–399.

Perlman, A.
 1975 *Neural Control of the Siphon in* Aplysia. Dissertation, New York University School of Medicine.

Peterson, L. R.
 1966 Short-term memory. *Sci. Am.*, **215**(1):90–95.

Peterson, R. P.
 1970 RNA in single identified neurons of Aplysia. *J. Neurochem.*, **17**:325–338.

Peterson, R. P., and D. Kernell.
 1970 Effects of nerve stimulation on the metabolism of ribonucleic acid in a molluscan giant neurone. *J. Neurochem.*, **17**:1075–1085.

Pieron, H.
 1913 Recherches expérimentales sur les phénomènes de mémoire. *Année Psychol.*, **19**:91–193.

Pinsker, H. M., R. Feinstein, and B. A. Gooden.
 1974 Bradycardial response in *Aplysia* exposed to air. *Fed. Proc.*, **33**:361, Abstract 877.

Pinsker, H. M., W. A. Hening, T. J. Carew, and E. R. Kandel.
 1973 Long-term sensitization of a defensive withdrawal reflex in Aplysia. *Science (Wash., D.C.)*, **182**:1039–1042.

Pinsker, H., and E. R. Kandel.
 1967 Contingent modification of an endogenous bursting rhythm by monosynaptic inhibition. *Physiologist*, **10**:279.
 1969 Synaptic activation of an electrogenic sodium pump. *Science (Wash., D.C.)*, **163**:931–935.
 UNPUBL. Short-term modulation of endogenous bursting rhythms by monosynaptic inhibition in *Aplysia* neurons: effects of contingent stimulation.

Pinsker, H., E. R. Kandel, V. Castellucci, and I. Kupfermann.
1970 An analysis of habituation and dishabituation in *Aplysia.* In *Biochemistry of Simple Neuronal Models,* edited by Costa and Giacobini. New York: Raven Press, pp. 351–373.

Pinsker, H., I. Kupfermann, V. Castellucci, and E. R. Kandel.
1970 Habituation and dishabituation of the gill-withdrawal reflex in *Aplysia. Science (Wash., D.C.),* **167:**1740–1742.

Pitman, R. M., C. D. Tweedle, and M. J. Cohen.
1972 Branching of central neurons: Intracellular cobalt injection for light and electron microscopy. *Science (Wash., D.C.),* **176:**412–414.

Pitts, W., and W. S. McCulloch.
1947 How we know universals. The perception of auditory and visual forms. *Bull. Math. Biophys.,* **9:**127–147.

Postman, L., and T. H. Alper.
1946 Rectroactive inhibition as a function of the time of interpolation of the inhibitor between learning and recall. *Amer. J. Psychol.,* **59:**439–449.

Pribram, K. H.
1971 *Languages of the Brain: Experimental Paradoxes and Principles in Neuropsychology.* Englewood Cliffs, N.J.: Prentice-Hall.

Prosser, C. L.
1934a Action potentials in the nervous system of the crayfish. I. Spontaneous impulses. *J. Cell. Comp. Physiol.,* **4:**185–209.
1934b Action potentials in the nervous system of the crayfish. II. Responses to illumination of the eye and caudal ganglion. *J. Cell. Comp. Physiol.* **4:**363–377.
1935 A preparation for the study of single synaptic junctions. *Amer. J. Physiol.,* **113:**108.
1936 Rhythmic activity in isolated nerve centers. *Cold Spring Harbor Symp. Quant. Biol.,* **4:**339–346.
1946 The physiology of nervous systems of invertebrate animals. *Physiol. Rev.,* **25:**337–382.

Prosser, C. L., and W. S. Hunter.
1936 The extinction of startle responses and spinal reflexes in the white rat. *Am. J. Physiol.,* **117:**609–618.

Pumphrey, R. J., and A. F. Rawdon-Smith.
1937 Synaptic transmission of nervous impulses through the last abdominal ganglion of the cockroach. *Proc. R. Soc. London* (B), **122:**106–118.

Quartermain, D., R. M. Paolino, and N. E. Miller.
1965 A brief temporal gradient of retrograde amnesia independent of situational change. *Science (Wash. D.C.),* **149:**1116–1118.

Quinn, W. G., W. A. Harris, and S. Benzer.
1974 Conditioned behavior in *Drosophila Melanogaster. Proc. Natl. Acad. Sci. U.S.A.* **71:**708–712.

Rahamimoff, R.
1974 Modulation of transmitter release at the neuromuscular junction. In *The Neurosciences: Third Study Program,* edited by F. O. Schmitt and F. G. Wordon. Cambridge, Mass.: M.I.T. press, pp. 943–952.

Rahamimoff, R., and E. Alnaes.
1973 Inhibitory action of ruthenium red on neuromuscular transmission. *Proc. Natl. Acad. Sci. U.S.A.,* **70:**3613–3616.

Raisman, G.
1969 Neuronal plasticity in the septal nuclei of the adult rat. *Brain Res.,* **14:**25–48.

Raisman, G., and P. M. Field.
1973 A quantitative investigation of the development of collateral reinnervation after partial deafferentation of the septal nuclei. *Brain Res.,* **50:**241–264.

Rall, W.

1959 Branching dendritic trees and motoneuron membrane resistivity. *Exp. Neurol.,* **1:**491–527.

1976 Core conductor theory and the cable properties of neurons. In "Cellular Biology of Neurons" (Vol. 1, Sect. I, *Handbook of Physiology, The Nervous System*), edited by E. R. Kandel. Baltimore: Williams and Wilkins.

Rasmussen, H.

1975 Ions as second messengers. In *Cell Membranes: Biochemistry, Cell Biology and Pathology,* edited by G. Weissmann. New York: Hospital Practice, pp. 203–212.

Remler, M., A. Selverston, and D. Kennedy.

1968 Lateral giant fibers of crayfish: Location of somata by dye injection. *Science (Wash., D.C.),* **162:**281–283.

Renshaw, B.

1946 Central effects of centripetal impulses in axons of spinal ventral roots. *J. Neurophysiol.,* **9:**191–204.

Rescorla, R. A., and R. L. Solomon.

1967 Two-process learning theory: Relationships between Pavlovian conditioning and instrumental learning. *Psychol. Rev.,* **74:**151–182.

Retzius, G.

1888 Ueber myelinhaltige Nervenfasern bei Evertebraten. *Verh. Biol. Ver. Stockholm,* **1:**58–62.

1891 Zur Kenntniss des centralen Nervensystem der Würmer. Das Nervensystem der Annulaten. *Biol. Untersuch.* (Neue Folge), **2:**1–28.

1902 Weiteres zur Kenntniss der Sinneszellen der Evertebraten. *Biol. Untersuch.* (Neue Folge), **10:**25–33.

Roberts, A.

1968 Recurrent inhibition in the giant-fibre system of the crayfish and its effect on the excitability of the escape response. *J. Exp. Biol.,* **48:**545–567.

Robison, G. A., R. W. Butcher, and E. W. Sutherland.

1971 Adenyl cyclase as an adrenergic receptor. *Ann. N.Y. Acad. Sci.,* **139:**703–723.

Roe, A., and G. G. Simpson, eds.

1958 *Behavior and Evolution.* New Haven: Yale University Press.

Roeder, K. D.

1948 Organization of the ascending giant fiber system in the cockroach (*Periplaneta americana*). *J. Exp. Zool.,* **108:**243–261.

1955 Spontaneous activity and behavior. *Sci. Mon.,* **80:**362–370.

1963 *Nerve Cells and Insect Behavior.* Cambridge, Mass.: Harvard University Press.

Roeder, K. D., L. Tozian, and E. A. Weiant.

1960 Endogenous nerve activity and behaviour in the mantis and cockroach. *J. Insect Physiol.,* **4:**45–62.

Romanes, G. J.

1883 *Animal Intelligence.* New York: Appleton.

1888 *Mental Evolution in Man.* London: Paul.

Roper, S.

1976 An electrophysiological study of chemical and electrical synapses on neurons in the parasympathetic cardiac ganglion of the mud puppy, *Nectures Maculosusi.* Evidence for intrinsic ganglion innervation. *J. Physiol.,* **254:**427–454.

Rose, R. M.

1972 Burst activity of the buccal ganglion of *Aplysia depilans. J. Exp. Biol.,* **56:**735–754.

Rosenblatt, F.
  1958 The perceptron: A probabilistic model for information storage and organization in the brain. *Psychol. Rev.,* **65**:386–408.
Rosenthal, J.
  1969 Post-tetanic potentiation at the neuromuscular junction of the frog. *J. Physiol. (London),* **203**:121–133.
Rosvold, H. E.
  1959 Physiological psychology. *Annu. Rev. Psychol.,* **10**:415–454.
Rovainen, C. M.
  1967 Physiological and anatomical studies on large neurons of the central nervous system of the sea lamprey (*Petromyzon marinus*) I. Müller and Mauthner cells. *J. Neurophysiol.,* **30**:1000–1023.
Russell, W. M. S., A. P. Mead, and J. S. Hayes.
  1954 A basis for the quantitative study of the structure of behaviour. *Behaviour,* **6**:153–205.
Russell, W. R.
  1959 *Brain, Memory, Learning; A Neurologist's View.* Oxford: Clarendon Press.
Schacter, S.
  1971 *Emotion, Obesity and Crime.* New York: Academic Press.
Schleidt, W. M.
  1961a Ueber die Auslösung der Flucht vor Raubvögeln bei Truthühnern. *Naturwissenschaften,* **48**:141–142.
  1961b Reaktionen von Truthühnern auf fliegende Raubvögel und Versuche zur Analyse ihrer AAM's. *Z. Tierpsychol.,* **18**:534–560.
Schneider, G. E.
  1970 Mechanisms of functional recovery following lesions of visual cortex or superior colliculus in neonate and adult hamsters. *Brain Behav. Evol.,* **3**:295–323.
Schneirla, T. C.
  1929 Learning and orientation in ants. Studied by means of the maze method. *Comp. Psychol. Monogr.,* **6**(4):1–143.
  1934 The process and mechanism of ant learning. The combination problem and the successive-presentation problem. *J. Comp. Psychol.,* **17**:303–328.
  1953a Basic problems in the nature of insect behavior. In *Insect physiology,* edited by K. D. Roeder. New York: Wiley, pp. 656–684.
  1953b Insect behavior in relation to its setting. In *ibid,* pp. 685–722.
  1953c Modifiability in insect behavior. In *ibid,* pp. 723–747.
Schwartz, J. H., V. F. Castellucci, and E. R. Kandel.
  1971 Functioning of identified neurons and synapses in abdominal ganglion of *Aplysia* in absence of protein synthesis. *J. Neurophysiol.,* **34**:939–953.
Segaar, J.
  1962 Die Funktion des Vorderhirns in Bezug auf des angeborene Berhalten des dreidornigen Stichlingsmännchen (*Gasterosteus aculeatus* L.)–zugleich ein Beitrag über Neuronenregeneration im Fischgehirm. *Acta Morphol. Neerl.- Scand.,* **5**:49–64.
Seligman, M. E. P., and J. L. Hager, comps.
  1972 *Biological Boundaries of Learning.* New York: Appleton-Century-Crofts.
Sharpless, S. K., and H. Jasper.
  1956 Habituation of the arousal reaction. *Brain,* **79**:655–680.
Sherrington, C. S.
  1906 *The Integrative Action of the Nervous System.* New Haven: Yale University Press. (Reprinted 1947.)
Shimahara, T., and L. Tauc.
  1972 Mécanisme de la facilitation hétérosynaptique chez l'aplysie. *J. Physiol. (Paris),* **65**:303A–304A.

1975a Multiple interneuronal afferents to the giant cells in *Aplysia. J. Physiol. (London),* **247:**299–319.

1975b Heterosynaptic facilitation in the giant cell of *Aplysia. J. Physiol.,* **247:**321–341.

Sidman, M.

1960 Normal sources of pathological behavior. *Science (Wash., D.C.),* **132:**61–68.

Simpson, G. G.

1949 *The Meaning of Evolution.* New Haven: Yale University Press.

Skinner, B. F.

1938 *The Behavior of Organisms; An Experimental Analysis.* New York: Appleton-Century.

1953 *Science and Human Behavior.* New York: Macmillan.

1956 A case history in scientific method. *Am. Psychol.,* **11:**221–233.

Smith, T. G., Jr., J. L. Barker, and H. Gainer.

1975 Requirements for bursting pacemaker activity in molluscan neurons. *Nature (London),* **253:**450–452.

Sokolov, E. N.

1963a Higher nervous functions: the orienting reflex. *Annu. Rev. Physiol.,* **25:**545–580.

1963b *Perception and the Conditioned Reflex,* translated by S. W. Waydenfeld. New York: Pergamon.

Solomon, P., and V. D. Patch, eds.

1971 *Handbook of Psychiatry,* 2nd ed. Los Altos, Calif.; Lange Medical Publications.

Sourkes, T.

1966 *Nobel Prize Winners in Medicine and Physiology 1901–1965.* New York: Abelard-Shuman.

Spencer, W. A.

1976 Physiology of supraspinal neurons in mammals. In: "Cellular Biology of Neurons" (Vol. 1, Sect. 1, *Handbook of Physiology, The Nervous System*), edited by E. R. Kandel. Baltimore: Williams and Wilkins.

Spencer, W. A., and R. S. April.

1970 Plastic properties of monosynaptic pathways in mammals. In *Short-Term Changes in Neural Activity and Behaviour,* edited by G. Horn and R. A. Hinde. Cambridge, Eng.: Cambridge University Press. pp. 433–474.

Spencer, W. A., and E. R. Kandel.

1967 Cellular and integrative properties of the hippocampal pyramidal cell and the comparative electrophysiology of cortical neurons. *Int. J. Neurol.,* **6:**266–296.

Spencer, W. A., R. F. Thompson, and D. R. Neilson, Jr.

1966a Response decrement of the flexion reflex in the acute spinal cat and transient restoration by strong stimuli. *J. Neurophysiol.,* **29:**221–239.

1966b Decrement of ventral root electrotonus and intracellularly recorded PSPs produced by iterated cutaneous afferent volleys. *J. Neurophysiol.,* **29:**253–274.

Spencer, W. A., and R. Wigdor

1965 Ultra-late PTP of monosynaptic reflex responses in cat. *Physiologist,* **8:**278.

Sperry, R. W.

1951 Mechanisms of neural maturation. In *Handbook of Experimental Psychology,* edited by S. S. Stevens. New York: Wiley, pp. 247–252.

1963 Chemoaffinity in the orderly growth of nerve fiber patterns and connections. *Proc. Natl. Acad. Sci. U.S.A.,* **50:**703–710.

1965a Discussion of papers by J. C. Eccles. In *The Anatomy of Memory,* edited by D. P. Kimble. Palo Alto, Calif.: Science and Behavior Books, pp. 75–76.

1965b Selective communication in nerve nets: Impulse specificity vs connection specificity. *Neurosci. Res. Program Bull.,* **3**(5):37–43.

Spira, M. E., and M. V. L. Bennett.

1972 Synaptic control of electrotonic coupling between neurons. *Brain Res.,* **37:**294–300.

1973 Function of synaptic modulation of electrotonic coupling between neurons. *Program Abstr. Soc. Neursci., 3rd Annu. Meeting.*

Spray, D. C., and M. V. L. Bennett.

1975 Pharyngeal sensory neurons and feeding behavior in the opisthobranch mullusc *Navanax. Neuroscience Abstracts,* **1:**570.

Stefani, E., and H. M. Gerschenfeld.

1969 Comparative study of acetylcholine and 5-hydroxytryptamine receptors on single snail neurons. *J. Neurophysiol.,* **32:**64–74.

Steiner, D., S. Cho, P. E. Oyer, S. Terris, J. D. Peterson, and A. H. Rubenstein.

1971 Isolation and characterization of proinsulin C-peptide from bovine pancreas. *J. Biol. Chem.,* **246:**1365–1374.

Stent, G. S.

1973 A physiological mechanism for Hebb's postulate of learning. *Proc. Natl. Acad. Sci. U.S.A.,* **70:**997–1001.

Stephens, C. L.

1973 Progressive decrements in the activity of *Aplysia* neurones following repeated intracellular stimulation: Implications for habituation. *J. Exp. Biol.,* **58:**411–421.

Stevens, C. F.

1966 *Neurophysiology: A Primer.* New York: Wiley.

1969 Voltage-clamp analysis of a repetitively firing neuron. In *Basic Mechanisms of the Epilepsies,* edited by H. H. Jasper, A. A. Ward, and A. Pope. Boston, Mass.: Little, Brown, pp. 76–82.

Steward, O., C. W. Cotman, and G. S. Lynch.

1973 Re-establishment of electrophysiologically functional entorhinal cortical input to the dentate gyrus deafferented by ipsilateral entorhinal lesions: Innervation by the contralateral entorhinal cortex. *Exp. Brain Res.,* **18:**396–414.

Stinnakre, J., and L. Tauc.

1969 Central neuronal response to the activation of osmoreceptors in the osphradium of *Aplysia. J. Exp. Biol.,* **51:**347–361.

1973 Calcium influx in active *Aplysia* neurones detected by injected aequorin. *Nature New Biol.,* **242:**113–115.

Stretton, A. O. W., and E. A. Kravitz.

1968 Neuronal geometry: Determination with a technique of intracellular dye injection. *Science (Wash., D.C.),* **162:**132–134.

1973 Intracellular dye injection: The selection of procion yellow and its application in preliminary studies of neuronal geometry in the lobster nervous system. In *Intracellucar Staining in Neurobiology,* edited by S. B. Kater and C. Nicholson. New York: Springer, pp. 21–40.

Strickberger, M. W.

1968 *Genetics.* New York: Macmillan.

Strumwasser, F.

1962 Post-synaptic inhibition and excitation produced by different branches of a single neuron and the common transmitter involved. *Int. Congr. Physiol. Sci., 22,* **2:** No. 801.

1967a Neurophysiological aspects of rhythms. In *Neurosciences: A Study Program,* edited by G. C. Quarton, T. Melnechuk, and F. O. Schmitt. New York: Rockefeller University Press, pp. 516–528.

1967b Types of information stored in single neurons. In *Invertebrate Nervous Systems: Their Significance for Mammalian Neurophysiology. Conference on Invertebrate Nervous Systems,* edited by C. A. G. Wiersma. Chicago: University of Chicago Press, pp. 291–319.

1968 Membrane and intracellular mechanism governing endogenous activity in neurons. In *Physiological and Biochemical Aspects of Nervous Integration,* edited by F. D. Carlson. Englewood Cliffs, N.J.: Prentice-Hall, pp. 329–341.

1971 The cellular basis of behavior in *Aplysia. J. Psychiatric Res.,* **8:**237–257.

1974 Neuronal principles organizing periodic behaviors. In *The Neurosciences: Third Study Program,* edited by F. O. Schmitt and F. G. Worden. Cambridge, Mass.: M.I.T. Press, pp. 459–478.

Strumwasser, F., and R. Bahr.

1966 Prolonged in vitro culture and autoradiographic studies of neurons in *Aplysia. Fed. Proc.,* **25:**512, abstract 1815.

Strumwasser, F., J. W. Jacklet, and R. B. Alvarez.

1969 A seasonal rhythm in the neural extract induction of behavioral egg-laying in *Aplysia. Comp. Biochem. Physiol.,* **29:**197–206.

Stuart, A. E.

1970 Physiological and morphological properties of motoneurones in the central nervous system of the leech. *J. Physiol. (London),* **209:**627–646.

Takeuchi, A., and N. Takeuchi.

1966 On the permeability of the presynaptic terminal of the crayfish neuromuscular junction during synaptic inhibition and the action of γ-aminobutyric acid. *J. Physiol. (London),* **183:**433–449.

Tanzi, E.

1893 I fatti e le induzioni nell'odierna istologia del sistema nervoso. *Riv. sper. Freniat. Med. leg Alien. ment.,* **19:**419–472.

Taub, E., and A. J. Berman.

1963 Avoidance conditioning in the absence of relevant proprioceptive and exteroceptive feedback. *J. Comp. Physiol. Psychol.,* **56:**1012–1016.

1968 Movement and learning in the absence of sensory feedback. In *The Neuropsychology of Spatially Oriented Behavior,* edited by S. J. Freedman. Homewood, Ill.: Dorsey Press, pp. 173–192.

Tauc, L.

1958 Processus post-synaptiques d'excitation et d'inhibition dans le soma neuronique de L'*Aplysie* et de L'*Escargot. Arch. Ital. Biol.,* **96:**78–110.

1960 Diversité des modes d'activité des cellules nerveuses du ganglion déconnecté de l'Aplysie. *C. R. Seances Soc. Biol. Fil.,* **154:**17–21.

1962a Site of origin and propagation of spike in giant neuron of *Aplysia. J. Gen. Physiol.,* **45:**1077–1097.

1962b Identification of active membrane areas in the giant neuron of Aplysia. *J. Gen. Physiol.,* **45:**1099–1115.

1965 Presynaptic inhibition in the abdominal ganglion of Aplysia. *J. Physiol. (London),* **181:**282–307.

1966 Physiology of the nervous system. In *Physiology of Mollusca,* Vol. 2, edited by K. M. Wilbur and C. M. Yonge. New York: Academic Press, pp. 387–454.

1967 Transmission in invertebrate and vertebrate ganglia. *Physiol. Rev.,* **47:**521–593.

1969 Polyphasic synaptic activity. *Prog. Brain Res.,* **31:**247–257.

Tauc, L., and J. Bruner.

1963 Desensitization of cholinergic receptors by acetylcholine in molluscan central neurones. *Nature (London),* **198:**33–34.

Tauc, L., and R. Epstein.

1967 Heterosynaptic facilitation as a distinct mechanism in Aplysia. *Nature (London),* **214:**724–725.

Tauc, L., and H. M. Gerschenfeld.

1961 Cholinergic transmission mechanisms for both excitation and inhibition in molluscan central synapses. *Nature (London),* **192:**366–367.

1962 A cholingeric mechanism of inhibitory synaptic transmission in a molluscan nervous system. *J. Neurophysiol.,* **25:**236–262.

Terrace, H. S.
  1973 Classical conditioning. In *The Study of Behavior: Learning, Motivation, Emotion, and Instinct,* edited by J. A. Nevin. Glenview, Illinois: Scott, Foresman, pp. 71–112.
Thomas, R. C.
  1969 Membrane current and intracellular sodium changes in a snail neurone during extrusion of injected sodium. *J. Physiol. (London),* **201:**495–514.
  1972 Electrogenic sodium pump in nerve and muscle cells. *Physiol. Rev.,* **52:**563–594.
Thompson, E. B., C. H. Bailey, V. Castellucci, and E. R. Kandel.
  1976 Two different and compatible intracellular labels: A preliminary structural study of identified sensory and motor neurons which mediate the gill-withdrawal reflex in *Aplysia californica. Neuroscience Abstract* 2.
Thompson, E. B., J. H. Schwartz, and E. R. Kandel.
  1976 A radioautographic analysis in the light and electron microscope of identified *Aplysia* neurons and their processes after intrasomatic injection of H-L-fucose. *Brain Res.,* **191**(1).
Thompson, R.
  1969 Localization of the "visual memory system" in the white rat. *J. Comp. Physiol. Psychol. Mongr.,* **69:**4 pt. 2.
Thompson, R. F., and W. A. Spencer.
  1966 Habituation: A model phenomenon for the study of neuronal substrates of behavior. *Psychol. Rev.,* **73:**16–43.
Thorndike, E. L.
  1898 Animal intelligence. An experimental study of the associative processes in animals. *Psychol. Rev. Ser. Monogr.* (Suppl. 2), **4:**1–109.
Thorndike, E. L., and R. S. Woodworth.
  1901a The influence of improvement in one mental function upon the efficiency of other functions. I. *Psychol. Rev.,* **8:**247–261.
  1901b The influence of improvement in one mental function upon the efficiency of other functions. II. The estimation of magnitudes. *Psychol. Rev.,* **8:**384–395.
  1901c The influence of improvement in one mental function upon the efficiency of other functions. III. Functions involving attention, observation and discrimination. *Psychol. Rev.,* **8:**553–564.
Thorpe, W. H.
  1943 Types of learning in insects and other arthropods, I. *Brit. J. Psychol.,* **33:**220–234.
  1944a Types of learning in insects and other arthropods, II. *Brit. J. Psychol.,* **34:**20–31.
  1944b Types of learning in insects and other arthropods, III. *Brit. J. Psychol.,* **34:**66–76.
  1956 *Learning and Instinct in Animals.* Cambridge: Harvard University Press.
Tinbergen, N.
  1951 *The Study of Instinct.* Oxford: Clarendon Press.
Toevs, L. A. S., and R. W. Brackenbury.
  1969 Bag cell-specific proteins and the humoral control of egg laying in *Aplysia californica. Comp. Biochem. Physiol.,* **29:**207–216.
Tolman, E. C.
  1932 *Purposive Behavior in Animals and Men.* New York: Appleton-Century-Crofts. (Reprinted 1967 by Meredith.)
  1938 The determiners of behavior at a choice point. *Psychol. Rev.,* **45:**1–41.
Tomita, T.
  1970 Electrical activity of vertebrate photoreceptors. *Q. Rev. Biophys.,* **3:**179–222.
Truex, R. C.
  1959 *Strong and Elwyn's Human Neuroanatomy,* 4th ed. Baltimore: Williams and Wilkins.
Uexküll, J. von.
  1957 A stroll through the worlds of animals and men: A picture book of invisible worlds. In *Instinctive Behavior,* edited by C. H. Schiller. New York: International Universities Press, pp. 5–80.

Verworn, M.

    1889 *Psycho-physiologische Protistenstudien. Experimentelle Untersuchungen.* Jena.

Villee, C. A., W. F. Walker, Jr., and F. E. Smith.

    1963 *General Zoology,* 2nd ed. Philadelphia: Saunders.

Von Baumgarten, R. J.

    1970 Plasticity in the nervous system at the unitary level. In *The Neurosciences: Second Study Program,* edited by F. O. Schmitt, G. C. Quarton, T. Melnechuk, and G. Adelman. New York: Rockefeller University Press, pp. 260–271.

Von Baumgarten, R. J., and B. Djahanparwar.

    1967a Time course of repetitive heterosynaptic facilitation in *Aplysia californica. Brain Res.,* **4:**295–297.

    1967b Beitrag zum Problem der heterosynaptischen Facilitation in *Aplysia californica. Pflügers Archiv gesamte Physiol. Menschen Tiere,* **295:**328–346.

Von Neumann, J.

    1956 Probabilistic logics and the synthesis of reliable organisms from unreliable components. In *Automata Studies,* edited by C. E. Shannon and J. McCarthy. Princeton, N.J.: Princeton University Press, pp. 43–98.

Wachtel, H., and E. R. Kandel.

    1967 A direct synaptic connection mediating both excitation and inhibition. *Science* (Wash., D.C.), **158:**1206–1208.

    1971 Conversion of synaptic excitation to inhibition at a dual chemical synapse. *J. Neurophysiol.,* **34:**56–68.

Wald, F.

    1972 Ionic differences between somatic and axonal action potentials in snail giant neurones. *J. Physiol. (London),* **220:**267–281.

Wall, P. D.

    1970 Habituation and post-tetanic potentiation in the spinal cord. In *Short-Term Changes in Neural Activity and Behaviour,* edited by G. Horn and R. A. Hinde. Cambridge, Eng.: Cambridge University Press, pp. 181–210.

Watanabe, A., and H. Grundfest.

    1961 Impulse propagation at the septal and commissural junctions of crayfish lateral giant axons. *J. Gen. Physiol.,* **45:**267–308.

Watson, J. B.

    1913 Psychology as the behaviorist views it. *Psychol. Rev.,* **20:**158–173.

    1930 *Behaviorism.* New York: Norton.

Waziri, R.

    1969 Electrical transmission mediated by an identified cholinergic neuron of *Aplysia. Life Sci.,* **8:**469–476.

    1971 Electrotonically coupled interneurones produce two types of inhibition in *Aplysia* neurones. *Nature New Biol.,* **232:**286–288.

Waziri, R., W. T. Frazier, and E. R. Kandel.

    1965 Analysis of "pacemaker" activity in an identifiable burst generating neuron in *Aplysia. Physiologist,* **8:**300.

Waziri, R., and E. R. Kandel.

    1969 Organization of inhibition in abdominal ganglion of *Aplysia.* III: Interneurons mediating inhibition. *J. Neurophysiol.,* **32:**520–539.

Waziri, R., E. R. Kandel, and W. T. Frazier.

    1969 Organization of inhibition in abdominal ganglion of *Aplysia.* II. Posttetanic potentiation, heterosynaptic depression, and increments in frequency of inhibitory postsynaptic potentials. *J. Neurophysiol.,* **32:**509–519.

Weber, E. H.

    1834 *De Pulsu, Resorptione, Auditu et Tactu: Annotationes Anatomicae et Physiologicae.*

Leipzig: Koehler. (English translation by B. Herrnstein in *A Source Book in the History of Psychology,* edited by R. H. Herrnstein and E. C. Boring. Cambridge, Mass.: Harvard University Press, pp. 64–66.)

Weight, F. F.

1974 Synaptic potentials resulting from conductance decreases. In *Synaptic Transmission and Neuronal Interaction* (Society of General Physiologists Series. Vol. 28), edited by M. V. L. Bennett, New York: Raven Press, pp. 141–152.

Weight, F. F., and J. Votava.

1970 Slow synaptic excitation in sympathetic ganglion cells: Evidence for synaptic inactivation of potassium conductance. *Science (Wash., D.C.),* **170:**755–758.

Weinreich, D.

1971 Ionic mechanism of post-tetanic potentiation at the neuromuscular junction of the frog. *J. Physiol. (London),* **212:**431–446.

Weinreich, D., S. A. Dewhurst, and R. E. McCaman.

1972 Metabolism of putative transmitters in individual neurons of *Aplysia californica:* Aromatic amino acid decarboxylase. *J. Neurochem.,* **19:**1125–1130.

Weinreich, D., M. W. McCaman, R. E. McCaman, and J. E. Vaughn.

1973 Chemical, enzymatic and ultrastructural characterization of 5-hydroxytryptamine-containing neurons from the ganglia of *Aplysia californica* and *Tritonia diomedia. J. Neurochem.,* **20:**969–976.

Weiskrantz, L.

1966 Experimental studies of amnesia. In *Amnesia,* edited by C. W. M. Whitty and O. L. Zangwill. London: Butterworth, pp. 1–35.

1970 A long-term view of short-term memory in psychology. In *Short-Term Changes in Neural Activity and Behaviour,* edited by G. Horn and R. A. Hinde. Cambridge, Eng.: Cambridge University Press, pp. 63–74.

Weiss, K. R., J. Cohen, and I. Kupfermann.

1975 Potentiation of muscle contraction: a possible modulatory function of an identified serotonergic cell in *Aplysia. Brain Research,* **99:**381–386.

Wells, M. J.

1968 *Lower Animals.* New York: McGraw-Hill.

Werman, R.

1966 Criteria for identification of a central nervous system transmitter. *Comp. Biochem. Physiol.,* **18:**745–766.

1969 An electrophysiological approach to drug-receptor mechanisms. *Comp. Biochem. Physiol.,* **30:**997–1017.

Wernig, A.

1972 Changes in statistical parameters during facilitation at the crayfish neuromuscular junction. *J. Physiol. (London),* **226:**751–759.

1975 Estimates of statistical release parameters from crayfish and frog neuromuscular junctions. *J. Physiol.,* **244:**207–221.

Whitman, C. O.

1899 *Animal Behavior. Biological Lectures.* Boston, Mass.: Marine Biological Laboratory, Woods Hole.

1919 *The Behavior of Pigeons: Posthumous Works of Charles Otis Whitman,* edited by H. A. Carr. Washington, D.C.: Carnegie Institution of Washington.

Whittaker, V. P., and H. Zimmermann.

1974 Biochemical studies on cholinergic synaptic vesicles. In *Synaptic Transmission and Neuronal Interaction* (Society of General Physiologists Series, Vol. 28), edited by M. V. L. Bennett. New York: Raven Press, pp. 217–238.

Wickelgren, B. G.
  1967 Habituation of spinal interneurons. *J. Neurophysiol.*, **30**:1424–1438.
Wickelgren, W. A.
  1973 The long and the short of memory. *Psychol. Bull.*, **80**:425–438.
Wiersma, C. A. G.
  1938 Function of the giant fibers of the central nervous system of the crayfish. *Proc. Soc. Biol. Med.*, **38**:661–662.
  1947 Giant nerve fiber system of the crayfish. A contribution to comparative physiology of synapse. *J. Neurophysiol.* **10**:23–38.
  1952 The neuron soma; neurons of arthropods. *Cold Spring Harbor Symp. Quant. Biol.*, **17**:155–163.
  1966 Integration in the visual pathway of Crustacea. *Symp. Soc. Exp. Biol.*, No. 20, pp. 151–177.
Wiersma, C. A. G., and J. L. M. Roach
  1976 Principles in the organization of invertebrate sensory systems. In "Cellular Biology of Neurons" (Vol. 1, Sect. 1, *Handbook of Physiology, The Nervous System*), edited by E. R. Kandel. Baltimore: Williams and Wilkins.
Wiersma, C. A. G., and K. Ikeda.
  1964 Interneurons commanding swimmeret movements in the crayfish *Procambarus clarki* (Girard). *Comp. Biochem. Physiol.*, **12**:509–525.
Wiesel, T. N., and D. H. Hubel.
  1965 Comparison of the effects of unilateral and bilateral eye closure on cortical unit responses in kittens. *J. Neurophysiol.*, **28**:1029–2040.
Wilcoxon, H. C., W. B. Dragoin, and P. A. Kral.
  1971 Illness-induced aversions in rat and quail: Relative salience of visual and gustatory cues. *Science (Wash., D.C.)*, **171**:826–828.
Williams, D. R., and H. Williams.
  1969 Auto-maintenance in the pigeon: Sustained pecking despite contingent non-reinforcement. *J. Exp. Anal. Behav.*, **12**:511–520.
Williams, L. W.
  1907 *The Anatomy of the Common Squid* Loligo peallii *Lesueur.* Leiden: E. J. Brill.
Willows, A. O. D.
  1967 Behavioral acts elicited by stimulation of single, identifiable brain cells. *Science (Wash., D.C.)*, **157**:570–574.
  1968 Behavioral acts elicited by stimulation of single, identifiable nerve cells. In *Physiological and Biochemical Aspects of Nervous Integration*, edited by F. D. Carlson. Englewood Cliffs, N.J.: Prentice-Hall, pp. 217–243.
  1971 Giant brain cell in mollusks. *Sci Am.*, **224**(2):68–75.
  1973 Learning in gastropod mollusks. In *Invertebrate Learning*, edited by W. C. Corning, J. A. Dyal and A. O. D. Willows. New York: Plenum Press, pp. 187–273.
Willows, A. O. D., D. A. Dorsett, and G. Hoyle.
  1973a The neuronal basis of behavior in *Tritonia.* I. Functional organization of the central nervous system. *J. Neurobiol.*, **4**:207–237.
  1973b The neuronal basis of behavior in *Tritonia.* III. Neuronal mechanisms of a fixed action pattern. *J. Neurobiol.* **4**:255–285.
Willows, A. O. D., and G. Hoyle.
  1969 Neuronal network triggering a fixed action pattern. *Science (Wash., D.C.)*, **166**:1549–1551.
Wilson, D. L.
  1971 Molecular weight distribution of proteins synthesized in single, identified neurons of *Aplysia. J. Gen. Physiol.*, **57**:26–40.

Wilson, D. L., and R. W. Berry.
    1972 The effect of synaptic stimulation on RNA and protein metabolism in the R2 soma of *Aplysia. J. Neurobiol.,* **3**:369–379.

Wilson, D. M.
    1961 The central nervous control of flight in a locust. *J. Exp. Biol.,* **38**:471–490.
    1965 The nervous co-ordination of insect locomotion. In *The Physiology of the Insect Central Nervous System,* edited by J. E. Treherne and J. W. L. Beament. New York: Academic Press, pp. 126–140.
    1966 Central nervous mechanisms for the generation of rhythmic behaviour in arthropods. *Symp. Soc. Exp. Biol.,* **20**:199–228.
    1968 The flight-control system of the locust. *Sci. Am.,* **218**(5):83–90.

Wilson, W. A., Jr.
    1971 *A Voltage Clamp Analysis of Slow Waves and Prolonged Synaptic Potentials in Bursting Neurons.* Dissertation. Duke University.

Wilson, W. A., and H. Wachtel.
    1974 Negative resistance characteristic essential for the maintenance of slow oscillations in bursting neurons. *Science (Wash., D.C.),* **186**:932–934.

Wine, J. J., and F. B. Krasne.
    1972 The organization of escape behaviour in the crayfish. *J. Exp. Biol.,* **56**:1–18.

Winlow, W., and E. R. Kandel.
    1976 The morphology of identified neurons in the abdominal ganglion of *Aplysia californica. Brain Res.,* **191**(1).

Winograd, S., and J. D. Cowan.
    1963 *Reliable Computation in the Presence of Noise.* Cambridge, Mass.: M.I.T. Press.

Woodworth, R. S.
    1948 *Contemporary Schools of Psychology,* rev. ed. New York: Ronald Press.

Woodworth, R. S., and H. Schlosberg.
    1954 *Experimental Psychology,* rev. ed. New York: Holt, Reinhart, Winston, pp. 786–794.

Woollacott, M. H.
    1974 Patterned neural activity associated with prey capture in *Navanax* (Gastropoda, Aplysiacea). *J. Comp. Physiol.* **94**:96–84.

Wyers, E. J., H. V. S. Peeke, and M. J. Herz.
    1973 Behavioral habituation in invertebrates. In *Habituation,* Vol. 1, edited by H. V. S. Peeke and M. J. Herz. New York: Academic Press, pp. 1–57.

Yasargil, G. M., and J. Diamond.
    1968 Startle-response in teleost fish: An elementary circuit for neural discrimination. *Nature (London),* **220**:241–243.

Yonge, C. M.
    1949 *The Sea Shore.* London: Collins.

Yoon, M. G.
    1973 Retention of the original topographic polarity by the 180° rotated tectal reimplant in adult goldfish. *J. Physiol. (London),* **233**:575–588.

Young, J. Z.
    1939 Fused neurons and synaptic contacts in the giant nerve fibres of cephalopods. *Phil. Trans. R. Soc. London* (B), **229**:465–503.
    1971 *The Anatomy of the Nervous System of Octopus vulgaris.* London: Oxford University Press.

Younkin, S. G.
    1974 An analysis of the role of calcium in facilitation at the frog neuromuscular junction. *J. Physiol.,* **237**:1–14.

Zilber-Gachelin, N. F., and M. P. Chartier.

1973a Modification of the motor reflex responses due to repetition of the peripheral stimulus in the cockroach. I: Habituation at the level of an isolated abdominal ganglion. *J. Exp. Biol.,* **59**:359–381.

1973b Modification of the motor reflex responses due to repetition of the peripheral stimulus in the cockroach. II: Conditions of activation of the motoneurones. *J. Exp. Biol.,* **59**:383–403.

Zucker, R. S.

1972a Crayfish escape behavior and central synapses. I. Neural circuit exciting lateral giant fiber. *J. Neurophysiol.,* **35**:599–620.

1972b Crayfish escape behavior and central synapses. II. Physiological mechanisms underlying behavioral habituation. *J. Neurophysiol.,* **35**:621–637.

1972c Crayfish escape behavior and central synapses. III: Electrical junctions and dendrite spikes in fast flexor motoneurons. *J. Neurophysiol.,* **35**:638–651.

1973 Changes in the statistics of transmitter release during facilitation. *J. Physiol. (London),* **229**:787–810.

Zucker, R. S., D. Kennedy, and A. I. Selverston.

1971 Neuronal circuit mediating escape responses in crayfish. *Science (Wash., D.C.),* **173**:645–650.

Indexes

Name Index

An *italic* page reference refers to a figure legend on that page.

Adey, W. R., 213
Adrian, E. D., 47ff, 56, 64, l09, 644
Advokat, C., 659
Agranoff, B. W., 527, 588, 607, 642
Alexander, R. D., 35
Alnaes, E., 584f
Apler, T. H., 606
Alvarez, R. B., 407
Alving, B. O., 241, 243f, 255f, 260
Apáthy, S., 215
April, R. S., *483*
Arch, S., 408
Aristotle, 11
Armstrong, C. M., 163
Armstrong, D., 509f, *512*
Arvanitaki, A., 56f, 64, 67, 73,
    221, 238
Ascher, P., 275f, 304, 311, 313, 325,
    330, 342, 491
Ashby, W. R., 212n
Atkinson, R. C., 607
Attardi, D. G., 475
Atwood, H. A., 325
Atwood, H. L., 647
Atz, J. W., 58
Axelsson, J., 533n

Bahr, R., 615
Bailey, C. H., 354, 359, 365, 560,
    584, 611
Baker, P. F., 146
Baldessarini, R. J., 203n
Barker, J.L., 260, *267*
Barlow, G. W., 18, 33n
Barnett, S. A., 18
Barondes, S. H., 527f, 531, 588f,
    607, 642

Barrett, J. N., 258
Baylor, D. A., 58, 235, 237, 280
Beach, F. A., 13, 46
Bennett, M. V. L., 103, 277n, 291n,
    309, 394n, 404, 469f, 491,
    494, 584n
Bentley, D. R., 62, 65
Benzer, S., 60f, 63, 65, 661f
Berman, A. J., 645
Bernard, C., 653
Bernstein, J., 99
Berry, M. S., 328
Berry, R. W., 528
Beswick, F. G., 480
Beurle, R. I., 211f
Bezanilla, F., 163
Binstock, L., *165*
Bittner, G. D., 485
Bjork, R. A., 622
Blankenship, J. E., 229, 273, 299ff,
    303, 309
Blodgett, H. C., 20
Bloedel, J., 179
Bodian, D., 109
Bohdanecky, Z., 609
Boistel, J., 596f
Bon, C., 324
Boring, E. G., 11, 27, 212
Bottazzi, F., 221
Bower, G. H., 213, 606ff
Boycott, B. B., 47
Boyd, I. A., 187n
Brackenbury, R. W., 408
Breckenridge, B. McL., 589n
Brenner, S., 61, 65
Bridgman, P. W., 21
Brink, J. J., 527

Brinley, F. L., Jr., 119, *152*
Brock, L. G., 285, 484
Brodwick, M. S., 251
Brogden, W. J., 20
Bronk, D. W., 48, 55, 480f, 485
Brown, A. M., 313
Brown, J. S., 658
Brown, P. L., 19
Brunelli, M., 589
Bruner, J., 309n, 491, 506, 519, *521,
    524*, 564
Bryant, H., 250, 298, 306, 594f
Bryant, J. S., 594f
Bullock, T. H., 47, 52, *73, 219,* 354n
Bürger, O., 215
Burrows, M., 58, 235, 645
Butcher, R. W., 529n
Byrne, J., 230, *357, 366, 370,* 372ff,
    399, 416, 548f, 552, 560, *562*

Cajal, S. R., 37, 212, 282, 476, 479,
    536, 596f
Calabrese, R. L., 455f, 595
Callec, J. J., 596f
Campbell, C. B. G., 41
Cardot, H., 57
Carew, T. J., 74, 354f, 359, *361, 364,*
    387, *389, 391,* 401, *403, 435,*
    494, *542,* 546, 548f, 552, 554ff,
    560, 577, 580, *610, 613,* 622ff,
    626, 632, 637, 659
Carpenter, D. O., 241, 243f, 250,
    257, 260
Case, J., 235
Casola, L., 527
Castellucci, V. F., 74, 194, 229f, *366,
    369,* 372, 374, 491f, 498, 528,

Castellucci, V. F., *continued*
    530, 542, *544*, 548f, 552, 554ff,
    *558*ff, 562, 565, *567*, 572, 575,
    577, 580f, 583, 585f, 589, 616,
    *618, 626, 629*f, *632*
Ceccarelli, B., 195
Cedar, H., 529, 531, 588f
Chalazonitis, N., 57, 223, 238
Chartier, M. P., 596
Chen, C. F., 255f
Chiarandini, D. J., 301
Cho, S. (*see* Steiner, D. S.).
Chorover, S. L., 609
Christoffersen, G. R. J., 244n
Coggeshall, R. E., 57, *79*, 82f, 225,
    229, 273, 276, 295ff, 299, *407*
Cohen, H. D., 527f, 607
Cohen, J., 460ff, 647
Cohen, L. B., 538, 661
Cohen, M. J., 58, 235, 276, 368n
Cole, K. S., 56, 99f, 138, 143, 146,
    153, 244n
Coleman, J. C., 654
Connor, J. A., 257f
Conroy, R. T. W. L., 480
Cooke, I. M., 241, 244
Coombs, J. S., 98, 165, 285, 484
Cooper, D., 204
Cotman, C. W., 532
Cottrell, G. A., 328
Courtney, K., 532
Cowan, J. D., 211f
Craig, W., 16, 26
Craik, F. I. M., 609
Crill, W. E., 258
Culbertson, J. T., 211f
Curtis, D. R., 330
Curtis, H. J., 56, 99f, 138, 143,
    146, 153

Dagan, D., 596
Dale, H. H., 102, 283, 285ff, 342
Dallenbach, K. M., 606
Dane, B., 647
Darwin, C., 3ff, 40, 46, 64
Davis, R. E., 527
Davis, W. J., 52n, 65, 359, 438, 440,
    468ff, 473, 652
De Robertis, E. D. P., 205n
De Weer, P., 244n
Dees, V., *621*f
Del Castillo, J., 98, 187ff, 194,
    197n, 488
Dennis, M. J., (*see* Kuffler, S.W.)
Deschênes, M., 291n
Dethier, V. G., 658, 662
Dewhurst, S. A., 276, 304

Diamond, J., 215, 329
Dieringer, N., *87*, 309, 399, 420, 656
Djahanparwar, B., 498
Dodge, R., 11
Dorsett, D. A., 58, 234, 449
Doty, R. W., 448
Dragoin, W. B., 19
Dudel, J., 200, 485, 491, 524, 531
Duncan, C. P., 607

Eales, N. B., 69, 74n, *90*, 92
Eccles, J. C., 98, 102, 135, 165, 202
    209 285ff, 342, 477, 484, 488,
    524, 536, 650n
Eckert, R., 266n
Ehrenberg, C. G., 215
Eisenstadt, M., 274ff, 327n
Elliot, T. R., 102
Epstein, R., 499, 504
Ervin, F., 528
Evarts, E. V., 488f
Evoy, W. H., 647

Faber, D. S., 260, *264, 267,* 276
Fabre, J. H., 6
Farel, P. B., 600, 620, 651
Fatt, P., 98, 102, 163, 165, 185, 187,
    283, 285f
Fechner, G. T., 52ff
Feinstein, R., 420
Fentress, J. C., 488
Field, P. M., *532*
Finer, R. L., 631
Fischbach, G. D., 494
Flexner, J. B., 527, 607
Flexner, L. B., 527, 607
Forbes, A., 478
Fowler, H., 539
Frank, K., 524
Fraser Rowell, C. H., 235, 462,
    466, *468*
Frazier, W. T., 58, 225ff, *231,* 238,
    256, 260, *263,* 266f, 273, 276,
    295ff, *299,* 330, *332, 409, 449,*
    491, 509, 523
French, M. C., 58, 235
Frenk, S., 523
Freud, S., 4
Friedländer, B., 215
Frings, C., 457n
Frings, H., 457n
Frisch, K. von, 20, 46, 67n
Fulpius, B. W., 204
Fuortes, M. G. F., 524
Furshpan, E. J., 102f, 277n, 494
Furukawa, T., 103

Gainer, H., 235, 260, *267,* 270, 408
Gallagher, J. E., 540
Garcia, J., 18f, 26
Gardner, D., 58, 232, 304, 317f, 320f,
    336, 338, 340
Geduldig, D., 244n, 246f, 249
Geffen, L. B., 195
Gerard, R. W., 100
Gerschenfeld, H. M., 57, 67, 174n,
    270ff, 276n, 299, 301, 303, 319,
    327f, 330, 342, 491
Getting, P., 394, 452, 494
Giller, E., Jr., 273, 275, 304
Ginsborg, B. L., 209
Glanzman, D. L., 600, 651
Glaser, E. M., 600, 620
Glassman, E., 527f
Gola, M., 25l, 258, 260, 265, *267*
Goldberg, A. L., 53l, 589n
Goldberg, N. D., 588
Goldman, D. E., 142
Goldman, J. E., 276n, 427
Goldman, L., 165
Goldschmidt, R., 215, *219*
Goldstein, A. J., 620f
Gooden, B. A., 420
Gorman, A. L. F., 243ff
Goronowitsch, N., 215
Graham, C. H., 52f
Graham, J., 100
Grant, D. A., 638n
Graubard, K., 83
Graziadei, P. P. C., 533
Green, K. F., 18
Greengard, P., 529
Grether, W. F., 11f, 26, 499, 625
Grillner, S., 645
Grindley, G. C., *621*
Grinnell, A., 532
Groves, P. M., 599, 605n, 635, 642
Gruener, R., 246, *249*
Grundfest, H., 52, 109, 246, 593n
Guillet, J. C., 596f
Gunn, R., 257
Guthrie, E. R., 605n

Hagbarth, K.-E., 625, 631, *633, 635,*
    651, 656, 662
Hager, J. L., 18f, 27
Hagiwara, S., 246, 257, 645f
Hamilton, G. V., 540
Hamilton, W. J., III, 346, 418
Hanawalt, J. T., 647
Hardy, A., 74
Harlow, H. F., 540f, 654, 660
Harlow, M. K., 654, 660
Harris, A. J. (*see* Kuffler, S. W.)

Harris, J. D., 11, 540
Harris, W. A., 63
Hartline, H. K., 47, 52ff, 64, 109
Hebb, D. O., 31f, 477ff, 536, 607
Heinroth, O., 16, 20, 26
Hening, W. A., *637*
Herrnstein, R. J., 11, 27
Herz, M. J., 540, 607
Heuser, J. E., 195, 197, 533
Highstein, S. M., 584n
Hilgard, E. R., 12, 22, 24, 213, 478, 498, 606, 620
Hille, B., 139, 165n
Hinde, R. A., 18, 22, 347, 418, 603, 609, 620f, 651
Hodgkin, A. L., 56, 98ff, 138, 143, 146f, 153, 157ff, 206, 209, 244n
Hodos, W., 41
Holmgren, G., 523
Horn, G., 603
Horridge, G. A., 73, 219, 354n, 645
Hoy, R. R., 62, 65, 645
Hoyle, G., 58, 67n, 235, 394n, 449
Hubel, D. H., 54, 214, 475, 532, 660
Huber, F., 645
Hughes, G. M., 223, 233, 645ff
Hughes, J. R., 488f
Hull, C. L., 605n
Hultborn, H., 286
Humphrey, G., 11, 540, 611
Hunter, W. S., 599
Hurlbut, W. P., 195
Huxley, A. F., 56, 98ff, 146f, 157ff, 206
Hydén, H., 527

Ihering, H. von, 57, 221
Ikeda, J., 645f
Ikeda, K., 52n, 438
Iles, J. F. (*see* Pearson, K. G.)
Immelmann, K., 20

Jacklet, J. W., 58, 235, 354, 407, 649
Jahan-Parwar, B., 457n, 498
James, W., 3, 15n, 606
Jankowska, E., 286
Jansen, J. K. S., 251
Jarrott, B., 195
Jarvik, M. E., 609
Jasper, H., 625, 651
Jeffrey, W. E., 538, 603, 609, 661
Jenkins, H. M., 19
Jennings, H. S., 6f, 15, 26, 46, 64
John, E. R., 212f
Johnson, G. E., 646
Jordan, H., 457n
Junge, D., 246f, 249, 251, 260, 265

Kado, R. T., 246
Kandel, E. R., 41, 57, 68n, 74, 80, 84, 92, 194, 221, 223, 225ff, 232, 244, 260, *263*, 266f, 273ff, *291*, *293*, 295ff, 303ff, 311ff, 317f, *320*, 324, *332*, 335ff, 352ff, *361*, *366*, *369*, 372, 374, 378, 381, 387, *389*, *391*, 401, *403*, 411ff, 420f, *435*, *449*, 457n, 477, *490*, 494, 497ff, *501*, 506ff, *514*, *516*, *521*, 523ff, 528ff, 542, *544*, *546*, 548f, 552ff, 554ff, *559*, 562, 565, *567*, 572, 575, *577*, 580ff, 585f, 588f, *610*, *613*, *618*, 622ff, *626*, *629*, 632, *637*, 649, 659
Kao, C. Y., 593n
Karlin, A., 204f
Katayama, Y., 235
Kater, S. B., 58, 235, *458*, 462, *464*, 649
Katz, B., 98, 102, 138f, 143, 146f, 153, 157, 163, 179, 181n, 183ff, 187ff, 191n, 194f, 197f, 209, 30l, 488, 584f, 591
Kaufman, I. C., 654
Kebabian, J. W., 529
Kehoe, J. S., 58, 174n, 233, 273, 291, 298, 301, 304, 311, 322ff, 342, 506, 519, *521*, *524*
Kerkut, G. A., 58, 235, 241, 244n, 301, 593
Kernell, D., 528
Kety, S. S., 662
Keynes, R. D., 163, 244n
Kimble, G. A., 498
Kimmel, H. D., 620f, 638
Klee, M. R., 260, *264*, *267*, 276
Klett, R. P., 204
Kling, J. W., 622
Koester, J. D., *87*, *226*, 267n, *293*, 298, 306, 309, 399, 420f, 427, 429ff, *433*, *437*, *439*, *441*, 656
Koike, H., 273f, 304
Koketsu, K., 285, 330
Konorski, J., 9, 477f, 536, 605n, 625n
Kopp, R., 609
Kral, P. A., 19
Krasne, F. B., 473, 591ff, 601, 603
Kravitz, E. A., 58, 232, 235f, 276f, 280
Kretch, D., 23, 31n
Kriegstein, A. R., 74
Kristan, W. B., Jr., 235, 455ff, 645

Kuba, K., 531
Kuffler, S. W., 54, 96, 102, 200, 215f, 485, 491, 524, 533n
Kugelberg, E., 625, 631, *633*, *635*, 651, 656, 662
Kukenthal, W., 215
Kuno, M., 194, 485, 488
Kunze, D. L., 313
Kupfermann, I., 57, 74, 84, 86, 89, 221, 225, 270, *293*, 352n, 354f, *357*, *360*, 363ff, *371*, 378f, 381f, 385n, 394n, 407, 411ff, 421, 457n, *459*, 542, *544*, *546*, 548, 551, 556, 638, 647, 649, 658f
Kusano, K., 179, 257

Landgren, S., 285f
Lange, P. W., 527
Langley, J. N., 102
Larrabee, M. G., 480f, 485
Lashley, K. S., 3, 24, 46, 212f, 602, 650n
Leblanc, G., 241
Lehninger, A. L., 584
Lehrman, D. A., 18, 22, 347
Lettvin, H. Y., 54, 475
Levinthal, C., 61
Levitan, H., 58, 234, 319, 324, *327*, 470, 589
Levitan, I. B., 531
Leydig, F., 215f
Libet, B., 330
Lickey, M. E., 457n
Liebeswar, G., 276n, *293*, 420f, 427
Liley, A. W., 187n, 191n, 485f
Lim, R., 527
Lindström, S., 286
Livengood, D. R. (*see* Kusano, K.)
Llinás, R., 184
Lloyd, D. P. C., 283, 385, 480, *482*, 485, 488, *490*
Locke, J., 11
Lockhart, R. S., 609
Loeb, J., 6f, 15, 46, 64
Loewi, O., 102, 328
Loh, Y. P., 408
Lopresti, V., 61
Lorento de Nó, R., 478f
Lorenz, K., 16f, 20, 26, 346
Lubbock, J., 6, 46, 64
Lugaro, E., 476
Lukowiak, K., 354n, 649
Lux, H. D., 266n
Lynch, G. S., 532

Macagno, E. R., 61
McCaman, M. W., 276n

McCaman, R. E., 276, 304
McCleary, R. A., 213
McCulloch, W. S., 211n
McDougall, W., 607
McGaugh, J. L., 607
McGeoch, J. A., 606
MacGinitie, G. E., 406
McGowan, B. K., 18
McMahan, U. J., 277, 649
MacPherson, S. J., 621f
Maier, N. R. F., 20
Manning, A., 15, 18, 27
Marler, P., 346, 418
Marmor, M. F., 243, *245*
Marquis, D. G., 12, 22, 478, 498, 620
Marshall, W. H., 488
Martin, A. R., 103, 187n, 194, 209,
    485, 491, *493*, 570n, 651
Martini, E., 220
Mathieu, P. A., 265
Matthews, B. H. C., 51
Mauthner, L., 215
Mayeri, E., 255, *293*, 420ff, *425*ff, 436
Maynard, D. M., 235, 325, 473, 645f
Mayr, E., 16
Mazzarelli, G., 57, 221
Mead, A. P. (*see* Russell, W. M. S.)
Meech, R. W., 250, 301
Mendel, G., 59
Mendelson, M., 440
Metcalf, J. F., 533
Michaelson, D. M., 204
Miledi, R., 98, 179, 181n, 183f, 188,
    191n, 194, 197f, 533n, 566,
    584f, 591n
Miller, J., 246
Miller, N. E., 21f, 609
Miller, P. L., 645
Miller, S., 9
Minami, H., 606
Mirolli, M., 244n
Mittenthal, J. E., 593
Miyahara, J. T., 194
Miyamoto, M., 589n
Model, P. G., 584n
Morgan, C. L., 6
Morgan, T. H., 59
Morris, D., 17
Moruzzi, G., 523
Mott, F. W., 644
Mountcastle, V. B., 203n, 214
Mpitsos, G. J., 468, 470
Müller, G. E., 606f
Muller, H. J., 60
Muller, K. J., 276, 649
Mulloney, B., 235, 325, 473, 645f

Murray, M. J., Jr., 471
Myers, R. E., 213

Nabias, B. de, 221, 223
Nakajima, S., 246, 251
Nakamura, Y., 246
Navratil, E. (*see* Loewi, O.)
Neher, E., 257f
Neilson, D. R., Jr., 596, 599, 603,
    625, 631, 642, 651
Nernst, W., 141
Newby, N. A., 354n, 385n
Nicholls, J. G., 43, 54, 58, 66, 96,
    235, 237, 251, 280, 491, 532, 585
Nicholson, C., 184, 649
Niklaus, R., 596
North, K. A. K., 191n

Ort, C. A., 235, 238, 455f
Otsuka, M., 58, 232, 235f, 280
Oyer, P. E. (*see* Steiner, D. S.)

Paine, R. T., 470
Paolino, R. M., 609
Parmentier, J., 235
Parnas, I., 491, 509f, *512*, 596
Patch, V. D., 654
Paupardin-Tritsch, D., 174n,
    276n, 327f
Pavlov, I. P., 7, 9, 11f, 26, 32,
    605n, 625
Pearson, K. G., 645
Peckham, E. G., 11
Peckham, G. W., 11
Peeke, H. V. S., 540
Peretz, B., 352n, 354f, 385n,
    542n, 649
Perlman, A., 354, 359, 365
Peterson, J. D. (*see* Steiner, D. S.)
Peterson, L. R., 624
Peterson, R. P., 528
Pichon, Y., 596f
Pieron, H., 11
Pilar, G., 103, 485, 491, *493*
Pilzecker, A., 606f
Pinneo, J. M., 470
Pinsker, H. M., 84, 244, *263*, 266f,
    304, 311ff, 324, 355, *361*, 420,
    457n, 509f, *514*, *516*, 542, *544*,
    548f, 556, *610*, *613*, 636ff, 655
Pitman, R. M., 276
Pitts, W. H., 211n
Postman, L., 606
Potter, D. D., 54, 58, 102, 105, 215f,
    232, 235f, 277n, 280, 494

Pribram, K. H., 650n
Prosser, C. L., 47, 50f, 56, 64, 599
Pumphrey, R. J., 595
Purves, D., 491, 585

Quartermain, D., 609
Quinn, W. G., 45, 63

Raftery, M. A., 204
Rahamimoff, R., 198, 584f
Raisman, G., 532
Rall, W., 114
Ratliff, F., 52, 54
Rawdon-Smith, A. F., 595
Reese, T. S., 195, 197, 533
Reich, E., 204
Remler, M. P., 43, 58, 235, 280, 593n
Renshaw, B., 285
Rescorla, R. A., 9, 605n
Retzius, G., 54, 215f, 237
Roach, J. L. M., 648
Roberge, F. A., 265
Roberts, R. B., 527, 646
Robison, G. A., 529n
Roe, A., 16
Roeder, K. D., 47, 66, 596, 645
Rojas, E., 163
Romanes, G. J., 5f, 40, 46, 64
Roper, S., 328f, 532
Rose, R. M., 460
Rosenblatt, F., 211f
Rosenblum, L. A., 654
Rosenthal, J., 485, 488
Rosvold, H. E., 620
Rovainen, C. M., 651
Rubenstein, A. H. (*see* Steiner, D. S.)
Rubinson, K., 355, *361*, 548f
Russell, W. M. S., 33n
Russell, W. R., 607
Ryall, R. W., 330

Saito, N., 257
Sarne, Y., 408
Schacter, S., 659
Schiller, P. H., 609
Schleidt, W. M., 539
Schlosberg, H., 498
Schmidt, R. F., 524
Schneider, G. E., 532
Schneirla, T. C., 13f, 46, 540
Schwartz, J. H., 273ff, 304, 355, 491f,
    528ff, 585f, 588f, 649
Sealock, R., 324
Segaar, J., 213
Segundo, J.P., 58, 234
Seligman, M. E. P., 18f, 27

Selverston, A. I., 43, 58, 235, 280, 325, 473, 593, 645f
Sharpless, S. K., 625, 651
Shaw, T. I., 146
Sherrington, C. S., 11, 32, 283, 285n, 384f, 596f, 644
Shiffrin, R. M., 607
Shimahara, T., 330f, 499, 589
Sidman, M., 662
Siegler, M. V. S., 468
Simpson, G. G., 16, 40f
Singer, J. J., 531, 589n
Skinner, B. F., 4, 9, 21, 24, 31f, 34, 605, *639*
Skyles, M., 229
Smith, F. E. (*see* Villee, C. A.)
Smith, T. G., Jr., 260, 265, *267*
Sokolov, E. N., 600, 603, 638n
Solomon, P., 654
Solomon, R. L., 9, 605n
Spencer, W. A., 354, 480, *483*, 518, *542*, 596, 599, 603, 625, 631, 642, 649, 651
Sperry, R. W., 213f, 475, 638n
Spira, M. E., 404, 469, 494
Spray, D. C., 469f
Standen, N. B., 250
Stefani, E., 301, 319
Steiner, D. S., 408
Stellar, E., 527, 607
Stent, G. S., 235, 455, 457, 625
Stephens, C. L., 251, 260, 265
Stevens, C. F., 135, 157, 257f
Steward, O., 532
Stinnakre, J., 84, 246, *250*, 591n
Stretton, A. O. W., 276f
Strickberger, M. W., 60
Strumwasser, F., 86, 250, 256, 260, *263*, 265, 295, 407, 491, 509f, 512, 615
Stuart, A. E., 58, 235, 238
Sutherland, E. W., 529n

Takahashi, K., 246, 251
Takeda, K., 593
Takeda, R., 255f
Takeuchi, A., 202
Takeuchi, N., 202
Tanzi, E., 476, 536
Taub, E., 645

Tauc, L., 57f, 67, 84, 223, 225n, 233f, 238, 241, 246, *250, 253*, 270ff, 280, 299, 303, 309n, 319, *322*, 324, *327*, 330f, *490*, 498f, 501ff, 506f, 509, 519, *521*, 524, 526, 589, 591n, 638
Tchou, S. H., 57, 221
Terrace, H. S., 10n
Thesleff, S., 533n
Thomas, R. C., 241ff, 244n
Thompson, E. B., 83, 584, 649
Thompson, R. F., 213, 518, 596, 599f, 603, 605, 625, 631, 635, 642, 651
Thorndike, E. L., 7ff, 11f, 26
Thorpe, W. H., 20, 22, 27, 47, 64, 66, 540, 603, 609
Tinbergen, N., 15ff, 26, 346, 539
Toevs, L. A., 408
Tolman, E. C., 13, 23, 31f, 34
Tomita, T., 174n
Tozian, L., 645
Truex, R. C., *73*
Tweedle, C. D., 276

Uexküll, J. von, 15

Van Essen, D., 43, 54, 66
Vaughan, J. E., 276n
Verworn, M., 6, 46, 64
Villee, C. A., *49*
Von Baumgarten, R. J., 255f, 498
Von Neumann, J., 212n
Votava, J., 174n, 330

Wachtel, H., 265, 273, 300, *303*, 311, *332, 449*
Wald, F., 246
Walker, R. J., 58, 235
Walker, W. F., Jr. (*see* Villee, C. A.)
Wall, P. D., 600
Watanabe, A., 593n, 645
Watson, J. B., 6f, 23, 31f
Waziri, R., 58, 225, 260, *264*, 266f, 273, 276, 295ff, *299*, 313, *315*, 333, *337*, 381, 491, 509, 523, *525*
Weber, E. H., 52
Weiant, E. A., 645
Weight, F. F., 174n, 330

Weinreich, D., 250, 276, 298, 306, 327n, 485, *487*
Weiskrantz, L., 609
Weiss, K. R., 84, 270, 460f, 647
Wells, M. J., *71*, 638
Werman, R., 301
Wernig, A., 191n, 195
Whalen, R. E., 539
Whitman, C. O., 16, 26
Whittaker, V. P., 194
Wickelgren, B. G., 600
Wickelgren, W. A., 609
Wickelgren, W. O., 651
Wiersma, C. A. G., 47, 52, 54, 368n, 438, 645ff, 652
Wiesel, T. N., 54, 214, 475, 532, 660
Wigdor, R., 480
Wilcoxon, N. C., 19
Williams, D. R., 19 ﹨
Williams, H., 19
Williams, L. W., 54f
Willis, W. D., 524
Willows, A. O. D., 35, 43, 58, 66, 92, 234f, 394n, 449, *451*, 473, 494
Wilson, D. L., 268, 270, 311, 528
Wilson, D. M., 52, 67n, 440, 473, 644ff, 652
Wilson, V. J., 488, *490*
Wilson, W. A., Jr., 265
Wine, J. J., 593, 595
Winlow, W., *80*, 277f, 295, 298f
Winograd, S., 211f
Winsbury, G. J., 285n
Wollberg, Z., 270
Woodsmall, K. S., 591ff
Woodworth, R. S., 8, 30n, 498
Woollacott, M. H., 234f
Wyers, E. J., 540

Yasargil, G. M., 329
Yonge, C. M., 74
Yoon, M. G., 475
Young, J. Z., 47, 54ff, 64
Younkin, S. G., 488

Zilber-Gachelin, N. F., 596
Zimmermann, H., 195
Zotterman, Y., 48
Zucker, R. S., 195, 485, 593f

Subject Index

An *italic* page reference refers to a figure on that page.

Acetylcholine (ACh), 187 (*see also* Synaptic transmission; Transmitters)
Action potential
  definition, 100
  beating cells, 238, 256ff
  bursting cells, 238, 260ff
  calcium role in generation, 246ff
  vs. electrotonic potential (conduction comparison), 123ff
  generating (ionic) mechanisms, 143ff, 151ff, 246ff
  hyperpolarizing afterpotential, 102
  initiation site, 105, 251ff
  sodium inactivation, 144
  triggering by synaptic potential, 100, 527
  undershoot, 102
Afterpotential
  definition, 100
  generating mechanism, 251
Annelida (segmented worms)
  giant fiber system (*Lumbricus*), 215
  nervous system (*Hirudo*), 455
  neuronal sprouting, post-denervation (*Hirudo*), 532f
  neurons, identified (*Hirudo*), 215, 235
  swimming (*Hirudo*), 455ff
Aplysia (marine snail)
  body plan, 74, 76f
  central nervous system, 77ff
    analytic advantages, 67f
  circulatory system (*q.v.*)
  complex behavior, 85ff, 419ff, 436ff, 441ff
  development, 74f
  elementary behavior, 84, 350ff, 378, 385ff, 406ff
  egg-laying, 84, 406ff (*see also* Fixed acts)
  escape response, 87
  feeding behavior, 89

  components, 457f
  neural mediation, 460ff
fixed-action patterns, 419ff, 436ff, 441ff
fixed acts, 84, 87, 378, 385, 406, 441 (*see also* Fixed acts)
foot, 76f
ganglia
  abdominal, 77ff, 80, 225, 226f, 229ff
  buccal, 77f, 232, 460
  cerebral, 77f, 460
  pedal, 77f
  pleural, 77f, 233, 506, 519
gill movements (classification), 355
gill-withdrawal reflex (*see* Reflexes)
glandular fixed act, 84, 385f
head, 74, 76, 457ff
higher-order behavior, 80
homeostatic adjustments, 87f, 419f
inking, 84, 385f (*see also* Fixed acts)
life cycle, 74f
locomotion, 85f
mantle organs, 76f, 350
  defensive withdrawal of, 84, 350ff (*see also* Reflexes)
metacerebral cells, 327, 460ff, 506
motivation, 658ff
neuroendocrine fixed acts, 84, 406ff
neuroendocrine reflexes, 84
neurons, identified, 225ff
reflex acts, 84 (*see also* Reflexes)
respiratory pumping, 378, 441ff
sexual behavior, 90
somatic-motor reflex act (example), 84, 350
visceral mass, 76f
visceral-motor fixed-action patterns, 87f, 378ff
visceral-motor fixed act (example), 378

Arousal
  components, 635
  and dishabituation, 585
  functions of motivational state, 659
  and sensitization, 635

Behavior. *See also* Fixed-action patterns; Fixed
    acts; Reflex acts; Reflex patterns
  definition, 30f
  abnormal behavior (stress-induced)
    anxiety (pathological), 654f
    and development, 660f
    disturbance of motivation, 658ff
    extreme sensitization (as model of
      anxiety), 655ff
    neuroses and psychoses, 654
  action system, 15
  aggressive behavior (habituation of), 540
  cellular approach, 37f, 63f (goals); 650 (problem
    of emergent properties)
  classification, 32ff
  complex behavior (behavioral patterns), 32
  consummatory component, 16
  defensive behavior (*see* Fixed-action patterns;
    Fixed acts)
  effector systems
    multiple controls, 384f
    types, 350
  elementary behavior (behavioral acts), 32
  emergent properties, 650
  hierarchy of responses, 470f
  higher-order behavior, 33
  intervening variables, 23, 34
  invertebrates as simple models of, 29, 35f
  lower-order behavior, 33
  modification (*see also* Conditioning; Dishabitua-
    tion; Habituation; Learning; Memory;
    Sensitization)
    capability (and common neural
      mechanisms), 39f
    and development, 660f
    by heterosynaptic facilitation, 498f
    neural analogs, 494f
    neural dynamic-change hypothesis, 478f
    neural plastic-change hypothesis, 476ff
    short-term (*see* Conditioning, *etc.*)
    short-term vs. long-term, 606ff, 625ff, 636ff
  and motivational state, 658ff
  psychological models, 5ff
  quantitative neural analysis, 374, 377f, 650f
  sexual behavior (*q.v.*)
  stimulus–response relationships, 348f
  units, 346ff
Brain
  effector systems controlled by, 350
  functioning (conceptual approaches), 212f

Calcium
  in action potential generation, 246ff
  role in plasticity, 485f, 588ff
  in transmitter release, 183f, 197f, 584f
Cellular connectionism, 212
Cholinergic transmission. *See* Neurons, multiac-
    tion; Synaptic transmission; Transmitters
Circulatory system (*Aplysia*)
  anatomy, 421f
  antagonist fixed acts (coordination), 443ff
  behavior, 419f
  command elements, 438ff, 443, 445
  heart rate (central program)
    decrease, 436ff
    increase, 429ff
  hormonal control, 436
  interneurons, 428f
  motor neurons, 421ff
  and respiratory system (coordination), 441ff
Conditioning
  alpha, 495ff
  classical, 7ff, 18f, 496ff
    models, 638f
    neural analog, 497
    and sensitization, 636, 638
  extinction, 11, 496
  and habituation, 11, 22, 518f
  instrumental, 8, 19
    associative changes, 506ff
    neural analog, 497, 506
    nonassociative changes, 506ff
  operant (*see* Conditioning, instrumental)
  pseudoconditioning, 12, 495
  reinforcement, 8, 496
  taste-avoidance (bait-shyness), 18f
Crustacea
  command cells (crayfish), 646f
  escape-swimming habituation (cellular analysis
    in crayfish), 591ff
  excitatory synapse (electrical, crayfish), 102f
  locomotion, neural control, 645f
  neurons, identified, 215, 235f
    multiaction (lobster), 325
  sensory system, functional organization
    (crayfish), 647f
Cyclic adenosine monophosphate (cAMP), 529n
Cyclic nucleotides
  as intracellular messengers for hormone
    action, 588
  presynaptic facilitation (possible role in), 591
  stimulation by synaptic transmitters, 589
  synthesis (neuronal control of), 529f

d-tubocurarine (*d*-tbc), 272, 325
Defensive behavior. *See* Fixed-action patterns;
    Fixed acts

Definitions:
  anomalous rectification, 115
  behavioral acts, 32
  behavioral patterns, 32
  binomial distribution, 189ff
  cable properties (membrane), 32
  capacitance, 115
  capacitive current, 118f
  capacitor, 115
  conditioned stimulus (CS), 9
  coupling ratio, 132
  delayed rectification, 115, 144
  gating current, 163
  ionic current, 118f
  leakage current, 157
  length constant (membrane), 121
  membrane battery, 113, 147ff
  mobilization (transmitter), 195
  noise, 126
  overshoot, 102
  quantal hypothesis, 189
  rectification, 115
  resistance, 111
  resistive current, 118f
  retroactive interference, 606
  space constant (membrane), 121
  spatial summation, 130
  spike, 100
  temporal summation, 127
  time constant (membrane), 121
  unconditioned stimulus (US), 9
  undershoot, 102
Depolarization, 100
Dishabituation. *See also* Sensitization
  definition, 11, 625
  and arousal, 585
  cyclic AMP role in, 588f
  vs. habituation, 625ff
  molecular model, 585ff
  as sensitization, 627, 631, 636
  short-term (cellular model), 583ff
Dynamic change, 478

Escape responses. *See* Fixed acts; Reflex acts

Facilitation. *See* Synapses
Feed-forward substitution, 334
Feed-forward summation, 338
Feeding behavior
  Aplysia, 89, 457f
    consummatory response, 457ff
    neural mediation, 460, 462
    orienting response, 457, 460
  Helisoma, behavior and neural control, 462ff
  Navanax, 469f
  Pleurobranchaea, 468f

Fixed-action patterns, 6, 16ff, 33, 346ff
  definition, 16, 32ff, 346ff
  antagonist (coordination, *Aplysia*), 443ff
  in circulation control (*Aplysia*), 419f
  consummatory behavior, 16
    in feeding, 457
  escape responses
    Aplysia, 87
    habituation, 595f (cockroach); 591ff (crayfish)
  escape swimming (*Tritonia*), 449f
    central program, 452f
    neural mediation, 449, 451ff,
  vs. fixed acts, 454f
  swimming (*Hirudo*), 455
    central program, 456
    neural mediation, 455ff
  visceral motor (*Aplysia*), 87f
Fixed acts. *See also* Behavior; Reflexes
  definition, 33, 346ff
  common features (*Aplysia*), 415
  egg-laying (*Aplysia*), 84, 406ff
    neuronal circuitry, 408ff
    stimulus–response relationship, 411
  inking (as glandular response, *Aplysia*), 84, 385f
    central program, 397ff; 399ff (triggering)
    neuronal circuitry, 387, 391; 393ff (properties)
  integration into fixed-action pattern
    (*Aplysia*), 441ff
  neuroendocrine (*Aplysia*), 84, 406ff, 415
  vs. reflex acts, 415f
  respiratory pumping (*Aplysia*), 87f, 378;
    380f (neuronal circuitry)
  visceral motor (*Aplysia*), 378

Gamma-aminobutyric acid (GABA), 202, 235
Gastropoda. *See* Mollusca
Gill-withdrawal reflex. *See* Reflexes
Goldman equation, 142

Habituation. *See also* Conditioning; Dishabituation; Learning
  definition, 11, 538
  adaptive functions, 540
  of aggressive behavior, 540
  at cellular level, 538
  cellular analyses
    Aplysia, 542ff, 609ff
    cockroach, 596
    crayfish, 593ff
    vertebrates, 596ff
  as central synaptic change, 548ff
  in complex behavior, 540
  vs. dishabituation, 625ff
  as elimination of inappropriate responses, 538f

Habituation, *continued*
  of escape responses
    birds, 539
    cockroach, 595f
    crayfish, 591
  as familiarity with environment, 538
  features, 518f
  high-order, 600f
  homosynaptic depression as common
      mechanism, 600f, 617, 631
  isolated test systems, 552ff
  of learning (most elementary type), 538
  locus, 594, 616f
  long-term, 609
    acquisition (relation to transmitter
        release), 616f
    and complex learning, 620
    learning curve, 620ff
  monosynaptic analog (frog spinal cord), 600
  motor neuron input resistance in, 555
  neural analogs, 519ff
    heuristic value, 524, 527
  in perceptual organization, 538
  quantal analysis, 565ff
  of reflexes (man), 631
  retention (effects of training schedule), 622ff
  of sexual responses, 539f
  short-term
    cellular electrophysiological model, 583ff
    and long-term (comparison), 617, 619
    transition to long-term (neural studies), 611ff
  spontaneous recovery, 538
  subcellular mechanisms, 617, 619, 641
  synapses (vs. usual synapse features), 564
  synaptic changes (and behavioral change), 620
  synaptic input changes, 555ff, 574f
  synaptic mechanisms, 519, 555ff
  training schedule effects on, 622ff
Hyperpolarization
  definition, 100
  posttetanic, 251

Insects
  escape response, habituation (cockroach), 595f
  locomotion (neural control), 644f
Invertebrates
  behavior
    simplicity of, 35
    study of (strategy and limitations), 38ff
  nervous system (simplicity of), 36f
  neuron identification, 215ff, 234ff
  neuronal polarity, 80, *82*
  synapses (usage effects) 491
  and vertebrate behavior evolution, 40f

Learning. *See also* Conditioning; Habituation;
    Memory; Sensitization

  definition, 22
  associative (*see* Conditioning)
  complex learning
    habituation, 540f
    and long-term habituation, 620ff
  distributed training (effects on habituation), 622ff
  equipotentiality (of cortical neurons), 212
  imprinting, 20
  latent learning, 20
  law of effect, 8
  mass action, law of, 212
  massed training (effects on habituation), 622f
  neural analogs, 494ff
    heuristic value, 524, 527
  neuronal dynamic-change hypothesis, 478f
  neuronal plastic-change hypothesis, 476f
  paradigm, of stimulus patterning, 494f
    neural analogs, 496
  pseudoconditioning, 12, 495
  sensory preconditioning, 20
  from synaptic usage-induced change, 480
  transfer of learning, 8
  trial-and-error, 8
Leech (*Hirudo*). *See* Annelida
Lobster. *See* Crustacea
Locomotion
  Aplysia, 85f
  Crustacea, 645f
  Hirudo (swimming), 455ff
  insects, 644f
  neural control 644ff
  Tritonia, 449ff

Macromolecular synthesis. *See also* small-molecule
    synthesis
  neuronal control (and neuronal short-term
      functioning) 528f
  in neurons, 268ff
Membrane
  active responses, 111
    vs. passive properties, 123ff
  cable properties, 110, 125
  capacitance, 115ff
  conductance, 113
    conductance channels, 147
  electrical equivalent circuit, 112, 120
  external medium resistance, 110
  input capacitance, 117
  input resistance, 111
  ionic channels, 138
  ionic current, 118f
  length constant, 121
  noise, 126
  passive properties
    vs. active properties, 123ff
    and neuronal integration, 126ff
    and synaptic transmission, 130ff

resistance, 111ff
   rectification, 115ff, 144
resistive current, 118f
sodium–potassium pump, 145
space constant, 120ff, 127
time constant, 120ff, 127

Membrane potential
   definition, 98
   afterpotential, 100, 251
   depolarization, 100
   equivalent circuit, 146ff
   gating current, 163
   generating mechanisms, 140ff, 239ff
   hyperpolarization, 100
   hyperpolarizing afterpotential, 102, 251
   ionic hypothesis, 137
      and electrical equivalent circuit, 146ff
      testing of, 145f, 153ff
   leakage current, 157
   overshoot, 102
   of presynaptic terminals, 198ff
   rectification, 115ff, 144
   resting membrane potential (*see* Membrane
      potential)
   spike (*see* Action potential)
   synaptic potential (*q.v.*)
   threshold (for action potential), 100
   undershoot, 102

Memory. *See also* Learning
   definition, 606
   acquisition, 606
   consolidation, 607
   long-term
      store, 607f
      from synaptic plastic change, 477
   multiprocess models, 607ff
   retention, 606
   retroactive interference, 606
   short-term
      to long-term (transition), 606ff
      reverberatory circuits in, 478f
      store, 607f
      from synaptic plastic change, 477
   synaptic processes, 617ff

Mollusca (*except Aplysia*), 69ff
   autoactive cells, 257
   escape-swimming (*Tritonia*), 449
   feeding behavior (neural control)
      Helisoma, 462, 464ff
      Navanax, 469f, 494
      Pleurobranchaea, 469f
   giant axon (squid), 99f, 138ff, 143ff
   metacerebral cells, (*Helix*), 221, 223ff
   multiaction cells (*Navanax*), 324f
   multiaction dopaminergic cell (*Planorbis*), 328
   neurons, identified, 221, 223ff, 234f

Motivational state
   as intervening variable, 23f
   and stress, 658ff
Motor neurons
   in circulation control (*Aplysia*), 421ff
   in effector system (multiple and hierarchical
      controls), 384f
   in egg-laying (*Aplysia*), 408ff
   in feeding
      Aplysia, 460ff
      Navanax, 494
   final common motor neurons, 385
   in gill-withdrawal reflex (*Aplysia*), 354ff
      quantitative individual contributions, 362ff
   in inking (*Aplysia*), 387ff
   input resistance (in response habituation), 555
   rebound excitation, 380
   in respiratory pumping (*Aplysia*), 380f
   in siphon-withdrawal reflex (*Aplysia*), 364f, 611n

Nematoda (threadworms)
   head ganglia (invariance, *Ascaris*), 215ff
   larval organs (invariance), 220
Nernst equation, 141
Nervous system
   cellular elements, 96f
   central vs. peripheral in molluscs, 80
   central program, 17
   glial cells, 96
   information storage mechanisms, 535
   organization (common patterns), 39f
Neurons. *See also* Motor neurons; Sensory
     neurons; *and individual organisms*
   action potential (*q.v.*)
   autoactive rhythms, 238, 255, 506, 518, 609
      contigent stimulation effects on, 510, 514
      mechanisms, 256ff, 260ff
      noncontingent stimulation effects on, 509f
      signaling functions, 256
   axon, 96
   beating cells, 238, 256ff
   bursting cells, 238, 260, 265f
   cell body, 96
      isopotentiality, 123
   circuitry
      convergent aggregates, 281, 330
      convergent high-order connections, 331ff
      divergent aggregates, 282f
      feedback by follower neurons, 333
      feed-forward connections, 334ff
      feed-forward substitution, 334, 445
      feed-forward summation, 338ff
      reverberatory aggregates, 478f
   command elements, 438, 440, 646f
   commissures, *81*
   components, 107f
   connectives, *81*

Neurons, *continued*
  current–voltage relationship, 114ff
  cytoplasmic resistance, 113
  dendrites, 80, 82n, 96
  electrical signals (*see also* Action potential;
      Membrane potential)
    measurement techniques, 98
    saltatory conduction, 126n
  electrotonic potential and propagation, 100
    vs. action potential, 123ff
  equipotentiality for learning (cortex), 212
  excitability (changes due to), 477
  firing patterns, 238f
  functioning
    dynamic-change hypothesis, 478f
    plastic-change hypothesis, 476ff
    probabilistic features, 212
    single function (at synapse), 286
  identification, 215
    criteria, 225, 228
  identified
    in Annelida, 215, 235
    in Crustacea, 235f
    in gastropods, 221ff
  interconnection (*see* Synapses)
  interneurons (*see also below,* multiaction cells)
    in circulatory system control (*Aplysia*), 428f
    differential response depression, 599f
    dual-action mediation
      to multiple cells, 295ff
      to single follower cell, 304ff
    and sensory neurons (input potential
        similarities), 108
    synaptic potential (and trigger zone), 108
  invertebrates vs. vertebrates (common plan),
      80, *82,* 96f
  macromolecular synthesis (control of), 528
  membrane potential (*q.v.*)
  model neuron, 109f
  morphology, 276ff
  motor neurons (*q.v.*)
  multiaction cells
    cholinergic, in *Aplysia,* 295ff (abdominal
        ganglion), 317ff (buccal ganglion), 322
        (pleural ganglion); in lobster, 325; in
        Navanax, 324f; in vertebrates, 328ff
    dopaminergic, 328
    serotonergic, 327f
  neuronal integration (and passive membrane
      properties), 126ff
  neuropil, 80
  nomenclature, 228
  pacers, 238
  plasticity, 475f
    biochemistry, 527ff
    chemical vs. electrical synapses, 491ff
    depression, 488f
    heterosynaptic facilitation, 498ff
    in learning, 476f

  morphology, 531ff
  posttetanic facilitation, 480
  polarity, 80, *82*
  polarization, 98
  presynaptic terminal, 96
  protein synthesis, 268ff
  receptor surfaces, 96
  repetitive firing (rate-limited), 251
  resting potential (*see* Membrane potential)
  reverberatory circuits (in short-term
      memory), 478f
  sensory neurons (*q.v.*)
  single follower, dual action mediation to
      (*Aplysia*), 304ff
  single function at synapse, 286
  spatial summation (postsynaptic cell), 128ff
  specialization, *284f*
  sprouting (into denervated regions), 531ff
  synapse (*see* Synapses)
  synaptic field, 80
  temporal summation (postsynaptic cell), 127ff
  transmitter biochemistry and pharma-
      cology, 270ff
  trigger zone, 107f, 130
  vertebrates vs. invertebrates (common plan),
      80, *82,* 96f

Plastic change, 476
Poisson distribution, 191ff
Postsynaptic potential (PSP). *See* Membrane
    potential
Potentials
  action potential (*q.v.*)
  afterpotential (*q.v.*)
  electrotonic potential, 100
    vs. action potential, 123ff
  end-plate potential, 185ff
  equilibrium potential, 141
  excitatory postsynaptic potential (EPSP) (*see*
      Synaptic potential)
  generator potential, 107
  inhibitory postsynaptic potential (IPSP) (*see*
      Synaptic potential)
  membrane potential (*q.v.*)
  pacemaker potential, 258
  resting membrane potential (*see* Membrane
      potential)
  reversal potential, 155
    for synaptic excitation, 166, 175
    for synaptic inhibition, 174, 177
  spike (*see* Action potential)
  unit synaptic potential, 188
Protein synthesis. *See* Macromolecular synthesis
PSP (postsynaptic potential). *See* Membrane
    potential.

Quantal release of transmitter
  at vertebrate nerve-muscle synapse, 184ff
  at *Aplysia* sensory-motor synapse, 565ff

Reflex acts
    definition, 11, 33, 347
    and fixed acts, 415f
    sequence of neural events, 106f
    somatic motor, 84, 350ff
Reflex patterns
    definition, 17, 346
    appetitive behavior, 16
    and fixed-action patterns, 17, 33, 346ff
Reflexes
    abdominal, habituation (man), 631
    all-or-none (relation to fixed acts), 347ff
    extinction, 11
    flexion withdrawal reflex (cat), 105ff
        habituation, 596ff
        neuronal circuit, 106
    gill-withdrawal reflex (*Aplysia*), 84, 350
        behavioral analysis (quantitative, reduc-
            tionist), 374, 377f
        dishabituation (locus and mechanism), 575
        features, 545
        ganglion test system (isolated), 552ff
        habituation, 542ff; 625, 627 (vs. dishabitua-
            tion); 555 (input resistance vs. synaptic
            drive); 546 (locus of change); 546, 548
            (motor fatigue and receptor adaptation);
            611ff (neural studies); 548ff, 555, 557ff
            (synaptic changes in CNS)
        neural circuitry, 546f, *626*
        reflex test system (isolated) 552ff
    neural circuitry, 105ff
    neuroendocrine reflex (*Aplysia*), 84
    orienting reflex, 11
    responsiveness (long-term enhancement), 636
    siphon-withdrawal reflex (*Aplysia*), 351
        neural mediation, 611n
        sensitization, 637
    spinal reflex
        neural circuitry, 105ff; 596ff (mechanism)
        posttetanic facilitation, 480ff
Respiratory pumping (*Aplysia*), 87f, 378
    neuronal circuitry, 380f
Resting potential. *See* Membrane potential
Reverberatory circuits. *See* Neurons

Sensitization
    definition, 12, 625
    as arousal, 635
    cellular mechanisms, 575
    and classical conditioning, 636, 638
    dishabituation as special case, 627ff
    extreme form (from chronic or repetitive
        stress), 655ff
    role of cyclic AMP, 588f
    short-term vs. long-term, 636
Sensory neurons
    components, 107
    in effector system (multiple and hierarchical
        controls), 384f

    functional organization (crayfish), 647f
    generator potential, 107
    in somatic motor reflex act (*Aplysia*), 365ff
        quantitative individual contributions, 372ff
Serotonin. *See* Synaptic transmission; Transmitters
Sexual behavior
    Aplysia, 90
    habituation, 539f
Sign stimulus, 15, 648
Small molecule synthesis (neuronal control), 529f
Squid. *See* Mollusca
Symbols and abbreviations:
    ACh (acetylcholine), 187
    C_{input} (input capacitance), 117
    C_m (membrane capacitance), 117
    cAMP (cyclic adenosine monophosphate), 529n
    CV (coefficient of variation), 194
    d (diameter), 114n, 121
    Δx (change in x), 111
    e (constant, 2.71828), 121
    E (excitatory synaptic potential amplitude), 565
    E_m (membrane battery), 113
    E_x (battery or equilibrium potential of x), 141
    EPSP (excitatory postsynaptic potential), 105
    F (farad), 118
    F (Faraday constant), 142
    G_x (conductance of x), 113
    g_x (specific ionic conductance of x), 147
    GABA (gamma-aminobutyric acid), 202, 235
    I (current), 111
    I_C (capacitive current), 118
    I_m (membrane current), 115
    I_R (resistive current), 118
    IPSP (inhibitory postsynaptic potential), 105
    λ (membrane space constant), 121
    m (quantal output), 189, 565
    N (number of stimuli in series), 193
    N_0 (number of failures in N), 193
    $\bar{v}$ (amplitude of synaptic potential), 193
    Ω (ohms), 113
    p (probability of success), 189
    q (amplitude of unit synaptic potential), 565
    q (charge in coulombs), 117
    q (probability of failure), 189
    R_i (resistance of intracellular medium), 113f
    r_i (intracellular resistance to axial current flow
        per unit length), 114n
    R (universal gas constant), 142
    R_m (membrane resistance), 114
    r_m (membrane resistance per unit length of
        membrane cylinder), 114n
    r_0 (resistance (per cm) of extracellular medium
        to axial current flow), 114n
    t (time), 120
    τ_m (membrane time constant), 121
    TEA (tetraethylammonium) 163, 166, 291
    TTX (tetrodotoxin), 162f, *165,* 181, 183f
    T (temperature), 142
    V_m (voltage or potential of the membrane),
        111, 113n

Synapses. *See also* Synaptic potential; Synaptic transmission
  definition, 97
  central nervous system
    depression in habituation, 564ff
    habituation of reflex paths, 596ff
  chemical
    common connections, 292
    comparison to electrical synapse, 103ff
    direct connections, 289ff
    plasticity, 491ff
  convergent aggregates, 330
  coupling ratio, 132
  divergent aggregates (elementary), 282ff
  dual action mediation
    to multiple follower cells, 295ff
    to single follower cell, 304ff
  electrical (electrotonic), 132
    action by multiaction cell, 313ff
    comparison to chemical synapse, 103ff
    connections, 294
    plasticity, 491ff
  end-plate potentials, 185ff
    frequency changes, 485f
  end-plate region, 187
  excitatory
    depression (in habituation), 519, 555ff
    equivalent circuit, *167, 176*
    and inhibitory actions (dual mediation), 304ff
    onto interneurons (plastic change), 599f
    posttetanic facilitation, 480ff
  gap junction, 103
  habituated changes (relation to behavioral changes), 620
  in habituating pathway (vs. usual synapses), 584
  heterosynaptic facilitation, 498ff
  homosynaptic depression (as habituation mechanism), 564f, 600f, 617
  homosynaptic facilitation, 484
  inhibitory
    equivalent circuit, *172, 178*
    posttetanic facilitation, 491
    two-component mediation, 311ff
  inputs to motor neuron (in response habituation), 555
  ionic mechanism (single) 287
  low-frequency depression, 488, 490f
  nerve-muscle
    motor fatigue and receptor adaptation in habituation (*Aplysia*), 546, 548
    posttetanic facilitation, 485
  multiaction interneurons, 295ff
  postsynaptic
    cell, 97, 127ff (spatial and temporal summation)
    fiber, 102
    inhibitory buildup (in habituation), 555ff

  potentials, 105; 165ff, 175ff (excitatory); 171ff, 177ff (inhibitory)
  receptors (*see* Transmitters, receptors)
  sensitivity decrease (in habituation), 565ff, 575
  posttetanic
    depression, 488f
    facilitation, 480ff, 491; 499 (vs. heterosynaptic facilitation)
  potentials (*see* Synaptic potential)
  presynaptic
    cell, 97
    facilitation, 202, 589 (molecular hypothesis, and serotonin-stimulated), 591 (cyclic nucleotides in)
    fiber, 102
    inhibition, in habituation, 557ff, 619 (role of transmitter release in buildup)
    vesicles, 195f
  quantal release, 184ff, 565ff
  reverberatory aggregates, 478f
  single-function neurons, 286
  summation effects, 127ff
  synaptic cleft, 102
  tight junction, 103
  usage effects
    in invertebrates, 490
    in vertebrates, 480ff
  zone of apposition, 102
Synaptic potential
  excitatory postsynaptic potential (EPSP)
    definition, 105
    depression (in habituation), 519
    due to ionic-conductance decrease, 175ff
    due to ionic-conductance increase, 165ff
  inhibitory postsynaptic potential (IPSP)
    definition, 105
    buildup (in habituation), 519
    due to ionic-conductance decrease, 177ff
    due to ionic-conductance increase, 171ff
Synaptic transmission, 102ff, 130ff (*see also* Membrane potential; Synaptic potential; Transmitters)
  chemical, 102ff
  cholinergic, 270ff
    multiaction neurons, in *Aplysia*, 295ff (abdominal ganglion), 317ff (buccal ganglion), 322 (pleural ganglion); in lobster, 325; in *Navanax,* 324f; in vertebrates, 328ff
    receptors (pharmacology), 324f
  conduction time, 289
  cyclic nucleotide role in, 531, 588f
  dopaminergic, 276
    multiaction cell (*Planorbis*), 328
  electrical, 102ff
  excitatory (in habituation), 519
  exocytosis, 195
  ionic mechanism (single), 287

latency, 289
postsynaptic mechanisms
    excitatory, 165ff, 175ff
    inhibitory, 171ff, 177ff
presynaptic facilitation
    in dishabituation, 627
    relation to sensitization, 631
presynaptic inhibition, 200
    in habituation, 557
presynaptic mechanisms, 179ff
quantal hypothesis, 189
serotonergic, 274, 276, 327f, 462
    multiaction neurons (*Aplysia*), 327f
    role in dishabituation and sensitization, 589ff
synaptic delay, 289

Tetraethylammonium (TEA), 163, 166, 291
Tetrodotoxin (TTX), 162f, *165,* 181, 183f
Threshold. *See* Membrane potential
Transmitters
    acetylcholine, 270ff
    biochemistry and pharmacology, 270ff
    chemical nature, 202f
    dopamine, 276
    gamma-aminobutyric acid (GABA), 202, 235
    mobilization, 195
    quantal release, 184ff
        in habituation, 565ff
        in dishabituation, 575
    receptors
        chemical nature, 204
        cholinergic (pharmacology), 324f
        ionophore, 204

receptor site, 204
    sensitivity decrease (in habituation), 565ff, 575
releasable pool, 194, 208, 617, 624
release, 195
    calcium-mediated, 183f, 197f
    in central synaptic depression, 565ff
    and long-term habituation acquisition, 616f
serotonin, 274, 276
    presynaptic facilitation stimulation, 589
    role in habituation and dishabituation, 589ff
storage pool, 195, 208, 617, 624
Transport systems
    electrogenic, 241
    sodium–potassium exchange, 241
    sodium pump (electrogenic), 241ff
d-Tubocurarine (*d*-tbc), 272, 325

Vertebrates
    behavioral evolution (relation to higher inverte-
        brates), 40f
    escape responses (birds), 539
    habituation
        cellular analysis, 596ff
        vs. dishabituation, 631
        monosynaptic analog (frog spinal cord), 600
    neurons
        polarity, 80, *82*
        sprouting in olfactory system, 533
        sprouting into denervated ganglion (frog), 532
        synapses (usage effects), 480ff
    reflexes
        habituation, 596ff, 631
        neural circuitry for, 105ff
        posttetanic facilitation, 480ff